A COMMENTARY
ON THE
NEW TESTAMENT

Prepared by

THE CATHOLIC BIBLICAL
ASSOCIATION

under the patronage of

THE EPISCOPAL COMMITTEE
OF THE
CONFRATERNITY OF CHRISTIAN DOCTRINE

The Catholic Biblical Association, 1942

NIHIL OBSTAT

RT. REV. MSGR. HENRY J. GRIMMELSMAN, S.T.D.

REV. JOHN F. MC CONNELL, M.M., S.T.L., S.S.L.

REV. JOSEPH J. TENNANT, S.T.D., S.S.L.

IMPRIMATUR

✠EDWIN V. O'HARA, *Bishop*

KANSAS CITY, MISSOURI

FEAST OF THE ASSUMPTION, 1942

"The parts of the text of the New Testament contained in this book are reproduced by license of Confraternity of Christian Doctrine, Washington, D. C., the owner of the copyright of 'The New Testament of Our Lord and Savior Jesus Christ—a Revision of the Challoner-Rheims Version'."

Used with its permission. All rights reserved.

The text of the Commentary and all other material is

COPYRIGHT, 1942, BY

WILLIAM H. SADLIER, INC.

International Copyright under International Copyright Union
All Rights Reserved under Pan-American Copyright Convention

FIFTH PRINTING

PRINTED IN THE UNITED STATES OF AMERICA

PREFACE

The revision of the Challoner-Rheims New Testament has now been published as the Confraternity Edition. Our experience made us appreciate the real need of a text of the Scriptures which could be readily understood by the average Catholic reader. From the outset it was acknowledged by all that even the clearest version of the Bible must fail to eliminate all difficulties, some of which are inherent in the profound spiritual teachings of the Scriptures, that at times seem to exceed the power of human reason. Certain obscurities in the Bible are of human origin, resulting from an ancient manner of writing history, or from oriental modes of thought, or more generally from the unfamiliar or even lost culture that served as a background for the sacred books.

To aid the reader of the Scriptures, each collaborator working on the revision of the New Testament was asked to revise a portion of the text and, at the same time, to prepare a simple commentary on it. At first it was hoped that the revised Confraternity Edition and its Commentary would be published simultaneously. But many modifications adopted by the editors made it advisable to postpone the publication of the Commentary.

The revised text has now been very widely circulated. The Commentary, therefore, will be all the more welcome to those readers in whom a new and livelier interest in the New Testament has been awakened. Since the Episcopal Committee of the Confraternity of Christian Doctrine urged the publication of the Commentary and the Scriptural collaborators gave assurance of their co-operation, the Catholic Biblical Association entrusted the publication of this Commentary to a committee of its members.

This Commentary, which is intended as a supplement to the Confraternity Edition of the New Testament, presupposes that the reader has the revised edition before him. The same logical order as indicated by the marginal headings of the Confraternity Edition is followed. The Commentary bases its explanation on the revised text and often makes reference to its footnotes.

This Commentary has in view the needs of the average person who reads the New Testament in English and of organized lay members of Scripture discussion clubs. By clarifying certain difficult passages it makes possible a fuller appreciation of the divine

Preface

message. The purpose of this simple Commentary excludes an extensive discussion of difficulties, a thorough consideration of controverted questions, and lengthy citations of the opinions of different schools. Citation of modern scholarship is restricted to Catholic authors who have written in English. A bibliography of such works is provided for the convenience of those who may desire to pursue an investigation beyond the limits of this Commentary. The authors cited in this bibliography indicate the vast Catholic literature on the New Testament.

Every effort has been made to avoid repetition. Thus, in dealing with the Synoptic Gospels the parallels are explained but once, that is, on their first occurrence. An event, for example, which is found in all three Gospels is explained in Matthew and merely mentioned in the others, unless fresh details in the other Gospels call for attention. This is also true when a subject is found more than once in other books, for example, in Ephesians and Colossians.

Since the Commentary is intended for the non-professional reader, only translated citations from the Latin or Greek texts are given.

The typographical norms are in nearly every respect those of the Confraternity Edition of the New Testament. The bold face headings, the verse numbers, and the spelling are those of the revised text. Thus, Prophet and Gospel, when capitalized, refer to the canonical books; otherwise, prophet indicates an individual, and gospel means the New Testament message. Church, when capitalized, is to be understood of the Church universal; otherwise, church signifies a local group of Christians, as the church at Corinth or Ephesus. Both the revised text and the Commentary follow the traditional order of the books. By way of exception, the Epistles of the Captivity are grouped together in the Commentary because they are so closely related historically. The enumeration of the Psalms is that of the Vulgate.

With one or other exception, each part of this Commentary was written by the same author who revised the corresponding book of the New Testament. The nature of this work, however, and the desire for uniformity called for more than customary intervention on the part of the editors. The use of technical terms, as a rule, has been avoided. The style and general manner of expression of the various writers have to some extent been altered. In some matters of little moment divergent views of individual commentators have been harmonized.

Preface

This effort is particularly evident in the chronology adopted in the Commentary. It is largely conventional, taking the view that the public ministry of our Lord extended over three years. Similarly, uniform dates are given for the missionary journeys of St. Paul, with the year 63 A. D. being accepted as the end of his first Roman imprisonment. All parts of the Commentary have been made to agree with this chronology. The individual authors, therefore, are not necessarily responsible for the dates given in their portion of the work.

In another important matter the editors prevailed. They desired to limit the Commentary to one volume, which made it necessary to abridge the work of the individual writers, who previously had consented to this abridgment. The editors consequently share the responsibility for the separate parts of the Commentary to the extent of their intervention, and they take this occasion to express to the authors appreciation for their generous co-operation.

An expression of gratitude is due the members of the Episcopal Committee of the Confraternity of Christian Doctrine for their encouragement and support.

E. H. Donzé, S. M.
L. Hartman, C. SS. R.
W. L. Newton
C. H. Pickar, O. S. A.
W. S. Reilly, S. S.

CONTRIBUTORS

Matthew:	Rev. Mark Kennedy, O.F.M., M.A., S.S.Lect. Gen.
	Rev. Louis Hartman, C.SS.R., S.S.L., Ling. Or.L.
Mark:	Rev. Aloys H. Dirksen, C.PP.S., S.T.D.
Luke:	Rev. John E. Steinmueller, S.T.D., S.S.L.
John:	Rt. Rev. Msgr. William L. Newton, M.A., S.S.D.
Acts:	Rev. William A. Dowd, S.J., M.A., S.T.D., S.S.L.
Romans:	Rev. Joseph L. Lilly, C.M., S.T.D., S.S.L.
1 Corinthians:	Rev. Edward H. Donzé, S.M., S.T.D., S.S.L.
2 Corinthians:	Very Rev. Charles J. Callan, O.P., ST.M., Litt.D.
Galatians:	Rev. John F. Rowan, S.T.D., S.S.L.
Ephesians:	Rev. Wendell S. Reilly, S.S., S.T.D., S.S.D.
Philippians:	Rev. John J. Collins, S.J., M.A., S.S.L.
Colossians *Philemon*	Rev. Maurice A. Hofer, S.S.L.
Thessalonians:	Rev. Charles J. Costello, O.M.I., S.T.D.
Timothy *Titus*	Rev. Leo P. Foley, C.M., S.T.D.
Hebrews:	Rev. Edward A. Cerny, S.S., S.T.D.
James:	Rev. Charles H. Pickar, O.S.A., S.T.L., S.S.L.
Peter:	Rev. Charles G. Heupler, O.F.M.Cap., S.T.L.
John:	Rt. Rev. Msgr. Albert G. Meyer, S.T.D., S.S.L.
Jude:	Rev. Raymond F. Stoll, S.T.D.
Apocalypse:	Rev. J. S. Considine, O.P., S.T.Lr.
Articles:	Rev. Richard T. Murphy, O.P., S.T.D., S.S.L.
	Rev. John F. McConnell, M.M., S.T.L., S.S.L.
	Rev. Wendell S. Reilly, S.S., S.T.D., S.S.D.
	Rev. John E. Steinmueller, S.T.D., S.S.L.
Index of Texts:	Very Rev. John A. McHugh, O.P., S.T.M., Litt.D.

TABLE OF CONTENTS

Preface	iii
List of Contributors	vi
The New Testament Background	1
The Parables of the Gospels	12
The Literary Relations of the First Three Gospels	20
Matthew	25
Mark	200
Luke	227
John	292
Acts of the Apostles	365
The Life and Epistles of St. Paul	406
Romans	410
1 Corinthians	447
2 Corinthians	482
Galatians	502
Colossians	518
Philemon	532
Ephesians	534
Philippians	546
1 and 2 Thessalonians	556
The Pastoral Epistles	570
Hebrews	589
James	606
1 and 2 Peter	616
The Epistles of St. John	635
Jude	649
Apocalypse	655
Index of Scripture Texts	685
Additional Reading	715
Glossary of Terms and Names	722

ABBREVIATIONS OF BOOKS OF THE BIBLE

Abd.	Abdias	Jud.	Judith
Acts		Kgs.	Kings
Ag.	Aggeus	Lam.	Lamentations
Amos		Lev.	Leviticus
Apoc.	Apocalypse	Luke	
Col.	Colossians	Mach.	Machabees
Cor.	Corinthians	Mal.	Malachias
Dan.	Daniel	Mark	
Deut.	Deuteronomy	Matt.	Matthew
Eccles.	Ecclesiastes	Mich.	Micheas
Ecclus.	Ecclesiasticus	Nah.	Nahum
Eph.	Ephesians	Num.	Numbers
Esd.	Esdras	Os.	Osee
Est.	Esther	Par.	Paralipomenon
Ex.	Exodus	Pet.	Peter
Ezech.	Ezechiel	Philem.	Philemon
Gal.	Galatians	Phil.	Philippians
Gen.	Genesis	Prov.	Proverbs
Hab.	Habacuc	Ps.	Psalm
Heb.	Hebrews	Pss.	Psalms
Isa.	Isaias	Rom.	Romans
Jas.	James	Ruth	
Jer.	Jeremias	Soph.	Sophonias
Job		Thess.	Thessalonians
Joel		Tim.	Timothy
John		Titus	
Jon.	Jonas	Tob.	Tobias
Jos.	Josue	Wisd.	Wisdom
Jdgs.	Judges	Zach.	Zacharias
Jude			

NOTE: In citing passages from the Bible, the chapter numbers are always given in italics, the verse numbers in usual arabic type.

THE NEW TESTAMENT BACKGROUND

I

The Background. The true background of the New Testament is that of Palestine at the beginning of the Christian era—a background which is made up of many foreign elements. Previous to our era, this tiny land, alternately battle-ground and buffer-state, trembled often to the sound of marching legions from Egypt, or Assyria, or Persia, or Greece. But in the end it was Rome's turn to play an important part in the land where Christ was born.

Rome definitely entered into the politics of Palestine when Pompey marched southwards from Syria and took Jerusalem after a siege lasting three months (63 B.C.). The coming of Rome was a kiss of death for the Machabean or Asmonean dynasty founded a century before by the great Judas Machabeus. The Jewish state became a vassal of Rome and national independence was a thing of the past.

The Herod Family. Some few years later an unexpected character, Antipater of Idumea, appeared on the scene. This ancestor of the Herods, a favorite of the Romans and very ambitious, introduced his sons, Phasael and Herod, into local politics, the first being created governor of Jerusalem, the other of Galilee. During the last decades of the first century B.C., Asia changed hands with startling rapidity. Antony, master of Asia after the battle of Philippi (42 B.C.), made Herod and Phasael tetrarchs of the province of Judea. Originally "tetrarch" meant a "ruler of a fourth part." In New Testament times the title was used of a "petty prince" or "ruler of a district." The remainder of the country was attached to Egypt and Syria.

Shortly thereafter, one of the last descendants of the Machabean family, Mattathias or Antigonus, captured Jerusalem. Herod escaped and made his way to Rome, where a judicious use of money won him great favor; the Senate in formal session declared him king of Judea. It was not, however, until 37 B.C. that Jerusalem fell into Herod's hands.

The reign of Herod "the Great," so-called in comparison with his rather inferior descendants, is divided into three periods:

1

the consolidation of his power (37-25); the period of prosperity, Roman friendship, and great building activity (25-13); and that of domestic troubles (13-4 B.C.).

Antony's star suffered eclipse at Actium, 31 B.C. With characteristic adroitness, Herod passed over to the camp of the victorious Octavian (Augustus) at just the right moment. Making the most of Rome's favor, he initiated a great building program. The most important undertaking of this sort was the rebuilding and enlarging of the temple of Zorobabel, begun in the 18th year of his reign (20-19 B.C.), and completed shortly before the Jewish revolt and the fall of Jerusalem. "He who has not seen Herod's building has never seen anything beautiful," was a common saying of the day (cf. John *2*, 20).

The domestic troubles which marred the third period of Herod's life were of such gigantic proportions as to overshadow all else. He had in all ten wives and almost as many sons, and so many of the latter were murdered at his command that a saying became very popular to the effect that it was better to be Herod's pig than his son. How very little human life mattered to Herod is reflected in Matthew's account of the massacre of the Holy Innocents (*2*, 16-18).

Upon the death of Herod, his kingdom was divided among his three sons, no one of whom bore the title of king. Archelaus reigned as ethnarch of Judea (4 B.C. – 6 A.D.). The exact meaning of "ethnarch" (translated "governor" or "ruler of a people"), is not certain, but it appears to indicate a petty ruler subject to Rome, and higher in rank than a tetrarch. His reign was marked by tyranny and despotism. By far the worst of Herod's sons, he was summoned to Rome and exiled to France; his territory was incorporated by Cyrinus into the Roman province of Syria, and was put under the direct rule of a local procurator.

Philip ruled as tetrarch over northern Transjordan. The southern border of his territory was, roughly speaking, the river Yarmuk and neighboring Decapolis. Philip ruled (4 B.C. – 34 A.D.) justly and peacefully; but at the end of his long reign, his territory was also annexed to Syria. Two of the cities built by him (Cæsarea Philippi and Julia or Bethsaida) are mentioned in the Gospels.

Herod Antipas was the last son of the great Herod to occupy a position of authority. This is the New Testament Herod so fearlessly criticized by John the Baptist; the same Herod whom Jesus Christ stigmatized as "that fox" (Luke *13*, 32); before whom

Jesus stood silent during the Passion (Luke *23, 7*). He ruled as tetrarch over Galilee and Perea (4 B.C. — 39 A.D.), but the intervening territory of Decapolis was under the direct supervision of the governor of Syria. Towards the end of his reign, Antipas learned that the Emperor, Caligula, had presented the former territories of Philip and the title of king to Herod Agrippa, brother of Herodias. This ambitious woman who had already been the cause of much trouble, was so envious of her own brother's good fortune that she prevailed upon her husband, Antipas, to seek the same title for himself. Together they went to Rome, hoping to realize this ambition; but there they found themselves outwitted by Agrippa, and were rewarded, not with royal honors, but with degradation and exile.

Herod Agrippa I, grandson of Herod the Great and Mariamne, ruled briefly (41-44) over a territory as extensive as that of his grandfather, for the new Emperor, Claudius, added to his possessions (those which had formerly belonged to Philip and Antipas) both Judea and Samaria. He is mentioned in the Acts *(12)* as having killed St. James the Greater and imprisoned St. Peter. Rome grew suspicious of this petty king when he began to build the third wall around northern Jerusalem, and it is likely that his early death saved him from an official investigation. His son, Agrippa II, was too young to fill the important position held by his father, and so did not succeed him. Instead, after a few years, he was made king of a tiny principality in Chalcis, succeeding his uncle, Herod. Like him, Agrippa was guardian of the temple and its funds, and had the right to appoint the High Priests. Then, in exchange for Chalcis, Agrippa obtained the tetrarchy of northern Transjordan, to which parts of Galilee and Perea were in time added. He ruled until about 85 A.D. It was before this last representative of the Herod family that St. Paul appeared during the Cæsarean captivity (Acts *25,* 13 ff).

II

Roman Provincial Rule. In 27 B.C., Augustus divided into two classes the provinces which were under Roman rule. Those which he reserved to himself because they were difficult to manage, as Syria, for example, were called imperial, and those which were pacified and more or less accustomed to the domination of Rome were called senatorial provinces. The provinces were either consular or prætorian, depending on the rank of the governor in charge. The governors of the senatorial provinces were called proconsuls; while the imperial provinces were ruled either by legates *pro prætore,* as was the case of Syria, or by prefects, as in Egypt, although nominally the actual head was the Emperor. A few imperial provinces were ruled in an exceptional manner by governors of equestrian rank—procurators—when (as in Mauretania and Thrace) the rudeness of customs, or (as in Judea) the special tenacity of the natives in adhering to national customs rendered the usual methods of imperial government impossible.

The Procurators. From 6-41 A.D., Judea was ruled by seven procurators, one of whom was Pontius Pilate (26-36). The procurator was directly responsible to the Emperor, who could remove him at will. He was dependent upon the legate of Syria for serious military aid, and the legate could in some cases remove the procurator and send him to Rome to answer for his administration of the province. The residence of the procurator was maritime Cæsarea. Experience, however, had taught the governors to be in Jerusalem with troops ready for action on the occasion of the great religious feasts, when smouldering national resentment was most apt to burst out in flames. Whenever in the Holy City, the procurator resided on the western side of the city in what was formerly Herod's palace—now known as the Citadel.

In the Roman empire, the Jews were exempt from military service in Roman armies. The procurator had at his disposal five cohorts of mercenaries, recruited from the Greeks, Samaritans and Syrians; it seems that these cohorts (numbering about three thousand men in all) were supplemented by a wing of cavalry (cf. Acts *23, 23*). Each cohort (four hundred to six hundred men) was commanded by a *chiliarch,* and one of them was always quartered at the Antonia, the fortress or garrison overlooking the temple area. St. Paul was to owe his life to this fact (cf. Acts *21-22*).

One of the chief functions of the procurator was the administration of provincial finances. In fact, the name "procurator" comes from this function: to "procure" money in the form of taxes. The revenue obtained from Judea, an imperial province, went into the imperial treasury, called the *fiscus* to distinguish it from the *ærarium,* which was the treasury of the Senate. In literal truth, Judea rendered to Cæsar . . . taxes (Matt. *22,* 17 f). The actual collection of taxes was carried out by local agents, Jews, called "publicans" in the Gospels; to these men the procurator farmed out the taxes for a fixed sum, as was the common practice in antiquity. Whatever the collectors obtained over and above the fixed sum paid for the right of collecting the taxes, was to their own profit; if they failed to obtain the same amount they paid for the privilege, it was their own loss, and they could blame no one. Much room therefore was left for individual injustice and extortion, depending upon the rapacity of the collectors. This as a rule was great, and explains the cordial dislike felt by the people for the publicans. This dislike is reflected in the Gospels, where publican and sinner are regarded as synonymous. St. Matthew was a collector of the customs tax in Galilee (Matt. *9, 9*), and Zachæus was chief of the publicans who collected like taxes in Jericho (Luke *19*, 1-2), but both were recipients of Jesus' favor.

The Sanhedrin. Operating side by side with, but more or less independently of the Roman governors, were the local courts, and in the case of Judea, the *Sanhedrin*. The Sanhedrin or Council of the New Testament corresponds to the *gerousia* or Senate (cf. *1* Mach. *12*, 6; *14*, 28; *2* Mach. *1*, 10; *4*, 44), and is hardly much older than the times of Antiochus the Great (223-187 B.C.). According to the Talmud, there were seventy members of the Sanhedrin, presided over by a President. This supreme native court was empowered to decide on secular affairs such as civic improvements; of much more importance was its almost absolute authority to deal with violations of the Law, idolatry, blasphemy, false prophets and the like (cf. Matt. *26*, 65; John *19*, 7; Acts *4-6.23*). Only one galling restriction did Rome impose upon this court: the procurator alone possessed the right to exercise the *jus gladii*, that is, pronounce sentence of death. John *18*, 28 ff is a good illustration of this. Aside from this restriction, the Sanhedrin had independent authority in policing the city and temple. It could order the arrest of individuals guilty of disturbing the peace—such orders were carried out by its own corps of police (Matt. *26*, 47; Acts *4*, 3)—and it had power to mete out corporal punishment less than death (Acts *5*, 21-40; *2* Cor. *11*, 24-25).

All things considered, the Sanhedrin enjoyed a rather extensive jurisdiction under Roman rule. That this jurisdiction was valid only within limits is evident from the fact that the Roman authorities could at any time take the initiative themselves and proceed in trials independently of the Jewish court. This right they sometimes exercised, notably in the case of St. Paul (Acts *21,* 33 ff). Jews who were Roman citizens could in these circumstances exercise the right of appeal to the Emperor (Acts *25,* 10-12).

Clashes between Jewish and Roman authorities were normally to be expected. Valerius Gratus, procurator from 15-26 A.D., removed three High Priests from the presidency of the Sanhedrin in as many years, and then, in 18 A.D., appointed as High Priest that son-in-law of Annas, Joseph Caiphas, who conducted the religious trial of Christ. Caiphas remained in office until 36 A.D.

Sects: Pharisees. A word must here be said about the various sects existing at the time of Christ. These were principally two: the Pharisees and the Sadducees; and of the two, the Pharisees were the more influential and important. Referred to in Machabean times as "Assideans," they were avowedly "'pious," and their piety took the form of separation (whence their name derives) from heathen practice and from all contact with the common people. The Pharisees were a tremendous power in the life of Judaism. They taught belief in personal immortality and judgment after death, angels, resurrection of the just, and maintained the doctrine of free will and divine Providence. Moreover, they were men of vivid, Messianic faith, although this became so confused with national hopes and aspirations that it made no provision for a suffering Messias and a spiritual kingdom. By the time of Christ both their faith and its observance had degenerated into a sterile devotion to the externals of religion almost to the exclusion of an inner spirit of piety. Thus it is that while the people looked upon the Pharisees as the champions of the faith, and venerated them because of their zeal for the Law of Moses and for the traditions that had grown up around it, nevertheless Christ on many occasions condemned their narrow spirit of Sabbath observance (Luke *13,* 15), their many washings in the name of legal purity (Mark *7,* 1-8), and their scrupulous payment of tithes (Matt. *23,* 23), even to the exclusion of the more important duties towards their neighbor: justice, mercy and charity.

A large part of the sect of the Pharisees was made up of the Scribes; although the terms are not synonymous. The Scribes were the lawyers, the interpreters of the Law of Moses, the men who sat in the chair of Moses (cf. Matt. *23,* 1-2) and who were called Rabbi. Their decisions on custom and traditions, as well as on questions directly flowing from the Law, inevitably drew them into conflict with the second great religious sect of the times, the Sadducees.

Sadducees. The Sadducees were, in contrast to the Pharisees, a sect whose members were drawn chiefly from the sacerdotal and lay aristocracy. They claimed descent from Sadoc, a famous High Priest who lived in the time of Solomon (cf. *3 Kgs. 4, 2; 1 Par. 6,* 9-15); but not all of the Sadducees were priests; some of them were representatives of the better families. The sect had known varying fortune, but at the time of Christ the High Priest was invariably a Sadducee, and he was in the eyes of the Romans the official representative of the Jewish nation. The Sadducees were the inflexible conservatives, admitting nothing but the letter of the Law of Moses, and they were legal rigorists, demanding the full penalties of the Law in criminal trials. In this they are again contrasted to the Pharisees, whose acceptance of tradition mitigated to some extent the severity of the old *lex talionis;* this more humane attitude in court won for the Pharisees the friendship of the people—something the Sadducees apparently never succeeded in doing. The Sadducees did not believe in personal immortality, personal judgment, resurrection of the body, or in angels (cf. Acts *23,* 8). Comfortable in a material sort of way, they were little interested in the coming of the Messias and showed no serious concern about Christ until His popularity threatened to get out of hand. Then they forgot their century-old hostility towards the Pharisees, and joined hands with these born enemies in a successful effort to destroy Christ. They were themselves destroyed by the fall of Jerusalem, 70 A.D.

Pilate. The year 36 marked the recall of Pontius Pilate. He had been a man of an unbending and hard character, whose administration Philo describes as one of "corruptibility, violence, robberies, ill-treatment of the people, continuous executions without trial, endless and intolerable cruelties." Pilate's career as a petty governor of Judea was characterized by imprudence and bad judgment. There was first of all the incident of the military standards. The procurator had his soldiers march into Jerusalem by night carrying these flags (upon which was the figure of the Emperor), but the wholly violent opposition of the people compelled him to order their removal. (Pilate's action was looked upon as a flagrant violation of the first commandment, cf. Ex. *20,* 4; Deut. *5,* 8.) Next he confiscated the contents of the temple treasury in order to finance the aqueduct from Solomon's Pools to Jerusalem. The popular opposition to this high-handed procedure was crushed brutally with clubs, which only served to augment the growing dislike for the governor. Frequent outbursts of popular feeling against the government are only hinted at in the Gospels (Luke *13,* 1; *23,* 12.19), but it is impossible to read John's account of the civil trial of Jesus without sensing something of the intense, reciprocal hostility between governor and people.

During his later days as procurator, Pilate unwisely introduced a number of golden votive shields into his residence, the prætorium. The shields were without figures, and bore only the name of the Emperor. Upon receiving the protests of the Jews, Tiberius himself ordered the removal of the offending shields. Pilate's final blunder was to overestimate the importance of a group of Samaritans gathered on Mt. Garizim to see the vessels of the temple, supposedly buried there by Moses. He commanded his soldiers to disperse the crowd, and the order was carried out with much shedding of blood. Vitellius, legate of Syria, ordered Pilate to Rome to give an account of his conduct to the Emperor. Two procurators, Marcellus and Marullus, filled out the five years which then ensued until the brief reign of Herod Agrippa I, 41-44 A.D.

From Pilate to 70 A.D. The average Roman governor of a colony was not remarkable for his considerateness and tact in dealing with his subjects. Nor were the procurators, men of the equestrian order, often gentle or greatly concerned about the feelings and prejudices of their subjects. It sometimes seems that they deliberately acted in such a manner as to arouse the people. Those who ruled Palestine until the outbreak of the revolt were for the most part indifferent to right or wrong, and seldom listened to any voice save that of money.

Of the seven remaining procurators who ruled Palestine from 44 to 66 A.D., two deserve mention by reason of their connection with St. Paul. Felix (52-60) held Paul a captive at Cæsarea for two years (58-60), hoping no doubt that Paul would attempt to buy his freedom (Acts *24,* 26). Festus (60-62) fell heir to the chaos left by the miserable government of his predecessor, a chaos so complete and far-reaching that his well meant efforts in the interests of justice were quite hopeless. After a hearing, however, Festus commanded Paul to be sent to Rome for trial, for Paul had appealed to the Emperor (Acts 25, 10). Festus arranged to have Paul heard by Agrippa II and Bernice (Acts 25, 13 ff) before his departure for Rome.

Festus died in office, and there ensued a period of wild anarchy and violence, during which St. James, head of the Church in Jerusalem, was put to death by Scribes and Pharisees acting on orders of the High Priest, Ananus. It is quite possible that the stoning of St. Stephen occurred during a similar inter-regnum after the dismissal of Pilate in 36 A.D.

The final, bloody conclusion was soon to be written to this tale of continued misrule and brazen injustice. The glorious temple of Herod was completed under Albinus (62-64) some eighty years after its inception. This procurator was a grasping scoundrel, but a paragon of virtue in comparison with his successor, Gessius Florus. The tyranny of Florus (64-66) was unbounded; he plundered villages and communities as well as individuals, and went so far as to afford protection to bandits and robbers on condition that they share their spoils with him.

Thus the stage was set for the final conflagration. Only a spark was needed to set it off, and the spark was struck when Florus laid hands upon the temple treasury. In bitter protest and with mock solemnity, the people took up a collection of anything and everything for "poor Florus." Florus sought savage revenge. The destruction, plundering, imprisonment, and crucifixions which followed were indications of his wrath, but the popular temper was so aroused that the procurator was speedily compelled to withdraw to Cæsarea. The Revolution had begun.

After five months of siege, on September 8, 70 A.D., Jerusalem fell and was razed to the ground, and the story of Israel's quasi-independence came to an end. *Judæa capta* became a separate province of the Empire, and was henceforth distinct from that of Syria.

Importance of Rome. Rome and the Roman world were the bridge built by divine Providence for the spreading of the good tidings of salvation. The administrative unity of the empire which was the product of Roman genius, made possible and greatly facilitated the labors of the Apostles. Humanly speaking, the missionary career of St. Paul would have been of short duration, had he not traveled constantly through territories policed by Roman soldiery and managed by ever-watchful representatives of the Emperor. More than once his Roman citizenship saved him.

But Rome and particularly Roman education and belief were a formidable obstacle to the gospel. The upper classes, trained from youth to admire strength and beauty, and taught to venerate and love their own city and the gods, found it next to impossible to look favorably upon a religion whose hero died upon a gibbet without striking a single blow in self-defense. Traditional paganism, moreover, sometimes combined with local patriotism and private interests to resist the heralds of Christ (cf. Acts *19*, 23 ff). In the end, however, paganism fell because it was fundamentally incapable of providing adequate satisfaction to the aspirations and longings of the human mind and heart; and Christianity was able to provide such satisfaction.

Perhaps the greatest obstacle to the spread of the gospel, however, was the happy, prosperous condition of the world, inaugurated in the reign of Augustus and carried on by his successors. So much indeed was happily brought about by Rome, that some of the provinces were led to look upon the Emperor as the cause of this peace and prosperity. The cult of the Emperor sprang up; his divinization even during life soon followed. He was to be a potent rival and competitor of Christ until the time of Constantine, when the two rivals clashed in a battle for supremacy; then the battle was decided forever in favor of Christ the King who had risen from the dead. With the fall of the cult of the Emperor, paganism fell also, and Christianity's triumph was complete.

CHRONOLOGICAL TABLE

Palestine	Apostles	Procurators	Rome
Birth of Christ, 9-6 B.C.		Administer Judea from 6-41 and 44-66 A.D.	Augustus, 27 B.C.-14 A.D.
Public Life of Christ, 27-30		Coponius, 6-9 Ambivius, 9-14 Ann. Rufus, 12-15 Val. Gratus, 15-26	Tiberius, 14-37
	Stoning of Stephen, 36	Pon. Pilate, 26-36	

The New Testament Background

Palestine	Apostles	Procurators	Rome
	Conversion of Paul, 36	Marcellus, 36-37	
		Marullus, 37-41	Caligula, 37-41
			Claudius, 41-54
Herod Agrippa I, 41-44	Martyrdom of James, 42	Cuspius Fadus, 44-46	
		Tib. Alexander —48	
	Paul's 1st missionary journey to Cyprus, Asia Minor, 45-49	Ventidius Cumanus, 48-52	
Apostolic Council of Jerusalem, 49/50	Paul's 2d missionary journey to Galatia, Macedonia, Greece, 50-52		
		Felix, 52-60	
	Paul's 3d missionary journey to Asia Minor, Macedonia, Greece, 53-58		Nero, 54-68
	Paul imprisoned for two years in Cæsarea, 58-60		
	Roman captivity of Paul, two years, 61-63, followed by acquittal	Festus, 60-62	
	Martyrdom of James the Less, 62		
		Albinus, 62-64	
	Martyrdom of Peter, 64/67	Gessius Florus, 64-66	
Jewish insurrection, 66-70	Paul again arrested, 66		
	Martyrdom of Paul, 67		Vespasian, 69-79
Jerusalem falls, and the temple is destroyed, 70			
	Death of John, c. 95-100		

RICHARD T. MURPHY, O.P.

THE PARABLES OF THE GOSPELS

Name and Nature. The term *parable* is a Greek loan-word which, in the proper sense, meant a juxtaposition, in the metaphorical sense, a comparison. But in the Synoptic Gospels the Greek word has a wider significance than comparison. Elsewhere in the New Testament it occurs only in Heb. *9,* 9 and *11,* 19; in the former case it means a "type" or "symbol" certainly, in the latter case probably. Our Lord, speaking in Aramaic, probably used a term akin to the Hebrew word *mashal* which was ordinarily translated into Greek by the word *parable.* It is understandable that the Evangelists should choose the same Greek word to translate the Aramaic expression used by Jesus.

Now, *mashal* could be used in Hebrew to describe almost any saying which departed from the plain, prosaic, pedestrian; almost any manner of imparting a lesson by indirection, although the element of comparison is seldom completely lacking. *Mashal* was employed indiscriminately for those distinct figures of speech which we call similitude, metaphor, allegory, riddle, proverb and maxim, as well as for "parable" in the restricted modern sense of a fictitious but plausible narrative used to impart or illustrate a religious truth by means of an expressed or implied comparison.

It is, however, generally agreed that to deserve the name of parable a saying must contain a complete thought or narrative in figurative language which imparts or illustrates a religious truth by means of an expressed or implied comparison between the figurative example and the religious reality. It is not necessary that the saying be long; in fact very short sayings are called parables in the Gospels. But the briefer parables may at times be abbreviations of more developed sayings of our Lord. The parable differs from the *allegory* by the fact that the element of comparison is absent from the latter; in the allegory one thing is described on the surface but the reference is really to another as in the beautiful allegories of St. John's Gospel (*10,* 1-16; *15,* 1-8). It should be noted that these distinctions were not rigid in ancient writings or in the Gospels; the parable sometimes contained allegorical elements and the allegory parabolic features. The successful combination of these two forms was considered in antiquity to be an achievement which proved more than ordinary artistic power. The parable is also distinct from the *fable* which is a fictitious but implausible narrative in which animals and plants may be made to speak and the lesson conveyed is ordinarily one of purely human wisdom. There are two fables in the Old Testament (Judges *9,* 8-15; *4* Kings *14,* 9), none in the New.

This parabolic manner of teaching, so closely associated in our minds with our Lord who used it with such incomparable mastery and artistic perfection, was not His invention. The parable has been used among almost all peoples and has from time immemorial enjoyed great popularity in the Orient where the imagination is more highly developed. The very mystery and apparent obscurity of this method of instruction makes it much more attractive and stimulating there than the plainer and more direct manner of speech to which we of the West are more inclined.

Interpretation. The parable is basically a comparison and to discover the lesson of the parable we must know both terms of the comparison. Usually to find the principal lesson it suffices to reduce the parable to these two terms, e.g., *just as* a prudent man builds his house on rock and not on sand, *so* an aspirant to the Kingdom must not only listen to Christ's message but also put it into practice. When the comparison is not explicit we must seek in the introduction or conclusion the clue to the nature of the truth which the parable is meant to illustrate. Not every detail in the story need be taken up in the application; some of these are at times without significance. So it is not necessary in the parable of The Good Samaritan to give a meaning to the road to Jericho, the inn, the innkeeper, the ass, the oil and the wine: these details are added to the story mainly to make it more lively, attractive and interesting. But it is lawful to try to detect a spiritual significance in these details and parabolic exegesis of this kind has been very helpful in the development of the tradition of Christian spirituality. Thus, in the parable of The Prodigal Son it is permissible and useful to consider the fatted calf as the symbol of all the graces, including the Eucharist, which God showers on the repentant sinner. But it would be wrong to maintain that this is the *meaning* of the fatted calf which is introduced into the story only to show in a vivid manner the joy of the father over the return of his son, which joy is but a pale image of the happiness of God over the repentance of a sinner.

At times, it may be found difficult to make the parable fit the application which seems to be contained in the conclusion. In this case the Commentary should be consulted. But it may be pointed out here that the Evangelists have sometimes added to the parables, as to other sections of our Lord's teaching, sayings originally uttered in other circumstances and placed alongside the parables only because of some similarity of subject.

Most of the parables are not difficult. This is true especially of those which contain no admixture of allegorical traits and of those which teach a moral lesson. No one could fail to grasp the point in the parable of the Miser (Luke *12,* 16-21), of the Godless Judge (Luke *18,* 1-8) or of the Two Houses (Matt. 7, 24-27), for it is immediately obvious that they are splendid illustrations of the vanity of riches, the power of persevering prayer and the absurdity of a superficial and insincere attachment to God's message. In the majority of the parables, the main idea is clear enough to be grasped by the attentive reader, especially if he is a Christian well-grounded in his religion which is the realization of the Kingdom on earth.

Two pertinent questions might be asked concerning certain of the parables: Did Jesus ever describe a real incident in a parable? Did He ever repeat a parable? The first is commonly, and probably correctly, answered in the negative, although there are still some who think that Jesus may have been describing real incidents in the parables of The Good Samaritan and The Rich Man and Lazarus. To the second it may be replied that it is antecedently probable that our Lord did repeat some of His parables and did not hesitate to change either details or application of the story. This is probably the case with the two parables contained in Luke *19,* 12-27 and Matt. *25,* 14-30 and also with the other two in Matt. *22,* 1-14 and Luke *14,* 16-24.

Number and Classification. In many recent studies on the parables, there is a noticeable tendency to limit their number to thirty odd. Such a list really includes no more than the longer parables, but it is arbitrary to make length an essential factor. Among those authors who do not limit the concept of parable by length there is no agreement on the exact number of parables. Likewise no agreement exists on the principle according to which the Gospel parables should be classified.

Happily, these questions are of slight importance. Our Lord

has left us no scientific definition of the parable and there is no evidence that the Evangelists had any idea of arranging them systematically when they recorded them in the Gospels. The accompanying list is rather more inclusive than is customary today but there is justification for calling every saying in it a parable unless the concept be needlessly restricted. The parables in the list are divided into complete and condensed, and the former are classified according to their principal lesson, a system which is reasonably satisfying. The division and arrangement were suggested by Father Prat's *Jesus Christ,* I, 553 f, but the list has been enlarged in accordance with the suggestions of Father Holzmeister, *Biblica,* 14 (1933) 367 f; 15 (1934) 548.

Purpose. One reason why our Lord adopted the parable method of teaching may have been His wish to awaken His followers to the revelation of God in nature, and to form, in them and in His Church, an awareness of that mysterious harmony with the unseen which God had implanted in nature when He made it good. He seems to have acted thus when He pointed out that the lesson of God's loving care for us was to be learned in the splendor of the lilies of the field and the carefree existence of the birds of the air. Certainly, the lessons thus drawn by Him from the natural sphere to illustrate the supernatural have contributed largely to the formation of the Christian habit of seeing the eternal through the transitory and the invisible wonders of God in the visible world, a habit which reached a surpassing perfection in Christians so far apart in time, dispositions and training as St. Francis of Assisi for whom the birds were brothers and Cardinal Newman from whom the grass could not hide the angels. St. Thomas *(Summa Theol.,* 3, 1, 1) says that the world was made for this purpose: "It is highly fitting that the invisible realities of God should be made known by visible things: *for this end the whole world was made* as is clear from the words of the Apostle (Rom. *1,* 20)."

But this purpose is, at the most, only accessory. It seems obvious that our Lord used parables, as teachers have always used them, to enlighten His hearers, and, to be more specific, to enlighten them on the nature, future and destiny of the Kingdom of God and on the dispositions which were needed in those who wanted to become or remain its members. It was observed above that the very mysteriousness of the parables was stimulating for the Oriental mind but this mystery was present only as long as the point of comparison was undisclosed. In other words, the obscurity ended with the removal of the suspense. The lesson once revealed, the figure served mainly to fix it permanently in the memory.

There would be need for no further comment on the purpose of the parables if our Lord had not once uttered words, recorded in slightly different terms by three of the Evangelists (Matt. *13,* 10-17; Mark *4,* 10-12; Luke *8,* 9-10). The words, "That . . . hearing they might hear, and *not* understand" (Mark *4,* 12), do appear incompatible with the obvious intent of the parables. Our Lord seems to mean that He adopted the parabolic method not to enlighten but to darken the minds of the multitude; not to complete their religious formation but to hinder it and thus to fulfill the terrible words in which God announced to Isaias the tragic result of the mission which the prophet was about to begin (Isa. *6,* 9 f). The Commentary should be consulted for an explanation of this difficult passage in the Gospels. Here the following observations may be of use. (a) Parables are not of their nature obscure and the point of many of the parables of Jesus is readily grasped. (b) The words are used only with relation to parables treating of the nature of the Kingdom: these present special difficulties of interpretation, not so much for us who have a clear idea of the Kingdom but for the Jews whose ideas concerning it were not only obscure and imperfect but also false. (c) The disciples and "those outside" are not to be considered as definitely fixed groups: individuals could pass from one to the other so that there can be no question of consigning certain people to a condition in which it would be impossible for them to receive the religious instruction necessary for eternal salvation. (d) The Apostles did not understand the parables without Christ's explanation which they received for the asking as the multitudes might have, had they not been content to applaud the popular orator without seeking the full truth which the divinely accredited Master was willing to give them. (e) Catholic interpreters are agreed that our Lord did, in fact, continue to the very end trying to do good to the people by instructing them: His failure to speak openly about the mysteries of the Kingdom was part of a general plan not to arouse their false hopes and yet not to turn them away abruptly by directly presenting ideas so opposed to all their Messianic expectations. (f) If He foresaw that they would abuse the greater light given to the disciples, this reticence was a merciful disposition which saved them from a greater sin.

THE PARABLES OF THE GOSPELS

In the final analysis the real difficulty lies in the text of Isaias (*6*, 9 f) which must be interpreted in the light of God's character, the idiom of the Hebrew language and the Hebrew concept of divine causality. Then it becomes clear that any hardening or blinding of men begins in men and that this is the result of the refusal of men to embrace the light and grace which God offers in abundance. It is not only the God of the New Testament who is "rich in mercy" (Eph. *2*, 4), throughout the Old also His mercy is to His justice as a thousand to four (Ex. *20*, 5 f), and His will is the conversion of the sinner to life (Ezech. *18*, 23). It may also well be that the text is in some degree the ironical expression of the disappointed love of God provoked by His knowledge that Isaias' contemporaries were going to reject the clearly divine message of the prophet.

LIST AND CLASSIFICATION

I. *Complete Parables*

(a) Dogmatic

	Matthew	Mark	Luke
1. The Sower	*13*, 3-23	*4*, 3-20	*8*, 4-15
2. The Weeds	*13*, 24-30		
3. The Mustard Seed	*13*, 31-32	*4*, 30-32	*13*, 18-19
4. The Leaven	*13*, 33		*13*, 20-21
5. The Treasure	*13*, 44		
6. The Pearl	*13*, 45-46		
7. The Net	*13*, 47-50		
8. The Seed Growing of Itself		*4*, 26-29	

(b) Moral

	Matthew	Mark	Luke
9. The Two Houses	*7*, 24-27		
10. The Two Debtors			*7*, 41-43
11. The Good Samaritan			*10*, 29-37
12. The Persistent Friend			*11*, 5-8
13. The Rich Fool			*12*, 16-21
14. The Barren Fig Tree			*13*, 6-9
15. The Lost Sheep	*18*, 12-14		*15*, 4-7
16. The Lost Coin			*15*, 8-10
17. The Prodigal Son			*15*, 11-32
18. The Unjust Steward			*16*, 1-13

The Parables of the Gospels

		Matthew	Mark	Luke
19.	The Rich Man and Lazarus			*16*, 19-31
20.	The Godless Judge			*18*, 1-8
21.	The Stubborn Children	*11*, 16-19		*7*, 31-35
22.	The Pharisee and the Publican..			*18*, 9-14
23.	The Unmerciful Servant	*18*, 21-35		

(c) Prophetic

		Matthew	Mark	Luke
24.	The Laborers in the Vineyard..	*20*, 1-16		
25.	The Two Sons	*21*, 28-32		
26.	The Wicked Vine Dressers......	*21*, 33-46	*12*, 1-12	*20*, 9-19
27.	The Marriage Feast	*22*, 1-10		
28.	The Great Supper			*14*, 16-24
29.	The Wedding Garment	*22*, 11-14		
30.	The Ten Virgins	*25*, 1-13		
31.	The Gold Pieces			*19*, 11-27
32.	The Talents	*25*, 14-30		

II. *Condensed Parables*

		Matthew	Mark	Luke
33.	Physician, Cure Thyself........			*4*, 23
34.	The Salt	*5*, 13	*9*, 49	*14*, 34-35
35.	The Lamp on the Lamp-stand..	*5*, 14a-15	*4*, 21	*8*, 16; *11*, 33
36.	The City on a Mountain.......	*5*, 14b		
37.	The Opponent on the Way.....	*5*, 25-26		*12*, 58-59
38.	The Lamp of the Body.........	*6*, 22-23		*11*, 34-36
39.	Serving Two Masters...........	*6*, 24		*16*, 13
40.	Pearls Before Swine............	*7*, 6		
41.	Son Asking His Father	*7*, 9-11		*11*, 11-13
42.	As the Tree, so the Fruit.......	*7*, 16-20; *12*, 33-37		*6*, 43-45
43.	The Clients of the Physician....	*9*, 12-13	*2*, 17	*5*, 31-32
44.	Bridegroom and Wedding Guests	*9*, 14-15	*2*, 18-20	*5*, 33-35
45.	Old and New Garments.........	*9*, 16	*2*, 21	*5*, 36
46.	Old and New Wine-skins.......	*9*, 17	*2*, 22	*5*, 37-38
47.	Old and New Wine............			*5*, 39
48.	Harvest and Laborers..........	*9*, 37-38		*10*, 2
49.	Speaking from the Housetops..	*10*, 26-27	*4*, 22	*8*, 17; *12*, 2-3
50.	Disciple and Teacher..........	*10*, 24-25		*6*, 40
51.	Servant and Master............	*10*, 24-25		
52.	Household and Master.........	*10*, 25		
53.	Divided Kingdom	*12*, 25-28	*3*, 23-29	*11*, 17-23

The Parables of the Gospels

		Matthew	Mark	Luke
54.	Attacks of the Unclean Spirit...	*12*, 43-45		*11*, 24-26
55.	The Prudent Householder	*13*, 52		
56.	True Defilement	*15*, 10-20	*7*, 14-23	
57.	Uprooted Plant	*15*, 13		
58.	Blind Guides of the Blind	*15*, 14		*6*, 39
59.	Children and Dogs	*15*, 26-27	*7*, 27-28	
60.	The Last Place at the Supper...			*14*, 7-11
61.	Building a Tower			*14*, 28-30
62.	Going to War			*14*, 31-33
63.	King's Son Free from Tribute...	*17*, 24-25		
64.	The Watchful Servants		*13*, 34	*12*, 35-40
65.	The Faithful Steward	*24*, 45-51		*12*, 42-48
66.	The Closed Doors			*13*, 25-30
67.	Unprofitable Servant			*17*, 7-10
68.	The Body and the Eagles	*24*, 28		*17*, 37
69.	The Thief	*24*, 43-44		*12*, 39-40
70.	Fig Leaves a Sign of Summer...	*24*, 32-35	*13*, 28-29	*21*, 29-31
71.	Shepherd and Sheep	*26*, 31	*14*, 27	

III. *Gospel of St. John*

(a) Allegories

72. The Good Shepherd *10*, 1-16
73. The Vine and the Branches *15*, 1-11

(b) Condensed Parables

74. The Mysterious Wind *3*, 8
75. The Light of the World *3*, 19-21; *8*, 12; *9*, 5; *12*, 35-36
76. The Living Water *4*, 10-14
77. Sowers and Reapers............ *4*, 37
78. Walking in the Day........... *11*, 9
79. The Grain of Wheat........... *12*, 24-25
80. Washing after a Bath.......... *13*, 10
81. Joy of Motherhood............ *16*, 21

JOHN F. McCONNELL, M.M.

THE LITERARY RELATIONS OF THE FIRST THREE GOSPELS

The Question. To even a cursory reader of our first three Gospels their many resemblances are immediately evident. Nor are these resemblances due merely to that similarity which is natural in several treatments of the same subject. On the contrary, there is often a likeness and even identity of order and detail of language which demand a much fuller explanation than simple sameness of theme. The existence of these resemblances has been recognized since earliest Christian times and, in more recent centuries, the very name given to the Gospels of St. Matthew, St. Mark, and St. Luke expresses a general appreciation of the fact. They are called the Synoptics, because they can be so arranged as to permit their respective accounts of the same Evangelical fact to be taken in with a single glance *(synopsis)*. It is this fact of resemblance, coupled with the no less obvious fact of some striking differences, which raises what is known as the "synoptic question."

A solid tradition almost as old as the books themselves tells us the order of their composition, an order which is still preserved in our modern New Testament. St. Matthew wrote first; St. Mark wrote second; and St. Luke wrote third. This same tradition tells us something else, however, which profoundly affects any conclusions that may ultimately be drawn from this first piece of information. While St. Matthew did write first, the tradition says his Gospel was originally composed in Aramaic, the current language of the Jews. It was not until some years later that it would have been translated into Greek, whether by Matthew himself or by some other inspired writer. By that time, it may be supposed, St. Mark's Greek Gospel had already appeared. Therefore, the order of composition of the Greek Gospels may have been (1) Mark, (2) Matthew, (3) Luke. This sequence is generally accepted today by scholars and must be kept constantly in mind in any treatment of this question of the literary relationship of the Gospels.

Oral Theory. Some scholars have answered the question of literary dependence with a blanket denial. They say, and rightly, that the oral method of instruction was in particular honor among the Jews and that a stereotyped, oral gospel would doubtless have come into use shortly after the beginning of the Apostolic ministry. St. Peter, for example, would surely have been called upon on many occasions to give his eye-witness account of the Master's life, and, like almost all story-tellers, he would in time have settled upon a fixed group of events and sayings for his narrative. Many Palestinian teachers would have followed his "catechesis." Then, when St. Matthew, St. Mark and St. Luke at length had begun their written accounts, they would have used this fixed oral source, each in his own way and independently of the others.

This explanation, however, seems to demand too much of the primitive oral gospel; it demands indeed that such an oral source be fixed not in the language of the Jews (Aramaic) but in the language of the Gospels (Greek), and that, furthermore, it be not the gospel of one Apostle but that of all the Apostles taken collectively. The likelihood that either of these conditions was fulfilled is extremely small. Finally, it is a striking fact that the most important words of Christ—on the Eucharist, for instance— differ considerably in the different Synoptics; yet, the oral theory demands an identity of language on much less important points. As a matter of fact, this theory is not generally held today as a single solution; but it does stress a fact which must be noted, namely, that there was an oral gospel of a certain fixity.

Literary Interdependence. If, then, the use of a common oral source is not sufficient explanation of the problem, the answer must be sought in the dependence of one Gospel on another, or on other documents. There is no other possible solution. But taking for granted the fact of such dependence, how can we determine what is its precise nature? That St. Mark wrote the first Greek Gospel nearly all critics agree. There is a graphic freshness, a vividness and strength which mark it as an original work. It is a Gospel of action, not of long discourses. Like St. Peter preaching at Cæsarea (Acts *10*) it begins with the baptism of John. Though it is the shortest of the three Synoptics, it is rich in detail, something not characteristic of works dependent on other writings. Its theology has a more primitive note than that found in St. Matthew or St. Luke. For instance, St. Mark (*3*, 1-6), recounting one of Christ's tilts with the Pharisees, unqualifiedly attributes anger to our Lord, whereas St. Matthew (*12*, 9-14) and St. Luke (*6*, 6-11), telling the same story, refrain from any such attribution. Again, St. Mark (*6*, 5 f) tells us that Christ *"could not work any miracle"* at Nazareth, while St. Matthew (*13*, 58) much more carefully tells us that "because of their unbelief, he *did* not work many miracles there."

St. Luke on St. Mark. Likewise, all critics (except those few who hold exclusively to a common oral source) are agreed that St. Mark's Gospel is a source of St. Luke's. The resemblances found in all three Synoptics are most striking in the case of these two, and there seems to be no sufficient explanation other than that of literary dependence. Nor are the divergencies such as to make this dependence unlikely. St. Luke does make considerable additions, notably the "great interpolation" (Luke *9*, 51—*18*, 14) and the "lesser interpolation" (Luke *6*, 20 — *8*, 3), together with his Infancy narrative; but these and other smaller bits of new information are usually fitted into the Marcan framework without disturbing its order. Certain transpositions of incidents can be accounted for by didactic reasons or by fuller information. Lastly, his omissions are quite in keeping with what we know of the author and his purpose. Out of his abundant material he would select typical cases of our Lord's activity and avoid repeating similar anecdotes. Furthermore, having in mind an audience different from that of St. Mark, he would be led naturally to the omission of such things as might be suitable for Jewish readers but quite incomprehensible and sometimes offensive to those of Gentile background. An example of this is found in his failure to recount our Lord's long journey to Tyre and Sidon and his return through the Decapolis, a trip over ground unfamiliar to Greeks and one marked by an incident in which Gentiles are compared to dogs. In this same section St. Mark gives an account of the second multiplication of the loaves, easily omitted by St. Luke on the grounds that it provided no new instruction. The resemblances give a solid positive argument which the divergencies do not invalidate.

St. Luke reproduces about three-fourths of St. Mark's Gospel but this does not mean necessarily that he borrowed all of that from his predecessor. He might well have used the oral gospel, and, in any case, his additions prove his use of some other good and abundant source. The precise extent of his dependence on the Gospel of St. Mark cannot be determined. We can only conclude from a comparison of the two Gospels that he was influenced by the order of St. Mark, by his selection of material, by his ideas, and by his wording in those parts where they go together over the Public Life.

St. Matthew on St. Mark. Turning now to the Gospel of St. Matthew, we find that its resemblance to the Gospel of St. Mark, at least in order and wording points clearly to some dependence of one upon the other. The Greek Gospel of St. Matthew contains over nine-tenths of Mark's material, dealt with in the same way and in very nearly the same language. The comparatively primitive quality of St. Mark's writing indicates the priority of his work; and a comparison of the two does nothing to shake that assurance but rather adds to it. To give but one instance: St. Matthew (*8*, 16), after reporting the cure of St. Peter's mother-in-law, tells us that the people waited till evening to bring their sick to Christ. Under the circumstances this waiting is strange, and the fact of St. Matthew's stating it without explanation is also strange. On the hypothesis, however, that he borrowed the incident (at least, the language of it) from St. Mark, the thing is easily understood; for St. Mark's account tells us that it was the Sabbath, a detail which St. Matthew simply neglected to carry over into his Gospel.

It would be a great mistake, however, to think that in the case of every parallel between these two Gospels only dependence of St. Matthew on the work of St. Mark can account for the resemblances. It must be remembered that the Aramaic Gospel of St. Matthew was written even before St. Mark's account. Furthermore, tradition makes it certain, according to the Biblical Commission, that the Aramaic and Greek of St. Matthew are substantially identical. Therefore, we may not hold that the Greek Gospel of St. Matthew incorporated any considerable amount of material which was not already contained in the Aramaic. Then, too, as in the case of St. Luke's Gospel, there is no reason to believe that this Greek Gospel of St. Matthew did not profit from the same sources as did St. Mark's. The literary dependence on St. Mark's Gospel need only be such as to account for the sameness of order, the common Greek wording and formulas of transition, and perhaps a small amount of common subject matter.

St. Luke on St. Matthew. We have considered the Gospels of St. Luke and St. Matthew in connection with that of St. Mark. What now of their relation with each other? Some very good scholars have held that St. Luke borrowed from St. Matthew. For instance, Dom Butler in *The Harvard Theological Review* for October, 1939, devotes a long and very instructive study to this question; and he concludes that in the two hundred places in which St. Matthew and St. Luke are parallel, and not dependent on St. Mark, the resemblance or identity may best be explained by the dependence of St. Luke on St. Matthew. The usual objection to this position is well known; the differences between the Infancy Gospels, the Resurrection narratives, the reporting of the Sermon on the Mount, etc. Father Butler suggests that while St. Luke used St. Matthew's Gospel, he did so only after his own Gospel was nearing completion. The learned Benedictine reminds us that good historical method requires that we avoid creating an hypothetical common source for St. Matthew and St. Luke if we can get along without it.

"Q." Those who think such a source necessary call it "Q" (from the German *Quelle,* "source"). This "Q" is not so objectionable as it used to be when it was thought of as one of the two documents (St. Mark was the other) which accounted for St. Matthew and St. Luke. Many who formerly postulated such a document do not now speak of a Two Source theory; they speak of four or of many sources, realizing that both St. Matthew and St. Luke had abundant information apart from that which came to them from St. Mark and "Q". But even if "Q" is less objectionable than it used to be, there does not appear to be sufficient reason to admit its existence. There is no insuperable objection to the simpler view that St. Luke utilized the Gospel of St. Matthew.

WENDELL S. REILLY, S.S.

THE HOLY GOSPEL OF JESUS CHRIST ACCORDING TO ST. MATTHEW

INTRODUCTION

Throughout the Christian ages the Gospel according to St. Matthew has probably been the best known and the best loved of all the four Gospels. When the same words of our Lord are found in this and another Gospel, they are undoubtedly quoted more often, both by preachers and by the faithful, in the form in which they occur in this Gospel. Almost half of our "Sunday Gospels" are selections from this Gospel.

While this popularity may be explained in part by the leading position of the First Gospel at the beginning of the New Testament, still it must be rather the very nature and character of this Gospel that have endeared it in a special way to all Christians. The spirit of St. Matthew's Gospel is so typically Palestinian and the Semitic character of its original language is still so clearly seen beneath the veil of its Greek or Latin or even English translation, that the reader naturally feels quite close to the very days when our Savior went about the hills and fields of Galilee doing good.

The Author. Such a Gospel must have been written by one who lived in the same country and during the same time as Christ Himself. Catholic tradition has always been unanimous in ascribing its authorship to the Apostle St. Matthew. Our First Gospel was known, and indeed used as inspired Scripture, by the Fathers of the post-apostolic Church at the end of the first century and the beginning of the second—Clement of Rome, Pseudo-Barnabas, Ignatius of Antioch, Polycarp and Justin. We also have the testimony of the second-century Fathers, Papias, Irenaeus and Pantaenus, as well as that of Clement of Alexandria, Origen and Tertullian of the third century, explicitly ascribing our First Gospel to St. Matthew the Apostle.

Among these St. Papias is the earliest and therefore the most important. Eusebius of Cæsarea, the Church historian of the fourth century, to whom we are indebted for almost all that we know of Papias, has preserved for us a few quotations from this early ecclesiastical writer. From these passages we know that Papias was a disciple of St. John the Apostle and that, about the year 120, he wrote a commentary on "The *logia* of the Lord." Concerning this latter work Papias said, "Matthew composed the *logia* of the Lord in the Hebrew language and each one translated it (into Greek) as well as he was able."

From this statement some critics have concluded that Matthew was indeed the author of a work called "The Sayings of the Lord," but that this contained nothing or very little of the life, miracles, etc., of Christ. According to them Matthew's work on the Sayings of the Lord was later on, by some unknown writer, combined with the life of Christ as recorded in St. Mark's Gospel, to form our First Gospel.

This theory has been condemned by the Pontifical Biblical Commission. And rightly so. For there is not the slightest evidence that the early Church knew of the existence of such a work consisting chiefly, if not solely, of the words or discourses of Christ. All the early Fathers who were acquainted with this statement of Papias understand it as referring to our canonical Gospel according to St. Matthew. Moreover, the word *logia*, as used elsewhere in Greek literature, does not mean merely "sayings" but also "oracles, sacred records," whether of someone's words or deeds. Thus, Papias himself, in describing the origin of the Second Gospel, uses the expression, "the *logia* of the Lord" as synonymous with "what was said and *done* by the Lord." (For his full statement see the Introduction to the Gospel according to St. Mark.) Finally, even if one insist that *logia* means "sayings, discourses," the term would not be inaptly used for our First Gospel, since, as a matter of fact, it is largely made up of the sayings and discourses of Christ. (For numerous examples of the use of the word *logia* as the technical term for "Scripture, Gospel," in the early Church see *Biblica*, VII [1926] pp. 301-310.)

St. Matthew. Of St. Matthew's own life we know very little. His father's name was Alpheus (Mark 2, 14), certainly not to be identified with Alpheus, the father of St. James the Less (Matt. *10*, 3 and parallels). He was probably a native or at least a resident of Capharnaum, for he held the position of publican or tax-gatherer in that town. Called from the tollbooth to the apostolate by our Lord's simple command, "Follow me," Matthew celebrated his call with a feast at which many "sinners and publicans," his former friends, were present. Christ's presence at this banquet gave occasion to the accusation of the Pharisees that he ate and drank with sinners and publicans. As a tax-collector, a minor public official whose office called for some knowledge of writing, he may have had more literary culture than any other of the Twelve. He was thus well fitted for the task of drawing up the first written account of the Master's life and teachings.

A comparison of Matthew's own account of his call (*9*, 9-13) with the parallel accounts in Mark *2*, 14-17 and Luke *5*, 27-32 shows that he also bore the name of Levi. It was quite usual for the Jews of this time to have two or more names. Possibly Mark and Luke wished to conceal the identity of the chosen publican by calling him here by his less known name. In all the lists of the Apostles, except his own, he is simply "Matthew." In his own list he is "Matthew the publican"

ST. MATTHEW   INTRODUCTION

(*10, 3*). In all three lists of the Synoptic Gospels his name is coupled with that of Thomas; in Acts *1, 13* with that of Bartholomew. His name in its original form of *Matthai* probably means "gift of the Lord." Some, however, explain it as equivalent to the Hebrew name *Amathi,* "faithful."

That is all that we know of him from the New Testament. Later traditions concerning his apostolic labors and his death are uncertain and in part contradictory.

Original Language. Matthew wrote his Gospel in "Hebrew," according to the constant tradition of the Church. The testimony of Papias to this fact is given above. St. Irenaeus states, "Matthew published the writing of his Gospel among the Hebrews in their own language." All the later Fathers agree on this. It is possible, but not very probable, that this "Hebrew" was the Hebrew of the Old Testament or the Rabbinical neo-Hebrew of the Talmud. It is much more likely that the original language of St. Matthew's Gospel was Palestinian Aramaic, the ordinary language of the Hebrews of that time and therefore not inaptly called "Hebrew." Thus, according to Acts *21,* 40, St. Paul addressed the people of Jerusalem "in Hebrew": but he certainly used Aramaic on this occasion or they would not have understood him, as they did (Acts *22,* 22). Moreover, the few words left untranslated in the Greek text, as *raka* (*5,* 22), *mamona* (*6,* 24), and *korbona* (*27,* 6), are all Aramaic.

The text of this Gospel in its original language is completely lost. Apparently it disappeared at a very early date. Perhaps it was still in existence at the time of Papias. (Cf. his words cited above.) The so-called "Gospel according to the Hebrews" or "the Nazarenes," which a few of the Fathers, as St. Jerome, believed to be the original Aramaic Matthew, was perhaps a badly corrupted form of it. In any case, the few fragments of this strange Gospel that have been preserved in translation, show a text that is quite different from our Greek Matthew.

Greek Translation. Concerning the Greek translator, even St. Jerome confessed his ignorance: "Who later on translated the First Gospel into Greek is not quite certain." We must admit our ignorance on this point. But it is certain that this translation was made before the end of the first century, for several quotations from our Greek Matthew are found in the writings of this period—the Didache, Pseudo-Barnabas, Clement of Rome. Judged by the standard of the classics, the Greek of the First Gospel is quite good, better in fact than the original Greek of Mark or John. But that does not prove that our Greek Matthew is not a translation. For the translator need not have followed the original too slavishly, even though we hold, according to the decision of the Biblical Commission, that his translation is "identical in substance" with the original Gospel of St. Matthew. Moreover, even in the Greek the sentence structure and rhythm are typically Semitic. The great

INTRODUCTION — ST. MATTHEW

similarity between the Greek of the First Gospel and the Greek of the other Synoptic Gospels may be explained by supposing that in his work the translator followed a Greek oral catechesis parallel to the Aramaic oral catechesis used by Matthew and similar to those followed by Mark and Luke, or that he used the Gospels of Mark and Luke as his guide, or that this translation is older than these two Gospels and was used by these Evangelists. (See article on The Literary Relations of the First Three Gospels.)

Time of Composition. The tradition of the Church from the time of the earliest Fathers is that our four canonical Gospels were written chronologically in the order in which they appear in the Bible. Now it can be demonstrated that St. Luke wrote his Gospel not later than the year 63. (See the Introduction to the Gospel according to St. Luke.) St. Matthew therefore must have written his Gospel some years before that date. The intrinsic evidence shows that it was certainly written before the capture and destruction of Jerusalem by Titus in 70 A.D. On the other hand we cannot date it too soon after the Resurrection, for the author presupposes that the Church is fairly well established, and such expressions as "even to this day" (27, 8), "even to the present day" (28, 15), point to a considerable time after the Crucifixion. Some scholars have proposed the year 42 for the Aramaic Gospel and 62 for its Greek translation. The statement of Irenaeus, that "Matthew published the writing of his Gospel among the Hebrews in their own language, while Peter and Paul preached the Gospel and founded the Church in Rome," was probably meant to contrast the difference in method and locality between these Apostles rather than give the time when the First Gospel was written. At any rate, "this testimony of St. Irenaeus, the interpretation of which is uncertain and controverted, must not be considered of such authority as to necessitate the rejection of the opinion of those who consider it more in conformity with tradition that the First Gospel was completed even before the arrival of St. Paul at Rome" (Biblical Commission.)

Destination. The teaching of the early Fathers, that St. Matthew wrote his Gospel in Palestine for the people of that country, is amply confirmed by the intrinsic evidence of the book itself. Matthew generally supposes that the morals and customs of the Palestinians as well as the topography of the Holy Land are well known to his readers, as will be evident to any one who compares the First Gospel with the other two Synoptic Gospels. Thus, while Mark describes at length the frequent ablutions of the Jews (7, 3 f) and explains the meaning of any Aramaic word or typically Jewish expression that he may use, as "Corban" (7, 11) and "the Preparation Day" (15, 42), in all the parallel passages of the First Gospel (15, 1 ff; 27, 6.62), such explanations are omitted as unnecessary. In like manner Luke inserts geographical notes on Nazareth (1, 26), Bethlehem (2, 4), the country of the Gerasenes (8, 26),

Arimathea (*23*, 51) and Emmaus (*24*, 13). Matthew, in similar passages, considers such notes as superfluous. He alone uses the expression, "the holy city," as synonymous with Jerusalem (*4*, 5; *27*, 53). Unless the customs and history of Palestine at that period were known to the readers, many things would not have been understood which occur in the Sermon on the Mount (*5*, 22-26.34 f; *6*, 2.5.16), in the parables (*22*, 11 ff; *25*, 1 ff) and in the narratives, as "the flute players" at a funeral (cp. Matt. *9*, 23 with Mark *5*, 38 and Luke *8*, 52), the magnificence of the temple structure (cp. Matt. *24*, 1 with Mark *13*, 1 and Luke *21*, 5), etc. Finally, the special emphasis in this Gospel placed upon the relation of the New Law with the Old, as in the Sermon on the Mount (*5*, 17-48) and the frequent citations from the Old Testament (more than seventy in this Gospel against scarcely fifty in all the other three Gospels together) show clearly that the readers whom the Evangelist had in mind were either Jews or Christians of Jewish extraction, whether in or outside of Palestine. Although St. Matthew undoubtedly hoped that his work would be read by the unconverted Jews, still he must have intended it directly for the Christians converted from Judaism.

Purpose of the Gospel. This special group of readers also determined the scope that the Evangelist had in mind in writing his Gospel. He intended to provide for the religious needs of the Christians in Palestine. The first and most important of these was the defense of the Faith against the attacks and falsehoods of the unbelieving Jews. St. Matthew's Gospel is therefore the first *apologia* of Christianity. The Evangelist fully achieves his purpose by proving that:

1. Jesus of Nazareth is the Christ, the Messias foretold in the Old Testament. The First Gospel mentions explicitly the fulfillment of the prophecies concerning His birth and infancy (*1*, 22 f; *2*, 5 f.15.17 f.23), His precursor (*3*, 3; *11*, 10), His Galilean ministry (*4*, 14 ff), His use of parables (*13*, 14 f.35), His miracles (*8*, 17; *12*, 18 ff), His triumph at Jerusalem (*21*, 4 f.16), His rejection by the Jews (*21*, 42), the flight of the Apostles (*26*, 31) and the blood-money of Judas (*27*, 9 f). Implicit references to Old Testament prophecies are also numerous, e.g., *9*, 36 (cf. Ez. *34*, 5); *11*, 5 (cf. Isa. *35*, 5 f; *61*, 1); *12*, 40 (cf. Jon. *2*, 1); etc. Consequently the Kingdom of Heaven preached by Jesus is the spiritual Kingdom of the Messias promised in the Old Testament. The First Gospel therefore offers numerous descriptions of this Kingdom in the parables of Christ and gives much of the teachings of Christ concerning the spiritual qualities that the members of this Kingdom must have.

2. The New Law, promulgated by Jesus, does not really abolish the Old Law of Moses but fulfills and perfects it; the interpretation of the Law, however, belongs no longer to the Scribes and Pharisees—"the hypocrites," but to Christ. ("You have heard that it was said to the ancients . . . But I say to you . . .". Cf. especially *5*, 17-48 and *23*, 1-36.) Therefore the Evangelist recounts the frequent conflicts between Christ

and these false teachers concerning the observation of the pharisaical traditions, which were based upon their wrong interpretation of the Law.

3. Christ's sufferings and death on the Cross, "to the Jews indeed a stumbling-block" (*1 Cor. 1*, 23), were preordained by God and foretold in the Scriptures (*16*, 21; *17*, 12.22; *26*, 2.22. 42, 54.56), for the Messias had to "give his life as a ransom for many" (*20*, 28). The Jews were indeed to be the first to share in this Redemption (*10*, 5; *15*, 24); but the fact that the majority of the Jews failed to accept Jesus as the Messias was due to the false ideas that the people had of the expected Messias ("They took offense at him;" *13*, 57), and especially to the conceit of their leaders, the Scribes and Pharisees, "blind guides of blind men" (*15*, 14), who spread false rumors concerning Christ's resurrection "even to the present day" (*28*, 15). For rejecting their Savior the unbelieving Jews themselves are rejected by God (*11*, 20-24; *21*, 18 f.28-32), and indeed they called down upon themselves and their children the blood of the Messias (*27*, 25). Therefore it is clear why the Gentiles are to be admitted so freely into the Kingdom without the necessity of observing the Mosaic Law (*8*, 11 f; *28*, 19), because the Kingdom of God is taken away from the Jews and given to a people yielding its fruits (*21*, 43).

4. The Kingdom of Christ is a true society, a Church, having authority to correct abuses even with the power of excommunication (*18*, 17); its government is committed to the Apostles, to whom Christ gave special instructions (*10*); their decisions are ratified in heaven (*18*, 18). This Church is founded upon Simon Peter, "the Rock," to whom alone Christ gave the keys of the Kingdom of Heaven (*16*, 17 ff).

Structure. St. Matthew, of course, does not develop his thesis as outlined above. He presents this teaching mainly as spoken by the Master during His public ministry. But since the scope of the Evangelist is principally apologetic and polemic, he selects that material which best suits his special purpose and therefore does not intend to give a complete life of Christ. In a sense, then, his Gospel is not strict history, although every statement that he makes is strictly historical. Furthermore, in order to present his thesis in an attractive form, he skillfully arranges the words and deeds of the Savior in certain well-balanced groups, often departing thereby from the strictly chronological order, although retaining the broad outlines of the natural sequence of events. Hence the particles and phrases which in themselves are temporal, such as "then," "at that time," "on that day," etc., in this Gospel are often not much more than mere indications of transition from one topic to another.

One reason for this artificial arrangement might also have been the desire to assist the memory of the faithful in learning the Gospel by heart. For in those days of expensive, hand-made books the memorizing of important books played a much more important rôle than it does today. This use of memory-aids is found more or less in all three Synoptic Gospels, where often the only link between various sayings or deeds of Christ is some similar word occurring in two entirely different contexts. The use of different links in different Gospels results in a different order in each Gospel.

Probably to this same purpose of memory-aids is due St. Matthew's fondness for certain definite numbers in his artificial groupings. Thus, he divides all the generations from Abraham to Christ into three groups of fourteen names in each group (*1*, 17), even though he realized that his readers knew that this did not correspond exactly with the facts. So also he groups most of the teachings of Christ into five long discourses, each of which ends with almost the same phrase, "And it came to pass when Jesus had finished these words . . ." (*7*, 28; *11*, 1; *13*, 53; *19*, 1; *26*, 1). These form the Evangelist's own division of his work, and every true schema of his Gospel must take them into consideration.

Between these five great discourses the Evangelist has drawn up four groups of various deeds and sayings of Christ. Some of these groupings too are skillfully devised arrangements. Thus, between the first and the second of these great discourses there are three groups of three miracles each; between each group of miracles there is an intermediate group of two incidents.

COMMENTARY

PRELUDE: THE COMING OF THE SAVIOR *1-2*

This whole section wherein Jesus is shown by his descent and birth to be the Son of David and the Messianic King, is proper to Matthew. Luke has a similar prelude at the beginning of his Gospel (Luke *1-2*). But these accounts in the First and the Third Gospel are entirely independent of each other, although in no way contradictory. Mark begins his Gospel with the preaching of the Baptist, as did the original oral gospel of the Apostles. We do not know where Matthew received his information for the history of our Lord's Infancy. Since Luke's account of Christ's birth and childhood very probably has our Lady as its ultimate source, it may be that Matthew's account of these early events is based upon a tradition whose ultimate source is St. Joseph. (Cf. Matthew's account of the Virgin Birth (*1*, 18-25) which is entirely from the viewpoint of St. Joseph.) In any case, the events here narrated are true history and unbelievers lack all objective ground for dismissing them as legendary.

1, 1-17: **Genealogy of Jesus.** Genealogical records, a compendium of one's family history, have always been highly esteemed by all peoples. This was especially true among the Jews, because these records showed the degree of relationship in marriage, enabled individuals to prove their possible priesthood, and preserved the Messianic hope within the family of David. The sources for the genealogies were to be found in the Old Testament, the public archives, private documents and tradition. Matthew's source of information for his list of names from Abraham to Zorobabel (1-12) was the Old Testament. (Cf. the foot-note reference in the text.) The following names, Abiud to Jacob (13-15), do not occur in the Old Testament; Matthew must have taken them from oral or written tradition. For the period from Abraham to David the Evangelist could find only fourteen names in his source, although these are obviously too few for this long period. This number, however, being thus determined, Matthew gives only fourteen names in each of the next two groups also, although this necessitated the dropping of three names—Ochozias, Joas and Amasias (cf. *4 Kgs. 8,* 24; *11,* 2; *14,* 1)—which should appear between Joram and Ozias. Likewise the fourteen names (there seem to be only thirteen; cf. below) in the third group are far too few for this period of almost six centuries. Luke has twenty-three names for this period. Therefore we are justified in concluding that the Evangelist could not have intended the word, "begot," to have its usual sense here; it seems he meant it rather to signify, "had as a descendant," or "was succeeded by," directly or indirectly, in the royal line. Luke also gives a list of Christ's ancestors *(3,* 23-38). The main differences between these two lists are these: (a) Matthew's list is descending; Luke's, ascending; (b) Luke gives the names from Adam to Abraham; Matthew has no corresponding group; (c) the names from Abraham to David are common to both lists; from David to Salathiel the lists are different; Salathiel and Zorobabel are on both lists; from Zorobabel to Joseph the lists are again different. Since both lists are inspired, there can be no real contradiction in this apparent discrepancy.

Various theories have been proposed to reconcile these two independent genealogies. (a) Some scholars suggest that Matthew aims at giving the genealogy of Joseph and Luke that of Mary. However, this theory is largely abandoned for philological reasons and because it was not customary for Jews to trace their ancestry on the maternal side. Besides, this theory does not explain why the divergent lines should have met for two generations in Salathiel and Zorobabel and then have separated again. (b) According to others, the key is to be sought in the Jewish custom of the "levirate marriage." This theory was first proposed by Julius Africanus (d. after 240 A.D.) in his Epistle to Aristides (in Eusebius *E. H., 1,* 7). In short it is this. According to Deut. *25,* 5-10, when a man died without issue, his brother was to marry the widow and raise up children for his deceased brother. The firstborn son of this "levirate" or brother-in-law marriage was considered the heir and legal son of the deceased brother. Hence, according to Julius Africanus, Jacob and Heli were uterine half-brothers, but Jacob of the line of Solomon according to Matthew was the natural father of Joseph, whereas Heli of the line of Nathan according to Luke was the legal father of Joseph; that is, Heli died without issue and his half-brother married the widow. This theory lacks probability, because there is no proof that the "levirate" law applied to uterine half-brothers, since the purpose of the law was apparently to transmit property in the male line. Moreover, this same dubious process must be invoked again in order to explain the different fathers of Salathiel—a rather remarkable coincidence. (c) According to a third theory, Luke gives the actual ancestors of Joseph, while Matthew gives the royal or dynastic table that lists the true heirs to the throne through the centuries even though one line of dynasty may die out. The line of Solomon would then cease with Jechonias (cf. Jer. *22, 30*), and Salathiel, a descendant of David through Nathan according to Luke, succeeded to the royal rights; Salathiel transmitted these rights to his son Zorobabel and the latter in turn to his son Abiud. The line of Abiud became extinct with Jacob, whereupon Joseph (or one of his ancestors) of the line of Resa, another son of Zorobabel according to Luke, could lay just claim to the throne of David. This is merely a theory, of course, since no direct proof can be adduced to verify it. But it is quite in keeping with the normal human transmission of royal power, and the loose use of the word "begot" both in Matthew and in the Old Testament makes it at least possible. No serious objection can be raised against it.

1. *The book of the origin* is a Hebrew expression meaning "the document showing the genealogy." V.1 is therefore the title not of the whole Gospel or even of the whole first chapter but only of 2-17. Still, in this first verse Matthew presents the thesis of his whole Gospel: that Jesus is the Messias. For "the promises were made to Abraham and to his offspring" (Gal. *3, 16*; cf. Gen. *12, 3; 22, 18*). And all orthodox Jews at the time of our Lord held that the Messias would be a descendant of David (cf. Matt. *22, 41* f and parallels; John *7, 42*) and *Son of David* had become a Messianic title (Matt. *9, 27; 12, 23*).   **3-6.** The only women mentioned in the list are *Thamar, Rahab, Ruth,* and Bethsabee, *the former wife of Urias.* There is a note of humility in recalling among the ancestors of Christ Thamar, Rahab and Bethsabee whose lives were not always exemplary. Rahab and Ruth were not native Israelites; perhaps one reason why the Evangelist mentions them here is to signify the call of the Gentiles, who "will come from the east and from the west and will feast with Abraham and Isaac and Jacob in the kingdom of heaven" (*8, 11*). There is no record in the Old Testament of the marriage of Salmon and Rahab.   **11.** *Josias begot Jechonias* is also the reading of the oldest Greek MSS; but several Greek MSS read: *Josias begot Joakim and Joakim begot Jechonias.*

This latter reading is rejected by all the textual critics as a later correction of some copyist, but it may possibly represent the original reading. According to our text *Jechonias* must be counted twice to get the required fourteen names in each group, or the *Jechonias* of 11 must be considered as standing for "Joakim." **16.** The Evangelist words this sentence very carefully to show that Joseph was only the legal and not the actual father of Jesus.

1, 18-25: The Virgin Birth. **18.** *The origin of Christ:* His conception and birth. *Betrothed:* much more than our "engaged" but less than "married." The Jewish marriage ceremony consisted of two parts: the first was the sealing of the marriage contract whereby the bridegroom gave a certain sum of money, "the purchase price," to the father of the bride, and the bride received her dowry, usually equal to "the purchase price," from her father; the second ceremony, separated by several months (usually a year) from the first, was the solemn, formal induction of the bride into the bridegroom's house, the blessing of fruitfulness invoked upon the consummation of their union, and the joyful wedding feast. Between the two ceremonies the bride was said to be *betrothed*. But since the first ceremony effected a valid, though unconsummated, marriage, even before the second ceremony the bride and bridegroom were spoken of as *husband* and *wife* (19 f), and any unfaithfulness on the part of the bride during this period was considered adultery and punishable with death (cf. Deut. 22, 23 f). *Before they came together:* before the second ceremony had taken place. *She was found to be with child:* Joseph learned of her pregnancy either by his own observation or by being informed of the fact through Mary or one of her relatives; the peculiar passive construction used by the Evangelist would favor the latter opinion. The words *by the Holy Spirit* were added by the Evangelist to forestall any wrong ideas on the part of the readers: Joseph himself apparently did not know of the supernatural character of the conception until the mystery was revealed to him by the angel. **19.** *A just man:* one who conscientiously observed the Law; hence this does not give the reason why *he did not wish to expose her to reproach;* but his desire to save her from public shame and punishment flowed from his conviction of her innocence. Therefore he *was minded,* made up his mind, decided, to separate from her legally but quietly, by giving her a bill of divorce (cf. Deut. 24, 1) before two witnesses in private without stating the motive. **20.** *Joseph, son of David:* the angel thus addresses him to recall to him his dignity and to signify that through him the son of his virginal wife and therefore his son before the law, would also be a son of David. *To take one's wife* to oneself was the technical term for the performance of the second part of the marriage ceremony (cf. Deut. *20, 7*). **21.** *He shall save his people from their sins:* the angel alludes to the meaning of the name *Jesus,* "The Lord is salvation." **22.** The citation of a prophecy fulfilled is one of the characteristics of the First Gospel (see Introduction), and serves to show the intimate connection between the Old and the New Testament. St. Matthew wishes to demonstrate that the facts which he narrates have their cause in the free will of God who, disposing the events according to a pre-established plan, revealed at times to the prophets, thus brought them into actuality. The coincidence, then, between the prophecy which announces the fact and the fulfillment of the same is not by chance but depends on the providential disposition of God. **23.** The prophecy referred to is that of Isa. 7, 14, pronounced at a time of calamity for Juda when Achaz, the head of the House of David, refused to ask God for a sign. The Greek translator of the Aramaic Gospel of St. Matthew does not follow the Septuagint exactly in his version of these words of Isaias, yet both independently render the Hebrew word *almah* as *parthenos, virgin* (in the strict sense). *And they shall call:* in the Septuagint: "And thou shalt call"; in the Hebrew (according to the Massoretic Text): "And she shall call." But the meaning of all three variants is substantially the same. *Emmanuel; which is, interpreted, "God with us."* The Gospel understands the name as meaning not merely, "God is with us by His aid," but "God is with us personally by His Incarnation." **25.** On the meaning of this verse see the note to the text. In Luke *1,* 31 the angel Gabriel tells Mary to call her Son's name *Jesus;* here Joseph *called his name Jesus:* the Old Testament shows that both the father and the mother had the privilege of naming the child.

2, 1-12: The Magi. **1.** *Bethlehem of Judea,* or *Bethlehem of the land of Juda* (6): to distinguish it from Bethlehem of the tribe of Zabulon in Galilee (cf. Jos. *19,* 15). It was about five miles south of Jerusalem and famous as the birthplace of King David. On *King Herod* (the Great) see chapter on The New

Testament Background. *Magi* were Persian pseudo-scientists, devoted especially to astrology and medicine. They were the lineal descendants of the Babylonian astrologists and soothsayers; hence they came *from the East to Jerusalem,* that is, from the Parthian empire. Within the Roman empire the name was used for a less reputable class of men who were skilled in magic, such as Simon Magus of Samaria (cf. Acts *8,* 9 ff) and Elymas the magician of Cyprus (cf. Acts *13,* 6 ff). The Magi of Matt. *2* were not kings; this false notion originated from the liturgical use of Ps. *71,* 10 in the applied sense on the Feast of the Epiphany. The common idea of making them three in number probably arose from the consideration of the three types of gifts that they offered (11). Their names, "Gaspar, Melchior and Baltassar," date from the Middle Ages.   **2.** *In the East* probably means merely "while we were in the East"; it is less likely that it means to specify the position of the star when they saw it. We do not know why they concluded from the appearance of the star that a *king of the Jews* was just born. There are hundreds of Babylonian astrological tablets which pretend to foretell events from celestial phenomena, but no known cuneiform tablet contains any forecast like this. If the Magi understood from the beginning that the newly born *King of the Jews* was the Messias, they must have acquired this knowledge from the Jews of the Dispersion. The prophecy of Balaam (Num. *24,* 17) probably had nothing to do with it. The Gospel does not say that the star was visible to them all the time and guided them on their way; from 9 we would conclude that the opposite was the case.   **4.** *All the chief priests and the Scribes of the people:* a meeting of the Sanhedrin, but probably not an official, plenary session.   **6.** This quotation from Mich. *5,* 2 is cited somewhat freely according to the sense rather than strictly according to the wording of the Prophet, but the Scribes understood it correctly as foretelling that Bethlehem would be the birthplace of the Messias.

7. It would seem from 16 that the Magi told Herod that they had first seen the star two years before their arrival in Jerusalem. God, however, might have revealed the star to them some time before the birth of Christ.   **8.** *That I too may go and worship him:* obviously a lie, for the context shows Herod's real intention.   **9.** *The star . . . went before them:* either, "anticipated their arrival," or, as seems more natural, "preceded them along the way." That the star could point out an individual house shows that it must have been some luminous object very near the earth and therefore entirely miraculous. **11.** *The house:* hardly the stable where our Lord was born but a regular dwelling-place. Joseph probably intended to make Bethlehem his permanent home (cf. 22). *They worshipped him:* they bowed down in homage before Him. This verb in itself does not necessarily signify divine adoration; but it is evident that God must have revealed to the Magi something of the nature of the Child whom they worshipped or they would not have acknowledged Him even as king of the Jews in such lowly surroundings. *Gifts of gold, frankincense and myrrh:* evidently the most valuable exports of their native land which they thought any foreign king would be glad to receive. The Fathers of the Church have interpreted them mystically to typify Christ's Kingship, Divinity and mortal Humanity.

2, 13-15: The Flight into Egypt. **13 f.** The Jews had numerous colonies in northern Egypt, especially at Alexandria, where Joseph would be welcomed and where he could find a livelihood. Since this part of Egypt bordered on Palestine, many Jews found refuge there beyond the jurisdiction of Herod. **15.** The fulfillment of the prophecy of Osee *(11,* 1), interpreted in a typical sense. The Evangelist sees in the history of the people of Israel a figure of the life of Christ (cf. *1* Cor. *10,* 1-11), both having sojourned in Egypt, and therefore he takes the Exodus of Israel, the adopted son of God, as a prophetic type of the return of Jesus, the true Son of God.

2, 16-18: The Innocents. **16.** Bethlehem and its outlying farms probably had about two thousand inhabitants at that time; therefore the number of boys slain by Herod would have been about twenty. **17 f.** The words of Jeremias *(31,* 15) are interpreted by the Evangelist in the typical sense. At Rama, a village about five miles north of Jerusalem, within the ancient borders of the tribe of Benjamin, the Babylonians first assembled the captives from Jerusalem (cf. Jer. *40,* 1). The prophet poetically represents Rachel, the mother of Joseph and Benjamin, mourning for her descendants who are to be taken away into exile. St. Matthew sees in this a figure of the mothers of Bethlehem weeping for their slain children. Perhaps he also has in mind the ancient tradition which identifies Ephrata, the burial place of Rachel, with Bethlehem (cf. Gen. *35,* 19).

2, 19-23: The Return to Nazareth. **19.** The text seems to imply that the *angel appeared to Joseph* shortly after the death of Herod. Herod died in 4 B.C. Christ was born about 8 B.C. (See Commentary on Luke 2, 1 f) If our Lord was about one year old when the Magi came to Bethlehem (cp. 7 with 16), then the holy Family stayed about three years in Egypt. **20.** *Those . . . are dead:* the plural signifies perhaps Herod and his accomplices. **22.** On *Archelaus* see The New Testament Background. **23.** The prophecy to which Matthew refers cannot be identified with certainty. Most commentators follow St. Jerome who thinks that the Evangelist is alluding to Isa. *11, 1*, where the Prophet in a passage that is certainly Messianic says, "And there shall come forth a rod out of the root of Jesse, and a flower shall rise up out of his root." Now the word which is here translated as "flower" is in Hebrew *neser,* identical with the root of the words *Nazareth, Nazarene.* But since Matthew speaks in general of the *prophets,* other commentators think that the reference is to all those passages of the Prophets (as Isa. *53, 2* ff) where the Messias is spoken of as lowly and despised; for Nazareth was in fact a lowly and despised village (cf. John *1*, 46).

I. THE PUBLIC MINISTRY OF JESUS 3-25

1. THE PREPARATION 3, 1 — 4, 11

3, 1-12: John the Baptist. The life of Christ as first told to the people by the Apostles began with the preaching of John the Baptist (cf. Acts *1*, 21 f; *10*, 37; *13*, 24). This oral catechesis of the Apostles forms the basis of the three Synoptic Gospels. From here on therefore Matthew usually has parallel passages in Mark or Luke or both. The mission of John was twofold: (a) to reform the moral life of the Jews, so that they would be spiritually fit to receive the Messias (cf. Luke *1*, 16 f.76 f); this he did by the preaching of repentance, accompanied by the symbolic rite of baptism or bodily washing; (b) to point out the Messias to the people (cf. John *1*, 31). **1-6.** This description of the appearance and general activity of the Baptist parallels Mark *1*, 1-6 and Luke *3*, 1-6. **1.** *In those days:* This vague indication of time is typical of Matthew; Luke *3*, 1 f gives the exact year. (See Commentary there.) *The Baptist:* this word, taken over from the Greek, means, "The Baptizer." *The desert of Judea:* the region northwest of the Dead Sea. *Desert* in Scripture usually signifies not a barren sand waste but a region unfit for agriculture although used for pasturage, especially after the winter rains.   **2.** *Repent* and "repentance" are important words in the preaching both of the Baptist and of Christ; they signify not regret for the past or the performance of "penance" but rather a change of mind and heart, a new outlook on life in keeping with the will of God. *The kingdom of heaven* is almost always used in Matthew for Mark's and Luke's "kingdom of God." Both expressions are therefore synonymous although either of them may have slightly different nuances according to the various contexts, such as the reign of God in the hearts of men, the Messianic kingdom, the visible society established on earth by the Messias, God's reign with the angels and saints in heaven, etc. Here the sense is "the establishing of the Messianic kingdom, the coming of the Messias."   **3.** These words from Isa. *40, 3* refer to the custom of having a herald precede a king when the latter is on a journey, to forewarn the inhabitants of his arrival so that they can repair their ill-kept roads. The king in Isaias is the Lord leading back the Jewish captives across the desert from Babylon. To Matthew the king is Christ and His herald is the Baptist.   **4.** The garb of John recalls the appearance of Elias (cf. *4* Kgs. *1,* 8). Certain species of *locusts* could be eaten legally (cf. Lev. *11,* 21 f) and today the Bedawin still relish them. The *wild honey* would be either true honey from wild bees or the sweet sap of certain shrubs or trees which was also known by this name. It was a sign of great mortification and trust in God's providence to subsist on such fortuitous food.   **7-10.** This sample of John's preaching of repentance is parallel to Luke *3, 7-9.*

7. On the *Pharisees and Sadducees* see The New Testament Background. *Brood of vipers:* the same invective against the same class of men is used by Christ in *12,* 34; *23, 33.* Worthy offspring of their father, the devil (cf. John *8,* 44), "the ancient serpent, who is called the devil and Satan, who leads astray the whole world" (Apoc. *12,* 9). These hypocrites also, like a *brood of vipers,* deceive men to kill their souls with the poison of their false morality. *The wrath to come:* eternal damnation (cf. *23, 33*), God's judgment on the last day, or its foretype, the slaughter at the destruction of Jerusalem.   **8.** *Fruit befitting repentance:* conduct in keeping with pretended conversion.   **9.** Cf. John *8,* 33.39 for this boasting of the Jews in their descent from Abraham. It is only spiritual descent from Abraham, achieved by grace, that counts in God's eyes (cf. Rom. *9,* 6 ff).   **10.**

Christ also employs the figure of the worthless tree that is cut down and burned up, to signify God's rejection and punishment of the wicked (cf. 7, 19). **11 f.** This sample of John's preaching about the coming Messias parallels Mark *1, 7* f and Luke *3, 16* f. Cf. also John *1, 26* f.33. These words of the Baptist were also known to St. Paul (cf. Acts *13, 25*). **11.** *Baptize with water:* literally, dip into water, immerse. John's baptism was not a sacrament but a mere symbol giving external expression to the *repentance* of his converts. This rite was merely preparatory: Christ on the other hand gives the spiritual reality, the *baptizing* or immersion of His faithful in the *fire* of the *Holy Spirit*. **12.** *Winnowing fan:* the shovel with which the threshed grain is thrown into the air so that the wind blows aside the lighter chaff while the heavier kernels fall back to the threshing floor. Christ thus separates the good from the wicked not only on the Last Day but also during life according as men accept or reject His gospel (cf. Luke *2, 34* f; John *9, 39*).

3, 13-17: The Baptism of Jesus. Parallels in Mark *1*, 9-11 and Luke *3*, 21 f; cf. also John *1*, 32-34. **13.** *Jesus came to be baptized* primarily that John might thereby recognize Him and make Him known to Israel (cf. John *1*, 31 ff). Other reasons are also given why Christ wished to be baptized: "that He might cleanse the waters and bestow upon them the power of sanctifying" (St. Thomas, P. 3, q. 39, a. 1); that He who took upon Himself the sins of the world might give us an example of humility and repentance. We might also consider His baptism as a solemn inauguration of His public ministry, but not in the sense of the rationalists who falsely interpret this scene as the awakening of the Messianic consciousness in Christ. This latter opinion has not the slightest foundation in the Scriptures and is a denial of the divinity of Christ. **14.** Even before John baptized Jesus he recognized Him at least as a holy person. For the harmonization of this verse in Matthew with the Baptist's statement in the Fourth Gospel, "And I did not know him," see Commentary on John *1*, 31. **15.** *To fulfill all justice:* to carry out the will of the heavenly Father according to which Christ's baptism was a part of the divine plan of Redemption. **16 f.** Note that this manifestation from heaven takes place not during the baptism of Jesus but only after He had left the water. The Holy Spirit comes down *as a dove* upon Christ, for the dove is the traditional symbol of peace and love. The dove and the voice from heaven were perceived by the Baptist (cf. John *1*, 32-34), and probably also by the bystanders, for this manifestation was for our benefit, not for Christ's. The soul of Jesus possessed the plenitude of grace from His conception as man. It would therefore be wrong to suppose that on this occasion Jesus received an increase of grace or that He was then chosen for His mission. A special reason for this manifestation is the intimate connection between the Blessed Trinity and Christian baptism (cf. Matt. *28*, 19). **17.** Cf. the words of the heavenly Father at the Transfiguration (*17*, 5).

4, 1-11: The Temptation. Since "it was right that he should in all things be made like unto his brethren" (Heb. *2*, 17), Christ wished to prepare Himself in a human way for His public ministry by prayer, fasting and the overcoming of temptation. Various reasons are suggested why our Lord should have allowed the devil to tempt Him. (a) That just as the first Adam brought sin upon us all by succumbing to the temptations of Satan, so Christ, the Second Adam (Rom. *5*, 14), should redeem us all from sin by overcoming the temptations of Satan (St. Ambrose). (b) That He might teach us by His example how to resist temptation (St. Augustine). (c) "For in that he himself has suffered and has been tempted, he is able to help those who are tempted" (Heb. *2*, 18). "For we have not a high priest who cannot have compassion on our infirmities, but one tried as we are in all things except sin" (Heb. *4*, 15). These last words also show the nature of Christ's temptations: He was tempted in all things as we are but without sin; therefore (a) He did not consent in the least to these temptations, for that would be sinful; (b) His temptations did not arise from a sinful nature, as happens with us, but solely from outside Himself, from Satan. The devil tempted Christ principally for two reasons: (a) to discover whether He was the Messias, for he had observed the scene at the baptism of Jesus (Satan would not have tempted Him at all if he had known that He was the Son of God in the strict sense of the term); (b) to check the power of this holy Man, whether He were the Messias or not, at the very outset by making Him his slave through sin.
1. A rather late tradition identifies the scene of Christ's temptation with the desolate stretch of mountains a few miles west of Jericho, now known as Mount Quarantal, that is, the Mount of the Forty Days (Fast). **2.** The sacred number *forty* is of

frequent occurrence in the Bible. Moses (Ex. *34,* 28) and Elias
(*3* Kgs. *19,* 8) also fasted for forty days and forty nights. The
nights are mentioned as well as the days to show that it was an
absolute fast (cf. Luke *4,* 2). It does not necessarily follow from
this verse that our Lord suffered no pangs of hunger during His
fast; the Evangelist emphasizes His hunger *after fasting forty days*
to show the nature of the first temptation.   3. The first
temptation was not primarily one of gluttony, for Christ, having
completed the prescribed period of fasting, was entirely at liberty
to eat; it was rather a temptation to lose confidence in God's
providence. It was not the Father's will that Jesus should work
miracles merely to relieve His own necessities. *The Son of God:*
this term was used by Satan not in the theological sense but
merely as a synonym for "the Messias." Satan re-echoes the
words that he heard at the baptism of Jesus.   4. Our Lord
rejects each of the three temptations in such a manner that He
does not satisfy the curiosity of the devil concerning His Messiasship. To each suggestion He replies by quoting Sacred Scripture. It is rather remarkable that all three quotations are from
the same Book, Deuteronomy, and indeed all from the same
section of that Book. The words quoted in this verse, from
Deut. *8,* 3, were spoken by Moses in reference to God's gift of
the miraculous manna in the desert. The sense intended by
Christ therefore is, "If God so wills, He can sustain my life
miraculously without food." These words may also be understood in an applied sense as "The food of man's true (spiritual)
life is the Word of God" (cf. John *4,* 34).

5. *The holy city:* Matthew alone of the Evangelists uses this
term as a synonym for Jerusalem. The native Arabic name of
modern Jerusalem is likewise *El Quds,* "The Holy." *The pinnacle of the temple* was not a steeple on the house of God but a
wing-like projection of one of the porches which surrounded
the court of the temple; here is probably meant the southeast
corner at the juncture of the Royal Porch and the Porch of
Solomon (cf. John *10,* 23), where there is a sheer drop of several
hundred feet to the valley of the Cedron below.   6. *Throw
thyself down:* a temptation to presume on God's providence just
as the first had been a temptation to distrust divine providence.
It was also a temptation to vain ostentation and to a manner of
manifesting oneself as the Messias such as God had not intended.
Many of the Jews indeed thought, as we know from the Talmud,
that the Messias would manifest himself in this way. Since Jesus

had quoted Scripture to refute the first temptation, so also the devil now abuses Scripture in support of his second temptation. The words which Satan cites are from Ps. *90*, 11 f. The Psalmist indeed assures the just man of God's special protection but certainly not if he should rashly and without reason expose himself to danger. **7.** *Thou shalt not tempt the Lord thy God:* the sense of these words from Deut. *6*, 16 is not, "Satan, thou shalt not tempt me," but, "No one should try God or put Him to a test by foolishly asking for a miracle." The Israelites at the time of Moses had thus "tempted the Lord" by demanding water in the desert as a proof that the Lord was with them (cf. Ex. *17*, 2-7).

8. *A very high mountain:* we do not know what mountain this was; it would be impossible, of course, to see *all the kingdoms of the world* at one time from any mountain on earth. It would seem therefore that even though we hold, as the more natural sense of the text, that Satan really appeared in bodily form and that the temptations were really presented externally and not merely internally to the phantasy of Christ, still in this case it must have been only to the eye of Christ's mind that *the devil showed him all the kingdoms of the world and the glory of them.* **9.** This temptation is the climax of Satan's craftiness and audacity. Instead of saying as before, "If thou art the Son of God," Satan now cleverly assumes that this is the Messias in order to trick Jesus into confirming this assumption and thereby revealing His dignity. The temptation consisted in offering the political dominion of the whole world, which was the best Satan could even pretend to offer, to the Messias to whom God had promised primarily the spiritual lordship of the whole world (cf. Pss. *2*, 8; *71*, 8-11); and the price that Satan demanded for this bargain was the abominable sin of devil-worship. **10.** Jesus hides from the devil the fact that He is the Messias by again merely quoting from the Book of Deuteronomy (*6*, 13) the divine prohibition of all forms of idolatry. But because the devil had now played his trump and been beaten, Christ banishes him with the command, *Begone, Satan!*

2. THE INAUGURATION OF THE MINISTRY IN GALILEE *4,* 12-25

4, 12-17: Jesus in Capharnaum. **12.** This verse does not give the reason for Christ's withdrawal from Judea into Galilee, for if fear of Herod who had just arrested the Baptist had been Christ's motive, He would certainly not have left Judea where Herod had no authority, in order to return to Herod's own tetrarchy of Galilee. Matthew here merely points out the time of Christ's departure, as Mark *1*, 14a more clearly states. John *4*, 1-3 gives the immediate reason why he withdrew from Judea into Galilee: the incipient hostility of the Pharisees, which would increase now that the Baptist was out of the way. Luke *4*, 14a gives the fundamental motive: the inspiration of the Holy Spirit, for it was by divine plan that most of the ministry of Jesus should be in Galilee. **13.** *And leaving the town of Nazareth:* these words probably refer to Christ's first preaching at Nazareth as told in Luke *4*, 16 ff. But it is doubtful whether John *4*, 44 refers to this event. (See Commentary on this v.) *He came and dwelt in Capharnaum* (parallels Mark *1*, 21a and Luke *4*, 31a): Christ chose this town for the headquarters of His Galilean ministry because its situation was ideal for this purpose. Its name, *Kaphar-nahum,* means "Village of Nahum" (as a personal name) or "Village of Consolation." It is now almost universally identified with the mound of ruins known as *Tell Hûm.* The Franciscan Fathers made excavations here and partly restored its synagogue, which had probably been built a century or two after Christ but undoubtedly on the same site where the previous one had been. *Which is by the sea,* that is, of Galilee; also known as the Sea of Tiberias and the Lake of Genesareth. The territory of *Zabulon and Nephthalim* corresponds with Lower and Upper Galilee; Capharnaum was about on the border between these two regions. **14-16.** Some consider this prophecy of Isaias (*9,* 1 f) as spoken solely and directly of Christ's preaching in Galilee; others consider these words as referring directly to the deliverance of this territory from the oppression of the Assyrians, which in turn was a type of the spiritual deliverance which Christ brought here. The first part of this prophecy is quoted somewhat freely by St. Matthew. **17.** Christ begins His preaching at Capharnaum (referred to in a general way in Mark *1*, 21b and Luke *4*, 31b) by repeating the words of the Baptist (Matt. *3*, 2; see Commentary there). Similar words of Christ are recorded in Mark *1*, 14b-15. From this, one should not conclude that Jesus derived His teaching from that of John. He was rather following a sound pedagogical principle of continuing an instruction where it had been left off.

4, 18-22: **The First Disciples Called.** Mark *1*, 16-20 narrates this event in almost exactly the same words. But Luke *5*, 1-11 is so different that its harmonization with Matthew and Mark is disputed. See Commentary there. John *1,* 35-51 tells of Christ's meeting with these first disciples on a previous occasion. As narrated in John *1* they were called in a general way to be the disciples of Christ; in the narrative of the Synoptic Gospels they are called to be His intimate followers who will accompany Him on all the journeys of His ministry.   18. *Casting a net into the sea:* in the Greek text this net is called by a specific name, the "casting-net." This consisted of a circular piece of lightweight mesh around the ends of which were attached small weights. The fisherman gathered all the mesh carefully in his hand and standing on the shore or in a boat over shallow water, would hurl the net in such a manner that the weights spread out over the spot where he saw fish. A different Greek word (the "drag-net") is used in the parable of The Net (Matt. *13,* 47-50); this net was of an entirely different type and was probably also the type of net referred to in Luke *5,* 4-6 and John *21,* 6.

4, 23-25: **Mission of Preaching and Miracles.** These verses give a general summary of Christ's ministry in Galilee and serve at the same time as an introduction for the Sermon on the Mount (*5-7*). A similar general summary of the Galilean ministry after the first miracles at Capharnaum is given in Mark *1,* 39 and Luke *4,* 44. On the *synagogues* see under this word in the Glossary appended to the text, p. 758f.   24 f. Parallel accounts of miraculous cures in general and of the large crowds that gathered around Jesus are given in Mark *3,* 7-12 and Luke *6,* 17-19; cf. also Matt. *12,* 15-21. *Decapolis:* see Glossary at end of text, p. 747.

3. Second Period of the Ministry in Galilee and Across its Lake *5,* 1 — *15,* 20

St. Matthew begins this section of our Lord's ministry in Galilee with his "Sermon on the Mount" (*5-7*), one of the longest and undoubtedly the most famous of the Master's discourses. While it is generally admitted that St. Matthew has arranged his material in groups which do not always correspond with the actual chronology of Christ's life (see Introduction) still it would be rash to assert that this whole discourse is an entirely artificial arrangement of various short sermons delivered on various occasions. It seems quite certain that on this occasion (probably in the spring or summer of the second year of the Ministry) our

Lord did preach an outstanding sermon at least substantially the same as that given by St. Matthew. For the Sermon on the Mount as recorded in the First Gospel possesses on the whole too clear a unity and development of thought to be a mere artificial collection of various sayings of Christ. Moreover, St. Luke independently of St. Matthew (for the third Evangelist has too fine an artistic sense not to appreciate the beauty of Matthew's Sermon on the Mount had he known it), has recorded a sermon of Christ that is substantially the same as Matthew's even though it is much shorter (Luke 6, 20-49). Therefore we conclude that everything that is common to both these Sermons was certainly spoken on this occasion. Likewise, there is no reason for placing elsewhere the passages which occur only in Matthew's Sermon. The difficulty however consists in those passages of Matthew's Sermon on the Mount which occur in various parts of the Third Gospel other than Luke 6, 20-49. Undoubtedly Christ often repeated His doctrine. But it would be rather strange that He should have done so in exactly the same words on different occasions. Besides, some of these passages, as the "Our Father," fit in much better in the context of the Third Gospel than they do in the great Sermon of the First Gospel. In many of these cases we cannot know for certain whether the order of Matthew or of Luke or of both is correct chronologically. Sometimes the same words have a somewhat different meaning as they occur in a different context. This is especially the case with those few passages of the Sermon on the Mount which are also found in St. Mark's Gospel, sometimes in an entirely different context. The general theme of this great Sermon is that the morality of the Kingdom of Heaven is something interior, spiritual, as contrasted with the more external justice of the Old Law and the mere hypocrisy of the Pharisees. 1. We do not know what exact site is meant by *the mountain*, but because the article is used in this Greek expression, the reference seems to be to the hills in general around the lake as opposed to the level stretch immediately bordering on the lake. The oldest tradition (from the Byzantine period) localized this sermon on a hillside near the shore of the Lake of Galilee not far from the modern Ain Tabga, about one mile south of Tell Hûm (Capharnaum).

5, 3-12: The Beatitudes. If the Sermon on the Mount may be called the Constitution of the Kingdom of Heaven, the

Beatitudes form its worthy preamble. The eight (really nine) Beatitudes of Matthew differ not only in number and arrangement but also largely in form and substance from the four Beatitudes and the four Woes of Luke *6,* 20b-26. In the First Gospel the Beatitudes are concerned primarily with the interior dispositions of the members of the Kingdom, whereas in the Third Gospel the external circumstances of these members are stressed. It is simpler, therefore, to consider the Beatitudes in Luke not as identical with but as supplementary to those in Matthew. Thus the first Beatitude in Matthew need not be interpreted as referring to the same virtue as the first Beatitude in Luke; the same holds true of the fourth Beatitude in Matthew as compared with the second in Luke.

3. *Blessed:* in Greek literally "happy, fortunate, enviable," the equivalent of the Hebrew congratulatory phrase, "O happiness of those who . . . ." *Poor in spirit:* literally, "poor of spirit," since the Greek phrase here is of the same type as the "pure of heart" of the sixth Beatitude. Just as the latter phrase means "the pure-hearted," so the former phrase means "the poor-spirited," that is, those who admit that they are spiritually poor, who are not self-conceited, the humble; the opposite are "the rich-spirited," i.e., the Pharisees against whose spirit and teaching the Sermon on the Mount is primarily addressed; cf. Luke *1,* 51.53. The first Beatitude therefore strikes the keynote of the whole Sermon. However, since Luke's first Beatitude, "Blessed are the poor," certainly refers to those who are in want of the goods of this world, Matthew's first Beatitude is more commonly understood of those who are detached from material wealth.

4. In the Greek text the order of the second and the third Beatitude is inverted. The wording of this Beatitude is based upon Ps. *36,* 11, "The meek shall inherit the land." In the Psalm "the land" is the land of Canaan; Christ takes this as a type of the Kingdom promised to His disciples.   **6.** *They who hunger and thirst for justice* are they who ardently long for spiritual perfection. This Beatitude need not have the same meaning as Luke's "Blessed are you who hunger now," which refers to those who are in want of material food.   **7.** By *the merciful* are not directly meant they who give alms but rather they who forgive injuries. Christ often insists on this disposition in His disciples (cf. Matt. *6,* 14 f; *18,* 21-35; Mark *11,* 25 f; Luke *17,* 4; cf. also Jas. *2,* 13).   **8.** *The pure of heart* are they whose mind (according to the Hebrew usage the heart is considered the organ of the reason and the will) is free from duplicity; *they shall see God* because the eye of their mind is clear (cf. Matt. *6,* 22 f). **9.** *The peacemakers shall be called the children of God* because He is "the God of peace *(1* Thess. *5,* 23). **12.** *Your reward,* literally, "your wages," that is, the recompense due to you in justice. This verse is rightly used by theologians as a proof for the Catholic doctrine of merit and reward hereafter. Our Lord speaks at length elsewhere of the persecutions that the prophets of the Old Law endured as a type of the persecutions which His own disciples would have to suffer (cf. Matt. *23,* 29-37; Luke *11,* 47-51; cf. also Jas. *5,* 10 f).

5, 13-16: The Disciples Compared to Salt and Light. These verses serve as a transition from the introductory Beatitudes to the body of the Sermon. Having just spoken of the persecutions that His disciples were to suffer, Christ now encourages them by reminding them of the great role they are to play in furthering the spiritual welfare of mankind. They are to be *the salt of the earth,* to give the world its spiritual tone and to preserve it from moral corruption; they are to illumine the world by the light of their good example. But they must have the true interior spirituality of the Kingdom if they are to fulfill this high office (transition to the next section). **13.** Salt improves the flavor of food, makes it wholesome and preserves it from spoiling. The natural salt that was often used at that time in Palestine was chemically impure and therefore liable to undergo a chemical change which would render it worthless. The application of this figure to the disciples of Christ is obvious. The same thought is apparently intended in the parallel passage of Luke *14,* 34 f; but the very similar words in Mark *9,* 49 occur in such a different context that their exact meaning is doubtful. See Commentary there. **14-16.** If the disciples let men see their good works with the intention that their Father in heaven thereby receive glory, they are not disobeying Christ's command that they should do their good works —their almsgiving, praying and fasting—in secret *(6,* 1-6.16-18), for in the latter case it is the evil intention of self-glory that is condemned. However, the main thought here is not so much that the disciples should go out of their way to give good example but rather that because of their high office they cannot help being seen and therefore must be careful not to give bad example. *A city set on a mountain cannot be hid:* this sentence as well as the application of the figure of a lamp of good example is peculiar to Matthew. The figure of the lamp also occurs in Mark *4*, 21; Luke *8*, 16; *11*, 33; but in these passages the applications of the figure are different.

5, 17-20: The Old Law and the New. This whole paragraph is peculiar to Matthew with the exception of 18 which has a parallel in Luke *16*, 17, although in a different context. Here Matthew gives the theme of the Sermon on the Mount: that the morality of the New Law of the Kingdom is higher and more spiritual than that of the Old Law especially as interpreted by the Scribes and Pharisees, but at the same time it is not something entirely new but rather the natural development and perfection of the Old Law. **17.** *The Law, the Prophets:* the first two of the three divisions of the Old Testament; therefore the entire Old Dispensation is meant here. *To fulfill:* to bring the Law to its complete perfection by insisting not only on the external act but also on the interior dispositions. If we understand *the Prophets* here not as part of the Old Law in general but as the Messianic prophecies in particular, then *to fulfill* them would be to accomplish them by carrying them into execution according to His Father's will. **18.** The *jot* was the smallest letter in the Hebrew alphabet as written at the time of Christ; it corresponds to the Greek iota and the I of our alphabet. By a *tittle,* in Greek "a little horn," was probably meant the small projections by which one letter of the Hebrew alphabet is distinguished from another otherwise very similar letter. Therefore the sense of Christ's saying is, that not until the end of the world shall any essential part, even the smallest, of the Old Law be abrogated, but it shall rather be perfected. **20.** This is the thesis of the Sermon on the Mount which is then developed by several individual applications in the following paragraphs.

5, 21-26: Against Anger. Peculiar to Matthew except 25 f which occur in a similar form in Luke *12*, 58 f, where however they may have a somewhat different sense, due to a different context. **21 f.** In the Old Law the act of murder was condemned (Ex. *20*, 13; Deut. *5*, 17), but Christ teaches that according to His New Law the interior acts of anger which lead to the external act of murder are likewise condemned. Cf. *1* John

3, 15. The *judgment* mentioned here probably refers to the local tribunal. On the meaning of *Raca, fool* and *Gehenna* see the foot-note to the text; on the *Sanhedrin,* see Glossary at end of text, p. 757. Since there seems to be an ascending scale of punishment here, it is generally assumed that, whatever be the meaning of the term, to call one's brother *Raca* is worse than being merely angry with him, and to say to him, *"Thou fool"* is worst of all. But it is difficult to see such immense malice in these opprobrious expressions. Perhaps there is rather a descending scale here as in 39-42. In this case the sense would be: to be angry with one's brother is so obviously sinful that even a local tribunal can handle such a case; to insult him by calling him "empty-headed" is not considered as bad as murderous anger, yet there is guilt in this also which the highest tribunal, competent to judge the more difficult cases, will perceive; finally, even such a seemingly slight insult as the common expression, "Thou fool," is not free from all guilt in the eyes of God to whom alone belongs the right of condemning to Gehenna (Matt. *10, 28).* **25 f.** Christ refers to a well-known custom of human prudence: that it is better to settle a litigation equitably out of court rather than have a plaintiff who has some right on his side appeal to the law against you; the spiritual sense intended by Christ is: *"If thy brother has anything against thee, go first to be reconciled to thy brother* while thou art still *on the way* of life, for if thou art not yet reconciled when death summons thee it will go hard with thee when the case comes up before the divine Judge."

5, 27-30: Chastity of Mind and Body. **27 f.** Only in Matthew. The Old Law condemned not only the act of adultery (Ex. *20,* 14; Deut. *5,* 18) but also the desire of adultery (Ex. *20,* 17; Deut. *5,* 21). Christ perfects this law by teaching that even an act which is indifferent in itself when done with an evil intention (in this case the act of looking at a woman with the intention of thereby arousing sensual desires) is also sinful. *Who even looks with lust at a woman:* literally in Greek and Latin, "Whoever looks at a woman in order to lust after her." **29 f.** These same words on avoiding the occasions of sin are given in Mark *9,* 42.46 as spoken by Christ on another occasion. Hence one might argue that Matthew incorporated them into his Sermon on the Mount where they may not have been spoken originally. But since Matthew himself *(18,* 8 f) has the passage which strictly speaking parallels Mark *9*, 42 ff, it seems much more reasonable to suppose that Christ spoke them more than once. Therefore one may hold that they are original in the Sermon on the Mount where they form a slight, though natural, digression. These words need not be taken literally; yet they are not a mere counsel but a strict precept in the sense intended by Christ, that is, that we must give up the things that are dearest to us if they prove an occasion of sin. The *right* hand is mentioned specifically because normally it is more valuable than the left hand; this then induced a singling out of the *right* eye although each eye normally is of equal value. Note that in Mark and Matt. *18*, 8 f the hand is mentioned before the eye; probably the original order.

5, 31-32: Divorce. Although Christ treated the question of divorce on at least one other occasion (Matt *19*, 3-9; Mark *10*, 2-12) still this passage is undoubtedly original in the Sermon on the Mount, for it is introduced by the typical phrases, *It was said . . . but I say to you*. The similar statement of Luke *16*, 18 stands there in no clear connection with its context and probably represents the same saying of Christ as Matt. *5*, 32. **31.** Christ quotes (somewhat freely) Deut. *24*, 1 where Moses regulated the custom of divorce which had already been of long standing among the Israelites. Jesus here perfects the Old Law on this point by completely abrogating this custom which had been merely tolerated in the Mosaic Law. The teaching of Christ on divorce as recorded by the other Evangelists and understood by St. Paul (*1* Cor. *7*, 10 f.39; Rom. *7*, 2) makes it perfectly clear that His prohibition of divorce with the right to remarry is absolute. Therefore the seeming exception mentioned in the First Gospel, *save on account of immorality* (*5,* 32), *except for immorality* (*19,* 9) cannot be understood in the sense that the innocent partner of an unfaithful spouse may divorce the guilty one and marry another person. The traditional interpretation of these words is undoubtedly correct: unfaithfulness justifies separation from bed and board, but the bond of marriage remains unbroken. St. Matthew records the words of Christ in their entirety. Our Lord had reason to mention this partial exception lest His absolute prohibition of divorce seem to imply that the injured party is obliged to continue to live with the unfaithful spouse. The other Evangelists however, omit these words, probably intentionally, to forestall a false interpretation of them in the sense of permitting a divorce with the right to remarry. On the Jewish custom of divorce and Christ's attitude toward it see Commentary on Matt. *19*, 3 ff where our Lord speaks at greater length on this question.

5, 33-37: Concerning Oaths. Only in Matthew. **33.** Christ refers to the laws of Moses which prohibit perjury and the breaking of vows (cf. Lev. *19*, 12; Num. *30*, 3; Deut. *23*, 21). **34-37.** He then perfects the old Law in this regard: (a) by condemning the Pharisees' false interpretation of these laws according to which only those oaths in which the divine name was expressly mentioned were binding, while simulated oaths in which a person swore by something having some special relation to God, as heaven, the earth, Jerusalem, one's head, were not binding; (b) by prohibiting absolutely the use of oaths in private conversation: His disciples should have such a reputation for honesty that there should be no need for them to strengthen their affirmations or negations by calling on God as a witness. Therefore when they say yes or no it should be simply yes or no: adding an oath to this is instigated by the devil. Jas. *5*, 12 states this teaching of Christ with perfect clarity, probably re-echoing the words of the Master even more exactly than Matthew has done.

5, 38-42: The New Law of Talion. **38.** Christ quotes these well-known phrases from the Mosaic Law (Ex. *21*, 24; Lev. *24*, 19 f; Deut. *19*, 21) according to which willful damage was punished by inflicting the same damage on the offender. Other ancient nations, as the Babylonians and the Romans, had similar laws. Such laws were not dictated by a spirit of cruelty but rather a desire of restraining undue revenge. One may rightly doubt whether these laws were frequently carried out to the letter. **39-42.** Parallel in Luke *6*, 29 f. Christ perfects the Old Law on this point by prohibiting all private retaliation. He counteracts the spirit of revenge by inculcating the opposite spirit of charity by which His disciples should willingly give up their own just rights for the sake of peace. But Christ is not merely giving a counsel. His teaching on this matter "is partly of precept and partly of counsel. It is of precept: (a) that we should not seek revenge; (b) that we should rather suffer a second injury than revenge the first, *(turn to him the other cheek also);* (c) that we be willing to give up our rights whenever charity or the glory of God seem to demand it. It is of counsel that even though neither charity nor God's glory demand it, we nevertheless obey Christ's words literally for the sake of Christian self-denial" (Menochius). The words of St. Paul in *1* Cor. *6,* 7 f may be taken as an authentic commentary on Christ's teaching in this matter. In the application of this principle our Lord gives various examples in a descending scale of violated rights: physical violence, unjust litigation, pressing into public service, asking for a gift, asking for a loan. *Whoever forces thee:* the Greek and the Latin words used here are derived from the Persian word for "messenger, courier," and are based on the Persian custom according to which the public couriers had the authority to demand the necessary means of transportation from any citizen. The same word is used in Matt. *27,* 32 and Mark *15,* 21 of Simon of Cyrene being "forced" to carry the cross of Jesus.

5, 43-48: The Love of Enemies. **43.** Lev. *19,* 18 commanded the Israelite to love his neighbor. *Thou shalt hate thy enemy:* these words are not found in the Old Testament. Christ is probably citing the teaching of the Scribes who drew this false conclusion from the Old Testament teaching that the Israelites should not be contaminated by the surrounding pagan nations.   **44-47.** Parallel in Luke *6,* 27 f.32-35. Our Lord corrects this false notion by ordaining that we must be charitable to all alike, whether they love us or hate us.   **48.** This verse is often understood as a call to absolute, spiritual perfection in all respects, but this is correct only in an accommodated sense. According to the context *perfect* means "complete, all-embracing" in our charity just as God is all-embracing in His charity. That this is the direct literal sense is clear from the parallel, Luke *6,* 36.

6, 1: Purity of Intention. Only in Matthew. Having shown how the commandments of His New Law demand a higher morality and a greater heroism than those of the Old Law, our Lord now comes to the second section of His sermon. The general principle is stated in this verse: our acts of piety must be performed with a pure intention. The next three paragraphs apply this to three typical acts of piety: almsgiving, prayer and fasting.

6, 2-4: Almsgiving. Only in Matthew. **2.** To *sound a trumpet before* oneself is a metaphor for ostentation. There is no reason to believe that the Pharisees, here called the *hypocrites,* did so literally. **3.** Another metaphor, the sense of which is: "Do not even take self-satisfaction in thy almsgiving." **4.** *And thy Father, who sees in secret, will reward thee:* these words are repeated with powerful effect at the end of these three paragraphs (cf. 6.18).

6, 5-15: Prayer. **5 f.** Only in Matthew. Perhaps only these two verses formed part of the original Sermon on the Mount, while the remaining verses on prayer may have been added here from some other discourses of our Lord. The form of this paragraph as compared with those on almsgiving and fasting would seem to favor this view. Still it is quite possible that Christ Himself made this digression by adding some other thoughts on prayer. Prayer in common is not forbidden; in fact our Lord Himself recommends it elsewhere *(18,* 19 f). What is here condemned is praying in public for the purpose of ostentation. There is no emphasis on the word *standing,* for this was the normal posture at prayer (see note to text). **7 f.** Only in Matthew. Christ does not condemn vocal prayer. What is here condemned is the superstitious notion that certain prayer-formulas repeated a great number of times even though with no attention have a special efficacy. *Multiply words:* the meaning of the Greek verb that is rendered by this phrase is uncertain; perhaps it means "to mumble unintelligible words." A good example of the pagan custom of frequently repeating the same prayer-formula is seen in the incident of Elias and the priests of Baal *(3* Kgs. *18,* 26 ff). On the other hand, our Lord taught the usefulness and even the necessity of frequently repeating the same prayer as long as it comes from the heart, as can be shown both by His words (Luke *11,* 5-13; *18,* 1-8) and His example (Luke *6,* 12; *22,* 41-43; Mark *14,* 36.39). *Your Father knows what you need:* therefore we do not pray as if God had to be informed of our needs, but rather that we ourselves may have a more vivid realization of the truth that all good gifts come from Him.

9-13. The Lord's Prayer, in a slightly shortened form, is also given in Luke *11,* 2-4 in a context which seems more original than that of Matthew. Still there is nothing improbable in supposing that our Lord taught His disciples this beautiful prayer on more than one occasion. Despite its depths of pure piety, one of the chief characteristics of the Our Father is its clarity and simplicity, so that it should be meditated on rather than commented on. After its opening invocation it contains three petitions concerned with God's glory followed by four petitions concerned with our needs. Several of these petitions may be compared to similar words in certain Jewish prayers which can be traced back to about the time of Christ, but even when the words are the same they have a new and much deeper import as spoken by Christ and His disciples.

9. *In this manner,* therefore, not necessarily with these very words are we to pray. Christ wished to give us a model upon which we are to form our own prayers. But His followers rightly feel that they cannot improve upon the prayer of the Master and therefore they love to use the very words that He taught them. *Our Father who art in heaven:* in Greek simply, *Our Father in heaven;* but since the Greek must use the definite article before an attributive phrase, the Latin tries to imitate this by using a relative clause. We address God as our *heavenly* Father to distinguish Him from our earthly father. The Israelites frequently addressed the Lord as their Father, just as He often called them His children (cf. Deut. *32,* 6; Isa. *43,* 6; *63,* 16; *64,* 8; Ecclus. *23,* 1.4; etc.), and more was meant by this than the common fatherhood of God as Creator of all mankind. But this term has still deeper significance as used both by Jesus, the Son of God in His divine nature, and by the members of His Mystical Body who, being "led by the Spirit of God, are the sons of God," having "received a spirit of adoption as sons, by virtue of which we cry, 'Abba! Father!'" (Rom. *8,* 14 f). Even when a Christian prays alone he should say *Our* Father and not *My* Father, for he should be conscious of the spiritual union with his brethren. By the *name* of God is meant God Himself as known and spoken of by men. *Hallowed:* glorified, honored, reverenced, etc. **10.** *Thy kingdom come:* may God's reign be acknowledged by all. In this petition we pray that the Messianic kingdom, the Church of Christ, be spread more and more throughout the world and that Christ the King may reign in the hearts of all men. In the third petition we pray that all men on earth may obey God's will as perfectly as do the angels in heaven (cf. Ps. *102,* 20 f). *Thy will be done* is not merely a prayer of resignation. Although the words *on earth, as it is in heaven,* are usually understood as referring only to the third petition, perhaps Christ intended them to qualify all three preceding petitions.

11. *Bread* is used here, as part for the whole, for all that is necessary for our sustenance. It is well to note that this is the only one of the seven petitions that is concerned with the temporal needs of man: a good norm for our own prayers. The Greek word that is here translated as *daily* is of uncertain meaning, but *daily* is at least as probable as any of the other translations that have been suggested, such as, "supersubstantial," "for sustenance," "for tomorrow," "necessary." This petition by which we express our confidence that God will provide for our temporal life, is in perfect accord with Christ's teaching that we should not be anxious about our food and clothing (cf. *6,* 25-34). **12.** *Debts:* used figuratively for "sins," the word used by St. Luke in this petition, as is also evident from 14 f where *offenses* is used as a synonym. *Our debtors* therefore means those who have offended us. On sin as a "debt" cf. *18,* 21 ff. *As we also forgive:* from the Greek it is evident that this does not mean, "In proportion as we forgive," but "since we also forgive." With childlike simplicity we mention this as an inducement for God also to forgive us. But it is also a necessary condition for our pardon by God (cf. 14 f). **13.** For our own good God allows us at times to be tempted, but conscious of our weakness we humbly ask Him to spare us this trial as far as possible. Finally, even though we are tempted, we ask God to save us from falling into the evil of sin. However, according to the Greek where the article is used before the adjective *evil,* the more probable translation would be, "Deliver us from the evil one." There is no clear example where this adjective when used with the article in Greek signifies "evil" in the abstract sense, while it is frequently used of the devil, rendered either "the evil one" or "the wicked one" in our translation; cf. Matt. *5,* 37; *13,* 19.38; Eph. *6,* 16; *1* John *2,* 13 f; *3,* 12; *5,* 18 f; so also it can be understood in John *17,* 15. In the Our Father the traditional translation is retained because it is so well known in this form. The last petition then is essentially the same as the preceding one; perhaps for this reason it is omitted by St. Luke (Origen's explanation). **14 f.** The thought contained in the fifth petition (12) is elaborated here. Mark *11,* 25 f, in a different context on prayer, parallels this. Cf. also Matt. *18,* 35.

6, 16-18: Fasting. Only in Matthew. The third type of pharisaical piety to be avoided: ostentatious fasting. Far from condemning the practice of fasting as such, our Lord in this passage presupposes that His disciples will fast; but their works of mortification are not to be done in a spirit of vainglory. The Law of Moses prescribed only one general fast-day each year, on Yom Kippur, the great Day of Atonement (cf. Lev. *16,* 29-31; *23,* 27-32; Num. *29,* 7). In the Prophets there are many references to the proper spiritual dispositions that should accompany an acceptable fast. At least after the Babylonian Exile we find instances of special public fasts (cf. *1* Esd. *8,* 21.23) and private fasts (cf. *2* Esd. *1,* 4). At the time of our Lord pious Jews fasted often (cf. Matt. *9,* 14 and parallels), even twice every week (cf. Luke *18,* 12). According to the ancient practice of doing penance in sackcloth and ashes, the Jews, when fasting, denied themselves the pleasure of anointing their head, and instead of oil they sprinkled ashes over their head and face. These symbolic acts were not wrong in themselves but are here condemned by Christ because they were done from motives of ostentation.

6, 19-24: True Riches. Here begins a new section of the Sermon, which however flows naturally from the preceding considerations. Christ has just emphasized the truth that our works of piety must be done with the pure intention of pleasing God. Therefore we must serve God with an undivided heart. But the chief obstacle in the way of this whole-hearted service of God is man's preoccupation with the things of this world, especially the pursuit of earthly wealth. Man's excuse is that he must provide for the future. Christ answers that man's anxiety for the future is excessive because he lacks genuine trust in God. Therefore, since all this is so logically developed in the Sermon on the Mount, there is no good reason for assuming that it is not original here, even though almost all the words that Matthew attributes to Christ on this occasion are given by Luke in various parts of his Gospel other than in the Sermon on the Mount. Our Lord no doubt treated these important matters on more than one occasion. **19-21.** Parallel in Luke *12,* 33 f. People at the time of Christ hoarded up their wealth in the form of expensive clothing or precious metals and jewels. The former would be consumed by moths, the latter corroded by rust, all alike stolen by thieves. Therefore even from a natural viewpoint this was folly. Only treasures stored in heaven are secure. **19.** *Break in:* literally, "dig through," for the mortar used in the houses was not much more than mud. **21.** This is the fundamental reason why we must avoid the engrossing pursuit of earthly riches; the profound truth of experience that a man's interests, his *heart,* all his thoughts and desires are centered in the things which he values most highly, his *treasure.* Therefore if a man's treasure consists in earthly riches, he will be interested in the things of earth and not in God; if his sole treasure consists in the merits of his good works which are laid up for him in heaven, he will be interested only in God and the things of God.

22 f. Almost the same words in Luke *11,* 34-36, where the different context gives them a somewhat different meaning. In Matthew the sense of this passage seems to be this. Since God demands a whole-hearted service (preceding context) and man cannot combine His service with the service of mammon (following context), therefore we must be single-minded in His service. Christ teaches this truth here in the form of a little parable. It is through the eye that the light guides all our bodily actions, so that in a sense the eye is the lamp of the body. When the eye is sound and healthy the whole body is in the light and can adapt itself to the world round about. But when the eye is diseased or blind the whole body is partially or totally in the dark even though the sun be shining brightly. The spiritual sense of this parable is this. The eye of the body is the mind or purpose of the spirit. If the mind is spiritually healthy and "single," that is, directed solely towards the service of God, the whole spirit will be filled with God's light. But if the mind is evil, directed mostly or entirely to the service of earthly wealth, the whole spirit is in spiritual darkness. If then because of the excessive pursuit of wealth the spiritual light that should be in us is darkened, how great is the spiritual darkness within us! As Christ says in the Beatitudes, only the pure of heart, that is, the single-minded can see God. *If thy eye be sound:* literally "simple," that is, in the physical sense "sound and healthy," in the spiritual sense "single in purpose, pure-intentioned." *If thy eye be evil:* in the physical sense "diseased, blind," in the spiritual sense "evil, directed towards the things of earth."

24. Parallel in Luke *16,* 13. Christ illustrates the same truth by this little parable of the *two masters,* the application of which in the spiritual sense is obvious. *Mammon:* from the Aramaic word for "wealth," *mamona. Hate the one and love the other . . . stand by the one and despise the other:* probably referring indifferently to God and mammon, although it is possible that the sense intended is: those who *love* mammon *hate* God whose laws interfere with the unrestricted pursuit of wealth, while those who *stand by* God despite the allurements of wealth, *despise* mammon.

6, 25-34: Trust in God. Parallel in Luke *12,* 22-31, spoken on a different occasion but in the same sense. The common excuse for undue pursuit of worldly goods is desire to provide for an uncertain future, in modern language the desire for "social security." Christ, with a deep sympathy for the sufferings and needs of the poor, teaches them the folly of this excessive worry about the future. He argues (a) from natural reason, that worry is useless since it cannot change the future, and harmful since there are sufficient troubles in the present without adding to these by worrying about future troubles; (b) from faith, that God who takes such good care of His irrational creatures, such as the birds and the wild flowers, will certainly take care of us, His rational creatures, whom He loves much more than these. If we attend to the thing of primary importance, the service of God, He will attend to such secondary things as the preservation of our earthly life. Therefore we should not be preoccupied with the things of earth as the heathens are who do not know God.   25. If God gave us our life and body He will also give us the lesser gifts, the food and clothing that are necessary for our life and body.   26. Note that Christ does not tell us to imitate the birds and the lilies of the field in their freedom from work. The birds and the flowers are merely pointed out as examples of God's providence. Every creature of God must live according to its nature: the way that man's nature is constituted he must sustain himself by work. The birds and the flowers are especially chosen as illustrations because by them Christ can refer to man's two main worries, food from agricultural labor and clothing from the work of spinning and weaving.   **27.** *Stature* renders a Greek word which can mean either height (of the human body) or length (of human life). Since a lineal measure, the cubit which equals about half a yard, is mentioned, the Latin from which our translation is directly made, understood this Greek word in the sense of *stature,* height of the body. But the context would seem to demand the sense of length of life. For men are not normally worried about food and clothing in order to become a foot and a half taller, but they are anxious about their span of life. For although we must take a reasonable care of our health, the hour of our death is already determined for us by God, and anxiety about it will not change it. Worry, if anything, will rather shorten our span of life. **28.** *The lilies of the field:* probably the crimson anemones are meant, for these are a common wild flower of Galilee in the springtime and their color naturally suggests the royal robes of Solomon. **30.** *The grass of the field* here means all the wild verdure which when withered in the heat of summer is used as fuel in Palestine.

7, 1-6: Avoiding Judgments. Here begins the last section of the Sermon on the Mount. This section, the whole seventh chapter of Matthew, consists of the teachings of Christ on various topics which are arranged in such a way that they seem to lack all logical unity. From this however we should not conclude that St. Matthew has gathered various sayings of Christ from several different occasions and strung them together at the end of this sermon. For most of these sayings of our Lord occur also in Luke's Sermon on the Mount. Various explanations may be offered for this seeming lack of logical unity. (a) Some of these sayings of Christ belong in the previous parts of the Sermon where they would fit in very naturally. This is probably true of those sayings which St. Luke does place in such a more logical order. (b) St. Matthew has preserved for us in this section only fragments of the original Sermon, so that the connection of thought is often lost. (c) At this part of the Sermon Jesus permitted His hearers to ask Him questions. The Evangelist would then have preserved only the answers of these disconnected queries. In their teaching the Rabbis followed such a method (cf. Luke 2, 46).

1-5. Parallel in Luke's Sermon on the Mount *(6, 37 f.41 f)*, where it occurs after his Rules of Charity; in Matthew therefore these words belong logically at the end of 5. **1.** *Do not judge,* etc.: to be understood in the sense of "Do not condemn, that you may not be condemned." Christ forbids uncharitable criticism by private persons of their fellow-man. He does not forbid those in authority to pass judgment or condemn, nor does He forbid the individual the use of discernment in spiritual matters (cf. 15-19). What is here condemned is the hypocritical self-deceit of thinking oneself morally better than others. **2.** *With what measure you measure,* etc., is a Jewish proverb known also from the Talmud. In Mark *4,* 24b Christ uses this proverb in a somewhat different sense. **3.** *The speck:* literally in Greek, a particle of chaff, a splinter of wood; here figuratively for a small moral fault. In hyperbolic contrast our Lord speaks of a *beam* of wood to signify a much greater sin. **6.** Only in

Matthew and apparently in no connection with the context, unless this verse be considered with the preceding as part of a special instruction for the teachers of Christ's doctrine. Probable sense: the teachings of Christ which are *holy* and as precious as *pearls* (cf. *13,* 45 f) should not be rashly proclaimed to the unbelievers and scoffers, for such men will not only not appreciate them but will turn them as arguments against you to *rend you.* *Dogs and swine,* unclean animals to the Jews, were used as epithets for the Gentiles in Jewish literature. Cf. Matt. *15,* 26. These words of Christ therefore form the basis for the early Christian practice, known as the *disciplina arcani,* of not revealing all the truths of the Faith to the unbaptized. But even in the early Christian centuries (cf. the *Didache*) these words of Christ were applied in an extended sense to the prohibition of administering the sacraments, especially the Eucharist, to the unworthy.

7, 7-11: Power of Prayer. Parallel in Luke *11,* 9-13, where this section follows the Lord's prayer and a little parable on perseverance in prayer. In Matthew these verses have no connection with the immediate context but they would fit well after *6,* 15. These words on prayer consist of an affirmation in general terms of the efficacy of prayer (7 f) followed by an *a fortiori* argument in confirmation of the statement.   **9 f.** *For a loaf . . . a stone, . . . for a fish . . . a serpent:* the substitution would be suggested by the external resemblance.   **11.** *If you, evil as you are:* even though you are selfish and unkind.

7, 12: The Golden Rule. Parallel in Luke *6,* 31. In Matthew this verse stands entirely apart from its context; it would fit in much better after "The New Law of Talion" (*5,* 42) as in the Third Gospel. The Golden Rule states very wisely that true charity would never be violated if every one would treat others as he himself wishes to be treated. The same thought is stated negatively in Tob. *4,* 16. Love of others should therefore be on a par with love of oneself: "Thou shalt love thy neighbor as thyself" (Lev. *19,* 18). All the commandments of *the Law and the Prophets,* that is, of the whole Old Testament, are summed up in the law of charity. Cf. Matt. *22,* 39 f; Rom. *13,* 8-10; Gal. *5,* 14.

7, 13-23: Obstacles to Virtue. A general statement on the difficulty of salvation (13 f) followed by warnings against two dangers to salvation: (a) being deceived by false teachers, (b) deceiving oneself by a presumptuous self-confidence. Largely paralleled in Luke's Sermon on the Mount and therefore undoubtedly a part of the original discourse. **13 f.** The narrow gate and way. Similar words in Luke *13,* 24, where they form the answer to the question, "Are only a few saved?" Under these common biblical figures of the gate and the road Christ teaches that His doctrine is not at first sight easy and attractive. **15.** By false prophets are here meant not erroneous forecasters of the future but teachers of false doctrine who pretend to speak in God's name. They are *ravenous wolves in sheep's clothing,* that is, they deceive men by the pretense of teaching men an easy way to happiness but in reality they are ruining men's souls and destroying their true happiness. **16-18.** Parallel in Luke *6,* 43-45. Since a man's doctrine and philosophy of life affect his outward conduct, from the latter we can judge the former. The nature of the plant determines the nature of its fruit. Christ uses the same figure in *12,* 33. By a *good tree* and its *good fruit* are meant a cultivated tree and its edible fruit, while the *bad tree* with its *bad fruit* is the wild tree and its inedible fruit. **19.** The same figure of the destruction of the useless tree was used by the Baptist (cf. *3,* 10). Cf. also John *15,* 2.6. **21-23.** Lip-service is not enough; faith must be alive and effective. The same doctrine is taught in the parable of the Two Sons (*21,* 28-31). **21.** Similar words in Luke *6,* 46. **22 f.** Similar words in Luke *13,* 26 f. *That day* is a common biblical expression for the Last Day. For His own special reasons God may sometimes grant the gifts of prophecy and miracles to unworthy persons (cf. *1* Cor. *13,* 2). *I never knew you:* cf. *25,* 12. *Depart from me:* cf. *25,* 41. The words used here occur also in Ps. *6,* 9.

7, 24-27: Conclusion of the Sermon. Parallel in Luke *6,* 47-49, but the description of the storm is much more vivid and powerful in the First Gospel. In Palestine the torrential rains of the winter months, November to March or April, have largely denuded its soil and often destroy houses that are not solidly built.

7, 28-29: Epilogue. *And it came to pass when Jesus had finished these words:* substantially the same words are used at the end of the other four great discourses of Christ in the First Gospel (cf. *11,* 1; *13,* 53; *19,* 1; *26,* 1). *The crowds were astonished at his teaching,* etc.: the same words in Mark *1,* 22 and Luke *4,* 32 in connection with our Lord's first teaching at Capharnaum. *Authority:* literally, "power." The words *and Pharisees* are not in the Greek. The contrast between the teaching of Christ and that of the Scribes or Rabbis, the official teachers of Judaism, consisted principally in this: (a) the Scribes based their teaching on the opinions of their predecessors, "Rabbi So-and-so said this . . ."; Christ taught in His own name, *I say to you* . . .; (b) the teaching of the Scribes was largely confined to subtle points of casuistry; Christ taught the basic truths of religion and morality; (c) the teachings of the Rabbis as preserved in the Talmud make the driest and most uninteresting reading imaginable; the beauty of Christ's style and imagery thrilled His hearers and all succeeding generations with wonder and delight.

8, 1-4: A Leper. Parallels in Mark *1,* 40-45 and Luke *5,* 12-16. Many of the events in *8-9* took place before the Sermon on the Mount. The arrangement of this section in the First Gospel is largely artificial (see Introduction). Our Lord probably worked this miracle during His missionary journey in Galilee that followed His first ministry in Capharnaum (cf. Mark *1,* 39 f). The place was some town of Galilee other than Capharnaum (cf. Mark *2,* 1; Luke *5,* 12).   **2.** The leprosy for which Moses gave detailed regulations as well as the affliction of the *lepers* mentioned in the Gospels need not always be identified with the disease now known as leprosy. Apparently several diseases of the skin of a more or less serious character were called by this name in the Bible. But the immediate cure of these afflictions that Jesus achieved simply by His command was in any case certainly miraculous.   **3.** *Jesus touched him:* some such physical contact was the usual procedure in most of the miraculous cures worked by our Lord. He acted in this way probably to show His sympathy for the afflicted person and more particularly to conceal the entirely miraculous nature of these cures by using certain customary remedies (e.g. spittle or mud on sore eyes) or ceremonies.   **4.** *See thou tell no one:* most of the miracles of Jesus were not done primarily to prove to His immediate audience that He was the Messias and the Son of God but rather out of sympathy for the afflicted. On the other hand, during most of His ministry, at least in Galilee, He endeavored to conceal His miracles. For He did not wish to be proclaimed the Messias as long as the people had a false idea of the Messias as a mighty temporal prince (cf. John *6,* 14 f). For similar instances of our

Lord's efforts to conceal His true nature until the proper time cf. Matt. *9, 30; 12,* 16; Mark *1, 34; 3,* 12; *5,* 43; *7,* 36; Luke *4,* 41. By commanding the leper to carry out the prescriptions of Lev. *14,* 2-32 for cured lepers Christ wished to show His compliance with the Mosaic Law and to avoid all unnecessary hostility with the priests. The same command in Luke *17,* 14. *Offer the gift ... for a witness to them:* the leper is to bring his sacrifice to the priests so that they, in accepting it, would be witnesses of his cure and he in turn would receive a testimonial from them guaranteeing the genuineness of his cure. This seems more probable than the opinion which holds that the priests are thereby to have a proof that Christ worked a miracle. For this would hardly be in conformity with our Lord's command, which is still more strict as given in Mark *1,* 43, that the cured leper should conceal the miracle. Moreover, the priests need not have seen anything miraculous in the cure, for the prescriptions of the Mosaic Law evidently presuppose that the cure of some types of so-called leprosy is fairly common and entirely natural.

8, 5-13: **The Centurion's Servant.** Parallel in Luke *7,* 1-10, where this event is also placed in connection with the return to Capharnaum after the Sermon on the Mount. A *centurion* was a Roman military officer of inferior rank in command of a "century," a company of one hundred soldiers; hence the title. But because the Roman legions were seldom kept at the maximum strength of six thousand men, the century often consisted of only fifty or sixty men. A centurion may therefore be compared to a captain in modern military terminology, although he rose from the ranks and would properly be considered a non-commissioned officer. It is uncertain why this centurion resided at Capharnaum. Possibly there was a Roman garrison there; or perhaps he belonged to the little army of Herod Antipas, if we may suppose that Herod's soldiers used Roman titles. In any case he was a Gentile (cf. 10).   **5.** *There came to him a centurion:* not personally but by means of the Jewish elders who came as his intermediaries, as we learn from Luke.   **8.** *Lord, I am not worthy that thou shouldst come under my roof:* these words, made memorable as the Christians' humble prayer before Holy Communion, show the centurion's understanding of the Jewish custom of not entering the house of a Gentile (cf. John *18,* 28). *But only say the word:* the centurion believes that Jesus has such supernatural powers that He can work miracles at a distance.   **9.** The general sense

is clear: just as the soldiers who are subject to the centurion obey him, so also must nature obey the command of Jesus. But the exact meaning of the first part of the sentence is somewhat doubtful: either, (a) Even though I myself have higher superiors, still I can give commands; how much greater right hast thou to command who hast no higher superior; or (b) Just as my right to give commands rests on my own obedience to higher superiors, so also thou canst command nature, for thou also art obedient to thy higher Superior, thy heavenly Father; or (c) I also am a man placed in authority, etc. It is doubtful whether the centurion would have known enough of the nature of Jesus to conceive the ideas expressed in the first two opinions. According to the third opinion, *subject to authority* means "set in an authoritarian system."
11 f. This marvelous faith of the centurion, which was so much greater than that of the Jews, prompts Jesus to prophesy the great spread of the Faith among the Gentiles as well as the reprobation of the Jews for their lack of faith in Him. Luke has this prophecy in a different context *(13,* 28 f). Faith, not merely carnal descent, entitles one to share in the promises made to the Patriarchs. The blessings of the Messianic kingdom are pictured here, as frequently, under the image of a banquet. The figures of *the weeping and the gnashing of teeth* and the *darkness outside* to represent eternal perdition, are frequent in the First Gospel (cf. *13,* 42.50; *22,* 13; *24,* 51; *25,* 30), and with the exception of Luke *13,* 28 occur only in this Gospel.

8, 14 f: Peter's Mother-in-law. Parallels in Mark *1,* 29-31 and Luke *4,* 38 f, where this event is told with more details. It took place during Christ's first ministry at Capharnaum on a Sabbath (cf. Mark *1,* 21) after He and His disciples had returned from the synagogue. *Peter* is called "Simon" here by the other two Evangelists and with greater exactness, for our Lord had not yet given him his new name. According to John *1,* 44 Bethsaida was "the town of Peter and Andrew." Either this means merely their birthplace or they had a second residence at Capharnaum. Since Peter had a mother-in-law he was certainly married. That she *began to wait on them* at table would hardly imply that his wife was already dead. This item is added to show the completeness of her miraculous cure.

8, 16-17: Other Miracles. Parallels in Mark *1,* 32-34 and Luke *4,* 40 f. The people waited until *it was evening* before *they* *brought to him* their sick, because the Sabbath was over at sunset. In keeping with the teaching of the Scribes they would not ask Him to cure the sick on the Sabbath, or perhaps they considered the carrying of the sick a forbidden work. **17.** Only in Matthew. The words of Isa. *53,* 4 which are cited here by the Evangelist, refer directly to the Passion of Christ and are so understood in *1* Pet. *2,* 24. The literal sense of the prophecy is that Christ took upon himself the guilt of our mortal infirmities and bore the punishment for our sins. Matthew takes the words *infirmities* and *ills* in the physical sense. Some would therefore conclude that the Evangelist is using this text merely in an accommodated sense. But it is difficult to see how he could say of a merely accommodated sense, *that it might be fulfilled.* He could, however, say this rightly of a prophecy in the consequent or extended sense. That is, Matthew argues thus: Isaias prophesied that Christ would bear the punishment of our sins and would take away our sins and their consequences; but one of the punishments and consequences of sin is sickness; therefore, concludes the Evangelist, Isaias also prophesied that Christ would take away our corporal infirmities and would bear away the burden of our physical ills.

8, **18-22: Sacrifice to Follow Christ.** These sayings of Christ on the sacrifices that are demanded of His followers are placed by Matthew in connection with the first crossing of the Lake of Genesareth and the subsequent storm; Luke places them on the last journey to Jerusalem. On general principles the order of the Third Gospel is usually preferred here. Both Gospels certainly narrate the same events. **18.** Mark *(4,* 35) and Luke *(8,* 22) place this crossing of the lake after Christ's discourse in parables. According to Mark it was on the evening of the same day on which Jesus spoke in parables. **19-22.** Parallel in Luke *9,* 57-60. We may rightly suppose that the dispositions of these men who asked to follow Christ were good. Nothing in the text can help us decide whether these aspirants found the conditions too hard or not. **19.** This *Scribe,* since he belonged to an influential and respected class, would no doubt have been a valuable asset to the group of Christ's disciples; but our Lord is in no hurry to accept him. His rank connoted a certain amount of wealth and therefore Christ reminded him that His followers were expected to share the poverty and homelessness of their Master. Out of respect the Scribe addresses him as *Master,* literally in Greek, teacher, the equivalent of the Hebrew title Rabbi. **20.** This is the first time in this Gospel that the expression *Son of Man* is used. Our Lord used it often in speaking of Himself. One reason for this was probably the Semitic custom of polite speech which discouraged the too frequent use of the first person singular. But the reason why Jesus chose this specific title for Himself was because it was definitely Messianic yet at the same time not clearly understood by the people as a synonym for Messias. An examination of the numerous places where it is used (not only in the Synoptic Gospels but also in John) shows that Jesus used it of Himself in His office of Messias but with special emphasis on His human nature. No doubt it is ultimately based on Dan. *7,* 13-14, for even the enemies of Jesus understood this passage as Messianic (cf. Matt. *26,* 64 ff; Mark *14,* 62 ff). But that it was ordinarily not understood as such by the people seems clear from John *12,* 34. Therefore by calling Himself the Son of Man Jesus both proclaimed Himself the Messias and at the same time concealed this truth from the people because He did not wish to be taken for such a Messias as the people were expecting. (See Commentary on *8,* 4). Since this humble title seemed to stress the humanity of Jesus at the expense of His divinity, with the exception of Acts *7,* 56, where there is a reference to the prophecy of Dan. *7,* 13, it is never used in the New Testament by any one but Jesus Himself.   **21.** In Palestine at this time the dead were ordinarily buried on the same day on which they died. This man therefore wished to postpone his following of Jesus until the conclusion of the lengthy period of mourning, the settling of the estate, etc. Or perhaps his father was not yet dead but was expected to die soon.   **22.** The exact meaning of these words of Christ is somewhat disputed. (a) Some hold that *the dead* is to be taken in the same sense in both instances, i.e., let the dead take care of themselves for it is better that thy father remain unburied than thou shouldst fail to follow me. (b) Others take the first *dead* to mean the spiritually dead, i.e., let thy relatives who have not received the life of divine grace through faith in me, bury thy father.

8, 23-27: **The Storm on the Lake.** Parallels in Mark *4,* 36-40 and Luke 22b-25.   **23.** *His disciples followed him:* i.e., accompanied Him in the same boat, as is clear from the context and from the parallel passages.   **24.** This *great storm on the sea* of Galilee was undoubtedly of the same nature as those storms which often rise on this inland sea even today. The surface of this lake is about six hundred eighty-two feet below sea-level, while the mountains which surround it on almost all sides tower from one thousand to nearly two thousand feet above it. Consequently the air near the water becomes much warmer than that on the mountains only a few miles away. Sometimes, especially towards evening, this warm air begins to rise very rapidly and the cool air, rushing down from the mountains, causes sudden and extremely violent storms.   **25 f.** The disciples really showed faith in Jesus by calling on Him to save them from danger, but He tells them that they had but little faith in Him or they would not have become afraid in the first place; for even though He was sleeping, He knew of the danger.   **27.** The *men* can only be the disciples mentioned in the preceding verses. This is certain from the parallel accounts, nor is there reference anywhere to an additional crew. They *marvelled* because they knew that only God can rule the wind and the waves (cf. Pss. *17,* 16; *88,* 10; *106,* 23-30). The Fathers of the Church see in this episode a picture of the Church and of individual souls on the stormy sea of persecutions and trials. But Jesus is in the ship and therefore they should trust in Him without fear.

8, 28-34: Expulsion of the Devils in Gerasa. Parallels in Mark *5,* 1-20 and Luke *8,* 26-39, where this event is told with many more details. All three Gospels agree in placing this episode immediately after the first storm at sea. But while Matthew speaks of two demoniacs, Mark and Luke mention only one. Probably the second one in Matthew's account was rather unimportant and was therefore passed over in silence by the other two Evangelists.   **28.** *Gerasenes:* the people of Gerasa. From the whole context it is certain that this event took place somewhere quite near the shore of the eastern side of the lake. But there is much confusion in the MSS concerning the name of this site. For each of the three Gospels there are MSS reading Gadarenes or Gerasenes or Gergesenes. The last variant is generally explained as a mistaken emendation of some early scholar, possibly Origen. The testimony of the best Greek MSS favors as the original reading of the First Gospel "Gadarenes," of the Second and the Third Gospel "Gerasenes." Gadara was an important city of Decapolis about six miles southeast of the lake of Galilee and its suburban territory probably included the southeastern segment of the shores of the lake. Gerasa was the most famous city of Decapolis and its great ruins still testify to its former magnificence. But it was some twenty-six miles southeast of the lake. Since this seems too far for it to be mentioned in connection with the shores of the lake, it is generally assumed that there was an otherwise unknown village by the same name on the eastern shore of the lake. Some identify this village with the modern site of Kursi, directly opposite Magdala; others with modern Geradi, about five miles further south on the same shore; still others think that just as Matthew seems to name the site from its nearest city, Gadara, so the other two Evangelists seem to speak of the whole eastern shore of the lake as "the country of the Gerasenes" from the most important city in Decapolis, Gerasa.

29. These words were spoken by the demons who used the possessed men as their mouthpiece. *What have we to do with thee:* literally "What to us and thee," a Hebrew idiom which here signifies the demons' displeasure at being told to leave the possessed men. *Son of God* in the mouth of the demons probably signifies no more than Messias, for they would hardly dare to oppose Him at all if they had known His full divinity. The same title is given Him by the demons in Mark *3,* 11. *To torment us before the time:* to confine us in hell (cf. Luke *8,* 31) before the Last Judgment. Till then they have a certain liberty to roam about the world (cf. 2 Pet. *2,* 4 with *1* Pet. *5,* 8). For all the words spoken here by the demoniacs cf. the demoniac in the synagogue (Mark *1,* 24; Luke *4,* 34). 30. Even though the Jews were not permitted to eat pork, it should cause us no difficulty to read here of *a herd of many swine,* for this region was inhabited mainly by Gentiles. 32. *Down the cliff into the sea:* almost the whole eastern shore of the Lake of Galilee consists of high precipitous banks. There was no injustice done to the owners *when the whole herd ... perished in the water.* For in allowing the demons to cause this damage, Christ as the God-Man exercised His sovereign rights of dominion for some higher purpose known to Himself. 34. *They entreated him to depart from their district* because they stood in awe of the power of Jesus (cf. Luke *8,* 37) and probably also because they feared more temporal losses.

9, 1-8: A Paralytic at Capharnaum. Parallels with more details in Mark *2,* 1-12 and Luke *5,* 17-26. Both Mark and Luke place this event as the first of the five disputes with the Pharisees before the Choice of the Twelve and the Sermon on the Mount. **1.** *His own town* is Capharnaum (cf. Mark *2, 1*). However this return to Capharnaum after the expulsion of the demons in Gerasa is not to be connected with this cure of the paralytic. **2.** The *pallet* mentioned here was a sort of mat on which a sick person could be carried as on a modern stretcher. These men showed *their faith* not merely in bringing the paralytic to be cured by Jesus but in the extraordinary means that they used to achieve this end, as narrated in the other Synoptic Gospels. Instead of merely curing the paralytic Jesus first bestowed a greater boon upon him, the forgiveness of his sins. He did so primarily in order to prove to the Scribes that, if He had the power to work miraculous cures, He also had the distinctly divine prerogative of forgiving sin. **3.** If Jesus were not divine, the Scribes would have been entirely justified in concluding *"This man blasphemes."* **4.** Jesus gives the Scribes further proof of His supernatural power by showing them that He is able to read their hidden thoughts. **5 f.** Not only are these words of Christ almost exactly the same in all three Gospels but, what is more extraordinary, in all three Gospels the words of Christ are interrupted by the clause, *then he said to the paralytic,* which we might without irreverence call a sort of "stage-direction." Such a remarkable agreement cannot possibly be explained except by the assumption that here all three Synoptic Gospels clearly depend on one original source. Christ's argument is: it is no easier for one to work the sudden cure of a paralytic than it is to forgive sin; if He can do the one He can also do the other. Christ as the *Son of Man,* in His human nature, has this power of forgiving sin. Therefore He could also communicate it to His Apostles (cf. Matt. *18,* 18; John *20,* 22), but just as they worked miracles only in His name (cf. Acts *3,* 6), so also they and their successors can forgive sins only in His name, by His authority. **8.** *The crowds . . . were struck with fear:* i.e., they were filled with reverential awe at the sight of the manifestation of such divine power.

9, **9-13: The Call of Matthew.** Parallels in Mark *2,* 13-17 and Luke *5,* 27-32. This is the occasion of the second conflict with the Scribes and Pharisees, which Mark and Luke place before the Sermon on the Mount. **9.** Concerning St. Matthew and his call to the apostolate, see Introduction. Since Matthew was *sitting in the tax-collector's place,* the site of his call was probably a custom-house at the border. According to Mark this was near the seaside; therefore probably where the Jordan entered the lake, the border between the domains of Herod Antipas and Philip. **10.** *In the house,* i.e., of Matthew (cf. parallels). He held this banquet to celebrate his call. Among the friends whom he invited were *many publicans* (tax-collectors) *and sinners.* Since the Pharisees considered all publicans to be sinners, the second term really means "and other sinners." **11.** Oriental banquets are semi-public functions; thus the Pharisees saw this one. Their adverse criticism, which they did not dare to make directly to the Master, was most likely uttered only after the celebration was over. They endeavored to seduce the disciples from the Master by implying that no holy Rabbi would so act. Cf. the same accusation in Matt. *11,* 19; Luke *15,* 2; *19,* 7. **12 f.** Jesus answers their objection by three arguments. (a) By comparing His spiritual ministry to the corporal administration of a physician; just as the latter cannot avoid contact with the sick, so neither can He avoid the spiritually sick. (b) By showing the true nature of religion as taught by the prophet Osee (Os. *6,* 6). Only in Matthew, here and in *12,* 7. Mercy shown to one's fellow-man is more pleasing to God than sacrifice offered directly to Him. (c) By stating the essential nature of His Messianic office, the salvation of sinners. These words of Christ not only justify His consorting with sinners but at the same time show why the rigorist Pharisees can have no part in His kingdom, for they exclude themselves from His aid by pretending to be just and spiritually healthy. The first requisite for admission into the Kingdom of Heaven is humble acknowledgment that we are sinners and stand in need of Christ's salvation (cf. Rom. *3,* 23; *1* John *1,* 8-10). *Go and learn what this means* is a Rabbinical formula frequently used in the Talmud to introduce an argument from Scripture.

9, 14-17: The Question of Fasting. Parallels in Mark *2,* 18-22 and Luke *5,* 33-38; the third of the five disputes which took place before the Sermon on the Mount. On the Jewish custom of fasting, see Commentary on Matt. *6,* 16-18. **14.** Now that the Baptist was in prison, his disciples were probably somewhat jealous of the success of Jesus and His disciples. But their question on fasting seems to have been prompted by the Pharisees: *"Why do we and the Pharisees often fast . . . ?"* This was another attempt on the part of the latter to discredit Jesus as being too lax. Jesus answers their question by explaining that: (a) in the future His disciples will indeed fast but such penance is inopportune during the present happy period of His short sojourn with them on earth; (b) He does not wish His disciples to fast at present lest they seem to imitate the spirit of the Pharisees. The Christians' motive in fasting is fundamentally different from that of the Pharisees or even that of the disciples of John. These two different spirits cannot be combined without doing damage to both. This He illustrates by two simple comparisons. **15.** Christ demonstrates the inopportuneness of fasting at present by comparing the period of His public ministry to a wedding feast. The Baptist had already compared Jesus to a bridegroom (cf. John *3, 29*). The Apostles also liken the union of Christ with His Church to the espousals of a bridegroom with his bride (cf. 2 Cor. *11,* 2; Apoc. *21,* 2. 9 f; *22,* 17). *The wedding guests:* literally in Greek, "the sons of the bridal chamber," which expression reproduces the Hebrew title for the chosen companions of the bridegroom who had a special part to play at a Jewish wedding. These are the Apostles. This term is to be distinguished from that of "the friend of the bridegroom" (John *3, 29*) who was something like our "best man." This part was played by St. John the Baptist. *The bridegroom shall be taken away from them:* the verb implies a violent separation, a reference to Christ's death on the cross. *Then they will fast* points not only to the time but also to the reason of their fasting: that is, the penance and fasting of Christ's disciples have a special relation to His Passion and Death. Hence the Lenten fast of the Church. Christian penance receives its special efficacy from its union with the sufferings and death of Christ. Therefore it differs essentially from the fasting of the Old Dispensation which sought justification in the works of the Law (cf. Rom. *3-5*). **16 f.** To try to combine these two forms of penance would be as harmful as putting a *patch of raw cloth on an old garment* or pouring *new wine into old wine-skins. Raw cloth* is the newly woven material that has not yet been properly treated against shrinkage. To cut a piece from this for a patch not only damages the new cloth (cf. Luke *5, 36*) but the later shrinkage rends the old cloth to which it is attached. In the Near East the skins of animals are often used as containers for liquids. *New wine* is still subject to some fermentation; *old wine-skins* are not as strong as new ones and might not be able to stand this pressure. A similar figure in Job *32, 19*.

9, 18-26: The Ruler's Daughter; the Woman with a Hemorrhage. Parallels with more details in Mark *5,* 22-43 and Luke *8,* 41-56. These two Gospels place these events immediately after Christ's return from the country of the Gerasenes. The site was one of the towns with a synagogue on the western shore of the lake; perhaps Capharnaum. All three Gospels insert the account of the cure of the woman with a hemorrhage within the narrative about the ruler's daughter; while the order of events was certainly such as described, still this peculiar literary arrangement is undoubtedly due to the original source from which all three Gospels are derived.   **18.** *A ruler:* from the other Gospels we know that he was a ruler of the synagogue and that his name was Jairus. *Worshipped him:* paid homage to him, i.e., "he fell at his feet" (Mark *5,* 22). *"My daughter has just now died":* Matthew summarizes; at the first request of her father the girl was at the point of death, but she was dead before they reached the house (cf. parallels).   **20.** This woman was suffering from a pathological condition of what is otherwise the normal function of menstruation. Her disease rendered her legally unclean (cf. Lev. *15,* 25-30). *The tassel of his cloak:* at each of the four corners of their cloak the Jews attached a tassel by means of a blue cord to remind themselves of the obligation of keeping the commandments of the Lord (cf. Num. *15,* 38 f). To show their zeal for the Law the Pharisees wore very large tassels (cf. Matt. *23,* 5). Besides this woman other people were cured by touching one of the tassels of our Lord's cloak (cf. Matt. *14,* 36). There was nothing superstitious in this. It was merely a way of expressing one's faith in the wearer of the cloak. This woman chose this means probably from humility and embarrassment. She was cured at the moment she touched the tassel even before Christ spoke to her (cf. parallels).   **23.** *The flute players and the crowd making a din:* i.e., the professional mourners, the usual accompaniment of a funeral in the Near East.   **24.** Christ said, *"The girl is asleep, not dead,"* partly to conceal the miracle and partly to teach the true nature of death, because for a Christian death is only sleep in which he who has fallen asleep in the Lord awaits the resurrection. But this girl's death was a reality, as the mourners knew. All the people except the girl's parents and the three most intimate of the Apostles were put out of the room, so that the miracle might remain a secret (cf. parallels).   **26.** Matthew alone mentions this fact that the secret naturally could not be kept.

9, 27-31: Two Blind Men. Only in Matthew. *Son of David:* one of the typically Messianic titles. **28.** *The house:* probably our Lord's own house at Capharnaum. Christ waited until He was away from the public gaze before curing the blind men, in order to keep the miracle a secret (cf. 30). This miracle stresses the necessity of faith on the part of those who seek miracles from Jesus. Such faith requires not only that one firmly believe that He is the Christ and has the power to work the miracle but also that one sincerely trust that He will do it.

9, 32-34: A Dumb Demoniac. Only in Matthew. There is a very similar account in Luke *11,* 14 f, but since the latter passage is paralleled by Matt. *12,* 22-24, this event must be distinct. **32.** *As they were going out:* i.e., from the house mentioned in 28. **33.** It is clear from this that the man's speechlessness was the result of his being possessed by the devil. **34.** See Commentary on *12,* 24.

9, 35 — 10, 15: The Mission of the Apostles. **35-38.** Reason for this mission: the great amount of spiritual work to be done for the people and the impossibility, humanly speaking, for Jesus to do it all alone during the short time at His disposal. **35.** The same preaching tour of Jesus throughout Galilee is referred to in Mark *6,* 6b. Similar words, but referring to an earlier journey, in Matt. *4,* 23. **36.** Similar words in Mark *6,* 34 introducing the first multiplication of the loaves. The Israelites, abandoned by their leaders, are likewise compared to *sheep without a shepherd* in Num. *27,* 17; Ezech. *34,* 5. **37 f.** The same words of Christ in Luke *10,* 2 introducing the mission of the Seventy-two Disciples. For a similar thought, souls ripe for the kingdom of God likened to the white grain at harvesttime, cf. John *4,* 35b. The figure here is taken from the custom of hiring extra help at the harvest. **10, 1.** The bestowal upon the Apostles of the power to work miraculous cures. Parallels in Mark *6,* 7 and Luke *9,* 1 f. Christ allowed them to share in this divine power of His in order that they might thereby attract the people more easily and render their preaching more efficacious. *Having summoned his twelve disciples:* Matthew takes for granted that the Twelve already form a distinct group among the more general disciples of Jesus. They were in fact chosen much earlier in the ministry, immediately before the Sermon on the Mount, as we know from Mark *3,* 13-15 and Luke *6,* 12 f. **2-4.**

The list of the Apostles. Parallels in Mark *3*, 16-19; Luke *6*, 14-16; Acts *1*, 13. These four lists are substantially the same but have several minor differences both in the order of the Twelve and in their names and titles. In all the lists except Mark's the Apostles are grouped in pairs, probably because Christ "sent them forth two by two" (Mark *6*, 7), but possibly also as a mere memory-aid in the oral catechesis. Brothers are not always grouped together in the same pair. The order of the names is largely based on the relative importance of each of the Twelve. The order of the names of the Apostles in the Canon of the Mass differs somewhat from all the lists in the New Testament, not only by inserting Paul immediately after Peter but also by placing Thomas and James the Less much higher on the list.

2. The number *twelve* corresponds intentionally with the number of the tribes of Israel (cf. Matt. *19*, 28). The word *apostles* comes from the Greek and signifies literally "they who are sent," i.e., delegates, ambassadors, messengers. In the New Testament there are a few passages where the title is not restricted to the Twelve (e.g., Acts *14*, 13). First, not only in all the lists but also in importance because Christ bestowed on him the primacy, is *Simon, who is called Peter. Simon* is the Greek form of the Hebrew name *Simeon,* the name by which James the Less calls him in Acts *15*, 14 (according to the Greek and the Vulgate MSS). For being the first to acknowledge the divinity of Jesus he received from Him the appellation of "the Rock," in Aramaic, *Kepha*. However, during the mortal life of Jesus no one ever addressed him by any other name than Simon. (For the sole exception in Luke *22*, 34 see Commentary on that passage.) After the Ascension people commonly spoke of him as "the Rock" (Kepha) or "Simon, the Rock." But by the time of St. Paul this word had already lost its appellative force and was considered merely as another personal name of the Prince of the Apostles, receiving a Greek case-ending as *Kephas* or being translated into Greek as *Petros*. The translation however was not quite felicitous, for this Greek word with its masculine ending signifies merely a "stone," not the "bedrock" (petra). The Evangelists call him more or less indifferently "Simon," "Simon Peter," or simply "Peter." *Andrew,* a Greek word meaning "manly," is put in the fourth place in the lists of Mark and Acts in order to keep the three leading Apostles together. **3.** *James* always precedes his brother John in the Gospels, probably because he is older. Only in the list of Acts is the order reversed. On the call of these four

Apostles, see Matt. *4*, 18-22 and parallels. *Philip,* a Greek name meaning "lover of horses," occurs in the fifth place in all the lists. The Fourth Gospel alone gives us some further information about him. (cf. John *1*, 43-46; *6*, 5-7; *12*, 21 f; *14*, 8 f). *Bartholomew* follows Philip in all the lists except Acts where Thomas is placed between them. For this reason he is usually identified with Nathanael (cf. John *1*, 43 ff). Bartholomew, literally in Aramaic, "son of Tolmai," is a sort of "family name." *Thomas,* the Aramaic for "twin" as the Fourth Gospel correctly translates it (cf. John *11*, 16; *20*, 24; *21*, 2), is associated with *Matthew* in the lists of all three Synoptic Gospels. Only in the First Gospel is Matthew placed after Thomas and only here in all four lists is he called the *publican:* an act of humility which affords us a valuable confirmation of the authorship of this Gospel. *James, the son of Alpheus:* thus and in the ninth place in all four lists. He is very probably the same as "James the Less" of Mark *15*, 40 (cf. Matt. *13*, 55; Mark *6*, 3) and "James, the brother of the Lord" (Gal. *1*, 19), an important figure in the early Church at Jerusalem. Mark agrees with Matthew in the order of the last three names, but Luke, both in his Gospel and in the Acts, places *Simon* immediately after James and has "Jude the brother of James" instead of *Thaddeus.* Jude is therefore the same as Thaddeus; the latter name is really an epithet, "stout-hearted," from the Aramaic word for "breast." Instead of Thaddeus some Greek MSS have "Lebbeus" in the same sense, from the Hebrew and Aramaic word for "heart." 4. The appellation *the Cananean* or "the zealot" is added to Simon's name to distinguish him from Simon Peter. James, Jude and Simon were relatives of Jesus (cf. Matt. *13*, 55). *Iscariot* is usually explained as "man of Carioth" (a village in the extreme south of Juda, cf. Jos. *15*, 25) but this derivation is not entirely certain. In John *6*, 72 according to the best Greek and Latin MSS his father Simon is called Iscariot. *Who also betrayed him* is said by way of anticipation.

5-42. Instructions to the Apostles concerning their ministry. This is the second of the five great discourses of the First Gospel. It is intended in part for this present mission and in part for all their future ministry. As recorded in the First Gospel this instruction serves as a sort of early Canon Law for apostolic laborers. That Christ gave some such instruction on one certain occasion is placed beyond dispute by the partial parallels in Mark *6*, 8-11 and Luke *9*, 3-5. But Matthew's instruction is

10, 5b-15   St. Matthew

much longer than that of the other two Gospels. It incorporates almost all of the instruction to the Seventy-two Disciples as given in Luke *10,* 1-16 and in the section entitled "Opposition Foretold" duplicates many of the sayings of Christ on persecution as given in the so-called Eschatological Discourse or in other parts of the Synoptic Gospels.

5b-6. Only in Matthew. According to the plan of God in keeping with His promises to the Patriarchs, the Kingdom of God was to be preached first to the Jews alone. Therefore Jesus Himself limited His ministry to them (cf. Matt. *15,* 24). But after His resurrection He instructed His Apostles to preach the gospel to all nations (cf. Matt. *28,* 19; Mark *16,* 15). Consequently these words of Christ, forbidding the Apostles to preach to the Samaritans or the Gentiles, applied only to the period of our Lord's public ministry. Still, even after the Ascension the Apostles always endeavored to announce the glad tidings first to the Jews and only after these had heard the gospel did they preach directly to the Gentiles (cf. Acts *13,* 46; Rom. *1,* 16).

7 f. Christ instructs the Apostles to preach and to cure the sick. Parallels in Luke *9,* 2; *10,* 9. The summary of their preaching is to be the same as that of the Baptist and that of Christ Himself at the beginning of His ministry (cf. Matt. *3,* 2; *4,* 17). The purpose is to prepare the people for the Master Himself. *Freely you have received, freely give:* only in Matthew. Freely means "without cost." These words may rightly be used as an argument against simony.   **9 f.** Parallels in Mark *6,* 8 f and Luke *9,* 3; *10,* 4.7b. Apostolic poverty is inculcated in the spirit of the Sermon on the Mount, according to which the disciples of Christ are not to worry about future provisions (cf. *6,* 19-34). God will see to it that the people for whom they labor will supply them from day to day. Therefore they are not to acquire for themselves before the journey an extra tunic or an extra pair of sandals or an extra staff. This seems to be the sense of Christ's words in Matthew and Luke if they are to be harmonized with His words in Mark where He allows the Apostles one staff and one pair of sandals. By a *wallet* is here meant the cloth bag or knapsack in which travelers used to carry their provisions; the Apostles have no need of one for they are not to supply themselves with provisions for their journey.

11-15. Rules concerning the acceptance of hospitality. Parallels in Mark *6,* 10-11 and Luke *9,* 4 f; *10,* 5 f.10-12. These words presume that during the time the Apostles are engaged in their min-

istry in any town they accept free board and lodging in one of the
houses of the town. This was probably our Lord's own method
during His ministry. Hospitality has always been one of the most
highly cherished virtues in the Near East. Having found a respectable home where their good name will not suffer injury, the
Apostles are to accept hospitality there and make this their headquarters until they leave the town. To change residences would
only lead to jealousies and contentions among the townsfolk.
They are to be polite and use the ordinary greetings of all Semitic
peoples, "Peace." For them this is not to be an empty salutation,
as the world gives its greeting "Peace" (cf. John *14*, 27), but a
heart-felt blessing. If the person thus greeted is kind and good,
their *peace will come upon him*, i.e., their good wishes will be
efficacious; but if the person is unkind and unresponsive to their
preaching, their *peace* will return to them, i.e., their good wishes
will not be effective. According to the Hebrew idiom a word
"returns" to the speaker when it does not have its intended effect
(cf. Isa. *45*, 23). The symbolic act of shaking the dust from one's
feet to signify that one absolves oneself of all further responsibility toward an unreceptive community was actually carried
out by Paul and Barnabas at Antioch in Pisidia (cf. Acts *13*, 51).
The wicked cities of Sodom and Gomorrah (cf. Gen. *13*, 13; *19*,
1-26) had at this time become proverbial (cf. Rom. *9*, 29; Jude 7;
Apoc. *11*, 8), yet here and in Matt. *11*, 23 f Christ compares these
cities favorably with the towns that reject His teaching.

10, 16-42: **Opposition Foretold.** **16a.** Parallel in Luke *10*, 3.
These words, like all the following, refer to the persecution of
the Apostles and disciples of Christ after the Ascension. **16b.**
Only in Matthew. These words are not to be understood in too
general a sense, but directly in regard to persecution, as the context demands. The disciples of Christ are to use all possible
prudence to avoid being slaughtered as sheep are by wolves, but
at the same time they must never use duplicity or compromising
trickery in order to avoid persecution. Therefore they must be
as *wise*, i.e., prudent *as serpents*, but as *guileless as doves*, i.e.,
entirely free from the craftiness and trickery of serpents. Serpents
and doves have always been considered proverbial examples of
such traits.   **17 f.** A description of the future persecution of
Christ's disciples. Parallels in Mark *13*, 9 and Luke *21*, 12 f.   **17.**
But beware of men: "But be on your guard" (Mark) lest you fall
into the snares of wicked men. By *councils,* literally "sanhedrins."

are here meant the local tribunals; these courts had the power of condemning a culprit to be scourged in the synagogue, i.e., to "receive forty lashes less one" (2 Cor. *11,* 24). **18.** *For a witness to them,* i.e., the Jews (17), *and to the Gentiles:* in presenting their defense in these courts the disciples of Christ shall give testimony to Him and to His teachings, as Stephen did before the Sanhedrin (cf. Acts 7) and Paul before the Roman governors Felix and Festus, before King Agrippa (cf. Acts *24-26*) and even before the Roman Emperor (cf. Acts *27,* 24). **19 f.** Parallels in Mark *13,* 11 and Luke *12,* 11 f; *21,* 14 f. While using all prudence in defending themselves, the disciples are not to be worried by their lack of worldly eloquence, for the Holy Spirit will inspire their defense (cf. Acts *4,* 8). This reference to the bestowal of Christ's Spirit upon His disciples shows that the first three Evangelists presume a knowledge of this doctrine on the part of their readers. **21 f.** Parallels in Matt. *24,* 9 f.13; Mark *13,* 12 f; Luke *21,* 16 f. The disciples of Christ must expect persecution from all who do not believe in Him, even from their unconverted relatives. **21.** The teaching of Christ concerning the conduct of His disciples towards their relatives (cf. 35-37) should be understood in the light of these words. **23.** Only in Matthew. Cf. also *23,* 34. Christ never exhorted His followers to expose themselves rashly to martyrdom. The Apostles understood this teaching very well and escaped several times from their persecutors (cf. Acts *9,* 23-25.29 f; *12,* 5 f; *17,* 10; *20,* 1 f). On the other hand one who has a fixed charge of souls in a definite locality should not leave his flock and flee (cf. John *10,* 11 f). **V.23b** may be interpreted in two ways: (a) you will not have converted all of Israel before the Second Coming of Christ; (b) you will not have preached the gospel in all the towns of Palestine before the destruction of Jerusalem, which is a type of the end of the world. **24 f.** Similar words in Luke *6,* 40, but in a different context and with a somewhat different meaning. Here the sense is: the disciples of Christ cannot expect an easier fate than the Master has suffered. Cf. also John *13,* 16; *15,* 20. On the meaning of Beelzebub, see Commentary on *12,* 24.

26a. Only in Matthew. **26b-27.** A favorite saying of our Lord, spoken on various occasions, each time in a somewhat different sense. Here the meaning is: either (a) preach the gospel boldly and openly without fear, despite your persecutors; or (b) you may be reviled and persecuted now but on the Last Day you will be justified before the whole world. Luke *12,* 2 f records very similar words but with a somewhat different sense in regard to the hypocrisy of the Pharisees. A similar saying of Christ given in Mark *4, 22* and Luke *8, 17* refers to the clear explanation of the hidden truths of the parables.   **28-31.** Parallel in Luke *12, 4-7; 21,* 18. The thought of God should strengthen the disciples in time of persecution. They should remember that on the one hand, if they are unfaithful to Him, He will inflict much worse punishment than the persecutors can, but on the other hand, if they are faithful to Him, not the slightest temporal evil will befall them without His permission.   **28.** God does not *destroy* the existence of *both soul and body in hell,* but He does destroy their true life and happiness.   **29-31.** An argument *a fortiori* from God's care of such unimportant creatures as the little birds; the same argument in the Sermon on the Mount (cf. *6,* 26). The relative unimportance of sparrows is shown by their cheapness: *two sparrows for a farthing,* in Greek, "an assarion," about one cent. They are even cheaper when bought in larger quantities: "five sparrows sold for two farthings" (Luke *12,* 6).

32 f. Parallel in Luke *12,* 8 f. Similar words in Mark *8, 38;* Luke *9, 26.* The danger of persecution is that the disciples may deny Christ. In that case, when they appear before His *Father in heaven,* i.e., before the tribunal of Christ, He also will say that He does not know them (cf. *7, 23; 25, 12).*   **34-36.** Parallel in Luke *12, 51-53.* A prediction of the discord which His teaching will cause among men, even among the members of the same household.   **34.** Christ is speaking here of the results of His teaching: some will not believe and will hate those who do. But his intention was to bring peace to the world (cf. Luke *2,* 14). It would be an abuse of these words of Christ to quote them in favor of war. The *sword* is here a figure of speech for persecution. **35 f.** His doctrine will cause division even in such natural pairs as father and son, mother and daughter, etc. Both these verses are quoted from Mich. *7, 6.* It would be wrong to cite these words in the sense that normally in good Christian families the worst enemies of a man's salvation or of his religious vocation are the members of his own family.   **37-39.** Parallel in Luke *14, 26* f; *17, 33.* Similar words on the doctrine of the Cross for the followers of Christ in Matt. *16, 24* f and parallels.   **37.** That man would love his relatives more than Christ who would allow these to prevent him from following Christ.   **38.** Every true disciple of Christ must be willing to suffer even as his Master suffered. There is a clear reference here to the crucifixion of

Christ, even though He meant the *cross* for His followers mostly in the figurative sense. The Apostles should have understood His words without difficulty, for the Roman punishment of crucifixion, in which the condemned man had to carry his own cross to the place of execution, was only too well known to the Jews. **39.** *Life:* literally "soul," the principle of life, used here in a twofold sense: *He who finds his life,* i.e., he who saves his mortal life by being unfaithful to me and my doctrine, *will lose it,* i.e., his true eternal life; and vice versa. Cf. John *12,* 25.

40-42. The reward of those who are hospitable to the disciples of Christ. There is no close connection between this and the preceding; it is a return to the thought of hospitality in 11. **40.** Similar words of Christ in Matt. *18,* 5 and parallels; Luke *10,* 16; John *12,* 44; *13,* 20; cf. also Matt. *25,* 40.45. The thought is not merely that Christ will consider what is done to His disciples as if it were done to Him personally; it is done to Him and through Him to the Father because of the reality of the Mystical Body of Christ. **41.** Peculiar to Matthew; a further development of the preceding thought. *Receive a prophet's reward:* i.e., have a share in the reward of the prophet whom he befriends. **42.** Similar words in Mark *9,* 40. In the heat of the long dry summer of Palestine *a cup of cold water* is a welcome, if inexpensive, gift. It is not the material value but rather the spirit of kindness and above all the motive, "because you are Christ's" (Mark), that makes the act meritorious.

11, 1-6: The Baptist's Deputation. Parellel in Luke *7,* 18 f. 22 f. **1.** A general statement serving merely as a transition; the first part is the common conclusion to all the five great discourses of the First Gospel. **2.** *When John had heard in prison:* cf. *4,* 12; *14,* 3-12. According to Josephus, John was imprisoned in the frontier fortress of Macherus, situated on a high plateau near the river Zirka Main, to the east of the Dead Sea. *John heard of the works of Christ* from his own disciples (Luke). Hearing of His miracles would have been no cause to doubt the Messiasship of Jesus. *He sent two of his disciples to say to him:* literally in Greek, "having sent through his disciples, he said to him," i.e., he sent his disciples with this message. The reading *two of his disciples* is taken from Luke.

3. *He who is to come* is one of the regular titles of the Messias. Why did the Baptist have his disciples ask Jesus this question? According to a few Catholics and almost all non-Catholics Jesus did not measure up to the Baptist's expectations and idea of the Messias and John was now beginning to doubt whether he had really pointed Him out correctly. The traditional Catholic opinion holds that John never doubted for a moment that Jesus was the Messias; he asked this question not for his own sake but for the sake of the disciples who were somewhat scandalized by the conduct of Jesus (cf. *9,* 14), jealous of His success (cf. John *3,* 25 f), and who perhaps shared the ideas of the Messias common among the people. This opinion is much more probable not only for *a priori* reasons, that God would not allow the Precursor of the Messias to waver in his faith, but also by reason of the context, especially our Lord's magnificent eulogy of the Baptist, that he was not *a reed shaken by the wind* (7). Perhaps the situation was this: John's disciples in a spirit of jealousy told their master in prison of all the works of Jesus; John replied, "This is He of whom I said, 'He is to come after me.' No one can receive anything unless it is given to him from heaven. If you do not believe me, ask Him Himself." (Cf. John *1,* 15; *3,* 27-30.) **4-6.** Jesus cannot answer the question directly by saying, "Yes, I am the Messias," for the people have a wrong idea of the Messias. Therefore He lets John's disciples see the miracles that He was working at that very time (cf. Luke *7,* 21) and points out how these miracles fulfill the Messianic prophecy of Isa. *35,* 5 f; *61,* **1.** He who has not a false idea of the Messias *is not scandalized,* i.e., does not trip up and stumble spiritually, when he sees the true Messias.

11, 7-19: Christ's Witness Concerning John. **7-11.** Parallel in Luke *7,* 24-28. Christ's praise of the Baptist was spoken lest the people conclude from the question of his disciples that he himself had lost faith in Jesus. **7.** Christ waited till *these* disciples of John *were going away* before beginning His panegyric lest He seem to be currying the favor of the Baptist by flattering him before his disciples. *"What did you go out to the desert to see?"* is repeated three times for emphasis. On the crowds that were attracted to John's preaching, cf. *3, 5. To see:* literally in Greek, "to gaze upon" as a wonder. The sense is not "Did you go out to the desert just to see the scenery, the reeds by the bank of the Jordan?" but rather figuratively, "Did you go out to see a weak man who wavers in his opinions like a reed shaken by the wind?" **8.** John's rough garments were one of his characteristics (cf. *3,* 4). There is probably an allusion here to the immoral luxury in the palace of Herod. **9 f.** John was *more than a prophet,* for while the prophets of the Old Testament pointed out the Messias at a distance, the Baptist pointed Him out directly and physically. He thus fulfilled his office of messenger or precursor of the Lord spoken of by the prophet Malachias (Mal. *3*, 1). The prophet speaks in the person of Christ, "before my face . . . my way before me," but our Lord, by changing the pronouns, quotes the words of God the Father addressing His Son. In the Second Gospel these words are cited in the same manner by the Evangelist himself (cf. Mark *1*, 2). **11.** *A greater than John the Baptist* is not to be understood absolutely nor even as a comparison between him and all the others of the Old Dispensation but solely in regard to his prophetic office, i.e., "there is not a greater prophet than John the Baptist" (Luke). Yet the New Dispensation, in which every disciple of Christ is mystically united with the Son of God, is so far superior to the Old that even *the least in the kingdom of heaven is greater than he.* **12 f.** Similar words in Luke *16,* 16, but probably in a different sense. The exact sense of this difficult passage is disputed. Either: (a) Since the Old Dispensation came to an end with John, the kingdom of heaven, i.e., the New Dispensation, *has been enduring violent assault,* in Greek literally "is being violated," i.e., is being attacked by those who refuse to accept the change, *and the violent,* or "the violators," have *been seizing it by force,* in Greek literally "are plundering it." Or, (b) Because the Old Dispensation passed away with John, men are now rushing with violent enthusiasm to get into the kingdom of heaven, the New Dispensation, and they who rush through the crowd with force lay hold of the kingdom of heaven for themselves. The first opinion seems to fit the actual historical circumstances much better than the second. But it is to be noted that the Rabbis interpreted Mich. *2,* 12 f in the sense that, when the Messias would come, there would be a violent tumult of men jostling one another in an effort to get into His kingdom. The use of this text by ascetical writers in the sense that only they who do violence "to themselves" can enter the kingdom of heaven, would seem to be a mere accommodated sense. **13.** The whole Old Testament was a period of preparation which reached its climax and came to an end with John. **14.** Our Lord treats of the same question in Matt. *17,* 10-13; Mark *9,* 10-12. Mal. *4,* 5 f foretells the coming of Elias to prepare the people for the Messias. The Jews understood this prophecy in the literal sense that Elias would come in person before the coming of Christ. The reference to this prophecy in Luke *1, 17* shows that it was fulfilled in John the Baptist. Jesus therefore in this eulogy of the Baptist tells the people who John really is: "He is Elias who was to come," i.e., he is the Elias foretold by Malachias. Since a certain amount of good will is required to accept this interpretation of the prophecy, Christ says, "If you are willing to receive it," i.e., if you have the will to believe it. **15.** This phrase was frequently added by Christ to His sayings to emphasize their importance.

16-19. The Stubborn Children. Parallel in Luke *7, 31-35.* By means of a little parable Christ explains the reason why the men of His generation, especially the Scribes and Pharisees (cf. Luke *7,* 30), accept neither the Baptist nor Himself. **16a.** An oratorical question, as if Jesus were seeking for a suitable comparison. Cf. Mark *4,* 30. **16b.** *Who call to their companions:* their "call" is not the words of 17 but the words of wedding songs and funeral songs that they sing in "playing wedding" and "playing funeral." Luke reads, "calling to one another," i.e., alternately in two groups; this was the ordinary way of choral singing among the Jews at that time and has been preserved in the antiphonal singing of the Psalms in the liturgy. **17.** *And say:* the sense is, "This generation is like those peevish children who refuse to join in the games of the other children in the market place and to whom these children say, . . ." Such an elliptical construction is not uncommon to the popular parables of our Lord. The *market place* was the only open space in an ancient city and naturally served as the children's playground. *We have piped:* i.e., played joyful music such as was used for dancing at a wedding. For music and dancing at a feast, cf. Luke *15, 25. You have not mourned:* the children at play imitated the oriental custom of giving exaggerated signs of grief at a funeral (cf. Mark *5,* 38). **18 f.** Application of the parable. The men of this generation are like these peevish children: they are satisfied neither with John's asceticism nor with Christ's deep humanity, ignoring the stern preaching of the Baptist by saying, "He is demented," and rejecting the teaching of Jesus by saying, "He is a sinner like the rest of men." *And wisdom is justified by her children:* the best Greek MSS in Matthew read, "by her works;" the reading "by her children" has been taken over from Luke; but both expressions are synonymous, for "children" is a common Semitic figure for "works." The whole

11, 20-25 ST. MATTHEW

expression is evidently proverbial, as shown by use of the Greek aorist tense in the sense of a universal present, the so-called "gnomic aorist." The sense of the proverb is probably this: a wise man proves that he is wise by his actions; therefore the way the Scribes and Pharisees act proves that they are fools.

11, 20-24: The Impenitent Towns. Parallel in Luke *10*, 13-15. It is doubtful on what occasion Jesus spoke these Woes against the towns that rejected His teachings. **21.** *Corozain* was evidently a town of some importance, for this town and Bethsaida are here compared with the two important cities of Tyre and Sidon on the Phoenician coast. Today it is represented by a few ruins (Khirbet) known as Kerazeh, about two miles north of Tell Hûm (Capharnaum), a mute witness of God's punishment. This is the only place in which Corozain is mentioned in the New Testament, yet it was one of the *towns in which most of His miracles were worked* (20): a clear proof that our four Gospels give but a small part of our Lord's activity. On the site of Bethsaida, see Commentary on *14*, 13. *Tyre and Sidon* are large commercial cities whose riches and luxury undoubtedly made them corrupt with sin and vice. The frequent invectives of the Prophets (cf. especially Isa. *23;* Ezech. *26-28*) against these two cities had made them proverbial examples of wickedness. Christ passed near them on one occasion, but it seemed He did not enter the cities themselves (cf. Matt. *15,* 21; Mark *7,* 24.31). This region was almost entirely pagan, although there were probably small Jewish colonies there (cf. Mark *3,* 8; Luke *6,* 17); later on St. Paul found some Christians at Tyre (cf. Acts *21,* 3-6). *Sackcloth and ashes* are mentioned frequently in the Old Testament as the symbols of mourning and means of penance. For a whole city doing penance in sackcloth and ashes, cf. Jon. *3,* 5-8. **23.** "Shalt thou be exalted to heaven?" i.e., in pride. These words are reminiscent of the words of Isaias (*14*, 13-15) against the king of Babylon. **24.** See Commentary on Matt. *10,* 15.

11, 25-30: Jesus Draws Men Gently to Himself. 25-27. Parallel in Luke *10,* 21 f. The mystery of the Father and of the Son is revealed only to the humble. Luke connects these words with the return of the seventy-two disciples from their mission. **25.** God chooses the lowly. Cf. *1* Cor. *1,* 26-29. *I praise thee:* the Hebrew verb behind this expression signifies both praise and thanksgiving. The Gospels speak often of Jesus praying to His heavenly Father; this is a sample of His prayers. *Lord of heaven and earth* emphasizes the sovereignty of God as Creator of the universe; the same expression is used by St. Paul in his discourse at Athens (cf. Acts *17,* 24). *The wise and the prudent* are especially the Scribes and Pharisees who think themselves wise and clever. *The little ones:* literally "infants," the humble and lowly (cf. Pss. *8, 3*; *18,* 8). *These things that thou didst reveal* to them are the truths concerning the relationship between Jesus and His heavenly Father (27); these truths cannot be known by natural reason alone but require a divine revelation (cf. *16,* 17; Gal. *1,* 15 f). **26.** *Yes, Father* expresses the full accord of Christ's will with the will of His Father in this divine predilection of the lowly. **27.** *All things have been delivered to me by my Father,* especially all divine knowledge, for the verb used here refers especially to the handing down of knowledge. These words place the Son, Jesus, on an absolute par with God the Father: the clearest possible proof of the divinity of Christ. Since only the Father of Himself knows the Son and only the Son Himself knows the Father, it follows that man cannot have an adequate knowledge of either the Father or the Son except by revelation. The true nature of the Father is known to those men alone *to whom the Son chooses to reveal him.* But the Father also in His good pleasure reveals the Son to the little ones (25 f). God indeed offers this revelation to all men but the proud close their eyes to the light. Hence by an act of deliberate choice, of predilection, God gives the efficacious grace to believe only to the humble. On the whole passage, cf. John *1,* 18; *5,* 20a; *6,* 44-46.66; etc. This saying of Jesus that is preserved in two of the Synoptic Gospels bears a striking resemblance both in thought and in diction to the words of Jesus in the Fourth Gospel; an excellent argument for the authenticity of the words of Christ as recorded by St. John.

28-30. Only in Matthew. Since Jesus alone can teach men the true nature of God and His holy will in their regard, He now extends to all men this alluring invitation to be His disciples. For similar invitations given by divine Wisdom, cf. Ecclus. *24,* 26 f; *51,* 31.34 f. **28.** *All you who labor and are burdened,* with the sorrows and afflictions of life but above all with the burden of sin which the Law cannot relieve (cf. Acts *13,* 39; Rom. *3,* 28). The repose that Christ offers is spiritual *rest* and peace of heart. **29.** The Rabbis often spoke of the Old Testament as "the yoke of the Law"; Christ offers "His yoke" in its stead. *Learn from me,* be taught by me, be my disciples: directly this is the object of the whole invitation, but the passage can also be understood in the extended sense as an invitation to make Christ the center of our devotional life. *For I am meek,* etc.: this can also be translated, "that I am meek," etc. (so St. Augustine); but in the context the phrase is better understood as giving the reason. On the meekness and humility of Jesus, cf. *12,* 19 f. *You will find rest for your souls:* the same words in Jer. *6,* 16; cf. also Isa. *28,* 12.   **30.** Every commandment is a *yoke,* but Christ's yoke is easy to bear (cf. *1* John *5,* 3). The contrast is with the yoke of the Old Law, especially as interpreted by the Scribes, who made it really unbearable (cf. Matt. *23,* 4; Acts *15,* 10).

12, 1-8: The Disciples Pluck Grain on the Sabbath. Parallels in Mark *2,* 23-28 and Luke *6,* 1-5; according to these two Evangelists this is the fourth of the five disputes with the Pharisees before the Sermon on the Mount. The ripe grain in the field points to a season of the year near the Passover; apparently the second of the four Passovers of our Lord's public ministry. Here as often Matthew abandons the strict chronological sequence. This and the following dispute are concerned with the observance of the Sabbath by abstaining from labor. The Pharisees made much of this point of the Law in their opposition to Jesus. The Fourth Gospel agrees with the Synoptics in this regard (cf. John *5,* 9-18; *7,* 22 ff; *9,* 14.16). These controversies concerning the Sabbath show the typical narrow-mindedness of these hypocrites who quibbled over such fine points of the Law and ignored the basic principles of morality (cf. *23,* 23 f).   **1.** *Jesus went through the standing grain:* i.e., along a path with fields of ripe grain on each side. *His disciples . . . began to pluck the ears of grain:* in itself this was permitted by the Law (cf. Deut. *23,* 25).   **2.** Work on the Sabbath was forbidden by the Law (cf. Ex. *20,* 10). To safeguard this law, to "put a hedge around it" as the saying was, the Rabbis forbade anything having even the semblance of work on the Sabbath. Reaping on the Sabbath was indeed specifically mentioned in the Law as illicit (cf. Ex. *34,* 21), but Deut. *23,* 25 clearly distinguishes the plucking of a handful of grain from reaping with a sickle.   **3 f.** First argument of Christ in defense of His disciples: a positive law may rightly be dispensed with in a case of proportionate necessity. He argues this point by the example of David who, when hungry, ate the sacred bread which he should not otherwise have eaten (cf. *1* Kgs.

21, 6; on *the loaves of proposition which only the priests* could lawfully eat, cf. Lev. *24*, 5-9).   **5 f.** Only in Matthew. Second argument of Christ: the work done by the priests in the temple on the Sabbath was not considered a violation of the sabbatical rest; but He is greater than the temple; therefore the disciples who serve Him on the Sabbath are lawfully dispensed from the Sabbath law. On the sacrifices which the priests had to offer in the temple on the Sabbath, cf. Num. *28*, 9 f.   **7.** Only in Matthew. Third argument: God is more pleased with kindness and charity than with zeal for the Law; proved by quoting Os. *6*, 6 (cf. Matt. *9*, 13).   **8.** Final argument, or the conclusion of all these arguments (cf. "Therefore" in Mark *2*, 28): the Messias is the authentic interpreter of God's Law, including the law of the Sabbath.

12, 9-14: A Man with a Withered Hand. Parallels in Mark *3*, 1-6 and Luke *6*, 6-11. The last of the five disputes with the Pharisees before the Sermon on the Mount. There is a similar discussion in connection with the cure of the man with dropsy in Luke *14*, 1-6; cf. also the case of the stooped woman in Luke *13*, 10-16.   **9.** *And when he had passed on from that place*, not necessarily on the same Sabbath as the preceding; Luke says it was on another Sabbath. *Their synagogue:* the synagogue of the people here concerned, but the name of the town is not mentioned; perhaps Capharnaum.   **10.** *A withered hand* was one that was in some way shrunk, stunted or crippled; the popular language considered it deprived of its natural juices like a wilted plant. *They asked him:* they were Scribes and Pharisees (cf. Luke). In the other Gospels this question is asked not by these men but by Christ. However, since our Lord read their thoughts, (cf. Luke), Matthew summarizes by putting in their mouth the thought that was in their mind. *That they might accuse him:* perhaps they had brought this cripple into the synagogue intentionally as a test-case.   **11.** The opinions of the Rabbis as recorded in the Talmud differed on this case of an animal that falls into a pit on the Sabbath, but Christ's argument is evidently based on the actual practice at His time. The same argument is used in Luke *14*, 5 except that instead of a sheep Christ speaks of "an ass or an ox." Matthew is the only Evangelist that has this argument here; the other two have a more general question about doing good on the Sabbath.   **12.** An argument *a fortiori* such as Christ used so often. The conclusion shows that the specific case of the sheep was adduced as an argument to answer the general question as given in Mark and Luke. **14.** The earliest recorded instance in our Lord's public life of a plot to put Him to death. The Pharisees evidently tried to trump up a capital charge against Him, such as a flagrant violation of the Sabbath (cf. Ex. *31,* 15).

12, 15-21: The Mercy of Jesus. Peculiar to Matthew; but a similar account of Christ's miracles in general and of the crowds that followed Him is given before the Sermon on the Mount in Matt. *4,* 24 f; Mark *3,* 7-12; Luke *6,* 17-19. **17-21.** In this kindness and meekness of Jesus the Evangelist sees the fulfillment of the words of Isa. *42,* 1-4. This is the first of the famous "Servant of the Lord" hymns in Isaias. Since St. Matthew gives us the authentic interpretation of it, we may rightly conclude that all these hymns are Messianic even though some of them may not be explicity cited in the New Testament. For other references to the Messias as "the Servant of the Lord," cf. Acts *3,* 13.26; *4,* 27.30; out of respect for our Lord later copyists sometimes changed the word "Servant" into "Son." The quotation in Matthew follows neither the Hebrew nor the Septuagint text; it is probably based on an Aramaic translation or "Targum" that is otherwise not preserved. **18.** *I will put my Spirit upon him:* Jesus had God's Holy Spirit from the beginning; the words signify that He would show this forth publicly, or refer to the external manifestation of the Spirit at His baptism. *He will declare judgment to the Gentiles:* i.e., He will show the Gentiles the way of justice and holiness. **20a.** *A smoking wick* is the wick of a lamp that is dimly burning but in danger of going out completely. This figure and that of the *bruised reed* signify the slight evidence of good will in morally weak men; far from rejecting such men Christ encourages them to greater efforts. **20b-21.** Despite initial reverses final victory is promised the Messias in His efforts to teach men justice and holiness. *In his name,* i.e., in Him, *will the Gentiles hope:* so also the Septuagint; the Hebrew text reads, "The isles (i.e., the distant lands of the Gentiles) await his teaching."

12, 22-37: Blasphemy of the Pharisees. The accusation that Jesus drives out devils only by the power of Beelzebub and the refutation of this calumny (22-32) is paralleled in Mark *3,* 22-29 and Luke *11,* 14-23; *12,* 10. **22-24.** A distinct yet remarkably similar incident is given in Matt. *9, 32-34*. The Pharisees cannot deny the reality or the supernatural character of this cure. Therefore, in order to counteract the tremendous impression which this miracle makes on the people, they try to use this very miracle against Jesus by claiming that He Himself is possessed by the prince of devils and consequently has such control over the other devils. According to Mark these Pharisees were "Scribes who had come down from Jerusalem;" probably in a certain official capacity to investigate this Prophet from Nazareth. This also explains the importance of their diabolical calumny. *Beelzebub:* the form "Beelzebul" has far better support in the MSS. But "Beelzebul" cannot be explained satisfactorily as "the lord of dung," or "the lord of the dwelling." It is merely a corrupted form of the older pronunciation "Beelzebub." By assimilation the consonants of the last syllable were influenced by those of the first syllable. In Babylonian the "accuser" in a lawsuit was called the "beeldabab," i.e., "the lord of the word;" the Aramaic borrowed this word in the older form of Aramaic "beeldhebòb," which later became "beelzebūb." Therefore this word is the exact Aramaic translation of the Hebrew word "satan" (the accuser) just as the Greek "diabolos" (the devil) also means "the accuser." In *4* Kgs. *1, 2* the Jews likewise call the god of Accaron "Beelzebub," i.e., Satan.   **25 f.** Christ's first argument in refuting this calumny: the absurdity of supposing that Satan would destroy himself.   **27.** Christ's second argument: *your children,* i.e., the disciples of the Scribes, attempt exorcisms (cf. Acts *19, 13*) and believe that they occasionally succeed with God's help. *Therefore they shall be your judges:* i.e., the Scribes' own disciples will condemn their teaching that devils are cast out only by the power of Beelzebub.   **28.** The conclusion to the preceding argument.   **29.** Christ portrays His victory over Satan in the form of a little parable. *The strong man* is Satan; the "stronger than he" (Luke), who binds him is Christ. On the binding of Satan, cf. Apoc. *20, 2.*   **30.** The usual interpretation of these words is, "It is impossible for a man to be neutral in my regard; he must either be for me or against me." Essentially the same idea is expressed in Mark *9, 39,* but from the opposite viewpoint. Here however *he who is not with me* may refer directly only to Satan, i.e., since Satan is not helping me in casting out devils, he must therefore be opposed to me.

31. Blasphemy is used here in the original sense of the Greek word from which this English word is derived, i.e., "Speaking evil, slander, calumny," whether against God or man, as is clear from the following verse. *The blasphemy against the Spirit* is slandering the Spirit of God by saying that it is evil. But the Spirit of God is essentially good and holy. Therefore to call good evil and evil good is a complete inversion of all moral values. Such a sin is of its very nature unforgivable because it denies the very goodness of God, the only source of the forgiving of sins (cf. Rom. *2,* 4). These Scribes were guilty of such a sin, "for they said, 'He has an unclean spirit' " (Mark *3,* 30); i.e., they called the Spirit of God Satan. **32.** To *speak a word against the Son of Man* is to calumniate Christ in His human nature. Such a sin is not a deliberate blinding of oneself to the light and therefore of itself not unforgivable, as in the case of the Good Thief. This verse is often quoted as one of the proofs for the existence of Purgatory, but it is doubtful whether it has such probative force. For to state negatively that there is a certain sin that *will not be forgiven either in this world or in the world to come* does not necessarily imply the positive statement that certain other sins are forgiven in the world to come. Moreover, Purgatory is not a place where certain sins are forgiven but the place where the temporal punishment, still due to sins already forgiven, is suffered. However, this text may imply the existence of Purgatory (cf. 2 Mach. *12,* 46). **33-37.** Only in Matthew in this context. **33.** The same figure in the Sermon on the Mount (*7,* 17 ff; Luke *6,* 43 ff); but here the thought is: "Be consistent. If you pretend to be good internally, show this externally by good works." Or perhaps: "Be logical. If I produce good works, such as the casting out of devils, then admit that I myself am good." **34.** *Brood of vipers:* the same epithet for the Pharisees in *3,* 7; *23,* 33. A man's expressed thoughts show his mind and character. He cannot avoid this, for his speech comes forth *out of the abundance,* literally in Greek "the overflow," *of the heart,* i.e., of of his mind and sentiments. **35.** Luke *6,* 45 has the same words in the Sermon on the Mount. The same thought as the preceding, but here the mind is pictured as the *treasure,* i.e., the storehouse, of either good or evil things. **36 f.** *Every idle word* means every seemingly unimportant but consciously spoken word. Every deliberate human act necessarily has moral value for which we will either be rewarded or punished.

12, 38-45: The Sign of Jonas. Parallel in Luke *11,* 16.29-32. Cf. the similar but distinct incident in Matt. *16,* 1-4; Mark *8,* 11 f.

38. *Certain of the Scribes and Pharisees:* according to Luke they were different from those who accused Him of working miracles by the power of Beelzebub; but it was on the same occasion. It was probably out of sarcasm that these Scribes addressed Him as their equal, *"Master,"* i.e., Rabbi. *"We would see a sign from thee:"* not any miracle in general, such as curing the sick, but "a sign from heaven" (Luke), i.e., from the sky, such as Moses wrought in bringing down manna from heaven. They expected this of the Messias (cf. John *6,* 30). **39.** *Adulterous generation* (cf. also *16,* 4; Mark *8,* 38) is not to be understood in the literal sense but as equivalent to "perverse generation" (*17,* 16; Luke *9,* 41). This figure was often used by the Prophets who compared Israel's faithlessness to God's covenant to a wife's infidelity (cf. Os. *2;* Ezech. *16,* 15 ff; *23*). Cf. also Jas. *4,* 4. Such a generation does not deserve a sign. Therefore *no sign shall be given it but the sign of Jonas the prophet.* The same words in Matt. *16,* 4 are paralleled in Mark *8,* 12 as "Amen I say to you, a sign shall not be given to this generation." This agrees with the explanation of the "sign of Jonas" in Luke *11,* 30: "For even as Jonas was a sign to the Ninevites, so will also the Son of Man be to this generation." The whole point is that Jonas worked no miracle for the Ninevites; they simply believed his preaching. Just as Solomon gave no "sign" to the queen of the South except the wisdom of his words, so also Jesus will give no sign to the men of His generation except the wisdom of His teaching. Directly therefore Christ refuses them a sign. **40.** Nevertheless He does give them a sign, but one which because of their perverseness can only be verified too late, the sign of *Jonas in the belly of the fish.* All the parallel passages omit these words, but there is no reason to question their authenticity, for the Pharisees knew even before the Crucifixion that Jesus had prophesied His resurrection after three days (cf. *27,* 63). Christ actually rose on the morning of the third day, but according to the Hebrew way of reckoning, any part of a day (of twenty-four hours) at the beginning and at the end of a period of time was considered a whole day. In Jon. *2,* 1, which our Lord quotes literally, the whole or a part of a period of twenty-four hours is called "a day and a night." Therefore Christ truly fulfilled this prophecy of "three days and three nights" even though He was actually in the tomb only one full day and two nights. By *the heart of the earth* is meant not directly the tomb but the netherworld, the abode of the dead, which according to the ideas of the time was situated in the

center of the earth. **42.** *The queen of the South,* the queen of the Sabeans in southern Arabia (cf. *3 Kgs. 10).*
43-45. The return of the unclean spirit. Parallel in Luke *11,* 24-26. In the form of a little parable Jesus teaches that *this evil generation,* for failing to correspond to the grace that He offers, will in the end be much worse than if the grace had not been offered. *Dry places:* according to the popular notion, to which our Lord does not hesitate to accommodate Himself, the normal dwelling place of evil spirits is in the desert (cf. Isa. *34,* 14; Bar. *4, 35;* Tob. *8, 3).* In the application of the parable the soul that has once received God's grace but not fully corresponded with it, is indeed *swept and adorned* but *unoccupied,* i.e., God does not possess it fully. *Seven* is symbolic, signifying a large number; cf. the seven devils that had gone out of the Magdalene (Luke *8,* 2), the seven times in a day to forgive injuries (Luke *17,* 4), etc.

12, 46-50: Jesus and His Brethren. Parallels in Mark *3,* 31-35 and Luke *8,* 19-21. The first two Gospels agree in connecting this event with the Beelzebub dispute; Matthew has a definitely temporal phrase joining the two events. Luke on the other hand places this event, with no indication of time, after the Sermon in Parables. The order of the first two Gospels is therefore to be preferred here.   **46.** *His brethren:* this expression also occurs in *13,* 55 f; Mark *6, 3* (in both these passages there is mention also of "his sisters"); John *2,* 12; *7,* 3.5.10; Acts *1,* 14; *1* Cor. *9,* 5. Many non-Catholics, reviving the heresy of Helvidius (about 380 A.D.), consider these "brethren and sisters" of Jesus to be the later children of Joseph and Mary. This is contrary to the teaching of the Catholic Faith concerning the perpetual virginity of the Mother of God. The common opinion in the Eastern Church and also among Protestants regards these "brethren" as the children of Joseph by a previous marriage. This theory, although held by several Fathers of the Church, is ultimately based on apocryphal gospels. After being vigorously attacked by St. Jerome, it was no longer held by any Western Father. It is rightly rejected by Catholics, not only because it offends against the pious sentiment of the Church concerning St. Joseph, but also, because it contradicts the data of the New Testament in regard to at least some of these "brethren." For among them are James, Jude and Joseph, the sons of Cleophas (or Alpheus) and Mary, the "sister" of the Mother of Jesus (cf. Matt. *13,* 55; *27,* 56; Mark *6, 3; 15,* 40.47; Luke *24,* 10; John *19,* 25; Gal. *1,* 19; Jude 1). They were therefore not His brothers in the strict sense of the word but His cousins (in what degree of relationship is not quite certain). In Hebrew the use of the word "brethren" for any blood relatives was common (e.g. in Gen. cp. *13,* 8 with *14,* 12; *29,* 15 with *24,* 29). *Seeking to speak to him:* we do not know what they wished to tell Him. **48-50.** By these words Jesus showed: (a) that the bond of spiritual relationship, founded on the union of the true children of the heavenly Father, is more important than that of blood relationship; (b) that He Himself carried out His own precepts concerning the demands of the kingdom of God and the demands of family ties (cf. *8,* 21 f; *10,* 37). But there is no disrespect shown here to His mother and His brethren.

13, **1-23: Parable of the Sower.** *13,* 1-52 forms the third of the five great discourses of Jesus in the First Gospel. It consists of seven parables, the explanation of two of these, and other words about the reason for such a method of teaching. Since all but the first parable begin with the words, "The kingdom of heaven is like . . . ," this discourse is commonly known as "The Parables of the Kingdom." In the corresponding sermon in the Second Gospel Mark *(4,* 1-34) gives only two of these parables, but he has an extra one peculiar to himself. Luke *(8,* 4-18) has but one parable on this occasion, but he records two others of Matthew's seven on another occasion (Luke *13,* 18-21). This discourse in the First Gospel may therefore be a partially artificial arrangement; still it would seem from Mark *4,* 33 f that on this day in particular Jesus "spoke many such parables."

1-3a. The setting. All three Gospels agree that this discourse was spoken to *great crowds gathered about him.* The sloping shore offered a natural auditorium, while the water kept the crowd from pressing too closely around the Master. For other occasions on which Jesus taught from a boat anchored near the shore, cf. Mark *3,* 9; Luke *5,* 3. See article on The Parables of the Gospels.

3-9. The Parable of the Sower. Parallels in Mark *4,* 3-9 and Luke *8,* 5-8. From the context, the *seed* which the *sower went out to sow,* is obviously grain. The commonest grain in Palestine was barley, but wheat, rye and other grains were known. **4.** *Some seed fell by the wayside:* this is the literal translation of the Greek and Latin in all three Gospels. But in Aramaic the same expression, literally "along the road," means both "by the side of the road" and "on the road." The latter sense is clearly demanded here, for the seed "was trodden under foot" (Luke) and it fell to the ravages of the birds more readily than that which sunk into the soft ploughed earth.   **5.** *Rocky ground* means the place where there are only a few inches of soil above the bed-rock. In such ground the rain does not seep away so quickly and therefore the seed that fell there *sprang up at once.*   **7.** *Thorns* are plentiful in the more arid parts of Palestine and their seeds, blown on the arable land, are a constant source of trouble for the farmer (cf. Gen. *3,* 18).   **8.** Some parts of Palestine, when given the proper care, are very fertile.

10-15. Why Jesus taught in parables. Parallels in Mark *4,* 10-12 and Luke *8,* 9 f.   **11.** The very expression *the mysteries of the kingdom of heaven* implies that there is something hidden and secret about these truths. God reveals them to whom He will (cf. Col. *1,* 26 f).   **12.** This verse is given in Mark *4,* 25 and Luke *8,* 18b after the explanation of the parable of The Sower; cf. also Matt. *25,* 29; Luke *19,* 26. *Even that which he has shall be taken away:* i.e., "even that which he seems to have" *(25,* 29). See note to the text.   **13-15.** Matthew alone cites in full the words of Isa. *6,* 9 f. He quotes the Septuagint text exactly. These words of the prophet are important in explaining the failure of the Jews to recognize Jesus as the Messias and are therefore cited frequently in the New Testament. Cf. John *12,* 40; Acts *28,* 26 f; Rom. *11,* 8; cf. also John *9,* 39 ff; *2* Cor. *3,* 14. The natural purpose of a parable is to make clearer the doctrine which it illustrates. But without the key to its moral it may often be obscure in itself. Christ therefore chose this form of teaching in order that the people who were not yet well disposed, might not grasp the full import of His teaching, while those who were well disposed, as the Apostles, might understand it. However, this is not to be considered so much as a punishment directly intended by God but rather as the natural consequence of the rejection of the sufficient light which all the Jews received from God. On the purpose of the parables, see article on The Parables of the Gospels. *Hearing you will hear, but not understand,* etc.: a Hebrew idiom meaning, "Even though you hear, you v ill not understand; and even though you see, you will not perceive." *The heart of this people has been hardened:* in Hebrew, "has become fatty," i.e., insensible, dull, slow to perceive.   **16 f.** The disciples' privilege in learning these mysteries. Only in Matthew here, but the same words are given in Luke *10,* 23 f after the words about the revelation of the Father and of the Son.

18-23. Explanation of the Parable of the Sower. Parallels in Mark *4,* 13-20 and Luke *8,* 11-15. "The seed is the word of God" (Luke). The sower is therefore Christ, and the different types of soil upon which the seed falls are the different kinds of men. Therefore when Christ speaks of the different seeds as the different kinds of men, the wording is not to be pressed too strictly. Such loose construction is common in the parables (cf. Commentary on *11,* 17). Our Lord distinguishes four different types of souls to whom the word of God is preached. (a) The unreceptive type upon whom the word makes hardly any impression, *he does not understand it;* he is like the trodden path on which the seed cannot sink in. It is easy for the devil to make him soon forget all about it. (b) The enthusiastic but shallow and unstable type. His soul is like the fertile but shallow soil upon the rock. (c) The man who in himself possesses the proper dispositions for spiritual growth but is prevented from producing fruit because of external circumstances, i.e., *the care of this world and the deceitfulness of riches,* the great spiritual danger that Christ warned His disciples against in the Sermon on the Mount (cf. *6,* 19-34). Such a man is like the soil that is good in itself but encumbered with thornbushes. (d) The good man who is free from all three obstacles just mentioned. Like the good earth he bears fruit according to his capacity.

13, **24-30: The Weeds.** Only in Matthew. If this discourse of Christ consisted of several parables, as seems probable, then all of them were delivered consecutively and only after the sermon to the people did Christ explain these parables privately to the Apostles. But evidently the original oral gospel inserted the explanation of the parable of The Sower immediately after the parable itself; Matthew follows this order even though it interrupts the discourse as such. This Evangelist also gives Christ's explanation of the parable of The Weeds, but in this case the explanation is postponed till after the discourse in correct chronological order (36-43). However, for the sake of convenience the explanation will be treated here in conjunction with the parable itself. The story itself is clear and simple. However, it is important to note that the *weeds* of the parable are not just any kind of undesirable growth among the wheat but a definite species of weed is named, the *zizanium,* a kind of wild wheat, known variously as cockle, tare, or darnel, which can hardly be distinguished from true wheat until its ears of grain ripen. Therefore, even though these weeds undoubtedly hinder the proper growth of the wheat, to try to root them out would do more harm than good, not so much because the roots of the weeds might be entwined with those of the wheat but because the servants of the householder might not know how to distinguish the weeds from the wheat and wrongly root up the latter. It is necessary to wait until the harvest time; only then will both the weeds and the wheat receive their due treatment. This is clearly the main point of the story and therefore also the principal lesson intended by Christ in His explanation of the parable (36-43). Nevertheless, various secondary details of the story receive their proper interpretation from Christ, all in keeping with the leading thought. Hence we may learn from His manner of interpreting His parables that, while each parable has only one main idea, many of the details of the parables also are to be understood symbolically, but always as part of the principal thought. On the other hand, however, many details of the parables are merely added to make the story interesting. To see in every detail an allegorical meaning would be excessive and beyond Christ's intention. Although our Lord says in His explanation of this parable that *the field is the world,* the parable is not directly concerned with the general problem of evil in the world but rather with the special problem of evil in *his kingdom* (41).

The parable is prophetic, viewing all future ages of the kingdom until *the end of the world* (40). *The Son of Man* is *the householder,* the Lord of the kingdom, who commands not only *his servants* who labor in His kingdom but also *His angels, the reapers.* In His kingdom are not only good men, *the sons of the kingdom,* but also wicked men, *who work iniquity* and *cause scandals,* i.e., cause the good to stumble. The kingdom is therefore the Church, a visible society of good and evil men. Elsewhere He foretold that not every branch on His vine would be good (cf. John *15,* 1-6). But it is not His fault that the members of His kingdom do not always live up to His high ideals. For *while men were asleep,* Satan, *his enemy came and sowed weeds among the wheat.* These evil men therefore have only the appearance of belonging to His kingdom; they are really *the sons of the wicked one.* But the servants of Christ cannot know the dispositions of men's hearts; hence they cannot always distinguish the weeds from the wheat. It is right for them to root out the thorns and thistles, to excommunicate the obviously unworthy, but the weeds must continue to grow with the wheat until the harvest.

Only then will justice be done to both the good and the wicked. *The just shall shine forth as the sun* (cf. Wisd. *3, 7*; Dan. *12,* 3), for their true merits are now hidden (cf. Col. *3, 3* f). This will be *in the kingdom of their Father,* for this blessedness was prepared for them by the Father (cf. *25,* 34) and at the consummation Christ "delivers the kingdom to God the Father" *(1* Cor. *15,* 24).

13, 31-35: The Mustard Seed and the Leaven. 31 f. parable of The Mustard Seed. Parallels in Mark *4,* 30-32 and Luke *13,* 18 f. The mustard seed is not, absolutely speaking, *the smallest of all seeds;* but it is so in popular estimation and as such was often used as a proverbial figure of something very small (cf. *17,* 19). There are various kinds of mustard plants. The one referred to here is a fast growing annual herb. *It becomes a tree:* i.e., it grows as large as a tree. Under this image Christ foretells the marvelous growth of the Church, from its small lowly beginnings to the vast society it was soon to become. *The birds of the air come and dwell in its branches,* "beneath its shade" (Mark), i.e., mankind finds spiritual refuge in the Church. 33. The parable of The Leaven. Parallel in Luke *13,* 20 f, where it is also joined to the parable of The Mustard Seed. The preceding parable stressed the external growth of the Church. This parable emphasizes its intrinsic vivifying force. Just as a small quantity of leaven or yeast soon penetrates a large batch of dough, so the seemingly insignificant teaching of Christ, "the mysterious, hidden wisdom of God" *(1* Cor. *2,* 7), soon makes its benign influence felt throughout the whole world. It is true that "leaven" is considered elsewhere as the symbol of the influence of evil (cf. *16,* 6; *1* Cor. *5,* 6 ff; Gal. *5, 9*), but that is no reason for rejecting the traditional and obvious interpretation of this parable. *Three measures,* literally "three seahs" i.e. one ephah, the usual batch of dough (cf. Gen. *18,* 6; Jdgs. *6,* 19; *1* Sam. *1,* 24). One ephah was the equivalent of a little more than a bushel.

34 f. The Evangelist's first conclusion to this discourse. Parallel in Mark *4,* 33 f.   34. *Without parables he did not speak to them:* the sense is either that on this particular day Christ's discourse consisted solely of parables, or that all His teaching was combined with parables.   35. Only in Matthew. The quotation is from Ps. *77,* 2. This Psalm was written by Asaph the Seer, therefore correctly called by Matthew *the prophet.* Asaph reviews the history of Israel and sees in it hidden lessons for the people of his time which he presents in the form of enigmatic sayings

or "parables." So also the parables of Christ contain the hidden truths of the mysteries of the kingdom (cf. also Col. *1*, 26).

13, 36-43: Explanation of the Parable of The Weeds. See Commentary on *13*, 24-30.

13, 44-46: The Treasure and the Pearl. Only in Matthew. These two parables together with the following parable of The Net were apparently spoken to the Apostles alone after they had entered the house (cf. 36); but possibly Matthew intended these three parables as part of the discourse to the people, but added them here as an afterthought. **44.** Hoards of gold, jewels, etc., have often been hidden in the earth in times of danger. Evidently according to the Palestinian laws of that time the mere finding of such a cache did not entitle the finder to it unless he also owned the property where it was. This *hidden treasure* is the kingdom of heaven. He who discovers its great worth gladly renounces all else to possess it. **45 f.** The kingdom of heaven is also a *pearl of great price*. Its possession is likewise worth every sacrifice. In general therefore these two parables point the same moral. But in the first parable the treasure is found accidentally; in the second, the pearl is found only after diligent search.

13, 47-50: Parable of the Net. Only in Matthew. In the story, the net is the "drag-net," a very long net used near the shore; the *bad fish* are the inedible fish that have no market value. The moral is the same as that of the parable of The Weeds: in the kingdom, the Church, there are both bad and good men; their final separation and retribution is made only *at the end of the world*. Since the Apostles were called to be fishers of men (cf. *4,* 19), they are not to be discouraged if some of the men whom their preaching brings to the kingdom, are not good men.

13, 51 f: Conclusion. Only in Matthew. This is really the second conclusion to this discourse; cf. 34. Even this conclusion is in the form of a little parable. Just as a *householder brings forth from his storeroom things new and old,* so *every Scribe,* i.e., the scholar and teacher, who is an expert in the doctrine of the *kingdom of heaven* knows how to unite the teachings of the Old and of the New Dispensation and knows how to present the old, eternal truths in a new attractive form, as Christ did by means of the parables.

13, 53-58: **Jesus at Nazareth.** Parallel in Mark *6,* 1-6a and partial parallel in Luke *4,* 16-30. The first two Evangelists agree in placing this scene towards the close of the Galilean ministry. Luke places it at the very beginning of this ministry. All three Evangelists agree in giving a twofold reaction on the part of the Nazarenes: (a) they are amazed at His wisdom; (b) because of their knowledge of His relatives and of His previous life among them, they refuse to believe in Him and consequently He cannot work miracles among them as He had done at Capharnaum. Various solutions at harmonizing these accounts are proposed. The most probable seems to be that Jesus visited Nazareth twice during His Galilean ministry. At the first visit He is, on the whole, favorably received, "and all bore him witness, and marvelled at the words of grace that came from his mouth" (Luke). At the second visit they are angry at Him and even try to put Him to death. Each of the three Evangelists combines both visits into a single account, the first two Gospels placing the combined account where only the second visit properly belongs, the Third Gospel placing its combined account where only the first visit properly belongs. Matthew probably refers to the first visit in *4,* 13. Some commentators also see a reference to the rejection of Jesus at Nazareth in John *4,* 44.

53. Matthew's usual phrase at the end of each of the five great discourses. **54.** *By his own country,* literally "his fatherland," is here meant Nazareth. *In their synagogues:* the Greek text has the singular, "synagogue," here as in the parallel passages. There was only one synagogue in the little village of Nazareth. *They were astonished:* the same Greek verb is used to describe the reaction of the people at the end of the Sermon on the Mount (7, 28) and at our Lord's argument against the Sadducees (22, 23); it signifies glad amazement, joyous wonder. This probably refers to the first visit of Jesus. **55.** *The carpenter's son:* Mark has, "Is not this the carpenter?" A son usually followed his father's trade. The Greek word, like its Latin translation, signifies merely an "artisan," and is used of workers in stone (masons) and in metal (smiths) as well as in wood (carpenters). *His brethren:* see Commentary on *12,* 46. **56.** *And his sisters, are they not all with us* here at Nazareth? The other relatives had changed their residence with Jesus to Capharnaum, but these female relatives who were married at Nazareth could not well accompany them there. The word *all* implies more than two. **57.** See note to text. The words of Jesus are evidently a well-known proverb with the same general sense as our modern saying, "No man is a hero to his valet."

14, 1-12: **Death of the Baptist.** Parallel in Mark *6,* 14-29, which gives a more complete account. Luke has no account of the death of the Baptist, but he mentions his imprisonment (*3,* 19 f) and parallels the first two Gospels about Herod's opinion that Jesus was John risen from the dead (*9,* 7-9). Both Matthew and Mark narrate the imprisonment and death of the Baptist only after they have mentioned Herod's idea that Jesus was the Baptist come back to life. Such a peculiar arrangement can only be explained as a trait that both Gospels inherited from their common source, the original oral gospel.

1 f. On the popular opinions concerning the identity of Jesus cf. also *6,* 13 f. On Herod Antipas see The New Testament Background. Herod's great superstition combined with his evil conscience to produce this strange opinion about Jesus. He was strengthened in this opinion by the rumors of the people (cf. parallels). **3-5.** The imprisonment of John, cf. also *4,* 12; *11,* 2. **3.** *Because of Herodias* may mean either "because of Herod's marriage with Herodias which John condemned," or, "for the sake of Herodias," i.e., in order to please her. The latter meaning seems the more probable from Mark's account. *Herodias* had already been married to Herod Philip, a half-brother of Herod Antipas. This Philip lived as a private citizen at Rome and is not to be confused with another half-brother, Philip the Tetrarch, who is mentioned in Luke *3,* 1. Herodias herself was a daughter of still another half-brother, Aristobulus, and a sister of Agrippa I. All these half-brothers were the sons of Herod the Great, each by a different wife. Antipas had also been married previously to the daughter of Aretas, the Nabatean king, who is mentioned in 2 Cor. *11,* 32. While Antipas was in Rome on political business he met and fell in love with Herodias. Having divorced their previous partners, they married each other. This cost Antipas a disastrous war with Aretas. **4.** All these descendants of Herod the Great were at least nominally Jews and bound by the Law of Moses. But such an incestuous marriage was an abomination to the Jews and a direct violation of the Law (cf. Lev. *18,* 16; *20,* 21). The Baptist condemned it openly before Antipas and was consequently imprisoned. **5.** From Mark we learn that the real enemy of the Baptist was not Herod but Herodias. Herod "protected him" against Herodias who "laid snares for him," i.e., tried to murder him. The tetrarch therefore considered the imprisonment of John a mere "protective custody." When Matthew then says that Herod "would have liked to put him to death, but he feared the people," we must understand this in the sense that Herod would have acceded to his wife's desire, even though he himself was favorably inclined to John, if it were not for fear of the people. Added to this was Herod's own superstitious awe of the Baptist.

6-12: The Death of the Baptist. Like the well-meaning but weak Pilate, Herod also succumbed to the demands of less scrupulous persons. **6.** *Herod's birthday* was either the anniversary of his birth or the anniversary of his accession to the throne. We do not know in what time of year either of these events would have fallen. In any case the celebration had a thoroughly pagan character. Women were usually not present at such banquets, but professional female dancers and acrobats were often presented for the amusement of the guests. That *the daughter of Herodias danced* before them was degrading to her dignity. She evidently did so solely to further her mother's murderous scheme. Josephus tells us that Herodias had a daughter named Salome by her first marriage. This is most probably the same daughter who danced before Herod. Soon afterwards Salome was married to Philip the Tetrarch. **8.** Herodias was not present at the feast (cf. Mark). **9.** Herod was *grieved* at the request. Thus Matthew also shows that the tetrarch himself did not desire the death of John. People often have the false idea that rash oaths bind even though their execution involves the committing of sin (cf. the vow of Jephte, Jdgs. *11,* 30 f.35 ff). **10.** *He sent and had John beheaded in prison:* according to Josephus this was in the fortress of Macherus (see Commentary on *11,* 2). The banquet evidently was held in the same fortress, which was on the border between the domains of Herod and Aretas who were at this time at war with each other. **12.** It was customary to allow the relatives or friends of an executed man to have the body for burial (cf. *27,* 58). Herod would readily have granted this favor, since he was well inclined towards the Baptist, whose disciples had previously had access to their master in prison (cf. *11,* 2 f).

14, **13-21: Jesus Feeds Five Thousand.** This is the only important event in our Lord's public ministry, before the triumphal entry into Jerusalem, that is narrated by all four Evangelists. Parallels in Mark *6,* 31-44; Luke *9,* 10b-17; and John *6,* 1-15.

This event is of the greatest importance for two reasons. (a) The extraordinary character of the miracle admits of absolutely no natural explanation and aroused the people to enthusiasm more than any other miracle of Christ. (b) The special relation of this miracle to the Blessed Eucharist is evident not only in the following discourse in the Fourth Gospel but also in the narrative of the first three Gospels. A somewhat similar miracle is narrated in *4* Kgs. *4, 42* ff, but this miracle of Eliseus, which was a figure of the miracle worked by Christ, was in much smaller proportion.

13. *When Jesus heard this,* i.e., that the Baptist had been beheaded: this phrase merely states the time and is most probably not to be understood as giving one of the reasons why Jesus *withdrew by boat to a desert place apart.* Fear of Herod could hardly have been the motive (cf. Luke *13,* 31 f), for at this time Herod was not in Galilee but in southern Perea, and moreover, a day later Jesus is still in Herod's domain. The real reason for the departure is given in Mark and Luke, that the Apostles might rest after their missionary labors. *A desert place* means merely an uninhabited place, not a sandy waste. John says that this place was on "the other side of the sea of Galilee," i.e., on the eastern shore. Luke says, "to a desert place which belongs to Bethsaida" (this seems to be his meaning, although there is much confusion in the MSS). But Mark complicates matters by stating that after the miracle Jesus "made his disciples get into the boat and cross the sea ahead of him to Bethsaida" (*6,* 45). Outside of the Gospels only one Bethsaida is known: a small village on the northeastern shore of the sea of Galilee, about one mile east of the place where the Jordan enters the lake. Just north of this Bethsaida Philip the Tetrarch had founded a new city which he named Julias. According to the Third and the Fourth Gospels this miracle seems to have taken place somewhere near this Bethsaida-Julias. Mark's statement is explained in various ways. (a) There is another Bethsaida, on the western shore of the lake a few miles south of Capharnaum. This would be the Bethsaida of Galilee mentioned in Matt. *11,* 21 (Luke *10,* 13); Mark *8,* 22; John *1,* 44; *12,* 21. It is quite possible that there was more than one town by this name on the lake, for the word means "the house of fishing," i.e., fishermen's village. (b) There was only one Bethsaida, the one in the territory of Philip, which he renamed Julias. Mark would then mean that the disciples were told to row from the "desert place" of the miracle along the shore of the town itself. But then they could hardly be said to "cross the sea." Or the Greek text of Mark is to be translated, "and cross the sea ahead of him (to the land) opposite Bethsaida." The Greek may admit this meaning but it is rather forced.   **14.** If the boat had a head-wind against it, the *crowds who followed on foot from the towns* around Capharnaum (about four miles to Bethsaida-Julias) could easily have arrived there before the boat.   **15.** *When it was evening:* i.e., towards evening, the late afternoon; cf. Luke, "The day began to decline."   **17.** These words were spoken by Andrew (cf. John). He, like Philip, whom Jesus had asked about the matter, was a native of Bethsaida (John *1*, 44), and would therefore know more of the circumstances of the place.   **19.** The words used to describe our Lord's action is remarkably similar to words used to describe the institution of the Holy Eucharist (cf. *26*, 26 and parallels). This similarity is not accidental. Jesus intended this miracle to be a figure of the Eucharist. *He blessed the loaves:* i.e., He recited over them a prayer, blessing and thanking God for His gifts. He did the same at the Last Supper. The Canon of the Mass, beginning with the Preface, is also essentially a thanksgiving blessing said over the Sacramental Bread and Wine, and the whole service is therefore known as the Eucharist, i.e., "thanksgiving." *He broke the loaves* at the Last Supper also. This symbolic action is such an important part of the Eucharistic Sacrifice that the Mass in the Apostolic Church was generally called "The Breaking of the Bread" (cf. Acts *2*, 42.46; *20*, 7.11; *1* Cor. *10*, 16; perhaps also in Luke *24*, 35). The share which Christ made the Apostles have in the miracle prefigured the share they would have in the administration of His sacraments. *The crowd* reclined *on the grass* which was "green" (Mark) and plentiful (John), i.e., it was the season of the year after the winter rains and before the heat of summer had withered the grass. According to John the miracle took place shortly before the Passover, i.e., just one year before the institution of the Holy Eucharist.   **20.** Note the number of the *baskets full of fragments:* one for each of the twelve Apostles. **21.** The size of the crowd is naturally given in round numbers.

14, 22-33: **Jesus Walks on the Water.** Parallels in Mark *6*, 45-52 and John *6*, 16-21; but only Matthew has the episode of Peter's attempt to walk on the water. Cf. also the earlier miracle of the stilling of the storm on the lake (*8*, 23-27 and parallels).

22. Jesus *made his disciples get into the boat:* literally He "forced" them to do so; the word implies strong reluctance on their part, no doubt because they shared in the desires of the people to make Jesus their king (cf. John *6, 15*). *He dismissed the crowd:* i.e., He sent them away in orderly fashion. Perhaps this was His custom at the end of each day of religious instruction, but He had special reasons on this occasion, to calm the popular enthusiasm.   **23.** *The mountain* is the hilly region that surrounds the lake on almost all sides. The Gospels make special mention of Jesus praying alone at night before important stages of His life (cf. *26, 36;* Luke *5, 16; 6, 12; 9, 28*). On this occasion He faced a crisis in His Galilean ministry. On the morrow He was to reject the requests of the people for their type of Messias and was to preach instead a purely spiritual kingdom in which He would give them Himself as the Bread of Life (cf. John *6, 22-72*).   **24.** *The boat was in the midst of the sea:* i.e., it was still some distance from any shore, but not necessarily in the very middle of the lake. John says, "They had rowed some twenty-five or thirty stadia," i.e., about two-and-a-half or three miles, when our Lord appeared to them. From Bethsaida-Julias to Genesar would be about four or five miles.   **25.** At this time the period from sunset to sunrise was divided into four "watches"; the fourth watch of the night would be about from three to six A.M.   **26.** In the darkness and storm it would have been difficult for the disciples to recognize their Master. It is entirely understandable why they were overwhelmed with fear.

28-31. Matthew alone has the account of Peter walking on the water, but that should not be considered as weakening the authenticity of the fact. Mark follows the preaching of St. Peter and the latter evidently suppressed this item about himself intentionally. It is difficult to say just why Peter should have wanted to walk on the water. It would hardly have been from a desire to make sure that the apparition was really Jesus. In any case the incident is perfectly in keeping with the impetuous character of Peter who often said or did things without sufficient reflection (cf. *16,* 22; *26,* 69 ff; Luke *9, 33; 22, 33;* John *18,* 10). When he finally reflected on the stupendous nature of the miracle in which he himself was sharing, he lost faith more in himself than in Jesus, and *began to sink.* He did not doubt that Jesus could do the miracle but he doubted that He would do it for him. His cry for help shows his faith in Christ.   **32.** The fact that *the wind* fell and the storm ceased just at the moment when Jesus *got into the boat* was also a miracle. **33.** *They who were in the boat:* cf. the equally strange expression in *8,* 27, "the men." It seems to be solely the Apostles who were meant, for there is not the slightest inkling of a "crew" other than the Apostles. This is the first time in the Synoptic Gospels that Jesus is called the *Son of God* by any man. The exclamation of the Apostles, however, was based on an incipient faith in His divinity. For the whole context of *16,* 16 f seems to make it certain that Peter's profession of faith at Cæsarea Philippi was the first full acknowledgment of the divinity of Christ. Mark has a rather different conclusion to this miracle; see Commentary there.

14, **34-36: Other Miracles.** Parallel in Mark *6,* 53-56. After the miracle of the first multiplication of the loaves the boat of the Apostles had probably headed for Capharnaum (cf. John *6,* 17), but the storm had driven them somewhat off their course (cf. John *6,* 21). This episode is probably to be placed before the discourse on the Eucharist at Capharnaum which was held on the day following the miraculous feeding of the five thousand (cf. John *6,* 22). These miracles of curing the sick at Genesareth could easily have taken place on the morning of the same day as the discourse at Capharnaum while Christ and the Apostles walked back along the shore-road to that town. **34.** *Genesar:* the more common form is "Genesareth" as in Mark and in most Greek MSS of both Gospels. Both forms occur in the Talmud. On the shorter form cf. *1* Mach. *11,* 67. It is probably derived from the Hebrew "gan-hassar," "Garden of the Prince." The longer form arose under the influence of the ancient name of that district, "Kinnereth" (cf. Num. *34,* 11; Deut. *3,* 17; Jos. *12, 3*). This was the name of the littoral plain that extended for about four miles along the western shore of the lake and about two miles inland from Capharnaum on the north to Magdala on the south. It also gave its name to the lake (cf. Luke *5,* 1). It is well watered and at the time of Christ was famous for its fertility. **36.** *The tassel of his cloak:* see Commentary on *9,* 20.

15, **1-20: Jesus and the Pharisees.** Parallel in Mark *7,* 1-23. This is a very important controversy, for here Jesus shows that he has broken clearly and definitely with Judaism. One of the most characteristic features of Judaism, both as it was at the time of Christ and as it is today, is the insistence on numerous minute prescriptions concerning food, the so-called dietary or

15, 1-5   St. Matthew

"kosher" laws. These laws are partly based on the laws of Moses concerning clean and unclean food (cf. Lev. *11;* Deut. *14*) and partly on *the tradition of the ancients,* i.e., the interpretation of these laws and the new regulations which were given by the Rabbis ever since the Babylonian exile. Cf. Paul's "zeal for the traditions of the Fathers" (Gal. *1,* 14). The main purpose of these laws and regulations was to keep the Jews, who had lost their political independence, as a distinct people. Here Jesus rejects not only the Rabbinical traditions as not sanctioned by God and even at times contrary to God's basic law of morality, the Ten Commandments (1-9), but also the law of Moses itself concerning clean and unclean food (10-20).

1. Not the local Scribes of Galilee but the more learned *Scribes and Pharisees from Jerusalem* raise this basic issue. Perhaps they had come from the Sanhedrin in a more or less official capacity. From the Fourth Gospel we know that Jesus had already aroused the leaders at Jerusalem to hostility.   **2.** These wily Rabbis do not accuse Jesus Himself directly of violating the tradition. Their charge is that He permits and thereby encourages His disciples to do so. The particular example that they cite is the Apostles' disregard of the regulations concerning ritual washings. Cf. Mark for a description of these customs which were based on Lev. *15.*   **3-6.** Jesus ignores this point of their traditions to attack the importance that they place on their traditions in general. He shows that on at least one point their traditions are contrary to the fundamental law of God. **4.** Our Lord quotes the fourth commandment (cf. Ex. *20,* 12; Deut. *5,* 16). The Hebrew verb *honor* means more than "show respect to"; it is also used in the sense of "repay a favor with gifts" (cf. e.g., Num. *22,* 17.37; Jdgs. *13,* 17): in the fourth commandment it certainly includes the idea of "support." Christ immediately cites the penalty for cursing one's parents (cf. Ex. *21,* 17; Lev. *20,* 9). The reason why He adds this is because the case which He now mentions involves a virtual curse of one's parents.

5. *Dedicated to God:* Mark gives the original Hebrew word that was used in this vow, "Corban." Some commentators have understood the passage to mean, that when a man dedicated to the temple the money that would otherwise have been used in supporting his parents, he could offer up the spiritual merits of this gift to the temple in favor of his parents and was thereby freed from all further obligations in their regard. This would then be a trick of the Scribes for the purpose of enriching the temple. But the "Corban" vow is clearly explained in the Talmud, and it has absolutely nothing to do with the temple. Whatever be the origin of the custom, at the time of Christ the word "Corban" had become a mere oath-formula signifying that a man swore that such and such a thing would be used for such and such a purpose. Thus a man could swear to abstain from wine by saying, "Any wine that may be offered to me is Corban." Only by a fiction of law was the object considered as dedicated to God. The Rabbis considered such a vow binding, even though the act itself that was vowed was sinful. The very case referred to by Christ is given in the Talmud. Only the most subtle casuistry of the Rabbis was able at times to circumvent the force of the vow. Our Lord does not accuse the Rabbis of approving this vow as such. But He condemns their traditions which consider the vow as binding.   6. Note that the Scribes not only say that a man who has taken such a vow *does not have to honor his father or his mother;* they also "do not allow him to do anything further for his father or his mother" (Mark 7, 12). 7. Jesus stigmatizes such teaching as hypocrisy.   8 f. The very appropriate words of Scripture which He quotes are from Isa. *29,* 13. The passage is quoted in exactly the same way in Mark, although it differs slightly from both the Hebrew and the Septuagint texts. Cf. also Ps. *77,* 36 f; Col. *2,* 22.

10 f. Turning from the Pharisees to the crowd, Jesus considers the original question of eating with unwashed hands. But He treats it on the general principle, that all externals as such have no moral value, but it is solely the evil dispositions of the soul that *defile a man.*   12. This statement of Christ went further than the rejection of the traditions of the ancients: it abrogated the Mosaic distinction between clean and unclean food. It is therefore hardly surprising that *the Pharisees* should *have taken offense at hearing this saying.* We need not suppose, however, that the Apostles grasped the full import of our Lord's teaching at that time. In this matter, as in others, they were slow of understanding (cf. *16,* 9) and it was not until several years after Christ's ascension that the Apostles under the guidance of the Holy Spirit definitely settled the moot question of the obligation of the Mosaic Law.   13. The traditions of the Pharisees have no divine approval, for the development of this spirit is a spurious growth *that the heavenly Father has not planted.* For God's work among men under the figure of His

care for a plant, especially a vine, cf. Ps. *79,* 9-16; Isa. *5,* 1-7; Jer. *2,* 21; *12,* 2; Matt. *21,* 33. If a work is not from God it will not prosper (cf. Acts *5,* 38). **14.** *Blind guides:* cf. *23,* 16. 24. A learned Jew boasted of being "a guide to the blind" (Rom. *2,* 19). "The blind leading the blind and both falling into a pit" was evidently a proverbial saying, for it is quoted in almost the same words in Luke *6,* 39. **15-20.** The third and final part of this episode took place privately between Christ and His disciples (cf. Mark *7,* 17). Even the disciples found it hard to understand this revolutionary teaching that abolished the distinction between clean and unclean food. Therefore Christ repeats in clearer language what He had already said in 11. Even after Pentecost St. Peter still needed a special revelation from heaven to free him from this racial prejudice against unclean food (cf. Acts *10,* 9-16. 28; *11,* 1-10). A remnant of the Mosaic legislation (the prohibition of partaking of "anything strangled" or of "blood," cf. Lev. *3,* 17; *17,* 10 ff) was still preserved by the Apostles assembled at the Council of Jerusalem (cf.Acts *15,* 20. 29), lest they unnecessarily offend the Jews (cf. *1* Cor. *10,* 32). But St. Paul insisted on the full force of Christ's teaching in this matter (cf. Rom. *14,* 14; *1* Cor. *10,* 25-30; *1* Tim. *4,* 3 f; Titus *1,* 14 f). By these words Christ did not wish to deprive the Apostles or the Church of the right to make positive precepts concerning food; but a Catholic who breaks the laws of the Church on fast and abstinence commits sin not by eating the food as such but by his disobedience. **18.** Cf. *12,* 34b.

4. Ministry Mostly in the Regions Bordering on Galilee *15,* 21 — *18,* 35

15, 21-28: The Canaanite Woman. Parallel in Mark *7,* 24-30. The two accounts are independent but harmonize perfectly. Mark gives only the conversation between Jesus and the woman after "He entered a house;" Matthew also tells us what was said while Jesus and His disciples were on the way to this house (22-24).

21. These words imply a definite departure of Christ from Galilee, most probably in order to avoid the growing hostility of the Pharisees. Hereafter we find Him but rarely in Galilee proper. His instructions are henceforth mostly limited to the Apostles, to whom He now begins to reveal the mystery of His approaching passion, death and resurrection. *The district of Tyre and Sidon:* the region near the coast of the Mediterranean sea northwest of Galilee. Here Jesus was also out of the jurisdiction of Herod, for this district belonged to the province of Syria. On these two cities see Commentary on *11, 21*. 22. *A Canaanite woman:* in true Jewish fashion Matthew uses the Old Testament name for this people; but the Canaanites as such had long since disappeared from history. Mark calls her a "Syrophœnician" woman, i.e., a Phœnician of Syria. The Phœnicians were descended from the Canaanites (cf. Gen. *10, 15*). *Out of that territory:* according to Mark *7, 31* this event took place nearer Tyre than Sidon. In the neighborhood was Sarepta where Elias had aided the widow and restored her son to life (cf. *3 Kgs. 17, 9* ff; Luke *4, 26*). Jesus wished His presence in that region to remain secret (cf. Mark), but the woman recognized Him and acknowledged Him as the Messias by the typical Jewish title, "Son of David." Possibly she had been among "the large crowd of those about Tyre and Sidon" who came to hear His preaching (cf. Mark *3, 8*; Luke *6, 17*). Matthew places together the various cries which the woman uttered as she followed Christ and the Apostles, *crying after them.* 23. *"Send her away":* from *besought* and from Christ's response the sense seems to be, "Grant her request and dismiss her." The Apostles probably felt embarrassed by the scene she was making and were losing their patience with her. 24. According to the divine plan the gospel was first to be preached to the chosen people, Israel, before it would be announced to the Gentiles. See Commentary on *10, 5 f.* 25. *She came:* i.e., she came into the house where He had entered (cf. Mark); *and worshipped him:* i.e., fell on her knees before Him, the natural and traditional posture of a suppliant. 26. *The children* are the members of the house of Israel, "who have the adoption as sons" (Rom. *9, 4*); *the dogs* are the Gentiles (cf. *7, 6*). The Greek has the diminutive "little dogs," i.e., household pets as distinct from the stray dogs so common in eastern cities. Perhaps our Lord intentionally softens the opprobrious term that the Jews used for the Gentiles. 27. The woman then understands the hint and keeps up the figure by pointing out that the little dogs of the house receive the leavings of the children's food. However, at this time the Greek diminutive had often lost its original force. 28. The woman's *faith* was *great* not only because she acknowledged Jesus as the Messias who could and would grant her request, but also because she had such marvelous perseverance in her prayer despite the first rebuffs and humiliations. Cf. our Lord's words to the centurion at

Capharnaum (*8,* 10. 13) whose request Christ granted more readily because of the different circumstances of his case.

15, 29-31: Jesus Heals the Suffering. There is no strict parallel in the other Gospels, but Mark *7,* 31-37 mentions the same journey and gives one example in detail (the healing of a deaf-mute) of the miracles which Matthew refers to in general. **29.** Jesus first went further north along the coast of the Mediterranean (cf. Mark), then turned east, crossing the high mountain ranges of the Libanon and the Antilibanon (or Hermon), and finally journeyed towards the south, so that He went along the sea of Galilee on the heights above its eastern shores, until He arrived in "the midst of the district of Decapolis" (Mark). The purpose of such a circuitous route was to avoid Galilee completely. **30.** For other occasions when the miraculous cures performed by our Lord are mentioned only in general cf. *4,* 23 f; *8,* 16; *9,* 35; *12,* 15; *14,* 14. 35 f; *19,* 2; *21,* 14; similar summary accounts in all the other Gospels. **31.** The expression the *God of Israel* shows that the people who witnessed these miracles were mostly Gentiles who worshipped many gods. This agrees with Mark who places our Lord's activity at this time in Decapolis.

15, 32-38: Jesus Feeds Four Thousand. Parallel in Mark *8,* 1-9. Since no new indication of place is given in either Gospel, the scene is probably still in Decapolis. See Commentary on *14,* 13-21, the very similar miracle of the first multiplication of the loaves. St. Jerome sums up the differences between the two miracles as follows: "There were five loaves and two fishes, here seven loaves and a few little fishes; there they reclined upon the grass, here upon the ground; there five thousand are fed, here four thousand; there twelve baskets were filled; here seven large baskets." Most non-Catholic critics consider the two miracles as two varieties of one original story. But both Matthew and Mark give both accounts, and both of these Evangelists cite the words of our Lord in reference to these miracles as two distinct events (cf. Matt. *16,* 9 f; Mark *8,* 19 f). To admit that they were mistaken in this matter is contrary to the Catholic doctrine on the inspiration and inerrancy of Sacred Scripture. Nothing could prevent Christ from working two similar miracles, and the early oral tradition naturally recounted both events in very similar words. **32.** *Three days:* At the first multiplication of the loaves the people were with Christ only one day and were miraculously fed because it was too late in the day for them to buy food. On this occasion the people were not necessarily without food for all three days, but they had already finished the food that they had brought with them. **33.** *In a desert,* for large regions of Decapolis were sparsely inhabited, despite its dozen fair-sized cities. **35.** *On the ground,* for it is now later in the season and the plentiful grass of early spring is already withered. **36.** *He gave thanks:* in Mark, "He blessed;" both terms are synonymous. **37.** *Seven baskets:* the Greek and Latin have a different word here than is used for the "twelve baskets" of the former miracle. In *16,* 10 and Mark *8,* 20 it is more exactly translated "large baskets." It was in one of the "large baskets" of this type (perhaps similar to our "wash-basket" as distinct from our "market-basket") that St. Paul was let down over the wall of Damascus (cf. Acts *9,* 25).

15, 39 — 16, 4: **The Pharisees and Sadducees Ask a Sign.** Parallel in Mark *8,* 10-12. Cf. also the earlier incident of the "Sign of Jonas" (Matt. *12,* 38-40; Luke *11,* 16. 29 f). **39.** *Magedan* is the reading of the best MSS of Matthew, while the best MSS of Mark have "Dalmanutha." Neither name is known outside of this passage and therefore the site of this incident is uncertain. All that can be said is that it is somewhere near the sea of Galilee, since Christ and His disciples go there by boat. Either Magedan is the name of a town and Dalmanutha the name of the region in which this town lay, or vice versa. The identification of Magedan with Magdala is improbable. *16,* **1.** We do not know why the *Sadducees* should have united forces on this occasion with their inveterate enemies the *Pharisees,* especially since the Sadducees, the rationalists among the Jews of that time, were hardly interested in a Messias at all. Perhaps it was their affiliation with the pro-empire political party of the High Priests that made them hostile to Jesus. *A sign from heaven:* see Commentary on *12,* 38. **2-4b.** The two best Greek MSS omit the words, *"When it is evening . . . signs of the times."* But since the vast majority of the MSS have these words, they are generally considered authentic. Luke *12,* 54-56 has a similar saying of our Lord, but it cites a different weather forecast. The two passages are independent of each other. In Matthew our Lord refers to a weather rule that is known to almost all peoples; cf. our "Red in the morning, sailors take warning; red at night, sailor's delight." *Signs of the times:* the fulfillment of the prophecies in Jesus. **4b. On the**

16, 5-13　　　　　　St. Matthew

adulterous generation and *the sign of Jonas* see Commentary on *12,* 39.

16, 5-12: The Leaven of the Pharisees and Sadducees. Parallel in Mark *8,* 13-21. Cf. also Luke *12,* 1.　　**5.** *When his disciples crossed the sea:* the sense seems to be, "While they were crossing the sea"; cf. Mark, where this incident takes place in the boat before they reached their destination. They sailed from Magedan (Dalmanutha) to Bethsaida-Julias (Cf. Mark *8,* 22) on their way towards Cæsarea Philippi. But "to cross the sea" does not necessarily mean to sail from the western shore to the eastern or vice versa. Mark says that they had only one loaf of bread with them in the boat, but since this would be insufficient for the thirteen of them, Matthew makes no mention of it.　　**6.** Our Lord's thoughts are still on the Pharisees and Sadducees who had demanded a sign from heaven. Therefore under the figure of *leaven* He warns His disciples to beware of their evil influence and of the evil principles of their teachings. On the figurative use of the word "leaven" see Commentary on *13, 33.* Luke *12,* 1 has this saying of Christ in a different context. Instead of *Sadducees* Mark has "the leaven of Herod," perhaps in the sense of "the leaven of the Herodians" (cf. Matt. *22,* 16; Mark *3,* 6; *12,* 13). The leaven of these men, like that of the Sadducees, was the spirit of worldliness and rationalistic skepticism.　　**7.** The sense seems to be: They began to argue among themselves what He meant by this saying, and came to the conclusion that He said this because they had brought no bread with them (cf. 11). Forgetting the teaching of their Master that no food was unclean (cf. *15,* 11 ff), the disciples think that Jesus in true rabbinical fashion is giving them new precepts about unclean food.　　**8-12.** Christ's rebuke is even more severe in Mark. For the two multiplications of the loaves of which He now reminds them, cf. *14,* 13-21; *15,* 32-38; and parallels.

16, 13-20: Peter's Confession. Parallels in Mark *8,* 27-30 and Luke *9,* 18-21. This passage is of prime importance for the Messiasship and divinity of Jesus, the primacy of Peter and the nature of the Church.

13. The scene is near the ancient Paneas, modern Banias, which Philip the Tetrarch had rebuilt and named *Cæsarea* in honor of the Emperor. It had the added title *Philippi,* i.e., of Philip, to distinguish it from several other **Cæsareas,** notably the

Cæsarea in Palestine on the Mediterranean which is mentioned frequently in Acts. Cæsarea Philippi was situated in the southern foot-hills of Mount Hermon at one of the sources of the Jordan, about thirty miles north of the sea of Galilee. *The Son of Man* is not synonymous with "Messias" here, or the answers of the disciples would be unintelligible. Mark and Luke have given the question in a more exact form. Or perhaps the original form of the question was, "Who do men say that I am, I who call myself the Son of Man?" Jesus was not seeking information on this point; this question was merely an introduction for the important question in 15. 14. It is rather remarkable that the disciples do not report that one of the popular opinions about Jesus considered Him to be the Messias. Previously at least some of the people had acclaimed Him as the Son of David, a Messianic title (cf. *9, 27; 12, 23; 15, 22*). But ever since Jesus had refused a temporal kingship after the first multiplication of the loaves (cf. John *6, 15*), the people were disappointed in the Messianic hopes that they once had in Him. *And others, Jeremias:* only in Matthew. Although there was no foundation for the opinion in the Old Testament, many Jews at this time thought that Jeremias had been taken up into heaven without dying and would come back to earth.

15. Christ's second question is addressed to all the Apostles. Although they all no doubt, with the exception of Judas (cf. John *6, 65. 71 f*), shared to a certain extent in the faith of Peter, and in a sense he acted as their spokesman, still he alone actually made the profession of faith in Jesus on this occasion and to him alone were the words of 17-19 spoken. 16. *Simon Peter:* Matthew and Luke who refer to him here by his full name, stress the fact that it was on this occasion that Simon received the epithet of "the Rock" (Peter). On the history of his name, see Commentary on *10, 2*. Mark gives the confession of Peter simply as, "Thou art the Christ." Luke has, "The Christ of God." Only the First Gospel records the additional words, "The Son of the living God" together with Christ's reply to Peter. All the MSS of Matthew's Gospel contain these words; therefore no objection against their authenticity can be raised on textual grounds. Nor have the objections of the higher critics any value here. The other two Evangelists fail to record these words because their account of this incident is ultimately based on the preaching of St. Peter who omitted this encomium of himself out of humility. The words *Son of the living God* might not

16, 17-18   St. Matthew

of themselves prove that Peter was here proclaiming his faith in the divinity of Jesus. The expression "Son of God" had been used before this time in Jewish apocryphal literature as a mere synonym for "Messias" with the simple meaning of "God's favorite, God's special envoy," etc. The word *living* distinguishes the true God from the false gods of the pagans, mere lifeless idols; this word adds special solemnity to Peter's confession but does not affect the meaning of the word *Son.*

17. However, from our Lord's answer to Peter it is certain that Peter meant these words in the strict theological sense of divine filiation. For Jesus says that Peter learned this truth not from any human being or from any natural source but by a special revelation from the Father (cf. *11,* 25-27). Any one who had seen the miracles of Jesus could and should have been led by his natural reason to recognize Him as the Christ. But not even from a knowledge of the Old Testament would one have known clearly that the Christ would be the very Son of God. It also follows from these words of praise and reward that Jesus spoke to Peter, that this was the first occasion on which any one had publicly proclaimed his faith in the divinity of Jesus; therefore similar expressions of faith spoken on previous occasions (cf. *14,* 33) were uttered without full realization of the truth of the words. *Blessed:* literally, "Happy, fortunate, lucky." *Simon Bar-Jona:* Jesus calls him by his full name to emphasize the importance of what He is about to say. *Bar-Jona* means "son of Jona." In John *1,* 42 he is called "Simon, the son of John." Either Simon's father had more than one name, a not uncommon custom of that time, or more probably Jona and Johannes are but variant forms of the original Hebrew name of Johannan. *Flesh and blood:* a Hebrew expression meaning "mortal man" as contrasted with God.   18. *And I say to thee:* the original may likewise be translated, "I also say to thee," with the sense that as Peter has given testimony of Jesus, so now Jesus also gives testimony of him, or more probably, as the Father has given Peter this revelation, so also Jesus gives him a special gift. Our Lord said in Aramaic, "Thou art a rock (or, the rock), and upon this rock," etc. The Greek translator of the Aramaic Gospel of St. Matthew, by translating the first "kepha" as a proper name, greatly weakened the force of Christ's words. On the word *Church,* see note to the text. There is emphasis also on the word *my:* this is the Church of the Son of God, a divine, not human institution. The rock upon which Christ builds His Church is not

merely such faith as Peter professed but Peter himself. *The gates of hell:* cf. Job *38,* 17; Pss. *9,* 15; *106,* 18; Wisd. *16,* 13; Isa. *38,* 10. In all these passages the Hebrew has "the gates of Sheol" which the Greek renders either "the gates of Hades" or "the gates of Death." Sheol or hades was the name of the abode of the dead. The word *gates* is used in the Scripture not only for the large doors in the walls of a city but also for the open space at these doors where the authorities of a city held their meetings (cf. Deut. *21,* 19; *25,* 7; *2* Par. *32,* 6; 2 Esdr. *8,* 1; Job *31,* 21; Isa. *29,* 21; Amos *5,* 12. 15; Zach. *8,* 16; etc.) The whole expression in our passage means "the power of death." Christ therefore predicts that His Church will be immortal. Since it is principally Satan who would endeavor to destroy the Church, in an extended sense "the gates of hell" means the "power of Satan."

19. He who has *the keys* of a house is the steward or administrator of that house (cf. Isa. *22,* 22; Apoc. *3,* 7). *The kingdom of heaven* is here synonymous with the Church, God's kingdom on earth. Peter is therefore made Christ's viceroy in governing His kingdom. The common idea that St. Peter is the doorkeeper of heaven is based on a misunderstanding of this passage. To *bind* is to declare authoritatively that something is obligatory, to *loose* is to declare authoritatively that something entails no moral obligation. These Aramaic expressions are used several times in the Talmud in such a sense. Christ therefore gives Peter the power of making official statements on matters of faith and morals; his decisions will be ratified by God. According to the context of *18,* 18 these same expressions signify also the power to admit into and to exclude from the Christian community. Therefore Peter is here given wide legislative, administrative and judicial powers, including and extending beyond the power to forgive and to retain sins (cf. John *20,* 22 f). Note that all the verbs in this passage are in the future, for the Church as such was not founded till after the Resurrection; only then did Christ actually bestow the primacy on St. Peter (cf. John *21,* 15-17). That Peter really exercised the primacy in the Apostolic Church is perfectly clear in Acts. The few instances that are raised as objections by non-Catholic critics (cf. Acts *11,* 2 ff; *15;* 2 Cor. *11,* 5; Gal. *2,* 11; *1* Pet. *5,* 1) can be explained as due to Peter's spirit of humility and conciliation, or to the special personal privileges that the other Apostles received from Christ. Our Lord built His Church not on Peter alone but on all the

Apostles (cf. Eph. *2,* 20; Apoc. *21,* 14). To all of them He gave the power of "binding and loosing" and the power of forgiving and remitting sins (cf. Matt. *18,* 18; *19,* 28; John *20,* 21-23). But a group cannot act without a head. St. Peter alone was made this head, and to him alone was given "the keys of the kingdom of heaven." **20.** For the reason of this strict command, see note to the text.

16, 21-23: Passion and Resurrection Foretold. Parallels in Mark *8,* 31-33 and Luke *9,* 22. **21.** *From that time:* i.e., after the Apostles have shown firm faith in Jesus; they should not now be scandalized by being told the full truth about Christ. Our Lord had previously hinted at His coming death (cf. *9,* 15; *12,* 40; John *2,* 19; *3,* 14; *6,* 52), but now "what he said he spoke openly" (Mark). On two later occasions Jesus gave special instructions to His disciples about His passion, death and resurrection (cf. *17,* 21 f; *20,* 17-19; and parallels), on each occasion emphasizing some point in particular. Here the emphasis is on His rejection by the leaders of the people, the very ones who should have given Him official recognition. **22 f.** Peter remonstrates with Jesus and is rebuked by Him. *Far be it from thee:* literally in Greek, "God be good to thee," i.e., God forbid! Peter is still so attached to the popular idea of the Messias as a triumphant temporal king that he does not want to understand the words of Jesus in their literal sense. His endeavor to persuade Jesus to comply with the popular idea of the Messias and take His own crown without a cross is very similar to the temptation of the Devil in the desert (cf. *4,* 9). Therefore our Lord rejects this temptation of Peter in almost the same words that He spoke to Satan himself. **23.** *He turned:* most probably to be understood in the sense of Mark's "Turning and seeing his disciples"; i.e., Peter had taken Him aside in order to persuade Him in private, but Jesus rebukes him publicly before the other disciples. They all had to learn this lesson. *Get behind me:* i.e., "Get out of my way; thou art blocking my path towards the cross." There is irony in the fact that the *Rock* on which Christ was to build His Church, should try to be a *scandal,* i.e., a *"stone* of stumbling" to Christ. *Dost not mind:* dost not have a mind for it, an inclination and sense of perception for. *The things of God:* Jesus had said that He *must* suffer; it was His Father's will as revealed in the prophecies of old (cf. Ps. *21;* Isa. *53).*

16, 24-28: The Doctrine of the Cross. Parallels in Mark *8,* 34-39 and Luke *9,* 23-27. This passage is intimately connected with the preceding, for not only must Christ suffer first in order to enter into His glory, but His disciples must do the same.

24. *If anyone wishes:* no one is forced; it must be an act of free will aided by Grace. *To come after me:* to be my follower, my disciple. Christ does not say directly that His disciple must deny certain pleasures "to himself," but that he must *deny himself* (direct object). Just as Peter "denied" Christ, so also the disciple must say that he does not know himself, his sinful self has no claim on his allegiance. *Take up his cross:* figurative language borrowed from the custom of the time according to which a man condemned to crucifixion had to carry his own cross to the place of execution (see Commentary on John *19,* 17). That our Lord is using a metaphor is clear from the word "daily" in Luke. The *cross* that the follower of Christ must willingly embrace is any suffering, even martyrdom itself, that results from being a true disciple of Christ. Cf. *10, 38.* **25.** Similar sayings of Christ in *10,* 39; Luke *14,* 26; *17,* 33; John *12,* 25. In the original the word that is translated here twice as *life* is the same word that is translated twice in the next verse as *soul.* This word is used both of the life-principle and life itself and indeed both in regard to the natural life and in regard to the supernatural life. Therefore the sense is, "He who would save his mortal life by being unfaithful to me will lose his immortal life (his soul); but he who loses his mortal life for my sake will save his immortal life (his soul). **26.** *In exchange for:* literally "as barter-value for"; nothing in the world has as much value as one's immortal soul.

27. The sequence of thought here would not be clear, had we not the connecting verse in Mark and Luke which Matthew has in *10, 33.* The sense is: if a man disowns Christ in order to save his own life, Christ will disown him when He comes to judge the world. **28.** To *taste death:* an Aramaism, not occurring in the Old Testament but common in the Rabbinical writings, meaning simply "to die." *The Son of Man coming in his kingdom:* cf. the variant expressions in Mark and Luke. It is not certain what this refers to. Some of the Apostles will see it; others will be dead before that time. It cannot refer to Christ's coming at the end of the world when all the Apostles will long since have been dead. Nor can it refer to some event in the near future, such as the Transfiguration, Resurrection or Ascension. For all of the Apostles lived to see these events. which, moreover do not contain the element of retribution demanded by the context. It probably refers to the destruction of Jerusalem in which Christ vindicated His honor by punishing the city that slew Him. The destruction of Jerusalem is a type of the destruction of the world on the last day; as such, the expressions which refer properly to the one are also used analogously of the other. See Commentary on *24*, 15-35.

17, 1-8: Jesus Transfigured. Parallels in Mark *9*, 1-7 and Luke *9*, 28-36. **1.** *After six days:* so also in Mark; this is most probably according to the Hebrew way of reckoning and means "on the sixth day after the preceding." See Commentary on *12*, 40. Therefore Luke's "About eight days after these words" is only a rough calculation meaning "about a week later." All three Evangelists indicate the interval of time in order to show that this event was intimately connected with the events at Cæsarea Philippi: Peter's Confession, the Passion Foretold and the Doctrine of the Cross. "The principal purpose of the Transfiguration was to remove from the hearts of the disciples the scandal of the Cross" (St. Leo the Great). Therefore the same three Apostles who were to witness Christ's hour of humiliation in the garden of Gethsemani, were now chosen to be the special witnesses of His divine glory. Cf. also Mark *5*, 37.

A high mountain: St. Cyril of Jerusalem († 386) was the first to identify this mountain with Mount Thabor. Since that time tradition has known no other site for the Transfiguration. However, certain objections, not all of equal value, are raised against this traditional identification which render it somewhat doubtful. The main objections are: (a) Thabor was considered a part of Galilee; but according to Matt. *17*, 21; Mark *9*, 29 it would seem that Jesus and the disciples did not return to Galilee until after the Transfiguration. (b) None of the Evangelists give a name to the mountain. This would seem to exclude both Thabor and Hermon, for Matthew especially would have pointed out the connection with the well known words of Ps. *88*, 13, "Thabor and Hermon will rejoice in thy name," if the event had taken place on either of such famous mountains. It is possible, moreover, that these words might have given rise to the fourth century tradition. (c) The place was *a high mountain.* Thabor is an isolated peak rising some one thousand six hundred and fifty feet above the surrounding plain of Esdraelon and somewhat less than two thousand feet above sea-level. But among the mountains of Palestine this would not be considered as extraordinarily *high*. Hermon is some nine thousand feet in height. (d) Both at the time of Antiochus the Great (218 B. C.) and during the Jewish rebellion against Rome (67 A. D.) the top of Thabor was occupied by a town and a fortress. The top of Hermon on the other hand is snow-clad almost all the year. But none of the Evangelists speak of the Transfiguration as taking place on the "top" of the mountain. **2.** He *was transfigured:* more than a merely external change in appearance is indicated; the Greek word used here (from which we get our word "metamorphosis") means "His form was changed." This is to be compared with Phil. *2,* 6 f (literally according to the Greek): "Being in the form of God . . . he emptied himself, taking the form of a slave." That is, at the Incarnation Christ hid His divine glory; at the Transfiguration He allowed this divine glory to shine forth for a brief moment. St. Peter considered the Transfiguration of Jesus one of the strongest proofs of His divinity (cf. *2 Pet. 1,* 16-18). Possibly St. John, another witness of the event, is also referring to the Transfiguration when he writes, "We saw his glory—glory as of the only-begotten of the Father" *(1,* 14b). *White as snow:* the Greek has, "White as light"; some early Latin copyist confused the two Latin words *"lux,* light" and *"nix,* snow." All the Evangelists stress the dazzling brightness of the transfigured Christ. **3.** *Moses and Elias* represent the Law and the Prophets, i.e., the Old Testament. Luke adds that they "spoke of his death." This was to teach the Apostles that Christ's sufferings and death were foretold in the Old Testament. For the other interesting items given only by St. Luke, see Commentary there. **4.** *It is good for us to be here:* this is most probably the sense of the Greek which is literally, "It is good that we are here"; the sense is hardly, "It is a good thing we are here for now we can quickly put up three huts of boughs," etc. (Lagrange). It is rather futile to inquire what purpose Peter had in mind in making this proposal, "for he did not know what he said" (Mark and Luke). **5.** The *bright cloud* symbolizes the special presence of God, known technically as the "Shekinah"; cf. the cloud which covered Mount Sinai (Ex. *24,* 15 ff), the Ark of the Covenant (Ex. *40,* 32 ff), and Solomon's temple *(3* Kgs. *8,* 10-12). Compare the *voice* from heaven at the Baptism of Jesus *(3,* 17 and parallels). Both in classical literature and in the Septuagint the Greek word that is here translated as *beloved* means "only"; the Greek

17, 8-12 ST. MATTHEW

Fathers rightly used this text against the Arians, arguing that if Jesus was the "only Son" of God He was not merely "a son of God" in the sense of "a man especially chosen by God" but was truly divine, having the same nature as the Father. *In whom I am well pleased:* only in Matthew here, but all three Evangelists have the expression at the Baptism of Christ. *Hear him,* with the emphasis on *him:* Christ now supplants Moses and Elias, i.e., the Law and the Prophets, as the teacher of God's will. **8.** *Jesus only:* Moses and Elias disappear to signify the passing of the Old Covenant.

17, 9-13: On the Coming of Elias. Parallel in Mark *9,* 8-12. **9.** Probably one of the reasons why Jesus wished the Transfiguration to remain a secret for the time being, was the same as the reason why He wished most of His miracles to remain secret. See Commentary on *8,* 4. But it is to be noted that not even the other Apostles are to be told about the Transfiguration *till the Son of Man has risen from the dead,* i.e., until Jesus has permanently assumed the glory of His divinity which He manifested briefly on the mountain. Luke, who has nothing else of this paragraph, mentions the fact that these three Apostles carried out this command of our Lord (cf. Luke *9,* 36b). Cf. also Mark *9,* 9a. *The vision:* literally "the thing seen"; cf. Mark, "what they had seen." The word *vision* in no way implies a denial of the objective reality of the Transfiguration. **10.** What induced the disciples to ask this question? Evidently the sight of Elias at the Transfiguration started the train of thought, but the exact purpose of the question is uncertain.

11. The common belief that Elias will come in person before the Second Coming of Christ at the end of the world is based on this verse; it receives no other support in the New Testament. But the sense of this verse is probably, "The Scribes say truly that Elias is to come and restore all things before the (first) coming of the Messias, *but I say to you,* i.e., I (as contrasted with the Scribes) explain the true meaning of this prophecy to you, *that Elias has come already."* Our Lord clearly means that the prophecy of the coming of Elias was already completely fulfilled in the ministry of the Baptist. See Commentary on *11,* 14. **12.** *They did to him whatever they wished:* i.e., they killed him. On the expression itself, cf. Luke *23,* 25b. The *they* need not refer solely to the Scribes; it is the indefinite "they," i.e., people in general, here meaning Herod in particular. The reason why

Jesus then refers to His own coming sufferings is clear from Mark who mentions the fact that the disciples had also asked Christ what He meant by saying, "Till the Son of Man has risen from the dead." See Commentary on Mark *9*, 9-12. **13.** This correct explanatory note is added only by Matthew, not by Mark.

17, 14-20: **A Possessed Boy.** Parallels in Mark *9*, 13-28 (where a much more complete account of this event is given) and Luke *9*, 37-44a. **14.** The scene need not have been at the very foot of the mountain of the Transfiguration. *Lunatic* is a literal translation of the original Greek and Latin word used here, but at that time lunacy was the name given to any mental or nervous disorders which occurred at more or less regular intervals and were therefore attributed to the changes of the moon (*luna*). From the description of the boy's symptoms, especially in the detailed account of Mark, it seems certain that he was suffering from epilepsy. In this case the nervous disease was accompanied or even caused by diabolical possession. **15.** Although Christ had given His disciples the power to cast out devils (cf. *10*, 1. 8), still their faith was not strong enough to effect a cure in this case (cf. 19). **16.** Christ's cry of impatience was addressed to all present, the crowd, the Scribes, the father of the epileptic and the disciples. *Unbelieving,* at least in regard to the disciples, means "of little faith" (19). **18.** *Privately:* according to Mark, Christ and the disciples had entered a house in the neighborhood. **19.** Cf. *21*, 21; Mark *11*, 23; Luke *17*, 6. On the *mustard seed* as the symbol of something very small, cf. *13*, 31. **20.** This verse is missing from some of the best Greek MSS; perhaps it was taken over into the First Gospel from the Second, where, however, these same MSS omit the words *and fasting*.

17, 21-22: **The Second Prediction of the Passion.** Parallels in Mark *9*, 29-31 and Luke *9*, 44b-45. The little band assembles in Galilee for the last time prior to the final journey to Jerusalem. For the first and the third predictions of the Passion, cf. *16*, 21-23; *20*, 17-19; and parallels. The new element in this prediction is that *the Son of Man is to be betrayed into the hands of men.* The disciples grasped the general sense that great afflictions were awaiting their Master, so that *they were exceedingly sorry;* but the exact meaning of Christ's words about the betrayal "was hidden from them" and "they did not understand the saying, and were afraid to ask him" (cf. Mark and Luke).

17, 23-26: **Paying the Temple Tax.** Only in Matthew. **23.** This is the last visit of Jesus to *Capharnaum;* apparently all the events up to *19,* 1 take place in that town (cf. Mark *9,* 32; *10,* 1). The *didrachma,* i.e., the coin worth two drachmas (two denarii), was considered equivalent to the Old Testament half-shekel. The half-shekel was the annual poll-tax imposed on every Jewish man who was twenty years old or older, the money being used for the maintenance of the temple. To pay the *didrachma* was therefore a popular expression meaning "to pay the temple tax." The earliest mention of half-shekel tax for the sanctuary is in Ex. *30,* 11-16. But apparently it was not a regular feature of Jewish life until the time of Nehemias (cf. *2* Esdr. *10,* 32 f), when, however, the impoverished Jews could pay only a third of a shekel each year. This tax fell due shortly after the Passover. But since Christ had left Galilee at about the Passover, i.e., shortly after the first multiplication of the loaves (cp. *14,* 13 ff; *15,* 21 with John *6,* 4), He and His disciples had not yet paid the tax and now it was long overdue. When the tax-collector at Capharnaum reminds Peter of the obligation, Peter answers without hesitation that his Master certainly keeps this law and pays the tax.   **24.** Jesus was apparently in *the house* while this conversation between the tax-collector and Peter was taking place outside, but He knew of it by His supernatural knowledge. *Tribute or customs:* in Greek the first of these words signifies the excise tax or custom-duty paid on merchandise; the second word, the direct tax on persons (poll-tax) and on property. **25 f.** Since a king's son does not pay taxes to his father, so also Jesus, the Son of God, is under no obligation to pay taxes for the upkeep of His Father's house.   **26.** This is the only reference in the New Testament to angling with a hook and line. Fish have been found with almost every conceivable small object in their stomachs. Therefore there need not be anything miraculous in the fact that a fish has a stater in its mouth. But the fact that this extraordinary thing should have occurred just when Jesus predicted it, is certainly supernatural, and shows a special disposition of divine Providence. A *stater* was a coin worth two didrachma; therefore just enough to pay the tax both for Jesus and for Peter. The actual carrying out of Christ's command is not mentioned, but of course there would be no point to telling this story unless Peter really caught the fish with the stater.

18, 1-4: **Against Ambition.** Parallels in Mark *9,* 32-35 and Luke *9,* 46 f. Beginning with this incident our Lord gives an instruction to His disciples which covers various points but possesses a certain unity of subject-matter. The stereotyped phrase at its end (*19,* 1a) shows that Matthew considers this the fourth of the five great discourses of Christ in his Gospel. For a title we may call this an "Instruction on Fraternal Charity" or "Duties towards Believing Brethren." Under such a heading all of its topics show a certain logical unity: Humility versus Ambition (1-4), the Evil of Giving Scandal to humble souls (5-9), The Great Value of Every Soul as shown by the parable of The Lost Sheep (10-14), Fraternal Correction (15-18), Prayer in Common (19-20), Forgiving Injuries as exemplified by the parable of The Unmerciful Servant (21-35). Most of these sayings of Christ are found only in the First Gospel. The scene is at their home in Capharnaum (cf. Mark). Therefore this discourse was spoken to the disciples alone; Christ no longer taught the people publicly during His last short stay in Galilee.

1. Actually the disciples were ashamed to ask Jesus about this question which they had discussed among themselves on the way to Capharnaum (cf. Mark and Luke). "But Jesus, knowing the reasoning of their heart" (Luke), of His own accord gave them the answer. Matthew, as is his custom, summarizes the whole situation by representing the disciples as asking Christ the question. The word *then* implies the previous discussion. The reason for the argument was probably the promise of the primacy to Peter at Cæsarea Philippi. It could hardly have been the favor shown to the chosen three at the Transfiguration, for the rest of them did not know of this event. For a similar discussion about the first place in the Kingdom, cf. *20,* 20-28; Luke *22,* 24-30.   **2.** According to a ninth-century tradition this *little child* was St. Ignatius of Antioch. But it is very improbable that St. Ignatius was a native of Capharnaum. Since the event took place in Peter's house, the child was more likely a member of his household.   **3.** *Unless you turn:* you are headed in the wrong direction; on the road of ambition you will not even reach the kingdom of heaven, to say nothing of being the first in it. Christ's disciples must *become like* little children in humility and simplicity.   **4.** The most probable sense of these words seems to be, "True greatness in the kingdom of heaven, i.e., in God's sight, consists in humility." On the whole episode cf. the similar incident in *19,* 13-15 and parallels. According to Mark and Luke, St. John interrupted Christ's discourse with an entirely extraneous subject; see Commentary on Mark *9, 37-40.* Matthew omits this episode lest it spoil the unity of the discourse.

18, 5-9: Avoiding Scandal. Parallel in Mark *9,* 36. 41-47. Luke *17,* 1 f has a similar saying of Christ but in a different context. **5.** *These little ones who believe in me* should not be understood solely of children; it refers rather to the true disciples of Christ who are children in spirit, cf. also *11,* 25; *1* Cor. *14,* 20. *Receives me:* see Commentary on *10,* 40. **6.** *Causes one . . . to sin:* in Greek, "scandalizes one," i.e., causes one to trip up: according to the context the particular sin that is meant is to cause a humble disciple of Christ to lose this spirit of humility or to cause him to lose faith in Christ. *A great millstone:* literally, "an ass-millstone," i.e., the large millstone turned by an ass, as distinct from the small stone turned by hand. Death by drowning was not a common penalty among the Jews but it was inflicted by several of the surrounding nations for the most heinous crimes. The terror of this punishment was due to the great importance attached by all ancient peoples to a decent burial. But such a punishment is less an evil than scandalizing the innocent. **7.** *Woe to the world:* some take this to mean, "Alas, poor world which suffers so much because of scandals!" But in keeping with the following sentence it means more probably, "Woe to the world, the cause of so much scandal!" *It must needs be,* morally speaking, because of the wickedness of men, *that scandals come.* **7-9.** See Commentary on *5, 29* f.

18, 10-14: The Lost Sheep. Luke *15,* 4-7 presents the same parable but with a somewhat different application. In the Third Gospel this parable illustrates the truth that "there will be joy in heaven over one sinner who repents, more than over ninety-nine just who have no need of repentance." Here the parable shows the great care that the Son of Man, "the Good Shepherd" (cf. John *10,* 11-18), has for the soul of even one of His least disciples that goes astray. **10.** The first argument for the importance of every human soul: God "has given his angels charge over" them (cf. Ps. *90,* 11). This text can rightly be cited as a proof for the doctrine of the guardian angels, a doctrine which was held by the Jews at that time and is simply taken for granted in the New Testament (cf. Acts *12,* 15). *Their angels in heaven always behold the face of my Father in heaven:* these words are intended to show the great dignity of the guardian angels who are not deprived of the beatific vision, even though they are fulfilling their office on earth. Others take these words to mean, "Do not bring evil to these little ones, for their angels will avenge them before God's throne." **11-14.** The second argument for the importance of every soul: Christ's zeal in seeking even one soul that goes astray. **11.** This verse is missing in the best Greek MSS, and is generally considered to be taken over from Luke *19,* 10 (cf. also Matt. *9, 13*). Yet it seems to form a natural and almost necessary introduction here to the parable. **12.** This is certainly not intended to show the proportion of the just and sinners as ninety-nine to one: it merely emphasizes Christ's care for each individual soul as well as for the flock taken as a whole.

18, **15-18: Fraternal Correction.** Only in Matthew; but cf. Luke *17, 3.* If administered from a purely unselfish motive for the good of an erring brother or for the common good, fraternal correction is a form of fraternal charity. But the good name of the sinner must be preserved as far as possible. Therefore correction has three progressive stages: private (15), before a few witnesses (16), and finally before the whole congregation (17). **15.** The words *against thee* are missing in the best Greek MSS, and rightly so, for they give an entirely false idea of the matter that is treated here. Personal offenses are considered in 21 ff. Here there is a question of sin in general, which from the context (cf. 17), is presumed to be of a very serious nature. *Thou hast won thy brother:* i.e., thou hast brought him back to the brotherhood, as the Good Shepherd brings back the erring sheep into the fold. **16.** *On the word of two or three witnesses every word may be confirmed:* a quotation from Deut. *19,* 15, cited also in 2 Cor. *13,* 1; *1* Tim. *5,* 19; Heb. *10,* 28. **17.** *The Church* cannot mean here the entire group of all Christ's disciples as in *16,* 18. On the other hand this word in itself does not mean "the authorities of the Church." Christ is no doubt referring to contemporary Jewish customs, so that the word is used here in its original meaning of the "assembly, congregation," i.e., the local Christian community. But since the context presupposes that *the Church* gives some decision in the matter *(if he refuse to hear even the Church),* and since the community as a whole cannot well do this, implicitly the word here signifies "the authorities of the Church." *As the heathen and the publican* are to the Jews, i.e., excommunicated. For certain forms of excommunication as a means of fraternal correction in the Apostolic Church, cf. *1* Cor. *5,* 9-13; *2* Thess. *3,* 6; Titus *3,* 10; *2* John 10.   **18.** The same power that was given to Peter as Christ's vicar on earth (*16,* 19) is here given to all the Apostles, to be exercised in harmony with his supreme authority. On the meaning of "to *bind*" and "to *loose*" see Commentary on *16,* 19. See also note to the text.

18, 19-20: The Power of United Prayer. Only in Matthew. This passage fits in well with the general theme of the whole discourse, the relation of the brethren towards each other. But *I say to you further* implies a more intimate connection with the preceding, i.e., just as the decisions of the Church are ratified in heaven (18), so also the united prayer of the Church has special efficacy before the throne of God. The ultimate source of both powers is due to the presence of Christ Himself in His Church (cf. *28,* 20). This presence of Christ is felt in the Church through the influence of the Holy Spirit, to whom are directly due both the infallibility of the Church and the efficacy of its prayers (cf. John *14-16;* Rom. *8,* 26 f). The prayers of the Mystical Body of Christ are the prayers of Christ Himself, who is our great Intercessor (cf. Rom. *8,* 34; Heb. *7,* 25).   **20.** Our Lord may be alluding to the well-known saying found in the Talmud, "When two or three are gathered together to study the Law, the Shekinah is in their midst." If so, He identifies His own presence with the divine presence.

18, 21-35: The Unmerciful Servant.   21 f.   Luke *17,* 3 f has similar words of Christ on the frequency of forgiving injuries; but these words were probably spoken on a different occasion.   **21.** *Sin against me:* a personal offense as distinct from the sin considered in 15-17; but Peter's question was probably prompted by our Lord's words in 15. *Seven times:* Peter thought he was generous. *Seven* is here a symbolic number meaning "often."   **22.** *Seventy times seven* is also symbolic and signifies "without limit." Cf. the same symbolic numbers in Gen. *4,* 24 in regard to vengeance.

23-35. The parable of The Unmerciful Servant is found only in the First Gospel.   **24.** *Ten thousand talents,* the equivalent of about nineteen million, two hundred thousand dollars in U. S. A. currency (see under word "Money" in Glossary at end of text, p. 752), an enormous sum at that time. **25.** In ancient times a man's wife and children were regarded as his property and were therefore forfeited by a defaulting debtor. This was the common practice among the pagan nations and was not unknown among the Israelites (cf. *4* Kgs. *4,* 1), although among the latter the Mosaic Law endeavored to mitigate its evils (cf. Lev. *25, 39*). **26 f.** Note that the servant, not trusting the goodness of his master, asks only for a moratorium on this debt which he could not possibly pay; but the master overlooks this insincerity and remits the whole debt of the unworthy servant. **28.** *A hundred denarii,* about sixteen or seventeen dollars, an insignificant amount compared with the other debt. He *throttled him,* whereas the master had used no physical violence. **29.** The first servant had made exactly the same petition (26), but here it was spoken with evident sincerity. **34.** Tyrants, especially in the Orient, made use of torture in order to wring from their victims the confession of a hidden source of wealth or to have their relatives and friends pay the required money out of compassion. Here, however, the master is entirely justified in inflicting torture as a punishment on the wicked servant. This torture is without end, for the immense debt can never be paid. **35.** The application of the parable is perfectly clear. On the necessity of forgiving our fellowmen, cf. *6,* 12 (where sin is also considered a "debt"); *6,* 14 f; Mark *11,* 25 f. The parable likewise teaches that our offenses against God are infinitely greater than the offenses we receive from our neighbor (24.28). If we refuse to forgive our brother, we not only make God *angry* (humanly speaking), but also make our fellow-Christians *very much saddened,* and it is their prayers to God to redress this wrong that will bring God's vengeance upon us (31). *From your heart:* our forgiveness of injuries must be sincere.

5. Ministry on the Journey to Jerusalem *19 — 20*

This account of the events of our Lord's last journey from Galilee to Jerusalem is closely paralleled in Mark *10.* But the corresponding section in the Third Gospel (Luke *9,* 51 — *18, 34*) is much longer and contains much material which Matthew has in the earlier sections of the Public Ministry. See Commentary on Luke *9,* 51. According to the first two Gospels this journey

19, 1-4   ST. MATTHEW

of Christ and His disciples was more or less direct, from Galilee through northern Judea to Perea and finally via Jericho to Jerusalem. Nevertheless it was not a hurried trip but a missionary tour.   **1.** *The district of Judea beyond the Jordan:* strictly speaking no part of Judea was on the east side of the Jordan; either Matthew speaks loosely of Perea as a part of Judea, which hardly seems probable, or his original text reads the same as the best Greek MSS of Mark *10,* 1, "the district of Judea and (the district) beyond the Jordan."   **2.** *And great crowds followed him;* therefore his journey is certainly distinct from the secret journey of Jesus to Jerusalem mentioned in John *7,* 9 f.

19, 3-12: The Question of Divorce. Parallel in Mark *10,* 2-12. See Commentary on *5,* 31 f.   **3.** *For any cause:* Mark omits this phrase, yet it is essential to the question. The second Evangelist, writing primarily for the Gentiles, considered the purely Jewish question about the grounds for divorce as of no interest to his readers. But all the Jews at the time of Christ thought that divorce at least on some grounds was licit. Only there was a difference of opinion among the Rabbis as to what grounds were sufficient for divorce. The liberal school of Hillel taught that a man could divorce his wife for almost any reason whatever; if she spoiled the cooking, if her beauty no longer pleased him, etc. The stricter school of Shammai permitted a man to divorce his wife only if she was guilty of some infidelity to her marriage vows. All the Rabbis agreed that after a divorce both parties could remarry but that the first husband could never take back the divorced wife. These Pharisees in Perea wished to learn what school of thought Jesus favored, Hillel's or Shammai's. There is no need to read a hostile intention into their *testing him.*

4-6. Jesus rejects both opinions and teaches that there are no licit grounds for divorce with the right to remarry. His arguments from the Scripture are: (a) at the creation God made but one man and but one woman; therefore even successive bigamy is wrong (4); (b) marital relations cause such an intimate union between man and wife, that the severance of this bond is against the natural law (5); (c) God Himself is the author of this bond; therefore its severance is not merely a question of expediency or inexpediency, but it is impossible for any man to break the bond that God has tied (6).   **4.** The reference is to Gen. *1,* 27. *From the beginning:* i.e., at creation; the phrase itself is an allusion to the first words of Gen. *1, 1. Male and female:* it is important to note that in the original both of these words are in the singular: God made only one man and only one woman; therefore the institution of monogamous marriage is from God himself.   **5.** A citation of Gen. *2,* 24; quoted also in Eph. *5,* 31, and in part in *1* Cor. *6, 16.*   **7.** To the Pharisees Christ's complete denial of the right of divorce seemed opposed to the Law of Moses. There is only one reference to divorce in the Mosaic Law: Deut. *24,* 1-4. It seems well to quote here in full a literal translation of this passage according to the original Hebrew: "When a man marries a woman and has relations with her, if she does not find favor in his eyes, because he finds in her some matter of shame, and he writes a document of divorce for her and gives it into her hand and sends her from his house, and she departs and goes forth and becomes the wife of another man, and the second husband also hates her and writes a document of divorce for her and gives it into her hand and sends her from his house, or if the second husband who married her dies, the first husband who sent her away cannot take her again as his wife after she has been defiled, for this would be an abomination before the Lord."   **8.** Christ gives the authentic interpretation of this law. Moses did not "command" a man to divorce his wife, as the Pharisees said. He merely *permitted,* i.e., tolerated, divorce because it was an abuse of long standing among the Israelites, although it was not according to God's institution of marriage at the creation of man. Moses merely takes cognizance of this abuse and forbids further evils that may result from it.   **9.** Conjugal infidelity justifies separation but not divorce with the right to remarry. See Commentary and notes on *5,* 31 f. Cf. also Mal. *2, 14* ff. Christ had to mention this apparent exception here on account of the nature of the question proposed by the Pharisees (3).

10-12. Only in Matthew. According to Mark *10,* 10-12 the disciples asked Jesus in private about this matter, and the Master repeated the words that He had spoken to the Pharisees; hence this conclusion of the disciples: "If a man cannot divorce his wife, it is better not to marry."   **11.** *This teaching:* the obvious meaning seems to be, "This teaching that I have just given on divorce." Others understand it as, "This teaching that you are giving by saying that it is not expedient to marry." The latter interpretation is more in keeping with the following words. *To whom it has been given:* a special grace of God is needed.

12. Besides the two classes of men who are physically incapable of begetting children, our Lord teaches that there is a third class who voluntarily abstain from marriage *for the kingdom of heaven's sake.* The last words show that Christ gives His approval to this third class. *Let him accept it who can;* therefore voluntary celibacy is proposed as a counsel, not as a precept, to His disciples. Not only would it be inexpedient but it would be against God's will to impose perpetual virginity as a precept. (The Church can, of course, prescribe celibacy as a precept in the clerical state of the Latin rite, but no one has an obligation to enter that state.) Marriage is from God (cf. 4-7) and therefore is something good in itself. Celibacy in itself is something negative and therefore of itself is not something better than the good from which it abstains. Celibacy, when embraced for purely selfish motives, is not as good as matrimony. But celibacy *for the kingdom of heaven's sake,* i.e., not only for the sake of facilitating the work of the divine ministry and the works of mercy but also for the sake of one's own personal sanctification, is better than the married state. This is also the clear teaching of St. Paul (cf. *1* Cor. 7).

19, 13-15: Jesus Blesses the Children. Parallels in Mark *10,* 13-16 and Luke *18,* 15-17. Cf. the previous incident of Jesus and a child in *18,* 1-5 and parallels. The two events are certainly distinct, even though the words of Christ on each occasion are very similar, and His words about receiving children seem to fit in better with the second occasion, while His words about children serving as a model for His disciples fit the first occasion better.   **13.** The word *Then* is probably a mere connective particle implying no immediate relation with the preceding. To *lay hands on and pray* for some one is an ancient symbolic gesture in bestowing a blessing, signifying the granting of the things asked for in the prayer. The interference of the disciples, while undoubtedly well meant, showed that they did not fully grasp the true nature of the kingdom of heaven; therefore our Lord *rebuked them.*   **14.** From these words of Christ one may rightly argue to the licitness of infant-baptism, for unless a child is baptized, it cannot come to Christ. The disciples of Christ must be distinguished by their practice of the virtues which are seen as the natural characteristics of all good children: humility, simplicity, docility, purity, etc.   **15.** *He departed from that place:* only in Matthew; but since we do not know what this place was nor where Christ went next, this remark is of no great help.

19, 16-30: The Danger of Riches. Parallels in Mark *10,* 17-31 and Luke *18,* 18-30. All three Gospels have the same three sections of this episode in the same order: (a) the incident of the rich young man (16-22); (b) the conversation on the danger of riches (23-26); (c) the reward promised by Christ to those who leave all for His sake (27-30).

16. *A certain man:* Luke calls him "a certain ruler," which may mean either "a ruler of a synagogue" or, more probably, "a member of the Sanhedrin." Matthew alone calls him a "young man" (20), but this Greek word was used for any man who had not reached middle age. That he was no longer young seems clear from the fact that he was a "ruler" and that he says, "From my youth" (according to the Greek of Mark and Luke). In Matthew the best Greek MSS have simply *Master;* the preceding word *Good* was probably taken over into the First Gospel from the other two.   **17.** The words *that is God* are not in the Greek text and represent a correct gloss added to the Latin text. In Matthew the sense of our Lord's answer is, "No action is morally good except in relation to the *One who is good,"* i.e., the mere observance of the Law as such cannot give justice and eternal life (cf. Gal. *3,* 21b); the formal reason why the observance of the commandments leads to eternal life is because they are the will of God and the norms by which man imitates God's own goodness. In the other two Gospels a somewhat different answer of Christ is recorded, on the meaning of which see Commentary on Mark *10,* 18. But these two answers are not at all contradictory, and undoubtedly both were given by Christ.

18. The order of the commandments as cited by Christ differs somewhat in each of the Gospels and there is still further confusion in the MSS. Cf. Ex. *20,* 12-16; Deut. *5,* 17-20. In all three Gospels the fourth commandment is placed after the others. In Matthew alone is Lev. *19,* 18 cited, while in Mark alone occur the words "Thou shalt not defraud," which probably refers to Deut. *24,* 14.   **20.** It seems clear from the context that the young man was sincere in what he said.   **21.** *Sell . . . and give:* cf. Luke *12, 33. Treasure in heaven:* cf. *6,* 20. *If thou wilt be perfect:* these words make a clear distinction between what is obligatory for all (18 f) and what is recommended by Christ to those who would be His closer followers, i.e., the counsels of perfection. This passage is rightly considered the principal basis for the three traditional vows of religious life: poverty (*sell what thou hast, and give to the poor*), obedience (*come, follow me*), and chastity, since any one pledged to absolute poverty cannot raise a family (cf. also *19,* 12).

23-26. It is very difficult for a rich man to save his soul; not that the possession of riches in itself is necessarily sinful, but riches are often unjustly acquired and therefore unjustly retained, they easily lead a man to commit sins of self-indulgence and, what is an important point in Christ's teaching, the pursuit of the things of this world keeps a man from the whole-hearted service of God and induces a certain self-confidence that is opposed to the humble trust that we must have in God's providence. Christ is emphatic on this matter (cf. *6,* 19-34; Luke *12,* 13-34; *16*), and any attempt to weaken this teaching is unworthy of His disciples. *A camel through an eye of a needle* is a proverbial expression meaning that something is impossible. Similar paradoxical expressions are found not only in the Talmud but also in Greek and Latin literature. To try to explain camel by a similar-sounding Greek word meaning "rope," or to interpret an *eye of a needle* as meaning a low gate in the walls of a city through which pedestrians, but hardly camels, can pass, are futile attempts to whittle down the force of Christ's words.   **25.** That the disciples understand this teaching literally is seen in the fact that they *were exceedingly astonished* and "amazed" (Mark).   **26.** *Looking upon them,* to stress the importance of this teaching, Jesus repeats His doctrine without figurative language: it is humanly *impossible* for a rich man to be saved, but by the grace of God it is *possible* for him to give away his wealth to the poor or at least to use it as a wise steward uses the property which his master entrusts to him (cf. Luke *16,* 1-13), which practically means ceasing to live as a rich man.

27-30. The reward Christ promises to those who renounce all things to follow Him.   **28.** This special reward for the Apostles alone is given in this context by Matthew only, but cf. the similar words in Luke *22,* 29 f. *Regeneration,* or "rebirth": the Greek word that is used here is found in only one other place in the New Testament, viz., Titus *3,* 5, where however it signifies the spiritual rebirth at baptism (cf. John *3,* 3. 5). Here it is a synonym for the "Last Day" when God will restore the world to the primeval happiness and harmony it enjoyed before the fall of our first parents (cf. Isa. *65,* 17; *66,* 22; Acts *3,* 21; Rom. *8,* 19-21;

2 Pet. *3,* 13; Apoc. *21,* 1-5). *Judging* probably signifies here not "condemning" but "obtaining justice for, avenging the wrongs of" (according to the common Hebrew usage). *The twelve tribes of Israel* are all the elect of both the Old and the New Covenants (cf. Apoc. 7).   **29.** *House* signifies not so much the material building as "household, home." *Wife:* not in Mark, and missing in the best Greek MSS of Matthew, but certainly authentic in Luke. But the disciple of Christ could not leave his wife without her consent. Perhaps what is meant here is "the prospect of marrying." *A hundredfold:* the other two Gospels add "in the present time." This promise need not be taken literally; but cf. *6, 33*; in place of the relations whom the disciple of Christ abandons (cf. *10,* 35-37), he receives many more new "brethren" in Christ.   **30.** A proverbial saying often used by our Lord; besides the parallel passage in Mark *10,* 31, it also occurs in Matt. *20,* 16 and Luke *13,* 30. In the present context the meaning is, "They who are rich and honored in the present life will be poor and without honor in the world to come, but they who give up all for Christ and are despised by the world will be the richest and the most honored in the future life." God's standards are not the standards of the world. However, this verse probably belongs logically to *20,* 1-16, and as such has a somewhat different meaning.

20, 1-16: Parable of the Laborers in the Vineyard. Only in Matthew. This parable taken merely as a story presents no special difficulties.   **2.** *A denarius a day* was the regular daily wage at that time.   **3-8.** The day, considered as the period of time from sunrise to sunset, was divided into twelve "hours." According to the varying length of daylight throughout the years these "hours" would be correspondingly longer or shorter. The sixth hour always began at midday. The third hour began at the middle of the forenoon, the ninth hour at the middle of the afternoon. These divisions formed four main periods of the day, so that "the third hour" meant the whole period up to the ninth hour, etc. This fact is to be remembered in regard to the mention of the various hours in connection with our Lord's crucifixion (cf. Matt. *27,* 45 f; Mark *15,* 25. 33; Luke *23,* 44; John *19,* 14). The wages were paid every evening (cf. Lev. *19,* 13; Deut. *24,* 14 f).   **15b.** *Art thou envious . . .?:* literally in the original, "Is thy eye evil . . .?" The expression "to have an evil eye" is a Hebraism meaning "to look with envy" (cf. the same expression in the original of Deut. *15,* 9; Prov. *23,* 6; *28,* 22; Ecclus. *31,* 14; Mark *7,* 22). Note the different agreements made between the householder and the laborers: the first group of laborers are promised a specific wage, the middle groups a less definite promise of "whatever is just," the last group no promise at all.

The principal lesson intended by our Lord in giving this parable is not certain. But it at least seems clear that the parable is given as an illustration of the truth that *the last shall be first and the first last.* For this truth is enunciated both at the beginning and at the end of the parable (*19,* 30; *20,* 16), and the conjunctions *For* (1) and *Even so* (16) show that the parable was meant to illustrate this truth. Moreover, the preceding context concerning the reward for the faithful disciples of Christ (*19,* 27-29) and the importance of the *wages* (8) in the parable seem to indicate that *the first* and *the last* signify those who are in some way more highly favored or less favored in regard to the spiritual reward that they receive from God. There is not the slightest reference to the distinction between Jews and Gentiles either in the parable or in the context, as there is in Luke, *13,* 28-30. Nor does it seem probable that the application of this parable is to be limited to the Apostles, the immediate hearers of these words, as if, e.g., it meant that St. Stephen would receive his reward before they would receive theirs. Christ is probably speaking in general terms of all His followers who labor in His vineyard. They all receive what is essentially the same reward, the beatific vision of God in heaven, even though there are different degrees of glory in heaven according to each one's merits (cf. *1* Cor. *15,* 41). In this regard God is not unjust to any one. But from a merely human viewpoint He seems to be more than just to some. Towards these He is extremely generous. They have not bargained with Him about wages and rewards but accept His call to work in His vineyard on faith in His justice and goodness. The main lesson therefore of the parable seems to be that God's bestowal of reward is not to be judged too strictly according to man's idea of justice.

In the application of the parable no special emphasis is to be placed upon the various hours of the day when the laborers were called. This is merely a device of the story to signify that those who may appear to men to be less deserving of reward will be more highly favored by God. Those who worked for only one hour not only received as much as those who worked longer but they received their pay before the others, an additional advantage. God gives His grace to whom He will. He is good to all; but if He is more generous to some, the others have no reason to complain. God's standards of judging even in purely spiritual matters are not necessarily the same as man's standards. **16b.** The words *For many are called but few are chosen* fit in with the interpretation given above, but they are not in the best Greek MSS here; they were probably taken over from *22, 14*.

20, 17-19: The Third Prediction of the Passion. Parallels in Mark *10,* 32-34 and Luke *18,* 31-34. Cf. the first two predictions of the Passion, *16,* 21-23; *17,* 21 f; and parallels. This third and last of our Lord's predictions of His death and resurrection is by far the most detailed and explicit of all. The Third Gospel adds Christ's words that all these sufferings had been foretold by the Prophets; Luke alone notes that the disciples did not understand this. For the circumstances in which this prediction was made, see Commentary on Mark *10, 32.*

20, 20-28: The Mother of James and John. Parallel in Mark *10,* 35-45.   **20.** Matthew had already mentioned that the sons of Zebedee were James and John (cf. *4,* 21); therefore here, as also in *26,* 37 and *27,* 56, he does not repeat their names. The name of their *mother* was very probably Salome (cp. *27,* 56 with Mark *15,* 40). In Mark the petition is presented directly by James and John, but Matthew is more exact here. No doubt this ambitious desire was shared both by the mother and by her sons. Perhaps they thought it would look less self-interested, if the request was made by the mother; or they had more reliance on her feminine powers of persuasion. *Worshipping:* cf. *15, 25. She made a request:* no doubt her words were essentially the same as those which Mark attributes to her sons, i.e., "Lord, grant me whatever I ask of thee." She wished Christ to promise to grant her request even before she told Him what it was, just as Herod pledged himself to grant the daughter of Herodias whatever she would ask.   **21.** *To sit at the right hand* of a king meant to have the highest place of honor after the king himself; the place at his left hand was the next best (cf. *3* Kgs. *2,* 19; Ps. *109,* 1). The petition of the Apostles, which seems prompted at least in part by Christ's promise in *19, 28,* really shows their unshaken faith in Jesus, for they firmly believe that, despite the humiliations and suffering and death that He was predicting for Himself, He would eventually come into His *kingdom*, "in glory" (Mark).

22 f. In Matthew as in Mark the conversation is now directly between Christ and the sons of Zebedee. It is not correct to say that Jesus rebukes them and refuses their petition. Their prayer is heard and granted but in a better way than they had intended. They had sought important positions in His kingdom on earth, His Church. But God the Father, the Creator, has already determined this by giving to some better natural qualifications for these positions than to others. Because Jesus had come, not to do His own will, but the will of Him who sent Him (cf. John *6,* 38), He will not change this divine decree. But He offers the sons of Zebedee a greater boon, an exalted position in His heavenly kingdom. This does not necessarily correspond to the position held in His kingdom on earth (cf. *18,* 4). The closer one imitates the sufferings of Christ (22) and His spirit of self-sacrifice in the service of others (27 f), the greater one will be in heaven. The *cup* is a metaphor for the lot in life to which God has destined one (cf. Ps. *22,* 5), especially the lot of affliction (cf. *26,* 39; Isa. *51,* 17; Jer. *25,* 15; *49,* 12; Ezech. *23,* 33 f). This prophecy of Christ was fulfilled in the fact that St. James was the first of the Apostles to suffer martyrdom (cf. Acts *12,* 2) and that St. John was scourged (cf. Acts *5,* 40), exiled to Patmos (cf. Apoc. *1,* 9), and according to Tertullian, immersed in a cauldron of boiling oil, from which he was miraculously delivered.

24-28. Jesus teaches His disciples the nature of true greatness in His kingdom. Luke *22,* 25-27 has very similar words of Christ following a contention among the Apostles at the Last Supper. Since the Church is a visible society there must be those who exercise authority in it. But they must not imitate the pagans in this. Those who have worldly authority use it to their own advantage and prestige. But those who have spiritual authority in Christ's kingdom must imitate His humility and self-sacrifice in the service of others. This alone constitutes true greatness in God's sight (cf. *18,* 4). Since the less authority a man has, the more easily he can practise these virtues, it is foolish to contend for higher authority. *Servant, slave:* cf. *1* Cor. *9,* 19. *The Son of Man has come . . . to give his life a ransom for many:* this and the words of Christ at the institution of the Holy Eucharist (cf. *26,* 28 and parallels) are the only references in the Synoptic Gospels to the redemptive value of Christ's Death. The Apostles were **not** fully prepared to receive this great truth which they understood clearly only after they had received the Holy Spirit at Pentecost. These authentic words of Jesus are of great value in refuting the heresy of the modern critics who claim that this doctrine was an invention of St. Paul. *Ransom:* the same Greek word (in the plural) is used in the Septuagint in the sense of the "price" paid for the freeing of slaves (Lev. *19,* 20) and captives (Isa. *45,* 13) and of the "fine" imposed instead of the death penalty (Ex. *21,* 30; Num. *35,* 31 f). Christ gave His life therefore for our spiritual freedom or "Redemption" (cf. 1 Cor. *6,* 20; *7,* 23; *1* Pet. *1,* 18 f). *For many:* literally in Greek, "instead of many"; for we of ourselves were unable to pay the price. In such a sense the so-called "substitution theory" of the Redemption is correct, but it should not be over-emphasized or it will give a distorted picture of how Christ really redeemed us. Our Redeemer won salvation for all men, but He uses the word *many* either: (a) in the sense of "all" but with special emphasis on the immense number of Adam's descendants (cf. Rom. *5,* 15, where however the article is used with this adjective in Greek); or: (b) in reference to the actual effect, for "many," but not all men, co-operate with the grace by which the merits of Christ's death are applied to themselves.

20, 29-34: **The Blind Men at Jericho.** Parallels with more details in Mark *10,* 46-52 and Luke *18,* 35-43. There are certain difficulties in harmonizing these three accounts. (a) Matthew speaks of two blind men, Mark and Luke of only one blind man. Cf. the similar difficulty in the expulsion of the devils at Gerasa (*8,* 28-34 and parallels). (b) Both Matthew and Mark place this event *as they were leaving Jericho,* Luke says it happened "as he drew nigh to Jericho." (c) The conversation between Jesus and the blind men is essentially the same in all three Gospels.

Various solutions are proposed. (a) One blind man was cured before Jesus entered Jericho (Luke), another blind man was cured after Jesus left Jericho (Mark), and Matthew combines the two cures as taking place after the departure from that town. But this seems unlikely, for the narrative of all three Gospels evidently treats of one and the same miraculous cure. (b) As Jesus entered Jericho a blind beggar implored His help (Luke), but Jesus at first ignores his plea. As our Lord and His companions pass through Jericho, the blind man follows them, continually crying out his prayer. Somewhere along the line of march he is joined by another blind man (Matthew). After leaving Jericho, Jesus stops and cures them (the conversation and cure as given by all three Gospels). (c) There were two Jerichos at that time: Old Jericho, on the site of the Canaanite city that was destroyed by Josue (cf. Jos. *6*) and rebuilt by Hiel (cf. *3* Kgs. *16,* 34); and New Jericho, also known as Phasael, built by Herod the Great, which was about two miles to the south of the old city. (The modern village of Jericho, er-Riha, is about two miles to the east; eight hundred and twenty-five feet below sea-level.) This miracle then took place between Old and New Jericho. The first two Evangelists, according to the Jewish custom, refer to the older city as Jericho; Luke, as a Gentile, refers to Herod's new city of Jericho. Mark tells us the name of one of the blind men, Bartimeus, i.e., son of Timeus. Probably he became a Christian and was known to the early Church (cf. Mark *15,* 21). Luke undoubtedly refers to the same man. The other blind man was probably of less importance and is mentioned by Matthew only.

29. *A great crowd followed him:* no doubt these people were mostly pilgrims from Galilee who were on their way to Jerusalem for the Passover. These same enthusiastic Galileans gave Jesus a triumphant welcome at Jerusalem (cf. John *12,* 12). **30 f.** *Son of David:* a Messianic title. **32.** Our Lord wishes that a prayer of petition be specific. **34.** *They followed him* not only "along the road" (Mark) but probably also as His disciples in the "Way" (cf. Acts *9,* 2; *19,* 9).

6. Last Ministry at Jerusalem *21-25*

21, 1-11: Triumphal Entry into Jerusalem. Parallels in Mark *11,* 1-11a; Luke *19,* 29-44; John *12,* 12-19. The only Evangelist who gives us definite information about the time when this occurred is John. According to the Fourth Gospel the Passover in that year fell on a Saturday (cf. *18,* 28; *19,* 14.31); "six days before the Passover," when Christ was anointed at Bethany (cf. *12,* 1), would seem to be the preceding Sunday; hence "the next day" after the anointing, when Jesus entered Jerusalem in triumph (cf. *12,* 12), would apparently be Monday. However, some scholars, arguing from the Synoptics' account of the Last Supper, conclude that the Friday on which Our Lord died was the Passover itself and that the anointing at Bethany "six days before the Passover" occurred on the preceding Saturday; the triumphal entry into Jerusalem on "the next day" would then have taken place on Sunday. The Church in her liturgy favors this second opinion.

1. From the Synoptic Gospels alone we might think that our Lord made the whole journey from Jericho to Jerusalem (about twenty miles of steep ascent) on the same day. But from John *12,* 1. 12 we know that He stayed at Bethany for at least one day before entering the Holy City. *Bethphage* is generally considered the name of a village that lay somewhere between Bethany and the Mount of Olives, although its exact site cannot be determined. But according to Luke, the only one besides Matthew who speaks of it, Christ apparently reaches Bethphage before He comes to Bethany. *Bethphage* means "the house of green figs," i.e., the grove of fig-trees. In the Talmud it is the name of the whole region immediately around Jerusalem. Perhaps the Gospels also use this word as the name of a region in which Bethany was situated. *The Mount of Olives* is the ridge of high hills immediately to the east of Jerusalem, separated from the city by the deep valley of the Cedron. It received its name from the "Olivet" (Luke), i.e., the olive grove, that grew on its lower slopes.   **2-4.** Jesus intentionally makes arrangements in order to fulfill the Messianic prophecy of Zach. *9,* 9. Although His followers still have a wrong idea of the Messias as a temporal prince, He now permits them to acclaim Him publicly as the Messias. His purpose was partly to reward their fidelity with this brief moment of joy and triumph and to strengthen their faith against the approaching hour of darkness, and partly to force the issue with His adversaries and thus bring about His sacrificial death according to the will of His Father.   **5.** This prophecy of Zach. *9,* 9 is acknowledged by all as directly Messianic. It is also cited in John *12,* 15. Both Evangelists give a somewhat free translation. In Matthew the first words are assimilated to Isa. *62, 11.*

6. Mark and Luke tell us in detail that everything happened just as Jesus had foretold.   **7.** Matthew alone mentions both animals. In the prophecy of Zacharias the *ass* is the same animal as the *colt,* synonyms being used for the sake of Hebrew parallelism. It is unfair to St. Matthew to say that he did not understand this. He simply narrates the fact that the colt on which Jesus rode was accompanied by its mother. The *cloaks* laid on the animals and spread out upon the road were the customary signs of a festive procession.   **8.** These branches were no doubt from the olive trees of the grove through which they were passing. John (*12,* 13) alone mentions "the branches of palms" that they carried in their hands as symbols of triumph.   **9.** *The crowds that went before him,* as distinct from the crowd that accompanied Him from Bethany (7; John *12,* 9) and *followed him,* were Galilean pilgrims who came out of Jerusalem to meet Him (cf. John *12,* 12 f); it is unfair to accuse them of fickleness, for the crowds that demanded His death on Good Friday were natives of Jerusalem who had always been opposed to Him (cf. Acts *13,* 27 f). The words *Blessed is he who comes in the name of the Lord!* are from Ps. *117,* 26. Perhaps these people meant them in the sense that "Christ comes in the name of the Lord," i.e., as God's Anointed. But in the original Psalm they mean, "Blessed in the name of the Lord is he who comes," i.e., the Lord's name is invoked in a blessing upon him who enters the temple. *Hosanna* occurs in the preceding verse of this Psalm in its original sense of "Do save!" But here it is used as a mere cry of cheering. Each of the four Evangelists records different exclamations of the people.   **10 f.** The excitement within Jerusalem which this triumphant entry of Jesus caused, is mentioned only by Matthew. But cf. Mark *11,* 11a.

21, 12-17: Cleansing of the Temple. Parallels in Mark *11,* 15-19 and Luke *19,* 45-48. John *2,* 13-17 records a similar but distinct event. See Commentary there. It was this final attack on the lucrative business of the High Priest's party in the temple that, humanly speaking, sealed the fate of Jesus. It was not so much His earlier adversaries, the Pharisees, as it was these new enemies, the party of the High Priest, who brought about the death of Christ. Charges of hostility to the temple played an important part in His trial (cf. *26,* 61; *27,* 40). There can be no doubt therefore that this last "cleansing of the temple" took place but a few days before the Crucifixion. Mark is explicit in stating that on the day of the triumphal entry into Jerusalem Jesus merely "looked round upon all things" in the temple, and that it was only on the next day that He drove the scandalous traffic from the temple (cf. Mark *11,* 11 f. 15). Matthew and Luke, who place this event on the same day as the triumphal entry, must therefore be considered as combining both visits to the temple into one narrative.   **13.** This quotation from Isa. *56,* 7 is cited in full in the Second Gospel. *A den of thieves:* cf. Jer. *7,* 11.   **14-16.** This account of the miracles in the temple and of the hosannas of the children is peculiar to Matthew. These events probably took place on the second day in the temple, for on the first day Christ spent very little time there (cf. Mark *11,* 11b).   **14.** *The blind and the lame* used to gather at the gates of the temple to beg alms of the worshippers (cf. Acts *3*, 2).   **15.** *The chief priests and the Scribes* are here probably not a formal delegation of the Sanhedrin as in 23. They were *indignant* at the acclamations of the children because the *Son of David* was a Messianic title.   **16.** By citing these words of Ps. *8, 3* Jesus shows that He accepts with approval the children's praise. Humble and simple souls recognize Him as the Christ, while the worldly-wise do not (cf. *11,* 25; *1* Cor. *1,* 26-29).   **17.** Each night except His last during Holy Week Jesus stayed at Bethany (cf. Mark *11,* 11. 19; Luke *21, 37*), probably in the house of Lazarus or of Simon, the leper (cf. *26,* 6). Bethany was about one and a half miles from Jerusalem, on the southeast side of the Mount of Olives.

21, 18-22: Jesus Curses a Fig Tree. Parallel in Mark *11,* 12-14. 20-24. Matthew again compresses two events into one. Mark gives us the exact order of events: on the morning after the triumphal entry into Jerusalem, i.e., on the same day as the cleansing of the temple, Jesus cursed the fruitless tree; on the following morning the disciples noticed that the tree was completely withered, and Jesus then gave them an instruction on the power of prayer that is accompanied by unwavering faith. Matthew's statement that *immediately the fig tree withered up* as soon as Jesus cursed it, must therefore be understood to mean that the curse was immediately effective and the tree died at once, but that the withering of its leaves was not noticeable until the next day. Unless we see a deeper meaning in this deed of Christ we can hardly understand why He acted as He did or why the Evangelists should have thought it worth while recording. Nor is it reasonable to suppose that Christ cursed the fruitless tree solely for the purpose of teaching the disciples about the power of prayer with unwavering faith. The latter lesson was taught merely because the disciples were more interested in knowing *how* Christ had worked the miracle rather than in knowing *why* He had acted as He did. Our Lord's action on this occasion is probably to be considered as purely symbolic like the frequent symbolic actions of the prophets (cf. *1* Kgs. *11,* 7; *3* Kgs. *11,* 29-32; *22,* 11; Isa. *20,* 2 ff; Jer. *13,* 1 ff; *27,* 2 ff; Ezech. *4; 5;* Acts *21,* 11; etc.).

The meaning of Christ's action is clear from His parable of The Barren Fig Tree (Luke *13,* 6-9). Whether Jesus actually was physically hungry or not is immaterial (in Palestine people ordinarily did not eat in the morning); He told the disciples that *he felt hungry* (if He had not said so, they would not have known of His hunger), to signify His ardent desire to see the good fruits of religion. The *fig tree* standing alone (in Greek 19 means, "And seeing a single fig tree by the wayside") signifies Israel whom God had planted and separated from the Gentiles. The appearance of leaves but no fruit symbolizes Israel's pretense to a righteousness which it should have possessed but did not. Near Jerusalem a fruit fig tree should be bearing unripe figs at the beginning of April. The small green figs appear before the leaves at the beginning of March and are fully ripe by the beginning of June. Mark's statement that "it was not the season for figs" refers to the fully ripened figs. Symbolically this signifies that Christ had the right to expect of Israel at least the imperfect justice of the Old Law even though it was not yet time for the full perfection of the New Law. Israel was cursed and died spiritually at the time of Christ's crucifixion, although for a while it still bore the appearance of spiritual life before it withered completely at the destruction of Jerusalem. **21 f.** Cf. *17,* 19; Luke *17,* 6; Jas. *1,* 6.

21, 23-27: The Authority of Jesus. Parallels in Mark *11,* 27-33 and Luke *20,* 1-8. **23.** *The chief priests and elders of the people:* the other two Evangelists also mention the Scribes; this was therefore an official delegation from the Sanhedrin. They were not primarily concerned about Christ's right to teach. These men were mostly Sadducees and up to the present had largely ignored His teaching. But by attempting to correct the abuses in the temple He had now come in direct conflict with them. To *do these things* therefore means "to drive the worldly business from the temple." Jesus would have no authority to do so unless He were the Messias. But it would be of no avail for Him to tell them this outright, for He had already given them more than sufficient proof of this truth and they would not accept it. **24 f.** Our Lord's question should not be considered merely as an irrelevant conundrum proposed simply for the sake of embarrassing His adversaries. The question about John's authority has a very intimate relation to the question of His own authority. John "was sent from God, . . . to bear witness concerning the light, that all might believe through him" (John *1,* 6 f). The Baptist had pointed out Jesus as the Messias (cf. John *1,* 29-34; *3,* 25-30), and Jesus had often appealed to this testimony (cf. especially John *5*, 31-35). **27.** By feigning ignorance about John's authority these men admit failure in their official duty, for according to the Talmud one of the chief functions of the Sanhedrin was to judge and condemn false prophets. For our Lord's refusal to enlighten them any further, cf. *13*, 12. Note that after the first "Cleansing of the Temple" Christ's authority was likewise questioned (cf. John *2*, 18).

21, **28-32: Parable of the Two Sons.** Only in Matthew. This parable, as well as the following one, was spoken to the same men who had questioned His authority (cf. 31 f. 45). **29-31.** Almost all of the critical editions of the Greek text follow Codex Vaticanus and a few other MSS in reading for the first son, *"But he answered and said, 'I go, sir'; but he did not go";* and for the second son, *"But he answered and said, 'I will not'; but afterwards he regretted it and went."* In this case the answer is of course that "the latter" son did the father's will. This order is more in keeping with Christ's other parables (cf. the older and the younger son in Luke *15*, 11 ff) and more in conformity with Christ's answer in 32. Otherwise the inversion in the Vulgate is unimportant. On obedience in deed rather than in mere word, cf. *7*, 21; Luke *6*, 46. **31.** *The publicans and harlots are entering the kingdom of God before you:* i.e., they have a better chance of being saved than you have (cf. Luke *18*, 14), for, although they at first say, "I will not" to God's commands, later on they repent and do God's will. **32.** *In the way of justice:* i.e., preaching justice (cf. Luke *3*, 10-14) and leading a just life (cf. *11*, 18), so that all should have acknowledged that his authority was from God (cf. 25). *The publicans and the harlots believed him:* cf. Luke *3*, 12 f. *Seeing it:* i.e., the repentance of the publicans; even this moral miracle did not move the chief priests to repentance —a clear sign of their own reprobation. Cf. Luke *7*, 29 f.

21, **33-46: Parable of the Vine-dressers.** Parallels in Mark *12*, 1-12 and Luke *20*, 9-19. As a story this parable is perfectly clear. **33.** *A hedge:* more exactly according to the Greek, "a fence, a wall" of stones. The present very common use of cactus hedges in Palestine began only after the discovery of America, whence the cactus plant was introduced into the Old World. The purpose of the fence about the vineyard was to protect it from marauders and wild animals. *A wine vat* for the juice of the pressed grapes: this is the literal meaning of the word in Mark, but in Matthew the word is more exactly "a wine press," i.e., the upper receptacle in which the grapes were crushed under foot (cf. Isa. *16,* 10; *63,* 2 f; Jer. *48,* 33; Joel *3,* 13). The *tower* in the center of the vineyard gave the watchman who guarded the vines against birds, animals and robbers, an elevated and sheltered observation post (cf. Job *27,* 18; Cant. *1,* 5; Isa. *1,* 8). These *vine-dressers* were "share-croppers"; cf. Luke: "That they might give him part of the fruit of the vineyard." **34-36.** According to Matthew the servants who are sent to collect the owner's share of the crop come in groups, according to Mark and Luke they come individually at various times. **37.** *His son:* in the other two Gospels, "His beloved son," i.e., his only son (see Commentary on *17,* 5); he was the only messenger that the householder had left (cf. Mark). **38.** It would be beside the point to look for the legal grounds upon which the vine-dressers hoped to gain possession of the vineyard, for the parable aims to show that they acted not only unjustly but insanely. **39.** Luke agrees with Matthew in having the son first cast out of the vineyard and then killed; according to Mark the son is first killed inside the vineyard and then his corpse is cast out; the latter would be the more natural order in the story itself, while the former is more in keeping with the application of the parable. **40 f.** In Mark and Luke Christ Himself answers the question. However, Matthew seems more exact here; for it was Christ's custom to make His adversaries add the final word to a parable before they are aware of the full implication of their own admission (cf. 31; Luke *10,* 36). In this case the delegates of the Sanhedrin probably answered the question in a general way and when Christ made clear the full meaning of the parable they exclaimed, "God forbid!" (cf. Luke).

The application of the parable is clear and certain. In it Christ proposed several important truths. **33.** The *householder* is God. His *vineyard* is Israel. The same figure is used in Isa. *5,* 1 f, from which passage Christ borrowed the description of God's care for the vineyard. This description need not be understood symbolically in all its details, although the *hedge* was probably intended to signify the moral and physical safeguards that God employed in order to keep His chosen people separated from the Gentiles (cf. "the intervening wall of the enclosure" between the Jews and the Gentiles which Christ broke down, Eph. *2,* 14—the same word in Greek in both passages). The vine-dressers are the rulers and the people of Israel who were to bring forth spiritual fruit in due season. That the householder *went abroad* signifies that after the bestowal of the Law on Mount Sinai God did not ordinarily intervene directly except through the assistance of His grace but let Israel work out its own salvation.   34. *His servants sent to the vine-dressers to receive his fruits* were the prophets whom God sent at various times either singly or in groups.   35. All of the prophets were more or less rejected by the people, several of them were physically maltreated and a few of them, as Zacharias the son of Joiada and John the Baptist, were put to death. Cf. *23,* 35-37.   37 f. The *son* and *heir* of the householder is Jesus. He clearly differentiates Himself from all the previous prophets: they were but the *servants,* literally the "slaves," of God; He is God's "only Son." This is the first instance in the Synoptic Gospels where Jesus publicly proclaims that He is not only the Messias but the Son of God in a unique sense. God's special love for Israel is shown by sending as His last messenger His own Son in the hope that they would respect at least Him.   39. That *they cast him out of the vineyard and killed him* signifies that Christ would be rejected by the Jews and handed over to the Gentiles to be crucified (cf. *20,* 18 f); others see in this a prediction that Jesus would "suffer outside the gate" of Jerusalem (Heb. *13,* 12), but this is rendered somewhat doubtful by the variant form in Mark. 40 f. The punishment of Israel is two-fold: their spiritual inheritance will be taken from them and given to the Gentiles (cf. 43), and they themselves will be destroyed—a prophecy of the destruction of Jerusalem, the authenticity of which is admitted even by the critics.

42. The quotation is from Ps. *117,* 22, cited also in Acts *4,* 11 and *1* Pet. *2,* 7; for a similar prophecy of Christ as the corner stone cf. Isa. *28,* 16; *1* Pet. *2,* 6; and as a stone of stumbling, cf. Isa. *8,* 14; *1* Pet. *2,* 8; both passages are combined in Rom. *9,* 33. *The corner stone:* literally "the head of the angle," a Hebrew expression which may also mean "the keystone" of an arch; in any case it refers to Christ as the bond between two rows of stones which are the Old and the New Covenants; cf. Eph. *2,* 20.   44. This saying of Christ is not in Mark and is of doubtful authenticity in Matthew, but it is certainly original in Luke. There is an allusion to the prophecy of Isa. *8,* 14 f and probably also to Dan. *2,* 34 f. 45. The "crucified Christ is a stumbling-block to the Jews" (*1* Cor. *1,* 23).   45 f. The rulers of the Jews were now determined to do away with Jesus because they understood

that He proclaimed Himself the true owner of God's vineyard and that He threatened to depose them and appoint new rulers of His people.

22, 1-14: The Marriage Feast. Only in Matthew, although there is a very similar parable of The Great Supper in Luke *14,* 16-24. **1.** *Addressed them:* literally, "answered them," i.e., in response to the anger and hatred that His preceding parable had evoked, Jesus reiterated His prediction of the rejection of the Jews in still stronger terms under the form of another parable. Logically this parable is closely connected with the preceding. As a story the whole scene is in keeping with the customs of a royal banquet in the ancient Orient. The first refusal of the invitation is not surprising, for according to oriental customs of politeness a man often waits for a second invitation before attending a banquet. *The crossroads* (9): literally "the roads by which the roads go out," are the main highways from the city which branch out into secondary country roads; the sense is therefore, "Go out into the country." The context presupposes that it is the man's own fault that he has entered without a wedding garment (11 f), otherwise the king would not have punished him.

In the application of the parable not all the details need to be understood symbolically.   **2.** The *King* is God. *His son* is Christ, whose union with His Church through the Incarnation is frequently compared to a *marriage* (see Commentary on *9,* 15); for the kingdom of heaven compared to a banquet, cf. *8,* 11.   **3 f.** Mention is made of two groups of *servants* who are *sent to call in those* who had been previously *invited:* perhaps we are to see in the first group the prophets of the Old Testament, and in the second group the Apostles. In this parable Christ Himself is not considered as one of these messengers; He is the royal bridegroom. The second call is more urgent, for now the kingdom of God is at hand.   **5 f.** Most of those who had been invited fail to heed the call because they are preoccupied with worldly affairs: Christ recognized that this was the fundamental reason why the vast majority of the Jews rejected Him. *The rest:* only a minority of the Jews were guilty of violent persecution of Christ and His Apostles.

7. A clear prophecy of the destruction of Jerusalem.   **8.** The *Then* need not be understood as meaning "after the destruction of Jerusalem."   **9.** Those outside the city are called to the marriage feast, i.e., the Gentiles. **10.** There are *both good and bad* in the Church; cf. *13,* 41. 48 f.

11-13. The parable of The Wedding Garment. This is really a distinct parable, but it was probably added to the preceding by Christ Himself, not merely by the Evangelist. This further thought was suggested by the mention of both good and bad at the marriage feast. **11.** The *wedding garment* does not signify faith, for this man had accepted the call and entered the Church, but charity and good works, or the preservation of the state of grace which makes a man acceptable to God. **12.** *Friend:* literally "comrade," a word peculiar to Matthew; cf. *20,* 13; *26,* 50; in all three cases the word is used in a friendly expostulation. The man has no excuse to offer; it was clearly his own fault that he had no wedding garment. Evidently the host had offered this guest a festive garment and the refusal to wear it constituted an insult to the host. **13.** According to the better MSS, *Bind his feet and hands.* On *the darkness outside and the weeping and the gnashing of teeth,* cf. *8,* 12. **14.** This verse does not refer to the parable of The Wedding Garment but to the parable of The Marriage Feast (2-10). The sense is: "Although the Jews who were invited to the kingdom of God were many, yet comparatively few of them answered the invitation and actually entered the kingdom." This saying of Christ therefore does not answer the question whether most or only a few of those who are baptized are actually saved. The *chosen* or "the elect" is a technical expression for the members of Christ's kingdom; cf. Apoc. *17,* 14. This term has nothing to do with the Calvinistic idea of predestination. Sometimes it is used as entirely synonymous with "the called." When the two terms are distinguished, as here, "the elect" are those who of their own free will co-operate with grace.

22, 15-22: Tribute to Cæsar. Parallels in Mark *12,* 13-17 and Luke *20,* 20-26. **15.** The ultimate purpose of Christ's adversaries in seeking to *entrap him in his talk* was that, if He would deny the political supremacy of the Emperor, they would "deliver him up to the ruling power and to the authority of the procurator" (Luke); if on the other hand Jesus would acknowledge this supremacy, they hoped that He would lose caste with the people who still expected Him to proclaim Himself their ideal Messias who was to give them political independence. **16.** That the plot might be less obvious, the leading *Pharisees* did not come in person but *sent their disciples,* who of course were also "Pharisees" (Mark), to ask the question. Why *the Herodians,* the supporters of Herod's dynasty, were in on the plot is not clear: either they had their own reason for opposing the payment of tribute to the Emperor, or as favorers of the imperial party they were brought along to be witnesses of any seditious statement that Jesus might make. The question is introduced with elaborate flattery in an unsuccessful attempt to conceal their treacherous intentions.   **17.** *Tribute* is the correct word, corresponding with the technical Greek word used by Luke; but Matthew and Mark use a more popular Greek word meaning "Poll-tax," since it was by such a tax that this tribute was collected. The emperor at this time was Tiberius. The questioners meant the word *lawful* in the sense of "morally licit."

19. The purpose of our Lord's question was to show that He was not ignorant of their wicked trickery, and perhaps also that He might awaken their conscience to the malice of their plot. He calls them *hypocrites* because they were simply pretending to seek information and because they themselves, at least in practice, approved of the tribute. Christ wished to have the coin of tribute shown in order to drive home the object lesson. The tribute had to be paid in the silver *denarius* of the Empire. Since this bore a human figure, the head of the emperor, which was something idolatrous for a good Jew, they did not have such a coin with them and had to fetch one (cf. Mark). The local authorities, such as Herod and the procurators, were permitted to mint only copper coins; out of regard for the scruples of the Jews these bore no human figures.   **20.** Jesus asked this question, not as if He Himself did not know whose image and inscription were on the coin, but in order to have His adversaries themselves admit that they used Cæsar's money. It was an axiom, admitted also in the Talmud, that if a people used the coinage of a certain king they also acknowledged his sovereignty. This denarius would have borne the portrait head of Tiberius in profile with the Greek words *KAISAROS SEBASTOU,* i.e., (denarius) "of the revered Cæsar."

21. *Render:* i.e., give back to Cæsar what is due to him. By this wise decision Christ avoided the trap set for Him. He placed the burden of responsibility upon His adversaries. If they are willing to accept the benefits of Roman rule—and these were many—then they also have the duty of contributing to the support of the Roman government. They do nothing morally wrong in this, provided that at the same time they render, i.e., give back, to God whatever is due in justice and gratitude to Him. By this answer Christ neither approved nor disapproved of either the Roman Empire or of Jewish nationalism. His kingdom was not of this world (cf. John *18,* 36) and He refused to become involved in questions of politics. Since this saying of Christ is stated as a general principle, it is valid for all times. Our Lord recognizes the State and the Church as distinct, each sovereign in its own sphere. Normally there should be no conflict between the temporal jurisdiction of the one and the spiritual jurisdiction of the other. A Christian has moral obligations to both authorities. As long as civil authority is not unjust and tyrannical and does not interfere in the legitimate sphere of religion, it must be accepted and obeyed as God's will (cf. John *19,* 11; Rom. *13,* 1-7; *1* Pet. *2,* 13-17); but if any conflict arises between the authority of the State and the authority of God, "we must obey God rather than men" (Acts *5,* 29).   **22.** Although they failed to trap Jesus into making a seditious statement, still they did not hesitate to accuse Him falsely before Pilate of "perverting our nation and forbidding the payment of taxes to Cæsar" (Luke *23,* 2).

22, 23-33: The Sadducees and the Resurrection. Parallels in Mark *12,* 18-27 and Luke *20,* 27-39.   **23.** *On the same day,* but not necessarily immediately after the preceding discussion. It is to be noted that the Sadducees were not concerned with the question of the immortality of the soul as such. For all Jews took a realistic view of human life. Man has a body as well as a soul and therefore they did not consider the mere existence of a disembodied soul as a truly human life. The Platonic concept of the immortality of the soul is entirely foreign to all the Sacred Scriptures, the New Testament as well as the Old. St. Paul teaches at least implicitly that if there is no resurrection of the body there is no future life (cf. *1* Cor. *15,* 12-19).   **24-28.** The argument of the Sadducees is based upon the Mosaic law of levirate marriage (cf. Deut. *25,* 5 f). The question is purely theoretic, for this law had already fallen into desuetude; moreover, it would be highly improbable that such a case would befall the same woman seven times. However, *seven* is merely a symbolic number here, signifying "several."

29-33. Jesus affirms the doctrine of the resurrection and refutes the objection of the Sadducees. The doctrine itself is proved from *the Scriptures* (31 f); the objection is removed by explaining how *the power of God* will give man "a spiritual body" (*1 Cor. 15,* 44) at the resurrection.   **30.** The man is said to *marry,* the woman is *given in marriage. At the resurrection the body* "rises in incorruption" (*1* Cor. *15,* 42); therefore, since the risen body will not die, as our mortal bodies do at present, there will no longer be any need to continue the existence of the human race by the procreation of offspring. In this respect men and women after the resurrection will be *as angels of God in heaven.* For the angelic existence is not continued by married life. From this passage arose the custom of calling sexual purity "the angelic virtue"; but strictly speaking this virtue cannot be predicated of the angels who have neither body nor sex. The comparison to the angels was probably added by Christ to show that He rejected also the fallacy of the Sadducees *which denied* the existence of angels.

31 f.   The scriptural proof. The quotation is from Ex. *3, 6.* Christ's argument from this text is usually explained by emphasizing the tense of the verb, i.e., "I *am* the God of Abraham," not, "I *was* the God of Abraham." But it is to be noted that in the original Hebrew, as also in the original Greek of Mark, no verb is used here at all. Moreover such an argument would at best prove only the immortality of the soul, which is quite a different thing from the resurrection of the body, which Christ wished to prove. Finally such an interpretation does not explain why Christ chose such a seemingly weak argument when He could have quoted texts from the Old Testament which directly affirm the truth of the resurrection (cf. Isa. *26,* 19; Dan. *12,* 2). It is not sufficient to say that the Sadducees would not have accepted these texts because they did not hold the Prophets as inspired Scripture; for there is no certain evidence that the Bible of the Sadducees consisted only of the Pentateuch.

The reason why Christ chose this text from Exodus was because He wished not only to prove the fact of a future resurrection but also to explain why there must be a resurrection. The force of Christ's argument is clearly explained in Heb. *11,* 8-16, where the author of this Epistle alludes to this same text of Exodus in his conclusion, "Therefore God is not ashamed to be called their God, for he has prepared for them a city" (Heb. *11,* 16). When God sent Moses back to Egypt, He reminded him of His promises to the Patriarchs. But these promises concerned not only their posterity but also them personally. These promises were made not only to Abraham (cf. Gen. *12*, 1-3; *13*, 14-17; *15; 17*, 7 f) but also to Isaac and Jacob (cf. Gen. *26*, 3-5; *28*, 13-15), "the co-heirs of the same promise" (Heb. *11*, 9). For Abraham the true "Land of Promise" was not Palestine where "he abode as in a foreign land, dwelling in tents . . . ; for he was looking for the city that has the foundations, of which city the architect and the builder is God" (Heb. *11*, 9 f). The "heavenly country" (Heb. *11*, 16) that was promised them, is not merely heaven itself where their immortal souls would dwell, but "the holy city, the New Jerusalem, coming down out of heaven from God" (Apoc. *21*, 2), i.e., the rejuvenated earth (cf. Rom. *8*, 21) where they will live a fully human life with body as well as soul. "In the way of faith all these died without receiving the promises" (Heb. *11*, 13); therefore God, who is faithful to His promises, must still raise up their bodies. The ultimate reason for this is that, if God is the God of the Patriarchs, they share in His life. Therefore *He is not the God of the dead, but of the living*, "for all live to him" (Luke), i.e., all, not only the Patriarchs, "who shall be accounted worthy of that world and of the resurrection from the dead," for they are "sons of God" and therefore "sons of the resurrection" (Luke).   **33.** Only in Matthew; but cf. Luke *20*, 39.

22, 34-40: The Great Commandment. Parallel in Mark *12*, 28-31. Luke *10*, 25-28 has a similar but distinct incident introducing the parable of The Good Samaritan.   **34 f.** The Pharisees were pleased at Christ's rebuttal of the Sadducees' errors; therefore they held a private meeting and sent a delegate to Jesus to see if He also agreed with them on the question of the greatest commandment in the Law. According to the Talmud, it was the common teaching of the Rabbis that the greatest commandments were precepts to love God and one's neighbor. This whole incident is quite amicable; there is no need to read a hostile motive here into their *putting him to the test* (cf. Mark).   **36.** *Great:* by popular usage the positive degree is used for the superlative; cf. 38 where the same Greek word is translated (also in the Latin) as *greatest*.   **37.** The quotation is from Deut. *6*, 5. *With thy whole mind* is the Septuagint translation of the first phrase, *with all thy heart;* in Hebrew the third phrase is "with all thy strength," i.e., intensely; in Mark and Luke both forms are given. All these phrases are more or less synonymous here, the same thought being repeated for the sake of emphasis; the sense is, "With all thy faculties." God must be loved and served with whole-hearted devotion, because He is the only God (cf. the introductory words of this precept as cited by Mark). **39.** The quotation is from Lev. *19,* 18. *As thyself:* the norm of love of neighbor should be the treatment that one would wish to receive from others. The same principle is stated in the Golden Rule; see Commentary on 7, 12. Although Jesus had not been asked about *the second* commandment, still He adds these words to show the intimate bond between these two commandments, and perhaps also to show that in this matter He was in agreement with the teaching of the majority of the Rabbis. Objectively considered, the precept of the love of neighbor is second in importance to the precept of the love of God, but it is nevertheless *like it,* i,e., for all practical purposes of equal importance because it is fundamentally the same precept (cf. *1* John). **40.** Only in Matthew. *The whole Law and the Prophets:* the whole Old Testament. All the other commandments of God are but elaborations and applications of these two fundamental commandments. Cf. 7, 12.

22, 41-46: The Son of David. Parallels in Mark *12,* 35-37 and Luke *20,* 41-44. **41.** After the Pharisees had used all their arguments in vain against Jesus, it was now His turn to take the offensive. But in proposing this question our Lord had a higher purpose than merely to put His adversaries to confusion. According to the Second Gospel Jesus spoke these words not only to the Pharisees, but also, "while teaching in the temple" . . . "the mass of the common people" who "liked to hear him" (Mark *12,* 35. 37). On His triumphal entry into Jerusalem a few days before, the people had acclaimed Him as "the Son of David" and they undoubtedly hoped that He would restore the temporal kingdom of David (cf. Mark *11,* 10). Jesus therefore now teaches them that, while He accepts the title of "Son of David," He has a far higher dignity and destiny than is implied in that title. Cf. also John *18,* 33-36.

42. This question does not, of course, mean directly, "What do you think of *me?* Whose son am I?" For the Pharisees did not accept Jesus as the Christ. The common opinion among the Jews was that Christ would be a descendant of David (cf. John *7,* 42). **43 f.** The quotation is from Ps. *109,* 1. Our Lord's words imply that His hearers consider this Psalm to be written by David about the Messias; else His argument would not be valid. Christ Himself certainly considered this Psalm to be

Messianic. Is He also teaching here that it is Davidic? According to almost all modern critics this Psalm was addressed *to* a king, not written *by* a king, and hence cannot possibly be written by David. It is true that the word "David" is sometimes used merely in the sense of "the Psalter" (cf. Heb. *4, 7*, literally "in David"). But Christ here says that "David himself says" this of Him (cf. Mark and Luke), and it is scarcely conceivable that He is merely accommodating Himself here to a popular but erroneous opinion. Therefore the Biblical Commission has rightly decided (May 1, 1910) that all Catholics must hold the Davidic authorship of this Psalm. *In the Spirit* means "under the inspiration of the Holy Spirit"; some commentators, however, take this to mean, "in the spiritual sense" intended by God. Literally in Hebrew, "The oracle of Yahweh for my lord," i.e., the words of God to the Messias.   **45.** Jesus does not answer this question which involves the whole doctrine of the mystery of the Incarnation. But immediately after Pentecost the Apostles proclaimed that these words were fulfilled in Jesus, who is both true God and true man (cf. especially Acts *2*, 34-36).   **46.** Matthew notes that this was the last discussion between Jesus and His adversaries. Mark (*12*, 34b) has the same statement after the Scribe's question about the Great Commandment, for this was really the last time that the Pharisees questioned Jesus. Luke, who omits the question about the Great Commandment, makes the same statement at the end of the preceding discussion about the resurrection from the dead (cf. Luke *20, 40*). All three Evangelists are right, because each speaks from a different viewpoint.

23, 1-12: Hypocrisy of the Scribes and Pharisees. Partial parallels in Mark *12*, 38-40 and Luke *20*, 45-47. This whole chapter of the First Gospel consists of a powerful denunciation of the Scribes and Pharisees. About half of these words of Christ are not recorded elsewhere, the remainder occur in the Third Gospel as spoken under different circumstances. No doubt our Lord could have repeated these sayings on various occasions, but it seems probable that Matthew has at least in part incorporated into this discourse certain sayings that were spoken on another occasion (cf. Luke *11*). However, some of these sayings seem much more suitable here than they do in Luke (cf. 37 ff).

1. This first section of the discourse (1-12) is addressed directly *to the crowds and to his disciples;* the second section (13-36) is spoken directly to the Scribes and Pharisees, while the concluding

paragraph (37-39) is an apostrophe to all Jerusalem. **2.** This may be understood literally, for there was a certain seat in the synagogue known as *the chair of Moses;* but it is more probably meant in the figurative sense, "they have taught in the name of Moses." In Matthew this whole discourse is directed against both the Scribes and the Pharisees considered as one group, but in the corresponding passages in Mark and Luke some sayings are against the Scribes alone, the other sayings against the Pharisees alone. While most of the Scribes were also Pharisees, only a small portion of the Pharisees were Scribes.
3. The usual interpretation of this passage is that our Lord recognizes the Scribes and Pharisees as the legitimate interpreters of the Law.
4. Cf. the same words in Luke *11,* 46. The teachings of the Scribes and Pharisees were an unbearable yoke (cf. Acts *15,* 10 and contrast Matt. *11,* 30). *With one finger . . . to move them:* they themselves do not make the slightest effort to bear this burdensome yoke.   **5-7.** On the ostentation and vanity of the Pharisees, cf. *6,* 2. 5. 16; and the parallel passages in Mark *12,* 38 f and Luke *20,* 46.   **5.** On the *phylacteries* see the foot-note to the text. On the *tassels* see Commentary on *9,* 20.   **6.** Almost the same words in Luke *11,* 43; cf. also Luke *14,* 7.   **8 f.** Christians are forbidden the use of titles for the mere purpose of ostentation; see foot-note to the text. *All you are brothers:* in the Apostolic age "brother" and "sister" were the only titles used among Christians, including the Apostles themselves (cf. Acts *9,* 17; *21,* 20; 2 Cor. *1,* 1; *2* Pet. *3,* 15; Apoc. *1,* 9). The expression "the brethren" was synonymous with "the Christians."   **10.** Cf. John *13,* 13.   **11.** The same words in *20,* 26.   **12.** The same words in Luke *14,* 11; *18,* 14.

23, 13-39: Woe to the Scribes and Pharisees.   **13.** Cf. the similar words of Luke *11,* 52 against the "lawyers," i.e., Scribes. This may be understood in the specific sense that the Scribes and Pharisees refuse to accept the teaching of Christ and so enter His kingdom, or more probably, as in Luke, in the general sense of eternal salvation. It is bad enough that they themselves will lose their souls, but much worse that by their false teaching, "by taking away the key of knowledge" (Luke), they prevent others from entering heaven.   **14.** This verse is probably not authentic in Matthew but an adaptation of Mark *12,* 40 and Luke *20,* 47, made by some copyist, possibly for the purpose of having "eight

Woes" to correspond with the "eight Beatitudes." See Commentary on Mark *12,* 40.   **15.** From the Greek word which is here translated as *convert* we have the English word "proselyte," which is used especially in the sense of a convert from paganism to Judaism. This missionary zeal of the Scribes and Pharisees was not in itself wrong, but these *hypocrites,* as Christ calls them in each of the "Woes," were animated less by a desire of promoting God's glory and the good of souls than by the desire of self-aggrandizement and of boasting of the increase of their sect (cf. Gal. *6,* 13). Since they lacked the true motive in making converts, their disciples imitated the worst features of the teachings and practice of their masters and became more wicked even than they.

16-22. A severe condemnation of the Pharisees' casuistry about oaths which encouraged dishonesty. Cf. the similar teaching of Christ on oaths in *5,* 33-37. *The gold of the temple* probably means "the gold that is offered to the temple," for this expression is treated here as entirely parallel with *the gift that is upon the altar.* Such gifts were "Corban" (cf. Mark *7,* 11), and therefore the Scribes taught that to swear by them was to swear the inviolable "Corban oath" (see Commentary on *15,* 5), but to swear merely by the temple or the altar had no binding force. Christ shows what *blind fools* the Scribes were in making such a distinction, for the temple and the altar are more important than the gifts which they, or rather God through them, *sanctifies.* Therefore all these oaths have a special relation to God and hence are all binding.

23. Similar words in Luke *11,* 42. The Law commanded that tithes, i.e., the tenth part, of all the produce of the fields should be offered each year to the temple (cf. Deut. *14,* 22). To show their zeal for this law, the Pharisees gave tithes even of the little garden herbs that were used as condiments. (The *anise* mentioned here is really the dill plant.) In itself this was not wrong—"they should not leave these things undone." Christ did not approve of the violation of even the least law (cf. *5,* 19). But even though every law of God is, in a sense, important (cf. Jas. *2,* 10), some laws are far more important than others. The hypocrisy of the Scribes and Pharisees consisted in this, that they pretended to be very scrupulous about these minute prescriptions, while they flagrantly violated the far more important laws of God that concerned *right judgment,* i.e., justice in general, *and mercy and faith,* i.e., fidelity to pacts and promises. Christ calls these the *weightier matters* of the Law. This may mean: (a) that He is alluding to the rabbinical distinction between "heavy" and "light" precepts, or as moral theologians would say, those which bind under grave or under venial sin; (b) "the basic principles of moral law"; (c) "the more burdensome precepts of the Law," for, while it is comparatively easy to give tithes on such inexpensive things as condiments, it is far more difficult to be consistently just, kind and faithful to one's neighbors.

24. A proverbial saying, probably with no allusion intended to the law against eating unclean animals and insects (cf. Lev. *11*, 4. 20. 23). The Pharisees, like all false moralists, lost sight of the just proportion that exists between more important precepts *(the camel)* and less important ones *(the gnat)*. **25 f.** Similar words in Luke *11,* 39 f. Our Lord is not referring directly here to the rabbinical traditions on the washing of cups and dishes (cf. Mark *7,* 4b). *The outside of the cup and the dish* is a metaphor for the externals of religion and morality, in regard to which the Pharisees were immaculate. *Within:* i.e., inside the cup and dish; the continuation of the same metaphor, signifying that interiorly the Pharisees are full of wickedness, despite their pretense of external piety and virtue. *They* refers directly to *the cup and the dish. Uncleanness* represents one of the various words found here in the Greek MSS; the best MSS read "intemperance"—the same word which is well translated "lack of self-control" in *1* Cor. *7,* 5. The externals of virtue should not be despised *(that the outside too may be clean),* but the interior dispositions are far more important *(clean first the inside).*

27 f. Christ continues the same thought by using the striking metaphor of the *whited sepulchers.* Lest the numerous strangers in Jerusalem at the Passover might become defiled by contact with tombs that they might not recognize as such, shortly before that feast (therefore at the very time while Christ was saying this) these tombs were whitewashed, not in order to beautify them but in order to warn strangers of the uncleanness within. The biting wit of our Lord's saying consisted in this, that while the Pharisees "whitewashed" themselves by their external pretense of piety (cf. "thou whitewashed wall" in Acts *23,* 3), this very "whitewashing" betrayed them, and showed them, as it did the sepulchers, to be full of uncleanness within. In Luke *11,* 44 the figure is somewhat different: there the Pharisees are portrayed as successful in their hypocrisy and defiling men who come in contact with them. **29-32.** Substantially the same thought is expressed in Luke *11*, 47 f. The Jews at the time of Christ had built or repaired the monuments and reputed tombs of the prophets near Jerusalem. They pretended thereby to honor the prophets and to atone for the guilt of their forefathers who rejected and persecuted them. But Christ says that this in itself *(Thus)* reveals their evil conscience. For in spirit they were the worthy descendants of the murderers of the prophets. It was as if they were the accomplices of their ancestors and said to them, "You kill the prophets and we will bury them for you." Through their hatred of Jesus they showed themselves to be even worse than their ancestors. Therefore Christ says, *"You fill up the measure of your fathers,"* i.e., "You supply what is wanting to the fullness of the wickedness of your ancestors; or perhaps the thought is, "You exhaust God's longanimity by adding to the sins of your fathers" (transition to the following thought).

33-36. The punishment that is to come upon these persecutors of God's prophets. **33.** Cf. *3*, 7; *12*, 34. *The judgment of hell:* the condemnation to eternal punishment. **34-36.** Parallel in Luke *11*, 49-51. Jesus, like the prophets of old, speaks in God's name, or rather as God Himself. Therefore in Luke the words of 34 f are attributed to divine Wisdom, but this is probably not to be understood as a quotation from some otherwise unknown prophetic writing. **34.** The Apostles and teachers of the New Dispensation are given the titles of God's ambassadors of the Old Dispensation. There may be a reference to the three main periods in Hebrew literature: that of the *Prophets,* that of the *wise men,* i.e., the writers of the "wisdom" literature, and that of the Scribes of the two or three centuries immediately before the time of Christ. *Persecute from town to town:* cf. *10*, 23. **35.** *That upon you may come all the just blood:* i.e., the responsibility for these deaths; cf. the similar phrase in *27*, 25; Apoc. *18*, 24. For the sense in which this generation also shared in the guilt of the preceding generations, cf. 31 f. *From the blood of Abel the just to the blood of Zacharias:* i.e., from the first to the last murder mentioned in the Old Testament (cf. Gen. *4*, 8; 2 Par. *24*, 21 f). According to the order of the present Hebrew Bible the Books of Paralipomenon or Chronicles are at the very end; from the words of Christ it would seem that this order was already established in His time. There can be but little doubt that Christ is referring to the murder of Zacharias the son of Jojada, who was stoned "in the court of the house of the Lord" (2 Par. *24*, 21). There was a Zacharias the son of Barachias in the days of Isaias (cf. Isa. *8*, 2), and the second last of the Minor Prophets was also called Zacharias the son of Barachias (cf. Zach. *1*, 1); but there is no reason to believe that either of these men was slain in the temple. It would be contrary to Christ's divinity and to the inerrancy of the Evangelist to admit that either of them made the mistake of confusing one Zacharias with another. The words *the son of Barachias* apparently formed part of the original Greek translation of the First Gospel, for almost all extant MSS contain them. If these words were also in the original Aramaic Gospel of St. Matthew, they may be explained by assuming that Jojada was known also as Barachias, or that Barachias was one of his ancestors. St. Jerome tells us that the Aramaic "Gospel of the Nazarenes" (cf. Introduction), read "Zacharias the son of Jojada," in Aramaic "Zachariah Bar Jodae." If this does not represent a later correction of the apparent error, we may perhaps explain the error of the Greek as due to the translator who mistook "Barjodae" for "Barachias." **36.** *This generation* is here to be understood of those then living, for Christ is referring to the destruction of Jerusalem.

37-39. The thought of the punishment that is soon to be inflicted upon the Holy City wrings from the Savior this touching cry of compassion. Luke *13*, 34 f records almost the very same words of Christ, but in a far less appropriate context. Here these words, spoken right in Jerusalem, form a magnificent conclusion to His ineffectual ministry in that city. **37.** *How often* refers directly to Christ's previous visits to Jerusalem, as recorded in the Fourth Gospel; but it may possibly be understood also in the sense that Christ is speaking here as God and referring to the numerous graces granted to the Holy City since the time of David. **38.** These words are from Jer. *22*, 5. Cf. also Jer. *12*, 7; *3* Kgs. *9*, 7 f. *Your house* may refer either to the temple or to the political power of Jerusalem. **39.** A prediction of Israel's rejection until Christ's Second Coming when He will be welcomed by the people in the words of Ps. *117*, 26. But this passage probably implies also the final conversion of Israel (cf. Rom. *11*, 25-29). For the meaning of Ps. *117*, 26 see Commentary on *21*, 9.

24, 1-14: Destruction of Jerusalem and End of the World. With this section Matthew begins the last of the five great discourses of Christ as recorded in the First Gospel (cf. the typical conclusion in *26*, 1). This so-called "Eschatological Discourse" is partially paralleled in Mark *13* and Luke *21,* 5-38; but in the First Gospel it is much longer, embracing two long chapters *(24-25).* Its divisions are: (a) Prophecies concerning the destruction of Jerusalem and the end of the world *(24,* 1-35); (b) The necessity of being always prepared for the unexpected coming of Christ *(24,* 36-51); (c) Two parables demonstrating this truth *(25,* 1-30); (d) The Last Judgment *(25,* 31-46).

1-3. The occasion of the discourse. Parallels in Mark *13,* 1-4 and Luke *21,* 5-7. **1.** The temple of Herod was justly famed for its grandeur. Herod had a mania for building on a gigantic scale. Some of the stones of his temple which still remain in Jerusalem are about fifteen feet long. The Talmud gives enormous dimensions for some of them. Cf. the exclamation of one of the disciples in Mark. **2.** *One stone upon another:* actually at least one section of the retaining wall of the temple's substructure is still intact. This is the well-known "Wailing Wall" of the Jews. But our Lord is speaking of the temple proper, of which not a vestige remains. **3.** From the summit of the *Mount of Olives* one can obtain a magnificent view of all Jerusalem and especially of the temple area. *Privately:* therefore this discourse was not spoken publicly to the people, nor was it spoken even to all the Apostles, for Mark tells us that Christ's audience consisted solely of Peter, James, John and Andrew. *These things* refer to the destruction of the temple which Christ had just foretold, undoubtedly to the great amazement of the disciples. *The sign of thy coming* seems to correspond to "the sign when all things (the destruction of Jerusalem) will begin to come to pass," as the question is recorded in Mark and Luke. Perhaps at this time the disciples considered that these two events would be more or less simultaneous. Christ therefore answers their question by speaking both of the destruction of Jerusalem and of His coming at the end of the world. But nothing in His discourse shows that He considered these two events as simultaneous. On the contrary, He clearly taught that there would be an indefinitely long period between them (cf. Luke *21,* 24b). The common opinion both of the Fathers of the Church and of modern exegetes holds that some parts of the discourse refer primarily to the destruction of Jerusalem and some parts primarily to the end of the world. But since the former event was intended by God as a type of the latter, the thought may seem to pass rather abruptly from one event to the other, and one cannot always be certain which event Christ is primarily referring to.

4-8. Parallels in Mark *13, 5-8* and Luke *21, 8-11*. These words probably refer primarily to the troubled years that were to precede the destruction of Jerusalem. But they can be understood also of the troubles on earth during all the following centuries. **4.** In times of grave crises men's minds are easily upset by false rumors. Therefore such warning phrases as *Take care,* occur frequently in this discourse. Christ told His disciples of these future events primarily in order to warn them against such dangers, not in order to satisfy our idle curiosity about the future. **5.** This was a time of great Messianic expectations. Therefore the Jews who refused to accept Jesus as the Christ were often deceived by men who falsely claimed to be the Christ. For the period between the death of Jesus and the destruction of Jerusalem we know the names of some of these false christs, such as Menander, Dositheus and Theudas. **6.** During this same period there were wars along the eastern borders of the Roman Empire, and several times before the actual outbreak of hostilities between the Jews and the Romans there were serious threats of war between them. **7.** These words were also fulfilled in the first generation after Christ, but they have been even more completely fulfilled in all the following centuries; the break in the narrative as given in Luke might be understood in the sense that Christ is now referring rather to these later times. **8.** *The beginnings of sorrows:* literally in Greek, "the beginning of the birth-pangs"; in the Jewish apocalyptic literature of that time "the birth-pangs" is a common expression referring to troubled times preceding the "rebirth" of the world (see Commentary on *19,* 28).

9-14. The persecutions which are to befall the followers of Christ. In Matthew these words seem to refer to all the persecutions and troubles of Christians until the end of the world. But the corresponding sections in Mark *13,* 9 f and Luke *21,* 12 f are perhaps to be understood only of the persecutions by the Jews before the destruction of Jerusalem. Cf. *10,* 17 f; John *15,* 20; *16,* 2. On the great defection in the last days when through the efforts of the *false prophets* of the Antichrist *many will fall away* and *the charity of the many will grow cold,* cf. 2 Thess. *2,* 3-10; 2 Tim. *3,* 1-5. **14.** *The whole world:* the Greek can also be understood in the sense of "the whole Roman Empire" (cf. Luke *2,* 1). *To all nations,* or "to all the Gentiles": the preaching of the gospel will be *for a witness* to them, i.e., that this fact may be used as testimony against them on the last day, if they will not receive the gospel. Mark *13,* 11-13 and Luke *21,* 14-17 add here other words of Christ concerning persecution which Matthew gives in *10,* 19-22.

24, 15-22: **Destruction of Jerusalem.**   15-20. Parallels in Mark *13,* 14-18 and Luke 21, 20-23a. All commentators are agreed that this section refers directly to the destruction of Jerusalem. *The abomination of desolation:* cf. note to the text. Christ gives this as a warning sign for the Christians to leave Jerusalem at once. Luke omits this sign but records another sign, "When you see Jerusalem surrounded by an army." These two signs were probably seen more or less simultaneously, but they were certainly not identical, for the *abomination of desolation* was to be seen in the *holy place,* i.e., in the temple (cf. Acts *6,* 13; *21,* 28). Dan. *9,* 27; *11,* 31; *12,* 11 prophesied that the *abomination of desolation* would be seen in the temple. According to *1* Mach. *1,* 57 this prophecy was first fulfilled in 168 B.C. when Antiochus Epiphanes placed the statue of Jupiter Olympius in the temple of Jerusalem. But this desecration was itself a type or figure of the desecration that was to take place there during the siege of Jerusalem by the Romans. This latter profanation of the temple must have happened about the same time as the first appearance of the Roman soldiers before the walls of Jerusalem (cf. Luke); therefore this prophecy cannot be understood as fulfilled in any earlier profanation, such as Caligula's attempt to place his statue there in 38 A.D., nor to any event after the fall of Jerusalem. The Holy City was first besieged by the Romans in 66 A.D. A few months later the party of Jewish fanatics known as the Zealots seized the temple and with the aid of the Idumeans slaughtered the priests and the people, even within the temple itself, and caused the daily sacrifice to cease. We know from Josephus that at this time many of the inhabitants fled from the doomed city, the Christians no doubt being among these fugitives. The *abomination of desolation* is therefore probably to be understood of these outrages of the Zealots and Idumeans. The word *standing* can be understood in the general sense of "being," i.e., taking place; in the fulfillment of the prophecy it need not be understood as referring to a person or a statue. The parenthetical phrase *let him who reads understand* occurs also in Mark where there is no preceding reference to Daniel, as there is in Matthew. Many commentators conclude from this, that this remark is made by the writer, not by Christ, and that both Evangelists are therefore using a common *written* source which contained this remark. The sense would then be, "Let him who reads these words of Christ understand His warning and flee from Jerusalem when they see these things come to pass." But even in Mark Daniel is implicitly referred to, and therefore we can just as well understand this parenthetical phrase as spoken by Christ, in the sense, "Let him who reads the prophecy of Daniel understand." The warning to the reader that is given in Apoc. *13,* 18 about the intentionally obscure allusion of the writer, is not quite parallel.

17 f. The Christians must flee from the doomed city with all possible haste. **20.** *Or on the Sabbath:* only in Matthew, the Gospel intended primarily for the Jewish Christians. By these words Christ is not implicitly inculcating an obligation concerning the Jewish prohibition of making long journeys on the Sabbath; He merely takes cognizance of the fact that many Jewish Christians will still feel bound by their ancient customs. **21 f.** It is not certain whether these words refer to 70 A.D. or to the terrible afflictions before the end of the world. The corresponding, but not parallel, passage in Luke *21,* 23b-24a is certainly to be understood only of the Roman conquest of Palestine. Although this passage in the First and Second Gospels may seem too strong for such a sense, still it is possible to understand it in such a limited sense. **21.** In the siege of Jerusalem, according to Josephus well over a million people perished—perhaps the most frightful siege in the history of the world. **22.** The sense is, "Unless God had decided for the sake of the Christians to curtail his wrath in those days, all men (in the limited sense of the passage, in Palestine) would perish."

24, **23-31: The Signs of the Last Day.** **23-25.** Parallel in Mark *13,* 21-23. Jesus refers here to *the false prophets,* i.e., the false teachers who at various times in the Christian era but especially in the last ages of the world, would induce men to believe in *false christs,* i.e., false saviors of mankind. *The great signs* and wonders may signify certain prodigies done by magic or with the help of the devil, or this may be understood in the more general sense of the illusions of earthly prosperity and happiness by which these deceivers will lead men astray. *If possible:* it is only too possible that *even the elect* may be led astray by these false allurements; but with the help of God's grace the elect can remain true to the Faith.   **26-28.** Only in Matthew; but cf. a similar saying of Christ in Luke *17,* 23 f. The coming of Christ at the last day will not be in some out-of-the-way place but will be known to the whole universe.   **28.** Cf. the same words in Luke *17,* 37. *The body:* literally in Greek, "the corpse" or "the carcass." *The eagles:* apparently "vultures" are meant. Probably a proverbial saying. Various interpretations are offered. (a) *The body* signifies the spiritually dead nation of the Jews in the doomed city of Jerusalem; *the eagles* are then the Roman soldiers who surround it and tear it to pieces. But the immediate context is not concerned with the destruction of Jerusalem. (b) *The body* represents the sinners whom Christ and His angels come from heaven to punish. This is a rather far-fetched interpretation. (c) Just as vultures are drawn from afar to the place where a carcass lies, so also will all men necessarily be drawn to the judgment-seat of Christ.   **29-31.** Parallels in Mark *13,* 24-27 and Luke *21,* 25-28. This passage is certainly to be understood of the coming of Christ at the last day.

29. If the word *immediately* is to be taken literally, *the tribulations of those days* must refer to the afflictions shortly before the end of the world, which however are not mentioned in the preceding verses. It is certain that Christ foretold an indefinitely long period between the destruction of Jerusalem and the end of the world (cf. *nor will be* in 21, and Luke *21,* 24b). But *immediately* may also be understood in the sense of "suddenly, unexpectedly" (cf. Apoc. *4,* 2). This description of the cosmic disturbance is in the language of the prophets (cf. Isa. *13,* 10; *34,* 4) and need not be understood literally.   **30.** *The sign of the Son of Man:* according to the traditional interpretation, the Cross of Christ. *Then will all the tribes of the earth mourn:* cf. Zach. *12,* 10 f; they will be overcome with fear, but the Christians should rejoice (cf. Luke *21,* 26. 28). This description of the *Son of Man* is based on Dan. *7,* 13 f.   **31.** The sound of the *trumpet* on the last day is also referred to in *1* Cor. *15,* 52; *1* Thess. *4,* 16; Apoc. *8,* 2; it is probably to be understood figuratively of God's power summoning the living and the dead to judgment, the figure being derived from the ancient Jewish custom of summoning the people to religious services by means of the ram's horn (cf. Lev. *25,* 9; Num. *10,* 1-10; *29,* 1; Joel *2,* 1. 15; and the figurative use in Isa. *27,* 13).

24, 32-35: Jerusalem's Impending Destruction. Parallels in Mark *13,* 28-31 and Luke *21,* 29-33. Having given a description of the destruction of Jerusalem with its preceding signs of warning and a description of the end of the world when Christ will come as unexpectedly as lightning, our Lord now speaks of the time when each of these two events will occur. The former event will take place in the near future and the disciples of Christ will have ample signs to warn them of its coming (32-35), but the time of the latter event cannot be foretold; the Second Coming of Christ at the last day will be sudden and unexpected, and therefore the disciples must always be prepared for it (36 ff). **32.** *Now* signifies a transition, i.e., from the thought of the last day back again to the thought of the destruction of Jerusalem; the same Greek particle is more accurately rendered as *but* in 29 and 36. *Parable* is used here in the more general sense of any comparison; just as men know from such natural signs as the budding of the trees that a new season is approaching, *even so* the disciples are to know from the signs of which Christ told them that the destruction of Jerusalem is fast approaching. **34.** According to the immediate context, *this generation* has its natural meaning of "the men now living"; Jerusalem was destroyed about thirty-seven years after Christ spoke these words. If this passage is understood of the end of the world, *this generation* would probably signify "the Jews" who will always exist as a distinct group (cf. Rom. *11,* 25). **35.** A general statement referring to the absolute truth of all of Christ's predictions. The present universe is finite and cannot last eternally, but must give way for "a new heaven and a new earth" (Apoc. *21,* 1).

24, 36-41: The Need of Watchfulness. **36.** Parallel in Mark *13,* 32. God alone knows when the last day will come. The better Greek MSS have in Matthew, as do all the MSS in Mark, the words *nor the Son,* which must mean "the Son of Man," i.e., Christ as man in the special circumstances under which His human nature operated during the period of His mortal life (cf. Phil. *2,* 6 f). Christ of course knew of the time of His Second Coming not only by His divine knowledge but also by His supernatural human knowledge. But during His mortal life Jesus ordinarily acted like other men according to His experimental knowledge and the knowledge derived from this by human reason. Since a knowledge of the time of His Second Coming surpassed all finite *intelligence,* Jesus could thus honestly say that

He was ignorant of it. Moreover, it was the Father's will that Christ should not communicate this knowledge to men (cf. Acts *1, 7*). **37-39.** Cf. the similar saying of Christ in Luke *17, 26* f. The end of the world is very aptly compared to the Deluge (cf. Gen. *7*), for (a) both events come suddenly; (b) only a few are prepared; (c) most men are preoccupied in worldly pleasures and are overwhelmed by the disaster. Cf. Luke *21, 34-36*; also Heb. *11, 7*; *1* Pet. *3, 20*; *2* Pet. *2, 5*. **40 f.** Cf. the similar words of Christ in Luke *17, 34* f. At Christ's coming people will be engaged in their ordinary occupations. Here to *be taken* means to be saved, to *be left* is to be damned. These words, as well as this whole section on Christ's coming at the end of the world, can very aptly be applied to Christ's unexpected coming to each person at the hour of death.

24, 42-51: Exhortation to Vigilance. **42-44.** In order to show the necessity of always being spiritually awake and ready for His coming, Christ compares Himself to a *thief* who comes in the night when man least expects him. The same little parable of the thief occurs in Luke *12,* 39 f; cf. also *1* Thess. *5, 2*; *2* Pet. *3,* 10; Apoc. *16, 15*. **45-51.** The parable of the faithful servant and the wicked servant. The same parable occurs in a different setting, in Luke *12, 42-46*. There is a similar parable on the unexpected return of the master in Mark *13, 33-36*; Luke *12, 37* f. These parables were intended especially for the Apostles and the others in authority in the Church, who are the stewards in the household of the Faith; but they were also intended for all Christians (cf. Mark *13, 37*). **48.** *That wicked servant:* but no wicked servant has been mentioned in the preceding verses; the sense is probably, "But if that servant, instead of being faithful and prudent, should be wicked and say to himself . . ." *My master delays his coming:* cf. 2 Pet. *3,* 3 f. 8 f. **49.** Those in authority in the Church must not be cruel to their subjects (cf. *20, 25*). **51.** *Cut him asunder:* an extraordinarily gruesome punishment that was unknown even among the wicked tyrants of Christ's time. The same verb in Hebrew and in Aramaic signifies both "to cut asunder" and "to cut off." If the Greek verb can be understood in the latter sense, it would fit in very well with the context, i.e., the unfaithful servant is "cut off" from his office and made to *share the lot of the hypocrites*. *The lot of the hypocrites* is the eternal punishment of hell, as is clear from the following words (cf. *8,* 12; *22,* 13; *25,* 30).

25, 1-13: Parable of the Ten Virgins. Only in Matthew; but cf. the little parable of the "master's return from the wedding" in Luke *12, 35* f. In order to understand this parable it is necessary to know something of the wedding customs of that time, which Christ supposes are well known to His hearers. On the day of the wedding the bridesmaids assembled at the house of the bride. After sunset the bridegroom, accompanied by his male friends (the "friend of the bridegroom" and the "sons of the bridal chamber"; see Commentary on *9, 15*), went to the bride's house, where they were greeted by the bride and her bridesmaids, and then both parties returned together in a joyous procession, that was illumined by lamps or torches, to the wedding feast in the house of the bridegroom.

1. In the parable *ten virgins* or bridesmaids are mentioned because this was the regular number at an ordinary Jewish wedding, as certain other accounts would seem to indicate. The words *and the bride* are not in the best Greek MSS; they were added by some copyist who did not understand the customs of a Jewish wedding. With her friends the bride awaited the arrival of the bridegroom. She, of course, was ready when the bridegroom arrived and therefore she is not mentioned in the parable. The words *who went forth to meet the bridegroom* must be considered either as a preliminary statement of the meeting which is told more in detail in 6-10, or in the sense, "They took their lamps and went forth from their own homes to the home of the bride, in order to meet the bridegroom when he would arrive there."   **2.** *Foolish* and *wise:* here in the sense of "improvident" and "provident."   **3 f.** The foolish virgins *did not take* extra *oil* with them; but *the wise took* extra *oil in* auxiliary *vessels* besides the lamps.   **6.** *A cry arose:* probably one of the friends of the bridegroom went a little ahead of the bridegroom's party in order to warn the bride and her maids that the bridegroom was approaching.   **7.** While the virgins slept, all of their lamps had burned low.   **10.** The foolish virgins probably tried to arouse some of the neighbors in order to buy oil from them. *Went in with him to the marriage feast:* in the house of the bridegroom, not of the bride.

13. This verse shows that the principal lesson intended by Christ in this parable is that His disciples must always be ready for His unexpected coming. But many of the details of the parable fit in so aptly with this leading thought that we may rightly conclude that these also were intended by our Lord to signify a spiritual truth. The *bridegroom* is Christ (see Commentary on *9, 15*). The bride is His Church (cf. especially Apoc. *22, 17* where the bride longs for His coming). The *virgins,* the friends of the bride, are the members of the Church. All Christians are virgins spiritually, inasmuch as they are not contaminated by heresy (cf. Apoc. *14,* 4 f); but in the parable no stress is laid on their virginity, for bridesmaids were always unmarried and presumed to be virgins. That half of them were not found ready at the coming of Christ should not be taken as an answer to the question in Luke *13,* 23; but this parable reaffirms the truth that not all the members of the Church are saved (cf. the parable of The Weeds and the parable of The Net, *13,* 24 ff. 47 ff). The *lamps* which all the virgins had, may signify faith which all Christians have; the extra supply of oil would then be perseverance in good works. *The bridegroom* was long in coming. He would normally be expected to arrive shortly after sunset, but it was only *at midnight* that he came: a warning to the early Christians who hoped for a speedy return of the Savior (cf. also *24,* 48). *They all became drowsy and slept:* even the good succumb to a certain amount of indifference. But the important thing is that they are found ready; they are in the state of Grace. The last-minute efforts of the others are of no avail. *The door was shut:* death and Judgment seal the fate of every one for all eternity. 11 f. Cf. *7,* 21-23; Luke *13,* 25-27.

25, 14-30: Parable of the Talents. The parable of the Gold Pieces in Luke *19,* 12-27 is similar but distinct. This parable, like the preceding, teaches that Christ's Second Coming may seem long delayed; however, it does not stress its sudden and unexpected character, but rather the judgment that will accompany Christ's return. This forms, therefore, a natural transition to the last section of the discourse, the Last Judgment. 14. The *man going abroad* is Christ, who deprives His disciples for a while of His visible presence. *His servants* are all Christians. *His goods* are the spiritual benefits and graces that He won by His Redemption. 15. A *talent* was not a coin but a definite sum of money, amounting to twelve thousand denarii or about one thousand nine hundred and twenty dollars. Many consider the talents of the parable to signify both the natural and the supernatural gifts that God bestows on us. In the sense of natural abilities, especially of mental endowments, the word "talents" has been taken from this parable into English and other modern languages. But this is not entirely correct. In the parable the *talents* signify certain supernatural gifts and responsibilities which Christ gives *to each according to his particular ability,* i.e., according to his natural endowments which he already has from the Creator. On the apparently unequal distribution of spiritual gifts, cf. Rom. *12,* 6-8; *1* Cor. *12,* 4-11; Eph. *4,* 7; *1* Pet. *4,* 10.   **16 f.** These gifts are "given to everyone for profit" (*1* Cor. *12,* 7).   **18.** Therefore this servant does wrong in merely taking care that his gift is not lost.

19. *After a long time:* cf. 5. The Judgment is a "settling of accounts."   **20-23.** Note that both scenes are identical. Each of the good servants receives the same reward, because this is measured not by the original gift but by the degree of co-operation. Contrast the similar parable in Luke, which considers the same truth from a different aspect. There is joy in the words of the good servants, "Look, Master, I have doubled thy money!" The reward consists in a sharing of Christ's own *joy;* but heaven does not consist in inactivity: *I will set thee over many things* (cf. also Luke *19,* 17. 19).   **24-27.** The *wicked servant* failed to use the talent that his master had entrusted to him: (a) because he was *slothful;* (b) because he had a wrong idea of his master as *a stern man;* (c) because he *was afraid:* entrusting the money to the bankers involved a certain amount of risk. Thus a Christian may fail to use the graces that Christ bestows upon him: (a) because of spiritual sloth; (b) because he has an exaggerated idea of God's justice and a mean idea of His mercy; (c) because he is spiritually timid and too proud to risk making a mistake. **28-30.** The punishment of the wicked servant.   **28.** God's grace is not given in vain (cf. Isa. *55,* 10 f), if some one neglects or misuses it, it is taken from him and given to another (cf. Apoc. *3,* 11).   **29.** Just as natural faculties become more perfect through use or become atrophied through disuse, so also grace that is used leads to an increase of grace, whereas grace that is neglected tends to be lost; cf. *13,* 12.   **30.** Cf. *8,* 12; *13,* 42. 50; *22,* 13; *24,* 51.

25, 31-46: The Last Judgment. Only in Matthew. While it is certain that all men, whether Christians or not, will be judged by Christ on the Last Day, still it is not certain whether this description was meant by Christ to be understood in such a sense, or whether this description is limited to a portrayal of the judgment that will be passed on Christians alone. All the preceding parables (from *24,* 42 on) are concerned with good and bad Christians only, and this would seem to be the case here also. Not only those on the right but also those on the left address Christ as "Lord" (37. 44), i.e., they speak to Him as only Christians would. On the other hand this is a judgment of *all the nations* (32); still this might mean only "the Christians of all the nations" (cf. *8,* 11; *28,* 19).   **31.** On the Last Day Christ *shall come in his majesty* as *king* (34; cf. also Apoc. *19,* 16). *And all his angels with him:* these words are from Zach. *14,* 5, where "the saints," i.e., "His holy ones," really means "His angels." Christ frequently mentions the angels as playing an important part in the Last Judgment (cf. *13,* 39. 41. 49; *16,* 27; *24,* 31). **32 f.** The little parable of the Sheep and the Goats. Because most of Palestine offers rather poor pasture land, goats are often as common as sheep. During the day they all graze together, but in the evening the shepherds put the sheep in folds apart from the goats. So also in the Church the good and wicked are not separated until the Last Day. The goat is a common symbol of the wicked, whether from its vile smell or from its "capricious" nature. The left has always been considered the evil, unlucky side.

34-46. The criterion is the exercise of Christian charity. This should not be surprising. The non-Christians are already condemned for not believing in Christ (cf. John *3,* 18). But the Christian's faith will not save him unless it is a living faith (cf. Jas. *2,* 14 ff) which proves itself in doing God's will. All the commandments of God can be summed up in the love of God and the love of neighbor (cf. *22,* 37-40). And the true test of sincere love of God is love of neighbor (cf. *1* John *4,* 20 f), which is shown in acts of charity (cf. *1* John *3,* 17 f). This argument is developed on these same lines in Jas. 2. From the particular acts of charity which Christ mentions here, is derived the list of "the corporal works of mercy"; but this list should not be limited to these. Note how similar yet diverse are these two judgments. The repetitions, interrogations, etc., are merely a literary device used by Christ to impress this truth on the minds of His disciples; we need not believe that such conversations will actually take place at the Last Judgment. Contrast *blessed of my Father* with *accursed ones;* the latter are cursed by the good God only because they have damned themselves. Contrast the *kingdom prepared for you* with *the everlasting fire prepared for the devil;* hell was not intended originally for men, for "God our

Savior wishes all men to be saved" (*1 Tim. 2*, 4). *My brethren* does not mean "all men" but "my disciples"; in the New Testament *brethren* in this sense always means "Christians" (cf. Heb. *2*, 11 f). While the New Testament teaches the unity of the human race (cf. Acts *17*, 26), the modern concept of "the brotherhood of man" is not found in it. We must, of course, "do good to all men, but especially to those who are of the household of the faith" (Gal. *6*, 10). *The least* of Christ's brethren are, in a sense, all of His disciples, who should all be humble and lowly, but here are directly meant the most afflicted and despised of His disciples who would on this account be more easily neglected. Christ does not say that He will *consider* what is done to these as though it were done to Him; He says simply that it *is* done to Him. This should be understood in the light of the Mystical Body of Christ, the intimate union of Christ Himself with each of His members.

II. THE PASSION, DEATH AND RESURRECTION *26-28*

1. THE LAST SUPPER *26*, 1-35

From here on, agreement between all three Synoptic Gospels is even greater than it is in the narrative of our Lord's Public Ministry. It seems probable, therefore, that in the early oral catechesis of the Apostles the account of Christ's passion, death and resurrection was considered as a special narrative, partly independent of the account of His preaching and miracles. Most of the events that are here narrated in the first three Gospels are paralleled, in a broad sense, in the Fourth Gospel also. St. John, however, rather presupposes the account of the Synoptic Gospels as known to his readers and supplements it with new items of his own. There are certain seeming discrepancies between the Synoptic Gospels on the one hand and the Fourth Gospel on the other, especially in regard to chronology. It seems that all four Gospels count time according to the Jewish method of reckoning, i.e., the day considered as a period of twenty-four hours begins at sunset. Therefore in all four Gospels the Last Supper takes place, strictly speaking, on the same day on which Christ died; the preparations for it are made on the previous day. All four Gospels agree that our Lord died on a Friday at about three o'clock in the afternoon. According to the Synoptic Gospels this Friday would seem to have been the great feast of the Passover, at the beginning of which, that is, on the preceding evening according to our way of reckoning, the Passover Supper was eaten. But according to the Fourth Gospel this Friday was the day before the Passover (cf. John *13*, 1 f; *18*, 28; *19*, 14. 31). Astronomical calculations (the date of the Passover was determined by the moon) confirm the correctness of the Fourth Gospel. During the whole period of Pilate's governorship (26-36 A.D.) no Passover could have fallen on a Friday; but there were two years when it fell on a Saturday: 30 A.D. (April 8) and 33 A.D. (April 4). We have good reasons from extra-biblical sources for believing that at the time of Christ, when the Passover fell on Saturday, many anticipated the Passover Supper by one day. (See Lagrange, *The Gospel of Jesus Christ*, Vol. II, p. 193 f.) According to the Synoptic Gospels Christ also seems to have done so.

1. The conclusion of the preceding discourse according to the usual formula of the First Gospel.   **2.** This brief prediction of the Crucifixion is only in Matthew. *After two days:* we would say, "The day after tomorrow." Since Christ is apparently speaking of *the Passover* as the day on which He would eat the Passover Supper, i.e., Friday according to the Jewish way of reckoning time, this prediction of His crucifixion was made on Wednesday.

26, 3-5: The Council. Parallels in Mark *14*, 1 f and Luke *22*, 1 f.   **3.** *Then* refers to the day mentioned in 2, as is certain from Mark. This meeting therefore took place on Wednesday. *The chief priests and the elders:* Mark and Luke mention the Scribes also; therefore members of all three sections of the Sanhedrin were present. However, this was probably not an official session of that body but rather a private meeting of its leaders.   **4.** They had already decided to put Jesus to death (cf. John *11*, 53). They were now discussing ways and means of arresting Him secretly, for an attempt to seize Him while He was publicly teaching in the temple would have caused a riot among the people.   **5.** Since there were so many Galileans at Jerusalem for the Passover, the Council finally decided, no doubt with reluctance, to postpone the arrest of Jesus until the feast and its octave were over. But the offer made by the traitor Judas led them to change their plans.

26, 6-13: The Anointing at Bethany. Parallels in Mark *14,* 3-9 and John *12,* 1-8. Luke omits this, probably because he has already (in *7,* 36 ff) given an account of a somewhat similar but certainly distinct anointing of Jesus. The anointing at Bethany took place on the day before Christ's triumphal entry into Jerusalem, as John *12,* 1. 12 explicitly states. Matthew and Mark place this event in connection with the meeting of the Council and with the treason of Judas probably in order to show that this incident completely embittered Judas against his Master and so led to his treason. But strange to say, it is only the Fourth Gospel that mentions Judas by name as one of those disciples who were indignant at the seeming waste of the precious ointment. Perhaps the intimate connection between this anointing and the burial of Jesus led to the transference of this event from its proper chronological place to its present incorporation in the account of Christ's Passion.

6. *Simon the leper:* he must have been cured of his leprosy or he could not have acted as host. "Simon" was a very common name at that time; it was by a purely accidental coincidence that the Pharisee in whose house Jesus was anointed by the penitent woman also bore the name of Simon (cf. Luke 7, 40 ff). **7.** We know from John that this *woman* was Mary of Bethany, the sister of Martha and Lazarus. *Precious ointment:* Judas estimated its value as worth three hundred denarii (cf. Mark and John), a very large sum indeed, since one denarius was the ordinary daily wage of a laborer (cf. *20, 2*).   **8 f.** John tells us that these words were spoken by Judas. But the other indignant disciples shared his sentiments, although they were no doubt sincere in their intention of having the money given to the poor. **10.** The disciples were evidently trying to stop the woman, at least by rebuking her; cf. Mark: "And they grumbled at her." What this *good turn* consisted in is explained in 12.   **11.** Cf. Deut. *15,* 11. Christ is simply stating a fact; His words should not be abused as an argument against honest efforts to banish poverty from the world as far as possible. The special circumstances, i.e., the close proximity of our Lord's death and burial justify this lavish expenditure.   **12.** The ancients considered a decent burial with proper anointing of the body to be of great importance. Because there was no time to wash and anoint the body of Jesus on Good Friday evening, the women postponed this act of piety until the end of the Sabbath (cf. Mark *15,* 46; *16,* 1). But Christ had then already risen without having received this intended anointing. Hence the great importance of this anticipated anointing of Christ's body by Mary of Bethany. **13.** Mary's deed merits immortal fame not merely because of its intrinsic goodness (any other good deed done to Christ would merit the same), but because of its prophetic nature, i.e., it foretold the fact that Christ was soon to be buried without being anointed after death. We need not suppose, however, that Mary understood all this.

26, 14-16: The Betrayal. Parallels in Mark *14,* 10 f and Luke *22,* 3-6. Judas made his offer to the chief priests probably on the same day that the Council met (3-5), i.e., on Wednesday of Holy Week. Some think that 16 implies a longer period. But the Council would hardly have come to its first decision (5) if Judas had already made his proposal. Matthew alone mentions the amount of money that Judas received (cf. also *27, 3*). By a "piece of silver" is here meant the Jewish "shekel," the equivalent of the stater which was worth four denarii. *They counted out* is correct according to the sense, for the money given to Judas was undoubtedly in coins; but the Greek means literally "they weighed"; for the whole sentence is a quotation from Zach. *11,* 12. According to Mark and Luke the money was merely promised to Judas on this occasion. Probably it was set aside for him on this day and handed over to him as soon as Jesus was delivered into their hands. Moreover, in Mark and Luke the money does not play a prominent role in the betrayal. Possibly Judas's question, as given in 15 was merely a cloak to hide his real motives. It is true that Judas was certainly avaricious (cf. John *12,* 6) and no doubt avarice played an important part in the betrayal. But the popular notion that Judas sold Christ merely out of avarice is difficult to reconcile with other known facts of the case. For if avarice were his sole motive, why should remorse have induced him to throw this money away when he saw that Jesus was condemned to death? Moreover, thirty pieces of silver were a comparatively *small* sum. As treasurer of the Apostolic band Judas could probably have purloined much more than that during every month of our Lord's Ministry. This was the amount of indemnity, determined by the Law, which a man had to pay who was responsible for the accidental death of another's slave (cf. Ex. *21,* 32). Probably the chief priests, out of hypocritical scruples, felt that this was the minimum they should pay Judas for the death of his Master. But Judas was invaluable to them. Without his assistance they could not have arrested Christ. If Judas were acting solely out of avarice, he could have demanded a much higher price and he would have received it. A few prefer to explain the treachery of Judas as motivated by sheer malice and hatred for Jesus. It is true that Judas had long since lost faith in Christ (cf. John *6*, 65. 71); but this theory also fails to explain his later remorse (cf. *27*, 3 ff). Another unsatisfactory theory seeks to lessen the malice of Judas by explaining his action as based on a desire to hasten the crisis and force Jesus to proclaim Himself the Messias according to the popular conceptions of that dignity; when Judas saw that this plan had failed, it was supposedly out of true but misguided affection for his Master that he committed suicide. But this theory, besides being entirely unknown to tradition, is opposed to Christ's strong condemnation of Judas's treason (cf. 24). One thing that is certain is that Judas acted under the instigation of the devil (cf. Luke *22*, *3*; John *13*, 2. 27). His true motives are known only to God and to Satan.

26, 17-19: Preparation. Parallels in Mark *14*, 12-16 and Luke *22*, 7-13.   **17.** *The first day of Unleavened Bread* was originally equivalent to the first day of the Passover and its octave, i.e., the fifteenth of Nisan, at the beginning of which the Passover meal was eaten; but in order to insure the absence of all leaven and leavened bread at the feast, the Rabbis ordained that all leaven had to be out of the house by the morning of the previous day, i.e., the fourteenth of Nisan, the day on which the Passover lambs were sacrificed. Thus at the time of Christ the fourteenth of Nisan became known as "the first day of the Unleavened Bread" (cf. Mark and Luke). It was sometime on Thursday when the Apostles spoke these words. *To eat the passover* means to eat the famous meal prescribed as a memorial of the flight from Egypt (cf. Ex. *12*).   **18.** *A certain man:* the Greek is equivalent to our "Mr. So-and-so"; i.e., Christ named or described this man but the Evangelist does not identify him; cf. the directions which Christ gave Peter and John on how they should find this man's house, as recorded by Mark and Luke. Evidently this man was a disciple of Christ. According to tradition his house, now known as "the cenacle," became the headquarters and first church of the Apostles in Jerusalem. Jesus meant *My time* as the time for Him to establish His kingdom by His death and resurrection. *Keeping the Passover* is equivalent to "eating the passover" of 17.

26, 20-25: The Betrayer. Parallels in Mark *14,* 17-21; Luke *22,* 14. 21-23 and John *13,* 21-30. In Luke the order of events at the Last Supper is different from that of Matthew and Mark. Thus, in the Third Gospel the denunciation of the Betrayer is mentioned after the institution of the Holy Eucharist. But, as is shown in the Commentary on Luke, the order of events at the Last Supper as recorded in the Third Gospel is not always strictly chronological. **20.** At their festive banquets the Jews at this time followed the custom of the Greeks and Romans who *reclined at table,* i.e., lay on their side upon couches, supporting the head with the left hand and eating with the right. From the description in the Fourth Gospel it seems that Jesus lay facing towards John, while John was perhaps in the same position in regard to Peter; Judas lay on the other side of Jesus, while some other Apostle lay on the other side of Judas. These five occupied the broad couch in the center, the other eight Apostles in two groups of four each occupying the two side couches. From this arrangement of the table in the form of a square U the dining-room at that time was known as the "triclinium," i.e., the "three-couch" room. **21.** *While they were eating:* this episode evidently took place near the end of the meal; the institution of the Holy Eucharist took place *"after the supper"* (Luke). **22.** Although each of the Apostles probably felt some apprehension of the possibility of his own personal guilt, yet their question had rather the sense of self-justification, "It is not I, is it, Lord?" **23.** Apparently this sign by which the traitor might be known, was given only to John and perhaps through him to Peter, for it seems from the Fourth Gospel that the other Apostles were ignorant of the identity of the traitor at the time that Judas left. *He who dips his hand with me in the dish* should be understood as the same sign as "He for whom I shall dip the bread and give it to him" (John); the latter gives the more exact description of the action. Cf. Ps. *40,* 10. **24.** *Goes his way, as it is written of him,* i.e., goes to His destiny, His death, as is foretold in the Prophets, especially Isa. *53.* **25.** Christ's answer must have been whispered to Judas, so that the other Apostles did not hear it. Judas left immediately, as soon as he was sure that Jesus knew of his treachery (cf. John *13,* 30).

26, 26-29: The Holy Eucharist. Parallels in Mark *14*, 22-25; Luke *22*, 18-20; and *1* Cor. *11*, 23-25. While all these four accounts are in substantial agreement, there are certain minor differences between them even in regard to Christ's words. These differences are probably due to the fact that this action of our Lord was repeated from the beginning at least once a week in every Christian community; thus there naturally arose certain minor variants which are reproduced in these four accounts. Luke's account is very similar to that of *1* Cor., while Mark's phraseology is much closer to Matthew's.   **26.** *While they were at supper:* i.e., "While they were eating" (Mark; the same Greek expression is used in both Gospels); contrast this with "after the supper" (Luke), "after he had supped" (*1* Cor.; the same Greek expression in both instances) in regard to the cup of wine. Either the consecration of the bread and that of the wine were separated by an appreciable interval at the Last Supper, their present close juncture in the Mass being a liturgical adaptation, or both temporal expressions are more or less synonymous, signifying "at the end of the meal." *Bread:* more exactly "a loaf." *Gave it:* i.e., the loaf of bread broken into pieces. The symbolism of "the breaking of the bread" is explained according to the reading of many MSS in *1* Cor., "This is my body that is broken for you," i.e., that is broken with suffering on the cross as a sacrifice for you; cf. the words of the consecration of the cup. The word *blessed* is synonymous with the phrase *gave thanks,* both expressions referring to the unrecorded prayer of Christ in which He thanked His heavenly Father for His gifts and called down the divine blessing upon this food and drink. This prayer of Christ is no doubt substantially repeated in the prayers of the Canon of the Mass, beginning with the Preface. The simple, direct statement, *"This is my body"* cannot possibly be understood in a figurative sense. Even many modern rationalists now admit that Christ meant these words in the sense in which they have always been understood in the Catholic Church, although these critics, of course, deny Him the divine power of actually doing what He said He was doing, i.e., truly giving His disciples His body as food under the appearance of bread. In theological language this change of the substance of bread into the substance of Christ's body is known as Transsubstantiation.   **27 f.** The *cup* contained wine, as is certain from 29. What has just been said of the bread applies also to the change of the wine into Christ's blood. Cf. also *1* Cor. *10,* 16.

The word *new* before *covenant* is missing in the best MSS of Matthew and Mark; it occurs, however, in all MSS of Luke and 1 Cor.; cf. Jer. *31,* 31; Heb. *7,* 22; *8,* 6; *12,* 24. Jesus is certainly alluding to Ex. *24,* 8; cf. Heb. *9,* 15-22. By these words Christ shows that this action of His at the Last Supper is a sacrifice, for He speaks of His blood as similar to the sacrificial blood with which Moses sealed the Old Covenant between God and Israel. Moreover, by the words *which is being shed* our Lord shows that this act is related in a special way to His sacrifice on the cross. *For many:* see Commentary on *20,* 28. *Unto the forgiveness of sins:* this is therefore primarily a propitiatory sacrifice. 29. Luke *22,* 18 records these words of Christ as spoken in connection with what is apparently a different cup of wine, partaken of before the Eucharist; but see Commentary on Luke. If these words are understood as referring to the Holy Eucharist (so they are apparently to be understood in Matthew and Mark), they signify not only that Jesus wished to tell the Apostles that He was about to depart from this world, but also the next time that He would give Himself to them in the Holy Eucharist He would be in the kingdom of His Father; whereas the present sacrificial banquet at the Last Supper is an anticipation of the Death on Calvary, all repetitions of the Liturgy of the Last Supper by the Church are memorial sacrifices, renewing the Sacrifice of the Cross in which Christ "offered himself up once for all" (Heb. *7,* 27).

26, 30-35: Peter's Denials Predicted. Parallel in Mark *14,* 26-31. Luke *22,* 39 and John *18,* 1 also record the walk of Jesus and His disciples to Gethsemani. But the Third and the Fourth Gospels mention the prediction of Peter's denials as taking place in the cenacle during Christ's discourse after the Last Supper. Their accounts are too similar to those of Matthew and Mark to suppose that this episode occurred twice. The First and the Second Gospels evidently forsake the chronological order here in order to place the prediction of Christ's abandonment by the Apostles in closer proximity to its fulfillment. 30. *After reciting a hymn:* literally in Greek, "having hymned"; several Psalms might have been sung. 31. *Be scandalized:* literally, "be tripped up," i.e., the Apostles will stumble on the path of fidelity to Jesus, in so far as they will abandon Him and scatter like frightened sheep. This scattering of Christ's little flock had been foretold in Zach. *13,* 7. Cf. also John *16,* 31 f.

32. Whenever Christ forewarned His disciples of His approaching passion and death, He also consoled them with the thought of the Resurrection. Just as Jesus had courageously led His disciples from Galilee to Jerusalem (cf. Mark *10,* 32), so also will He lead them back from Jerusalem to Galilee, where He will appear to a large group of His faithful followers (cf. *28,* 16; *1* Cor. *15,* 6). **33-35.** Peter's protestation of unwavering fidelity and Christ's prediction of his three denials are substantially the same in all four Gospels. Mark alone mentions the cock crowing twice, both in the prediction and in its fulfilment.

2. THE PASSION AND DEATH OF JESUS *26,* 36 — *27,* 66

26, **36-46: The Agony in the Garden.** Parallels in Mark *14,* 32-42 and Luke *22,* 40-46. There is also a brief anticipation of Christ's Agony in John *12,* 27. We are here in a dark sanctuary of deep mysteries which we should enter reverently, humbly confessing our ignorance. Although it was impossible for the human nature of Christ to be separated from His divine nature and all that this union entails, still in this agony Christ seems in a certain sense to suspend the spiritual support which His human nature normally received from His divinity. He is here the Second Adam, voluntarily undoing by His loving obedience to the Father the sins which the disobedience of the First Adam had brought upon mankind (cf. Rom. *5,* 12-19; Heb. *5,* 7 f). Therefore, since Jesus acts here as the representative of all mankind, the Man *par excellence,* He acts, as far as possible, purely according to His human nature, suffering all the natural, though sinless, frailties of human nature, the dread of suffering and death, the shrinking from degradation and revilement, the intense sorrow of being abandoned by His friends and betrayed by His own disciple, the overwhelming grief at the thought of men's black ingratitude in despising the cost and value of His Sacrifice, and perhaps, as far as it would be compatible with His sinless soul, a sense of guilt for all the shameful sins of the world that He was taking upon Himself to atone for (cf. *2* Cor. *5,* 21; Gal. *3,* 13; *1* Pet. *2,* 24).

36. *Gethsemani,* a Hebrew name meaning "olive-oil-press"; its traditional site at the base of the Mount of Olives is marked by century-old olive-trees. **37.** The same three Apostles who were witnesses of Christ's divine power in the raising of the daughter of Jairus and witnesses of His divine glory in the Transfiguration, were now chosen to be witnesses of the humiliating agony of His humanity. But Jesus sought human solace from them in vain. **38.** *Even unto death:* Christ's anguish was so great that of itself it could have caused His death. **39.** From this passage it is certain that Christ's human will is distinct from, but in conformity with, His divine will; cf. also John *5,* 30; *6,* 38. On the meaning of the *cup* see Commentary on *20,* 22; here, however, there may be the additional thought of "the cup of God's wrath" (cf. Ps. *74,* 9; Isa. *51,* 17; Jer. *25,* 15; Apoc. *14,* 10; *16,* 19). **40.** It was now near midnight and the Apostles were exhausted from the emotional strain of the evening (cf. Luke). **41.** To *enter into temptation* may mean either "to be tempted" or "to consent to temptation." Cf. *6,* 13. The temptation against which the Apostles should at present use all natural and supernatural means of help (vigilance and prayer) would be to forsake Christ and even deny Him. *The spirit indeed is willing:* even the best good-will of men is often overcome by the weakness of human nature. **45.** *"Sleep on now":* these words may be understood as a concession to the tired Apostles; but in view of the following statements, which were apparently spoken at once, they should rather be understood as ironical.

26, 47-56: Jesus Arrested. Parallels in Mark *14,* 43-50; Luke *22,* 47-53; and John *18,* 2-11. **47.** *A great crowd:* very numerous at least in comparison with the small band of the Apostles, although Peter apparently felt that he had some chance of successful resistance against them. John speaks of a "cohort" and its military "tribune," but these terms should not be understood in the strict sense, as if a body of Roman soldiers were present. This rabble no doubt consisted entirely of Jews, among whom were some of the leaders of the High Priest's party (but certainly not Annas and Caiphas) and the "captains of the temple" (Luke), i.e., Levites who served as the temple-police. That they were not a regular military force is seen from their weapons, *swords and clubs;* the Greek word that is here translated as "swords" is not the technical word for "soldiers' swords." **48 f.** This signal had been agreed upon, probably in order to avoid causing undue alarm, for it was customary for a disciple of a rabbi to greet his master with a kiss. **50.** *Friend:* see Commentary on *22,* 12. *"For what purpose hast thou come?":* the Greek phrase, which is of uncertain significance, is a relative clause, literally, "for which thou hast come"; perhaps the sense is, "Dost thou dare kiss me as a friend in this business for which thou hast come!" (cf. Luke). **51.** It is only from the Fourth Gospel that we know that it was Peter who *drew his sword* and that the name of the wounded man was Malchus. **52-54.** Christ was opposed to all physical violence on this occasion, but His words should not be overstressed, as if He forbade His disciples under any circumstance to have recourse to active resistance against unjust aggression. His arguments here are: (a) Physical violence always recoils upon him who uses it, and He wishes no unnecessary harm to befall His Apostles (52); (b) This action shows lack of confidence in God who could send *more than twelve legions of angels* instead of twelve weak Apostles, if it were His will to rescue Jesus from His enemies (53); (c) It is the will of the Father, expressed in His words to the prophets, that Jesus should be captured and put to death (54). Moreover, Peter's action would compromise Christ's cause (cf. John *18,* 36). **55.** In these words of calm majesty Jesus shows His enemies that He is freely allowing them to arrest Him and that the means they are using in order to capture Him are absurd. **56.** According to Mark, Christ Himself pointed this out to His enemies, i.e., He is allowing Himself to be arrested because it is His Father's will, as expressed in the Scriptures; God permits Christ's enemies and the power of darkness to have this brief hour of seeming triumph (cf. Luke).

26, 57-68: Jesus before the Sanhedrin. The only parallel in the strict sense is Mark *14,* 53-65. John *18,* 13 f. 19-24 gives a different account of the inquisition held by the High Priest during the night. Luke *22,* 54. 63-71 recounts the abuse heaped upon Jesus during the night in the house of the High Priest, but puts the trial by the Sanhedrin as taking place at dawn. However, the account of the Jewish trial in Luke is too similar to that of Matthew and Mark to be considered as a distinct repetition, held for the sake of legality in the morning. The most probable harmonization of the four accounts of the Jewish trial is as follows. After His arrest in Gethsemani Jesus is brought to the house of the High Priest, where, apparently, both Annas and Caiphas had their apartments (Matt. *26,* 57; Mark *14,* 53; Luke *22,* 54; John *18,* 13 f. 24). Here a preliminary investigation is held by the leaders of the Sanhedrin, in order to get sufficient matter for the formal charge that would be presented before the whole Sanhedrin in the morning; this inquisition was accompanied and followed by insults and physical violence, somewhat in the manner of a modern "third-degree" (Matt. *26,* 67 f; Mark *14,* 65; Luke *22,* 63-65; John *18,* 19-23). At dawn Jesus is brought before the whole Sanhedrin and two charges are raised against Him: of claiming to be the Messias and of claiming to be the Son of God. Upon His admission of the latter charge He is condemned to death as guilty of blasphemy (Matt. *26,* 63-66; Mark *14,* 61b-64; Luke *22,* 66-71). It is not clear whether the false witnesses (Matt. *26,* 59-63a; Mark *14,* 55-61a) appeared at the preliminary investigation or at the formal trial (cf. Luke *22,* 71). **61.** Mark records the false testimony somewhat differently; possibly Matthew records the words of one witness and Mark those of the other witness. No doubt these men were referring to the words of Christ as recorded in John *2,* 19. A charge of crime against the temple would have been fatal for Jesus if it could have been proved true. The fact that the leaders of the Sanhedrin did not accept this false testimony shows that they were anxious, in their hypocrisy, to observe all the customary legalities. If this is so, it would seem certain that the formal trial did not take place at night. **63.** Jesus had no need to answer charges upon which the witnesses themselves did not agree. *I adjure thee by the living God* is a formula by which Jesus is put under oath to answer truthfully. His acceptance of the oath shows that His prohibition against oaths (cf. *5,* 34) is not to be understood in regard to necessary legal oaths. **64.** Christ affirms His divinity by His allusion to the prophecy of Dan. *7,* 13; *sitting at the right hand of the Power,* i.e., of God Almighty, certainly implies that He is the Son of God in the strict sense. The Sanhedrin also understood it as such. **65.** The tearing of one's garments was a gesture of the Jews to express great grief or horror, here the horror at hearing what was considered a blasphemy. This was, however, a formality. If the claim which Jesus made were not true, His words would certainly have been blasphemous. By formally stating under oath the fundamental Christian truth of His divinity, Jesus placed the act for which He was condemned to death. **66.** *Liable to death:* more exactly according to the literal sense of the Greek and Latin, "guilty of death," i.e., guilty of a crime for which the penalty is death. In these words the members of the Sanhedrin cast their votes by which they condemned Jesus to death. **68.** *Prophesy to us:* i.e., tell us by divine knowledge; Jesus had been blindfolded (cf. Mark and Luke) and could not know naturally who struck Him.

26, 69 — 27, 2: Peter's Denial. Parallels in Mark *14,* 66-72; Luke *22,* 54-62; and John *18,* 17 f. 25-27. All four Gospels agree substantially, but there are many minor differences of detail among them. All four agree: (a) that there were three distinct denials; (b) that the charge on all three occasions was that Peter was a disciple of Jesus; (c) that the first denial took place at the fire in the courtyard and was occasioned by a maidservant's suspicions. But in the second and third denials there are differences among the four Gospels in regard to both the exact place in the courtyard and the accusers. We should probably consider each of these denials as really a group of several accusations and denials, one Gospel recording one part of it and another Gospel another part. Matthew and Mark alone mention the fact that Peter added oaths to his denials, and indeed with increasing vehemence. **69.** *A maidservant:* she was the portress who had admitted Peter (cf. John). Peter had special reasons for being afraid, because he had wounded Malchus. **70.** In Matthew and Mark Peter's first denial is mild enough, being merely an evasive answer, that he did not understand what she was driving at. But a little lie soon grows to big proportions. **73.** The Galileans spoke a dialect of Aramaic that was somewhat different from Judean Aramaic. **74.** *A cock crowed:* we cannot really determine the time of the third denial from this incident, for cocks crow occasionally at any hour of the night; however, at the first signs of dawn they all begin to crow lustily. **75.** *Peter remembered:* it was Christ's look that helped to remind him (cf. Luke). A cock crowed also after the first denial (cf. Mark) and apparently Peter had heard that one too, but it did not remind him of Christ's prediction.

27, 1. The official meeting of the Sanhedrin at dawn. Parallels in Mark *15,* 1a and Luke *22,* 66a. They *took counsel together* concerning the charges against Jesus that they would present to Pilate; but the original phrase here may mean simply, "They held a council." In any case it was at this meeting that Jesus was formally condemned to death by the Sanhedrin. **2.** Jesus is delivered to Pilate. Parallels in Mark *15,* 1b; Luke *23,* 1; and John *18,* 28. The Romans allowed the Sanhedrin the right to condemn a man to death, but required that the sentence be ratified by the procurator; if he thought that there was a miscarriage of justice he could review the case himself (cf. John *18,* 31). The Jewish penalty for blasphemy was stoning (cf. Lev. *24,* 23), and apparently the Jews at the time of Christ inflicted this penalty without taking the trouble to consult the procurator (cf. John *8,* 5 ff. 59; Acts *7,* 58). But the Sanhedrin preferred to have Jesus condemned by the Roman authorities on a civil charge and executed by the Roman punishment of crucifixion, in order to discredit His followers all the more.

27, 3-10: The End of Judas. Only in Matthew; but cf. Acts *1,* 18 f.   **3.** Either Judas had not expected Jesus to be condemned, or after the betrayal his feelings towards Jesus had changed. He *repented* in the sense of "suffered remorse"; his repentance was not "according to God" (cf. 2 Cor. *7,* 10).   **5.** *Temple:* the Greek word signifies the temple-building proper, not the courts of the temple. *Hanged himself:* see foot-note to Acts *1,* 18. **7.** *The potter's field:* a field which was worthless for agriculture, either because the potter had dug his clay there, or because he used it as a dump for his broken pots.   **8.** *Haceldama* is an Aramaic word. According to Matthew this field received its name because it was bought at the *price of blood.* According to Acts it was called "the Field of Blood" because the blood of Judas was shed there. These two explanations of its name are not mutually exclusive and undoubtedly both are correct.   **9 f.** This quotation is a somewhat free translation of Zach. *11,* 12 f. The word *Jeremias* may have been in the original Aramaic Gospel of St. Matthew or it may have been added by the Greek translator or some early copyist, the purpose being perhaps to call the reader's attention to somewhat similar passages in Jer. *18,* 2 ff; *32,* 6 ff.

27, 11-25: Jesus before Pilate.   **11-14.** The first hearing before Pilate. Parallels in Mark *15,* 2-5; Luke *23,* 2-4; John *18,* 29-38. The Jews at first try to have Pilate ratify their sentence without investigating the case (John 29-32). Failing in this, they enter the formal charge of sedition against the Roman authorities (Luke 2).   **11.** Christ answers affirmatively only in regard to His spiritual kingdom (cf. John 33-38). **12-14.** Only Matthew and Mark record that, when the Jews repeated these accusations, Jesus remained silent; but John *19,* 9 records Christ's later silence before Pilate. This first part of the Roman trial ends by Pilate declaring Jesus innocent of these false charges of sedition (Luke 4; John 38b). In order to evade the persistence of the Jews, Pilate tries in vain to pass on to Herod the responsibility for Christ's execution (Luke *23,* 5-15).

15-18. 20-21. Christ and Barabbas: parallels in Mark *15*, 6-11; Luke *23*, 17-19; John *18*, 39 f.   **15.** The custom of releasing prisoners at great national holidays was common in many countries.   **16.** Barabbas was *a notorious prisoner,* not only because he "was a robber" (John) but also because he "had committed murder in a riot" (Mark and Luke); the word for "riot" signifies properly an "insurrection" against the authorities. In one group of Greek MSS in Matthew he is called "Jesus Barabbas." This is possibly correct, for Barabbas was really his "last name" (literally, "son of Abbas") and this explains the contrast with *Jesus who is called Christ* (17).   **17.** *When they had gathered together:* at the beginning of the trial before Pilate only the chief priests and elders (12), i.e., Christ's enemies in the Sanhedrin, had been present; now the people assembled for the purpose of demanding the release of some prisoner (cf. Mark). The initiative in general therefore came from the people, but it is not likely that they took the initiative in demanding Barabbas. Apparently Pilate was free to offer them only a limited choice. He thought of Barabbas as the least welcome alternative to Christ.   **18.** Knowing that Jesus was innocent and that it was purely out of spite that the chief priests demanded His death, Pilate relied on the more generous nature of the common people to effect the release of our Lord.

19. Matthew alone mentions this incident of Pilate's wife. Nothing else is known of either her or the rest of Pilate's family, but this verse has offered the dramatizers of the story of Christ's Passion a rich field for speculation. Apparently Pilate received this message while he was waiting for the people to make up their mind about the release of Barabbas or Jesus. *Have nothing to do with:* i.e., do nothing against. *I have suffered many things:* whatever the nature of this *dream,* it must have greatly frightened her. No doubt this dream was caused by God as a special grace to save Pilate from committing this sin against Christ.

20. The *crowds* here were probably the rabble of Jerusalem who had their own reasons for disliking this Prophet from Nazareth. They acted mostly through ignorance (cf. Acts *3,* 17). In any case, they could not speak in the name of the whole Jewish nation. The latter rejected Jesus when the majority of the Jews throughout the world rejected the later preaching of the Apostles.   **21.** Although the people now express their clear preference for Barabbas, still Pilate apparently did not release him until the end of Christ's trial (cf. Luke v.25).

22 f. The mob clamors for the crucifixion of Jesus. Parallels in Mark *15,* 12-14; Luke *23,* 20-23a. A similar scene is recorded in John *19,* 6. 15 as taking place somewhat later in the trial, but no doubt these cries for Christ's crucifixion were repeated at every attempt that Pilate made to save Jesus. Pilate irritates them, perhaps intentionally, by repeatedly referring to Jesus as "He *who is called the Christ."* If the proposal to release Barabbas, considering him as a popular leader, had originated with the people and not with Pilate (Mark seems to suggest this, although the contrary opinion is expressed above), the procurator's repeated attempts to substitute Jesus for Barabbas must have infuriated the crowd all the more against our Lord.

24 f. Only Matthew records Pilate's handwashing and the self-curse of the people. This scene probably took place immediately before Pilate passed sentence on Jesus and is therefore to be connected with John *19,* 13-15.   **24.** *A riot was breaking out:* it would have been fatal for Pilate if the report reached Tiberius that he had caused a riot by opposing the Jews precisely in a matter where they appeared to be more anxious to preserve the Roman dominion in Palestine than he himself was (cf. John *19,* 12). Pilate finally had to choose between justice for Jesus or his own personal interests. He chose the latter. The symbolic act of washing one's hands in public, to signify that they were morally clean in some affair where innocent blood was shed, was a Jewish custom (cf. Deut. *21,* 1-9) which Pilate made use of here, apparently in order to accommodate himself to the native customs.   **25.** By these words the people probably merely wished to say that they would assume responsibility before Cæsar for this act. But since neither they nor Pilate could avoid the moral responsibility before God for the murder of Jesus, their words are at least implicitly a curse upon themselves. This curse was fulfilled at the destruction of Jerusalem. But it is unfair to lay this curse upon the whole Jewish race.

27, 26-30: The Scourging and Crowning. Parallels in Mark *15,* 15-19 and John *19,* 1-3. Luke omits this scene, probably because it seemed too repulsive, but he agrees with John *19,* 4 f that Pilate had tried to appease the Jews by this half-way measure (cf. Luke *23,* 16. 22).   **26.** Matthew and Mark seem to speak as if Christ were scourged after sentence of death had been passed on Him. Scourging was the normal preliminary of the Roman punishment of crucifixion, and for this reason the first two

Gospels bring Christ's scourging in closer proximity to His crucifixion. There is no reason for thinking that this cruel punishment was inflicted twice upon our Lord. The victim of a Roman scourging was stripped of his clothes, bound to a pillar and beaten with a whip, usually consisting of a couple of leather lashes, the ends of which were weighted with small leaden balls.   27. The prætorium was the Roman name for the soldiers' barracks. Its mention here gives probability to the opinion that the trial before Pilate took place at the fortress of Antonia, just north of the temple, rather than at the palace built by Herod the Great, at the western wall of the city, where the procurators usually resided while in Jerusalem. Christ was mocked in "the courtyard of the prætorium" (Mark). Note that he is brought in here only after the scourging. According to the Roman custom scourging took place in some public place. *The whole cohort* signifies "the whole band of soldiers that was stationed here." This was Pilate's bodyguard and almost certainly did not consist of a full cohort of a thousand men. The Greek word is often used in the general sense of any military unit (cf. John *18*, 3).   28. Since Matthew had not mentioned the stripping which preceded the scourging, he now mentions it, in order to show that the *scarlet cloak* was placed directly upon the bruised back of Jesus. There is no need to suppose that after the scourging our Lord was first clothed with His own garments and then immediately stripped again. The Greek word that is here translated as *scarlet cloak* signifies the cloak or large cape worn by the soldiers. It was ordinarily of a scarlet color. Here it is put on our Lord as a ridiculous substitute for the crimson garments of royalty. The *crown of thorns* and the *reed* for a scepter were also intended as the mock-emblems of royalty. Their words are a parody on the customary greeting to the emperor, *Ave Cæsar*. The soldiers no doubt indulged in this rough horse-play on their own initiative. For them Jesus was just a common rebel against the great power of Rome.   30. *The reed* was a sort of bamboo cane and blows with it upon Christ's thorn-crowned head must have been extremely painful.

27, 31-33: The Way of the Cross. Parallels in Mark *15*, 20-22; Luke *23*, 26-32; John *19*, 16 f. For the events between the mockery and the sentence of crucifixion read John *19*, 4-15.   31. *His own garments:* this is given special mention, because according to the Roman custom the condemned man was not clothed after the scourging but driven naked with whips to the place of execution, his arms tied to the cross-beam upon which he was to hang. The clothing of Christ before going to Calvary was probably a concession to the better sense of decency among the Jews.   **32.** At first Jesus Himself carried His own cross (John). No doubt it was because of His extreme weakness that it was laid upon Simon. There is good foundation therefore in the Gospels for the tradition that Jesus fell several times on the way to Golgotha. The Evangelists no doubt could have told us many more incidents of this sorrowful journey, but their very constraint adds all the more to the pathos of their narrative. *Cyrene* was at this time a flourishing city in the Roman province of Libya (west of Egypt); there was a colony of Jews there (cf. Acts *2,* 10). Simon, however, was probably a resident at Jerusalem at this time, for Mark and Luke add that he was "coming from the country," i.e., coming in from the fields, from his farm. Apparently he was coming home from work. If so, we have here a strong confirmation even in the Synoptics that this was the eve and not the day itself of the Passover. The Cyrenian Jews had a synagogue in Jerusalem (Acts *6,* 9). *Forced,* literally "pressed into public service," does not necessarily imply that Simon performed this act of charity unwillingly. He took up the whole burden for Christ, not merely assisting Christ by carrying the cross with Him. But whether this was the entire cross or only its horizontal beam is uncertain.   **33.** *Golgotha,* the word in the local Aramaic dialect for *Skull;* in good Aramaic really "Gulgoltha." From the Latin word for skull, *calvaria,* we have the name "Calvary." It was probably a small hill which in some way resembled a skull. It was outside the walls (Heb. *13,* 12) but "near the city" (John *19,* 20), at the side of the highway (30). The traditional site, the Church of the Holy Sepulcher, is certainly the true one.

27, 34-44: The Crucifixion. Parallels in Mark *15,* 23-32; Luke *23,* 33-43; John *19,* 18-24.   **34.** Before executing a criminal the Jews were accustomed to give him a drugged drink which would deaden his nerves and thereby lessen his sufferings. For this purpose they used wine mixed with incense or myrrh, which would not be unpleasant to drink. According to Mark the wine that was offered to Jesus was "mixed with myrrh." Possibly this was prepared by the pious women who followed Christ and ministered to His needs (cf. 55), for the Jews generally left this charitable office to women. But according to Matthew this wine was *mixed with gall,* which is extremely bitter. There can hardly be any doubt that both Evangelists are referring to the same wine. Probably Christ's enemies intentionally added the gall to the prepared drink in order to torment Him the more. The *gall* recalls Ps. *68,* 22. Jesus *tasted it,* perhaps to suffer in His sense of taste, or to show His gratitude to those who had meant well in offering the drugged drink. But *he would not drink it,* because He wished to remain fully conscious to the end. "This was not the chalice which He had promised His Father to drink" (Lagrange).   **35.** Jesus was then stripped of His garments. According to the Roman custom, at least outside of Palestine, a crucified criminal was completely naked; but it seems probable that in Palestine, out of regard for the Jews' sense of decency, the victim was permitted a loincloth. On the division of Christ's garments, see Commentary on John *19,* 23 f. The reference to the prophecy of Ps. *21,* 19 is not found in the best MSS of the First Gospel; it was taken over into Matthew's Gospel from the Fourth Gospel where all the MSS have it. After the stripping Jesus was crucified. If done according to custom, His arms were first fastened to the horizontal beam that He had carried to the place of execution; this was then attached to the vertical beam or large stake which stood permanently on the site. There was usually a sort of projecting horn which served as a seat to support the weight of the body. It is certain that Jesus was nailed through His hands and feet to the cross, and not merely fastened with ropes (cf. Luke *24,* 39 f; John *20,* 20. 25. 27).   **36.** The soldiers who carried out a sentence of crucifixion were obliged to see to it that no one rescued the victim but that he died on the cross. The crucifixion of our Lord was entrusted to four soldiers (cf. John *19,* 23) under the command of a centurion (cf. 54).   **37.** That this placard was *put above his head* shows that the cross on which Jesus died had four extremities, i.e., the form in which it is traditionally represented, although other forms and shapes of crosses were used for crucifixions. The four Evangelists agree in substance but differ in details on the exact words of this inscription. The full inscription seems to have been, "This is Jesus of Nazareth, the King of the Jews"; for it was customary to give both the name of the criminal and the crime for which he was being executed. Perhaps the inscription differed slightly in each of the three languages (cf. John) in which it was written. Only John mentions the dispute between the chief priests and

Pilate about this inscription. The latter could not have worded the charge other than he did, for Jesus was executed by the Romans not for *pretending* to be, but for actually being, the King of the Jews.
38. *Robbers:* more exactly, "brigands, highwaymen." Jesus is put in the prominent place between them, as if He were their leader. Mark alone mentions the fulfillment of the prophecy of Isa. *53,* 12 in this. **39 f.** The mockery by *the passers-by:* only in Matthew and Mark. These people were either strangers who were arriving at Jerusalem for the feast or the natives of the city who were coming or going on business. Perhaps they were not greatly interested in the execution but asked about it and were told the worst: that Jesus had threatened the temple and had blasphemed by saying He was the Son of God; such crimes would have seemed outrageous to any good Jew, so these people joined in mocking Christ. **41-43.** The mockery by the chief priests and their adherents: in all three Synoptics. This consisted principally in ridiculing our Lord's miracles and His claim to be the Messias. **43.** Only in Matthew. Cf. Ps. *21,* 9; Wisd. *2,* 13. 18. These men, no doubt unwittingly, used the words of this Psalm which certainly refers to the Passion of Christ: cf. also 8 of this Psalm. **44.** It seems fairly certain that at first both of the robbers reproached Jesus, although it is possible that Matthew and Mark use the plural merely in the sense of "one of the robbers"; at least later on one of them defends Jesus, expresses his belief in His Messiasship, and gains Paradise (cf. Luke *23,* 39-43).

27, 45-56: The Death of Jesus. Parallels in Mark *15,* 33-41; Luke *23,* 44-49; John *19,* 25-30. **45.** *From the sixth hour . . . until the ninth hour:* i.e., from twelve noon until three in the afternoon, when the sun would normally be at its brightest. We need not, however, understand this as meaning exactly three hours to the minute. The darkness apparently began after our Lord had been on the cross for some time; according to Mark *15,* 25 the crucifixion began somewhat before noon. But it was evidently not long after three o'clock when our Lord died. The *darkness* was certainly not due to an eclipse, for the Passover always occurs at the period of full moon when an eclipse of the sun would be impossible; therefore this darkness was miraculous, however it may have been brought about. We need not understand *the whole land* to mean "the whole earth"; it is sufficient to understand the Evangelists to mean "the whole land of Palestine" or even only "the whole region around Jerusalem." **46.** This is the only one of the "Seven Words" of Christ on the Cross which Matthew and Mark record. This cry of our Lord is the beginning of Ps. 21. In both Gospels these words are given in a mixture of Hebrew and Aramaic, but in a different mixture in each. Our Lord would probably quote this prayer in the sacred language in which it was originally written, and it seems that He really did so, for He must have said *Eli* rather than *Elohi* if the bystanders understood, or pretended to understand, this as "Elias" (in Hebrew *Eliyah*). In the Psalm these words meant merely, "Why hast thou left me to the fury of my enemies?" Our Lord probably meant them in the same sense. They are a prayer for help, coming from the human soul of Christ. Far from containing the slightest hint of despair, they are expressive of loving confidence in God. **47.** *The bystanders* who said this must have been Jews; otherwise they would not have known of Elias. **48.** This drink was offered to Jesus in answer to His cry, "I thirst" (cf. John *19*, 28 f). Since this *common wine,* which resembled vinegar (cf. Ps. *68*, 22), was the ordinary drink of the Roman soldiers, it was probably one of the soldiers who gave our Lord this drink. No doubt he acted out of genuine kindness. Luke *23*, 36 probably refers to some previous mockery. **49.** *The rest* who tried to prevent this charitable deed were most likely Jews (cf. the reference again to Elias). **50.** This *loud* cry at the very moment of death is generally taken to prove that Jesus died of His own free will (cf. John *10*, 18); but this might be wrongly understood in a sense resembling suicide. Christ's death was voluntary in the sense that He did not shirk the responsibilities which ultimately led to His crucifixion. But He died from the tortures inflicted on Him just as any one else would have died under the same circumstances. However, since this loud cry was no inarticulate shout but identical with the prayer recorded in Luke *23*, 46, we may at least conclude from it that our Lord was conscious to the very end.

51-56. Various witnesses to the death of our Savior. Only 51b-53 are peculiar to Matthew: nature itself is a witness to the death of the God-man. **51.** It is not certain whether this *curtain of the temple* was the "first curtain" at the entrance of the Holy Place (so St. Jerome), or the "second curtain" (cf. Heb. *9*, 3) which separated the Holy Place from the Holy of Holies; nor are Catholic interpreters agreed about the significance of this rending of the curtain: some see in it the visible token of God's displeasure with Judaism, the abolition of the temple as God's special abode (cf. *23, 38*), etc.; others prefer to understand this as a sign that God then took away the barrier between Jew and Gentile (cf. Eph. *2,* 14), that all men are admitted to the worship of the true God (cf. Heb. *6,* 19 f; *9,* 8; *10,* 19 f), etc. **52 f.** *The tombs were opened* at the moment of Christ's death, by the shock of the earthquake; but it seems that *the saints who had fallen asleep,* i.e., the holy persons who had died before this time, did not rise until Christ Himself arose from the dead. This event is to be connected with our Lord's descent into limbo (cf. *1* Pet. *3,* 19; *4,* 6). But this passage is somewhat obscure: some interpreters hold that these *saints* rose in their glorified bodies and ascended with Christ into heaven; others think that they appeared in the manner of ghosts and then returned to their tombs (so Lagrange); for the rest of the New Testament seems to imply that Christ alone is "the firstborn from the dead" (Col. *1,* 18) and that no one has risen or will rise in a glorified body before the Last Day (excepting, of course, our blessed Lady's Assumption into heaven). **54.** *They were very much afraid:* i.e., they were overwhelmed with awe at the terrible portents in nature, the darkness, the earthquake, etc. Their words need not be understood as a confession of the divinity of Christ; it is sufficient to understand them in the sense given in Luke, "Truly this was a just man." According to Mark the centurion said this because he had seen how Jesus had died and had heard His last words. Now from Luke *23,* 46 we know what His last words were; therefore it seems that it was the sight of the great moral courage that Jesus had shown in His death and of the confidence that He had shown in calling God His "Father" even as He expired, that made the centurion exclaim, "Truly this man was the *Son* of God." **55 f.** All three Synoptic Gospels refer to these holy women as being eye-witnesses of our Lord's death. No doubt these women together with St. John were the original source whence the Synoptics drew their information of what happened on Golgotha. **55.** Among the *many women* was also the Mother of Christ (cf. John *19,* 25). *Looking on from a distance:* the Synoptic account indeed seems to be based on the reports of persons who had viewed the scene from a distance; cf., e.g., the "loud cry" of Jesus which Matthew and Mark mention without apparently knowing what words were said. *Who had followed Jesus . . . :* cf. Luke *8,* 1-3. **56.** *Mary Magdalene:* i.e., Mary of Magdala, a town in Genesareth; she is most probably to be distinguished both from Mary of Bethany and the Penitent Woman of Luke 7, 36 ff (see Commentary there). *The mother of the sons of Zebedee* is apparently the same as "Salome" in the parallel passage of Mark (cf. *20,* 20).

27, 57-61: The Burial. Parallels in Mark *15,* 42-47; Luke *23,* 50-56; John *19,* 38-42. **57.** *When it was evening:* i.e., towards evening, the latter part of the afternoon. Jesus had died about three in the afternoon. Some time after this His side was pierced with a lance (John). Only then did Joseph of Arimathea go to Pilate to ask permission to take away the body of Jesus (cf. John *19,* 38). More time was consumed by Pilate's investigation (Mark), Joseph's purchase of the linen cloth, etc., so that it must have been near sunset before they were ready to bury our Lord. But since at sunset the Sabbath began, when no work could be done, this burial was necessarily of a rather hurried and provisional nature; cf. Mark *16,* 1. *A certain rich man:* Mark and Luke tell us that he was also "a councillor," i.e., "a member of the Sanhedrin"; and for this reason, lest the reader might think that he was one of Christ's enemies, all the Evangelists are careful to point out that he *was also a disciple of Jesus.* **58.** *He asked for the body of Jesus,* because according to Jewish Law the bodies of persons who had been condemned to death by the Sanhedrin were not to be given honorable burial but were to be thrown in a special tomb reserved for them. The Sanhedrists had already asked Pilate for this (cf. John *19,* 31), but the procurator was only too glad to offend them by granting Joseph his request. **59.** According to the burial customs of that time a corpse was placed on a winding-sheet or shroud, here called a *linen cloth,* which was then folded over it. This was a new and *clean* linen cloth, for Joseph had just bought it (cf. Mark). When John *19,* 40 speaks of "linen cloths" in the plural (but a different word), either merely this shroud of the Synoptics is meant or this with other minor cloths for some other special purposes; but there is no foundation for the assertion that this passage in John means that our Lord's body was swathed in bandages. The identification of the Holy Shroud which is now venerated in Turin with the linen cloth in which our Lord was buried has strong arguments in its favor. **60.** *His new tomb:* no one had ever been buried in it (Luke and John); it was near the place of the Crucifixion (John). Matthew alone mentions that this tomb belonged to Joseph. From the sepulchres of that period which are still preserved around Jerusalem, it seems that the tomb where Christ's body was laid was in the form of a cave; a large stone, resembling a millstone standing on its edge, was rolled in a slot that was left for this purpose in the hewn rock at the entrance. **61.** The holy women apparently did nothing at the burial of Jesus except to watch what was being done. The Evangelists refer to them as witnesses in proof of the fact that our Lord's body was laid in the same tomb which was found empty two days later. The prominent mention of Mary Magdalene by all the Evangelists shows that she played an important rôle in these events. *The other Mary* was the mother of James and Joseph (cf. 56).

27, 62-66: Precautions of the Chief Priests. Only in Matthew. The First Gospel alone makes mention of the guard at Christ's tomb, not only here but also in *28,* 4. 11-15. For this reason most of the rationalist critics reject this part of the Resurrection story as a later legend. But these men overlook the fact that it was only the First Gospel, the Gospel written for the Christians of Palestine, that had the necessity of explaining and refuting the lying report that was spread among the Jews of Palestine by the chief priests (cf. *28,* 13. 15). **62.** *The Preparation,* i.e., for the Sabbath, was the ordinary name for Friday among the Jews. The corresponding Greek word, *Parasceve,* is still the regular name for Friday in the Greek Church, but in the Latin Church this name is now limited to Good Friday. *The day after the Preparation* is a rather remarkable way of saying simply "the Sabbath," but Matthew wishes at the same time to point out that Christ died on Friday, the "Preparation Day," a fact which all the other Evangelists had already mentioned in connection with the death or burial of Jesus (cf. Mark *15,* 42; Luke *23,* 54; John *19,* 31. 42). The Sanhedrists sought authorization from Pilate, for they themselves really had no right to interfere with a man's private tomb in this way. **63.** *That deceiver:* not only the Master but also His disciples are given the same opprobrious name (cf. *2* Cor. *6,* 8). Christ had foretold His resurrection on the third day to His enemies as well as to His disciples (cf. *12,* 40; John *2,* 19). The critics object that, if the disciples themselves did not understand Christ's prediction (cf. Mark *9,* 31; Luke *18,* 34; John *2,* 22), the chief priests would not have understood it. But the latter could have understood Christ's words just enough to make them forestall any false claims of the fulfillment of this prediction; they were taking no chances.   **64.** *The last imposture:* the first deception, according to these men, was when "that deceiver" said that He was the Messias and the Son of God. This last claim would be *worse:* they foresaw in a dim way the importance of the Resurrection.   **65.** *You have a guard:* Pilate does not give them the use of any Roman soldiers, but permits them to use the Jewish temple-police (see Commentary on *26,* 47). *As well as you know how:* the sense is, "Do with your own guard whatever you want." Pilate is tired of the whole sorry affair and does not want to become involved any further in it.   **66.** Cf. Dan. *6,* 17. Probably clay or wax, impressed with a special seal, was put between the stone and the tomb. This was an additional precaution, to prevent even the guard from conniving at any one opening the tomb.

3. The Resurrection of Jesus 28

28, 1-10: The Women at the Grave. Parallels in Mark *16,* 1-8 and Luke *24,* 1-11. John *20,* 1-18 gives an independent account of the events on Easter morning. All four Evangelists are in complete agreement on all important points. In certain minor details they show some differences which cannot be called discrepancies in the strict sense. Various solutions of these minor differences are offered: the Evangelists view the same event from different angles; at times they may not be referring to the very same event, e.g., the various women who come to the tomb; each Evangelist selects certain details for emphasis and omits others, according to his own special purpose; etc. There are so many plausible proposals for harmonizing these various details that it is difficult to say which is the best. No attempt at the harmonization of all these details is made here. See the Commentary on each of the Gospels for the details proper to each.

1. Since the Sabbath ended at sunset, the sense here is obviously, *late in the night* after *the Sabbath. The other Mary:* cf. *27,* 56. 61. *To see the sepulchre:* more specifically, they came to anoint the body of Jesus (Mark); but Matthew has already mentioned that the tomb was sealed and a guard set there, and therefore he prefers not to interrupt his narrative by explaining that these women knew nothing of all this.   **2-4.** Only in Matthew. If we had only the First Gospel, we might be led to believe that the women witnessed this sight; but it is clear from the other

Gospels that they did not. These events, which evidently occurred shortly before the arrival of the women, are told in this place by Matthew because he must explain how the soldiers and the seal on the entrance to the tomb were no hindrance to the women, when they discovered the empty tomb. Probably the original source of Matthew's information here was the true report of the guards. Note that none of the Evangelists speaks directly of the Resurrection as such. Our Lord's glorified body was not seen by any mortal eye as He left the grave. But it is generally supposed that the moment of His resurrection was simultaneous with the *great earthquake. Sat upon it:* more exactly according to the Greek, "sat above it," i.e., on the upper part of the sepulcher. **3.** Cf. *17*, 2b. **4.** *Like dead men:* but they soon recovered their senses and fled to the city, (cf. 11) before the women arrived at the tomb. **5.** *Even as he said:* cf. *12*, 40; *16*, 21; *17*, 22; *20*, 19. **6.** *See the place:* The angel points to the empty grave as evidence of the truth of his statement. **7.** *Behold, I have foretold it to you:* perhaps this is to be understood as a "sign," i.e., when the Apostles and the holy women see Jesus in Galilee, they will know that the angel spoke the truth. But Jesus Himself had told the Apostles, that He would go before them into Galilee (cf. *26*, 32). Perhaps we should read here in Matthew as it is in Mark, "*Behold,* He has *foretold it to you.*" **8.** *In fear and great joy:* overcome with awe, they were afraid at first to tell the vision to anyone (cf. Mark *16*, 8), but reflecting on the stupendous fact of their Lord's resurrection, their joy then made them hasten to bring the glad tidings to the Apostles (cf. Luke *24*, 9). **9.** *Jesus met them* while they were on their way to the Apostles. Unless we understand Matthew here as giving a summary account of several apparitions of Christ including the apparition to Mary Magdalene (John *20*, 14-17), we must assume that she was not with "the other Mary" now but that the latter was accompanied by "the other women" (Luke *24*, 10); see Commentary on Mark *16*, 9. They *embraced his feet* and no doubt kissed them; cf. our Lord's words to Mary Magdalene in John *20*, 17, which really mean "Stop embracing me." **10.** Christ repeats the command already given to the women by the angel (7); cf. 16 ff.

28, 11-15: The Guards and the Chief Priests. Only in Matthew; see Commentary on *27*, 62 ff. **11.** *While they were going,* i.e., while the women were on their way to bring the good

28, 12-17   St. Matthew

news to the Apostles, these guards had already arrived in the city and made their report; it is clear therefore that they had left the tomb before the women arrived there.   **12.** Most likely not a full, formal meeting of the Sanhedrin.   **13.** The precautions taken by the chief priests had worked out to their own disadvantage. The very fact that a guard had been set at the tomb could be used by the disciples of Christ in confirmation of the truth of His resurrection, for the presence of the guard made it impossible for the disciples to steal His body. The story invented by the chief priests involves the absurdity of witnesses testifying to a knowledge of what happened while they were asleep.   **14.** This cannot be taken as an argument that this guard consisted of *Roman* soldiers, for anyone could be prosecuted and punished for spreading calumny.   **15b.** These words can rightly be used as an argument to prove that this Gospel was written for the inhabitants of Palestine.

28, **16-20: Commission of the Apostles.** Only in Matthew; but cf. Mark *16*, 15 f. Following his own special plan, Matthew makes no reference to the apparitions of our Lord to the Apostles which occurred in Jerusalem; Luke on the contrary, according to *his* special scope, limits his account to the manifestations of Christ in Judea; John tells of Christ appearing to His disciples both in Jerusalem and in Galilee.   **16.** They *went into Galilee* in obedience to Christ's command (7. 10). We do not know what particular mountain this was, nor whether it was before His passion (cf. 26. 32) or after His resurrection (by means of the women), that *Jesus had directed them to go* there. The repeated commands about this meeting of Christ and His disciples show the great importance that He attached to it; this importance consisted undoubtedly in the great commission given then to the Apostles.   **17.** *Some doubted:* it is hardly possible that any of the eleven Apostles doubted the truth of Christ's resurrection at this time; perhaps Matthew is again summarizing and referring to the doubt of Thomas (John *20,* 24 f); or some doubted not the truth of the Resurrection but whether this was really Jesus, for His countenance seems to have changed somewhat after the Resurrection (cf. Luke *24,* 16; John *20,* 14; *21,* 4b); and perhaps He stood at first at a distance (cf. the next words, *And Jesus drew near*); or perhaps the *some* refers not to the Apostles but to other disciples who might have been with them, for many interpreters with great probability identify this apparition with the one mentioned in *1 Cor. 15,* 6 (it is hardly likely that "more than five hundred brethren" would have been found at this time anywhere outside of Galilee). **18.** *Has been given to me:* cf. *11,* 27; this power or authority (as the Greek word also means) Christ possessed, even as man, from the first moment of His incarnation, but He voluntarily abstained from using it until He had also merited it by His obedient sufferings; therefore this power was *given* to Christ by the Father at the Resurrection (cf. Phil. *2,* 9-11; Eph. *1,* 20-23). *In heaven and on earth:* over the invisible as well as the visible world. **19 f.** Substantially the same command is given in Mark *16,* 15 f: perhaps both Gospels are here giving in summary the same words of Christ; cf. also Luke *24,* 47; Acts *1,* 8. *Therefore* expresses the consequence: since Christ has this power, He communicates it, necessarily in a limited degree, to the Apostles and their successors, in order that His work may be continued to the end of time. For this purpose He established a Church embracing *all nations,* no longer limited to the Jews (cf. *10,* 5 f). Christ then lays down the conditions of membership in His Church: (a) faith in Him, implicitly contained in the word *disciples* and explicitly stated in Mark *16,* 16; (b) Baptism; (c) the observance of His commandments. *In the name of the Father,* etc.: from the beginning the Church has used these words as the "formula" of Baptism, i.e., the words to be recited while this rite of "washing" is being performed (so already in the "Didache," c. 100 A.D.). The Greek is literally "Into the name of the Father, etc."; this may mean "into the possession of . . . ," or "into union with . . . "; this is probably the meaning of the phrase, "to baptize in the name of Jesus Christ" (Acts *2,* 38; *8,* 16; *10,* 48; *19,* 5). These words also clearly teach the doctrine of the Blessed Trinity: one God *(the name,* in the singular, signifies the unity of nature) and three Persons, distinct but equal to one another. Finally this passage may be used as a proof of the absolute divinity of Christ: *the Son,* by which word Christ certainly means Himself, is placed on a par with *the Father* and *the Holy Spirit.* Although Christ is soon to deprive His disciples of His visible presence by His ascension into heaven, still, He remains with them in an invisible manner both by His presence in the Holy Eucharist and by the presence of the Holy Spirit, who is "the Spirit of Christ" (Rom. *8,* 9). With these words, not his own, but the Master's, St. Matthew ends his Gospel. A more magnificent conclusion cannot be found in any other book.

Louis Hartman, C.SS.R.   Mark Kennedy, O.F.M.

THE HOLY GOSPEL OF JESUS CHRIST ACCORDING TO ST. MARK

INTRODUCTION

Until modern times the Second Gospel was but little studied and commented upon, since almost all its narratives are found in Matthew and Luke. But modern critical studies now attach great importance to this Gospel. Most non-Catholic critics consider Mark to be the first written Gospel, one of the sources used by the authors of the First and of the Third Gospels, and consequently our most reliable guide in judging the nature and mission of the historical Jesus. For the evaluation of these opinions the reader should consult the article on the Literary Relations of the First Three Gospels.

St. Mark. The author of the Second Gospel was known either by the Hebrew name of John (Acts *13,* 5.13) or the Latin name of Mark (Acts *15,* 39; Col. *4,* 10; *2* Tim. *4,* 11; Philem. 24; *1* Pet. *5,* 13). In Acts *12,* 12.25; *15,* 37 he is called "John who was surnamed Mark." All of the above-mentioned passages speak of one and the same person. Few modern critics question this identification, though the Roman Martyrology distinguishes "Mark the Evangelist and the Apostle of Egypt" (April 25) from "John Mark, disciple and cousin of Barnabas" (September 27). Identifying them, we find in the texts quoted considerable information concerning the life of St. Mark. Although his cousin Barnabas was a native of Cyprus (cf. Acts *4,* 36), Mark was probably born in Jerusalem, for his mother Mary resided there. She seems to have been fairly well-to-do. Her home served as a place of assembly for the Christians of Jerusalem—possibly the "upper room" of the Last Supper and of the Descent of the Holy Spirit. In 44 A.D. Mark left Jerusalem for Antioch in the company of Paul and Barnabas (Acts *12,* 25). In the following year he accompanied them on the missionary journey through Cyprus, but upon reaching the mainland at Perge in Pamphylia he returned home to Jerusalem. He seems to have been considerably younger than Paul and Barnabas. Hence, it was probably the dread of further physical hardships rather than a dispute over methods or principles that induced their assistant to abandon the missionaries. In any case Paul was greatly offended by Mark's desertion. In 50 A.D. Mark was again in Antioch, probably having returned with Barnabas after the Council of Jerusalem. In the same year he accompanied his cousin on a second missionary journey in Cyprus. Thereafter we lose sight of him for a few years. We find him, however, in Rome during both of St. Paul's imprisonments in that city (61-63, 66-67 A.D.). During this same period he assisted St. Peter, who, writing from Rome, calls Mark his "son." This may be either a term of affection or may signify that Peter had given him spiritual birth in Christ through baptism.

The Author. That St. Mark wrote the Second Gospel should be beyond question. Some fifty years ago the theory of a "proto-Mark," a writing upon which the present Mark was thought to be based, was current; but today the Marcan authorship of this Gospel is supported not only by Catholic scholars but by many non-Catholic critics. Tradition has never associated the authorship of the Second Gospel with anyone but Mark. Such ancient writers as Papias, Irenaeus, Clement of Alexandria, Origen, and Tertullian explicitly attest the Marcan authorship. Early Christian literature has much implicit testimony to its existence. If we read the Gospel itself, we soon see that the author was a Palestinian Jew, familiar with Palestine and with the customs and institutions of its inhabitants. His own Semitic mentality is plainly reflected in his thought and language. All this fits Mark admirably.

The oldest and by far the most important testimony concerning the origin of St. Mark's Gospel is the statement of Papias (c. 120 A.D.) which has been preserved for us by the Church-historian Eusebius (*Hist. Eccles.* III. 29, 15). The immense importance of this testimony lies in the fact that Papias himself is merely quoting the words of "the Elder," who is commonly believed to be none other than St. John the Apostle. The statement of Papias is as follows: "This also the Elder used to say, 'Mark, having become the interpreter of Peter, wrote down accurately, although not in an orderly arrangement, the sayings and deeds of the Lord, as far as he recalled them. For he himself neither heard the Lord nor was he His follower, but he was later on, as I have said, a follower of Peter. The latter used to deliver his instructions as circumstances required, but not like one who draws up an orderly arrangement of our Lord's activity. Hence Mark did nothing wrong in thus writing down certain things as he remembered them. For his sole purpose was to omit nothing of what he had heard and to falsify nothing in recording this.' "

From these words we might rightly conclude, not only that Mark wrote an account of Christ's ministry according to the preaching of St. Peter, but also that this account is identical with our Second Gospel. For (a) this Gospel lacks the "orderly arrangement," i.e., the artificial disposition of material according to subject-matter, that is seen so clearly in the First Gospel and in part also in the Third; (b) the choice of subject-matter and its general chronological arrangement are exactly the same in this Gospel, which was intended primarily for Gentile Christians, as they are in Peter's discourse to the Gentile centurion Cornelius (Acts *10*, 36-43); (c) the many graphic details that are recorded in this Gospel come from one who was an eye-witness of these scenes; (d) the most distinctively Petrine touch, a tribute to the humility of Peter's preaching, is seen in the fact that in this Gospel such things as might redound to his honor are lightly passed over by Mark, whereas those of little or no credit to him are mentioned, even where the other Evangelists omit them. Considering all these facts, we are not surprised to learn that St. Justin (c. 150 A.D.) called this Gospel simply "the Memoirs of St. Peter."

Nature and Characteristics. Since Mark's main purpose in writing was merely to give a faithful reproduction of St. Peter's preaching, his Gospel is by far the most primitive of all, even though the Aramaic Gospel of St. Matthew was written at a somewhat earlier period. Mark's language is the ordinary *koine* Greek of every-day life as spoken by the Jews of the Diaspora. He does not make the slightest pretense at composing a piece of "fine literature." Yet there is a certain great charm and beauty in this very simplicity. His style is direct, vigorous and characterized by a realism that one would expect to find in a primitive narrative. The uneven structure of the sentences, the frequent use of the historical present, the vividness and graphic touches that are seen throughout, all show that this is essentially an oral narrative of an eye-witness. Nevertheless, we should not conclude from this that St. Mark acted merely as an amanuensis who mechanically records the speeches of another. Mark and not Peter is the true author of this Gospel; he alone was inspired by the Holy Spirit to write down this story which Peter had told over and over again to the first Christians.

Purpose of the Gospel. From the nature and origin of this Gospel, it follows that Mark's scope and purpose are identical with Peter's scope and purpose in preaching these truths, i.e., to show that Jesus of Nazareth is the Christ, the Son of God. For this reason Mark emphasizes mainly the deeds and miracles of Christ to prove His divine mission. No other Gospel so forcefully depicts the divine power of Jesus, which is shown especially in His power over demons. But at the same time no other Gospel gives us such a vivid picture of the true humanity of Christ, for Mark is pre-eminently the historian of the earthly life of Jesus. Since an account of Christ's teachings would distract somewhat from this main purpose, Mark records very little of the words of Christ—not much more than a few parables and a part of the Eschatalogical Discourse. Hence his Gospel is much shorter than the others, even though it is generally longer and more complete than Matthew and Luke in the narratives which he has in common with them.

Circumstances of its Composition. Early tradition and modern criticism are agreed on this, that the Second Gospel was written before the Third. But it is certain that St. Luke wrote his Gospel before he composed the Acts of the Apostles, i.e., some time prior to 63 A.D. On the other hand we know that Mark did not write his Gospel until he had lived for some time with St. Peter at Rome, to which city St. Peter may have gone in 42 A.D. (cf. Acts *12*, 17).

The Second Gospel was written at Rome and intended primarily for the Christians of that city. Early tradition attests that it was written at the express request of the faithful of Rome. These consisted principally of Greek-speaking Gentile converts. The internal evidence confirms the truth of this tradition. For (a) Mark's language is typical of the Greek spoken at Rome, as its Latinisms and even words borrowed directly from Latin prove; its Semitisms and few Aramaic words are relics from the original oral Gospel of St. Peter; (b) Mark directs his Gospel principally to Gentile readers for whom he often explains Jewish customs and institutions.

Integrity. The end of the Second Gospel (chapter *16*) has come down to us in a fourfold form. (a) In the two best Greek MSS, Codex Vaticanus and Codex Sinaiticus, and in a few MSS of the oldest versions the Gospel ends with 8. (b) A few MSS give after 8 a short conclusion of about thirty words. (c) The Vulgate together with the vast majority of Greek MSS give after 8 the conclusion which we have in our text (9-20). (d) A reading known to St. Jerome, and found also in the recently discovered Washington Codex, has the same conclusion as (c) but also inserts in 15 before the words "Go into the whole world" a rather long statement of Christ about the overthrow of Satan's power.

What is to be said of each of these four readings? Most literary critics admit that Mark did not intend his Gospel to end with 8, and hence that the MSS which end here, form (a), are based upon an early copy in which the original conclusion of Mark's Gospel was accidentally lost. The short conclusion of form (b) is generally conceded to be a later composition of some scribe who noticed that the Gospel could not end with 8. The insertion of form (d) has all the characteristics of being an apocryphal addition. Form (c) is very well attested in the MSS. Since it forms a substantial part of the Vulgate, Catholics hold that this conclusion is a part of the canonical and inspired Scriptures. The view that this conclusion was added by some inspired author other than St. Mark after the original conclusion of his Gospel was lost cannot be demonstrated, according to the Pontifical Biblical Commission in its decision of June 26, 1912.

COMMENTARY

Note. Only those words and passages which have no parallel in the First Gospel are commented upon here. For the rest the reader should consult the Commentary on the parallel passages in Matthew.

I: THE PUBLIC MINISTRY OF JESUS 1-13

1. PREPARATION FOR THE PUBLIC MINISTRY 1, 1-13

1, 1-8: John the Baptist. V.1 is generally taken as the title of this Book. But many interpreters, following Origen and Basil, consider 2 f as a parenthesis and join 1 with 4 despite the grammatical difficulty in getting the sense, "The beginning of the Gospel of Jesus Christ was the preaching of John in the desert." *The Gospel of Jesus Christ* here means "the good news concerning Jesus Christ." *The Son of God:* see note to text. These words are wrongly omitted here in some critical editions on the authority of a few ancient MSS.; cf. Lagrange *ad loc.*   2-8. Parallels in Matt. *3*, 1-6. 11 and Luke *3*, 2b-4. 16. The first part of this quotation (2) is from Malachias but Mark refers all these words to Isaias, the better known prophet. The Second Gospel alone cites Mal. *3*, 1 in connection with the Evangelists' account of the activity of the Baptist, but in Matt. *11*, 10, and Luke *7*, 27 Christ Himself quotes these words in reference to His Precursor.

1, 9-11: The Baptism of Jesus. Parallels in Matt. *3*, 13. 16 f and Luke *3*, 21 f. *The heavens opened:* the Greek verb in Mark is different from that used in Matthew and Luke; here literally, "split open, rent asunder." Perhaps the sky had been completely overcast with clouds which now suddenly parted.

1, 12-13: The Temptation. Parallels in Matt. *4*, 1. 11b and Luke *4*, 1 f. Mark does not mention the three particular temptations, but he alone speaks of Christ being *with the wild beasts*. This item is mentioned in order to show what a desolate region Jesus was then in. The words *and forty nights* were inserted here by an early copyist from the First Gospel.

2. Inauguration of the Ministry in Galilee 1, 14 – 3, 19

1, 14-15: In Galilee. Parallels in Matt. *4*, 12. 17 and Luke *4*, 14a; cf. also John *4*, 1-3.

1, 16-22: The First Disciples Called. Parallel in Matt. *4*, 18-22; cf. also Luke *5*, 1-11. The only item peculiar to Mark is that Zebedee was also assisted by *hired men*.   **21.** Jesus teaches at Capharnaum. Parallel in Luke *4*, 31; cf. also Matt. *4*, 13 and John *2*, 12.   **22.** Parallels in Matt. *7*, 28 f and Luke *4*, 32. *Astonished* is a favorite word of Mark; the same Greek word occurs in *6*, 2; *7*, 37; *10*, 26; *11*, 18.

1, 23-28: The Cure of a Demoniac. Parallel in Luke *4*, 33-37. **24.** Literally, "What is to us and to thee"; the same idiomatic phrase commonly expressing dissent occurs also in *5*, 7; Matt. *8*, 29; John *2*, 4 (which see); and often in the Old Testament. *The Holy One of God:* the demon recognized, not indeed the divinity of Christ, but His special relationship to God and perhaps also His Messiasship.   **25.** *Hold thy peace:* literally, "Be muzzled." For the reasons why Jesus did not wish to be recognized as the Messias so early in the ministry, see Commentary on Matt. *8*, 4.   **26.** *Convulsing him:* this physical violence was without permanent effect, for the devil "went out of him without harming him at all" (Luke).

1, 29-31: Peter's Mother-in-law. Parallels in Matt. *8*, 14 f, and Luke *4*, 38.

1, 32-39: Other Miracles.   **32-34.** Parallels in Matt. *8*, 16 and Luke *4*, 40 f. Mark alone adds the graphic touch that *the whole town had gathered together at the door*, which, however, need not be taken too literally.   **35-39.** Jesus preaches and works cures in the synagogues of Galilee. Parallels in Matt. *4*, 23 and Luke *4*, 42-44. Mark alone speaks of Jesus praying on this occasion and of *Simon* finding Him; note the Petrine touch.

1, 40-45: A Leper. Parallels in Matt. *8*, 1-4 and Luke *5*, 12-16. **45.** Mark alone records the fact that this leper did not obey Christ's injunction to keep the miracle a secret. The cured man probably thought that Jesus desired the secrecy merely out of humility.

2, 1-12: A Paralytic at Capharnaum. Parallels in Matt. *9,* 2-8 and Luke 5, 17-26. **4.** *They stripped off the roof:* the flat roof of this house was covered with "tiles" (Luke), i.e., large slabs of brick or stone, which could be removed without causing too much inconvenience to the people who were with Christ in the room immediately below.

2, 13-17: The Call of Levi. Parallels in Matt. *9,* 9-13 and Luke 5, 27-32. 13 is peculiar to Mark.

2, 18-22: The Question of Fasting. Parallels in Matt. *9,* 14-17 and Luke 5, 33-38. **18.** *Were fasting:* i.e., "were accustomed to fast" on certain days; an explanation necessary for Mark's Gentile readers.

2, 23-28: The Disciples Pluck Grain on the Sabbath. Parallels in Matt. *12,* 1-8 and Luke *6,* 1-5. **26.** *When Abiathar was high priest:* these words of Christ are found only in Mark and cause some difficulty, since, according to *1* Kgs. *21,* 1 ff, Achimelech was high priest at this time. See note to text. **27.** Mark alone records this important principle which Christ enunciates here: the positive laws of God are intended for man's good; therefore, if under certain circumstances they conflict with man's greater good, they are no longer obligatory.

3, 1-6: A Man with a Withered Hand. Parallels in Matt. *12,* 9-14 and Luke *6,* 6-11. **4.** To omit a good deed that can be done and that should be done, at least from an obligation in charity, is virtually the same as *to do evil.* **5.** Mark alone mentions Christ's *anger* on this occasion. *The blindness of their heart:* cf. *6,* 52; *8,* 17; literally, "the callousness of their heart," i.e., moral insensibility, spiritual stupidity, the failure to perceive true spiritual values. *With the Herodians:* cf. Matt. *22,* 16.

3, 7-12: The Mercy of Jesus. Parallels in Matt. *4,* 24 f and Luke *6,* 17-19; cf. also Matt. *12,* 15 f. In Matthew and Luke this description of the crowds that followed Jesus serves as an introduction to the Sermon on the Mount, but in Mark the scene is at the shore of the Sea of Galilee. **8.** *Idumea:* the only place in the New Testament where this region is mentioned. See Glossary at end of text.

3, 13-19: The Choice of the Twelve. Paralleis in Matt. *10*, 1-4 and Luke *6*, 12-16.   **17.** This parenthesis is peculiar to Mark. *Boanerges* should really be pronounced "Bwanērges," since it represents in Greek letters the Hebrew or Aramaic phrase "B'nē-regesh." *Sons of Thunder* is a Semitism meaning "the Thunderers." The sons of Zebedee won this epithet from their vehement disposition or from their eloquence. Some interpreters see a connection between this title and the incident mentioned in Luke *9*, 54. But it is to be noted that the Semitic word *regesh*, which means "thunder" in Arabic, is used in Hebrew and Aramaic only in the figurative sense of "uproar, tumult." Whatever the exact explanation of this title may be, it certainly stands in sharp contradiction with the false picture that is so often drawn of St. John the Apostle, as if he were almost an effeminate character.

3. SECOND PERIOD OF THE MINISTRY IN GALILEE AND ACROSS ITS LAKE *3*, 20 — *7*, 23

3, 20-30: Blasphemy of the Scribes.   **20 f.** These two verses are peculiar to Mark and their interpretation is very uncertain. One interpretation is that which is given in our translation and explained in the foot-note. This explanation presupposes that there is some connection between these verses and 31 ff; but this is not certain. *They went out:* from where? Hardly from Nazareth, which is not mentioned in the context. Moreover, *his own people* is a very dubious translation of the Greek which means literally "those by Him," i.e., those who were with Him in the house. Finally, it is to be noted that the Greek word for *crowd* is masculine, singular; therefore the *him* and *He* may just as well be translated as "it," referring to the *crowd*. Hence it is possible to translate 21 as follows: "But when they who were with Him (in the house) heard it, (i.e., the enthusiasm of the crowd outside), they went out (of the house) to lay hold of it (the crowd, to keep it in check), for they said, 'It (the crowd) has gone mad.'" This is the interpretation of the group of ancient MSS known as the "Western Family" of MSS.   **22-30.** Parallels in Matt. *12*, 24-29 and Luke *11*, 15. 17-22; *12*, 10.

3, 31-35: Jesus and His Brethren. Parallels in Matt. *12*, 46-50 and Luke *8*, 19-21.

4, 1-20: Parable of the Sower. Parallels in Matt. *13*, 1-15. 18-23 and Luke *8*, 4-15. Only 13 is peculiar to Mark; the other Evangelists suppressed this reference to the Apostles' lack of spiritual understanding.

4, 21-25: Purpose of This Teaching. Parallel in Luke *8*, 16-18. This is a group of proverbial sayings which Christ apparently uttered on various occasions. Some of these sayings have a somewhat different meaning when joined to other contexts. **21.** The sense of these words here is: "Just as a lamp is intended to give light, so these parables are intended to be understood and give spiritual light." For the same words used with other applications, cf. Matt. *5*, 15; Luke *11*, 33. **22.** The same thought as in 21. For the same words with different connotations, cf. Matt. *10*, 26; Luke *12*, 2. **24b.** *With what measure . . . :* Mark alone has these words in connection with the parables; here the sense seems to be, "Be generous in sharing these spiritual truths with others," or, "Your spiritual profit will be in proportion to the understanding and practical application that you give to these truths." Cf. the same words with different meaning in Matt. *7*, 2b; and Luke *6*, 38b. **25.** Cf. also Matt. *13*, 12; *25*, 29; Luke *19*, 26.

4, 26-29: Seed Grows of Itself. This is the only parable that is found solely in the Second Gospel. Commentators are not agreed on what our Lord intended as the principal lesson of this little parable. By the *seed cast into the earth* is certainly meant the word of God, or divine grace, which is given to man. But beyond that, this parable should not be interpreted allegorically; for there is no comparison here between the sower of the seed and God, since we may not say that "the seed grows without *His* knowing it." It would not be right to say that the soul "of itself" has the power to give increase to the seed of God's grace just as *of itself the earth bears the crop.* Nothing in this parable speaks of the seed having the power to grow of itself. The emphasis of the parable seems to be on the *gradual* but *persistent* growth of the seed. A farmer cannot stand and watch his seed growing, but the *seed grows without his knowing it,* as he goes about his daily work, sleeping and rising. But the growth of the seed is steady: *first the blade, then the ear, then the full grain in the ear,* and finally *the harvest.* So also the growth of grace, the growth of *the kingdom of God.* It may be hardly perceptible to the disciples of Christ, nevertheless that growth is always there, a slow but steady development, until the Kingdom of God comes to full maturity at *the harvest* which is the end of the world. The moral of the parable: patience and trust in God!

4, 30-34: The Mustard Seed.   30-32. Parallels in Matt. *13,* 31 f and Luke *13,* 18 f.   **33 f.** Parallel in Matt. *13,* 34.   **34b.** Only in Mark.

4, 35-40: The Storm on the Lake. Parallels in Matt. *8,* 18. 23-27 and Luke *8,* 22-25. Mark gives more details than the other Synoptics.   **39.** *Peace, be still:* literally in Greek, *shut up, be muzzled;* these are words that one would use to a barking dog.

5, 1-20: Expulsion of the Devils in Gerasa. Parallels in Matt. *8,* 28-34 and Luke *8,* 26-39.   **9.** *Legion* was the designation of the basic division of the Roman army; the Latin word is used here even in the Greek text; evidently at this time the word had come to signify "a large group."   **13.** Mark alone, in his love of details, gives the number of the swine.   **18-20.** At this time the Decapolis (see Glossary at end of text, p. 747, and Map) was inhabited almost entirely by Greek-speaking Gentiles. Hence the cured demoniac also was probably a pagan. This explains why Christ did not permit him to become one of His intimate followers during the public ministry; but He did allow him to publish the fact of his cure among his own people and so, in a sense, to become the first Christian missionary of the Decapolis. This permission was contrary to our Lord's practice in Galilee, for there was no danger that the pagans of the Decapolis would proclaim Jesus the political Messias of the Jews.

5, 21-43: The Daughter of Jairus; the Woman with a Hemorrhage. Parallels in Matt. *9,* 1. 18-26 and Luke *8,* 40-56.   **30.** *Perceiving in himself:* Jesus knew not only by His supernatural knowledge but also by His natural, experimental knowledge, that He had worked this miracle; for the miraculous power flowed in a certain mysterious way from His human nature, so that He could *perceive,* i.e., feel even in His body, that He was working a miracle.   **31 f.** Although Jesus knew by His supernatural knowledge who it was whom He had cured, nevertheless, according to the conditions that He voluntarily imposed upon Himself during His mortal life (cf. Phil, *2,* 5-8), He normally acted solely from His experimental knowledge like other men.   **37.** The same three Apostles who were to witness the Transfiguration and the Agony in the Garden.   **41.** Mark alone (as he learnt from St. Peter, of course) has preserved for us the very words which Jesus spoke in Aramaic. *Talithá* is the usual Aramaic word for "girl." But the form *cūmi* is really Hebrew, for no Aramaic dialect at this period had preserved in pronunciation the archaic feminine ending of the imperative; the best Greek MSS give the correct Aramaic form *cūm*, "arise."   **42.** A rather peculiar place to mention the girl's age. Luke gives it in a more natural place at the beginning of the narrative. But lest the reader think that the revived girl was but an infant and that there was also something supernatural in her walking, Mark immediately adds her age here; the Greek reads, *"For* she was twelve years old." **43b.** Our Lord did not work miracles unnecessarily. Therefore, in order to keep the girl alive, He *directed that something be given her to eat.*

6, 1-6a: Jesus at Nazareth. Parallel in Matt. *13,* 53-58 and partial parallel in Luke *4,* 16-30.   **3.** *Is not this the carpenter?* This is the only place in the Gospels from which we learn that Jesus followed the trade of His foster-father. *The son of Mary* implies that Joseph was already dead. For these relatives of Jesus see Commentary on Matt. *12,* 46.

6, 6b-13: The Mission of the Apostles. Parallels in Matt. *9,* 35a; *10,* 1. 9. 11. 14 and Luke *9,* 1-6; cf. also Luke *10,* 4. 10 f.   **7.** Mark alone mentions that the Apostles were sent forth *two by two,* although Luke says the same thing of the seventy-two disciples.   **13.** They *anointed many sick people with oil:* peculiar to Mark; this anointing was not the sacrament of Extreme Unction but merely a symbolic ceremony insinuating, however, the sacrament.

6, 14-29: Death of the Baptist.   14-16. Herod thinks that Jesus is John the Baptist come back to life. Parallels in Matt. *14,* 1 f and Luke *9,* 7-9.   **17-20.** The imprisonment of John. Parallels in Matt. *14,* 3-5 and Luke *3,* 19 f.   **19.** *Laid snares for him:* the Greek means "had it in for him."   **20.** *He did many things:* thus the Latin with many Greek MSS; but this is hardly intelligible in the context; the best Greek MSS read, "He was much disturbed."   **21-29.** The Baptist is beheaded at

Herod's birthday-party. Parallel in Matt. *14,* 6-12. **21.** *A favorable day:* i.e., a day which gave Herodias her opportunity. *Officials, tribunes and chief men of Galilee:* in modern terms, "the nobility, army officers and the local aristocracy."

6, 30-33: Return of the Disciples. Parallels in Matt. *14,* 13 and Luke *9,* 10-11a.

6, 34-44: Jesus Feeds Five Thousand. Parallels in Matt. *14,* 14-21; Luke *9,* 11b-17; and John *6,* 1-15. **34.** *Because they were like sheep without a shepherd:* peculiar to Mark in this context, but cf. Matt. *9, 36.* **37.** Mark, in his fondness for details, is the only one of the Synoptics to mention the Apostles' estimate of the amount of bread required, but the Fourth Gospel agrees with the Second on this point. **39 f.** Mark's original expressions are much more picturesque. The first word for *groups* (39) signifies properly "groups that eat and drink together"; the second word for *groups* (40) means "garden-plots." The Greek is literally, "And he directed them to make all the people recline by banquet-parties and banquet-parties upon the green grass. And they lay down like flower-beds by flower-beds, some of a hundred each and some of fifty each."

6, 45-52: Jesus Walks on the Water. Parallels in Matt. *14,* 22-27. 32 f and John *6,* 16-21. **48.** *And he would have passed by them:* literally, "And he was willing to pass by them," i.e., He acted as if He intended to pass by them. **50.** *For they all saw him:* hence they could eliminate the possibility that one or the other was having a merely subjective illusion. **51 f.** Matt. *14,* 33 records their cry of *astonishment.* Mark explains here why they were so surprised and astonished: because *they had not understood about the loaves,* i.e., because they had not drawn the proper conclusion from the miraculous multiplication of the loaves regarding the nature of Jesus. *Their heart was blinded:* their spiritual insight had been dull; see Commentary on *3, 5.*

6, 53-56: Other Miracles. Parallel in Matt. *14,* 34-36.

7, 1-23: Jesus and the Pharisees. Parallel in Matt. *15,* 1-20. **2-4.** Peculiar to Mark, who felt that an explanation of these Jewish customs was necessary for his Gentile readers. **2.** *Defiled hands:* literally "common hands"; cf. Acts *10,* 14 f where the same Greek word is used; "common" is used here in the sense of "profane," i.e., "non-kosher." **3.** *Without frequent washing of hands* represents the reading of only a few Greek MSS and is hardly correct, since there is no evidence that even the strictest Pharisees demanded a *frequent* washing of hands before each meal; the reading of almost all the Greek MSS is "unless they wash their hands with the fist," the sense of which is uncertain. **4.** When they come from *the market:* the original is equivalent to our modern expression "from business." *Without washing first . . . washing of cups:* etc.: literally, "unless they are baptized . . . baptisms of cups," etc. This use of the word "baptize" in its original sense of "to dip into water" shows the primitive character of this Gospel: this word had not yet become limited to its technical, religious meaning. Normally it was sufficient to pour a little water on the hands before eating (3), but after returning "from business," where contact with Gentiles could not well be avoided, the hands had to be "dipped" into water.

4. Ministry Mostly in the Regions Bordering on Galilee 7, 24 – 9, 49

7, 24-30: The Canaanite Woman. Parallel in Matt. *15,* 21-28. **30.** *Lying upon the bed:* i.e., resting peacefully but still weak.

7, 31-37: Healing of a Deaf-Mute. This miracle is peculiar to Mark. But Matt. *15,* 29-31 mentions the same journey and speaks in general of numerous miracles of our Lord in this region. **32.** *Dumb:* the Greek means literally "speaking with difficulty," i.e., with an impediment in his speech, which may be the sense here, since this man, when cured, "began to speak *correctly*" (35); but this rare word seems to have been used in the Greek of this period (e.g., in the Septuagint of Isa. *35,* 6) simply in the sense of *dumb.* **33.** Jesus took the deaf-mute *aside from the crowd* in order to keep the miracle a secret (cf. 36); for the reason of this secrecy, see Commentary on Matt. *8,* 4. The elaborate means used by Christ in working this miracle should not make one think that it cost Him a certain effort. The reason for the unusual procedure in the present case seems to be that Jesus required a certain amount of faith on the part of the recipients of His miracles (cf. *6,* 5 f); but this man had been brought by others and had given no signs of faith; since he was deaf, this sign-language was necessary in order that our Lord could inform him that, if he believed, He would open his ears and loose the bond of his tongue. **34.** On Christ's prayer to His heavenly Father before working a miracle, cf. John *11,* 41. *He sighed:* i.e., in prayer; for the sighs or "groans" in the prayers of Christians cf. Rom. *8,* 26, where the same Greek word is used. Mark has again preserved for us one of the very words used by our Lord in speaking Aramaic (cf. *5,* 41); in standard Aramaic the form would be more like *Ethphetach;* the form *Ephpheta* may represent its pronunciation in the Galilean dialect, or more likely be a corruption due to Greek-speaking Christians in handing down this Aramaic word in the early oral catechesis. **37b.** The omission of the article in Greek before the word *dumb* gives the sense, "He has made the deaf-mutes both hear and speak."

***8,* 1-9: Jesus Feeds Four Thousand.** Parallel in Matt. *15,* 32-38.

***8,* 10-12: The Pharisees Ask a Sign.** Parallel in Matt. *15,* 39; *16,* 4; cf. also Luke *12,* 54-56, and the earlier occasion when a sign was demanded (Matt. *12,* 38-40; Luke *11,* 16. 29 f). **11.** *The Pharisees came forth:* perhaps the sense is, "out of their own region"; if so, then *the district of Dalmanutha* is not to be sought in Galilee. **12.** The *sighing deeply in spirit* signifies here Christ's weariness with this unbelieving generaton. Mark's account is important in making Christ's refusal of a sign absolute; see Commentary on Matt. *12,* 39.

***8,* 13-21: The Leaven of the Pharisees.** Parallel in Matt. *16,* 5-12.

***8,* 22-26: A Blind Man at Bethsaida.** Peculiar to Mark. **22.** On the site of Bethsaida see Commentary on Matt. *14,* 13. Jesus and the Apostles were now on their way north from Decapolis to Cæsarea Philippi. **23.** As in the cure of the deaf-mute (7, 31 ff), so also here the miracle is performed *outside the village* in order to keep it secret (cf. 26). Some commentators explain the use of spittle in this case as a means of increasing the faith of the blind man; the cure would then be gradual because the man's faith was at first imperfect. Other commentators see both in the use of spittle and in the gradual nature of the cure means used by our Lord in order to conceal the miraculous nature of the cure. It was commonly believed at that time that spittle was an effective remedy, especially if it were the spittle of one who had not yet eaten on that day. **24.** The man, not blind from birth, recognized these objects but vaguely. **25.** The translation of the Vulgate, *and he began to see,* is rather surprising in view of the fact that the text states that the man had already partially recovered his sight; the meaning of the Greek is, "and he looked straight ahead." **26.** There are several varieties of readings for this verse; the best MSS read, "And he sent him into his house, saying, 'Do not go into the village.'"

8, 27-30: Peter's Confession. Parallels in Matt. *16,* 13-20 and Luke *9,* 18-21.

8, 31-33: Passion and Resurrection Foretold. Parallels in Matt. *16,* 21-23 and Luke *9,* 22.

8, 34-39: The Doctrine of the Cross. Parallels in Matt. *16,* 24-28 and Luke *9,* 23-27. **38.** Cf. Matt. *10, 33.* **39.** The full expression, *the kingdom of God coming in power* is peculiar to Mark and is best understood in the sense of "the manifestation of divine power in the establishing of God's Kingdom on earth," such as would be visible even during the lifetime of the Apostles.

9, 1-7: Jesus Transfigured. Parallels in Matt. *17,* 1-8 and Luke *9,* 28-36. **3.** The words *as snow* are not in the best Greek MSS. **7.** The sense is not that the vision departed *suddenly* but that the three Apostles, upon *looking around,* i.e., lifting up their faces from the ground (cf. Matt.), saw "at once" that the vision had meanwhile vanished.

9, 8-12: On the Coming of Elias. Parallel in Matt. *17,* 9-13. The sequence of thought in this conversation between Christ and the Apostles, as recorded here, may not at first sight seem quite clear. But it is necessary to remember that these Apostles had two difficulties. (a) Christ spoke of His resurrection from the dead, thereby implying that He must die before He entered into His glory; but why should this be so, if they had already seen Him in His glory? (b) The sight of Elias at the Transfiguration had reminded them that the Scribes taught *that Elias must come first,* i.e., before the coming of the Messias, yet now it seemed to them that Elias had come at best only after the coming of Christ. Although the Apostles asked Jesus directly only about the second difficulty, our Lord answered both difficulties at the same time, since both were intimately connected and both were caused by the false interpretation which the Scribes gave concerning the Messias and His Precursor.   **9.** The original is, "And they kept the word to themselves"; but "the word" does not mean here *what he said* but is a Semitism meaning "the thing, the affair," i.e., the vision that they had seen (cf. Luke *9, 36*b). Therefore, instead of talking about the Transfiguration, "they discussed with one another, (saying,) 'What is this "rising from the dead"?' " (this is the literal translation of Mark's Greek text).   **11 f.** Since Mark no doubt gives only a summary of Christ's words, the sense will be clearer for us if we join the first sentence, *Elias is to come first and will restore all things* immediately with the third sentence, *But I say to you,* etc. This is the order in the First Gospel. On the meaning of this passage, see Commentary on Matthew. *How then is it written:* our Lord is undoubtedly referring to Isa. *53. As it is written of him:* Christ may be referring to the persecution of Elias by Jezabel (*3* Kgs. *19*) considered as a type of the persecution of the Baptist by Herodias.

9, 13-28: A Possessed Boy. Parallels in Matt. *17,* 14-20 and Luke *9,* 37-44a. Mark's account is much longer and more detailed.   **20-23.** This conversation between Jesus and the father of the boy is peculiar to Mark.   **22.** The Greek reads, "This 'If thou canst'! All things are possible to him who believes."   **23.** *Help my unbelief:* the sense is, "Supply what is still lacking to my faith."   **24.** Apparently the possessed boy, besides being an epileptic, was also a deaf-mute (cf. 16); Mark alone mentions this aspect of the effects of his diabolical possession.   **25 f.** We should not consider this an additional miracle, as if the boy had been dead and was brought back to life by our Lord. The last violent effort of the devil had so exhausted the boy that he collapsed and lay in an unconscious and motionless state *like one dead.*   **28.** A few of the best MSS omit the words *and fasting.* Since prayer and fasting are often mentioned together (cf. Luke *2,* 37; Acts *14,* 22), possibly some early copyist added these words here, just as they are added in most Greek MSS in *1* Cor. *7,* 5. According to Matthew, the Apostles were unable to drive out this devil because of their "little faith." These two reasons are, of course, not mutually exclusive.

9, 29-48   St. Mark

9, 29-31: The Second Prediction of the Passion. Parallels in Matt. *17*, 21 f and Luke *9*, 44b-45.   **30.** The best Greek MSS here, as always in Mark, have "after three days"; but this is synonymous with Matthew's "on the third day."

9, 32-40: Against Ambition and Envy.   **32-36.** Parallels in Matt. *18*, 46-48; cf. also Matt. *20*, 20-28; *23*, 11; Luke *22*, 24-30. **37-39.** Parallel in Luke *9*, 49 f.   **37.** St. John the Apostle is not named alone elsewhere in the Synoptic Gospels. This incident well illustrates his fiery temperament that may have won for him the title of "Son of Thunder" (*3*, 17). *A man who was not one of our followers:* i.e., not one of our group; but this man clearly believed in Jesus, for he would not have succeeded in driving out the devils in His name unless he believed in Him (cf. Acts *19*, 13-16).   **38.** Christ rebukes John for his intolerance. Even though this man did not belong to the "group," he cannot be opposed to Christ if he has such good dispositions; and every one who in such or similar circumstances is not opposed to Christ is really for Him. This lesson given to John has permanent value for us also; we must be intolerant of error but tolerant of those who in good faith are outside the body of the Church: in as far as they do good work, we should encourage rather than hinder them.   **40.** Cf. Matt. *10*, 42. Christ returns to His main theme of doing good *in my name* (so the phrase is in 37 also, where our text has *for my sake*); John had interrupted our Lord's discourse because this phrase had recalled to his mind the man who cast out devils *in thy name.*

9, 41-49: Avoiding Scandal.   **41-46.** Parallel in Matt. *18*, 6-9; cf. also Luke *17*, 1 f and Matt. *5*, 30 f.   **43. 45. 47.** This Quotation from Isa. *66*, 24 is peculiar to Mark, but in the best Greek MSS it occurs only once, i.e., in 47. The figure of speech is taken from the custom of dumping the refuse of Jerusalem in "Gehenna," i.e., the valley of Hinnom south of the city, translated *hell* in our text, where it was burnt. Since this *fire* burnt on indefinitely, while the unburnt material was always full of worms and maggots, these things became symbols of the eternal punishment of the wicked; cf. also Jud. *16*, 21; Ecclus. *7*, 19. Since "Gehenna" (42. 44. 46) is contrasted here with eternal *life,* it is correctly translated by the word *hell.*   **48 f.** A very difficult passage which some commentators consider as a mere collection of various sayings of Christ in which the word *salt* occurs, without other logical connection between them; these sayings, according to this theory, were added in the early oral catechesis to the preceding words of our Lord solely because the word *fire* occurs in the first of them.   **48.** The words *For everyone shall be salted with fire* are peculiar to Mark. If understood with the preceding context, the *fire* would be the eternal fire of hell, which unlike the fires on earth, does not destroy but on the contrary has a preservative effect like salt. Other commentators understand *fire* here of the fire of tribulation and temptation by which the just on earth are purified from the dross of their faults and preserved unto life everlasting, just as salt purifies and preserves. The words *and every victim shall be salted* are missing in the best Greek MSS and are probably a gloss added by some early copyist who took these words from Lev. *2,* 13.   **49a.** The same words occur in Luke *14,* 34 with no apparent logical connection with the context. The similar words in Matt. *5,* 13 refer to the power of good example. Perhaps that is the meaning here also, since the preceding context is on the avoidance of scandal. **49b.** *Have salt in yourselves:* etc., peculiar to Mark. In ancient times salt was considered the symbol both of wisdom and of friendship. Christ seems to be referring to this latter meaning of the word here. The words *be at peace with one another* probably refer back to the dispute of the Apostles (32) which was the cause of this discourse.

5. Ministry on the Journey to Jerusalem *10*

10, **2-12: The Question of Divorce.** Parallel in Matt. *19,* 1-9; cf. also Luke *9,* 51; *16,* 18.   **12.** Peculiar to Mark. Among the Jews it was not possible for a wife to divorce her husband; but Christ spoke these words in the land *beyond the Jordan* (1), where there were many pagans who allowed this right to women. Mark, writing for Gentile Christians, thought it well to preserve these words of Christ.

10, **13-16: Jesus Blesses the Children.** Parallels in Matt. *19,* 13-15 and Luke *18,* 15-17. Mark alone, both here and in *9,* 35, mentions that Jesus embraced the children.

10, **17-31: The Danger of Riches.** Parallels in Matt. *19,* 16-30 and Luke *18,* 18-30.   **18.** The man had addressed Jesus as *Good Master,* merely using a title of respect or perhaps a flattering greeting in order to win a favorable answer. Our Lord does not refuse absolutely the attribute implied in the word *good*, but He obscurely inculcates the true reason why He has a right to this attribute. **21.** The words *looking upon him, loved him*, are peculiar to Mark. The Greek word for *looking upon* signifies rather "gazing fixedly at"; cf. the same word in *8, 25*. Jesus must have given some external sign, or St. Peter, from whom we ultimately have this narrative, would not have known that Jesus *loved him;* according to Origen the meaning here is, "He kissed him." **23 f.** Note the repetition of Christ's words, which is peculiar to Mark. It was undoubtedly for the purpose of emphasizing the importance of these words that our Lord repeated them. **30.** In Mark alone have we Christ's words by which He included among these blessings *persecution* also, itself really a blessing in disguise.

10, 32-34: **The Third Prediction of the Passion.** Parallels in Matt. *20,* 17-19 and Luke *18,* 31-34. **32.** Mark alone mentions the dismay and fear of Christ's disciples. For one of the reasons for their fear, cf. John *11,* 8. 16. 56b. *On their way:* more exactly according to the original, "on the road." Jesus walks alone in front of His disciples like a resolute leader. The text seems to imply that there were two groups behind Him, those who *were in dismay* and those who *were afraid;* but the Greek is so strange that some commentators consider the text corrupt here.

10, 35-45: **Ambition of James and John.** Parallel in Matt. *20,* 20-28; cf. also Luke *22,* 24-27. **38 f.** The reference to the baptism of suffering is peculiar to Mark's account here, but the same figure of speech is used in Luke *12,* 50. In the Greek of this period the expression, "to be baptized," meant originally "to be immersed" (see Commentary on *7, 4*). It is used here in the metaphorical sense of "to be plunged deep into misfortunes, to be overwhelmed with afflictions."

10, 46-52: **The Blind Bartimeus.** Parallels in Matt. *20,* 29-34 and Luke *18,* 35-43. **50.** This verse, peculiar to Mark, shows the vividness of his narrative. **51.** *Rabboni* is the Aramaic equivalent of the Hebrew word "Rabbi"; both words have the same meaning, i.e., literally "My great one," but they were used as synonymous with "teacher"; the only other occurrence of *Rabboni* in the New Testament is in John *20,* 16.

6. LAST MINISTRY IN JERUSALEM 11 — 13

11, 1-11: **Triumphal Entry into Jerusalem.** Parallels in Matt. *21*, 1-11; Luke *19*, 29-44; John *12*, 12-19. **10.** The cry of *Blessed is the kingdom of our father David that comes* is recorded by Mark alone. No doubt most of the people meant this in the literal sense and were cheering for the re-establishment of David's temporal kingdom. **11.** See Commentary on Matt. *21*, 17.

11, 12-14: **Jesus Curses a Fig Tree.** Parallel in Matt. *21*, 18 f.

11, 15-26: **Cleansing of the Temple.** **15-19.** Parallels in Matt. *21*, 12-17 and Luke *19*, 45-48; cf. also John *2*, 13-17. **16.** Peculiar to Mark's account. Apparently the people were accustomed to use the temple as an ordinary thoroughfare. **20-24.** Parallel in Matt. *21*, 20-22; cf. also Matt. *17*, 19 and Luke *17*, 6. **25 f.** Cf. Matt. *6*, 12. 14 f; *18*, 35.

11, 27-33: **The Authority of Jesus.** Parallels in Matt. *21*, 23-27 and Luke *20*, 1-8. **32.** Christ's enemies leave this sentence unfinished, for they hate to think of the consequences.

12, 1-12: **Parable of the Vine-dressers.** Parallels in Matt. *21*, 33-46 and Luke *20*, 9-19.

12, 13-17: **Tribute to Cæsar.** Parallels in Matt. *22*, 15-22 and Luke *20*, 20-26.

12, 18-27: **The Sadducees and the Resurrection.** Parallels in Matt. *22*, 23-33 and Luke *20*, 27-39. **26.** *About the bush:* the sense of the Greek is "in the passage called 'The Bush' "; at the time of Christ there were no chapter or verse divisions in the Books of the Bible, or at least they were not numbered; but the various sections of each Book bore subtitles, usually taken from some prominent feature of that section, and these subtitles were then used as we now use chapter and verse number for the purpose of indicating more accurately a citation from the Scriptures. In the Third Gospel this same Greek expression is rendered more accurately in our English text.

12, 28-34: **The Great Commandment.** Parallel in Matt. *22*, 34-40; cf. also Luke *10*, 25-28. **29 f.** Mark gives the quotation from Deut. *6*, 4 f in its longest form, including the introductory phrase; these words have been used by the Jews from ancient times as a morning and evening prayer, which is known from its first word ("Hear") as the *Shema*. **32-34.** Peculiar to Mark. Matthew evidently considered the long repetition unnecessary. *A greater thing than all holocausts and sacrifices:* cf. *1* Kgs. *15*, 22; Osee *6*, 6; Matt. *9*, 13; *12*, 7. *Not far from the kingdom of God:* this Scribe had a correct concept of true religion and would make a good Christian. *And no one after that,* etc.: see Commentary on Matt. *22*, 46.

12, 35-37: **The Son of David.** Parallels in Matt. *22*, 41-46 and Luke *20*, 41-44. **37.** Mark alone notes here the large popular following that Jesus still had in Jerusalem.

12, 38-40: **Hypocrisy of the Scribes and Pharisees.** Parallel in Luke *20*, 46 f. In Matthew this discourse of Christ makes up the whole of chapter *23*, yet despite Mark's brevity, one of his verses (40) is not found in the original text of the First Gospel (cf. Matt. *23*, 14). **38 f.** Cf. Matt. *23*, 5 ff. From the Greek word which is here translated as *long robes* our word "stole" is derived; it signified "a festive garment"; hence the emphasis here is not on the length of these robes but on their unusual character which attracted attention and pleased the Scribes' vanity. **40.** Those who *shall receive a heavier sentence* are not all the Scribes in general but those in particular *who devour the houses of the widows:* this sin is worse than that of vanity. To *devour the houses of widows* certainly means "to steal their property." But it is not quite clear how the Scribes did this. Some commentators take *making pretense of long prayers* as expressing the means used by the Scribes; *the widows* would then be those pious women who fall victims to the pietistic pretenses of the Scribes and voluntarily donate money to them in order to have the benefit of their prayers. Other commentators consider the Scribes as acting in their capacity of judges and unjustly defrauding the widows and orphans of their property, while they seek to cover their wickedness by seeming to be very pious in their long public prayers; since this is a sin that cries to heaven for vengeance, Christ says that a heavier penalty is due to it.

12, 41-44: **The Widow's Mite.** Parallel in Luke *21*, 1-4. **42.** The *lepton,* here translated *mite,* was the smallest copper coin in circulation; see Glossary at end of text, p. 752. Since the *lepton* was not used in the West, Mark explains its value for his Roman readers by stating that it was worth only half as much as the Roman *quadrans,* the smallest coin used in the West. **43 f.** Christ points out the important truth that the value of an offering in God's sight does not depend upon its intrinsic value but upon the relative amount of sacrifice that the offering costs the one who makes it. It is not the gift itself but the generosity behind the gift that is pleasing to God.

13, **1-13: Destruction of Jerusalem and End of the World.** All of Chapter *13* is given to this so-called "Eschatological Discourse," by far the longest discourse of Christ that is recorded in the Second Gospel. **1-10.** Parallels in Matt. *24,* 1-14 and Luke *21,* 5-13; cf. also Matt. *10,* 17 f. **11-13.** Parallel in Luke *21,* 14-18. Substantially the same words but in a different context are given in Matt. *10,* 19-22; cf. also Luke *12,* 11 f.

13, **14-20: Destruction of Jerusalem.** Parallels in Matt. *24,* 15-22 and Luke *21,* 20-24.

13, **21-27: The Signs of the Last Day.** **21-23.** Parallel in Matt. *24,* 23-25; cf. also Luke *17,* 23. **24-27.** Parallels in Matt. *24,* 29-31 and Luke *21,* 25-28.

13, **28-31: Jerusalem's Impending Destruction.** Parallels in Matt. *24,* 32-35 and Luke *21,* 29-33.

13, **32-37: The Need of Watchfulness.** **32.** Parallel in Matt. *24,* 36. **33.** Parallels in Matt. *24,* 42 and Luke *21,* 36a. **34-37.** Peculiar to Mark, at least in this context. But there are very similar little parables in Matt. *24,* 45-51 and Luke *12,* 36-38. With Luke's corresponding parable of the Watchful Servants there is combined the picture of a wedding feast; here the master merely returns from a journey abroad at an unexpected hour. Under the image of the *man* who *leaves home to journey abroad* is meant Christ who departs from the earth to go to His heavenly Father. **35 f.** Just as the servants do not know the hour of their master's return, so we also do not know the time of Christ's coming at the end of the world or to each soul at death. He will come *suddenly,* i.e., at an unexpected time. The important thing is that we always remain awake, i.e., spiritually prepared

14, 1-16   St. Mark

for His coming. Our Lord refers to the four divisions or "watches" of the night, which were used throughout the whole Roman Empire at that time.

II: THE PASSION, DEATH AND RESURRECTION *14-16*

1. The Last Supper *14*, 1-31

14, **1-2: The Council.** Parallels in Matt. *26*, 1-5 and Luke *22*, 1 f.

14, **3-9: The Anointing at Bethany.** Parallels in Matt. *26*, 6-13 and John *12*, 1-8. **3.** *Genuine nard:* see Commentary on John *12*, 3. The ointment was in the form of an oil which could be *poured* out of the alabaster jar. The *alabaster jar,* if we judge by the specimens found by archeologists, was shaped somewhat like a bottle. Its rather narrow opening was sealed with wax, and it was this wax and not the jar itself that Mary broke; the Greek word which is here translated *breaking* signifies properly "crushing, pushing down."

14, **10-11: The Betrayal.** Parallels in Matt. *26*, 14-16 and Luke *22*, 3-6.

14, **12-16: Preparation.** Parallels in Matt. *26*, 17-19 and Luke *22*, 7-13. **13 f.** *Two of his disciples:* Luke tells us that they were Peter and John. These strange directions were probably due to Christ's desire to keep the exact location of the house a secret from Judas and thereby ensure safety from His enemies during the Last Supper. *There will meet you a man carrying a pitcher:* the sense seems to be, "Follow the first man that you meet who carries a pitcher," etc. Perhaps the sight would be rather unusual, since the task of fetching water from the public fountains was generally performed by women. This man is certainly distinct from *the master of the house.* The latter was no doubt a disciple of Christ. His house seems to have been the same as that in which the Apostles assembled after the Ascension (Acts *1*, 13) and identical with "the house of Mary, the mother of John who was surnamed Mark" (Acts *12*, 12). If that is so, it adds probability to the opinion that the "certain young man" mentioned in *14*, 51 f was St. Mark himself.

14, 17-21: The Betrayer. Parallels in Matt. *26,* 20-25; Luke *22,* 14. 21-23; John *13,* 21-30.

14, 22-25: The Holy Eucharist. Parallels in Matt. *26,* 26-29; Luke *22,* 18-20; *1* Cor. *11,* 23-25.

14, 26-31: Peter's Denials Predicted. Parallels in Matt. *26,* 30-35; Luke *22,* 31-34; John *13,* 36-38. Mark alone, both in the prediction and in its fulfillment, mentions the cock crowing *twice.*

2. THE PASSION AND DEATH OF JESUS *14,* 32 — *15,* 47

14, 32-42: The Agony in the Garden. Parallels in Matt. *26,* 36-46 and Luke *22,* 40-46.   **36.** *Abba* is the Aramaic word for *Father;* cf. Rom. *8,* 15; Gal. *4,* 6.   **41.** *It is enough:* this is the Vulgate's rendering of a Greek word which means literally "it holds off"; the exact significance of this word as used here is uncertain; possibly the sense is, "It is no use," i.e., of watching any more.

14, 43-52: Jesus Arrested. Parallels in Matt. *26,* 47-56; Luke *22,* 47-53; John *18,* 2-11.   **44.** Mark alone records the words of Judas, *And lead him away safely:* Judas knew that Jesus had previously slipped from the hands of His enemies on several occasions; cf. Luke *4,* 29 f; John *8,* 59; *12,* 36.   **51 f.** Peculiar to Mark. According to a probable opinion, which, however, was unknown to the Fathers, this *certain young man* was St. Mark himself. Assuming that the Last Supper took place in the house of Mary, Mark's mother (see Commentary on *14,* 13), the band led by Judas may have first sought Jesus there; Mark, aroused from sleep by this excitement, may have followed them to Gethsemani. The *linen cloth wrapped about his naked body* seems to have been a bedsheet, for he slipped out of it very easily in order to evade his captors. Mark mentions this otherwise unimportant incident to show the rough character of the mob with Judas, and perhaps, if he is the young man in question, to indicate this one little appearance he made in the drama of Christ's life.

14, 53-65: Jesus before the Sanhedrin. Parallel in Matt. *26,* 57-68; cf. also Luke *22,* 63-71 and John *18,* 13. 19-24.

14, 66-72: Peter's Denial. Parallels in Matt. *26,* 69-75; Luke *22,* 54-62: John *18,* 17 f. 25-27.   **72.** *He began to weep:* the Greek is, literally, "Having thrown upon, he wept"; but the meaning of this obscure expression is entirely uncertain. Besides the conjecture of the Vulgate, other suggestions are; "Reflecting upon it, he wept"; "Casting himself upon the ground, he wept"; "Covering his head, he wept."

15, 1-14: Jesus before Pilate. Parallel in Matt. *27,* 1 f. 11-23; cf. also Luke *23,* 1-4. 13-22; John *18,* 28-40; *19,* 4-7.

15, 15-19: The Scourging and Crowning. Parallel in Matt. *27,* 26-30; cf. also Luke *23,* 16. 25; John *19,* 1-3.

15, 20-22: The Way of the Cross. Parallels in Matt. *27,* 31-33; Luke *23,* 26. 33a; John *19,* 17.   **21.** Mark alone mentions the two sons of Simon of Cyrene; hence they were probably known to the Christians at Rome. The Rufus of Rom. *16,* 13 may be identified with this Rufus.

15, 23-32: The Crucifixion. Parallels in Matt. *27,* 34-44; Luke *23,* 33-38; John *19,* 18 f. 23 f.   **25.** Peculiar to Mark. This does not contradict John *19,* 14. Mark divides the day into four quarters as he does the night (*13,* 35); the second quarter, from nine in the morning until midday, he names after the hour with which it begins; hence our Lord was condemned by Pilate and crucified shortly before midday.   **28.** Mark alone in this context refers to the fulfillment of this prophecy of Isa. *53,* 12; but cf. Luke *22,* 37.

15, 33-41: The Death of Jesus. Parallels in Matt. *27,* 45-56; Luke *23,* 44-49; John *19,* 25-30.

15, 42-47: The Burial. Parallels in Matt. *27,* 57-61; Luke *23,* 50-55; John *19,* 38-42.   **44.** *Pilate wondered:* Jesus had evidently died in a shorter time on the cross than was usual; this is to be attributed to the extraordinarily violent treatment that He had received.

3. The Resurrection of Jesus *16,* 1-18

16, **1-8: The Women at the Grave.** Similar accounts in Matt. *28,* 1-10 and Luke *24,* 1-11. **1.** The burial on Friday evening had been so hasty that there had been no time to *anoint* Jesus then. **2.** *When the sun had just risen:* according to John *20,* 1 this happened "while it was still dark." Perhaps John refers to the moment of departure, Mark to the moment of arrival at the tomb; or, if both Evangelists are referring to the same event in general, Mark's expression would mean, "After the first streaks of dawn appeared in the east," while John's expression would mean, "Before it was full daylight." **4.** *For it was very large* gives the reason, not why *they saw that the stone had been rolled back,* but why *they were saying to one another, "Who will roll the stone back?"* etc. Mark has similar loose sentence-structure in other instances. **5.** *At the right side:* apparently Mark means, "At the right-hand side of the tomb-chamber," in order to point out that this, the most honorable place among the other unused graves in the chamber, had been accorded to Jesus. *Robe:* the same word as in *12,* 38; see Commentary there; on the "white robes" of the blessed in heaven, cf. Apoc. *7,* 9. **7.** Mark, like Matthew, stresses the prediction of the Galilean apparition. We should expect him to tell us further of this apparition in the original conclusion of his Gospel. **8.** *They said nothing to anyone:* we must understand this to mean that at first, from fright, they kept the fact of the empty tomb and the words of the angel a secret, but later told the Apostles, as the other Evangelists relate.

16, **9-13: Apparitions of Jesus.** On the authenticity of 9-20, see the Introduction to this Gospel. Mark records very little that seems original in these twelve verses. If we assume that his original conclusion was lost at a very early date, we might explain these verses by saying that he composed them by summarizing various events narrated in the other three Gospels. **9.** On the apparition to Mary Magdalene, cf. John *20,* 1-17. *Seven devils:* cf. Luke *8,* 2. **10.** Cf. Luke *24,* 10 f; John *20,* 18. **11.** Cf. Luke *24,* 22-25. **12 f.** Cf. Luke *24,* 13-35. *Even then they did not believe:* this refers probably only to the doubt of Thomas and perhaps of a few others, for in the meantime at least Simon had seen the Lord (cf. Luke *24,* 33 f).

***16*, 14-18: Commission of the Apostles.** **14.** Cf. Luke *24*, 36-49; John *20*, 19 f; *1* Cor. *15*, 5. **15 f.** Cf. Matt. *28*, 18 ff. We need not understand Mark to mean that this commission was given to the Apostles on the same occasion as that mentioned in 14. *Every creature:* every human being. **17 f.** Cf. *6*, 7. 13; Matt. *10*, 1. 8; Luke *9*, 1; *10*, 19.

4. THE ASCENSION OF JESUS *16*, 19-20

***16*, 19-20: The Ascension.** **19.** Cf. Luke *24*, 50-53; Acts *1*, 9-11. *After he had spoken to them* should not be understood as meaning "immediately after the preceding words." **20.** This brief summary of the activities of the Apostles implies that the Church was fairly well established throughout the Roman empire at the time when these words were written, but there is nothing said here that would not be in keeping with the traditional date of the composition of the Second Gospel.

ALOYS H. DIRKSEN, C.PP.S.

THE HOLY GOSPEL OF JESUS CHRIST ACCORDING TO ST. LUKE

Introduction

The Third Gospel and the Acts of the Apostles form two sections of what is really one single literary work. This is clear from the first verses of each book which serve as brief introductions. The unity of authorship of both Books is clear, and most of the observations which are made by way of introduction to one section of this larger work are equally true of the other. Since the Acts was written at Rome in the year 63 or 64 A.D., the Third Gospel must have been written before that date.

The Author. *External evidence.* Catholic tradition, from the second century on, is unanimous in attributing this work to St. Luke, the disciple and companion of St. Paul. Thus the Muratorian Fragment, St. Irenaeus, Tertullian, Clement of Alexandria, Origen, etc., all give explicit testimony to his authorship. Even on the part of most non-Catholic critics no objections are raised against the Lucan authorship of this work. Almost all scholars agree that if this very early tradition is so unanimous in attributing so important a document to a relatively unimportant member of the Apostolic Church, the only explanation for this can be that St. Luke is in truth its author.

Internal evidence. An examination of the Third Gospel tends to confirm this truth. For its author shows himself to be a Gentile Christian of no mean literary ability and a disciple of St. Paul, whose teaching on the universality of salvation is here presented in a manner that shows unmistakable signs of Pauline influence. Too much stress, however, should not be placed upon certain indications in this work which are cited as a proof that its author was a physician. For there is probably no medical term used here that would be so technical that it could not have been employed by another well-educated man who was not a physician. But knowing, as we do from other sources, that its author was a physician, we are justified in seeing the influence of his medical background in certain passages of this work.

St. Luke. The name of Luke, which is really an abbreviated form of the Latin name *Lucanus*, appears only three times in the New Testament. In all three places it is certainly the same man who is referred to. From Col. *4*, 14 and Philem. 24 we know that Luke was with Paul during the latter's first imprisonment in Rome (61-63 A.D.). In the first of these passages he is called by St. Paul "our most dear physician," while in the other passage he is numbered among the "fellow-workers" of the Apostle. From 2 Tim. *4*, 11 we know that Luke was Paul's only companion in Rome during a part of his second Roman imprisonment. It may be purely accidental, yet it is surprising, that in all three passages Luke's name is mentioned in close association with that of Demas, who, "loving this world," finally "deserted" St. Paul.

While St. Luke is not named in the Acts (which is in itself a confirmatory argument of its Lucan authorship), still in several passages of this Book the author speaks of St. Paul and his associates in the first person plural (the so-called "we" sections: Acts *16*, 10-17; *20*, 5-15; *21*, 1-18; *27*, 1 — *28*, 16). From these passages we learn that the author, St. Luke, came in contact with St. Paul at Troas, on the northwest coast of Asia Minor, near the beginning of the Apostle's second missionary journey (51 A.D.). The two went together to Philippi, where Luke may have remained until Paul returned there towards the end of his third missionary journey (57 A.D.).

According to a fourth century tradition (Eusebius, St. Augustine), St. Luke was a native of Antioch and a member of the Christian community in that city as early as 43 A.D. But this is perhaps only a conjecture based upon the reading in Codex Bezæ and a few other MSS of the so-called "Western family" of Acts *11*, 27 f, "Now in those days some prophets from Jerusalem came down to Antioch, and there was great rejoicing. But when *we* had assembled, one of them named Agabus spoke and revealed," etc.

Characteristics. Although we know so little of the life of this Evangelist, we are justified in concluding from his writings that he was a peaceful and gentle man with a fine aesthetic temperament. For one of the outstanding characteristics of this Gospel is the special stress that is laid upon the operations of God's mercy and Christ's compassion for sinners (e.g. the Penitent Woman, *7*, 36 ff; the Prodigal Son, *15*, 11 ff). Likewise notable in this Gospel is the reverence shown toward womanhood. Many types of admirable women are presented to the reader: the Blessed Virgin Mary, Elizabeth, Anna the prophetess, the widow of Naim, the penitent woman, Mary and Martha, the ministering women, Mary Magdalene.

Luke often emphasizes the social aspects of Christ's teaching; contrast, for example, the Beatitudes and Woes as recorded in his Gospel with the Beatitudes as recorded by Matthew. Greater prominence is given in this Gospel than in the others to Christ's love for the poor and to His teaching concerning earthly wealth.

This Gospel may also be termed the Gospel of prayer, since the subject of prayer is mentioned so frequently. St. Luke carefully notes the example of Christ Himself in this matter, and on several occasions he is the only Evangelist to record that Christ then prayed. Similarly we find Christ's instructions on prayer recorded here at greater length. The

Third Gospel alone has preserved the three hymns that the Church uses in her daily liturgy: the *Magnificat,* the *Benedictus* and the *Nunc Dimittis.*

This whole Gospel breathes a spirit of holy joy, the gift, no doubt, of that Holy Spirit of whom St. Luke speaks so often in the Acts. In fact, "the great joy to all the people" that overflows the hearts of the faithful every Christmastide, owes more than is usually realized to Luke's simple yet marvelously touching story of the birth of the Savior.

Even though the rather late tradition which makes St. Luke a painter may be very doubtful, still this Evangelist certainly shows that he was an artist in the more general sense of the word. With a few skillful strokes he delineates his brief biographical sketches and makes his portraits stand out as living beings. He is able to make the story that he is telling grip the very soul of his readers. He proves himself a master in portraying tenderness and sympathy for man's afflictions. This often reaches a climax of genuine pathos; note, for example, the manner in which he says that a sick or dead child was an *only* child (7, 12; 8, 42; 9, 38).

Luke is more of a "literary" man than are the other Evangelists. It is true that the language of his Gospel, much more so than that of the Acts, shows frequent traces of Semitisms, but its author apparently retained this Hebrew tone intentionally, either in order to remain faithful to his source or in order to produce a conscious imitation of the "Biblical" Greek of the Septuagint. But when the occasion allows, Luke shows that he can write as pure and faultless a literary Greek as any profane author of his time. In him genuine spirituality is not incompatible with secular learning. He alone of all the Evangelists links his Gospel account with the history of Syria and the Roman Empire. He is conscious of his responsibility as an historian and, even though he wrote under the inspiration of the Holy Spirit, he employs all human diligence "to follow up all things carefully from the very first," so that his reader "may understand the certainty" of the Gospel (*1, 3* f).

Sources. In his Prologue the author says that *many have undertaken to draw up a narrative concerning the things that have been fulfilled among us.* These "many" who preceded him in writing an account of Christ's life and teachings would seem to include more than the authors of the first two Gospels. In fact, it is not entirely certain that he meant to include Matthew and Mark among them. In any case he does not state that he intends to base his account on these previously written accounts. On the contrary, he says he *followed up all things carefully from the very first,* i.e., he investigated his sources to their very fountainhead. No doubt in this regard he followed the example of the other writers of Gospel accounts who preceded him and who, he says, drew up their accounts *even as they who from the beginning were the eyewitnesses and ministers of the word have handed them down to us.* From the Latin word that is used to translate this last expression we get our word "tradition." This, then, is Luke's primary source of information, the oral tradition handed down by the Apostles. By the *eyewitnesses and ministers* of the word the Evangelist means above all the Apostles. From Acts *21*, 17 f; *27*, 1 f it seems fairly certain that Luke remained with Paul during the latter's two years' imprisonment in Palestine. Probably it was especially during this time that Luke followed up this oral tradition to its original sources. For his first two chapters on the birth and childhood of Jesus Luke seems to be indebted either directly or indirectly to the Blessed Virgin Mary herself (cf. *2*, 19.51b). On the literary relationship between this Gospel and Matthew and Mark, see the article on the Literary Relations of the First Three Gospels.

Scope. Luke's purpose in writing his Gospel is stated in his words to Theophilus, to whom he dedicates this work. This, he says, is *that thou mayest understand the certainty of the words in which thou hast been instructed;* that is, his primary scope is to draw up in writing an entirely trustworthy account of Christ's life and teachings concerning which Theophilus had already received oral instruction. Luke's principal aim is therefore to confirm the faith of Theophilus and his other readers. In the Third Gospel there are some characteristic features in the life and teaching of Christ that are more accentuated than in the other Gospels. Many of these are traceable to the spirit and influence of St. Paul. Prominent in this regard is Luke's emphasis on the universality of salvation. The teaching of the Apostle of the Gentiles was that Christ is accessible to all men, regardless of color, race or previous relationship to the Mosaic Law. "It is not astonishing that this should go straight to the heart of Luke, the Gentile convert, the companion of Paul in his apostolate. And when he formed the plan of writing a Gospel, this doctrine of universal salvation shone before him as the guiding light of his work" (J. Huby, *The Church and the Gospels*, p. 131). This message is proclaimed throughout Christ's life: at His birth (*2*, 14), presentation in the temple (*2*, 32), at the beginning (*3*, 6) as well as at the end of His public life (*19*, 10). Divine forgiveness and salvation are offered to all: Jews, Samaritans, Gentiles, publicans and sinners.

Structure. In the general outline and development of his Gospel Luke does not differ greatly from Matthew and Mark. In this regard he undoubtedly shows his fidelity to his sources. Where his order of the events narrated is different from that in the Second Gospel, commentators are not of one mind as to which of these two Evangelists is to be followed to arrive at the original chronological order of the events. Luke indeed states in his Prologue that he intends to write *an orderly account* (*1*, 3). This Greek expression is in fact substantially the same as that which Papias used when he said that Mark did *not* write "an orderly account"; but the only "order" that is obviously missing from the Second Gospel is the artistic arrangement of its narrative. Luke, on the contrary, shows at least in a few places, that he has abandoned the strictly chronological order; e.g., in *1*, 80 and *3*, 19 f. The steady movement of events from Nazareth to Jerusalem, which can be noted in this Gospel, corresponding to a similar movement of events from Jerusalem to Rome in the Acts, points to an artistic arrangement of the contents of the book.

COMMENTARY

Note. Only those parts of the Third Gospel that are peculiar to it are commented on here. For all other parts the reader should consult the Commentaries on the parallel passages in the other Gospels.

Prologue 1, 1-4

In a single well-balanced sentence St. Luke gives an introduction to his book in a style that is very much like the classical elegance of Greek and Roman writers. In the rest of the Gospel Luke generally follows the ordinary style of the Synoptics. In this prologue the author states the occasion, sources, method and purpose of his history. For the sense of these verses, see the paragraphs on *sources, scope* and *structure* in the introduction above.

1. *Have undertaken:* literally, "have put their hand to, have taken in hand to, have tried their hand at"; this expression of itself says nothing as to the outcome of their efforts, but in Acts *9,* 29; *19,* 13 (the only other occurrences in the New Testament) St. Luke uses this word in the sense of unsuccessful attempts. *Fulfilled:* more than merely "occurred"; these events had been foretold by the prophets. **3.** *All things:* Luke aims at completeness. But the matter and form of the Synoptic tradition were already fixed. Hence his account is not really a complete life of Christ. *Most excellent:* a title of respect ordinarily used only of high officials in the Roman Empire, equivalent to our modern expression, "Your Excellency" (cf. Acts *23,* 26; *24, 3*; *26,* 25). *Theophilus* means "God's friend." Possibly this was not the man's real name but was used for the purpose of concealing his identity. A definite individual is meant, probably a noble Roman converted by St. Luke. But the Evangelist naturally has a much larger audience in mind. **4.** *Instructed:* the Greek verb, from which our word "catechesis" is derived, signifies "to teach orally by having the pupil 're-echo' the words of the teacher." Luke presumes therefore that Theophilus already knew by heart the basic facts of "all that Jesus did and taught" (Acts *1,* 1).

Prelude: The Coming of the Savior *1, 5 — 2,* 52

This account is entirely independent of the account of the birth and infancy of Jesus as recorded in the First Gospel. But there are no contradictions involved and the two accounts harmonize perfectly. There is not the slightest textual ground for questioning the authenticity of these first two chapters of Luke's Gospel. It is true that there is a more pronounced Hebrew tone to the Greek here than there is in the rest of the Gospel, but this can easily be explained as due to the original source, whether oral or written, that the Evangelist used for this section. It is very probable that Luke's ultimate source of information here was the Blessed Virgin Mary. Possibly the Evangelist met her personally during his stay in Palestine.

1, 5-25: Annunciation of the Baptist. **5.** On *Herod* the Great, see article on The New Testament Background; cf. also Matt. *2. Zachary* means "The Lord is mindful." *Abia:* cf. *1* Par. *24,* 10; *2* Esd. *12,* 4.17. *Elizabeth* was also the name of Aaron's wife (Ex. *6,* 23). **8.** *In the order of his course:* there were twenty-four courses; that of Abia was the eighth. Each course officiated twice a year for one week at a time. **9.** *The temple:* here the *naos,* the "holy place," the edifice proper which was the house of God. **10.** *Outside:* in the court of the temple. *The hour of incense:* either at the morning or at the evening sacrifice. **11.** Angelic apparitions play an important part both in the Third Gospel and in Acts. In all these apparitions the angel undoubtedly appeared in human form.
13. *Thy petition:* from the following words it would seem that this refers to the prayers which Zachary had made for a son while he still had hope to obtain one; others understand this of his prayers for the coming of the Messias. The child is to be called *John* because this name means "The Lord is gracious," i.e., he is God's answer to their prayers. **14.** The key-note of holy joy is struck at once. **15.** *Great:* cf. *7,* 28. *He shall drink no wine or strong drink:* cf. *7,* 33. This was one of the requirements for a Nazirite (cf. Num. *6*) and apparently John is to be a Nazirite for life (cf. Jdgs. *13,* 2-5), but there is no mention here of the other requirements for a Nazirite, that the hair of the head and the face must not be cut. *Strong drink* as distinct from *wine* means distilled liquor. *Filled with the Holy Spirit,* etc.: cf. 41.
16. John is to fulfill the prophecy of Mal. *4,* 5 f; cf. Matt. *11,* 14: *17,* 10-13. The *fathers* are the Patriarchs, who had turned away

from their descendants, the Jews, because of the unworthy conduct of the latter; but now the second Elias is to bring back the children of Israel to a way of life worthy of their ancestors (cf. the full prophecy in Malachias). *The incredulous:* literally in Greek, "those who cannot be persuaded," i.e., the disobedient. *To prepare for the Lord:* cf. Mal. *3,* 1. *A perfect people:* more exactly, "a people made ready," i.e., to receive Him worthily. **20.** Zachary's deprivation of speech was a punishment for his incredulity; it was also a sign that he had asked for (18), and was necessarily unpleasant to Zachary in order to insure his co-operation in John's conception.

21. *They wondered at his tarrying:* the priests were accustomed to fulfill their functions and then leave the Holy Place as soon as possible; if they delayed too long the people began to fear that they had been struck dead for not offering the incense according to the prescribed ritual (cf. Lev. *16,* 13). **24 f.** Since Elizabeth's sterility had been a *reproach* to her, we might expect her to glory in the fact that she had conceived; but on the contrary she *secluded herself,* perhaps "in order to prevent comments on the part of her neighbors" (Lagrange). No doubt her seclusion lasted during the whole period of her pregnancy, but Luke mentions "five months" because he is about to speak of what happened *in the sixth month* of her pregnancy (26).

1, **26-38: Annunciation of the Savior.** **26.** *Nazareth:* see Commentary on Matt. *2, 23.* **27.** Luke, like Matthew, emphasizes the fact that Mary was a *virgin;* the Greek word that is used here signifies a virgin in the strict sense, not merely "a young woman." *Betrothed:* on the Jewish marriage customs and the status of Mary and Joseph at the time of the Annunciation, see Commentary on Matt. *1,* 18. According to the structure of this sentence the phrase *of the house of David* refers to Joseph rather than to Mary, but according to the common opinion Mary also was a descendant of David. **28.** *Had come to her:* literally, "had entered to her"; this shows that Mary was at this time in a house. This was the house of her parents or her guardians, certainly not that of Joseph. Relying on the authority of the apocryphal gospels, the Greek Church places the scene of the Annunciation at the public fountain of Nazareth. *Hail:* the word used by Luke was the ordinary form of greeting among the Greeks, but perhaps it is used here in its original sense of

"Rejoice!" Many of the Greek Fathers understood it in this sense and compared the whole greeting of the angel with Soph. *3,* 14-17; cf. also Joel *2,* 21 f; Zach. *9,* 9. The phrase *full of grace* represents a single word in the Greek text but is a correct translation according to the sense. The angel explains this word in 30 as *Thou hast found grace with God.* The words *Blessed art thou among women* are not part of Luke's original text in this passage. According to the best Greek MSS they were spoken only by Elizabeth (42). Apparently the custom of combining the words of Gabriel with those of Elizabeth to form a prayer to our Lady goes back to a very early period in the Church and was the cause of some enlarging by an early copyist on Gabriel's greeting. Grammatically it is possible to translate, "The Lord be with thee"; but tradition has always understood this as a statement of fact, not a wish; cf. Soph. *3,* 14 f: "Rejoice, daughter of Sion . . . the Lord is in thy midst." **29.** The words *When she had seen him* are not in the best Greek MSS. Mary was disturbed not by the sight of an angel but by the strange nature of his greeting.

V.31 is almost identical with Isa. *7,* 14 and was undoubtedly intended to signify that this was the fulfillment of the great prophecy of the virgin-birth of the Messias. **32.** *Called:* i.e., acknowledged as. **34.** According to some commentators Mary understood the words *thou shalt conceive* as referring to the immediate future; hence, since several months were still to elapse before the final marriage ceremony between her and Joseph, she asks the angel for further instructions. But the traditional interpretation takes her last words in the sense of "since I shall not know man," and argues from this that Mary had a vow or at least a firm resolution to remain a virgin forever. In any case the fact of her perpetual virginity is absolutely certain, whatever might have been the state of her mind at this time.

35. Although the Incarnation, as something extrinsic to the Godhead, is a work common to all three Persons of the Blessed Trinity, it is appropriated to the Holy Spirit, perhaps because it is above all a work of love or because He is the living Spirit of God that "gives life" (Nicene Creed; cf. also Rom. *8,* 11; *1* Cor. *15,* 45; *2* Cor. *3,* 6). The word *overshadow* is used of the *Shekinah* or cloud of glory that indicated God's presence (Ex. *40,* 32 ff), and of the bright cloud that appeared during the Transfiguration (*9,* 34). It therefore signifies here the special presence and action of God in Mary. **36.** Although Mary did not ask for a proof, yet the angel gives her a sign to confirm his utterance, to increase her joy and to give her an opportunity of visiting her cousin Elizabeth.   38. In deep humility and loving obedience Mary accepts the dignity and responsibility of the divine maternity. *Handmaid:* literally "slave-girl." It is commonly believed that as soon as she pronounced these words the Son of God took upon Himself our human nature in her immaculate womb.

1, 39-56: **The Visitation.**   39. *In those days:* i.e., shortly after the Annunciation. Mary hardly went alone on this journey of several days but was probably accompanied by some older woman. It seems certain, however, that Joseph was not with her at the Visitation. *Arose and went:* a Hebraism signifying that a person starts on a journey only after some deliberation; cf. the use of the same expression in *15,* 18.20; Acts *10,* 19. *With haste:* because of her desire to share her joy with Elizabeth and also to be of aid to her elderly kinswoman. Since almost all of Judea was *hill country,* this item does not help us identify the *town of Juda* where Elizabeth lived. Luke probably noted the rugged nature of the country for the benefit of his Gentile readers, in order to show the arduous nature of Mary's journey. The nearest point in the region that was anciently assigned to the tribe of Juda was about ninety miles south of Nazareth. If the house of Zachary was in the priestly city of Hebron, as some of the Fathers thought, her journey would have been about twenty-five miles longer. A rather late tradition identifies this *town of Juda* with modern Ain Karem, a village about five miles west of Jerusalem. 41. *The babe in her womb leapt:* i.e., for joy (44). According to the common opinion John was sanctified in his mother's womb at that moment (cf. 15). This was the effect of Mary's greeting, and thus the first grace of God was poured out to man through the Blessed Virgin. *Filled with the Holy Spirit:* i.e., the following words were uttered under the special inspiration of God (cf. 67).   42. Note the parallelism of this Semitic poetry: *Blessed . . . blessed . . .*   44. The words *For behold* show that it was through the Holy Spirit's action upon the babe in her womb that God revealed to Elizabeth the fact of Mary's divine maternity.   45. *She who has believed* is Mary; apparently the contrast is with Zachary who doubted.

46-55: In this incomparable canticle of holy joy and thanksgiving, known from the first word of its Latin translation as the *Magnificat,* Mary reveals her intimate knowledge of the Books of the Old Testament, especially the hymn of Anna (*1* Kgs. *2,* 1-10) and the Psalms. On the main thoughts of Mary's hymn, see the note to the text.   **46 f.** *Soul* and *spirit* are here used synonymously in this poetic parallelism. With her entire being Mary joyfully praises God, her Lord and Savior.   **48-50.** God's goodness is shown not only in His regard for Mary's lowliness but also in His kindness to the poor and afflicted of all generations. **51-53.** The tense of the verbs used in these verses probably represents the Semitic use of the verb in general statements which are true of the past, present and future.   **56.** *About three months:* i.e., until the birth of Elizabeth's child. Because Mary's departure is mentioned here before the Evangelist speaks of John's birth, some commentators conclude that Mary was not present at this event. But if she visited her cousin in order to be of assistance to her, this was precisely the time when she was needed most of all. Moreover, Luke's knowledge of what took place at the birth of the Baptist seems to come ultimately from Mary, who probably therefore was present at these events. Luke sometimes abandons the strictly chronological order for the sake of finishing the narrative of one episode before beginning another (cf. *1,* 80; *3,* 19 f).

1, 57-80: Birth of the Baptist.   **59.** *On the eighth day:* cf. Gen. *17,* 12; Lev. *12,* 3.   **60.** It is possible that Elizabeth also had received a divine revelation concerning the name of her son, but it seems more likely that Zachary had made this known to her in writing.   **62.** From this it would seem that at the time Zachary was deaf and dumb. But perhaps, since he had to make signs in order to be understood, people were naturally led to answer him in a sort of sign-language even though he could have understood them if they spoke to him.   **63.** *A writing-tablet:* at that time the ordinary material used for writing down little items of no permanent value was a flat piece of wood painted or stained a dark color and covered with wax. The writing was done with a metal stylus. When covered with writing, the tablet was rewaxed.   **67.** *Prophesied:* i.e., spoke under the influence of the Holy Spirit.

68-79: Zachary's hymn of thanksgiving and joy is known from the first word of its Latin translation as the *Benedictus.*   **74.** *Delivered from the hand of our enemies:* if Zachary shared the popular idea of a Messias who would bring political independence, he at least thought of this merely as a means of serving

God more perfectly.   **77.** Here Zachary shows that he understands the true nature of the Messianic salvation, i.e., a spiritual salvation from the slavery of sin. *Forgiveness of sins* was later the theme of the Baptist's preaching (cf. *3,* 3).   **79.** *Peace* here is much more than earthly peace and freedom from wars; it is essentially peace with God, the only true basis of genuine peace on earth (cf. *2,* 14; *19,* 38).   **80.** Luke finishes his account of John's early life before beginning his account of the Savior's birth. Note the contrast between John's youth and that of Jesus (*2,* 40.52). The Baptist is a stern ascetic even as a boy.

2, 1-7: The Birth of Jesus.   1-5. The census of Cyrinus. Some years ago it was the custom for critics to attack the accuracy of St. Luke's statements regarding this census. But recent discoveries, especially of Egyptian papyri, have brilliantly verified the truth of these words of the Evangelist.   **1.** *Augustus* was the *Cæsar,* or emperor, of the Roman Empire from 31 B.C. to 14 A.D. During his reign a census of all the Roman citizens of the whole Empire was made on three occasions: in 28 B.C., in 8 B.C., and in 14 A.D. As to the enrollment of the non-citizens in the various provinces outside of Italy, we know from the Egyptian papyri that there was a regular periodic census of these people every fourteen years. We have records for all these enrollments for each fourteen-year period from 19/20 A.D. to 257/258 A.D. The census of 5/6 A.D. is mentioned by Josephus and referred to in Acts *5,* 37. Fourteen years before that would be 8 B.C. This was probably the first occurrence of this periodic census and coincided with the census of the Roman citizens, so that Luke could speak of *the whole world,* i.e., all the inhabitants of the Roman Empire, being enrolled.   **2.** Luke calls this the *first* census to distinguish it from the later census mentioned in Acts *5,* 37. However, the civil governor of Syria from 9 to 6 B.C. was Sentius Saturninus. Cyrinus did not become civil governor of Syria until 6 A.D. Moreover, Tertullian refers to the records of this census as accessible to those who might have had doubts about the birth of Christ and says that it was taken up in Judea by Sentius Saturninus. But on the other hand we have evidence that is just as certain, that at this time Cyrinus was military commander in Syria, and Luke clearly had good grounds for attributing this census to him. It is not quite certain what was the relationship between Saturninus and Cyrinus in regard to the census in Judea. Instead of *Cyrinus* the better Latin MSS have *Quirinus.* His full name was Publius Sulpicius Quirinus, (or, Quirinius). The Greek word for *governor* is often used in the sense of "military commander." **3.** *Each to his own town:* these fourteen-year enrollments were by "households," as is clear from the papyri; everyone had to go to the place where he had his official residence. But there is not the slightest evidence that people had to go to their ancient ancestral home. If that was the case all of Galilee would have been emptied by the census, for almost all of its inhabitants were Jews, i.e., descendants from the tribe of Juda, whose original home was in Judea.

4. Joseph must have considered Bethlehem his real home. Perhaps he owned a piece of property there, for one of the main reasons for the census was the assessment of taxes. From Matt. 2, 21 f it seems that Joseph had intended after the birth of Jesus to make Bethlehem his permanent home. **5.** *His espoused wife:* the best Greek MSS read simply, "his betrothed" (the same word as in *1,* 27). But it is certain that at this time Mary and Joseph were fully married (cf. Matt. *1,* 24), and the word, *wife,* gives the correct sense. Luke still speaks of Mary as merely "betrothed" even at Christ's birth, lest the reader might wrongly think that her marriage was consummated. Mary naturally accompanied her husband to Bethlehem, whether she herself was obliged to register in the census or not. **6.** *While they were there:* this would seem to imply that they had arrived in Bethlehem at least a few days before the birth of Jesus. **7.** The stupendous event of the appearance on earth of the Incarnate Son of God is told in a few touching words—a masterpiece of charming simplicity that is vastly superior to the vulgar exaggerations of the apocryphal Gospels. *Firstborn:* the Greek word corresponds with the Hebrew technical term of which neither the etymology nor the usage implies later births. The Evangelist wishes merely to stress the fact that the Mosaic Law of the firstborn, involving various duties and privileges (cf. Ex. *13,* 2), applied to Jesus also (cf. *2,* 22 f). Luke mentions that Mary wrapped him in swaddling clothes, not as if this were extraordinary, for it seems that all new-born infants were thus swathed, but because he is about to quote the angel's words in reference to this. The *manger* implies that our Lord was born in a place where cattle were kept; according to the traditional site in Bethlehem, this was a cave which was used as a stable. The Evangelist does not tell us why Joseph had sought for lodgings in the oriental *inn* or *khan,* nor do we know why he failed to find room there, but the commonly accepted explanation is that that town was overcrowded because of the census. Later on, the Holy Family lived in a regular house at Bethlehem (cf. Matt. *2,* 11).

2, 8-20: The Shepherds at the Crib. Just as Christ was born in a stable like the poorest of the poor, so it was fitting that His first visitors should be poor and simple peasants. **11.** *A Savior:* in the spiritual sense, saving man from his sins; an allusion to His name of Jesus (cf. Matt. *1,* 21). The only other instance in the Gospels of this title of our Lord is in John *4,* 42. *Christ the Lord:* perhaps the Greek word *christos* is used here as an adjective, i.e., "the Anointed Lord," or possibly the original reading was as in 26, "the Anointed of the Lord." In any case the shepherds would have understood the phrase merely in the sense of "the Messias." **12.** This *sign* was given to the shepherds not only that they might know that the angel had spoken the truth but also to assist them in finding the Child, for there would not be another infant in Bethlehem who would be so recently born that he would still be *wrapped in swaddling clothes* and at the same time lying in such an odd cradle as *a manger*. **14.** Cf. *19,* 38. *Among men of good will* is the reading of the best MSS, but many MSS read "among men good will." The Greek word that is translated here as *good will* is used especially of God's "benevolence" towards men. No particular group of men is meant here; the sense is "mankind to whom God is well-disposed." **17.** *They understood:* the Greek verb means "they made known." **19.** While the shepherds spread the good news, Mary meditated in silence on the deeper meaning of this mystery. We may rightly understand this verse as meaning, "It was Mary who remembered all these things," and infer that the Evangelist derived his information about the infancy of Jesus either directly or indirectly from Mary; cf. 51.

2, 21-40: Circumcision and Presentation. **21.** Jesus naturally observed the Jewish law of circumcision just as He observed all other laws of the Old Testament. The Evangelist mentions the circumcision merely in passing, and stresses rather the bestowal of the holy name of Jesus on this occasion (cf. *1,* 31b; Matt. *1,* 21.25b). The rite of circumcision was given to Abraham "for a sign of the covenant" between God and him (Gen. *17,* 11). No Jew considered himself a son of Abraham unless he was circumcized (cf. Rom. *4,* 11 f.); it was by circumcision that he became subject to the Law (Gal. 5, 3). But according to the divine will the Son of God not only became man, but also became a son of Abraham, subject to the Law of Moses, "born under the Law, that he might redeem those who were under the Law" (Gal. 4, 4 f). Therefore, according to the divine plan of Redemption it was necessary for Jesus to be circumcized.

22. *Her purification:* the Greek text has "their purification," which can only mean "the purification of Mary and Jesus." Two different laws of Moses were complied with on this occasion: the purification of the mother (Lev. *12*) and the presentation of the firstborn son (Ex. *13*, 2.12.15). The Evangelist, following perhaps the popular custom, combines both rites under the name of "purification." **24.** This *sacrifice* was connected with the rite of the mother's purification. It was really a double sacrifice: one dove or young pigeon as a sacrifice for "sin," i.e., ritual uncleanness resulting from childbirth, and a yearling lamb, for which the poor could substitute another dove or young pigeon, as a holocaust, i.e., a thanksgiving offering for the birth of a child. While Mary was not obliged, strictly speaking, to offer the former sacrifice, the thanksgiving offering was even more appropriate in her case than it was for ordinary mothers. Perhaps too much stress is placed upon the poverty of Mary in making the offering of the "poor," for from the way Luke cites the law it would seem that the substitution of a bird for the lamb was the ordinary practice at that time. No sacrifice was prescribed in connection with the presentation of the firstborn, but the infant son that was consecrated to the service of the Lord was to be bought back at the price of five shekels (Num. *18,* 16; cf. also Ex. *34,* 19; Num. *3,* 12 f; *8,* 16-18).

25. *Simeon* is a representative of that pious class of Jews who at this time were awaiting a spiritual Messianic kingdom (cf. 38b). He is generally spoken of as an "old man" but this is not stated in the text, unless it is perhaps implied in the promise that he had received, that he should not die until he had seen the Messias (26). *The consolation of Israel:* the coming of the Messias; cf. the same thought in different words, *the redemption of Jerusalem* (38). **28.** *He also received him:* from this phrase it seems fairly certain that Simeon was not the priest who officiated at the presentation of Jesus. *Blessed God:* gave grateful praise to God.

29-32: This hymn of Simeon, which is known from the first words of its Latin translation as the *Nunc Dimittis,* has been used in the Church since an early age as part of the liturgical night-prayers (Compline).   **29.** *Lord, servant:* more exactly, "Master, slave." *Dismiss:* properly, "liberate, enfranchise." *In peace:* i.e., contented, now that God's promise to him has been fulfilled (cf. Gen. *46,* 30).   **30.** *Thy salvation:* practically equivalent to "Him whom thou has sent for our salvation." **32.** *A light of revelation:* either, "a light that is to be manifested," or, "a light by means of which divine truth will be manifested"; cf. *1,* 78 f; Isa. *42,* 6; *49,* 6; *52,* 10.

33. *His father:* both before the law and in the eyes of the people Joseph was the father of Jesus; since the Evangelist has already stressed the virgin-birth of our Lord, he knows that the reader will not misunderstand this ordinary form of speech. **34 f.** The mission of Christ will have a twofold effect in accordance with the dispositions of men. Some will be scandalized by His humility, poverty and death on the cross, and hence refuse to acknowledge Him as the Christ, so that He will be the occasion of their eternal ruin. To others, however, who receive Him and His doctrine He will bring eternal life. Cf. Isa. *8,* 14; Rom. *9, 33; 1* Pet. *2,* 7. The sense would be clearer if the passage were punctuated: *And for a sign that shall be contradicted—and thy own soul a sword shall pierce—that the thoughts of many hearts may be revealed.* Christ makes every man take a definite stand in His regard; as a result every man thereby shows the dispositions of his heart. The prophecy concerning our Lady, which is expressed in the parenthesis, was fulfilled especially in Christ's Passion. Man's rejection of her divine Son was to her a sword of sorrow.   **36 f.** Presuming that *Anna* was at least fourteen years old when she was married, she would have been at least one-hundred-and-five years old at this time. But since the original reads "she was a widow until eighty-four years," many commentators understand this as meaning that she was eighty-four years old at this time.

39. According to the most probable opinion the visit of the Magi and the flight into Egypt, as recorded in Matt. *2,* must be placed between the presentation of our Lord and the return to Nazareth. Luke gives a mere summary of the events here, wishing simply to state that, although Jesus was born in Bethlehem, He passed His youth in Nazareth. Therefore, the phrase, *they returned into Galilee, to their own town of Nazareth* should most probably be understood as referring to the same return to Nazareth as is related in Matt. *2, 23*.   **40.** Cf. 52.

2, 41-52: The Child Jesus in the Temple. According to Ex. 23, 14-17 all male Israelites were obliged to visit the temple on the feasts of the Passover, Pentecost and Tabernacles. However, at this time the Rabbis interpreted this law of Moses as obliging only the Jewish men who lived in Judea. Women did not come under this law at all. But many who were not strictly obliged often attended these feasts at Jerusalem out of devotion. **42.** See note to text.

46. *After three days* is probably to be understood as "on the third day"; cf. the same phrase in regard to the Resurrection. Mary and Joseph spent one day in the caravan from Jerusalem, one day in the return journey, and on the third day they found Jesus. The Greek word that is here and often in the New Testament translated as *temple* signifies properly "the sacred enclosure," i.e., the courts and porches of the temple where the Rabbis and, later on, Christ Himself often taught the people. *Sitting:* the Jewish teacher sat on a chair while his pupils sat on the ground (cf. Acts 22, 3, where "a pupil of Gamaliel" is literally "at the feet of Gamaliel"). *Both listening and asking questions:* the regular rabbinical method of instruction. **47.** It is significant how this twelve-year-old Boy from Nazareth amazes the Rabbis with His wisdom. **48.** *They,* i.e. Mary and Joseph, *were astonished:* apparently the reason of their astonishment was the sight of Him in such a learned assembly.

49. *About my Father's business:* literally, "In the things of my Father"; many commentators, arguing from the usage of similar phrases in the Greek documents of that period, translate, "at my Father's," i.e., at my Father's house. The sense is "Why did you look elsewhere for me, since you should have known that I would be in the temple." **50.** Neither Joseph nor Mary understood at this time the relationship between this early public appearance in the temple and the great work of the Redemption. **51.** Luke stresses here the obedience of Jesus towards His parents, lest the reader might think that His words to them in the temple showed undue independence. *His mother kept all these things:* cf. 19. **52.** The sense is, "As Jesus grew older He constantly manifested greater wisdom and grace." *Grace* is not used here in the theological sense but in the original sense of the word, "graciousness, favor"; Jesus showed Himself universally attractive. On His advance in *wisdom,* see note to text.

I: THE PUBLIC MINISTRY OF JESUS 3-21

1. THE PREPARATION 3, 1 — 4, 14

Some authors think that the first draft of Luke's Gospel began here and that *1-2* were prefixed by the Evangelist before publishing the final draft. It would seem from Mark *1*, 1-4 and Acts *10*, 37 that the oral Gospel of the Apostles began with the preaching of the Baptist.

1 f. The first phrase of this elaborate introduction is meant by the Evangelist as a chronological note, the other phrases merely give the political background. *In the fifteenth year of the reign of Tiberius Cæsar:* unfortunately it is possible to reckon the first year of the reign of the Emperor Tiberius in two ways: either from his appointment as co-regent with Augustus in the autumn of 12 A.D., or from the death of Augustus and the beginning of Tiberius' rule as sole emperor on the 19th of August of the year 14 A.D. Accordingly, the fifteenth year of Tiberius would be either from the fall of 26 A.D. to the fall of 27 A.D. or from August 19th, 28 A.D. to August 18th, 29 A.D. The method of reckoning the years of the reign of Tiberius from the death of Augustus seems to have been the only one used by profane historians. But the other method is favored by many commentators because of other chronological considerations. It is to be observed that this chronological note is given by Luke as the date of the beginning of John's ministry. V.21 implies that the baptism of Jesus took place some months later. The latter event was about two months before the first Passover of Christ's public ministry. According to the two methods mentioned above, the first Passover would be either in 28 A.D. or in 30 A.D. Astronomical calculation (the day of the Passover being determined by the moon) makes it certain that the last Passover of Christ's life was either in 30 or in 33 A.D. *Lysanias, tetrarch of Abilina:* another Lysanias of Abilina, known from profane history, was put to death by Mark Antony in 36 B.C. The Lysanias mentioned here was his son or grandson. For the other rulers mentioned here by Luke, see article on The New Testament Background.

3. Parallels in Matt. *3*, 1-3 and Mark *1*, 1-4. *The region about the Jordan:* not the whole valley of the Jordan but the southern part only, the district about Jericho (cf. Gen. *13*, 10 f). **5 f.** Luke alone gives the full citation from Isa. *40*, 3-5. **7-9.** Parallel in Matt. *3*, 7-10: in Matthew, however, these words are

spoken to the Pharisees and Sadducees, in Luke *to the crowds.*
10-15. Peculiar to Luke. Advice to people in various states of life. Note the special emphasis laid upon social justice and charity, so typical of the Third Gospel.   **13.** The higher officials in the tax-office generally connived at the extortions of the lesser tax-gatherers.   **14.** *Plunder no one:* i.e., by physical violence; the Greek is literally, "Do not shake down anyone." *Neither accuse anyone falsely:* the Greek signifies, "Do not blackmail anyone," properly, "Do not threaten to inform on anyone."
16 f. Parallels in Matt. *3,* 11 f and Mark *1,* 7 f.

19 f. Cf. Matt. *14,* 3 f; Mark *6,* 17 f. It is rather remarkable that none of the Gospels puts the account of John's arrest in its proper chronological place. The first two Gospels speak of it only in connection with the death of the Baptist. The Third Gospel clearly anticipates this event, in order to finish all that it has to say of the Baptist before beginning the account of Christ's public ministry. Actually John was arrested by Herod about a year after our Lord's baptism (cf. Matt. *4,* 12; Mark *1,* 14; John *4,* 1 ff).

3, 21-22: The Baptism of Jesus. Parallels in Matt. *3,* 13-17 and Mark *1,* 9-11. *When all the people had been baptized:* this should not be understood in the sense that Jesus was the very last person baptized by John, but it does imply that many had been baptized before Him. Luke alone notes that our Lord was *in prayer* at His baptism.

3, 23-38: Genealogy of Jesus. See the Commentary on the genealogy that is given in Matt. *1,* 1-17, which is partly similar to and partly dissimilar from this one.   **28.** *About thirty years of age:* some take this to mean, "Perhaps somewhat more, perhaps somewhat less than thirty years old." But Luke, who gives the exact date of Christ's birth (*2,* 1 f) and the exact date of the call of the Baptist (*3,* 1 f), could have given Christ's exact age at this time, if he had wished to do so. The Greek is literally "as (one) of thirty years." According to Num. *4,* 3.47 the Levites had to be at least thirty years old before beginning their sacred ministry. The Rabbis applied this to other offices and did not allow a man to teach in public unless he was at least thirty years old. Luke merely wishes to point out that Jesus complied with this regulation, without stating how much time had elapsed since His thirtieth birthday.

4, 1-13: The Temptation. Parallels in Matt. *4*, 1-11 and Mark *1*, 21 f. The chief difference between the First and the Third Gospel is the order of the second and the third temptation. Matthew seems to have the better order here, for the temptation to worship the devil forms a fitting climax to Christ's words, "Begone, Satan!" It is also rather strange that Luke, who speaks so often of the angels, should have omitted all reference to the angelic administrations here (cf. Matt. *4*, 11; Mark *1*, 13b). **6.** *For to me they have been delivered,* etc.: peculiar to Luke. Because of Adam's sin all mankind is, in a certain sense, under the power of Satan, "who leads astray the whole world" (Apoc. *12*, 9); he is "the prince of the world" (John *12*, 31; *14*, 30). **13.** The phrase *for a while* implies that Jesus was tempted again by Satan later in His life. Although the Evangelists do not record any subsequent temptations, some commentators understand this of the Agony in the Garden, which was the hour of "the power of darkness" (*22*, 53). It is worthy of note that on that occasion Luke (*22*, 43) speaks of the angelic ministrations similar to those mentioned here by Matthew and Mark. Other commentators think that Luke merely refers to the later attacks on Christ by His enemies which were instigated by Satan (cf. *22*, 3; John *8*, 44).

2. The Inauguration of the Ministry in Galilee
4, 14 — 6, 16

4, 14-32: Jesus at Nazareth. **14a.** The return to Galilee; cf. Matt. *4*, 12; Mark *1*, 14a; John *4*, 1-3. **14b-15.** A general summary of Christ's preaching in Galilee; cf. Matt. *4*, 17; Mark *1*, 14b-15.

16-30. Christ's ministry at Nazareth; partial parallels in Matt. *13*, 53-58 and Mark *6*, 1-6a. On the relation of Luke's account with that of the other two Evangelists, see Commentary on the passage in Matthew. **16-20.** Luke gives a graphic picture of this part of the synagogue service (see under word *synagogue* in the Glossary at end of text). The ruler of the synagogue could invite any adult Jewish man to read the Scripture of the day and deliver a homily on it. Visitors from other Jewish communities were often called upon to perform this function, and Christ and the Apostles made frequent use of this opportunity in order to preach the gospel (cf. especially Acts *13*, 15 f). Perhaps on this occasion Jesus was called upon because He had just returned from His visit to John the Baptist.

16. Note that one *stood up to read* from the Law or from the Prophets but *sat down* (20) to deliver the sermon.   **17.** The reading of a definite Book was determined by the season of the year, but the passages to be read from it and commented on were left to the choice of the preacher.   **18 f.** This quotation from Isa. *61,* 1 f is given by Luke according to the Septuagint version. The words *To set at liberty the oppressed* are from Isa. *58,* 6; the preacher was permitted to combine texts from the same Book in such a manner. *And the day of recompense:* literally, "of vengeance"; but these words do not occur in the Greek text of Luke's Gospel. Christ probably did not read these last words, because His sermon on this occasion was one of consolation. Isa. *61* refers directly to the deliverance of the Jews from the Babylonian captivity, but this is a type of Christ's deliverance of mankind from the bonds of sin and spiritual ignorance. *To proclaim the acceptable year of the Lord:* the prophet refers to the custom of proclaiming the fiftieth year the "jubilee," when all debts were remitted and captives and slaves set free (cf. Lev. *25,* 10); these words do not in the least imply that Christ's public ministry lasted only one year.   **20.** *Closing the volume:* literally, "rolling up the volume," which was in the form of a scroll. **21.** These few words merely state the theme of Christ's sermon. **22.**   So far this account fits in very well with the very beginning of Christ's ministry in Galilee.

23-30. This rejection of Jesus at Nazareth according to some scholars took place at a later period in His ministry, for the people refer to the miracle that Jesus had already done at Capharnaum (23). Cf. John *2,* 12; *3,* 2.   **24.** Jesus answers the proverb of the Nazarenes with a proverb of His own.   **25-27.** Christ proves His statement that *no prophet is acceptable in his own country* by showing that the prophets Elias and Eliseus were more favorably received and did more miracles in foreign countries than in Israel (cf. *3* Kgs. *17,* 8 ff; *4* Kgs. *5*).   **25.** *Heaven was shut up for three years and six months:* so also in Jas. *5,* 17; but according to *3,* Kgs. *17,* 1; *18,* 1 the rain came in the third year. Some say that this third year is found in the story of Elias in Sarepta; before this time, however, the prophet had been at the brook Carith until it dried up. Others follow the Jewish tradition, i.e., since seven was the symbol of perfection, half of this, three-and-a-half, was considered an unlucky number. Three-and-a-half years was a symbol for a period of calamity (cf. Apoc. *11,* 2 f; *12,* 6.14; *13,* 5).   **30.** It was not yet

the hour determined by the Father for the death of Jesus (cf. John *8*, 59; *10*, 31.39). His escape need not have been entirely miraculous.
31. Jesus resides in Capharnaum: parallels in Matt. *4*, 13 and Mark *1*, 21. **32.** Jesus teaches with authority: parallels in Matt. *7*, 28b-29 and Mark *1*, 22.

4, 33-37: **The Cure of a Demoniac.** Parallel in Mark *1*, 23-28.

4, 38-39: **Peter's Mother-in-law.** Parallels in Matt. *8*, 14 f and Mark *1*, 29-31. **38.** *Great fever:* Luke the physician uses a technical term to distinguish a great fever from a slight one in accordance with the practice of the ancient physicians. **39.** *Rebuked the fever:* i.e., to restrain its violence; cf. the use of the same verb in regard to the wind (*8*, 24) and the demons (*4*, 41; *9*, 43).

4, 40-44: **Other Miracles.** **40 f.** Many cures at Capharnaum: parallels in Matt. *8*, 16 and Mark *1*, 32-34. **42-44.** Jesus in the synagogues: parallels in Matt. *4*, 23 and Mark *1*, 35-39. **44.** In Luke the best Greek MSS read *in the synagogues of Judea*. If this is the original reading here, *Judea* seems to be used in a broad sense, equivalent to "Palestine," including Galilee (cf. *7*, 17).

5, 1-11: **The First Disciples Called.** Matt. *4*, 18-22 and Mark *1*, 16-20 give an account of the call of these four Apostles which is in some respects very similar to Luke's account and in other respects strikingly different. In both accounts these disciples are fishing, but according to Matthew and Mark they are using the casting-net, according to Luke, apparently the drag-net. Both accounts end by saying *they left . . . and followed him,* but according to the first two Gospels they left their father and his fishing business, according to Luke they left *all*. Matthew and Mark place the call of these disciples at the very beginning of Christ's ministry at Capharnaum, while Luke places it at the end of the first sojourn at that town or even after Christ's first missionary journey through Galilee (*4*, 44). Luke alone mentions the miraculous catch of fishes. Many commentators think therefore, that the event mentioned by Luke happened on a different occasion from that mentioned by Matthew and Mark. But many other commentators, harmonizing the differences in various ways, consider that these are merely two variant accounts of one and the same call of the four disciples while fishing.

1-3. Cf. Mark *4,* 1 where there is a similar account of Jesus teaching the crowd from a boat on another occasion. (See on Matt. *8,* 27.)   **4.** The Greek word for *nets* is generic, including all varieties of fishing nets, but from the manner in which the net is used on this occasion it is hardly possible to identify the *nets* mentioned by Luke with the "casting-nets" mentioned by Matthew and Mark.   **6.** *When they had done so:* evidently the *they* refers to Simon and his brother Andrew, although the latter is not mentioned by name in this episode as recorded by St. Luke. *A great number of fishes:* cf. John *21,* 6, where a similar, but certainly distinct, miracle is recorded.   **7.** *Their comrades in the other boat:* evidently James and John are meant (cf. 10). **8.** Peter had witnessed many previous miracles of Christ but this was the first one that concerned him so intimately; in the presence of the supernatural he felt deeply the sense of his own unworthiness.   **10.** *Henceforth thou shalt catch men:* a prophecy of the many souls that the Apostles under the primacy of St. Peter would win for Christ.   **11.** *Followed him:* note that Luke does not record, as Matthew and Mark do, Christ's call, "Come, follow me."

5, 12-16: A Leper. Parallels in Matt. *8,* 2-4 and Mark *1,* 40-45. **16.** Luke alone records that our Lord was *in prayer* in the desert after this miracle, whereas Mark alone mentions that our Lord was praying in the desert shortly before this miracle (Mark *1,* 35).

5, 17-26: A Paralytic at Capharnaum. Parallels in Matt. *9,* 1-8 and Mark *2,* 1-12. V.17, which is peculiar to Luke, was clearly intended by the Evangelist as the introductory setting to the miracle that is narrated in the following verses. *Teachers of the law:* merely a synonym for the *Scribes* mentioned in 21. *Out of Jerusalem:* perhaps a delegation from the Sanhedrin as in Matt. *15,* 1 (Mark *7,* 1) and John *1,* 19. *And the power of the Lord was present to heal them:* the word *them* seems to refer to certain sick people who are not mentioned in the passage. According to this reading *the Lord* is generally understood as meaning Christ. But the best Greek MSS read, "And the power of the Lord was present for him to work cures." According to this reading *the Lord* may be understood as meaning God the Father.

5, 27-32: The Call of Levi. Parallels in Matt. *9*, 9-13 and Mark *2*, 14-17.   **30.** *Why do you eat,* etc.: the *you* includes Christ, for the question was intended as a slur upon Jesus in particular (cf. the wording of the question in the other Synoptics).

5, 33-39: The Question of Fasting. Parallels in Matt. *9*, 14-17 and Mark *2*, 18-22.   **33.** *And make supplications:* John had taught his disciples certain definite formulas of prayer (cf. *11,* 1).   **36.** *A patch from a new garment:* the other Synoptics speak merely of "a patch of raw cloth."   **39.** Peculiar to Luke. Jesus explains that it is the extreme conservatism of the Pharisees and the disciples of John that prevents them from embracing the new teaching of Christ. They are accustomed to the old and therefore prefer it to the new.

6, 1-5: The Disciples Pluck Grain on the Sabbath. Parallels in Matt. *12,* 1-8 and Mark *2,* 23-28.   **1.** *On the second first Sabbath:* the best Greek MSS read simply, "On a Sabbath," but this shorter reading may be due to a deliberate correction made by some early copyist who could not understand the strange expression. Some commentators explain this odd expression as a combination of two glosses: since 6 mentions *another Sabbath,* some annotator of the text may have marked 1 as the *first* Sabbath; another annotator, noting that Luke had already mentioned a Sabbath (*4,* 31), may have added the word *second* here. Since the first ripe barley was offered in the temple at the Passover, this event must have taken place shortly after the Passover, the second one in Christ's public ministry. Hence many commentators explain the *second first Sabbath* as meaning the second day within the Passover octave or the second Sabbath after the feast itself.

6, 6-11: A Man with a Withered Hand. Parallels in Matt. *12,* 9-14 and Mark *3,* 1-6.

6, 12-16: The Choice of the Twelve. Parallels in Matt. *10,* 1-4 and Mark *3,* 13-19; cf. also Acts *1,* 13.   **12.** Luke notes that Christ made a solemn preparation for the choice of His Apostles by spending the entire night in prayer.

3. Second Period of the Ministry in Galilee and Across Its Lake 6, 17 — 9, 17

6, 17-19: The Sermon on the Mount: The Scene. Parallels in Matt. *4,* 24 f and Mark *3,* 7-12; cf. also Matt. *12,* 15 f. See Commentary on Matt. *5,* 1 for the scene of this sermon.   **17.** *A level stretch* must be understood, not of the valley below the mountain, but of a sort of plateau, somewhat below the top of the mountain (12), yet high enough above the Lake of Genesareth that it could still be called by Matthew "the mountain."

6, 20-26: The Beatitudes and Woes.   **20-23.** The Beatitudes; cf. Matt. *5,* 3-12.

6, 27-38: The Rules of Charity.   **27 f.** Love for one's enemies: parallel in Matt. *5,* 44.   **29 f.** The new law of talion: parallel in Matt. *5,* 39-42.   **31.** The Golden Rule: parallel in Matt. *7,* 12.   **32-36.** Charity must be all-embracing: parallel in Matt. *5,* 45-47. The "publicans" and the "Gentiles" of the First Gospel, which was written for Jewish Christians, become simply *sinners* in the Third Gospel which was written for Gentile Christians.   **37 f.** Against uncharitable judgments: parallel in Matt. *7,* 1 f; cf. also Mark *4,* 24. *Forgive, and you shall be forgiven:* literally in Greek, "set free, and you shall be set free"; but this can rightly be understood of pardoning offenses (cf. Matt. *6,* 14 f; Mark *11,* 25).   **38.** God will give us an immense reward for our generosity towards our neighbor. The figure here is taken from the custom of distributing grain by means of a *measure:* God's measure for repaying our charity will be *good,* i.e., generously large; the grains will be *shaken together* in it, so that they will pack well; they will be *pressed down,* so that the measure may hold as much grain as possible; the grain will finally be heaped up above the level of the measure until it is *running over.*

6, 39-45: Self-Examination.   **39.** The figure of the blind leading the blind is used in Matt. *15,* 14 of the Pharisees. It was evidently a proverbial saying, which is the meaning of the word *parable* here. According to the following context, the sense seems to be, "It is fitting that one who teaches another how to be good, should be good himself."   **40.** There is a similar saying in Matt. *10,* 24 f.   **41 f.** One should not criticize the imperfections of others if he has greater faults himself: parallel in

Matt. 7, 3-5.   **43-45.** Internal dispositions reveal themselves in external actions.   **43 f.** Parallel in Matt. 7, 16-18.   **45.** Cf. Matt. 12, 34b-35, where there is a similar saying of Christ in a different context.

6, 46-49: Conclusion of the Sermon. Parallel in Matt. 7, 21.24-27.

7, 1-10: The Centurion's Servant. Parallel in Matt. 8, 5-13. Peculiar to Luke's account is the indirect appeal of the centurion by means of the *elders of the Jews* (3) and his *friends* (6). Likewise in Luke alone are certain details which add to the pathos of the scene: the servant *was dear to him* and *was at the point of death* (2).   **5.** *He loves our nation,* etc.: since it is certain that the centurion was not a Jew by birth (cf. 9), it seems that he was a proselyte, i.e., a convert to the Jewish religion.

7, 11-17: The Widow's Son. Peculiar to Luke.   **11.** In the Greek text the town is called *Nain*. No town by that name is mentioned in the Old Testament, but Nain is undoubtedly to be identified with the modern little village of the same name which lies at the eastern end of the plain of Esdraelon and at the northern base of the mountain which is now known as Little Hermon. A few miles to the east in the same valley of Jezrahel lies the ancient village of Endor (cf. Jos. 17, 11; *1* Kgs. 28, 7; Ps. 82, 11).   **12.** *The only son of his mother, and she was a widow:* note the skill that St. Luke shows in portraying the pathos of this scene.   **13.** Luke is the only Evangelist who speaks of Jesus before His resurrection as *the Lord:* so also in 10, 1; 11, 39; 22, 61; all of these instances are in passages peculiar to the Third Gospel.   **14.** Simply by the power of His word Jesus brings back to life the young man who was unquestionably dead. Contrast with this the symbolic rites and the supplications used by Elias to raise the son of the widow of Sarepta (*3* Kgs. 17, 19 ff) or by Eliseus to raise the son of the widow of Sunam (*4* Kgs. 4, 32-36). It is interesting to note that Sunam is only about three miles south of Naim.   **17.** *Judea* is used here probably in the general sense of all "Palestine" (cf. 4, 44).

7, 18-23: The Baptist's Deputation. Parallel in Matt. 11, 2-6. **20 f.** Peculiar to Luke.   **21.** *Afflictions:* literally, "scourges," as distinct from *diseases,* would seem from 22 to refer to such permanent misfortunes as lameness, deafness, etc.

7, 24-30: Christ's Witness Concerning John. Parallel in Matt. *11,* 7-11.   **29 f.** Peculiar to Luke. Some commentators take these two verses as a parenthetical note added by the Evangelist (so according to the punctuation of our text); others take these words as spoken by Christ about John. The sense of the Greek is, "And all the people who had listened to him and even the publicans justified God by being baptized with the baptism of John. But the Pharisees and the lawyers brought to naught God's purpose concerning themselves by refusing to be baptized by him." Since there is a direct contrast here, *justified God* seems to be directly opposite in meaning to *brought to naught God's purpose* and signifies "showed the wisdom of God's plan"; cf. the expression *wisdom is justified* in 35.

7, 31-35: The Stubborn Children. Parallel in Matt. *11,* 16-19.

7, 36-50. The Penitent Woman. Luke alone records this beautiful and touching example of Christ's mercy for sinners. The Evangelist introduces this narrative without any reference to time or place.   **36.** *One of the Pharisees:* his name was Simon (40.44). Since this was a very common name among the Jews at that time, it is merely coincidental that the similar anointing of Christ at a banquet in Bethany took place in the house of a certain Simon who was called "the leper" (cf. Matt. *26,* 6-13; Mark *14,* 3-9). These two events are certainly distinct. **37.** *A woman in the town who was a sinner:* i.e., she was known in that locality as leading an immoral life. The Evangelist does not identify her more definitely. Apparently St. Gregory the Great († 604 A.D.) was the first to identify her with Mary Magdalene and Mary of Bethany, the sister of Martha and Lazarus. Since his time this identification has been commonly accepted as true. The Gospel for the feast of St. Mary Magdalene is taken from these verses of St. Luke. But since this opinion, which originated in the sixth century and has always been unknown in the Eastern Church, cannot be considered as tradition in the strict sense, the affirmation or the denial of this identification has nothing to do with Catholic doctrine as such. Most modern Catholic Scripture scholars consider this penitent woman of Galilee and Mary of Magdala and Mary of Bethany to be three distinct women.

39. The Pharisee argues secretly as follows: Jesus naturally observes the traditions of the Pharisees according to which no good man would allow himself to be defiled by letting a sinful woman touch him; but since Jesus allows this woman to embrace and kiss His feet, He must not know what kind of a woman she is; hence He cannot be a true prophet, for a prophet would be expected to have an insight into the condition of her soul. **40-50.** Jesus therefore has to refute this false conclusion of the Pharisee by showing not only that He knows the true state of this woman's soul but that even from the Pharisees' viewpoint He is justified in letting her touch Him, because she is no longer a sinner. Christ makes Simon admit that the greater a man's sense of gratitude for a remitted debt, the greater will be the signs of love that this man gives (41-43). He then argues from the slight signs of love that Simon gave Him, that evidently Simon is but slightly conscious of any spiritual debts having been remitted him by Christ, whereas the great signs of love that this woman has given Christ show that she is deeply grateful for the great spiritual debt that He has canceled for her (44-47). Therefore, according to the context, the most probable and now commonly accepted interpretation of 47 is, "For this reason, I am justified in saying to thee that her numerous sins must have been forgiven, since she has given such great signs of her gratitude." *Shall be forgiven her:* the Greek has the perfect tense of this verb in the sense of, "have been forgiven her and now are forgiven." Of course, the woman must have had antecedent faith and love in order to have her sins forgiven, but this truth is not stated directly in the passage.   **48.** These words do not imply the actual moment of absolution, but rather Christ's encouraging assurance to the woman and a public announcement for the benefit of the others present, that her sins had already been forgiven.   **49.** Cf. *5,* 21.   **50.** Cf. the same words of Christ in *8,* 48; *17,* 19; *18,* 42. This *faith* signifies a firm, lively belief in Christ's divine power, combined with an unshaken confidence in His goodness.

8, 1-3: The Ministering Women. Peculiar to Luke.   **1.** Cf. *4,* 43 f.   **2.** *Mary, who is called the Magdalene* is mentioned here by the Evangelist in such a manner that he seems to be introducing her to the reader for the first time, so that she can hardly be the same penitent woman of whom St. Luke has just been speaking. She is called *the Magdalene* probably because of her origin from Magdala, a town on the western shore of the Lake of Genesareth, about six miles south of Capharnaum. Elsewhere she is mentioned only in connection with the crucifixion, burial and resurrection of Christ. *Seven devils*, perhaps merely in the sense of many devils; cf. the symbolic use of *seven* in *17*, 4; Matt. *18*, 21 f. Diabolical possession is an affliction of man's physical life and therefore implies nothing as to the moral life of the possessed person. Cf. Mark *16*, 9.   **3.** *Joanna, the wife of Chuza,* is mentioned only here and in *24*, 10. Judging from the office that her husband held, she was evidently an affluent and influential woman. *Susanna* is not named elsewhere.

8, 4-15: Parable of the Sower. The whole passage is much shorter in Luke than in Matthew or Mark.   **4-8.** The parable itself: parallels in Matt. *13*, 1-9 and Mark *4*, 1-9.   **9 f.** The reason why Jesus taught in parables: parallels in Matt. *13*, 18-23 and Mark *4*, 13-20.

8, 16-18: Purpose of This Teaching. Parallel in Mark *4*, 21-25; cf. also Matt. *5*, 15; *10*, 26; *13*, 12.   **18b.** Cf. *19*, 26; Matt. *25*, 29.

8, 19-21: Jesus and His Brethren. Parallels in Matt. *12*, 46-50 and Mark *3*, 31-35.   **21.** The description of our Lord's response is considerably longer in the other Synoptic Gospels.

8, 22-25: The Storm on the Lake. Parallels in Matt. *8*, 18.23-27 and Mark *4*, 25-40.   **23.** *They were filling:* grammatically the word *they* refers to the disciples, but according to the sense the subject of the verb is the *boat* (22).

8, 26-39: Expulsion of the Devils in Gerasa. Parallels in Matt. *8*, 28-34 and Mark *5*, 1-20. As usual, Luke's account is much more similar to Mark's than it is to Matthew's. But several of the Marcan details are omitted in the Third Gospel, which, however, has certain peculiarities of its own.   **26.** *Which is opposite Galilee:* this is added by Luke for the benefit of the Gentile Christians who would not be acquainted with Palestinian topography.   **27.** Luke alone mentions that the demoniac *wore no clothes.* But the Second Gospel presupposes this also, for we read there that after his cure the man was found "sitting clothed" (Mark *5*, 15).   **29b.** In Mark a similar description of the ferocious power of the demoniac is given when the man is first introduced.   **31.** *Not to command them to depart into the abyss* corresponds to Mark's "not to drive them out of the country." There is no contradiction in these two forms of the entreaty, but Luke is more exact here. The abyss, according to its original meaning, is any immeasurable depth. In the Old Testament this word is often used of the deep sea (in the Douay Version it is generally translated as "the deep"; cf. Gen. *1,* 2; *7,* 11; Ecclus. *1,* 2; *16,* 18; etc.), but also of the depths of the earth considered as the abode of the dead (cf. Ps. *70,* 20). In the New Testament it is used once in the latter sense (Rom. *10,* 7) and several times as the name of the abode of the demons (Apoc. *9,* 11; *11,* 7; *17,* 8; *20,* 1.3; etc.).

8, 40-56: The Daughter of Jairus; the Woman with a Hemorrhage. Parallels in Matt. *9,* 1.18-26 and Mark *5,* 21-43. **42.** Luke alone tells us that this was the *only* daughter that Jairus had. **43.** Contrast the harsher terms used by Mark in regard to the ill treatment that the woman had suffered from many physicians. Luke, the physician, speaks more gently of the men of his profession.

9, 1-11: The Mission of the Apostles. **1-5.** Commission and instruction to the Apostles: parallels in Matt. *10,* 1.9-11.14 and Mark *6,* 7-11. **6.** Ministry of the Apostles: parallel in Mark *6,* 12 f. **7-9.** Opinion of Herod and the people concerning Jesus: parallels in Matt. *14,* 1 f and Mark *6,* 14-16. From Luke's account alone one might be led to think that Herod himself did not accept the popular notion that Jesus was the Baptist come back to life. But from the other Synoptic Gospels it is certain that Herod himself believed in this idea of the people. This is the only time that Luke speaks of the death of the Baptist. **10a.** Return of the Apostles: parallel in Mark *6,* 30. **10b-11.** Across the lake: parallels in Matt. *14,* 13 f; Mark *6,* 31-34; John *6,* 1-3.

9, 12-17 Jesus Feeds Five Thousand. Parallels in Matt. *14,* 15-21; Mark *6,* 35-44; John *6,* 4-13. Luke has the shortest account of this miracle and records nothing that is not known from the other Gospels.

It is rather surprising that Luke has nothing at all of the following events of our Lord's life as recorded in the other two Synoptic Gospels: the walking of Jesus and Peter upon the water, the miraculous cures at Genesareth, the dispute about the traditions of the ancients, the cure of the daughter of the Syrophœnician woman, the cure of the deaf-mute and the other miracles in Decapolis, the second multiplication of the loaves, the warning against the leaven of the Pharisees, and the cure of the blind man at Bethsaida (Matt. *14,* 22 — *16,* 12; Mark *6,* 45 — *8,* 26). This is sometimes called "The Great Omission" of the Third Gospel. Those who hold that Luke used the Gospel of St. Mark as one of the sources of his Gospel have various ways of explaining why this whole section of Mark's Gospel was omitted by St. Luke.

4. Ministry Mostly in the Regions Bordering on Galilee *9,* 18-50

9, 18-22: Peter's Confession; Passion and Resurrection Foretold. **18-21.** Peter's avowal of his faith in Christ: parallels in Matt. *16,* 13-20 and Mark *8,* 27-30.   **18.** Luke does not mention the locality, but he alone tells us that Jesus offered up special prayers before this important occasion, as also before the choice of the Apostles (*6,* 12).   **20.** *The Christ of God:* i.e., the Anointed of God, a phrase peculiar to Luke (cf. *2,* 26). Peter's full confession and Christ's response to it are recorded in Matt. *16,* 16-19.   **22.** The first prediction of the Passion and Resurrection: parallels in Matt. *16,* 21 and Mark *8,* 31. Luke omits Peter's remonstrance and Christ's rebuke to him (cf. Matt. *16,* 22 f; Mark *8,* 32 f).

9, 23-27: The Doctrine of the Cross. Parallels in Matt. *16,* 24-28 and Mark *8,* 34-39.   **23.** The word *daily* is in Luke alone. This shows clearly that our Lord is speaking of the *cross* in a figurative sense, for a man cannot be crucified in the literal sense every day. We must take up the cross of self-denial anew every day, if we are to persevere in the following of Christ.

9, 28-36: Jesus Transfigured. Parallels in Matt. *17,* 1-8 and Mark *9,* 1-7.   **28.** *To pray:* Luke again notes the prayer of Christ before an important event.   **31.** *His death:* literally, "His departure," i.e., from this life. This shows the intimate connection between the Transfiguration and Christ's death: our Lord revealed His glory to these Apostles to prepare them for the scandal of the Cross. *To fulfill:* Christ's death was a fulfillment of the will of the Father as expressed in the prophecies of the Old Testament.

32. Peculiar to Luke. Note the similarity between this scene and that of the Agony in the Garden: in both places the same three Apostles are taken by Christ apart from the others *to pray* (28), and in both places they become drowsy while He prays alone. *When they were fully awake, they saw* . . . : It seems that the Apostles witnessed neither the beginning nor the end of the Transfiguration (see Commentary on Mark *9*, 7).   **33.** *As they were parting:* apparently the sight of the departure of Moses and Elias induced Peter to make the offer of erecting tents for them, as if he could thus detain them.   **36b.** These three Apostles kept the Transfiguration a secret because Christ had commanded them to do so (cf. Matt. *17*, 9; Mark *9*, 8 f). Luke has nothing of the following discussion about the coming of Elias (cf. Matt. *17*, 9-13; Mark *9*, 8-12).

9, 37-44a: A Possessed Boy. Parallels in Matt. *17*, 14-20 and Mark *9*, 13-28. Luke's account, which is the shortest of the three, is much more similar to Matthew's than to Mark's.   **38.** *He is my only child:* literally "my only-begotten one"; Luke alone strikes this note of pathos.   **44a.** *Majesty:* the only other occurrence of this Greek word in the New Testament is in 2 Pet. *1*, 16, the "grandeur" of Christ at the Transfiguration; literally it signifies "the manifestation of greatness or of great works."

9, 44b-45: The Second Prediction of the Passion. Parallels in Matt. *17*, 21 f and Mark *9*, 29-31.   **44b.** As given in our text, the sense seems to be, "Remember what I am now telling you, namely, that the Son of Man is to be betrayed . . ." But *words* may be used here in the sense of "things, events" as often in Greek; the sense would then be, "Remember these marvelous things which you have witnessed." In the Greek and the Latin text the following phrase is, *"For* the Son of Man . . ." Hence, not only the Transfiguration but also the miracles of Christ were to serve the Apostles as reminders of His divinity when they would see Him humiliated in the Passion.

9, 46-50: Against Ambition and Envy.   46-48. Against ambition: parallels in Matt. *18*, 1-5 and Mark *9*, 32-36.   **49 f.** Against envy: parallel in Mark *9*, 37-39.

5. MINISTRY ON THE JOURNEY TO JERUSALEM *9*, 51 — *18*, 34

This section, commonly called the Perean ministry and recorded to a large extent by St. Luke alone, gives the journey of

Jesus towards Jerusalem after he had left Galilee. John also mentions some important events belonging to this same period. Very little of this part of the Third Gospel has parallel passages in Mark, but there is a considerable amount in this section of Luke which is found in various parts of the First Gospel. Some authors do not take all these events, which are recorded in this section of Luke, as happening in chronological order on one long journey made continuously in the direction towards Jerusalem; for, even though indications of time and place are very rare in this section, we find our Lord successively in Samaria (*9*, 52), at Bethany (*10*, 38), in Herod's domain (*13*, 31), "passing between Samaria and Galilee" (*17*, 11), and finally at Jericho (*18*, 35), where Luke's narrative again joins that of Matthew and Mark. Perhaps it is best to consider our Lord's wanderings during this period as a general missionary tour in various directions, but outside of Galilee, and with the intention of arriving ultimately in Jerusalem (cf. *9*, 51; *9*, 57; *10*, 1; *13*, 22; *14*, 25; *18*, 31).

9, 51-56: The Unfriendly Samaritans. **51.** *To be taken up* seems to refer to the Ascension rather than directly to Christ's death. *He steadfastly set his face* expresses Christ's determined will to do, suffer and die at Jerusalem as His Father had decreed; cf. Isa. *50*, 7.   **54.** See Commentary on Mark *3*, 17. Some Greek MSS add to the Apostles' question the gloss, "As Elias did" (cf. *4* Kgs. *1*, 9-14).   **55 f.** On the testimony of the best MSS the critical editions of the Greek text reject this saying of Christ as unauthentic. The words *for the Son of Man . . . to save them* may indeed have been adapted here from Christ's words in *19*, 10 (cf. also John *3*, 17). If genuine, these words may be understood in the sense, "You should not seek personal revenge but should be willing to forgive injuries" (cf. the response of David in 2 Kgs. *16*, 10), or in the sense, "You are led rather by the spirit of Satan than by the Spirit of God" (cf. Matt. *16*, 23).

9, 57-62: Sacrifice to Follow Christ. The first two incidents mentioned here (57-60) are given in Matt. *8*, 19-22 as happening much earlier in the Ministry. But they seem more likely to have occurred, as Luke states, during the journey to Jerusalem, for now rather than during the Galilean ministry could Jesus well say, that *the Son of Man has nowhere to lay his head.*   **61 f.** Jesus demands an absolute detachment even from family ties in those whom He calls to be His most intimate followers (cf.

14, 25 f). But this is not demanded of all Christians, as is clear from the Epistles of the New Testament. **62.** The figure is taken from the agricultural life of the Holy Land. There the plowman must constantly keep his eyes on the furrow, lest his plow strike against the numerous rocks which often dot the Palestinian soil. So also the disciple of Jesus must give his undivided attention to his vocation and devote himself unreservedly to the Kingdom of God. These three incidents were probably a part of the sifting out of the candidates preparatory to the sending forth of the seventy-two disciples (cf. the words *Now after this* in the next v.).

10, **1-12: The Seventy-two Disciples.** Peculiar to Luke. But all the sayings of Christ, that are recorded here, are given in other contexts in the First Gospel, especially in Matthew's account of the Mission of the Apostles (Matt. *10,* 7-16). **1.** Some good MSS read *seventy* instead of *seventy-two,* but the latter number seems more probable. These disciples were certainly not on a par with the Twelve. After the Resurrection their early discipleship as such did not entitle them to an official position in the Church, but they undoubtedly formed the nucleus of the Church at Pentecost (cf. Acts *1,* 15). **2.** Cf. Matt. *9,* 37. **3.** Cf. Matt. *10,* 16. **4.** The same words on the poverty of Christ's missionaries are given in the instruction to the Apostles *(9, 3;* Matt. *10,* 9 f; Mark *6,* 8 f). The words of Christ, *and greet no one on the way,* are recorded here only; idle conversation on their journeys would only delay and distract the missionaries, but our Lord does not forbid Christians the ordinary greetings of social life. **5 f.** Cf. Matt. *10,* 11-13. The text may also be punctuated, "first say, 'Peace!' to this house." This seems more probable, for the ordinary Hebrew greeting was *Shalôm,* i.e., *Peace,* and Christ says that his customary greeting is to be spoken *to this house,* i.e., to the members of this household. *A son of peace:* i.e., one worthy of your "peace" or greeting (cf. Matt.). **7 f.** Cf. *9,* 4; Matt. *10,* 10b.11; Mark *6,* 10. Christian missionaries are to be contented with whatever food and drink they receive. **9.** Cf. Matt. *10,* 7 f. **10-12.** Cf. Matt. *10,* 14 f; *11,* 24; Mark *6,* 11.

10, **13-16: The Impenitent Towns. 13-15.** The same words of Christ are recorded in Matt. *11,* 20-23 in a different context. It seems from Matthew that our Lord addressed these Woes to Corozain and Bethsaida when He preached there for the last time. But it is unlikely that Christ returned to these towns of Galilee now that He was on His way to Jerusalem. **16.** Cf. Matt. *10,* 40; John *13,* 20.

10, **17-20: Return of the Disciples.** Only in Luke; but cf. the similar words of Christ after the Resurrection in Mark *16,* 17 f. **18.** Cf. Isa. *14,* 12; Apoc. *12,* 9. Some interpreters understand these words to mean, "At the beginning of the world I beheld Satan's fall from grace" (cf. John *8,* 44), with the implication, "Therefore, do not be too proud of your success, lest you fall as Satan did." But the use of the imperfect tense, *I was watching,* probably expressing concomitant time here, seems to point to the meaning, "While you were driving out devils during your mission, I was watching how Satan's power was overthrown." **19.** *The enemy* is Satan. **20.** The casting out of devils is a charismatic grace which does not necessarily imply the possession of sanctifying grace in the exorcist; Christ tells the disciples that they should rejoice less in the former grace than in the latter, or rather in their predestination. *Written in heaven* is the same as "written in the book of life" (Apoc. *21,* 27).

10, **21-24: Jesus Draws Men Gently to Himself.** **21 f.** The revelation of the Son and of the Father: parallel in Matt. *11,* 25-27. Luke's emphatic *In that very hour,* contrasted with Matthew's more general *At that time,* makes it probable that the Third Gospel has this very important saying of Christ in its proper context. **23 f.** The corresponding words in Matt. *13,* 16 f were spoken by Christ in regard to the Apostles' good fortune in learning of the mysteries of the Kingdom in parables. Here they are congratulated upon receiving the revelation of the relationship between Christ and His heavenly Father.

10, **25-37: The Great Commandment: The Good Samaritan.** **25-28.** There is a similar but distinct discussion of the Great Commandment of charity in Matt. *22,* 34-40; Mark *12,* 28-34. **29.** *Wishing to justify himself:* the sense seems to be, "Wishing to show that he had sufficient reason for asking this question and that he was not to be dismissed as a mere schoolboy with such a simple solution of the difficulty" (so Lagrange). Others explain these words in the sense of "moral justification" thus, "Wishing to show that he was a just man who observed this law." In any case, this doctor of the law, like the Rabbis whose opinions are recorded in the Talmud, probably limited the meaning of *neighbor* to fellow-Jews.

30-37. The Good Samaritan. This story is told in such a graphic manner that some have thought that it represents a definite historical occurrence. But it seems more probable that this, like the other parables, is simply the story of an imaginary but by no means impossible event. **30.** Possibly our Lord was actually on this road between Jerusalem and Jericho when He spoke the parable. **31 f.** One might have thought that the *priest* and the *Levite* would be more charitable than other men, but Christ tells us that even they failed on this occasion. **33.** On the contrary, the *Samaritan* would ordinarily be hostile to a Jew. **34.** Oil and wine were the common remedies for bruises and cuts among the Orientals. The wine was used for its cleansing, astringent properties, the oil for its soothing and healing effects. **36 f.** The lesson taught by Jesus in this parable is that we have an obligation of practising charity towards all men, even those who may hate us (cf. *6, 27-36*). The *neighbor,* whom, according to God's command, we must love as ourselves, is every man with whom we come in contact.

10, 38-42: Martha and Mary. **38.** *A certain village:* although Luke does not tell us its name, it was certainly Bethany. There is no reason to doubt the identity of the Martha and Mary who are mentioned here, with the sisters of Lazarus of Bethany. The characters of these women as drawn here agree perfectly with the busy, practical Martha and the quiet, contemplative Mary of the Fourth Gospel—a further confirmation of the historical accuracy of St. John's Gospel. **42.** *Only one thing is needful:* this is commonly understood as referring to the only essential thing in life, i.e., serving God and saving one's soul. But since the best Greek MSS read, "Few things are needed or only one," some commentators understand this of the meal which Martha was preparing. Christ could well have meant His words to have both this practical sense and the higher spiritual sense. Our Lord does not condemn Martha for her efforts to serve the Master in His temporal needs. But He does defend Mary's seeming idleness, for in listening attentively to His words of wisdom she is more pleasing to Him than Martha is. This is the classical text for proving the superiority of the contemplative life over the active life: cf. also *1 Cor. 7, 32-35*.

11, 1-13: **Lessons on Prayer.**   1-4. The *Our Father*. Matt. *6*, 9-13 has this prayer in the Sermon on the Mount. It can readily be granted that Jesus spoke to His disciples more than once on the important subject of prayer, although 1 f seem to imply that this was the first occasion on which Jesus taught them a definite formula of prayer. But even though the First Gospel may not have the *Our Father* in its proper chronological place, Matthew has preserved this beautiful prayer in a more perfect form.

5-8. Under the form of the little parable of the Persistent Friend, Jesus teaches us to persevere in our prayers of petition; cf. *18*, 1-8.   9-13. Parallel in Matt. *7*, 7-11 (the Sermon on the Mount).   11. Peculiar to Luke is the mention of the *egg* and the *scorpion*. The scorpion can roll itself up into a sort of ball and thus have a slight resemblance to an egg.   13. Note Luke's *Good Spirit* in place of Matthew's *good things*. We are not forbidden to pray for the things of this world, but it is far better to pray for spiritual gifts.

11, 14-26: **Blasphemy of the Pharisees.**   Matt. *12*, 22-37 and Mark *3*, 22-30 place this dispute as happening earlier in Galilee, and in a somewhat different order.   14 f. The accusation that Jesus casts out devils by Beelzebub: parallels in Matt. *12*, 22-24 and Mark *3*, 22.   16. The demand for a sign from heaven: parallel in Matt. *12*, 38.   17-23. Parables of the Divided House and the Strong Man: parallels in Matt. *12*, 25-30 and Mark *3*, 23-27. Luke has the longest and most detailed form of the latter parable.   24-26. The return of the unclean spirit: parallel in Matt. *12*, 43-45.

11, 27 f: **The Praise of Mary.**   27. Instead of saying simply, "Fortunate is thy mother," this woman singles out two characteristics of motherhood in the typical parallelism of Hebrew poetry.   28. Jesus does not forbid this woman to praise His mother, but He stresses the point that spiritual is above physical relationship and, at least implicitly, states that Mary is to be felicitated more for keeping the word of God than for being chosen to be His mother.

11, 29-32: **The Sign of Jonas.**   This is the answer to the demand made in 16. Parallel in Matt. *12*, 39-42; cf. also Matt. *16*, 4b; Mark *8*, 12.

11, 33-36: A Lesson from a Lamp. **33.** The lamp upon the lamp-stand: a proverbial saying, spoken by Christ in other contexts also; cf. *8,* 16; Mark *4,* 21 (in regard to the purpose of the parables); Matt. *5,* 15 (in regard to good example, in the Sermon on the Mount). Here, if this saying has a connection with its context, the meaning seems to be, "You should use the spiritual light that you have received, and you will recognize who I am." **34-36.** Similar words of Christ in Matt. *6,* 22 f (on singleness of purpose, in the Sermon on the Mount). For the meaning in this passage, see note to the text.

11, 37-44: Denunciation of the Pharisees. Matthew has these and the following sayings of Jesus (45-52) in his "Woes to the Scribes and Pharisees" as spoken by Christ during Holy Week (Matt. *23,* 13-36). In the First Gospel all these Woes are directed against both the Pharisees and the Scribes or lawyers together, whereas in the Third Gospel there are three Woes directly against the Pharisees (42-44) and three directly against the Lawyers (46-52). If spoken only once by Christ, both the form and the context favor the position in Luke as the original place.

37 f. For other banquets given to Christ by the Pharisees, cf. *7,* 36; *14,* 1. **38.** For a similar accusation made by the Scribes and Pharisees against the disciples, cf. Matt. *15,* 1; Mark *7,* 1-5. **39 f.** Cf. Matt. *23,* 25 f. **41.** See note to the text. **42.** Cf. Matt. *23,* 23. **43.** Cf. *20,* 46b; Matt. *23,* 6-7a; Mark *12,* 38b-39. **44.** Cf. Matt. *23,* 27, where the thought is somewhat different.

11, 45-54: Denunciation of the Lawyers. **45.** The Scribes or lawyers naturally felt offended at Christ's denunciation of the Pharisees, for the latter merely put into practice what was taught by the former. **46.** Cf. Matt. *23,* 4. **47-51.** Cf. Matt. *23,* 29-35. **52.** Cf. Matt. *23,* 13. *The key of knowledge:* i.e., the key that opens the door to knowledge. The only *knowledge* that Christ was concerned with is the knowledge of God and of His holy will. Worldly knowledge, like worldly wealth often tends to the greater harm of fallen man. **53 f.** Peculiar to Luke. But this was the typical attitude of the Scribes and Pharisees towards Christ throughout His ministry and especially during its last period.

12, 1-3: The Leaven of the Pharisees. **1a.** Under this indefinite introduction, which tells us nothing specific as to the time and place, Luke groups all the various sayings of Christ which are in this chapter. Much of this material is found in the other Synoptics also but in different contexts. Jesus undoubtedly spoke on the same or a similar topic on various occasions. Still, the mention of these *immense crowds* which thronged to hear the Master points to an earlier period of the Ministry in Galilee rather than to the last months of our Lord's life.

1b. Cf. Matt. *16,* 6; Mark *8,* 15. **2.** The same words are given in *8,* 17 and Mark *4,* 22 in the sense that the hidden truths of the parables will be made known. If they are to be understood here in connection with the preceding words, then the sense is, "The hypocrisy of the Pharisees is hidden now from men's eyes, but on the day of Judgment it will be disclosed to all." However, since 2-9 correspond almost exactly with Matt. *10,* 26-33, it seems more probable that these verses should be considered as a single unit with the same sense as in Matthew (see Commentary there).

12, 4-12: Encouragement in Persecution.   **4-9.** Cf. Matt. *10,* 28-33.   **10.** The blasphemy against the Holy Spirit: almost the very same words are recorded in Matt. *12,* 32 and Mark *3,* 28 f where they fit the context very well. Here these words seem to interrupt the context on persecution, but note that the expression *Son of Man* is used in the preceding context (8), while the *Holy Spirit* is mentioned in the following context (12).   **11 f.** Cf. Matt. *10,* 19 f.

12, 13-21: A Warning against Avarice.   **13.** According to Deut. *21,* 17 the firstborn son was to receive a double share of the paternal inheritance or twice as much as any of the other children. Apparently in the case that was presented to Christ, the eldest brother had refused to give the younger the share to which he was legally entitled.   **14.** There were courts of law for settling such cases. Jesus refuses to intervene in purely secular matters.   **15.** A warning based on great practical wisdom. Man's true happiness, not only in heaven but even here on earth, is not at all in proportion to the earthly wealth he possesses.

16-21. This forceful parable of the Rich Fool illustrates the truth that Christ has just enunciated (15) on the folly of accumulating wealth. This man gives himself no rest in his feverish pursuit of ever greater possessions. Just when he thinks that he can enjoy his wealth, death separates him from it all.   **19.**

I will say to my soul is a Semitic idiom equivalent to "I will say to myself." 20. We need not understand this to mean the rich man knew what *God said to him*. This is simply a literary device of the parable, representing God's judgment of the rich man, as if it were spoken directly to him. No doubt the world would have thought this man very wise in providing so well for the future. But God's judgment is entirely different. Knowing the infinitely greater value of the soul over the body, God laughs (cf. Ps. *2, 4*) at the folly of the poor rich man. *They demand thy soul:* i.e., thy life; it is immaterial who the *they* are. 21. Cf. 33; Matt. *6, 20*. This conclusion to the parable is developed at length in the following verses (22-34).

12, 22-34: **Trust in God.** Almost all of these words of Christ occur also in the Sermon on the Mount (Matt. *6,* 19-21.25-33). They fit well in the contexts of both the First and the Third Gospel. There is no reason why they could not have been spoken more than once by our Lord. 22-31. Cf. Matt, *6,* 25-33. 24. *The ravens* correspond to "the birds of the air" of Matthew. Some birds indeed store up seeds and nuts for the winter, but not the ravens and crows; cf. also Ps. *146, 9*. 26. Peculiar to Luke. The meaning of *a very little thing* depends on the interpretation of the preceding verse (see Commentary on Matthew). 29b. *And do not exalt yourselves:* only in Luke. The possession of great provisions for the future makes a man vain and gives him a sense of independence of God. However, the Greek is literally, "Do not be raised up, do not be suspended," which may have the sense, "Do not waver," i.e., in your trust in God. 32. Peculiar to Luke. Christ addresses the disciples as His *flock*, because He is their Good Shepherd (cf. *15, 3*; John *10,* 1-18) who provides for their wants. "He calls the flock of the elect *little* either from a comparison with the greater number of the reprobate or from His desire that they remain humble" (Venerable Bede). *To give you the kingdom:* cf. *22, 29*.

12, 35-41: **The Watchful Servants.** Similar words of our Lord occur partly in Matthew and partly in Mark. These two Evangelists record the words as part of the Eschatological Discourse, spoken during Holy Week, where indeed they seem more suitable than in the present context. 35-38. The little parable of the Master's Return from the Wedding. Mark *13*, 33-37 records a very similar parable of the Master's Return from a Journey

Abroad. But in Luke this parable has certain features in common with the parable of the Ten Virgins (Matt. *25,* 1-13). The general sense is perfectly clear: the hour of the Lord's coming is unknown and therefore we must always be ready. But certain details are uncertain. The *lamps* (35) are the festive torches as in the parable of the Ten Virgins. We might indeed consider this a sort of continuation of that parable. According to the oriental wedding customs, after the bridegroom and his male friends go to the house of the bride, they *return* with her and her female friends to the groom's house where the banquet is held (see Commentary on Matt. *25,* 1-13). Therefore the servants of the groom must be ready to receive the bridal party.   **37b.** Contrary to custom, the bridegroom himself now ministers to his servants at table (cf. *22,* 27b; John *13,* 4). *He will gird himself* (cf. John *13,* 4), just as Christ expects that His servants' *loins be girt about* (35); the long, loose garments worn in the East were a hindrance to activity, and therefore, they had to be tucked up under the girdle.   **38.** *The second watch* of the night was from 9 P.M. to midnight, *the third* from midnight to 3 A.M. Just as in Matt. *25,* 6, the bridegroom comes around midnight, much later than he was expected.

39 f. Parable of the Thief at Night: exact parallel in Matt. *24,* 43 f.   **41.** Luke alone records this question of St. Peter, but the answer, whether given on this occasion or on another, is recorded in Mark *13,* 37.

12, 42-48: Exhortation to Vigilance.   42-46. Parable of the Faithful Steward and the Wicked Steward: parallel in Matt. *24,* 45-51.   **47-48a.** The punishment of the disobedient servants will be in proportion to the knowledge that they had of their master's will. Cf. also Jas. *1,* 22; *4,* 17; 2 Pet. *2,* 21.   **48b.** The greater the spiritual gifts that a man receives from God, the greater is his moral responsibility.

12, 49-53: The Necessity of Struggle. The result of Christ's preaching will be persecution for Himself and for His disciples. The followers of Christ will be hated even by the members of their own household.   **49 f.** Peculiar to Luke. 49 has been interpreted in various ways: the *fire* is understood of the purifying fire of the gospel of Christ (see note to text), or of the love of God which Christ came to enkindle in the hearts of men; but if these words are to be understood as having a connection with the following context, they seem to refer to Christ longing for His own passion and death as well as for the persecutions which His disciples were to suffer; in that case, *fire* and "sword" (Matt.) would be symbols of persecution.   **50.** On the *baptism* of suffering, see Commentary on Mark *10,* 38.   **51-53.** Parallel in Matt. *10,* 34 f; although 52 is peculiar to Luke.   **51.** Luke has *division* where Matthew has "sword"; it is interesting to note that the Aramaic word for "sword" is also used in the figurative sense of "desolation, devastation," but not exactly in the sense of "disunion."

12, 54-59: Time for Reconciliation. In Matt. *16,* 2 f our Lord refers to a similar popular weather rule, but there the context concerns the Pharisees who demand of Jesus a sign from heaven to prove that He was the Messias. Here the sense is more general: the people are able *to judge the face of the sky and of the earth,* yet they are unable, not only to *judge this time,* i.e., to know that this is the Messianic time foretold by the prophets, but even among themselves to *judge what is right.*   **58 f.** If there is a more intimate connection between these words and the preceding beyond the external link of the same word *judge* in both passages, the logical sequence seems to be, "Learn from the signs of the time that the period given to you in this life to be reconciled with your neighbor is limited." Matt. *5,* 25 f records the same words of Christ in the Sermon on the Mount, where the context (a warning against anger) is more appropriate.

13, 1-5: The Necessity of Repentance.   **1.** We have no other record of this massacre of the Galileans in the temple, but this incident is entirely in keeping with what we know of Pilate's way of acting.   **4.** Likewise *the tower of Siloe* and its fall is mentioned only here. Perhaps this tower was a part of the structure of the old wall of the city near the pool of Siloe. This incident was evidently still fresh in the minds of Christ's hearers. The people readily saw in such misfortunes the punishment of God. On this occasion our Lord does not contradict this popular idea as He does in John *9,* 2 f, but He uses the opportunity to teach the people that they will all ultimately be punished by God if they do not repent and change their lives (3.5).

13, 6-9: A Barren Fig Tree. This parable is recorded here only, but there is a similar "parable in action" in Matt. *21,* 18 f;

Mark *11*, 12-14.20. Here, as in the preceding passage, the moral is, "The time for repentance is limited." **7.** *For three years:* this is merely a detail of the story and should not be cited as a proof that Christ's ministry lasted that long; actually the tree is given another year of grace (8).

13, 10-17: **A Stooped Woman.** This miracle is recorded here only, but cf. the similar disputes after other cures on the Sabbath in *6*, 9; *14*, 5; Matt. *12*, 11; Mark *3*, 4. **14.** The ruler of the synagogue does not dare to rebuke Jesus directly but satisfies his indignation by scolding the crowd for seeking cures on the Sabbath. Actually the woman had not asked for this cure, but Jesus *called her to him* (12).

13, 18 f: **The Mustard Seed.** Parallels in Matt. *13*, 31 f and Mark *4*, 30-32.

13, 20 f: **The Leaven.** Parallel in Matt. *13*, 33.

13, 22-30: **The Narrow Gate.** **22 f.** Only in Luke. **23.** The question about the proportion of the saved to the damned has intrigued men of all ages. **24-30.** Christ does not satisfy our curiosity on this point, but simply warns us in substance, "Let each one be concerned about his own salvation. Salvation is not easy and many will lose it." All these words of Christ are recorded in various other contexts in Matthew. **24.** Cf. Matt. *7*, 13 f. **25.** Cf. Matt. *25*, 10-12. **26 f.** Cf. Matt. *7*, 22 f. **28 f.** Cf. Matt. *8*, 11 f. **30.** Cf. Matt. *19*, 30; *20*, 16.

13, 31-35: **Jesus and Herod.** This event, which is recorded by Luke alone, took place most likely in Perea, which was under Herod's jurisdiction. **31.** These Pharisees hardly said this out of a desire to save our Lord's life. Whether their report was true or not, they hoped by this means to induce Jesus to leave their own district. **32.** *Fox,* more exactly "jackal," for the true fox is not a native of that part of the world. The jackal was a symbol not only of deceit and cunning but also of cowardice. Christ knew, even from a merely human viewpoint, that He had nothing to fear from Herod who was himself very much afraid of the opinions of the people (cf. Matt. *14*, 5). Jesus is not to die before the hour appointed by the Father. In the meantime he will continue to *cast out devils and perform cures* without fear of Herod's threats (cf. John *11*, 8-10).   **33.** It will not be in Herod's domain but at Jerusalem that Jesus, according to the divine plan, is to be put to death.   **34 f.** Jerusalem acted as if it had a special prerogative of killing the prophets. The same words of our Lord are recorded in Matt. *23*, 37-39 as spoken during Holy Week at Jerusalem.

14, 1-6: A Man with Dropsy.   **5.** For the subsequent dispute, cf. *6*, 9; *13*, 15; Matt. *12*, 11; Mark *3*, 4.

14, 7-11: The Last Seat.   **7.** For the Pharisees taking the first seats at banquets, cf. Matt. *23*, 6. This little parable is an elaboration of the practical advice given in Prov. *25*, 6 f.   **11.** The proud will be humiliated but the humble will be exalted. This proverbial saying occurs frequently in the New Testament; cf. *18*, 14; Matt. *23*, 12; Jas. *4*, 6.10; *1* Pet. *5*, 5 f.

14, 12-14: Poor Guests. This form of our Lord's saying about doing good to those from whom we can expect no natural recompense occurs here only; but cf. the similar words of Christ in *6*, 32-34; Matt. *5*, 46 f (the Sermon on the Mount).   **14.** Such charity is rewarded on the Last Day because it is done from a supernatural motive. Hence, when our Lord says, *Do not invite thy friends,* etc., He does not mean that there is anything wrong in doing so, but that such purely natural acts merit no supernatural reward.

14, 15-24: Parable of a Great Supper. Only in Luke; but there is a very similar, though distinct, parable of the Marriage Feast in Matt. *22*, 1-14.   **15.** Christ's mention of *the resurrection of the just* (14) reminds this man of the coming Messianic kingdom, which is often compared to a festive banquet in the rabbinical writings. He evidently says this in such a way as to imply that he will surely be present at this feast in the kingdom of God. In the following prophetic parable Christ foretells that many of the spiritual leaders among the Jews will not be at this feast.
    16. This *certain man* is God the Father. The *great supper* is the kingdom of God both as the Church on earth and in its crowning happiness in heaven.   **17.** *His servant* is Jesus Christ. Judging from 21-23, those first invited into the kingdom of God by Christ's preaching are the rulers and the upper classes of the

Jews. **18-20.** They neglect the invitation because they are preoccupied with the things of this world. **21.** The second group invited are also *of the city,* i.e., of Israel. They are the poor, uneducated Jews who were despised by the Pharisees. **23.** Those outside of the city along *the highways and hedges,* i.e., by-ways of the world, are the Gentiles. *Make them come in:* i.e., by moral persuasion through the preaching of the Apostles, not by physical force. **24.** The Jews who rejected Christ's gospel are rejected by God.

14, **25-35: Following of Christ.** **26 f.** The same words of Christ are recorded in Matt. *10,* 37 f. **26.** *Hate his father:* i.e., love his father less than me. Christ demands of His intimate disciples absolute renunciation from family ties; of all His followers in general He demands that they must be ready to resist the influence of relatives whenever this would be detrimental to their spiritual welfare.

28-33. Both parables, the first of which deals with private life and the second with political life, teach the same lesson. Temporary enthusiasm in following Christ is not sufficient. We must seriously reflect upon the sacrifices, duties and conditions to be fulfilled. Then courage and constancy in following Christ is necessary, otherwise we shall suffer a great loss and become an object of ridicule. Cf. *9,* 62.

34-35a. The same saying of Christ is recorded in Matt. *5,* 13, where its context is much more appropriate; cf. also Mark *9,* 49a. Here the sense seems to be, "A disciple of mine who refuses to make the sacrifices that I ask of him is as worthless as salt that has lost its strength."

15, **1-7: The Lost Sheep.** This whole chapter forms one logical unit. Vv.1 f state the occasion: the accusation of the Scribes and Pharisees that Jesus acts too friendly towards sinners (cf. the same accusation of consorting with sinners in *5,* 30; *19,* 7; Matt. *9,* 11; Mark *2,* 16). Christ defends His action in this regard by teaching that God rejoices over the conversion of one sinner and it is therefore pleasing to God to lead sinners to repentance, even though this may necessitate a certain consorting with them. Our Lord demonstrates His point by comparing God's joy in this to man's joy in finding some precious object that was lost. This He does in three parables, all of which teach the same fundamental lesson, which is directly stated only by the first two parables (7.10). But Christ goes further and states that God rejoices *more* over the repentance of one sinner than over the virtue of the just. According to some interpreters this is not to be taken too strictly but is merely based upon analogy with joy among men, with whom the greater the preceding sorrow, the greater the subsequent joy, as, for instance, the joy of a mother over the recovery of a sick child is greater than her joy would be if the child had never been sick. But perhaps in speaking of those *who have no need of repentance,* our Lord means those who *think* they have no need of repentance. Underlying our Lord's words is the thought that, because of Adam's fall, all men are sinners (cf. *1* John *1,* 10). The self-righteous who refuse to humble themselves before God for the sins they have committed, even though these be few and venial, are less pleasing to Him than those who have sinned much and grievously but have then repented sincerely.

3-7. This parable is also recorded in Matt. *18,* 12-14 but in another context to accentuate God's loving care for the "little ones."

15, 8-10: The Lost Coin. *Drachma* is the Greek word for the same coin which was called by the Romans a *denarius:* see under heading "Money" in the Glossary to text.

15, 11-32: The Prodigal Son. This is universally considered the most beautiful of all the parables of Christ. Its principal moral is the same as that of the two preceding parables. Hence in keeping with the titles given to these, this parable should be called more aptly "The Lost Son," for the prodigality of the son is not an important part of the story. But besides the main lesson of this parable, to show God's joy over the return of a repentant sinner and the unreasonable attitude of the self-righteous who are offended by this, many of the details of the parable were probably intended by Christ in an allegorical sense, especially to portray the fall and repentance of the sinner.

11. *And he said:* Christ's audience was undoubtedly the same as that mentioned in 3 where the *them* refers directly to the Scribes and Pharisees. But *the publicans and sinners* (1) were no doubt also present, and our Lord's portrayal of God's tender mercy for repentant sinners was also intended for them, to encourage them to return to their loving Father in heaven. **12.** On inheritance customs among the Jews see Commentary on *12,* 13.

The father was under no obligation during his lifetime to comply with his son's request. **15.** The feeding of swine was the lowest depths of degradation to which any Jew, especially the son of a respectable family, could sink. **16.** *Pods:* the fruit of the carob tree, which resembles the locust tree, whose long, fleshy pods are still used in the Orient for cattle fodder and even for human consumption by the poor. **17-20.** Misery brings the young man to his senses. He prepares his confession beforehand, lest his embarrassment prevent him from saying it aright to his father.

21. His father lets him begin but not finish his prepared confession. **22.** *Robe:* literally "stole," i.e., the festive garment; cf. Mark *12, 38; 16, 5.* The *ring* was not only a token of honor but if, as was usually the case, this was a signet-ring, used in sealing documents, this bestowal of the ring signified that the father made his son the administrator of his property (cf. Gen. *41, 42;* Est. *3,* 10). Sandals were the marks of freemen, since slaves went barefoot. **24a.** In his joy the father breaks forth into typical Semitic poetry, in balanced parallelism (cf. 32b). **30.** The elder son shows his meanness in his speech. He does not speak of "my brother" but of "this son of thine," throwing the blame for his wickedness upon the father. *With harlots:* Christ had not used this expression in describing the *loose living* of the younger brother. In any case, the older brother could not have known this. He was guilty of rash judgment, just as the self-righteous often are. **31 f.** The father overlooks this meanness but insists upon his right to receive back the prodigal son. No one, least of all a self-righteous hypocrite, has the right to murmur against the infinite mercy of God.

16, **1-13: The Unjust Steward.** This parable, which is found only in Luke, is often misunderstood. But Christ states its principal moral quite clearly in 9: a rich man may enter heaven if he uses his riches in charity to the poor. Certain details of the story may be somewhat obscure, since we are not fully acquainted with the economic customs of that time and place, but these details are unimportant. When the steward knew that he would soon lose his stewardship, he provided for his future by using his master's wealth, that had been entrusted to him, in such a way that he won the good-will and hospitality of his master's tenants. Christ tells us that we should imitate this prudence of the steward. God has entrusted certain earthly goods to us, but the time will soon come when we will be deprived of these by death. Therefore we should provide for our eternal future by being similarly generous towards the poor with our wealth.

1. This *steward* was the manager of a large estate. **5.** The *debtors* probably had not borrowed this oil and wheat but owed it as rent for their use of the master's lands. **8.** Not the dishonesty but the prudence of the steward is commended. *In relation to their own generation:* i.e., in their dealings with other worldlings in the affairs of this world. *The children of light:* i.e., the members of Christ's kingdom are not as far-sighted in providing for their eternal future. **9.** See note to text. *They may receive you:* it is immaterial who the *they* are, whether the friends among the poor whom you have won by means of your wealth, or the angels, or the good deeds themselves; the sense is simply, "That you may be received." Heaven is here called *the everlasting dwellings* because in the parable the steward spoke of being received into their *houses*.

Vv.10-13 contain other sayings of our Lord concerning the proper use of earthly wealth, but we need not seek a close logical connection between these words and the preceding parable. **10 f.** Wealth is *a very little thing* in God's sight. If we are not *faithful* in this regard, i.e., if we do not use the wealth, which God has entrusted to us, according to His wishes in generosity towards the poor, He will not entrust the things of *true* value, i.e., spiritual gifts, to us. **12.** *What belongs to another* is our earthly wealth, since this really belongs to God and has been merely confided to our care. *What is your own:* i.e., the spiritual wealth of the kingdom of heaven which was "prepared for you from the foundation of the world" (Matt. *25,* 34). **13.** Cf. the same words in Matt. *6,* 24 (the Sermon on the Mount).

16, **14-18: Pretenses of the Pharisees.** Vv.14 f are peculiar to Luke, but the words of our Lord in 16-18 are recorded in various other contexts in the First Gospel. Here Christ refers to these things as examples of the Pharisees' hypocrisy. **16.** Cf. Matt. *11,* 12 f. Here the sense seems to be, "Before the time of John the Baptist you Pharisees could, in a sense, boast of your scrupulous observance of the Law, for the Old Law was then still in force. But now the New Law is being promulgated, which good and sincere men are gladly embracing." **17.** Cf. Matt. *5,* 18. Christ adds these words to forestall a wrong idea of His attitude towards the Law. The New Law does not nullify but perfects the Old Law.   **18.** Cf. Matt. *5,* 32; *19,* 9. Apparently this saying is out of its proper context here, unless this is given by Christ as an example of His manner of perfecting the Old Law.

16, **19-31: The Rich Man and Lazarus.** In this parable, which Luke alone records, Christ returns to the theme of the right use of riches. A few of the Fathers thought that our Lord referred to a strictly historical fact here, but the common opinion holds that this is a true parable, i.e., an imaginary but entirely plausible story to point a moral lesson. The aim of this parable is clear: that there will be terrible punishment after death for those who live in riches and luxury and are callous towards the sufferings of the poor, whereas the latter, if they bear their poverty and sufferings patiently and with trust in God, may hope for happiness hereafter. Christ does not intend to teach directly the nature of the punishments and rewards after death. Much of the imagery in this regard is common in the rabbinical writings of that time. Some of these expressions must certainly be taken as merely parts of the literary device and not understood too strictly. Thus the rich man speaks as if he had a body after death but before the general resurrection, he converses with Abraham as if heaven and hell were but a stone's throw apart, he has feelings of charity towards his brothers on earth, which would be impossible in a damned soul, etc. But the fact that there are rewards and punishments awaiting the just and the wicked immediately after death is such an integral part of the parable, that we can rightly conclude that this truth is at least implicitly taught by our Lord here.

19. *A certain rich man:* As usual in the parables, Christ does not give him a name. But in English he is ordinarily known as "Dives," which is simply the Latin word for a *rich man.* No other sin is mentioned against him except his wilful failure to share his wealth with the poor.   **20.** *Lazarus* is the only name given to anyone in Christ's parables. We should not conclude from this, that our Lord had a definite historical individual in mind, or that this poor man has anything to do with Lazarus the brother of Martha and Mary of Bethany. Christ gave him this name in the parable because of its meaning: it is derived from the Hebrew "Eleazar" which means "God is (my) help." He wished to show thereby that the poor man went to heaven not merely because he was poor but because he bore his poverty out of love for God and with faith and trust in Him. From this passage we derive our English word *lazar* in the sense of "a leper."   **21.** It is implied that these *longings* were not satisfied. The *dogs* are mentioned, not to show that at least they had pity on Lazarus, but to show his extreme affliction, for dogs are considered as unclean animals in the East, and this only added to his affliction. **22.** See note to text. The *hell* mentioned here is not directly the place of the damned but the abode of the dead. **23.** *In his bosom:* this figure of speech is probably taken, not from the custom of reclining at table (cf. John *13,* 23), for in the language of the Rabbis the banquet with Abraham (Matt. *8,* 11) would not take place until the Messianic age, but from the image of a little son "on the lap" of his father (cf. John *1,* 18). **26.** The *great gulf* separating the abode of the damned from the abode of the blessed cannot be crossed, so that it would be impossible for Abraham to help the rich man even if he wished. We know from other sources that the states of the good and the wicked are eternally fixed and unchangeable after death, but because of the free use of figurative language in this passage it is doubtful whether this text can be used as a proof for this truth. **31.** If men do not believe God's inspired word in the Scriptures, they will find reasons for doubting the reality of someone's resurrection from the dead.

17, **1 f: Avoiding Scandal.** The same saying of Christ is recorded in a different context in Matt. *18,* 6 f; Mark *9,* 41.

17, **3 f: Forgiveness of Injuries.** There is a very similar saying of our Lord in Matt. *18,* 15a.21 f.

17, **5 f: Efficacy of Faith.** **5.** Cf. Mark *9,* 23. **6.** Similar words of Christ are recorded in Matt. *17,* 19; *21,* 21; Mark *11,* 23.

17, **7-10: The Unprofitable Servant.** Peculiar to Luke. By this little parable, taken from daily experience in life, Christ wished primarily to teach His disciples the necessity of being humble in their service of God. Unlike the Pharisees who boast of their good deeds, the lowly disciples of Jesus should recognize that they are only doing their duty towards God and should not pride themselves on their diligence and obedience. **7.10.** *Servant:* more exactly "slave," who has not the right of wages that a hired servant would have. So also God's creatures have no right to complain if at times during life He may seem to them to be a hard master. We should be glad for the privilege of serving the good God even without hope of a reward. But elsewhere Christ teaches expressly that a great reward awaits the just in heaven, (*6, 23.35*), and it is right and proper to use the hope of this reward as one of the motives for serving God.

17, **11-19: Ten Lepers.** Only in Luke, who mentions this incident not merely for the sake of recording another of Christ's miracles but also to teach a lesson on gratitude. **11.** Perhaps near Engannim where the road from Galilee running south towards Jerusalem enters the hill country of Samaria (see map at the end of text), or near Bethsan, if on this particular journey to Jerusalem Jesus passed through Perea rather than through Samaria. **14.** *Show yourselves to the priests:* see Commentary on Matt. *8, 4*. *As they were on their way:* hence Jesus delayed their cure in order to test their faith through obedience. **16.** *And he was a Samaritan:* in this typically Lucan style (cf. "And she was a widow" in *7,* 12) the Evangelist implies that from a Jewish viewpoint this despised *foreigner* would have seemed the least likely to thank God and Jesus who cured him. The text does not state explicitly that all the others who were cured were Jews, but this seems implied and is the common opinion. **19.** Cf. *7,* 50; *8,* 48; *18,* 42.

17, **20-37: Coming of the Kingdom of God.** Vv.20 f, concerning the first coming of Christ, are peculiar to Luke. The following verses, on His Second Coming, are largely paralleled in the eschatological discourse of the First Gospel.

20. The sense of the question, as proposed by the Pharisees, is, "When is the Messias coming to establish the Messianic kingdom?" *Unawares,* not in the sense of "unexpectedly" but in the sense of "without being easily noticed"; the original is literally, "The kingdom of God does not come with observation." The Pharisees expected certain clear and definite "signs from heaven" (cf. *11,* 16.29) which would unequivocally point out the Messias. **21.** *Within you:* see note to text. Some interpreters understand these words to mean, "Within your hearts." This indeed is a true concept, for the kingdom of God is essentially something spiritual. But this is hardly the sense in the present context, for Christ would not have said to the *Pharisees,* "The kingdom of God is in *your* hearts."

22. The sense is uncertain. Many understand *one day of the Son of Man* to mean "one day of my mortal life on earth"; our Lord would then say in substance, "The time is coming when you, my disciples, will long for the comfort of my visible presence, such as you are now enjoying, but you will not have it." However, in all the following context the expression *the day of the Son of Man* refers to the Second Coming of Christ and, to be consistent, it should also have the same meaning here. Therefore it seems more probable that our Lord means, "Times of persecution await you when you will long to see me; but you shall seek me in vain."   **23 f.** Cf. the same words in Matt. *24,* 26 f. **25.** These words are only in Luke. Christ speaks of the necessity of His sufferings and death before He can appear in glory (cf. *24,* 26).

26 f. Cf. the very similar words in Matt. *24,* 37-39.   **28-30.** Only in Luke, but the thought is the same as in the preceding verses. The account of the destruction of Sodom is given in Gen. *19.*   **30.** There are two points of similarity between these two early catastrophes and the coming of Christ at the end of the world: (a) all three events are sudden and unexpected; (b) men are engrossed in worldly affairs and found unprepared.   **31.** In Matt. *24,* 17 these words are to be understood literally, warning the Christians at the destruction of Jerusalem to flee at once from the doomed city. But here the sense is rather, "Do not be too attached to your earthly possessions."   **32.** Cf. Gen. *19,* 26. *Lot's wife* looked back towards Sodom because she was attached to her home there. "Beware, lest the same fate befall you for being too attached to the things of this world."   **33.** This saying was used frequently by our Lord, generally in regard to persecution (cf. *9,* 24; Matt. *10,* 39; John *12,* 25). Here the sense is more general, "He who loves the life of this world too much will lose the true life of heaven."   **34 f.** Cf. Matt. *24,* 40 f, where the thought is the same. V.34 is peculiar to Luke. In v.35 the words *Two men will be in the field,* etc., were incorporated here from Matthew, for they are not in the Greek text of Luke.

36. The sense of the question is, "Where, Lord, wilt thou appear on earth at thy Second Coming?"   **37.** Cf. Matt. *24,* 28.

18, 1-8: The Godless Judge. This parable, which is peculiar to Luke, teaches the same lesson as the similar parable of the Persistent Friend *(11,* 5-8), i.e., perseverance in the prayer of petition. But here the emphasis is on prayer to God for help in time of trials and persecutions (cf. 7 f), especially in the sufferings that will precede Christ's Second Coming (8). In both parables God allows Himself to be compared with man who must be importuned. Here, however, God is not so much compared as contrasted with the godless judge. The moral is: if such an unjust

judge will finally do justice for one who importunes him, then all the more the God of justice will surely hear the persistent prayers of His elect.
3. The *adversary* of the widow had probably defrauded her in regard to the estate of her deceased husband.   **5.** *Wear me out:* literally, "beat me black and blue"; this expression was taken from the language of pugilism (cf. the use of the same verb in *1* Cor. *9,* 27 where it is translated "chastise"), but here it is used in the figurative sense, "to cause intolerable annoyance to someone." A few interpreters, however, translate this passage, "lest she finally come (after losing all her patience) and beat me." **8.** The coming of Christ at the end of the world, when He will vindicate His elect, may seem to be delayed very long, but it will not be unnecessarily postponed. The *faith* spoken of here is especially a firm confidence in Christ's coming; but cf. Matt. *24,* 12.

18, **9-14: The Pharisee and the Publican.** This parable was intended by our Lord to teach the necessity of humility in prayer. **9.** *To some who trusted in themselves:* not necessarily to the Pharisees only but also to some of His disciples who may have been somewhat affected by the pharisaical spirit.   **10.** *The temple:* i.e., one of the courts of the temple, most likely the women's court. The temple was used not only for the public liturgy but also for private devotions.   **11.** *I thank thee:* thanksgiving is indeed one of the rightful forms of prayer, but this Pharisee was not really thankful to God, for he attributed all his virtues to his own merits. *Even like this publican:* it was bad enough for the Pharisee to make his prayer a hymn of self-praise but worse to use his prayer as a means of speaking ill of his fellowman.   **12.** *I fast twice a week:* cf. *5,* 33; Matt. *9,* 14. *I pay tithes,* etc.: cf. Matt. *23,* 23.   **13.** The very attitude of the publican is one of humility. *The sinner:* note the force of the definite article (so also in Greek); the publican singles himself out as the one in whom all sins are centered.   **14a.** *Justified rather than the other:* the sense is not that the Pharisee received some benefit from his prayer but less than the publican, but rather that the prayer of the latter alone was pleasing to God.   **14b.** Cf. *14,* 11.

18, 15-17: Jesus Blesses the Children. With this passage Luke again picks up the thread of the Synoptic narrative in common with Matthew and Mark. Parallels in Matt. *19,* 13-15 and Mark *10,* 13-16.

18, 18-30: The Danger of Riches. Parallels in Matt. *19,* 16-30 and Mark *10,* 17-31.

18, 31-34: The Third Prediction of the Passion. Parallels in Matt. *20,* 17-19 and Mark *10,* 32-34.

6. LAST MINISTRY AT JERUSALEM *18, 35—21,* 38

18, 35-43: A Blind Man at Jericho. Parallels in Matt. *20,* 29-34 and Mark *10,* 46-52.

19, 1-10: Zacchæus the Publican. Luke alone records this incident of Christ's mercy to a publican, since this aspect of our Lord's life is especially stressed in his Gospel. **1.** *Jericho* was at this time an important commercial city which did a good business in the exportation of balm. This fact as well as its position at the border on the main road between Judea and Perea accounts for the presence there of one of the higher tax-officials. **2.** The name *Zacchæus* is derived from a Hebrew word meaning "pure, innocent." From this Hebrew name, from the fact that he was called *a sinner* (8), and especially from Christ's designation of him as a true *son of Abraham* (9), there can be hardly any doubt that he was a Jew and not a Gentile. *A leading publican:* i.e., commissioner of taxes and customs for this district, having under his authority several ordinary publicans or tax-gatherers, such as St. Matthew had once been (Matt. *9,* 9). *He was rich,* no doubt from the income derived from his lucrative position. **4.** *Sycamore* means literally in Greek "fig-mulberry." This tree, which is related to the true fig-tree, has fruit resembling figs but leaves like the mulberry-tree. It is entirely different from the American buttonwood tree, which is also known as a "sycamore tree." The tree mentioned here does not grow to a great height and its lower branches are near the ground; hence it was easy to climb.

5. Jesus knew that Zacchæus was well-disposed. **7.** Although Gal. *2,* 15 uses the word *sinner* as designating a Gentile, the ordinary Jewish use of this term signified a fallen-away Jew. **8.** *I give:* not in the sense of "I am accustomed to give," but in the sense of "I here and now promise to give." *One-half of my possessions to the poor:* perhaps to make atonement for his past lack of almsgiving. *Fourfold:* i.e., out of the remaining half of his possessions. If he had been as unjust as the publicans were generally accused of being, he probably would not have much left.

But note that Christ does not demand an absolute renunciation of his office and all his possessions as He demanded of the publican Levi (cf. *5*, 27 f). The *fourfold* restitution which Zacchæus imposed upon himself was prescribed in Ex. *22*, 1 for a convicted thief.   **9.** *This house:* not the edifice but the household. The whole family of Zacchæus believed in Christ.   **10.** Cf. *9*, 56; Matt. *18*, 11; John *3*, 17.

19, 11-28: Parable of the Gold Pieces. In some respects this is very similar to the parable of The Talents (Matt. *25*, 14-30). There are, however, many differences between the two parables so that there is no reason to think that they represent merely two records of one and the same parable. Matthew places this parable as spoken by our Lord at Jerusalem during Holy Week, whereas Luke mentions explicitly that this parable was given on the way from Jericho to Jerusalem (11. 28). The principal lesson is the same as that of the parable of the Talents, i.e., that during the period between Christ's departure from this earth and His Second Coming Christians should make use of the spiritual gifts that He has entrusted to them; at His coming He will reward those who have made good use of these gifts but will punish those who fail to co-operate with His graces. However in the present parable a prophetic element is interwoven with this moral theme: Christ foretells the rebellion of the Jews against His spiritual Messiasship and their subsequent punishment (12. 14. 27). This latter theme is sometimes considered as a separate parable and called "The parable of The Rebellious Subjects."

11. The triumphant spirit that pervaded the Galileans who were accompanying Jesus on His last journey to Jerusalem (see Commentary on Matt. *21*, 1-11) led the people to believe that He was about to proclaim Himself the political Messias whom they desired.   **12.** Jesus spoke this parable to correct this false opinion which they entertained of Him: He had first to die and go to heaven before He would enter into His kingdom. At that time many of the petty kings of the Roman Empire had to go to Rome to obtain their crown from the Emperor. Our Lord may be borrowing this imagery of the parable from such historical circumstances.   **13.** *Ten gold pieces:* literally *"ten mnas";* the *mna* was a very ancient name of a certain definite weight. In regard to money it was the designation of a certain weight of gold or silver bullion. Actually no coin called a *mna* was ever minted. It was merely a fixed sum of money equivalent to one hundred shekels (two hundred denarii); sixty mnas made one talent. Here each of the servants receives one mna apiece. **14.** This very thing happened when Archelaus (cf. Matt. *2, 22*), the son of Herod the Great, went to Rome after the death of his father. The Jews sent a delegation to Rome protesting his appointment as king over them. Our Lord may here be alluding to this but His real meaning is, that the Jews will refuse to receive Him as their king.

16-19. In the parable of The Talents the servants receive different amounts to begin with, but finally each good servant receives the double amount as his reward. Here each receives the same initial amount of money, but the good servants increase their capital to different degrees. This is simply a different aspect of the same spiritual truth of man's co-operation with grace. Here the rewards vary according to each one's co-operation; in Matthew the reward is the same for all the servants, and all are pictured as co-operating to the same degree. **20-26.** Substantially the same as Matt. *25,* 24-29. **27.** A prophecy of the punishment of the Jews at the destruction of Jerusalem and of all the enemies of Christ at the end of the world.

***19,* 29-44: Triumphal Entry into Jerusalem.** **29-38.** Parallels in Matt. *21,* 1-9; Mark *11,* 1-10; John *12,* 12-15. **38.** Peculiar to Luke is the cry of the people, *Peace in heaven and glory in the highest:* cf. *2,* 14.

39-44. Only in Luke. **39 f.** Jesus defends the enthusiasm of His followers as a natural outburst which could not be checked; cf. the similar attempt of the chief priests to have Jesus stop the cheering of the children in the temple (Matt. *21,* 15 f). *The stones will cry out:* cf. Hab. *2,* 11.

41-44. Similar words of Christ are recorded in *13,* 34 f; Matt. *23,* 37-39; but this is on a different occasion and His words here are even more pathetic. **41.** *When he drew near and saw the city:* from the top of the Mount of Olives one beholds Jerusalem and its temple spread out in a magnificent panorama. Only on one other occasion does an Evangelist record that Jesus *wept* (cf. John *11,* 35). **42.** *If thou hadst known:* i.e., "O would that thou hadst known!" *In this thy day:* the time of Christ's ministry was Jerusalem's special day of grace. *For thy peace* may be understood in the general sense, "for thy happiness, for thy greatest good," but the following context seems to imply the meaning of *peace* in the limited sense of "freedom from wars." **43 f.** This prediction was fulfilled to the letter in the year 70 A.D., when the Romans under Titus completely destroyed Jerusalem. *The time of thy visitation:* i.e., "the time when God came to thee with His extraordinary graces."

19, 45-48: Cleansing of the Temple. Parallels in Matt. *21,* 12-14 and Mark *11,* 15-18; cf. also John *2,* 13-17.

20, 1-8: The Authority of Jesus. Parallels in Matt. *21,* 23-27 and Mark *11,* 27-33.

20, 9-19: Parable of the Vine-dressers. Parallels in Matt. *21,* 33-46 and Mark *12,* 1-12.

20, 20-26: Tribute to Cæsar. Parallels in Matt. *22,* 15-22 and Mark *12,* 13-17.

20, 27-40: The Sadducees and the Resurrection. Parallels in Matt. *22,* 23-33 and Mark *12,* 18-27.   **40.** Cf. Matt. *22,* 46; Mark *12,* 34.

20, 41-44: The Son of David. Parallels in Matt. *22,* 41-45 and Mark *12,* 35-37.

20, 45-47: Hypocrisy of the Scribes and Pharisees. Parallels in Matt. *23,* 1. 5-7. 14 and Mark *12,* 38-40; cf. also *11,* 43.

21, 1-4: The Widow's Mite. Parallel in Mark *12,* 41-44.

21, 5-19: Destruction of Jerusalem and End of the World. 5-13. Parallels in Matt. *24,* 1-9 and Mark *13,* 1-9; cf. also Matt. *10,* 17 f.   **14-19.** Parallel in Mark *13,* 11-13; cf, also *12,* 11; Matt. *10,* 19-21.   **18 f.** Peculiar to Luke; cf. *12,* 7.

21, 20-24: Destruction of Jerusalem. Parallels in Matt. *24,* 15-22 and Mark *13,* 14-20.   **24.** Peculiar to Luke. V.24b foretells a long period between the destruction of Jerusalem and the Second Coming of Christ. This is called *the times of the nations* or of the Gentiles, because during this period the Gentiles are predominant not only in worldly affairs but also in the kingdom of God.

21, 25-28: The Signs of the Last Day. Parallels in Matt. *24,* 29-31 and Mark *13,* 24-27.

21, 29-33: Jerusalem's Impending Destruction. Parallels in Matt. *24,* 32-35 and Mark *13,* 28-31.

21, 34-38: The Need of Watchfulness. The other Synoptic Gospels record somewhat similar words of Christ at the end of the Eschatological Discourse but nothing that is strictly parallel. **34-36.** Cf. the similar words of Christ, warning us not to be engrossed in the cares and pleasures of this world, in *17,* 26 ff; Matt. *24,* 36 ff; Mark *13,* 32 ff.   **35.** Cf. Isa. *24,* 17.   **37.** Cf. *19,* 47; Matt. *21,* 17; Mark *11,* 11.19.   **38.** Cf. *19,* 48; Mark *12,* 37b.

II. THE PASSION, DEATH AND RESURRECTION 22 – 24

1. THE LAST SUPPER 22, 1-38

22, 1-6: The Council and the Betrayal.   **1 f.** The Council: parallels in Matt. *26,* 1-5 and Mark *14,* 1 f.   **3-6.** The Betrayal: parallels in Matt. *26,* 14-16 and Mark *14,* 10 f.

22, 7-13: Preparation. Parallels in Matt. *26,* 17-19 and Mark *14,* 12-16.

22, 14-20: The Holy Eucharist. Luke's account here is somewhat different from that of the other two Synoptics; cf. Matt. *26,* 26-29; Mark *14,* 22-25.   **15.** Despite the chronological difficulties, the words *this passover* seem to prove conclusively that the Last Supper was the Jewish Passover meal. But some interpreters understand our Lord to mean by *this passover* simply the Holy Eucharist which is the true paschal meal of the New Testament. These authors then consider that in 15-18 Luke is referring in an obscure way, because of the "discipline of the secret," to the institution of the Blessed Sacrament, 15 f referring to the consecration of the bread and 17 f to the consecration of the chalice. It is indeed rather remarkable that the words of 18b, *for I say to you that I will not drink,* etc., occur in Matthew and Mark as spoken by our Lord after the consecration of the Eucharistic cup of wine. But Luke seems to distinguish clearly between the *cup* before the Eucharist (17) and the Eucharist *cup* (20). To this objection these authors answer that 19 f are of uncertain authenticity in Luke, since they are missing in the MSS of the so-called "Western Family" of the text (at least beginning with the words, *which is being given for you*). Nevertheless, the vast majority of the MSS, including the best and oldest, have these words in Luke; they are certainly authentic here. According to the much more common and more probable opinion the *cup* of 17 refers to one of the cups of wine drunk during the Passover supper before our Lord instituted the Holy Eucharist at the end of this meal.

19 f. The words of consecration as recorded by St. Luke are almost identical with St. Paul's account in *1* Cor. *11,* 23-25. **20.** Two metonomies are used here: *This cup,* i.e., the wine contained in this cup (the metonomy of the container for the thing contained); *is the new covenant,* i.e., the inauguration or cause of the New Covenant (the metonomy of the effect for the cause); *in my blood,* i.e., through the instrumentality of my Blood, *which shall be* (in Greek, "which is now being") *shed for you.*

22, 21-23: The Betrayer. Parallels in Matt. *26,* 21-25; Mark *14,* 18-21; John *13,* 21-30. In Matthew and Mark this incident occurs before the institution of the Holy Eucharist. Those who hold that Luke is always strictly chronological, consider that Judas was present at this Mystery and received Holy Communion. Many other scholars, however, prefer to follow the order in Matthew and Mark.

22, 24-30: Contention among the Apostles. The incident itself is peculiar to Luke in this place, but cf. *9,* 46 and parallels. The same words of Christ are recorded in Matthew and Mark in another context. This contention at the Last Supper seems to have risen in regard to the places at table, for *the greatest* among them would have the most honorable place at table. That this contention took place at the beginning of the Last Supper and not after the institution of the Holy Eucharist seems, according to some scholars, implied in Christ's action at the beginning of the meal whereby He gave them a lesson in humility (cf. John *13,* 1-15). **25-27.** Almost exactly the same words of Christ are recorded in Matt. *20,* 25-28 and Mark *10,* 42-45 in regard to the contention of the Apostles after the request of the sons of Zebedee. **28-30.** Cf. the very similar words of our Lord in Matt. *19,* 28.

22, 31-38: Peter's Denials Predicted. **31-34.** Jesus foretells the three denials of Peter. Matt. *26,* 30-35 and Mark *14,* 26-31 place these words of Jesus as spoken on the way to Gethsemani, whereas John *13,* 36-38 agrees with Luke in placing them at the Last Supper. The latter order seems the more probable, unless one prefers to consider these words as spoken twice by our Lord. **31 f.** Peculiar to Luke. **31.** Satan sought to destroy the work of Jesus by making the Apostles waver in their faith and fall away. This is illustrated by the metaphor of the sieve which is shaken so violently that the grains of wheat are lost with the chaff. In the providence of God, however, the actual result will be their complete cleansing and purification from evil. **32.** Christ shows His particular solicitude for St. Peter because of his singular position as head of the Apostles. *When once thou hast turned again:* i.e., when thou art repentant after thy fall. St. Peter really never lost faith in Jesus even for a moment. His denials were due merely to thoughtless cowardice, begotten of human respect. *Strengthen thy brethren,* since I have made thee their leader. **34.** This is the only instance in the Gospels, apart from the bestowal of this title (Matt. *16,* 18) where Jesus addresses Simon as *Peter.*

34b-38. Peculiar to Luke. **35.** Cf. *9,* 3; *10,* 4. **36.** It seems fairly certain that Jesus meant these words here in a figurative sense. Hitherto the Apostles had lived in peace and were without want, but now they will soon be confronted with all sorts of hardships and trials. To meet these dangers they must be prepared and armed with spiritual weapons. **37.** Cf. Isa. *53,* 12. **38.** Understanding the words of 36 in the literal sense, the Apostles missed the point completely. Jesus replied, *"Enough,"* i.e., let us drop the subject. The Apostles' misunderstanding of our Lord's words about the *sword* is shown in Matt. *26,* 51 f and parallels.

2. The Passion and Death of Jesus 22, 39 — 23, 56

22, 39-46: The Agony in the Garden. **39-42. 45 f.** Parallels in Matt. *26,* 36-46; Mark *14,* 32-42; John *18,* 1. **43 f.** Peculiar to Luke. A few MSS, some of them of first-rate authority, omit these two verses. But it is easier to explain why some early copyist may have omitted these words intentionally out of a mistaken reverence for our Lord rather than to explain why someone should have added them to Luke's Gospel. Hence there is no good reason to doubt their authenticity. **43a.** Cf. Matt. *4,* 11b; Mark *1,* 13b. Not knowing the full nature of Christ's agony, we are not sure what this strengthening by the angel consisted in. Possibly the angel reminded our Lord of the effect of His sufferings and death; that it would be for the glory of God, for the exaltation of His own human nature, and for the salvation of mankind. **44.** *As drops of blood:* our Lord's agony was so intense that it forced His very blood to flow from the pores of His body. Such a phenomenon is not unknown to medical science.

22, 47-53: Jesus Arrested. Parallels in Matt. *26,* 47-55; Mark *14,* 43-49; John *18,* 2-11. **49.** Luke alone records this question. Peter does not wait for an answer. **53b.** Peculiar to Luke. *This is your hour:* i.e., the time when, according to the will of the Father, you are allowed to enjoy an apparent triumph. *Power of darkness:* i.e., this is being done through the efforts of Satan.

22, 54-62: Peter's Denial. **54-57.** The first denial: parallels in Matt. *26,* 58.69 f; Mark *14,* 54.66-68; John *18,* 15-17. **58.** The second denial: parallels in Matt. *26,* 71 f; Mark *14,* 69-70a; John *18,* 18.25. *Someone else,* in the masculine, hence a man; there were apparently several denials on this second occasion, since Peter was standing in a group of people. **59-62.** The third denial: parallels in Matt. *26,* 73-75; Mark *14,* 70b-72; John *18,* 26 f. All three Synoptics refer to the same incident here, but John apparently refers to a somewhat different denial on the same occasion. **61a.** Luke alone mentions that *the Lord turned and looked upon Peter.* Jesus was probably being led at the time from one part of the High Priest's house to another part of it and thus passed through the courtyard where Peter was standing.

22, 63-71: Jesus before the Sanhedrin. **63-65.** Jesus is mocked by the Jews: parallels in Matt. *26,* 67 f and Mark *14,* 65. **66a.** The morning session: parallels in Matt. *27,* 1 and Mark *15,* 1a. **66b-71.** The Jewish trial. Almost the same account is given in Matt. *26,* 59-66 and Mark *14,* 55-64 as happening during the night. At this morning session (66) the Sanhedrin met to give their proceedings the semblance of legality, since Jewish law forbade trials at night. According to other scholars all three Synoptics combine the preliminary investigation which took place at night with the formal trial which took place at dawn, Matthew and Mark giving the combined account of what occurred at night, whereas Luke gives the combined account of what happened in the morning.

23, 1-7: Jesus before Pilate.   1. Jesus is delivered to Pilate: parallels in Matt. *27, 2*; Mark *15*, 1b; John *18, 28*.   2. The charge of sedition: peculiar to Luke; cf. *20, 20-26* and parallels.   3. Pilate interrogates Jesus about His claim to be the king of the Jews: parallels in Matt. *27, 11*; Mark *15, 2*; John *18, 33-38*a.   4. Conclusion of the first hearing before Pilate: parallel in John *18*, 38b.   5-7. Peculiar to Luke who here gives the reason why Jesus was sent to Herod (transition to the next section).

23, 8-12: Jesus before Herod. Only in Luke.   11. *Arraying him:* literally "throwing about him"; Jesus was not stripped of his own garments on this occasion. The *bright robe* was not necessarily "white." The Greek word (bright) is related to our word "lamp" and means "shining, flashy, brilliant, splendid." Apparently this was intended as a royal garment to ridicule Jesus as a mock king.   12. *Previously they had been at enmity:* we do not know the reason for their previous enmity, but it has been plausibly suggested that it was caused by Pilate's high-handed treatment of the Galileans in Jerusalem (cf. *13*, 1 f).

23, 13-25: Jesus Again before Pilate.   13-15. Peculiar to Luke.   16. *Chastise him:* i.e. have him scourged. Our Lord was scourged, not merely as the usual punishment preceding crucifixion, as one might conclude from Matt. *27, 26*; Mark *15, 15*, but, according to Pilate's original intention, as a substitution for crucifixion (cf. John *19, 1-5*).   17-22. Christ and Barabbas: parallels in Matt. *27, 15-23*; Mark *15, 6-14*; John *18*, 39 f. Pilate tried this expedient before ordering the scourging of Christ (cf. 22 and John *18*, 39 — *19*, 1).   23-25. Pilate consents to the condemnation of Jesus: parallels in Matt. *27*, 24-26; Mark *15*, 15; John *19*, 16a.

23, 26-32: The Way of the Cross.   26. Simon of Cyrene: parallels in Matt. *27*, 31b-32 and Mark *15*, 20b-22; cf. also John *19*, 16b-17.   27-32. Only in Luke.   28. The sense is, "Weep less for me than for yourselves and for your children." Jesus does not reject their sympathy but warns them of the terrible chastisements that shall soon befall Jerusalem.   29. The childless woman will be envied in those days, for the mothers will have additional sorrow in the sight of their suffering children.   30. They will wish to hide within the hills and mountains to escape from God's wrath.   31. The meaning of this figurative language is, "If the Romans, who know that I am innocent, treat me so severely, how much more severely will they treat you as a rebellious nation." The figure is taken from the fact that dry wood burns more easily than green wood.

23, 33-43: The Crucifixion. **33.** Parallels in Matt. *27*, 33. 35a. 38; Mark *15*, 22.25.27; John *19*, 17b-18. **34.** Luke alone records our Lord's first word on the Cross, whereby He taught in practice His own lesson on the forgiving of one's enemies (Matt. *6*, 14 f). Christ finds an excuse for them in the fact that they were ignorant of His true nature. On the dividing of Christ's garments, cf. Matt. *27*, 35; Mark *15*, 24; John *19*, 23 f.

35-37. Jesus is derided on the Cross: parallels in Matt. *27*, 41-44 and Mark *15*, 31-32a. **38.** The title over the Cross: parallels in Matt. *27*, 37; Mark *15*, 26; John *19*, 19. **39.** In Matt. *27*, 44 and Mark *15*, 32b both robbers are spoken of as reproaching Jesus. Either they both did so at first and then afterwards one repented, or, as seems more probable, Matthew and Mark merely use the plural number, "the robbers," in the sense of "one of the robbers." **42.** *Thy kingdom:* the penitent thief acknowledges Jesus to be a king who would enter his kingdom through the portals of death, i.e., the Messias in the true sense taught by Christ. Under the circumstances this was an extraordinary act of faith.

23, 44-49: The Death of Jesus. **44-45a.** Darkness for three hours: parallels in Matt. *27*, 45 and Mark *15*, 33. **45b.** The curtain of the temple is rent asunder: in Matt. *27*, 51a and Mark *15*, 38 this is placed more suitably as happening at the very moment of Christ's death. **46.** Luke alone records this prayer of Jesus, which is taken from Ps. *30*, 6, but the "loud cry" at His death is also mentioned in Matt. *27*, 50; Mark *15*, 37. **47-49.** The centurion and the women: parallels in Matt. *27*, 54-56 and Mark *15*, 39-41.

23, 50-56: The Burial. Parallels in Matt. *27*, 57-61; Mark *15*, 42-47; John *19*, 38-40.

3. THE RESURRECTION OF JESUS *24*, 1-49

24, 1-12: The Women at the Grave. This is similar to but independent of the account in Matt. *28*, 1-8 and Mark *16*, 1-8; see Commentary there. **4.** *Two men:* i.e., two angels in human form. Matthew and Mark speak of only one angel, i.e., the one who speaks to the women.   **9.** *The eleven:* i.e., the twelve Apostles minus Judas. V.11 is missing in the MSS of the so-called "Western Family" of the text and, since it seems to be a summary of John *20,* 3-10, is rejected by many textual critics as unauthentic here. But the vast majority of the MSS, including all the best and the oldest, have this verse in Luke.

24, 13-35: Emmaus. Peculiar to Luke, but cf. Mark *16,* 12 f. **13.** *Two of them:* i.e., two disciples who were among *all the rest* (9), distinct from the Apostles (cf. 33). The exact site of *Emmaus* is uncertain. *Sixty stadia:* about seven miles; hence some authors identify Emmaus with the modern village of Kubeibeh. But a few MSS read "one hundred and sixty stadia"; hence other authors identify Emmaus with Nicopolis, which is about twenty miles west of Jerusalem. Nicopolis was indeed one of the many places called "Emmaus" (literally "hot springs") in ancient times, but the context hardly allows for a forty-mile round trip on foot by these two disciples on the same day.   **16.** *That they should not recognize him:* Christ's glorified body assumed a somewhat different appearance after the Resurrection; cf. John *20,* 14; *21,* 4. **21.** *We were hoping:* these words do not necessarily imply that now they had lost all hope.   **24.** *Him they did not see:* apparently these two disciples had left Jerusalem even before our Lord appeared to the women (Matt. *28,* 9; Mark *16,* 9; John *20,* 11-17); Luke does not mention these apparitions.

30. *He took bread and blessed and broke:* these words are so similar to the account of the institution of the Holy Eucharist (*22,* 19 and parallels), that St. Jerome, St. Augustine, and many other commentators, both ancient and modern, hold that this *breaking of the bread* (35) refers to the Holy Eucharist. Though these disciples had not been present at the institution of the Blessed Sacrament, yet they had learned of it from the Apostles and by it immediately recognized Jesus. Others, however, hold that this refers merely to an ordinary meal.   **34.** *Appeared to Simon:* this apparition of our Lord to St. Peter is not mentioned elsewhere in any of the Gospels but is referred to in *1* Cor. *15,* 5.

24, 36-43: Jesus Appears to the Eleven. This is most probably the same apparition as that mentioned in John *20,* 19-23; cf. also Mark *16,* 14.   **41-43.** Jesus ate in the presence of the Apostles in order to prove to them the reality of His risen body.

24, 44-49: The Last Instructions of Jesus. It is not certain whether all of these words were spoken on the same occasion, nor whether this occasion was the evening of Easter Sunday as in 36-43, or shortly before His ascension into heaven. At least the last part, 46-49, was spoken in Jerusalem (cf. 49b).

4. THE ASCENSION OF JESUS *24,* 50-53

24, 50-53: The Ascension. Mark *16,* 19 also speaks of our Lord's ascension into heaven. In Luke these last verses of his Gospel form a natural transition to the second volume of his work, i.e., the Acts of the Apostles, which continues the narrative of the Gospel by repeating at greater length the story of Christ's ascension (Acts *1,* 1-14). **50.** *Toward Bethany:* i.e., they walked in the direction of Bethany but did not go as far as that town, for the Ascension took place on the Mount of Olives (Acts *1,* 12). **53.** *They were continually in the temple:* both before and after the descent of the Holy Spirit the Apostles and the early Christians at Jerusalem often went to the temple to pray (cf. Acts *2,* 46; *3,* 1; *5,* 20; *21,* 26). St. Luke ends his Gospel as he began it—with a reference to the temple.

JOHN E. STEINMUELLER.

THE HOLY GOSPEL OF JESUS CHRIST ACCORDING TO ST. JOHN

Introduction

In taking up the Gospel according to St. John, the reader will notice at once a difference from the viewpoint of the Synoptics. The first three Evangelists move among the plain realities and the simpler folk of Galilee. The fourth strikes a higher note in his account of our Lord's life and teachings. This is evident at once in the Prologue. Again, John the Baptist is introduced mainly as a witness, with little said of his ministry, which is known from the Synoptics. The public life of Jesus is given chiefly as it touches Jerusalem (though still more clearly in its chronological outline than in the other Gospels). For the Synoptics the ministry in Galilee is the chief concern. This field John seldom invades, and when he does it is with his own purpose in view. Even in the story of the Passion, where all four Gospels unite, John seems to follow his own path. He omits the institution of the Holy Eucharist, the Agony in the Garden, and even the trial before the Sanhedrin. He adds, however, many details throughout the Gospel which are passed over by the others.

The recognition of these differences leads to some considerations which are important for a better understanding of the Gospel of St. John.

First of all, the author presupposes that his readers know the story from the Synoptics. This is evident from an examination of any matter which John records in common with them. That this was his intentional method may be concluded from the fact that, while he certainly knew the other Gospels, he shows not the least trace of literary dependence on them. And yet his Gospel clearly grows out of the same experiences and traditions. In both John and the Synoptics the story is fundamentally the same: Christ is the incarnate Son of God, come into the world to accomplish man's redemption by His death on the cross. In all essential features this agreement will hold for the entire Gospel.

Secondly, the first three Gospels record those features of Christ's life which had commonly figured in oral teaching before any of the New Testament was written. St. John's message is rather the person of Christ as an object of belief, on whom depends our possession of light and life. This aim commits the author to a course not followed, though certainly supposed, by the others. John centers on the king, they on the kingdom; he upon the divinity of the messenger, they on the essential value of the message. The two positions are in no sense exclusive; they are rather correlative, and often they actually overlap.

Thirdly, the comparison reaches its high point in the modal differences apparent in St. John's picture of Christ. This must be expected from the author's point of view and the scenes in which he presents the

Master. Expounding His teachings before the mixed and unprejudiced folk of Galilee, employing stories and similes which abound in familiar, homely allusions, our Lord will appear in a different light. In Jerusalem, He not only had another audience, more sophisticated, narrower, but His instructions often took the nature of a debate, a polemic. The parables of Galilee there become allegories. The miracles are fewer, and are often selected because of their logical connection with the discourse the Evangelist is reporting. And yet there is no trait in John's picture of our Lord that opposes what we know of Him from the Synoptics. Both show us Christ, but under different conditions.

Finally, it will be noticed that the author of the Fourth Gospel has left upon his writings the impress of his own personality. This accounts for much of its difference in tone. The authors of the other Gospels were dependent upon formal oral teaching, not original with any of them. Their voices, therefore, blend with the common voice of the Church in Jerusalem. John, intent upon his own thesis, more uniformly colors both discourse and narrative with his own style and disposition. Thus, he is inclined to repetition, a trait he might have learned from Christ Himself. The discourses he recalls were spoken many years earlier, but were deeply impressed upon his mind; long and serious meditation had made them a part of his own thought. With a mind wholly conformed to the Master's, his expression of these discourses is largely his own.

These differences between the Fourth Gospel and the Synoptics were recognized in the early centuries of Christianity. Eusebius cites St. Clement as saying: "Last of all John, perceiving that the external facts had been made plain in the Gospels, being urged by his friends and inspired by the Spirit, composed a *spiritual* Gospel." St. Augustine warned that prayer was needed for reading the Gospel whose author, because of his sublimity, had merited the symbol of the Eagle among the Evangelists.

The Author. Catholic tradition has held unbrokenly from the second century that St. John the Apostle is the author of the Fourth Gospel. The same tradition attributes to him the writing of the Apocalypse and the three Epistles which bear his name in our New Testament canon. We can here but suggest the line of evidence upon which this tradition rests.

External evidence. In view of the fact that the Gospel was not written until about the year 100 A.D., the testimony of the second century is most interesting. This Gospel was known in all sections of the Church, and is often explicitly attributed to John the Apostle. Recently a papyrus dating from the first half of the second century was found in the John Ryland Library, Manchester, England, which proves to be a section of this Gospel in circulation in Egypt. St. Ignatius of Antioch, who died about the year 107 A.D., was probably influenced in his writings by the Gospel. Justin Martyr, who died at Rome in 165 A.D., certainly knew it. Tatian the Syrian (born 110 A.D.) took from it the chronological sequence for his narrative based on the four Gospels. The later ecclesiastical writers of this century and the first part of the third century explicitly ascribe the Fourth Gospel to John the Apostle. For this we can cite the Muratorian Fragment, Clement of Alexandria, Irenæus in Gaul, Tertullian in Africa, Polycrates in Ephesus.

Internal evidence. The Gospel itself confirms this tradition of its author's identity. His language, his knowledge of the geography and customs of Palestine, his acquaintance with rabbinical doctrine, prejudices and methods of argumentation, his familiarity with the local religious and political conditions, all indicate a native of Palestine. Further, he claims to have been an eyewitness of what he records (*1,* 14; *19,* 35). He apparently was present at the Last Supper, and hence was an Apostle. Still, he carefully hides his identity, referring to himself only as "the disciple whom Jesus loved." All of these conditions, as may be gathered from the other Gospels and the Acts, are fulfilled only in St. John the Apostle.

St. John. From what we know of the life of St. John, he appears especially qualified for the writing of the Gospel. This natural aptitude imparts a special value to his writings. He was the brother of the Apostle James. His father, Zebedee, was a prosperous fisherman (Mark *1,* 19 f). Salome is believed to have been his mother, and a sister of Mary, the mother of our Lord; she was among the holy women who followed Jesus (Mark *15,* 40 f; Luke *23,* 49). John's home was the village of Bethsaida, on the shores of Lake Genesareth. Though a relative of our Lord, John learned of His Messianic character first from John the Baptist, among whose disciples he was numbered. He was among the first to attach themselves to Christ (*1,* 35). Later he was called more permanently, with Peter, Andrew and James (Mark *1,* 20), and is named among the chosen Twelve (Mark *3,* 17). He was of an enthusiastic disposition, and shared with his brother the title of "sons of thunder" (Mark *3,* 17; *9,* 38; Luke *9,* 49.54). He may have shared the ambition of his mother, who asked for her sons important places in the Messianic kingdom of Jesus (Matt. *20,* 20; Mark *10,* 35 ff).

John was especially intimate with his Master. Together with Peter and James he was privileged to see the raising of the daughter of Jairus, the Transfiguration, and the Agony in the Garden. At the Last Supper he reclined at the right of Jesus, and rested his head on the Master's bosom. He was the only Apostle present at the Crucifixion, and there was honored by our Lord with the care of His Blessed Mother. Thereafter Mary resided in the home of John. He could therefore well speak of himself under the title by which Christian tradition knows him, "the disciple whom Jesus loved" (*21,* 20).

We know too that he was closely associated with Peter. He may have been the other disciple known to the High Priest (*18*, 15), though this is uncertain. He ran with Peter to the tomb on the morning of the Resurrection (*20*, 1 ff); he was again with Peter at the Sea of Galilee (*21*, 1 ff); he was close to him in Jerusalem in the early years of the Church (Acts *3*, 1; *4*, 13; *8*, 14; Gal. *2*, 9).

It is well attested from the second century that John later lived in Ephesus and died there at a ripe old age. When he first came to this city is not known, but it is generally thought to have been just before the destruction of Jerusalem—hence about 66 A.D. Of his life at Ephesus we know only that he was held in high respect because of his association with our Lord, and that he had some contact with the other churches of Asia Minor. Tradition has preserved a few other items. Tertullian relates that in the persecution of Domitian (89-96) John was cast into a cauldron of boiling oil at Rome. Clement of Alexandria, Origen and Eusebius tell of his more certain banishment to the island of Patmos, where he wrote the Apocalypse. During his residence at Ephesus he wrote the three Epistles which bear his name, and toward the end of his life, the Gospel. Irenæus is authority for the fact that John was still alive at the beginning of the reign of Trajan (98-117). His death at Ephesus is generally placed at about 100 A.D.

Conditions of the Times. More than the other Gospels the work of St. John reflects the conditions of the age in which it was written. In this respect it has further value as showing the trials through which the Faith had to pass at the end of the first century. In fact, it was written to protect the Faith against these difficulties.

Ephesus was one of the most important cities in the Roman Empire, and one of the great centers of the ancient world. We know something of its earlier character from the experiences of St. Paul (Acts *19*). The church at Ephesus and throughout the territory dependent on it owed its establishment to the Apostle of the Gentiles. At that time the obstacles to the Faith arose principally from the pagan worship of Diana, the tutelary goddess of the city, and from the cult of Rome and the Emperor. But the city also enjoyed an intellectual life which brought it fame. Several philosophers held forth there, and in one of their halls St. Paul had established himself to lecture on Christianity (Acts *19*, 9). While Paul was in prison in Rome the church at Ephesus and the vicinity was disturbed by false teachers who cast doubt on the power of Christ to meet the needs of mankind. (Cf. Epistles to Colossians and Ephesians. The continuation of these difficulties may be seen also in the Epistles to Timothy.) These doubts arose mainly from the influence of Judaism, from which sect many of the early converts had been won. The Baptist sects, which claimed John the Baptist as the author of their doctrines, were another disturbing element. Their presence may be learned from Acts *18*, 24-28; *19*, 1-7.

A later and graver danger to the Faith grew out of the philosophic traditions of the city. In Greek thought from the time of Plato there was a tendency to remove God from immediate contact with creatures, on the ground that spirit and matter are incompatible. This, applied to our Lord, became the basis of the teachings of Docetism, which denied the reality of His human nature. Others taught that the gulf between God and creatures was spanned by intermediate beings, of which Christ was one. Still others, like Cerinthus, thought that Christ was but a man upon whom the Deity came at the time of His baptism, and from whom the Deity departed just before the Passion. Thus the true nature of our Lord was obscured or denied, in the matter of either His divinity or His humanity.

These internal dangers were even more serious than the persecutions, which raged during St. John's last years. Inaugurated by the madman Nero, persecution of Christianity was, from the time of Domitian, the set policy of the Roman Empire. The Jews, who feared and hated the growing power of the Church, sought every opportunity to interfere with her progress. Their attack was both doctrinal (as we know from the Epistle to the Galatians) and political.

Purposes of the Gospel. St. John's office, therefore, was not to write once more the story of Christ and His teachings. These were everywhere available at the end of the first century, through the circulation of the Synoptics. He had rather to consider the new conditions faced by the Church, and to provide the instruction most needed.

    1. His main purpose he himself states in *20, 31*: "These (signs) are written that you may believe that Jesus is the Christ, the Son of God, and that believing you may have life in his name." In other words, he seeks to prove that Christ is God incarnate, and that only through faith in Him can we have life.

    2. Since this faith was endangered by the philosophical heresies then appearing about him, John also wished to show that our Lord possessed a real human nature while remaining the eternal Son of God. Of this he had already written in his first Epistle.

    3. Against the Baptist sects he wished to urge that the Precursor admitted the superiority of Christ, to whom he bore witness. Against the Jews he would point out their guilt in refusing to accept our Lord as the Messias.

None of these purposes in any way interferes with the historical accuracy of the facts John records. On the contrary, to be of value for his thesis, his evidence had to be most faithful. This thesis was of eminent importance: a faith in Christ which touched the eternal welfare of his readers. The detection of error or inaccuracy in his writing would have defeated his purpose. Hence, apart from our conviction that John's writing was inspired by the Holy Spirit, we have natural reasons for complete confidence in the historical worth of his Gospel.

COMMENTARY
Prologue *1*, 1-18

The opening passage of the Gospel is in the form of Hebrew poetry, as is evident from its verse and strophic arrangement, its diction, and conscious climactic structure. It touches upon the main thoughts that are later drawn from the life and teachings of our Lord. It is something more than the sort of preface with which Luke's Gospel begins. It rather states the author's thesis in terms familiar to the philosophies of the time. It declares that the true mediator between God and man is a pre-existent divine person, God become man to bring spiritual life and light into the world.

This thought unfolds progressively: the general concept of the Word in Himself (1-5); more particularly, the Word's mission in the world (6-13); still more particularly, the Incarnation of the Word (14-18).

1, 1-5: The Word in Himself. *Logos:* In Greek the term *Logos* means "word," "thought," "reason." In some Greek systems of Philosophy it denoted the principle of order in the world, and also the agency of that order. It might also designate those intermediate beings which, according to some philosophers, carried the divine influence into the world of creatures. In their paraphrases of the Old Testament, the Jews employed a similar expression, *memra,* to indicate the personified Wisdom of their didactic literature (e.g., Prov. *3,* 13-20; *8*; Wisd. *9*; *16*, 12; Ecclus. *24*). Such expressions may have influenced John to state his thesis in somewhat philosophical terms, but only for the purpose of setting his teaching in sharpest contrast to that of the pagan systems of thought, and of completing the partial revelation of the Old Testament. The idea which he conveys by the term is borrowed from neither. Foreshadowed in the Wisdom Books of the Old Testament (as in the texts cited above), it is clear in the teachings of St. Paul (*1* Cor. *8,* 6; Col. *1*, 16; Heb. *1*, 2). For John this Word is eternal, a person, and divine, and this cannot be said of the *Logos* of the pagan philosophers or the *memra* of the Targums.   **1.** *In the beginning* is an allusion to Gen. *1*, 1; before the creation of the visible world, the Word already existed, and existed continuously. He was *with God* in closest relationship; He is therefore not only coeternal, but also consubstantial with God. **V.2** reasserts this with some emphasis. **3.** The Word is the Creator of the world. This also is emphasized: nothing that exists could exist without His intervention. A Latin tradition slightly modifies this interpretation by connecting "what has been made" with the next words, reading: "What was made in him was life." Another interpretation would read: "Without him was made nothing that was made in him." The sense is, plainly, the absolute dependence of all created things upon the Word. The beautiful passage in the Proverbs (*8,* 22 ff) is evidently in the author's mind, and its theme is Wisdom's part in the creation of the world. The Word is not a mere instrument; His divinity is implied (*1* Cor. *8,* 6; Col. *1,* 16). **4.** The Word is also and especially the Creator of the supernatural order. *Life* and *light* belong to the nature of the Word, who therefore alone supplies the higher intellectual and spiritual needs of men. The terms are closely related, and express the ruling idea of the Gospel. *Life* is the life of God communicated to man here on earth through the Word, and to be enjoyed in full hereafter in heaven. *Light* is the revelation of supernatural truth, the enlightenment of men by the dispelling of intellectual and moral darkness. The further meaning of these will be unfolded in the Gospel. Cf. also *1* John *1,* 2 ff, and Ps. *35,* 10: "For with thee is the fountain of life, and in thy light we shall see light." By virtue of His eternal nature the Word has been the life and light of mankind through all its history. **5.** This is true in spite of the fallen nature of man. Under the image of the natural opposition between light and darkness, the author represents either active resistance to the light, or man's failure to appreciate it. Cf. Ps. *9,* 16 f; John *12,* 35.

1, 6-13: The Word's Mission. It is here treated in a general way, including His constant mission from the creation of man, of which the Incarnation is but one manifestation. This constant function of the Word in relation to the world and His special work in the Redemption are not separated here, since they are not separate in fact. **6-8.** At this point the witness of the Baptist is introduced to indicate that the light was not without testimony. This was the mission of the Baptist, to be God's agent, making the recognition of the light possible to all. *He was not himself the light:* those who exaggerated the significance of the Baptist's ministry are assured that, however venerable in person and achievement, his glory was but the reflection of the greater glory of the Word in whose service he was sent. **9-13.** This contrast leads to a definition of the function of the real light in actual experience. **9.** His work in general is to make spiritual life and light available to all men. "One coming into the world," as read by the Vulgate, may be understood as an Aramaism for "man." **10.** The thought is that of possible, not actual illumination; the light is not in fact accepted by all men. The Word was in the world in this capacity from the beginning, and although the world (all men) had its existence from Him, it did not recognize Him. The feeling with which this is stated now mounts still higher. **11.** The light is rejected, even by His own people. As in the Gospel, the people of Israel may be more properly meant, while the wider view of "all men" is not excluded (*13, 17-20*). **12 f.** The contrast between the well-disposed, who accept the light, and those who turn from it, again supposes that the light is made available to all. This will be most evident in the Incarnation. On the part of men there is required only acceptance of the light; and here, as throughout the Gospel, this is identified with faith. Given this condition, God accomplishes the rest; by a moral title the believer is empowered to become what of himself he could not be, a son of God, not in a natural but in a supernatural sense. This is the function of the Word— to elevate man to a divine sonship attainable not by carnal generation, but only through a divine regeneration. So in *3*, 1 ff and generally throughout the Gospel and *1* John.

1, **14-18: The Word Incarnate.** This strophe is closely connected with the one preceding: the possibility of that divine sonship is made actual through the Incarnation of the Word. This mystery the author simply affirms, and then declares its excellence and its evidence. **14.** *The Word was made flesh:* this last term is chosen as connoting the weakness of human nature. *And dwelt among us:* literally, "pitched his tent amongst us," an allusion to Old Testament representations of the residence of God with His people. The point is, this was so accomplished that the Word did not thereby cease to be God. The demand of evidence for this tremendous mystery is satisfied by John's personal testimony. *We saw his glory:* the divine nature of the incarnate Word as evidenced by His works. *Grace and truth* are divine attributes often mentioned in the Old Testament: God's loving-kindness and fidelity; their present connection with the Incarnation, however, gives them a fuller meaning, *grace* becoming the supreme favor of the Redemption, and *truth* the manifestation of the reality of God through the Son, as in 18.   15. Again the witness of John the Baptist is invoked, this time as if to support the author's witness, but also to re-emphasize the superiority of the incarnate Word. The Baptist not only pointed out the Word, but also confessed that, being prior to himself in nature, the Word both surpassed and superseded him. The author had first been led to Christ by this very testimony of the Baptist (*1,* 35 ff).   16 f. The meaning of this incarnation of the Word, both in itself and as related to the old dispensation. (a) It means abundance of grace for all Christians *(we all),* an abundance which results in the first from the *fullness* of the source, the infinite treasure from which all are enriched (cf. Eph. *1,* 23; *3,* 19; *4,* 13; Col. *1,* 19; *2,* 9). The same profusion is expressed by *grace for grace,* i.e., our grace corresponding to His plenitude of grace, or grace answering to grace. (b) Three points of comparison exalt the Incarnation above the old dispensation. The Law gives way to grace and truth; the one *was given,* the other *came;* the one was mediated by Moses, the other by the Word, God Himself.   18. The excellence of this mystery is summed up in the fact that the incarnate Word is a direct revelation of God to man. This again supposes that the Word, though incarnate, retains His divine nature and His relation to God as declared in 1. Man could not of himself know God, being unable to see Him. But the Word, the Son of God, hence of divine origin and nature, becoming man, has by His person and work revealed God to us. Cf. Col. *1,* 12-16; Heb. *1,* 1.

I: THE PUBLIC MINISTRY OF JESUS *1,* 19 — *12,* 50

1. Christ Reveals His Mission and Divinity *1,* 19 — *4,* 54

1, 19-34: **The Witness of John the Baptist.** Supposing the Synoptics' account of the Baptist (Matt. *3,* 1-12; Mark *1,* 1-8; Luke *3,* 1-20), John attends only to his witness to Christ. Two features of it serve as a further prelude to this account of Christ's self-manifestation.

John's witness before his official examiners. The scene presupposes the popular excitement over the Baptist's preaching, as described in the Synoptics, and also reflects the general expectation of the Messias.   19. *The Jews* in this Gospel are almost always the officials of Judaism, a fact which adds to the significance of this commission. They were to inquire into his authority both for baptizing and for preaching, since public teaching was a matter of authorization by the Sanhedrin. **20.** John's answer is emphatic, and recorded both affirmatively and negatively, in the Semitic fashion. *Not the Christ,* the expected Messias. **21.** The common opinion was that Elias would return before the coming of the Messias; this was inferred from Mal. *4,* 5-6 (cf. Ecclus. *48,* 10; Mark *9,* 9-12; Luke *1,* 17). *The Prophet* is the Messias, as foretold by Moses: Deut. *18,* 15-18. Cf. John *7,* 40; Acts *3,* 22 f; *7,* 37. **22-27.** Their insistence upon some answer shows again the official concern over the excitement caused by John's appearance. The latter's reply is a quotation from Isa. *40, 3,* and accords with the Synoptic account of his preaching of penance. **24.** The Greek text reads: "And there had been sent some of the Pharisees," which seems to support the opinion that there was a second deputation, distinct from that in 19. These do not claim the same authority and their question is of a different nature. In answering it the Baptist renders his testimony. Its essence is that Christ is present and is greater than the Baptist, the difference being not only in their respective baptisms (cf. 33) but also in their persons. This is illustrated by the figure of the sandal-strap, understood here as in Mark *1,* 7; Luke *3,* 16. **28.** The Evangelist interrupts the Baptist's testimony to identify the place. Bethany beyond the Jordan is mentioned also in *10,* 40, if not by name. The place cannot be identified today. Some MSS read "Bethabara"; a few others "Betharaba."

John's more direct witness. **29.** *The next day:* after the Pharisees' deputation. The place is the same but Christ has arrived, perhaps from the scene of His forty days' fast which followed His baptism (Matt. *4,* 1-11; Mark *1,* 12-13; Luke *4,* 1-13). *Behold the Lamb of God:* it is the common opinion, supported by the context, that the Baptist had in mind the prophecy of Isaias (*53,* 7) in giving our Lord this title. Since in his account of the Passion the Evangelist notes Christ's fulfillment of the type of the Paschal Lamb, we may also see a reference to it here. The title may also, though less probably, recall the *tamid,* the lamb sacrificed daily in the temple. See also Jer. *11,* 19. *Who takes away the sin of the world* is commonly interpreted by the Fathers as referring to Christ's immolation for sin. This again suggests Isa. *53,* from which the meaning can be drawn. In Jewish usage the expression "to take away sins" might mean (a) to answer for a crime, (b) to carry the sins of another, (c) to forgive another, (d) to take away the sins of another by securing God's pardon. The thought in Isa. 53 is clearly that the Servant "carried" our sins and infirmities in the sense of removing them. Cf. *1 John 3, 5*. *Sin of the world:* the singular in a collective sense.   **30-34.** He does not deny former acquaintance with Christ nor knowledge of His higher sanctity (Matt. *3*, 14), but only of His nature and mission. The sign by which he gained this fuller knowledge is described in the Synoptics. Cf. Matt. *3*, 13-17; Mark *1*, 9-11; Luke *3*, 21. John again emphasizes the superiority of Christ, basing his witness now on direct revelation. *This is the Son of God:* from the divine communication just described, the Baptist probably understood Son of God in its proper sense, of our Lord's divinity. His hearers, however, would not share this understanding.

1, 35-51: The First Disciples. There is a natural connection between the Baptist's witness and the beginning of Christ's public ministry; moreover, the Evangelist himself was one of these first disciples, having been a follower of the Baptist. The scene is again the Jordan valley, but not all may have received their vocation at this place or within one day. Our Lord was now probably in some temporary shelter, such as pilgrims to the Baptist would erect for themselves. The import of the episode is that from the first the disciples attached themselves to Christ as the Messias, and did so because of the Baptist's witness. **38.** Christ's question, *What is it you seek?* implies no harshness. *Rabbi* on their lips is a title of courtesy.   **39.** *Come and see,* from the sequel, must have invited them to more than a mere visit. *The tenth hour* would be about four in the afternoon, whence the events of 40-42 must probably be assigned to the next day.   The remainder of the chapter serves to identify the first five disciples.   **40.** The two who first acted on the Baptist's words were Andrew, and undoubtedly, John the Apostle, who here, as throughout the Gospel, conceals his own name.   **41 f.** The third follower is Simon, whose singular position in the Gospel is at once signified by the change of his name to *Peter*. "Kepha" in Aramaic and "Petros" in Greek mean "rock."   **44.** *Bethsaida*, situated on the shore of Lake Genesareth. To Philip, Christ merely says *Follow me*.   **45 f.** Philip summons Nathanæl, who is probably identical with Bartholomew, one of the chosen Twelve (cf. Mark *3*, 18; John *21*, 2). The basis of the disciples' Messianic expectations was the cumulative ideal of a personal

Messias drawn from the writings of Moses and the Prophets. Nathanæl's objection reflects a popular low opinion of Nazareth. *Come and see* announces something important, and implies that Nathanæl will be convinced at sight.   **47-49.** *A true Israelite:* one worthy of the name. Our Lord alludes to some circumstances connected with the fig-tree which Nathanael would understand. This evidence of preternatural knowledge awakened a faith so deep that it burst forth in the words, *Thou art the Son of God, thou art King of Israel.* The second title is Messianic, and probably defines the force of the first, which therefore is not a recognition of Christ's divinity. As the Gospel will show, this recognition had yet to grow upon the Apostles.   **50 f.** If this slight evidence begot faith, their future experience with Him would fully confirm it; they would witness His intercourse with heaven. Cf. Gen. *28,* 10-17. The assertion is most solemn; *Amen* (which only John thus twice repeats) is stronger than "truly." *Son of Man:* a Messianic title based on Dan. *7,* 13-14, and applied to Christ in the Gospels by Himself only. It occurs eleven times in John. Cf. Commentary on Matt. *8,* 20.

2, 1-12: The Marriage Feast at Cana. Cana was a village in lower Galilee, about five miles northeast of Nazareth; it is now generally identified with Kefr Kenna.   **1.** *The third day,* probably from the call of Nathanæl. *A marriage,* an occasion of great joy and feasting, the celebration lasting for seven days; hence the exhaustion of the wine can readily be explained.   **3.** When the wine failed, Mary confided the situation to her Son. The presence of the disciples suggests the nature of their attachment to Jesus, and provides His motive for the miracle, the first sign wrought to confirm their faith (cf. *1,* 50-51).   **4.** *Woman* was the customary respectful address, similar to our "Madam," though less formal; cf. *19,* 26. The question is a familiar Jewish phrase, its conciseness in Aramaic, "What to me and to thee?" is kept in the original Greek. For its use in ordinary situations see Judg. *11,* 12; 2 Kgs. *16,* 10; *19,* 22; 2 Par. *35,* 21; Matt. *8,* 29; Mark *1,* 24; *5,* 7; Luke *4,* 34; *8,* 28. It commonly implied dissent, though not always of the same intensity. Its true force in the present context is determined by the facts that Mary's suggestion could not involve a fault, that it could not be received with the least harshness, and that she at once understood that it had not been repulsed, as her order to the servants shows. *My hour has not yet come:* St. Augustine and many of the Fathers understand by this the hour of His Passion, but this hardly fits the circumstances, which rather suggest His hour for the performance of such miracles, i.e., the public ministry. **6.** *Manner of purification:* the Jewish custom of washing the hands twice during the course of a meal was of ritual significance. As the *measure* was about nine gallons, the quantity was considerable. **8.** The *chief steward* is a functionary not otherwise known from Jewish sources, but the Greek term indicates one in charge of the table arrangements. **11.** *First of his signs:* i.e., signs of His real nature, evidence of His divinity. So Christ's miracles are usually termed by John. In the later idiom of the Rabbis, "sign" meant "wonder," while in John it is almost a synonym for "evidence," a "manifestation of His glory." *His disciples believed in him:* while the miracle may have had other motives, such as regard for His mother, kindness to His host, approval of the occasion, insinuation of the transubstantiation of the Eucharist, yet the Evangelist implies that the first motive was the confirmation of the incipient faith of the disciples. **12.** Most of the five disciples hitherto named were from the neighborhood of Capharnaum on the shore of Lake Genesareth. This was to be the center of our Lord's activity in Galilee, of which John has little to tell. *Brethren:* near relatives.

2, 13-25: Cleansing of the Temple. In John this event is clearly associated with the first Passover in Christ's public ministry (13). A similar event is recorded in the Synoptics (Matt. *21,* 12 f; Mark *11,* 15 ff; Luke *19,* 45 f), but assigned to the last Passover just before the Passion. Some hold that all record the same occurrence, John giving it its proper place, while the Synoptics, which say little of Christ's visits to the temple, connect it with His last discourses there. Others feel that the differences in detail between John and the Synoptics demand two distinct occasions. But John places correctly the occurrence he records.

13. What is now described took place before the opening of the feast, when the traffic in preparation for it was in full progress. **14.** *In the temple:* i.e., in the Court of the Gentiles. Oxen and sheep were needed for the sacrifices, doves for the offerings of the poor (Lev. *5,* 7; *12,* 8; *15,* 14; Luke *2,* 24). The money-changers were at hand to exchange foreign coins, which bore forbidden images, for the shekel required for the temple tax. This traffic was generally under the supervision of the High Priest, and also for his profit. **15 f.** *A whip of cords,* probably for driving out the animals, the merchants being expelled by His authoritative command. The scene recalls the prophecy in Zach. *14,* 21. **17 f.** The event elicited different emotions in those who witnessed it. To the disciples, recently confirmed in their attachment to Christ, it recalled a quality described in Ps. *68,* 10, a psalm accepted as Messianic (John *15,* 25; Acts *1,* 20; Rom. *11,* 9; *15,* 3). This light on the event confirmed their belief. To the Jews, in this case the authorities of the temple, the act seemed a usurpation of their prerogatives which called for justification. Hence their challenge, *What sign dost thou show?* **19.** Jesus' answer was enigmatic. In Greek the words can mean more definitely, "When you have destroyed this temple," but still some ambiguity remains. Both *temple* and *raise* could have two meanings. **20-22.** The Jews, and probably at that time the disciples also, thought only of the edifice in which they were standing. *Forty-six years:* the conflicting dates for Herod's beginning of the building leave this passage of little value for the dating of Christ's ministry. The true meaning of Christ's prediction became evident to the Apostles only after the Resurrection. *The Scripture:* probably Ps. *15,* 10; Isa. *53,* 10-12.

The beginning of faith in Christ. **23.** The Greek text can be read "during the feast," i.e., within the eight days through which it lasted. John conveys to us the important fact of the motive of this faith, Christ's miracles. *In his name:* i.e., in Him. **24.** The Evangelist, however, hastens to add that this was but an incipient faith, too weak for the full truth. The sequel will prove it. *Jesus did not trust himself to them:* did not clearly reveal to them His Messianic character, as He had to the disciples. *He knew all men,* and understood the weakness of their present disposition toward Him. **25.** And He required no human assistance to this understanding, since He of Himself could read the hearts of men. Cf. Matt. *4,* 19-22; *9,* 4; *12,* 15.

3, **1-21: Nicodemus** gives an example of the imperfect faith just mentioned, showing how slow were even learned Jews to accept Christ's teaching. **1.** *Nicodemus* bore a Greek name, like many Jews of the time. He was among *the Pharisees . . . a teacher* (10) or rabbi, and a *ruler* or magistrate, probably a member of the Sanhedrin (*7,* 50). **2.** *At night:* a caution implying the rise in Jerusalem even now of opposition to Christ, and perhaps also revealing a trait of Nicodemus' character. The interview would take place in some "upper room." *Rabbi* here expresses more than the conventional degree of reverence, since miracles are acknowledged as evidence of a teaching commission direct from God. We have no account of these particular miracles, but Galileans also had been impressed by them (*4, 45*). Here John reflects the Synoptic traditions of Matt. *8,* 16 f and parallels; again, in *7, 31* he is in accord with Matt. *4,* 23 f and parallels. **3.** *Born again:* the adverb may also mean "from above," as usually in John (*3,* 31; *19,* 11), but here it is not determined to a local application by the context as in the other instances. The rebirth must come from God (Titus *3,* 5; Jas. *1,* 18; *1* Pet. *1, 23*). **4-6.** As Nicodemus fails to see that the new birth is spiritual, Christ, in reaffirming its necessity to salvation, specifies birth *by water and the Holy Spirit*. Without this rebirth man is excluded from the supernatural order (Council of Trent, Sess. VII, *de Baptismo,* Can. 2: Denzinger 858). Its necessity is finally emphasized in Mark *16,* 16. That Christ would baptize with the Holy Spirit had been foretold by John the Baptist, as Nicodemus might have recalled.   **7 f.** Christ commends the mystery to belief by a comparison (cf. Eccles. *11,* 5): if the operation of natural forces may be partly obscure, much more may that of supernatural powers. The Greek word for "spirit" is the same as for "wind," giving the illustration a force that is lost in English. St. Augustine and others read it as "Spirit"; we, with St. John Chrysostom, St. Thomas Aquinas and many others, understand "wind." The image suggests other properties of the regenerate soul, such as interior liberty with divine direction. It seems to apply in part to the mission of Christ in *8,* 14 and *9,* 30.

9-15. Nicodemus, still doubting what he cannot comprehend, is now instructed in the necessity and motives of belief: (a) As *a teacher in Israel* he should have expected that spiritual infusion of new life promised in Ezech. *11,* 19; *36,* 26 f; Joel *2,* 28 f. (b) Christ as *a teacher from God* merits belief as supreme witness (*1* John *1,* 1-5). The plural *we* is that of solemn attestation. (c) Even less would such be inclined to believe His revelation of still higher truths. The rebirth, being an induction into the Kingdom in this life, might be called an *earthly thing* if compared with such *heavenly things* as the nature and attributes of God. (d) Since *no one has ascended into heaven,* man must rely for divine truth on Him who alone *descended from heaven.* (e) The required faith is the means of eternal life for man. Here (as less explicitly in *8,* 28 and *12,* 32) the brazen serpent of the desert marches (Num.

21, 8 f) is appealed to as a type of the saving power of the Passion of Christ.

Vv. 16-21 may be understood as John's observations on the theme of the interview.   16. *Gave his only-begotten Son* measures the divine love by the completeness of its offering, in pursuance of the thought of 14 f.   17. To *judge the world* in these verses means to condemn it, since *he who believes is not judged*.   18. To refuse belief is to bring judgment on oneself. 19-21. The cause of unbelief (as in *1*, 10-11) is opposition to the light. One guilty of evil avoids the light as tending to expose him (*1* John *1*, 6), while the clear conscience welcomes its revelation. Jesus thus becomes the occasion of a division of men into two groups, those whose belief in Him wins them eternal life, and those whose rejection of Him destines them to judgment. Hereafter the Gospel keeps these two groups in view, and shows their gradual progress in opposite directions.

3, 22-36: The Witness of John the Baptist. This is the only record of a Judean ministry in the Gospels. How long it lasted will depend upon our view of the duration of Christ's public activity. This work in Judea probably extended from the Passover (*2*, 1) to the following autumn (cf. *4*, 35). Its omission by the Synoptics is explained by their primary interest in the Galilean ministry, while John's brevity on the subject is due to his exclusive use of features pertinent to his thesis. Not Christ Himself, but His disciples were baptizing (*4*, 2). As this baptizing is not mentioned by the Synoptics, it probably played no part in the Galilean ministry. With St. John Chrysostom we should hold that it was merely the penitential rite employed by the Baptist. St. Augustine, St. Bede, St. Thomas Aquinas, Maldonatus, a Lapide, Calmes, and others understand a baptism of sacramental efficacy, but this seems less probable (7, 39; *16*, 7).

22. *The land of Judea:* in distinction from Jerusalem, where Christ had just been preaching. The disciples took part in this mission, probably by preaching the call to penance which they had learned from the Baptist.   23. *Ænnon near Salim* is in the Jordan valley, about eight miles south of Scythopolis.   24. John here supposes Matt. *14*, 3 ff; Mark *6*, 17 ff; Luke *3*, 19 f. The arrest of John the Baptist probably occurred in the autumn of the year; there seems to be some relation of time between it and the matter recorded here.   25. It is not clear what this dispute was about, except that it involved our Lord, as is evident

from the sequel. Most Greek MSS read "with a Jew." It is evident that the disciples of John the Baptist (as later the disciples of Christ, cf. Luke *9,* 49 f) were jealous of their master's mission and resentful of the fact that "all the people" were coming to Christ. They had not weighed their master's previous testimony. **27-30.** His reply to their complaint touches first the cause of their envy: Jesus' influence over the people is given Him from God, and because He is the expected Christ. He illustrates his personal relation to Christ by a custom connected with marriage: *the friend of the bridegroom* may take his place for a time, but gladly yields to the bridegroom on the latter's arrival; so, now that the Christ has come, it can be but a joy to His precursor to retire in His favor.

Again in 31-36, as in 16-21, probably the author takes up the thought, applying it to faith in Christ. **31.** Not the Baptist only, but all human teachers, in virtue of their earthly origin, are here contrasted with Christ, whose heavenly origin raises Him above them in nature and doctrine. **32.** Christ can speak as eyewitness (*1,* 18; *3,* 11 ff), whence the rejection of His witness is matter for astonishment. **33-35.** *His seal:* cf. Est. *8,* 8. One cannot believe God without accepting Christ, His messenger (cf. *1* John *5,* 12). *For not by measure* recalls the Old Testament prophets who, as God's agents, were said to possess His Spirit. As applied here to Christ it has been variously understood: (a) as if referring to the Spirit given to Christ; (b) as a contrast of Christ's plenary knowledge with the limited revelations to the prophets; (c) as meaning that God is not confined by any law in the bestowal of His gifts. In any case, Christ's message is fully the message of God, and merits all the credence due to God Himself. **36.** The decisive reason for accepting Christ is that salvation rests with Him. He who believes has life (cf. *1,* 12); he who "is stubbornly disobedient" (Greek) will have instead of life the anger of God (cf. *17,* 2).

4, **1-45: The Samaritan Woman. 1-8:** The meeting at Jacob's well. **1.** The Pharisees, having questioned John the Baptist (cf. *1,* 24 ff), would be even keener to investigate Christ's activities. If we are now in the fall of the year, the Baptist has been imprisoned, and the Pharisees would be still more watchful of Christ because of His growing influence with the people. **V.2** implies that baptizing in particular was scrutinized. This is the first indication of a group opposed to Christ; the tendency will grow.

3-5. Cf. Matt. *4,* 12; Mark *1,* 14; Luke *4,* 14. The usual road from Judea to Galilee passed through Samaria. This would bring Jesus to Sichar, the present Askar, or Askaroth, both close to Sichem (Nablus). On Jacob's field see Gen. *33,* 19; *48,* 22; Jos. *24,* 32. The word for *field* may denote a small tract or a large territory. In rabbinical tradition the gift to Joseph included all of Sichem, the land between Mts. Ebal and Garizim.   **6.** *The sixth hour* was about noon. Jesus and His disciples, on the road probably from early morning, might well be weary. In their absence Jesus rests on the coping of the well. This picture of His human susceptibility, though deftly drawn, is perhaps not unstudied.   **7 f.** On the Samaritans see *8,* 48 and Luke *9,* 52 f.

9-15: Jesus the source of life. The woman's attention is first caught by the fact that a Jew (perhaps known by His dress and accent) should ask a favor of her. For the enmity between Jews and Samaritans see Ecclus. *50,* 27 f. Her next surprise comes with our Lord's offer to give her *living water*—the common phrase for "running water," but here symbolizing grace (*4,* 14; *7,* 38 f). Thus Jesus leads her from temporal to spiritual concerns. The misunderstanding expressed in her doubt leads to clearer revelation: the water of which He speaks is spiritual life leading to eternal life. (For the figure see Ecclus. *24,* 29, where thirst means increasing desire.) "To drink water" given by another is to become his pupil or disciple. Abundance of water was expected in the Messianic era. Though not perceiving these Messianic allusions, the woman is led to seek the offered gift. This is the first stage of her education.

16-26: Jesus the Messias.   **16-18.** The woman evinces a sincere desire. Proceeding to her further instruction, Jesus humiliates her, as He did Nicodemus (*3,* 10), and at the same time manifests His superior knowledge. Conscience lends force to this woman's perception of our Lord's knowledge.   **19-24.** Perhaps to justify herself or to support a doubt about Him, she raises the moot question of the central place of worship. *This mountain* was Garizim, the site of the Samaritan temple destroyed in 129 B.C. by John Hyrcanus. *Will you worship* includes all sincere worshippers.   **V.22** is parenthetic: the Jews had the true knowledge from fuller revelation and because of their place in the redemptive plan of God (Rom. *3,* 1 f; *9,* 4 f).   **23 f.** Returning to His theme, Jesus defines true worship as that which is offered *in spirit,* i.e., with internal devotion, and *in truth,* i.e., with a sincere disposition. Those who thus worship are sought by the Father, whom only sincere devotion can please. **25 f.** The Messianic hope was current in Samaria. Far advanced from her first attitude, the woman expresses in this hope of the Messias an implicit act of obedience, which merits the revelation *I am he.* This is our Lord's first open declaration of His Messianic character; it was possible here in Samaria.

27-38: Instruction of the disciples. **27.** *They wondered,* because rabbis seldom spoke with women, especially in public, since women were considered incapable of instruction in the Law; but reverence and confidence withheld the disciples from questioning. Christ now unfolds His plan for the Samaritans, and draws from it a lesson for the disciples. **31-34.** Christ prepares the disciples for the lesson by refusing the food they had bought. Away so short a time, they doubted if anyone could have provided food in their absence. He teaches them that His nourishment is the accomplishment of His mission, the will of the Father. **35.** The circumstances further permit the applying of this lesson to their part in His work. From where they stood they could look out over fertile fields, and see the people issuing from the city. The scene suggested a harvest of souls. *There are yet four months, and then comes the harvest.* Some, taking the time to be May or June, and the harvest already ripe, suppose our Lord to quote a popular proverb, merely meaning by it, "You will soon see the reaping of the spiritual harvest." Such a proverb, however, is not otherwise known, though the calendar discovered at Gezer allows four months between sowing and reaping. If, on the other hand, the words express a present fact, the time will be December to January. This is more probable, and in this case, the Judean ministry lasted from the Passover (*2, 1*) until the winter. **36-38.** The disciples are enlightened about their own future office. The harvest of souls brings joy alike to the sower, who sees the fruition of his labors, and to the reaper, who receives his reward. Christ is the sower, as often in the Synoptics; cf. Matt. *13 passim* and parallels. A proverb known to the Greeks, and reflected in Job *31,* 8; Isa. *65,* 22; Mich. *6,* 15, points out that two agencies combine to produce the fruits of harvest. This is eminently true in the case of the Apostles; the spiritual harvest they are to reap was sown by others. These were the prophets, John the Baptist and, above all, Christ Himself.

39-42: Jesus among the Samaritans. The success of our Lord's mission among the Samaritans is impressive. Contrast this with what is said of the Jews in *2,* 23-25. The difference is the more striking since the Samaritans received Him *as the Savior of the world,* a Messianic title.

43-45: The return to Galilee. Our Lord was well received by the Galileans, particularly because they had seen His miracles at the feast in Jerusalem (*2,* 23; *3,* 2). Hence 44 can imply no censure here, and yet the same proverb is turned against Galilee in Matt. *13,* 57; Mark *6,* 4; Luke *4,* 24. The apparent discrepancy is variously explained. St. Augustine takes *his own country* to mean "His own people," and concludes that John here contrasts the attitude of the Samaritans with that of the Jews. Others think the proverb to be differently applied, our Lord's country in John being Judea, in the Synoptics Galilee. Still others think that John merely recalls what Christ said of Galilee on another occasion, though now He was well received. In any case the passage introduces one of the few events of the Galilean ministry recorded by John, an illustration of our Lord's methods and of the effect of His miracles. Faith is again the central theme.

4, **46-54: The Official's Son.** There is no reason to identify this event with that of Matt. *8,* 5-13 and Luke *7,* 1-10. **46.** *Cana of Galilee:* cf. *2,* 1. This *royal official* was a Jew, probably in the service of Herod Antipas. From Cana, in the hills of Lower Galilee, to Capharnaum, on the shore of Lake Genesareth, was about sixteen miles. Some think the official to have been Chuza (Luke *8,* 3), others Manahen (Acts *13,* 1), but neither identification is at all certain. **48.** The criticism, *unless you see signs and wonders, you do not believe,* is aimed as the plural shows, at the Jews as a class. The official had some faith in Christ, but so limited as to demand His presence at the bed of the sick boy. We may note that, like Nicodemus and the Samaritan woman, the official is humbled by the criticism. *Wonders,* a frequent term in the Synoptics and the Acts, is nowhere else applied by John to Christ's miracles. **49.** In the official's repeated petition there is perhaps some impatience, but also willingness to believe. **50.** Jesus puts his faith to the final test of accepting His word without proof, and the test is endured. *Thy son lives* means that the boy is out of danger.

51-53. The event confirms the official's faith. *Seventh hour:* about 1 o'clock (see similar expressions in *1,* 39; *4,* 6; *19,* 14). If the official started at once for Capharnaum he would have met his servants the same day; hence *yesterday* seems difficult. But if the meeting with the servants was after sunset, as it may well have been, 1 o'clock of the same afternoon could be called "yesterday," as the Jewish day ended with sunset. A better explanation, however, would be that the official delayed his departure from Cana till the next day; and *departed* (50) does not exclude this. *He believed:* when he fully realized the miracle, his faith lost all reserve, and in this he was joined by the members of his household.   **54.** *Second sign* recalls the first at Cana, 2, 1 ff. It marks the inception of the Galilean ministry, to which John will refer again only at its conclusion (*6*, 1 ff).

2. Christ Confirms His Mission 5, 1 — 6, 72

5, 1-18: The Cure at the Pool of Bethsaida.   1-9: The cure. **1.** *After this* indicates no definite time. *A feast of the Jews:* this causes some discussion. Some important MSS read "the feast." Which feast is meant is important in determining the length of Christ's public ministry. If the Passover, we must (with *2*, 13; *6*, 4; *12*, 1) allow three full years and some months; if some other feast near the time of the Passover, such as Purim, two years and some months will suffice. Of those who think this feast some other than the Passover, the majority prefer Tabernacles (September-October). Admitting uncertainty, we may hold that the Passover accords better with *4*, 35 and *6*, 4. In this case the year between *5*, 1 and *6*, 4 is not recorded by John. *Jesus went up:* His disciples are not mentioned, but the same is true of 7.   **2.** The place-name, taken from the local idiom, varies with the MSS. *Bethsaida* is the form of the Vulgate and of the Latin tradition generally. Others read Bezatha, Bethzatha or Bethesda. Such a pool with porches may be seen near the Church of St. Anne, just north of the site of the temple. **3 f.** *Awaiting the movement of the water,* with all of 4, is wanting in many Greek and Latin MSS. We accept it as part of the Vulgate text, allowing the uncertainty of its origin. The passage only contributes to the setting, and has no bearing on the miracle.   **5.** Early interpreters considered the man a paralytic, which seems justified by 7. Though infirm for thirty-eight years, he may not have been so long at the pool.   **6.** Both our Lord's question and the subsequent miracle display His compassion.   **7.** Having always been disappointed, the man had lost hope. What he says of the stirring of the water requires the statement in 4. He may be only repeating a popular conviction.   **8 f.** *Pallet:* any cloak or blanket on which the man rested, a better term than "bed." As a work of mercy, not designed as public evidence, it is not called "a sign."

10-18: Discussion with the Jews. **10.** *Thou art not allowed,* etc. Among the thirty-nine classes of work forbidden on the Sabbath by their tradition was the carrying of an article from one place to another. The penalty was severe. The guilt in this case was fixed on the instigator of the act (11-12). **14.** Cf. Matt. *9,* 1-8; Mark *2,* 3-12; Luke *5,* 18-26. The remission of sin was also an element of this cure. The warning for the future is a lesson on the cause of the man's disease. **15.** His information was not necessarily prompted by malice. His very conduct shows great respect for our Lord's power. **16-18.** The effect on the Jews. *Were persecuting* denotes an attitude of which this is but one instance, and supposes other occasions, such as in Mark *2,* 23-28; *3,* 1-6 and parallels. **17.** The words convey a profound revelation; according to St. Augustine, they affirm the union of the Father and the Son in the creation and the government of the world: the Son does what the Father does (this more expressly in 19). **18.** *To put him to death* reveals the gravity of the persecution; cf. *7,* 20. It becomes their purpose henceforth.

5, 19-30: Christ's claim to Divinity is that the Father and the Son work together, even in judgment. It amplifies the claim of 17. The passage is of great importance to John's thesis. It is also a link in His demonstration of the Jews' lack of faith, and evidence of the reason for their opposition.

19. The action of the Father and that of the Son are ultimately one because Their nature is one, even though the Son be here considered as incarnate. **20.** *Greater works:* the power to give life and to judge. **21.** The raising of the dead, whether in the physical or the spiritual order, is a divine prerogative: cf. Deut. *32,* 39; Tob. *13,* 2; Isa. *26,* 19; Ezech. *37,* 1-14; Os. *6,* 2 f; etc. *The Son gives life to whom he will:* this power is sovereign in the Son as in the Father. **22.** Judgment is another divine prerogative, often announced by the prophets, especially in connection with the "day of the Lord." The Father's communication of this power to the Son is another note of equality between them. **23.** The practical consequence: the Son must be honored as is the Father. **Vv. 24-30** apply all this to the earthly mission of the Son. **24.** *He who hears:* i.e., believes. This faith is the condition of life, the only way of escape from punishment. **25.** Cf. *8,* 43; *9,* 27. The spiritually dead are meant, and their rising to spiritual life through faith in Christ. Revival from physical death (as in the case of Lazarus) is rather the theme of 28. **26 f.** Cf. *1,* 4; *5,* 21; etc. *Given to the Son* to dispense to others. *Son of Man:* a Messianic title, although here without the article. These attributes of the Word are inalienable from the Word Incarnate. **28.** *At this:* at all that has been said, but especially in 25. His power to give spiritual life will be evident when the physical resurrection, in which they believe, will take place at the sound of His voice. Cf. *1* Thess. *4,* 14 ff. **29.** Cf. Matt. *25,* 46. Judgment here is condemnation. The two groups are already distinguished by their faith or want of faith. **30.** Cf. *5,* 19; *8,* 15 f.

5, 31-47: Justification of Christ's Claims. The evidence He offers is not His own word, nor even that of the Baptist, but the Father, the miracles, the Scriptures. Refusal to accept this evidence is proof of their ill will. **Vv.** 31 f anticipate the objection that the foregoing claim of parity with the Father must have adequate support. Cf. *8,* 12-19. He can summon other witness. Sts. Augustine and Cyril and the Fathers generally hold that this witness is the Father Himself. **33-35.** Although they believed the Baptist, and he bore witness to Christ, Jesus assures them that He neglects that testimony for the present. *I do not receive:* I do not now make use of it. *He was the lamp:* cf. *1,* 7-9; though not himself the true light, yet the Baptist was a light. *Burning* and *shining* with the heat of zeal and the light of a divine message: Cf. Ecclus. *48,* 1. *For a while:* their adherence to the Baptist was not long-lived. **36.** The evidence of our Lord's miracles surpasses the witness of John. Cf. Matt. *11,* 4; Luke *7,* 22. **37a.** Cf. *1,* 18. Apart from the works, the Father has borne witness in the Scriptures (cf. 38-40). V.38 seems to require the sense: "but you refuse this testimony, you who never heard God's voice." *Search* is better taken as the indicative: you search the Scriptures, looking for guidance to life; they give it by bearing witness to me. **40.** Cf. *2* Cor. *3,* 14 f. The truth was not forced upon them. **Vv.41-47** develop the charge against the Jews in *8,* 42 f: they are ill disposed to Christ because ill disposed to God the Father. **V.41** begins by anticipating the objection, you seek to glorify yourself; cf. *1* Thess. 2, 6. **42.** Cf. *1* John 2, 15; *3,* 17; *4,* 12; *5,* 3. Wanting the love of God, how can you love me? **43.** The proof: you reject God's ambassador. *Another:* any false prophet or teacher; cf. Matt. *24*, 24; Mark *13*, 22; 2 Thess. *2*, 10 f. Some see in this a reference to antichrist. **44.** Cf. Matt. *23*, 5; Luke *11*, 43. Behind this unreasonable attitude stand vanity and pride. **Vv.45-47** recur to 39. No need that Christ accuse them; the Messianic prophecies of the Pentateuch, now fulfilled in Him, make them worthy of arraignment by Moses, in whose writings they profess to trust while remaining obstinate toward Him.

For the intervening development of the public ministry of Christ, read Mark 2, 23 — 7, 23.

6, 1-15: Jesus Feeds Five Thousand. Cf. Matt. *14*, 13-21; Mark *6*, 30-44; Luke *9*, 10-17. **1.** *After this,* as in *5,* 1, is indefinite. *Sea of Galilee:* here and in *21,* 1 named for Tiberias, a city on its shore built by Herod Antipas. The reason for going to the other side of the lake is given in the Synoptics. **2.** *Followed him* includes more than this occasion, and again implies the Synoptic account. **4.** It is in John's manner to mention the nearness of the Passover. On the same feast in the following year the Eucharist prefigured in this miracle will be instituted. **5 f.** The first detail of interest to John is the question put to Philip, which was probably our Lord's response to the disciples' question in the Synoptics. Why Philip? Cf. *14*, 8. He had been present at Cana when the water was changed to wine. *To try him:* to excite his confidence or to elicit its expression. **7.** *Two hundred denarii* (cf. Mark *6,* 37) would be a large sum for them (some thirty-eight to forty dollars in our money); and yet not enough to provide for such a crowd; this is a confession of powerlessness. **8 f.** Another confession: the five barley loaves and two fishes are certainly not enough. Hence something else must be done. They are ready for the miracle.

10. *Recline* expresses the posture at meals according to the fashion of the times. **11.** *Thanks:* the Synoptics say "he blessed," but the parallel in Matt. *15,* 36 and Mark *8,* 6 shows that both terms cover one act. The blessing began with the words, "We give thee thanks." *As much as they wished,* to be understood of both bread and fish, indicates the completeness of the miracle. **12 f.** The gathering of the remnants, here performed by the Apostles, calls their attention to the abundance of the provision. **14.** Cf. Deut. *18,* 15-18. In contrast to the disposition of the officials, the people accepted the evidence of the miracle. Cf. also *1* Mach. *4,* 44-46; *14,* 41. Associated with the Messianic hope then current was the idea of a bounteous repast, based on Cant. *5,* 1; Isa. *25,* 6, also reflected in the apocryphal Henoch *62,* 14; Syr. Baruch *29,* 8. Cf. Luke *14,* 15. **15.** The danger in this popular enthusiasm explains much of our Lord's reserve in His public ministry.

6, 16-21: Jesus walks on the water. Cf. Matt. *14,* 22-33; Mark *6,* 45-52. **16.** As in the Synoptics, the event must be placed later that evening. **17.** *They went:* i.e., were going. *Across the sea*: St. Luke places the multiplication of the loaves in a deserted spot near Bethsaida. Most commentators identify this with Bethsaida Julias, on the north shore of the lake, just east of the Jordan. *Jesus had not come to them:* He was still in hiding from the people; cf. 22. **18.** Though comparatively small, the lake is capable of violent storms, dangerous to the small fishing craft. Cf. Matt. *8,* 23-27; Mark *4,* 35-40; Luke *8,* 22-25. **19.** The lake was about 6 miles wide by 14 long. *Frightened:* thinking they saw a specter. **20.** John omits Peter's experience (Matt. *14,* 28-31) and the quieting of the storm. **21.** *Immediately* does not necessarily suppose another miracle; they were probably near the land. *Land toward which,* etc.: the part to which the storm had driven them, i.e., Genesar (Matt. *14,* 34; Mark *6,* 53), the littoral plain near Capharnaum.

6, 22-72: The Discourse on the Eucharist is progressive, leading from the symbol of the manna to the revelation of the full reality. We can distinguish the setting (22-25) and two main divisions, the bread of life (26-47) and the Eucharist (48-59), followed by certain effects of the discourse (60-72).

22-25: The setting. Next day the people, knowing that Jesus had not gone with the disciples, looked for Him near the scene of the miracle. (The note that other ships docked there is a parenthetic explanation of 24b.) Finding that He also was gone, they went by boat to look for Him at Capharnaum; and meeting Him there, expressed surprise at His having evaded them. Their motive, vitiated by material notions of the gifts of the Messias, is the starting point of His discourse.

26-47: The bread of life. The place is the synagogue at Capharnaum (60). In this part of the discourse our Lord leads His audience from their material aims to a knowledge of the true bread which gives spiritual life, and of the need of faith in His words. He addresses those who had witnessed His miracle, the people of Capharnaum, who had not, and the officials of the synagogue. *Which endures:* i.e., produces an enduring effect, eternal life. *Set his seal* by divine confirmation, through miracles. **28.** Their question means, what work, in addition to fulfilling the Law, can win us this divine reward? **29.** Faith in Christ, the constant lesson of the Gospel, is now required particularly to receive the coming revelation.

30-36: The true bread from heaven. **30.** The question is somewhat impatient, either reflecting the official temper, or defending the sufficiency of the Law against implied derogation. **31.** Yesterday's miracle is balanced in their minds by the manna of Moses, which was known as *bread from heaven:* cf. Ex. *16,* 13-15; Ps. *77,* 24 f; Wisd. *16,* 20. For the demand of a further sign cf. Matt. *16,* 1; Mark *8,* 11; Luke *11,* 16. **32.** The manna was *from heaven* only in the sense that it was given by God. In itself it was perishable, and it nourished bodily life only. It is not the *true bread.* **33.** This true bread not only *comes down from heaven,* but also *gives life,* the real life to mankind. **34.** *Give us . . . this bread* recalls the request of the Samaritan woman for the living water. **35.** Our Lord's answer leads them a step nearer His object. He is the heavenly bread; faith is necessary to make it available; and its effect is a life which knows neither hunger nor thirst. Cf. Isa. *49,* 10. **36.** Reminding them that He has already adverted to their lack of faith, Jesus now emphasizes this radical impediment.

37-47: The necessity of faith. **37.** The call of the Father is vital to faith. *Cast out:* cf. Matt. *22,* 13. **38-40.** Our Lord holds an essential place in the divine plan to give mankind eternal life, so that belief in Him is absolutely necessary. **41.** The Jews, without replying, whisper to one another an objection which reveals some misunderstanding on their part. **42.** They object that they know His parents: cf. Matt. *13,* 55; Mark *6, 3.* Knowing nothing of His virgin birth, they think His earthly origin incompatible with the claim of descent from heaven. **43-47.** Jesus does not reply by explaining His Incarnation, but repeats His relation to the divine plan, insisting that correspondence with the Father's invitation means adherence to the Son. The quotation in 45 is from Isa. *54,* 13 (cf. Jer. *31, 33*). *In the Prophets:* i.e., in the volume containing their writings. *Not that any one has seen:* cf. *1,* 18; *5,* 37. *Him who is from*

God, not only by mission but also by nature. The conclusion is that faith in Him ensures eternal life (*5,* 24; *6,* 40).

48-59: The Eucharist.   **48-51.** Christ is the bread of life. This returns to the direct line of the revelation to which Jesus is tending.   **48.** *I am the bread of life* definitely identifies with His own person the notion of the true bread which gives spiritual life.   **49 f.** This bread surpasses the manna in two respects: it gives life, and that life is eternal.   **52-59.** The climax, the revelation of the Eucharist.   **52.** The Eucharist explicitly announced: the bread which gives eternal life is the flesh of Christ.   **53.** The Jews understood Jesus to speak of the eating of His flesh, and the statement puzzled them.   **54-59.** Our Lord's reply is a solemn insistence upon this literal meaning of His words.   **54.** Spiritual life requires this eating of the flesh and drinking of the blood of the Messias.   **55.** In the Semitic manner, the thought is repeated in a more positive form, excluding any symbolic interpretation.   **56.** It is repeated a third time, with emphasis on the fact that this food is His flesh and blood.   **57.** The effect of this food is to communicate life by uniting him who eats of it with Christ; cf. *15,* 4-7; *17,* 23; *1* John *3,* 24; *4,* 16.   **58.** *I live because of the Father* probably signifies both His eternal generation and the Incarnation; further, it expresses Christ's devotion to the work of His Father, and His servant's devotion to the service of Christ.   **59.** The revelation concludes with a final comparison between the Eucharist and the manna; the comparison is that of death with life.

60-72: The effect of the discourse. This illustrates the crisis in faith which terminated the Galilean ministry as narrated in the Synoptics. Its value to John's thesis is patent.

60-67: The doctrine alienates many followers.   **60.** Cf. *18,* 20; Matt. *4,* 13; *11,* 23. This supposes that we know the circumstances of the Galilean ministry. What follows probably did not occur in the synagogue.   **61.** The crisis is the graver as chiefly involving disciples (in the general sense of the term); probably the people had already displayed coldness. Cf. Matt. *11,* 20-24; Luke *10,* 13-15. *Said:* they only thought this, or spoke of it quietly among themselves. *A hard saying:* from the Greek term "severe," or as the Septuagint Version uses it, "forbidding." *Who can listen to it:* i.e., believe it. This was the crisis; they could not accept our Lord's word for what seemed difficult to them.   **62-65.** He encourages them to believe by promising stronger evidence of His knowledge of things now hidden from them. *Knowing in himself:* by His supernatural knowledge. *Scandalize:* a word implying both surprise and consequent loss of confidence. **63.** Cf. *1,* 50; *3, 12.* Jesus promises an equally marvelous confirmation of the literal meaning of His words. His ascension will reveal His divine nature, though He remains *the Son of Man.* **64.** *Flesh* in John's usage generally denotes human nature unaided by grace; if that is its signification here, the meaning is that they cannot understand the revelation without the help of the Spirit. Another explanation: it is not the flesh as such, but the flesh united to the Divinity, that becomes a food of supernatural efficacy. His revelation, if accepted, will lead them to this spiritual life. **65.** *From the beginning* of their association with Him. *Who should betray him* points to Judas as the chief example of unbelief. **66.** *This is why:* i.e., lack of faith; implying that many must have followed our Lord for temporal reasons. **67.** *Turned back* from His society, and perhaps also returned to their former occupations; the word has both moral and physical import.

68-72: The Twelve are put to the same test: a final confirmation of the literal meaning of our Lord's words on the Eucharist. **69.** Simon Peter, speaking for them all, replies in the terms of 64. He acknowledges that Christ has no superior in their interests, and that they accept this doctrine on His word. This is true faith. **70.** *We have come to believe:* from witnessing His miracles and from their general experience with Him. The sense is, we now believe that thou art the Messias. *Son of God:* or, "the Holy One of God." Cf. *10,* 36; Mark *1,* 24. This is the better reading. Peter regards Christ as the one uniquely set apart and consecrated for the work of God. This is not a full confession of His divinity; which was revealed to Peter later: cf. Matt. *16,* 13-16; Mark *8,* 27-30; Luke *9,* 18-20. **71.** Cf. *13, 18; 15, 16. A devil:* cf. *13,* 2.27. Here the term is comparative: like unto the devil. Jesus shows that Peter's words of faith do not express the sentiments of all the Apostles, puts them on guard against the scandal of Judas's betrayal, points to the need of guarding their faith. **V.72** expresses John's own wonder at the betrayal.

3. Conflicts with the Jews 7, 1 — 12, 50

This aspect of the development of the public ministry is in perfect accord with the Synoptic narrative (cf. Mark *10,* 1-52). But John confines his attention to Jerusalem, and particularly to the officials there. All that he reports in this section can be dated from September to the following April, the time of the Passion. It comprises our Lord's final efforts to convince the Jews of His mission, and the gradual progress of their opposition to its culmination in the sentence of His death.

7, 1-15: Jesus Goes Secretly to the Feast. **1.** *After these things* is indefinite. Judea was now practically closed to Christ, a fact that reveals the attitude of the officials of Judaism. **2.** The Feast of Tabernacles, from the 15th to the 22nd Tishri (September-October), commemorated the desert wanderings after the Exodus. It was very popular, and not seldom an occasion of national uprisings. **3-5.** This may have suggested to our Lord's *brethren* (near relatives) the idea of a Messianic demonstration. Their words reveal a misconception of His mission which was probably shared by many Galileans, and which, as John points out, was due to want of faith. **6 f.** *My time:* the time of His Passion. *Your time:* as involving no critical issue in opposition to worldly aims. The *world* here refers especially to the officials of Judaism (1). Cf. Matt. *5,* 20; Mark *3,* 29; *7,* 6; etc. **8.** *I do not go up:* i.e., not with you; or, not as in solemn pilgrimage; or, not openly; or (with the Greek), "not yet." **9.** Jesus remained for the time in Galilee, but went later, perhaps with some of His disciples. Some believe Luke *9,* 51-56 a record of this journey. **11-13.** *The Jews,* the officials who sought to apprehend Him. The people, undecided, discussed Him secretly among themselves. **14.** This is the first notice of our Lord's formally teaching in the temple. The scribes or rabbis who taught there had the sanction of the Sanhedrin. **15.** His mastery of the expository methods of the rabbis, and His acquaintance with the text of the Law astonished them in one who was known not to have studied under any of the great rabbis. (Cf. Paul and Gamaliel, Acts *22,* 3.)

7, 16-24: The Source of Christ's Teaching. Cf. *5,* 16-18. **16.** His answer meets the implied criticism of His teaching (cf. *5,* 30): you cannot reject my teaching without rejecting that of the Father. Cf. *12,* 44-49. **17.** Cf. Num. *16,* 28-31. God's will is expressed in the Law and the Prophets; cf. Matt. *22,* 40; Gal. *5,* 14; Jas. *2,* 8. The observance of God's will enables one to recognize His teachings, as sin blinds one to the understanding of divine truth: cf. *1* Cor. *2,* 5 ff; *3,* 1 f; Heb. *5,* **11.** This implies criticism of their conduct. **18.** Another test of the divine origin of His teachings: He seeks neither popular recognition nor applause but only the glory of the Father. Cf. *5,* 41; *6,* 38. *No injustice:* since His doctrine is not of His own invention, as theirs often is (Matt. *15,* 3 f; Mark *7,* 7-13), He violates the rights neither of the Sender nor of those to whom He is sent. **19.** Cf. Matt. *23,* 15; Rom. *2,* 23. Our Lord's aim is to anticipate by this statement the objection based on His violation of the Sabbath. Another aim would be the argument: you who do not keep the Law of Moses can hardly invoke it against me. The Greek text reads, "Did not Moses give you the Law?" This more directly accuses them of its violation. Cf. Matt. *12,* 34. **20.** Cf. *5,* 18; *7,* 1. The crowd, ignorant of the officials' purpose, imagined that an evil spirit had prompted what seemed to them a foolish speech; cf. *10,* 21; Matt. *11,* 18; Luke *7,* 33. Jesus had been thus accused before, e.g., Matt. *9,* 34; *12,* 24; Mark *3,* 22; Luke *11,* 15. **21.** He does not at once reply directly to the charge, but continues His defense. *One work,* the healing of the man at the pool (see on *5,* 8 f). *You all wonder,* and that with displeasure; the charge includes the people with the officials. **22.** *For this reason:* because it is from the fathers. Many others explain: for this reason (given in 21) you wonder. Better, perhaps: Moses' reason for giving you the law of circumcision is also my motive in healing on the Sabbath. Circumcision was from Abraham (Gen. *17,* 10 ff), though the laws regulating it came from Moses (Lev. *12,* 3). Except in the case of non-Jews, all things necessary to circumcision were lawful on the Sabbath. **23.** The law of the Sabbath yields to that of circumcision; much more to the general precept of charity; cf. Mark *3,* 4. *A whole man:* whereas circumcision might be regarded as the healing of one member. **24.** *By appearances:* largely the case in Pharisaic practice. A *just judgment* must be in accord with God's will.

7, 25-52: Christ's Origin. Belonging perhaps to a later occasion. Taking advantage of a better disposition, He teaches the people the true origin of the Messias. **25 f.** *The people of Jerusalem* itself would be more aware of the official attitude. *Rulers:* here the sanhedrists, the leaders of Judaism. *The Christ:* these citizens recognize His claim, and it governs their present attitude; they seem to be in doubt. **27.** *Where this man is from:* His place of origin and His family. In the case of the

Messias, both were known; He was to be born in Bethlehem, of the house of David. Cf. Matt. *2,* 4-6; John *7,* 42. His advent, however, was to be sudden and from an unknown place: this opinion could find plausible support in Ps. *71,* 6; Isa. *53,* 8; Mich. *5,* 2; Mal. *3,* 2.   **28.** He answers the objection in the course of His teaching in the temple: they fail to recognize Him as Messias because they do not know the One who sent Him. *I have not come of myself:* like a pretender such as Judas the Galilean (Acts *5,* 37). *He is true:* having the right to send; perhaps also with reference to the authorization of rabbis. *Whom you do not know:* cf. *5,* 37 f; *6,* 45.   **29.** *I am from him* is much better supported than the reading "with him." In keeping with the context, our Lord is *from Him* by generation, in which aspect His origin is really unknown to the people.

Vv.30-36 specify two effects of this teaching: the officials were the more disturbed, and tried to apprehend Christ, whereas many of the people, moved by His miracles, accepted Him as the Messias.   **31.** This question implies some previous discussion, perhaps with the officials trying to dissuade them.   **32.** *Whispering:* still in fear of the officials (cf. 13). *Rulers:* the Greek text reads "chief priests and Pharisees." Cf. *7,* 45; *11,* 47; *18,* 3. Not all the Pharisees were members of the Sanhedrin. *Attendants:* servants whose office was to preserve order in the temple. **33 f.** Cf. *8,* 21; *13,* 33. Our Lord assures them they can do nothing until the time set by the Father; then they will seek Him with another motive, but will not be able to reach Him. The words convey a threat.   **35 f.** Will He go to the Jews scattered among the Gentiles, and to the Gentiles themselves? Cf. Isa. *11,* 11 f; *49,* 1.6.   **37-39.** *On the last day* occurred all that remains to this chapter. *If anyone thirst:* the symbol of *living water,* somewhat differently applied than in *4,* 10ff may have been suggested by the libation of water on each day of the feast. But abundance of water was one of the blessings expected of the Messias; Ezech. *47,* 6-12; Joel *3,* 18; Zach. *13,* 1; *14,* 8.

37 f. According to punctuation, either Christ is the source of the *rivers of living water,* or the one who drinks is the source. With the balance of patristic authority, we understand it in the latter way. The quotation in 38 has not been identified; various texts have been suggested, as Ecclus. *24,* 28-32; Isa. *44,* 3; *48,* 21; *55,* 1; *58,* 11; etc. Cyprian quotes *1* Cor. *10,* 4, and refers it to Isa. *48,* 21.   **39.** The water libation was a symbol of the Holy Spirit. Cf. Isa. *44,* 3; *55,* 1 ff; Ezech. *36,* 25; and John *16,* 7.

40-52: Diverse effects upon the hearers.   **40-44:** The people.   **42.** Cf. *7, 27:* they did not know of His birth in Bethlehem. The discussion was heated; some of the people sided with the officials in seeking an arrest, but were prevented by either awe or indecision.   **45-49:** The attendants. These had been listening while awaiting opportunity to carry out their orders, and had meanwhile been affected by His teachings. The Pharisees' angry reply again reveals the cleavage between the official and the popular attitude.   **49.** *Is accursed:* the unlearned in the Law were despised as belonging to Israel only in part. Ignorance was thought to induce violation, and violation to incur uncleanness, which must fall under the malediction of the Law.   **50-52:** The Pharisees. The intervention of Nicodemus, although justified and cautious, elicits only an intemperate reply. *At night:* the Greek text reads "lately." *One of them:* a sanhedrist; cf. Luke *23,* 50. Nicodemus as a doctor of the Law, objects that it is against them. Cf. Ex. *23,* 1 f; Lev. *19,* 15; Deut. *1,* 16 f. *Art thou a Galilean?:* a compatriot of Jesus might be expected to defend Him, but Nicodemus was a Judean. *Arises no prophet:* Scripture names no great prophet whose place of origin was or was to be Galilee. But cf. *4* Kings *14,* 25 on Jonas.

7, 53 — 8, 11: The Adulteress. This section is absent from many of the chief Greek codices. In some others it is found out of its present place, at the end of John, or after *7, 36,* or after Luke *21,* 38. The Greek interpreters generally seem not to know it; so also many Latin and Syrian commentators. Discussion of its origin should distinguish between canonicity and authorship. The passage is firmly fixed in the Vulgate from the beginning, and hence is an integral part of the Bible. However, the Council of Trent, in defining the canonical books, did not intend to decide their authorship, a question depending upon the available evidence. As to this section of 13 verses, several Catholic authors are now inclined to deny that St. John wrote it. Others think it derived from him, but that its present form is later. Because of its position in the Vulgate, we retain it here, without going further into the matter of its origin.

7, 53 — 8, 1. During the feast our Lord spent the night on Mt. Olivet or at Bethany (cf. Luke *21,* 37), returning each morning to the temple. Mt. Olivet is east of Jerusalem, beyond the Cedron.   **2.** *At daybreak:* the time of the morning sacrifice, at which many would be present. **3-5.** Stoning was prescribed for this crime in the case of an espoused girl: Deut. *22,* 23 f. For other women death was prescribed without the manner of execution: Lev. *20,* 10; Deut. *22, 22;* Ezech. *16,* 38-40. **6.** *To test him:* if He said "No," He would seem to disregard the Mosaic Law; if "Yes," He might be accused to the Romans. Jesus, probably seated on a cushion or mat, leaned forward to write on the ground. Jerome records an opinion of his day, that Christ wrote the sins of the accusers before them; but more probably His act was a mere gesture of indifference (cf. Luke *12,* 13 f). **7.** Cf. Deut. *13,* 9 f; *17,* 7. **9.** The scribes and Pharisees departed, not the people. **10 f.** Jesus manifests His mercy while refuting His adversaries.

The whole of 8, 12 — *10, 42 is a series of encounters with the Jews, closely related, and probably all belonging to the period from the close of the Feast of Tabernacles to the Feast of the Dedication, in December.*

8, 12-20: The Light of the World. Cf. *5,* 31-47. The place is the temple; the audience chiefly officials and Pharisees. The crowd is not mentioned again until *11,* 42. **12.** *Light of the world:* this might have been occasioned by the illumination of the temple on the Feast of Tabernacles; or some known features of the Messianic hope might have led to it: cf. Isa. *9,* 1 (Matt. *4,* 14 f); *42,* 6; *49,* 6 (Luke *2, 32*); Baruch *5,* 9. **13.** Our Lord had already answered this objection: *5,* 31 ff. **14-18.** Because of His divine origin, already revealed to them, our Lord can bear this witness in His own favor. They are hindered from accepting it by their own lack of spiritual knowledge. **15.** A double contrast: they judge, He does not; their judgment is false, His is true. *I judge no one:* cf. *3,* 17; *12,* 47. This in contrast to their temerity. **16.** But if He were to judge, His judgment would be true. This insinuates that later He will come to judge them. His judgment is true because it is divine. **17 f.** The required two witnesses are found in Himself and in the Father. Cf. Deut. *17,* 6; *19,* 15; Matt. *18,* 16; 2 Cor. *13,* 1; Heb. *10,* 28. **19.** Cf. *5,* 38. Our Lord again identifies Himself with the Father. It is their fault that they do not appreciate the Father's witness. **20.** *In the treasury:* at or near the room adjoining the Court of the Women. Cf. Mark *12,* 41-44; Luke *21,* 1 f.

8, 21-30: The Son of God. Cf. *7, 31-36.* The subject had been broached before, but the tone is now of more solemn warning: there is grave danger in not recognizing God's legate. This occasion is not necessarily the same as in 12-20, though the place is the temple.   **21.** To His former prediction our Lord adds *in your sins you will die:* cf. Deut. *24,* 16. They alone will be responsible for their failure to share in the Redemption.   **22.** Cf. *7, 35.* Suicide was considered a great crime. **23 f.** Ignoring the interruption, our Lord assigns a reason for His statement: their refusal to accept Him as the Messias. They have still time to believe. *I am:* cf. *8,* 28.58; *13,* 19; *18,* 5-8.   **25.** The meaning is uncertain. The Latin of the Clementine Vulgate is generally conceded to express the thought of Apoc. *1,* 17; *2,* 8. The Greek text is slightly different, and has been variously interpreted, the most probable sense being, "even that which I tell you from the beginning;" or, *Why do I speak to you at all?*   V.26 repeats the threat in 24, but again modified. The truthfulness of the Father is warrant for our Lord's message, therefore they ought to take His warnings seriously. **27.** The Evangelist's remark further reveals the disposition of the audience. *God* is wanting in the Greek text. In v.18 the allusion seemed to be understood; either the audience was now different, or in the distraction of the present subject they lost the import of our Lord's words.   **28.** Cf. *3,* 14; *12,* 32 f; Luke *23,* 48.   **29.** Cf. *5,* 19 ff. The union of the Father and the Son should be clear to them from His miracles and the nature of His teaching.   **30.** Many were convinced by this.

8, 31-47: The Children of Abraham.   **31.** *If you abide in my word:* i.e., persevere in observing it. For firm adherence He promises them full discipleship, knowledge of the truth He came to reveal, and consequent freedom.   **33.** The unfriendly reply seems little in keeping with the faith ascribed to them in 30. Two explanations are offered: (a) that the speakers were "not they who believed, but they in the crowd not yet believing" (St. Augustine); (b) that their faith was yet too weak to tolerate an apparent slight to their national honor. Both explanations may be true. Cf. Matt. *3,* 9.   **34-38.** Jesus corrects them on two points. (a) Sin induces spiritual slavery, from which only the Son of God can free them. For v.35 cf. Gal. *4,* 21-31; Heb. *3, 5* f. (b) They are physical descendants of Abraham, but their actions show them conscious only of the carnal relationship.   **39-41.**

Further proof of the same: one's actions reveal one's nature. Their conduct manifests that they are not true sons of Abraham. On one side stands Abraham's hospitality to angels and reverence for God's word: Gen. *12*, 4; *15*, 6; *18*, 2; *22*, 2. On the other is their wish to kill Him, who has proved His claim to be God's messenger. *Fornication:* cf. Os. *1—2*. The term implies idolatry or attachment to another god.   **42-45.** The argument carries to its proper conclusion their refusal to believe. They cannot be sons of God, since they refuse to love His envoy.   **43.** Again, because they cannot understand His meaning.   **44.** But since their father is not God, he must be God's adversary, the devil; and two things prove this to be the case. The devil was a murderer from the beginning; it was he who introduced death into the world: cf. Wisd. *2*, 24; Rom. *5*, 12 ff. Again, he is by nature a deceiver: cf. Gen. *3*. These Jews resemble him in both features: their purpose is homicidal, and they are unable to recognize the truth.   **46 f.** Cf. *18*, 37 f; *1* John *2*, 24. The argument culminates in a further contrast: if I speak the truth, there must be some reason why you do not believe; that reason must be, that you are not of God. In the whole argument we may read a plea on the part of Jesus to avert the heinous crime they are meditating, the murder of God's Son.

8, 48-59: Christ and Abraham. After humbling them and correcting their errors, Jesus seeks to favor them with fresh truths. But they offer further resistance, and end by attempting to stone Him.   **48.** The Jews show their spirit in obloquy. *Samaritan:* (cf. *4*, 9), a stranger and hostile to the chosen people. *Devil:* (cf. *7*, 20; Matt. *12*, 24-32), one who is demented.   **49.** That Jesus honors His Father is sufficient refutation of the first charge. His direct denial of the second accusation serves to connect His preceding with His following words.   **50 f.** Cf. v.15. Their attitude is an evil so grave that the Father will avenge it (cf. Deut. *18*, 19). This divine judgment is synonymous with *death,* avoidable only by adherence to Christ; cf. *3*, 16; *4*, 13; *5*, 24.   **52 f.** To Christ they now seek to oppose Abraham and their traditions. This is further evidence of their unwillingness to understand; they will not admit Him to the category of Abraham and the prophets.   **54.** Cf. *5*, 31. The Father glorifies the Son; see *2*, 11; *12*, 27 f.   **55.** Cf. *1* John *2*, 4. *Know* is here used in the Semitic sense of a practical knowledge manifest in conduct, almost equivalent to "love."   **56.** Generally among earlier writers, it is believed that Abraham saw Christ's day in figure, in faith or in prophetic vision. Cf. Heb. *11, 13.* The rabbis taught that God had revealed the future to Abraham. A later opinion is that Abraham saw the actual advent of Christ from his place in Limbo. *My day:* His life on earth; cf. Luke *10, 24.*   **57.** Again the Jews are captious. Jesus' age cannot be estimated from this statement; fifty years stood for the start of old age in the common estimation: Num. *4, 3; 8, 24* f.   **58.** *I am:* a deliberate assertion of Christ's continued existence, and contrasted with *come to be.* The phrase was the more arresting because of its similarity to the divine name, "Yahweh," derived from a form of the Hebrew verb "to be."   **59.** The Jews understood this allusion, or at least the implication of His words, and attempted to stone Him.

9, **1-41: The Man Born Blind.** Christ having been virtually expelled from the temple (*8, 59*), the following event can hardly have occurred just as He was leaving it—perhaps not even soon afterward. At least it preceded the date next given (*10, 22*), that of the Feast of the Dedication.

1-12: The miracle.   **1.** Beggars frequented the approaches to the temple. No other cure of congenital blindness is expressly recorded in the Gospels.   **2.** Whether an unborn child could sin was a disputed question, but all agreed that a parent's sin might be visited upon a child: cf. Deut. *5, 9; 2* Kgs. *12,* 14.   **3-5.** Neither supposition is true in this case; the physical evil was permitted by God in order to lead to a greater good.   **4.** The time for that good (the manifestation of divine glory and power in Christ) is fast elapsing; *night is coming,* a definite allusion to His approaching death, though expressed in the terms of a general truth.   **5.** *Light of the world:* the present function of the light is the revelation which our Lord has been giving, the evidence of His nature which is a call to faith.   **6 f.** He used clay in this instance, Chrysostom thinks, to make the blindness more evident before the miracle. The clay, at all events, was not the medium of the cure. *The pool of Siloe* was at the lower end of the Tyropœon valley, within the city walls. John suggests a symbolic relation of its name to the Hebrew verb "send": Christ is the sent of God. In some instances the pool was used for ritual purifications.   **8-12.** The manner in which the miracle became known is significant.

13-34: The Pharisees investigate the miracle. The affair illustrates the popular recognition of the Pharisees' authority in questions of the Law. The investigation of the case may have run into the following days. The episode also contrasts the Pharisees, unmoved by the miracle, with the man who was cured and believed. **15 f.** The first examination results in dissension among themselves. Cf. 7, 12.48. What had been violated was their traditional interpretation of the precept concerning the Sabbath. **17.** Their second inquiry, not wholly free from malice, shows the man's faith in contrast to their doubt. **18-23.** Their third question is put to the parents, in an effort to disprove the fact. *He is of age:* i.e., over thirteen years. The parents' fear again shows the official opposition to Christ. Exclusion from the community had been a penalty imposed since the time of the exile; Cf. *1* Esd. *10,* 8. Its character at the time of Christ is not known. **24-33.** The last question. *Give glory to God:* a solemn adjuration to tell the truth. Cf. Jos. 7, 19; *1* Kgs. *6,* 5. They hoped he would either deny the fact or withdraw his previous confession of faith. *We know that this man is a sinner* hints at the answer they await; the *we* is emphatic. **26.** As they repeat their first question, the man's surprise becomes indignation. The Greek text reads, "and you did not hear," meaning "you refused to hear." **28.** *Disciples of Moses* was a current designation of the Pharisaic scribes. They repeat the popular doubt already voiced in Jerusalem; cf. 7, 25 ff. **30.** The man's answer condemns their bad faith. **31.** *Does not hear sinners:* the sinner has no claim upon divine favor such as this miracle demonstrates. **32.** *From the beginning of the world:* a Semitic phrase meaning emphatically "never." **33.** For the cured man the evidence is cogent: if He were not from God, He could do nothing of this order. **34.** The result: the Pharisees repel the man from their presence, definitely displaying their own voluntary blindness. *Born in sins* alludes to his former blindness and his ignorance of the Law. *Dost thou teach us,* expresses a typically Pharisaic point of view.

35-41: Sight and blindness contrasted. He who had been blind received also the new vision of faith; they who thought themselves enlightened in the Law are declared blind. Cf. Isa. *6,* 9; *42,* 16; *52,* 10. **35.** *In the Son of God:* not His divinity, but His supernatural power and mission. Many Greek MSS read "the Son of Man," meaning the Messias. Thus the man's courageous faith is rewarded with the offer of still higher faith. **37 f.** *Worshipped him:* perhaps with religious devotion; he had confessed to the Pharisees the power of God in Christ. This is faith. **39.** *For judgment:* cf. Matt. *11,* 25; Luke *10,* 21; *1* Cor., *1,* 26-29; the sense is "discernment," "discrimination." Recall *3,* 17 f; *5,* 24; cf. Deut. *30,* 15-20; Luke *2,* 34. Mankind is divided into two groups, those accepting and those rejecting Christ, the one represented by the cured man, the other by the Pharisees. But observe that God gives both the same opportunity. **40 f.** *If you were blind:* i.e., ignorant of what you are doing. They pride themselves on their knowledge of the Law. *Your sin remains:* it is habitual, formal.

10, **1-21: The Good Shepherd.** The claims already made by our Lord are now urged under the image of the Good Shepherd. This symbol is frequent in the Old Testament: cf. Pss. *22; 78,* 13; *94,* 7; *99,* 3; Isa. *40,* 11; Jer. *23,* 1-4; Ezech. *34; 37,* 24; Mich. *7,* 14; Zach. *10,* 3; *13,* 7. As to the occasion of this discourse, the fact that some of the hearers knew of the cure of the blind man (*10,* 21) suggests a link between the two. Most interpreters are inclined to connect *10,* 1-21 with *9,* 40 f as having a somewhat polemic aim. Many others, however, both in ancient and in modern times, regard the theme of the Good Shepherd as a new one, neither polemic nor apologetic.

1-6: The parable, most apt in the pastoral country of Judea, supposes rather than describes its scene. A barrier of stones and briers for the protection of their flocks has been erected by a group of shepherds, one of whom stands guard over the entrance of this fold. **1 f.** A thief finds no access by the normal entrance, but climbs into the fold at some unguarded spot; the true shepherd enters by the gate. **3 f.** The practice of naming one's sheep is still observed in Palestine. The sheep run to their own shepherd's call. Having let out his sheep (Pss. *76,* 21; *77,* 52), he goes before; he leading, they obedient to his voice. **5.** The thief is met quite otherwise. **6.** *This parable:* John's term is more strictly "proverb," but rather to be understood as "parable." A parable is by nature somewhat obscure, Jesus now applies the image to Himself, selecting two features, the fold and the good shepherd.

7-10: Christ the door of the fold. **7.** Jesus alone can admit to His Kingdom. **8 f.** The contrast is between true and false shepherds. The thieves, failing to pass through this door, were not recognized by the sheep. *All whoever have come:* the context implies "not through me." The Greek text has "all who have come before me." It refers to those who came before Christ, but also without His authorization; not to legitimate agents of God. Cf. Matt. *23*, 4-13; Luke *11*, 46-52. It can however apply to normal teachers who have lost the true spirit of their doctrine. **9.** The true shepherds, who enter through Christ, enjoy the safety of His protection, free access to the fold, and pasturage for their flocks. Cf. Num. *27*, 17; *1* Par. *4*, 39 ff; Jer. *23*, 3; Ezech. *34*, 14.   **10.** The motive of the thief is forcibly contrasted with that of Christ, which is to give the sheep abundant spiritual life.

11-18: The Good Shepherd. In the Old Testament, God is spoken of as king and shepherd of Israel: Ps. *79*, 2. Cf. the good shepherd in Isa. *40*, 11; Jer. *31*, 10; Ezech. *34*, 11-16; *37*, 24. The shepherd David is a type of the Messias: Ezech. *34*, 23. Hence the image, though not claimed in proof of His Messianic mission, eminently befits our Lord.   **11.** *I am the good shepherd:* this element of the original parable is now reapplied with an aim of contrast. The good shepherd is faithful to his duties, and devoted to the sheep, even to giving his life for them. Cf. Gen. *31*, 40; *1* Kgs. *17*, 34 f.   **12.** The contrast is not now with the thief, but with the shepherd unfaithful to his duty, called *the hireling* because of his selfish motive; cf. Ezech. *34*, 2.5. He is not a shepherd because he has no devotion to the sheep; he does not own them, and is not concerned about them. Cf. *1* Cor. *13*, 5; Phil. *2*, 21; Titus *1*, 11; *1* Pet. *5*, 2.8.   **13.** The contrast lies in the hireling's conduct in time of danger: he thinks only of his own safety.   **14-16.** To such a disposition Christ opposes His death for the salvation of mankind. He knows His sheep with a practical and intimate knowledge, such as He described on the night of the Last Supper. The model of this relationship is the mutual knowledge of the Father and the Son. He is willing, nay, fully intends, to sacrifice His life for the sheep. Cf. Isa. *53*. **16.** *Other sheep,* not of the house of Israel; cf. Isa. *2*, 2 f; *11*, 9 ff; *42*, 4; etc.; Amos *9*, 11-12. *Bring:* more properly, "lead." The reading of the Greek text, "one flock and one shepherd," is better than that of the Vulgate. Cf. Ezech. *34*, 23; *37*, 22.24.   **17 f.** The death of Jesus is the design of the father. His Resurrection also is foretold as in *2*, 19. Cf. Mark *8*, 31; *9*, 20; *10*, 34 and parallels; Phil. *2*, 8 f. He teaches that redemption is consummated only in His death; that He freely accepts death for love of men and in obedience to the Father; that He has the power to resume His life, a power which proves that He is God, as His death proves that He is man. *For this reason:* because of His obedience.

19-21: Effect on the audience.   **19.** *Again* has *9,* 16 in view. One party in official circles obstinately accuses Christ of madness; the other is inclined in His favor by the miracle of *9.*

***10,* 22-39: At the Feast of the Dedication.** The time between Tabernacles and Dedication was spent by our Lord away from Judea, or at least out of Jerusalem. John here tells of another claim of Christ to equality with the Father (*10,* 22-39), and of another stage in the opposition of the Jews (*10,* 40-42).

22. The Feast of the Dedication, later known also as the Feast of Lights, fell in mid-December. It commemorated the consecration (*chanukah*) of the temple in 165 B.C., after its profanation by Antiochus Epiphanes; cf. *1* Mach. *4,* 30-59. Its celebration was similar to that of Tabernacles. Beginning on the 25th Kislev, it lasted eight days. *Winter* was reckoned from the middle of December to the middle of February.   **23.** *Solomon's porch,* a portico along the eastern wall of the Court of the Gentiles; cf. Acts *3,* 11; *5,* 12.   **24.** *Jews:* as elsewhere, the officials of Judaism. *Keep us in suspense:* literally, "take away our soul." The expectation of the Messias, then so general, was especially keen on festal occasions, and was now accentuated by the works of Christ. The Jews seem to challenge Him either to declare His kingdom or to renounce His claim. With merely worldly ideals, they could not understand His delay (St. Augustine).   **27.** Recurrence to the image of 1-6 emphasizes their refusal to believe; nothing prevents them but themselves.   **29.** The Greek text reads "My Father . . . is greater," though the text of the Vulgate has good support in the MSS. *What my Father has given me:* interpretations vary. Some hold that the gift is the divine nature; others, the divine power of Christ. The former is the general opinion. The Fourth Lateran Council (Denzinger 432) cites this text in defining that "the Father by generating the Son from eternity gave to Him His own nature."   **30.** This clear statement of equality with the Father follows logically from 29. The responsive action of the Jews (vv.31, 33, 36, 38) shows that they understood it.   **31.** Cf. *7,* 20; *8,* 59; *19,* 7.

32-39: As in *6,* our Lord is not deterred by opposition from maintaining the full meaning of His revelation.   **32.** His works were *good,* as being not only merciful, but also wonderful; they were evidence of a divine mission.   **33.** Blasphemy was punishable by stoning; see note on *8, 59*. It included any injury to the divine name. Our Lord is here understood to identify Himself with God. **34-36.** That their offense was unreasonable He proves from Scripture, arguing *a minore ad majus*. Quoting Ps. *81, 6*, He recalls that under the Old Law even the magistrates were called "gods" (cf. also Ex. *7*, 1; *22*, 8 f), though at times unworthy of their office. How much more right to the title has He who not only is sent by God, but has been made holy by Him. *Made holy:* set aside especially for God's work. The reply is against their insistence upon words, and their hasty action based upon the words of the Law. **34.** *In your Law:* as in *7, 49; 12, 34; 15, 25*, used of Scripture in general; a designation known also from rabbinical literature. **35.** *Called them gods:* the context of Ps. *81, 6* shows that unfaithful magistrates are meant. They were the ordinary interpreters of the law. *Scripture cannot be broken:* cannot give false information. **36.** However they may have understood this argument, His next words permit no qualification to the claim in 30. **37 f.** *The works of my Father:* the miracles, exceeding natural powers, could be recognized as from God. **39.** Cf. *8, 59*; Luke *4, 30*. This escape was not necessarily miraculous.

10, 40-42: **Jesus in Perea.** It is suggested that we are here again in contact with the Synoptic tradition as in Matt. *19*, 1; Mark *10*, 1; Luke *13*, 22 f. **40.** The place would be Bethany; cf. *1, 28*. **41.** *Many came to him:* for the details we might consult Luke *13, 22* and the chapters following. The popular belief is based upon the word of John the Baptist and the miracles of our Lord. **42.** The people's belief contrasts with the unbelief of the officials, just above.

11, 1-44: **The Raising of Lazarus.** The narrative has arrived at the final developments before the Passion. Opposition to Christ culminates in the official decree of His death. The faith of His adherents advances until He is popularly acclaimed the Messias. Details are supplied by the Synoptic Gospels.

1-16: Jesus is called to Bethany. **1.** *Bethany:* cf. Luke *10, 38-42*, a passage which John may have in view. The village lay on the eastern slope of Mt. Olivet. **2.** The reference is to John *12, 1-8*, not to Luke *7, 38*. **4.** *He said to them:* to the messengers, for conveyance to Martha and Mary. Our Lord foresaw the death of Lazarus, but also his return to life. Cf. *9, 3*. The glory of God is co-ordinate with that of the Son of God. **5 f.** Human attachment might have induced Him to hasten to their assistance; His respect for the known will of His Father made Him remain where He was. **7 f.** Recall *10, 32*. The concern of the disciples shows their real attachment. **9 f.** Cf. *9, 4; 12, 35* f. The parable involves His own experience. Conforming to the will of His Father, Christ is safe from harm; the Jews can do nothing before the hour set by the Father. **11-13.** Sleep was one of the six things good for a sick person. The disciples readily misunderstood in their eagerness to withhold our Lord from danger. *Lazarus sleeps:* "sleep" was an expression for death, though with the possibility of implying the coming revival. Cf. Matt. *9,* 24; *27,* 52 and parallels; Acts *7,* 60; *13,* 36. **14 f.** Their mistake demanded this explanation; so did the motive of further support for their faith. **16.** "Didymus" (the Twin) is the Greek form of Thomas. The Hebrew *To'am,* Aramaic *Te'oma,* meant "twin." Cf. *20,* 24; *21,* 2. Thomas expresses the feelings of the other disciples. In view of his conduct on a later occasion (*20,* 19-29), this is an important indication of his character.

17-32: Jesus and the two sisters. **17.** *Four days:* at least two days had intervened between the arrival of the messengers and His departure for Bethany. **18.** *Fifteen stadia:* about 1.7 miles. This distance explains the next verse. **19 f.** The mourners (cf. Gen. *50,* 10; *1* Kgs. *31,* 13) from Jerusalem were in keeping with custom, but one of the sisters must remain as hostess. **21 f.** Cf. v.32. Martha's greeting is not so much a complaint as a profession of faith in our Lord's power, though a faith not yet perfect. Cf. *4,* 46 ff. She indirectly asks for her brother's restoration (Cyril, Augustine). **23-26.** Cf. *5,* 28 f; *6,* 39 f; Dan. *12,* 2; *2* Mach. *7,* 9.14. To encourage the faith she expresses, Jesus leads her to surpass the ordinary Jewish belief in the resurrection, and to recognize in Him the Messias, the author of the true and spiritual resurrection. Cf. *1,* 4; *5,* 24; *8,* 51. He who believes in Christ enjoys a life that is proof against this natural death. Hence Christ is the true resurrection. Cf. Mark *12,* 24-27 and parallels. **27.** The Greek text reads "the son of God, who comes into the world," i.e., who is destined to come. Responding to the demand for a higher faith, Martha confesses that Jesus is the Messias. *Son of God* at least surpassed the popular Messianic expectation. It is in the manner of St. Peter (cf. *6,* 69). **28-30.** We are not told when Christ expressly asked for Mary; in any case, her response is prompt. **31.** The people were thus drawn to the scene of the miracle. **32.** Mary's words are the same as Martha's, but her attitude is different. Her faith, it would seem, required no further encouragement, for the miracle is at once related.

33-44: The raising of Lazarus. **33-35.** *Groaned in spirit* and *was troubled* describe deep human emotion and an effort of the will to control it. The Greek verb for *groaned* signifies the manifestation of some passion, e.g., anger. Cf. Matt. *9,* 30; Mark *1, 43; 14, 5.* *In spirit* is the same as *in himself* in 38. The simplest explanation is that our Lord, with the rest, was moved to sorrow. His emotion would be deeper. His weeping is defined by the Greek word as the shedding of silent tears. **37.** The reference is clearly to *9,* 7. The people's sentiment is not unbelief, but regret analogous to that of Martha and Mary. **38.** Lazarus was the occasion of our Lord's emotion, and the approach to the grave reawakened it. Cf. above; and Matt. *27,* 60; Mark *5,* 3 for the nature of the grave. The stone would be *laid against* or "upon" the tomb according to the position of the excavation. **39.** The thought of again exposing the corpse was repulsive to Martha, and another trial to her faith. *Decayed:* the methods of embalming then in use could not prevent decomposition. **40.** We are not told when our Lord had assured her that she would see the glory of God, it might be implied in 25 f, the Evangelist omitting a part of that conversation. *Glory:* the divine power. **41 f.** The prayer of thanks was uttered aloud for all to hear; the sequel in a lower tone. His purpose is to remind them again of His relation with the Father. The tone of humility in the prayer does not detract from equality between them. Cf. *14,* 28. It should be noted that Christ wrought this miracle on behalf of the faith of the people, calling the attention of all to His power of giving life. **44.** The appearance of Lazarus, still wrapped in the burial cloths, must have amazed them all, but the wonder increased when, on the removal of the cloths, they saw that he lived. Few of our Lord's miracles were accompanied with such demonstration; but the Passion, the last act, was now close at hand.

11, 45-53: **The Council.** **45.** *Many of the Jews* seems now to refer to the people from Jerusalem, a number of whom did not share the sentiments of the officials. For them the miracle confirmed a faith in Jesus as at least the agent of the Father. **46.** Why did they tell the Pharisees? Some think there was no malice in their report; cf. *9,* 13. The majority of interpreters, however, suppose ill will. The main point is that the news of the miracle reached the Pharisees. **47 f.** This was a council of men competent to discuss official action. It included *the chief priests,* i.e., members of the high priestly families, and the Pharisees, whose influence in the Sanhedrin was great. It was probably not a session of the Sanhedrin, for the Greek text calls it "a council." The fear of the officials was based on the obvious effect of our Lord's miracles upon the people. But since He had never betrayed political ambitions, this motive for taking action against Him was really unjustified. *Our place* was the temple, though the city and their own authority might be included. *Nation:* the Romans had forcibly intervened on other occasions. **49 f.** Caiphas gives this advice with some authority. *That year:* cf. *18,* 13: the great year of the Redemption (Origen). John does not mean that the tenure of office was annual; he certainly knew that Caiphas had held it from 18 to 36 A.D. **51 f.** Caiphas thus predicted, though unconsciously, the future event: Christ was to die for the nation indeed, and even for all mankind. *Being high priest:* chief official intermediary between God and His people. *Prophesied:* in the true sense, delivered a message from God. The gift of prophecy does not imply its possessor's holiness, nor need the prophet understand the full import of his words: cf. the case of Balaam, Num. *23,* 1 — *24,* 25. **53.** The voice of Caiphas decided the issue and sealed the fate of Christ by a formal resolution to put Him to death. This is the climax of their opposition. Cf. *7,* 30. 45; *8,* 59; *10,* 31; Matt. *12,* 14; etc.

11, 54-56: Jesus in Ephrem. **54.** *Ephrem* (Efrem, Ephraim, Ephren) was a village on the borders of the Judean desert, some twenty miles north of Jerusalem and five miles east of Bethel. The village gave its name to the whole district. Today it is et-Taiyibeh. For our Lord's further activity at this time see Luke *17,* 11 — *19,* 28. **55.** The Passover could be celebrated only by those who were "clean" according to the Law. This obliged pilgrims from other regions to reach the city a few days earlier, since contact with Gentiles exposed them more to the risk of defilement. **56.** Some commentators see two questions in the people's words: what do you think (of Christ)? has He not come? Possibly some were still in doubt.

12, 1-11: The Anointing at Bethany. This event is given a different connection by Matt. (*26,* 6-13) and Mark (*14,* 3-9), who

12, 1-18 ST. JOHN

place it immediately before the Last Supper. Their aim is to associate it more closely with the betrayal by Judas. See Commentary on Matt. **1.** Cf. Luke *18,* 35 (Matt. *19,* 1; Mark *10,* 1). *Six days* does not conflict with the "two days" in Matt. *26,* 2 and Mark *14,* 1; these rather date the betrayal. **2.** Matthew and Mark tell us that Simon the Leper was Jesus' host on this occasion. **3.** *A pound:* twelve ounces. *Genuine nard:* or spikenard, a fragrant ointment derived from an Indian plant. The great value of the ointment, the prodigality of its use, her hair serving as towel, all manifest Mary's ardent devotion. **4 f.** *One of his disciples:* Matthew and Mark say, "the disciples." The sentiment was probably general, but Judas expressed it. The contrast is between Mary's generosity and his disposition. **6.** Judas' motive became evident to John later. *The purse:* Judas kept what money the group possessed. *Used to take:* indicating that Judas had been dishonest for some time. His avarice was to lead to the betrayal of his Master. **7.** Our Lord's answer has been variously understood. A pause may be assumed after "Let her be!" The rest of the statement requires some completion; e.g., she did not sell it, that she might keep it; or, this has occurred that ... Mary had offered an act of homage such as might be associated with solemn burial; our Lord reveals its ultimate significance, hidden now both from her and from the disciples. Cf. Matt. *26,* 12; Mark *14,* 8 f. **8.** This calls special attention to His burial. **9-11.** The divergent attitudes are emphasized.

12, 12-19: Triumphal Entry into Jerusalem. Cf. Matt. *21,* 1-11; Mark *11,* 1-11; Luke *19,* 29-45. The Synoptics must be read for other details. John's purpose is to bring into stronger relief the two dispositions of belief and unbelief. **12.** Many of those present had not seen the miracle at Bethany. **13.** *Branches of palms:* cf. *1* Mach. *13,* 51; *2* Mach. *10,* 7. These were a symbol of victory and joy, brought from the Jordan valley as decorations for the feast. The cry was a recognition of Jesus as the Messias, and it expressed the hope that His kingdom might now be established. The words are from Ps. *117,* 25 f. *Hosanna:* an exclamation of joy and blessing, whatever its original meaning. **14 f.** With the Synoptics John sees here the fulfillment of prophecy (Zach. *9,* 9), and hence the justification of this Messianic cry. **16.** Cf. *2, 22; 20,* 9. Only after the Resurrection did the disciples recognize this fulfillment of the prophecy. Cf. *16,* 13. **17 f.** The raising of Lazarus was the cause of this popular demonstration. *Bore witness:* i.e., raised the jubilant cry of welcome to Jesus as the Messias. Cf. Matt. *21,* 9.   **19.** Cf. Luke *19,* 39-41. The Pharisees did not share the enthusiasm; rather, their purpose was confirmed.

12, 20-36: Last Words of Jesus to the People.   20-22. *Gentiles:* better, "Greeks," half-proselytes who could enjoy a limited share in the feast. Their interest suggests the extension of Christ's kingdom beyond the confines of Israel; cf. *10,* 16; Ps. *21.* It is likely that they approached Philip only by chance. His need of consulting with Andrew is a further indication of his character.

23-33: The real triumph of Christ will be His death.   **23.** *Jesus replied* may mean, in general terms, "continued" His discourse to the people. His real triumph is not the recent entry into the city; the hour of His glory is that of His death. Cf. *7,* 30; *8,* 20; Luke *24,* 26. For the present this thought transcends all else.   **24 f.** Under the symbol of the grain of wheat our Lord announces His death and resurrection. His guiding principle is the sacrifice of earthly life for that which is eternal. Cf. *1,* Cor. *15,* 36. This principle is also the guide for those who would share in His glory. Cf. Matt. *10,* 39; *16,* 21-25; Mark *8,* 31-35; Luke *9,* 22-24; *14,* 27; *17,* 33.   **26.** To be Christ's associate, one must be ready to follow Him through everything, even death, to share in the honor which the Father grants. Cf. Matt. *10,* 32; Mark *8,* 35; Luke *12,* 37.

27. *Father, save me from this hour* is by many interpreted as a question, equivalent to a negation: He will not make such a petition. The more general opinion, on the contrary, sees in the words a prayer for deliverance, which, however, is immediately withdrawn, Christ thus expressing His acceptance of the mission. In either case this is a brief anticipation, perhaps even the beginning, of the Agony in the Garden of Olives. *Troubled:* His human revulsion from this suffering and from the sins which occasion it, and the effort of His will to control the emotion. *Glorify:* His death is for the restoration of God's glory; and this is His motive for refusing to ask deliverance from *this hour.* **28.** Cf. Matt. *3,* 17. The voice answers this prayer. *I have glorified it,* in the miracles; *I will glorify it,* in the Passion and the Resurrection.   **29.** The people could not clearly distinguish the words. Some heard so indistinctly that they thought it thundered; others, recognizing a voice, thought of angels. Cf.

Gen. *16,* 9; *21,* 17; Judg. *2,* 1; *5,* 23; etc. To all it was a sign from heaven. **30.** As such it was a confirmation of our Lord's words, and this sufficed them.

31. Christ's victory in the Passion is the realization of that judgment often spoken of by the prophets. Cf. Isa. *2,* 12 ff; Rom. *16,* 20; Col. *2,* 15. *Prince of the world,* in John, and generally in rabbinical literature, is the devil. Cf. *16,* 8-11; 2 Cor. *4,* 4. **32 f.** Cf. *3,* 14; *8,* 28. His death is the salvation of all men; cf. *6,* 44. *All things:* the Greek text reads "all men." John explains the expression in 33. **34.** As the Messianic kingdom was to endure forever, so also was the Messias to live forever. The Jews later modified this, conceiving of the Messianic era as a prelude to "the world to come." *We have heard:* in the synagogue services; cf. Luke *4,* 16; Acts *13,* 15; *15,* 21. *Law:* again designating Scripture generally. The expectation could have been based on Isa. *9,* 7; Ps. *109,* 4; Dan. 7, 13 f; 2 Kgs. *7,* 16; Jer. *33,* 17 f. *Who is this son of Man?* i.e., a different one than we have heard of. **35 f.** Not an answer to their question, but to their hesitation. The light is Christ Himself, the symbol here recalling the prologue (*1,* 9) rather than *11,* 9 f. Cf. *7,* 34; Pss. *26,* 1; *118,* 105; Acts *13,* 47; *1* Thess. *5,* 4. This is probably our Lord's last exhortation to the people. To John it is significant as an exhortation to faith, the main theme of the Gospel. **36 b.** Our Lord probably retired to Bethany or to the Mount of Olives: cf. Matt. *21,* 17; Mark *11,* 11; Luke *21,* 37.

12, 37-50: Incredulity. This exhortation concludes John's evidence from the public ministry of Christ. Before proceeding to the Passion, he adds some remarks on faith (37-43) and supports them with sayings of our Lord on that theme (44-50).

37-43: The Evangelist's remarks; incredulity and human prudence. **37.** Cf. *6,* 9; *14,* 9; *20,* 30; *21,* 11. The unbelief of which he speaks is chiefly that of the officials, though the people are not wholly excused; they should have believed because of the miracles. **38-41.** Cf. Rom. *10,* 16-21. John sees in this the fulfillment of two sayings of Isaias. In the first, Isa. *53,* 1, the prophet, about to announce the sufferings of the Messias, declares the fact so marvelous as to challenge belief. In the other, from Isa. *6,* 9 ff, the prophet is told to blind the people (etc.) in punishment of their offenses; this text is cited as fulfilled. Cf. Matt. *13,* 14 f; Acts *28,* 26 f. In both cases John has in mind the guilt of the Jews' refusal to believe. St. Augustine observes,

"God predicted, but did not create this unbelief." **41.** Cf. *12*, 45. The reference is to the vision of Isaias in the temple (*6*, 1-6). John identifies Christ with God, who was seen in this vision. **42 f.** Recall the attitude of Nicodemus and of Joseph of Arimathea. The danger was expulsion from the synagogue. Cf. *4*, 1; *7*, 32; *7*, 47; *9*, 16.22. The Pharisees were dominant at the time. Cf. *7*, 13; *19*, 38.

44-50: Sayings of Christ on unbelief. It is probable that the following statements are collected here as pertinent to John's present theme, not as spoken on this occasion. **44 f.** Cf. *1*, 14.18; *3*, 11.13; *5*, 17 f. 36 f; *6*, 45; *7*, 28; *8*, 19.42; *10*, 30.38; *14*, 7-10. **46.** Cf. *1*, 4 f; *3*, 9; *8*, 12; *12*, 35 f. **47.** Cf. *3*, 17; *8*, 15. **48.** Cf. *3*, 18; *5*, 22.27.29; Mark *16*, 16. **49.** Cf. *5*, 19 ff; *7*, 16; *8*, 16.28 f. 55. **50.** Cf. *5*, 24.40; *6*, 40.47.69; *8*, 26.

II: THE PASSION, DEATH AND RESURRECTION *13-21*

1. THE LAST SUPPER *13*, 1 — *17*, 26

13, **1-20: The Washing of the Feet.** Comparison of this whole section with the Synoptics (Matt. *26*, 17-35; Mark *14*, 12-25; Luke *22*, 7-38) shows John's additions to their narrative, which he assumes as known.

V.1. is a general introduction to the scene of the whole chapter. *Pass out of this world:* an expression for death; *to the Father* has in view the divine nature of the victim. *His own:* primarily the Apostles, but also others who had come to Him in faith. *Loved them:* demonstrated His love. *To the end:* to the limit of love's perfection. The proximity of His death and the ardor of His love color all the events of the evening.

2. *When the supper was in progress:* it is difficult to choose between two readings in the Greek, one in the present, the other in the past. Vv.4 and 12 seem to imply that the act interrupted the supper. 1 Cor. *11*, 25 indicates that the institution of the Eucharist followed the regular supper; if so, the present action might have been between the two. *The devil,* etc.: contrasting our Lord's affection with the betrayal which was to lead to His death. Here John supposes Matt. *26*, 14; Mark *14*, 10; Luke *22*, 3; again in *12*, 4-6 he insinuates that the evil design has already been formed. **3.** Cf. Matt. *28*, 18; John *1*, 14; *3*, 31; *8*, 42; etc. Conscious of His dignity and office, He yet gives this humble sign of His love. **4.** *Rose from the supper:* probably after the eating of the lamb. *Garments:* a plural of generalization, referring to the usual outer garment or tunic. **5 f.** Washing of feet was the task of non-Jews, of slaves for their masters, of wife for husband, of children for father. It seems probable that our Lord began with Peter, though this is disputed. Peter's surprised question expresses his reverence as touched by the self-abasement of the act. **7.** *Thou shalt know hereafter* is generally understood of the instruction in 13 ff, though some prefer the fullness of knowledge to be infused at Pentecost. **8.** Peter's resistance is characteristic, but still impelled by reverence. *If I do not wash thee:* the warning corrects Peter's attitude, but it may mean more. Probably a cleansing is implied, but this supposition should not be carried too far. It definitely implies the acceptance of Christ's love, and this leads to the participation indicated in 10 and 13. **9.** Again the impetuous Peter, "disturbed by love and fear" (St. Thomas). **10.** A partial washing, either of hands or feet, satisfied the demands of both etiquette and ritual. The meaning is, it is enough that you submit to my wishes. *You are clean:* free of any defilement that might separate them from Him. *But not all:* the first mysterious allusion to the betrayer. **11.** Cf. *6,* 70.

12-17: Our Lord explains His action. **12.** *Reclined again* would indicate that the supper was not yet ended. The question prepares for the instruction to follow; it reflects the surprise of the Apostles, and assumes its answer to be "no." **13-15.** A rabbi expected honor and respect from his students; Christ was even more entitled to them. It was not the action itself on which He insisted, but the lesson it taught. **16.** Cf. Matt. *10,* 24; Luke *6,* 40. The proverb explains the mandate. Cf. *15,* 20. **17.** Some would place here the contention of the Apostles before the supper; cf. Luke *22,* 24. **18-21.** Cf. Matt. *26,* 21-25; Mark *14,* 18-21; Luke *22,* 21-23. This episode took place between the ritual supper and the institution of the Eucharist. Omitting the details recorded by the Synoptics, John specifies his own part in the scene. **18 f.** *I do not speak of you all:* the betrayer is excepted from the blessing in 17, though he still has time to repent of his plan. The indication of him has two purposes: it warns the Apostles of the scandal, and recalls to them another Messianic prophecy fulfilled in Christ (cf. Isa. *43,* 10; John *8,* 24). This quotation is from Ps. *40,* 10. To "lift the heel" is a figure of injury, such as tripping in a race; the thought is, a familiar friend has injured me. **20.** Cf. Matt. *10,* 40. The place of this verse in the present context is difficult; probably the sense is that in spite of this betrayal, our Lord's mission will be accomplished; in spite of His removal by death, His missionaries will represent Him. In their mission the Apostles will enjoy His protection.

13, 21-30: **The Betrayer.** 21. *Troubled in spirit:* cf. *12,* 27; the cause is the betrayal. *Said solemnly:* a clearer indication of the betrayal than in 18. 22. The surprise of the Apostles at this incredible thing was tinged with fear. Here should be placed Matt. *26,* 25; Mark *14,* 19; Luke *22,* 23. 23. He was on Jesus' right. See Commentary on Matt. *26,* 20. In his position John could easily lay his head back on the bosom of Jesus. This he may have done several times during the meal, but obviously on the occasion of Peter's request. Peter may have been next to John. 24. The Greek text reads "say, who is it . . ." Peter first got John's attention by beckoning. 25. The Greek reads "leaning back thus upon . . ." Cf. *21,* 7.20. This memory became attached to John's name. He repeated the question so as to be heard by our Lord alone. 26. The Greek reads, "it is he for whom I dip the morsel, and to whom I give." It was dipped in the *charoseth* (a mixture of nuts and fruits), or in salt water or vinegar. 27. *After the morsel:* some hold that Judas now realized that his crime was detected, and resolved to hand Jesus over to the officials at once. *Satan entered into him* to take more complete possession. *Do:* St. Augustine understands this as a permission, since only Christ could have given Himself up. *Quickly:* at once, there being no further need of delay. Cf. Matt. *26,* 5. 28. The company were still in the dark; we may doubt, from their attitude, that even Peter or John understood these final words. 29. *For the feast:* if this supper was the paschal meal, the "feast" is the solemn repast on the 15th Nisan. From *18,* 28 it is fairly certain that John did not consider the next day the 15th. *Give something to the poor:* alms were customary on the Feast of the Passover. 30. Judas' departure dissolved the intimacy involved in his eating the paschal supper with the group. Probably he left before the institution of the Eucharist. *It was night:* the meal could begin only after sunset. Symbolically, too, Judas was going forth into spiritual darkness.

13, 31-35: **The New Commandment.** A prelude to the farewell address *(14-17).* 31. Judas' departure was the beginning of the Passion (St. Thomas), and thus of our Lord's glorification and that of His Father. But this was His mission. Cf. *3,* 14; *8,* 28; *10,* 17; etc.; Rom *3,* 26; *5,* 8 f.   **32.** *In himself* is generally understood of the Father. Others, stressing *himself,* understand, through himself not through another (cf. Heb. *1,* 3).   **33.** *Children:* the usual address of a master to his pupils, but now with a special tenderness. For the rest see *7,* 34; *8,* 21; He would withdraw His bodily presence from the Apostles, as from the Jews, but for a different reason and with other results.   **34.** A *new commandment:* fraternal charity was already a precept of the natural law and also of positive divine law; cf. Lev. *19,* 18; *1* John *2,* 7-11. Its new element is that its norm and example is Christ: *as I have loved you.* Cf. Matt. *5,* 43-47; *22,* 34-40; Mark *12,* 28-34; Luke *10,* 27. He now removes from it all the limitations which custom and national prejudice had imposed. Christian charity is the charity of Christ, the bond of Christian unity. **35.** There is special reason why disciples of Christ should love one another. This is the theme of John's Epistles. It does not exclude the love of others. Here it becomes the badge of Christ, to be worn by any who claim to be His disciples.

13, 36-38: Peter's Denials Predicted. Cf. Matt. *26,* 31-35; Mark *14,* 27-31; Luke *22,* 31-34. John and Luke seem to record this announcement in its actual setting, though Matthew and Mark connect it with the walk to Gethsemani.   **36.** *Thou canst not follow me now,* either because of imperfect faith, or because Christ has further work for him to do; cf. *21,* 18 f.   **37.** Peter is too self-confident.   **38.** The cowardice that will lead to his denial is solemnly revealed.

14, **1-31: A Word of Comfort.** This is clearly one continuous discourse. It attaches to *13,* 35, the theme of Christ's departure. Some MSS even indicate this by a fresh introduction.

1. After He is gone their faith will keep Him present. *Believe . . . believe:* the Greek has both in the imperative. Another clear identification of Father and Son; cf. *5,* 17 f; *5,* 36; *8,* 18.28. 38; *12,* 44-50.   **3.** *I am coming again:* of personal reunion after death. Cf. 2 Cor. *5,* 8; Phil. *1,* 23; *1* John *2,* 28; *3,* 2. Some, however, would refer it to the *Parousia,* or coming of Christ at the Last Judgment.   **4.** Cf. *7,* 33. They knew of His return to the Father. *The way:* this was the burden of His instruction; cf. Matt. *16,* 21; *17,* 12; *20,* 18 f. In *6* Jesus had revealed Himself as the way to life: cf. *5,* 40; *6,* 35.39-40.47.

5. Thomas is prompted by a desire to understand more clearly. **6.** Our Lord's answer is a compendium of the New Covenant itself. *The way:* as *the door* in *10,* 7; He is the mediator between God and man; cf. Acts *4,* 12; Heb. *1-3. The truth:* the revelation of truth in both word and person; cf. *1,* 18. *The life:* the source of spiritual life; cf. *1,* 4; *5,* 21.26. No one can come to the Father except through Christ. **7.** In the second member the Greek text reads, "and henceforth you know." Though mildly reproachful, the thought is also consoling. The identity of Christ with the Father, implied in 6, is confirmed by both statements.

8. Philip looked for a more direct manifestation of the Father, perhaps as in the ancient theophanies. *It is enough:* we shall ask no further consolation. **9.** The doctrine of Christ's consubstantiality with the Father, together with their distinctness in person, is explicit in this verse, which reveals John's meaning in *1,* 18. **10.** Real faith in Christ will recognize His identity with the Father. Both His words and His miracles are evidence of it. **11.** This is addressed to all the Apostles (cf. *10,* 38) as a further rebuke to their slowness in faith. **12.** Their work is to be the extension of the Messianic kingdom. The fulfillment of this promise is the thesis of Acts. **13.** Cf. *15,* 4. *In my name:* cf. Mark *9,* 40; Acts *3,* 6. 16; *16,* 18. **14.** Our Lord's intervention is their consolation.

15-17: Further assistance is to come from the Advocate. In 15 the Greek text reads, "you will keep." This is the condition for receiving the Advocate, and it rests on love of Christ. The *commandments* are those of the New Law; cf. *1* John *2,* 3-5; *3,* 23 f; *5,* 2-4. **16.** The title *Advocate* is used only by John; cf. *14,* 26; *15,* 26; *16,* 7; *1* John *2,* 1. In classical Greek it denotes one called to the assistance of a person accused. In Job *16,* 2 it is rather one who consoles. The ancient Latin interpreters preferred *Advocatus,* implying an intellectual assistance. Our English term *Advocate* is more expressive than the "Paraclete" of the Douai-Rheims. Jesus Himself is such an Advocate (*1* John *2,* 1), and He will secure them another who will remain with them permanently. Moreover, this new Advocate is a person, distinct from the Father and the Son, identical with the Holy Spirit. **17.** *Spirit of truth:* proceeding from and teaching the truth.

18-21: This, again, is not the *Parousia,* or Second Coming of Christ. Nor is it the same as in *14,* 3. It is generally considered to be His return in the Resurrection. But this, if true, should not be restricted to visible appearances of His risen body; it is rather His mystical union with those who love Him: cf. Matt. *28*, 20. In addition, therefore, to the aid of the Advocate, Christ also will be with them.   **19.** *The world sees me no more,* etc.: after death He will be hidden from the world, but interiorly visible to the disciples in virtue of His union with them.   **20.** *In that day* of the fulfillment of this promise. *You will know:* their own perception will reveal to them fully the union of Christ with the Father, and their own union with Christ.   **21.** The observance of Christ's commandments is the condition of this manifestation. Cf. Wisd. *1,* 2. Through grace it is realized in the individual soul and in the collective life of the Church.

22-24: *Judas,* the brother of James and author of the Epistle which bears his name, perhaps the Thaddeus of the Synoptics. His question shows how much the Apostles shared the popular view of the Messianic manifestation. *The world,* here the Jewish world; but cf. Isa. *2,* 2; *11,* 10; *42,* 4; *49,* 6; *52,* 15; etc. Judas has in mind the *Shekinah,* the dwelling of God with His people; cf. Zach. *2,* 10. The condition of the true *Shekinah,* the dwelling of God in our souls, is the observance of His commandments. *We will come:* the three persons. *Make our abode:* cf. Rom. *8,* 9.11; *1* Cor. *3,* 16; Gal. *4,* 6; *2* Tim. *1,* 14. *He who does not love me* defines *the world. Words ... word:* the whole message brought by Christ.

25 f: Further assistance from the mission of the Advocate. The questions of the Apostles (*13,* 36; *14,* 5.8.22) have shown how imperfect was their understanding; our Lord now encourages them to expect enlightenment from the Advocate.   **26.** *Whom the Father will send:* (cf. *16,* 7), again the perfect union of Father and Son. *He will teach you:* the Advocate will take over the teaching office of Christ, instructing no longer by word of mouth, but by interior communication.

27-31: This is not a simple adieu, but includes assurances of serenity of mind, tranquillity of soul, simplicity of heart, fellowship in charity.   **27.** *My peace:* the interior peace which only God can give, the desire of the whole Old Testament, the great boon of the Messianic era; cf. Ps. *71,* 3.7; Isa. *9,* 6 f; *26,* 3; *27,* 5; *54,* 13; *57,* 19.   **28.** *The Father is greater than I:* this can only refer to Christ's inferiority in the Incarnation. The Church Fathers offer two explanations: (a) the Son has His divine nature from the Father, and in this sense is inferior; (b) Christ refers here to His human nature. The second is the better opinion. **30.** *The prince of the world* (cf. *12,* 31) is Satan. *In me has nothing:* he has no claim on me, as he had, e.g., on Judas. Cf. John *8,* 44; Rom. *6,* 23; Heb. *2,* 14; *2* Peter *2,* 19; *1* John *3, 8.*   **31.** The first words should be connected with 30; the thought is, I am to die, not because the author of death has any claim on me, but that the world may know, etc. *Arise:* cf. Matt. *26,* 46; Mark *14,* 42. If our Lord and the Apostles actually left the cenacle at this point, what follows in *15-17* either was delivered on the way to Gethsemani, or assembles thoughts expressed by our Lord on other occasions. But it is hard to conclude that they left the cenacle at this point; *18,* 1 implies the contrary.

15, **1-17: Union with Christ.** The theme of this figure is the corporate union of Christ with His disciples in the wider sense, as in the "Mystical Body" of St. Paul. It is a parable-allegory, somewhat in the manner of the Good Shepherd. The figure is familiar from the Old Testament. Cf. Gen. *49,* 22; Pss. *79,* 9; *127,* 3; Ecclus. *24,* 16.23; Isa. *5,* 1-6; *27,* 2 ff; Jer. *2,* 21; *12,* 10 ff; Ezech. *15,* 1 ff; *17,* 5 ff; *19,* 10 ff; Os. *10,* 1; also Matt. *21,* 33 and parallels.
 1. *The true vine:* i.e., in the highest sense of the figure, communicating real life, so *true light, 1,* 9; *true bread, 6,* 32. Thus Christ is the source of grace. The image prescinds from the relation of the divine persons.   **2.** In Greek the verbs of this verse are in present time. The fruitless branches are those not united to Christ in charity; those which bear fruit are capable of still closer union and richer yield. *Take away* expresses permanent separation. *Cleanse:* by removing obstacles to greater fertility.   **3.** Cf. *13,* 10; Acts *15,* 9.   **4.** The lesson of the imagery is their complete dependence upon Christ for success; they are also warned of the possibility of falling from grace. Cf. Col. *1,* 18; Eph. *2,* 20-22.   **5.** Cf. *10,* 9.14. Repeating the image for the sake of emphasis, our Lord offers a motive to continued union: without Him they can do nothing of supernatural value.   **6.** Another motive is that separation must mean destruction. Cf. Matt. *13,* 36-43. Dead branches and chaff were commonly used as fuel; cf. Ezech. *15,* 1-5.   **7.** Cf. Mark *11,* 24 and parallels. The warning is now repeated as a motive for adherence to Him in whom their prayers will be efficacious.

8. Cf. *14,* 13; *1* Pet. *2,* 12. The end of this fruitful union is that of Christ's mission, the glory of the Father.

9-17: The lesson of the image of the vine: union with Christ. **11.** *These things:* the exhortation to union with Him and its condition. True happiness can be had only through union with Christ; cf. *1* John *1,* 4; Acts *13,* 52; Rom. *15,* 13. **12.** Cf. *13,* 34; Mark *12,* 31 and parallels. **13.** The example of His love for them; cf. Rom. *5,* 6-10; *1* John *3,* 11 f. **14.** Cf. Matt. *12,* 50 and parallels; Luke *12,* 4. **15.** Cf. *16,* 12. The relation of master to disciple at the time was like that of lord and servant; whereas the revelation of intimate knowledge is a token of friendship; cf. Prov. *25,* 9. **16.** Cf. *6,* 70; *13,* 18. The student usually chose the teacher under whom he would study; this was not the case with Christ and His Apostles. *Appointed you:* i.e., destined you; cf. Acts *13,* 47; *1* Thess. *5,* 9; *1* Pet. *2,* 8. *That whatever,* etc.: co-ordinate with the two preceding clauses, and dependent on *appointed.* **17.** *These things I command you,* namely, *that you love,* etc.: cf. *13,* 34; *15,* 12. The precept of charity is thus the sum of Christ's teachings; all can be reduced to the commandment: Love one another.

15, 18-27: The World's Hatred. United with Christ, the Apostles share His fate in the world. The thought is most appropriate at the Last Supper, and we find no difficulty in accepting this passage as delivered on that occasion. **18.** *World:* men of worldy desires. Cf. Matt. *10,* 22; *24,* 9; Mark *13,* 13; Luke *21,* 17; *1* John *3,* 13 f. *If the world hates you:* implying that it will. **19.** The cause of the world's hatred; cf. *7,* 7. **20.** This emphatically predicts the persecution of the Apostles and the rejection of their teachings. In *13,* 16 the proverb is somewhat differently applied. Their fate will be the future fate of Christ in the world. **21.** This is because the world is ignorant of Christ's mission. The same division of men into two groups, so evident in this Gospel, will continue into the future. Cf. Matt. *10,* 22. The fulfillment will be seen in Acts *4,* 17; *5,* 40 f; *9,* 21; *12,* 1 ff; cf. *1* Pet. *4,* 14-16. **22-25.** This persecution is inexcusable, for, like the Jews, all others could know Christ if they would. Refusal to accept the evidence of His miracles makes their offense fully voluntary. The quotation in 25 is from either Ps. *68,* 5 or Ps. *34,* 19. This fulfillment of a prophecy is further support for the Apostles' faith. **26 f.** Cf. Matt. *10,* 20; Mark *13,* 11; Luke *12,* 12. In spite of the world's hostile reception, the Apostles with the Advocate will bear witness to their Lord. *Who proceeds from the Father:* the Catholic interpretation refers this to a divine emanation; this has been controverted by the Greeks, who understood it of the temporal mission of the Holy Spirit. *You will bear witness:* the verb in Greek is present in form, but in the sense "are about to bear witness." Cf. *1* John *1,* 1; Acts *1,* 21 f; *10,* 40 f. *From the beginning:* i.e., of the public ministry.

16, 1-5: Persecution Predicted. Some details are now given of the persecution already foretold. The section *16,* 5 ff is closely connected; the Advocate will support them in their trials. **1.** *Scandalized:* recall the circumstances of *6,* 61, when many of Christ's followers withdrew; cf. also *16,* 32. **2.** Cf. *9,* 22; *12,* 42; Acts *8,* 1-4; *9,* 1; Phil. *3,* 6. Sharers in the execution of one condemned by the Law naturally thought they were doing the will of God. Cf. Ex. *32,* 29; Deut. *13,* 8; *17,* 7. *Offering worship:* by performing a religious act. **3.** Cf. *8,* 19; *15,* 21. **4.** Cf. *13,* 19; *14,* 29. This prediction is to prepare them for the shock when it comes. *I did not tell you:* there had been some previous warning (Matt. *5,* 11 f; *10,* 14.16; *20,* 23; Luke *6,* 22; *12,* 4; *21,* 12), but the present one has two novel features: they are His representatives, and what is done to them is done as a religious act (St. Thomas). *Because I was with you:* this makes a great difference.

16, 6-33: The Rôle of the Advocate. This is the third main topic of the discourse, and complements the thought in *14,* 15-26. It falls into three parts.

6-11: The Advocate and the world. **5.** *No one of you asks me:* Peter and Thomas had done so earlier: *13,* 36; *14,* 5. 28; but the situation has changed. **7.** Cf. *14,* 16 f; *15,* 26; *7,* 39. The coming of the Holy Spirit is one of the providential fruits of our Lord's sacrifice. In Acts and the Epistles there is much evidence of the working of the Spirit in the Church, and it is accepted as the result of Christ's oblation. **8-11.** *Convict:* bring conviction relative to these truths. It implies the obstinate disposition of the world. The Advocate is pleading the cause of our Lord. *The world:* including more than the Jews. In the following three charges, *because* explains why the world is convicted, rather than directly specifies its offense. **9.** *Of sin:* the world is made to realize that, refusing to believe in **Christ,**

it remains in its guilt. **10.** *Of justice:* Christ's return to His Father proves that His cause was just, while His leaving the world is especially chargeable to its opposition. **11.** *Of judgment:* in the Passion a divine judgment was passed on the ruler of this world: cf. *12,* 31; Eph. *2,* 2-10; *6,* 12; Col. *2,* 14 f; Heb. *2,* 14. With him were judged his followers.

12-15: The Advocate will complete the Apostles' knowledge. **12.** *Many things: 15,* 15. The Acts and the Epistles contain new elements in the Christian revelation; see Acts *10,* 15; Rom. *11,* 25; *1* Cor. *7,* 12; *15,* 23; *1* Thess. *4,* 14; *2* Thess. *2,* 3; *2* Pet. *3,* 7 f. **13.** The Greek text reads, "He will guide you to the whole truth." Cf. Ps. *24,* 5. The teaching of the Spirit, as that of the Son, is the Father's. *Whatever he will hear* conveys the same thought. If this is addressed to the Apostles alone, it means that revelation closed with the Apostolic age (cf. Council of Vatican, Denzinger, 1836, 2021); if it extends beyond them, it can only apply to fuller knowledge of truths already revealed. *The things that are to come:* see predictions of the future in Acts *11,* 28; *21,* 11; *27,* 22; and cf. *1* Cor. *14,* 24 f; Eph. *4,* 11. Vv.14 f emphasize the identity of the Spirit's teaching with that of the Son. *He will glorify me* by completing Christ's teaching, as Christ had glorified the Father by delivering His message; cf. *5,* 19; *8,* 26. 40; *15,* 15. (See also Denzinger, 1787.) The passage indicates most clearly the Blessed Trinity: unity of nature, distinction and relation of persons. The Father communicates this nature to the Son; the Father and the Son to the Holy Spirit.

16-24: Joy at Christ's return. **16.** The Greek text lacks *because I go to the Father,* which probably found its way here from v.10. **17 f.** They could not reconcile His departure and quick return. **19.** The Greek text makes this a question: cf. *2,* 25; *6,* 62; Mark *9,* 31. Christ again reveals His supernatural knowledge. The next verses do not answer their question, but repeat the consoling promise of his speedy return. **20.** His death will bring sorrow to the Apostles, joy to the world; His resurrection will turn their sorrow to joy. **21.** This figure is a favorite simile with the Old Testament prophets (Isa. *13,* 8; *21,* 3; Jer. *4,* 31; *13,* 21; *22,* 23; *30,* 6; Os. *13,* 13; Mich. *4,* 9), and very apt here. **22.** Cf. Isa. *66,* 14; 2 Cor. *4,* 14. The fulfillment is recorded in John *20,* 20. **23.** *In that day,* primarily of the Resurrection, but also of subsequent time. *Ask:* either put a question or make a request; the former predominates in 23a, the latter in 23b. **24.** Cf. *14,* 13; Matt. *7, 7* and parallels; *1* John *5,*

14 f. Thus far they have not asked in His name: He now urges them to do so. Perfect joy is to be had in God alone.

25-33: Conclusion of the discourse. **25.** *Parables* in the sense of mysterious sayings. Most of the parables were obscure; cf. Matt. *13,* 10-15. At this time many of our Lord's statements were hardly understood by the Apostles, not yet enlightened by the Holy Spirit. *The hour is coming:* mainly that of the Resurrection, but also of the subsequent mission of the Holy Spirit. **26 f.** They will then ask in His name because they will know Him as their mediator. *I do not say:* Christ will not cease to be supreme mediator, but their own office too will be efficacious through Him. **28.** Cf. *6,* 46; *7,* 29; *8,* 42; *13,* 3. *Came forth* at least implies His eternal generation, though His temporal mission may be primarily intended. This is the final explanation of His leaving them. **29.** Cf. *16,* 25; *17,* 7. With joy the Apostles understood the meaning of this statement. **30.** *For this reason:* Christ has revealed His supernatural knowledge, and thus manifested His divinity. Cf. *1,* 48 f; *4,* 19.29; Ps. *7,* 10; Wisd. *1,* 6; Jer. *17,* 9 f. **31.** But their faith was not yet complete. The meaning of Christ's reply will vary with the location of the emphasis. Read as interrogative, it can mean, do you at last believe? or, are you sure you believe? If read as affirmative, it is, now indeed you believe. In either case they are warned against the coming trial of faith. **32.** To show the insecurity of their disposition, He reveals what they will do within a few hours. *To his own house:* cf. Matt. *26,* 31; Mark *14,* 27. *I am not alone:* cf. *8,* 16.19. **33.** Cf. *12,* 31; *13,* 31. *I have overcome the world:* in conquering its ruler. His grace raises them above all the trials of the world. Cf. *1* Cor. *15,* 57; *1* John *4,* 4; *5,* 4.

17, 1-26: Christ's Priestly Prayer for Unity. The following prayer is called "priestly" as being a prelude to that sacrifice in which Christ was both priest and victim. Even though a prayer, it is partly instructive in purpose. The association of the passage with the Last Supper is generally conceded.

1-5: Christ prays for Himself. **1.** *These things:* the discourse in the cenacle, therefore at least *13-14. Raising his eyes:* a formal attitude of prayer; cf. Luke *18,* 13; Acts *7,* 55. *The hour:* of the Passion. *Glorify:* the sacrifice of our Lord restored due glory to God, and its accomplishment is the glory of the Son; cf. *7,* 39; *12,* 16; *13,* 31. **2.** The motive of the prayer. Jesus is the divine agent of grace to men. Cf. Rom. *4,* 25. *Power over all* *flesh:* i.e., over all men. Cf. Pss. *2*, 7 f; *71*, 8 f; Isa. *49*, 6; etc. *To all* who accept the offer, cf. *3*, 35 f; *6*, 44; Matt. *22*, 2 ff. **3.** *That they know,* etc.: cf. *8*, 55; *10*, 15; *16*, 3; *1* John *2*, 3 f; *3*, 1; *4*, 8. The knowledge is a practical acquaintance, acceptance and service. This is a divine gift, effective through faith and love. Cf. *3*, 16; *1* John *5*, 20. Eternal life is the glorification of God, in which man can share even on earth by "knowing" Him. *Only true God:* in opposition to the false gods whom man serves. Cf. *5*, 24; *6*, 40.55. **4.** Cf. *3*, 16; *4*, 34; *10*, 17 f; *14*, 31. Christ here includes the accomplishment of all His mission. **5.** Cf. *1*, 1 ff; *2* Cor. *8*, 9; Phil. *2*, 5 ff. Our Lord is speaking as God-Man. As God, He had never ceased to be present in the bosom of the Father: as man He had "emptied Himself." Now He prays that His human nature be elevated to the right hand of the Father, the position eternally natural to the Son.

6-19: Christ prays for the Apostles. **6.** *Given me:* by calling them; cf. *6*, 37.44.66. *They were thine:* men ready to please God and to establish His kingdom; cf. *3*, 21; *8*, 47. *Thy word:* Christ's message; cf. *5*, 30; *7*, 16; *8*, 38. **7.** Cf. *10*, 25; *12*, 49 f; *14*, 10. **8.** The object of faith is Christ, sent by the Father and originating in him. This preliminary praise of the Apostles is the basis of the prayer. Vv.**9-11a** reveal the reasons for this special prayer. **9.** *I pray for them* who have thus shown themselves worthy; the reasons will follow. *Not for the world:* He does pray for the world in due time *(18,* 21), but now His petition is for the Apostles. *Because they are thine:* the first reason. **10.** The parenthesis, *all things mine are thine,* could be uttered only by Christ as the divine Son. I am glorified in them: this is the second reason, their faith, and their spreading of Christ's name. **11a.** The third reason: He is about to leave them in the world.

11b. The prayer itself begins. *Those whom:* the Greek reads "which," referring it to *thy name* "which thou hast given me" to be revealed, and which is to be the bond of their union. *Holy:* separated from the world. The model of the unity for which Christ prays is the unity of the Father and the Son: supernatural, perfect, founded on faith and love. Cf. Acts *2*, 42-47; *4*, 32; Eph. *4*, 4.12.13. While our Lord is with them He can foster this union, but His departure will expose them to danger. **12 f.** *No one of them perished:* cf. *10*, 28; *18*, 9. The fall of Judas was foreseen in Pss. *40*, 10; *108*, 4 ff. *Son of perdition:* cf. Matt. *26*, 24; it were better he had not been born. *I speak in the world:* now, in the presence and hearing of the Apostles. **14-16.** A new petition for protection against the world. *I have given them thy word,* to which the world is opposed. *World:* humanity that is hostile to God. **15.** *From evil* in the Greek can also be read "from the evil one" (as in *1* John *2,* 13; *3,* 12; *5,* 18) and is so understood by many recent interpreters. More generally, however, with Sts. Augustine, Chrysostom, Cyril, Thomas, the impersonal reading is maintained. **17-19.** The climax of the petition. **17.** *Sanctify them:* set them aside from the world, as devoted to the work of God. Or, more significantly: as their mission partakes in that of Christ, they need His holiness. *In truth:* in the service of this truth, and as transformed thereby. The "truth" in question is the word of God, the source of all truth. **18.** Their mission is analogous to that of the Son; cf. *4,* 38; *10,* 36; *20,* 21; Matt. *10,* 16 f. **19.** *I sanctify myself,* devote myself as victim in this sacrifice; cf. Ex. *13,* 2; *28,* 38; Deut. *15,* 19. *That they may be sanctified,* as above. *In truth* also as above, although some take it here adverbially, "veritably," and not in an external way.

20-23: Christ prays for future believers. Some take this as merely an extension of the above prayer, and applicable to the Apostles only. **20.** Cf. Acts *1,* 8; Rom. *10,* 14.17. **21.** The unity petitioned for the Apostles: one among themselves, one with God, united in faith and love. *That the world may believe:* the visible unity of Christians is a motive of faith in Christ. This supposes an external manifestation of such unity, e.g., in the Church. **22.** *Glory:* either divine power, or the divine sonship. Cf. *1,* 12; Rom. *8,* 9-16; *2* Cor. *13,* 5; Gal. *2,* 20. Union with Christ permits us even here to enjoy eternal life (cf. *3,* 15. 36; *5,* 24), and hence it can be spoken of as "glory." Others stress more the gifts which this union confers. Thus St. Augustine speaks of immortality; St. Thomas, of the glory of the resurrection; Maldonatus, of love. **23.** Cf. Eph. *3,* 17. A fuller statement of the manner and effect of this union. Christ is in the faithful, He is also in the Father; thus through Him unity is perfected. Cf. *2* Pet. *1,* 4; *1* John *3,* 9; *4,* 12 f.

24-26: The conclusion of the entire prayer. **24.** *Whom thou hast given me:* some restrict this to the Apostles; but the reward seems rather destined for all whom the Father draws to the Son. *My glory:* His glory as the Son, enjoyed from all eternity. *That they may behold:* in the full force of seeing, appreciating, enjoying. **25.** *Just Father:* who decides the cause of faithful and unbeliever. The discrimination is based on their "knowing" or not "knowing" God.   **26.** The full significance of this knowledge: it is the knowledge of the Father. Its practical effect is the enjoyment of God's love. Faith makes this possible. St. Augustine: we are Christ's members, and in Him we are loved, since He is loved in entirety.

2. The Passion and Death of Jesus *18-19*

18, 1-11: Jesus Arrested. Cf. Matt. *26,* 47-56; Mark *14,* 43-52; Luke *22,* 47-53. The Synoptics record the Agony, which is omitted by John.   **1.** After the above discourse, our Lord and His Apostles left the cenacle. *Cedron* means "muddy." It can be called a torrent, although its bed is dry except in the rainy season. We know the name of the garden, Gethsemani, from Matthew and Mark.   **2.** Cf. Luke *21,* 37; *22,* 39. This may have been the lodging of our Lord and the Apostles during the feasts. Thus it would be known to Judas.   **3.** *Cohort* here probably means a "group." The danger of popular interference (Mark *14,* 2) may have required a large number of soldiers. The arrest was made officially by the attendants of the High Priest.   **4.** Cf. *13, 3.* The Greek text reads "that were coming upon him." Our Lord made it clear that He could not be taken against His will.   **5 f.** In His manner or aspect there was something awe-inspiring, which caused them to draw back and fall to the ground. Perhaps only those nearest Him and ready to apprehend Him were thus affected.   **7-9.** It would seem that the officials also had in mind the arrest of the disciples. Note *(17,* 12) our Lord's solicitude for His Apostles. There the danger is moral; here, physical; but both are provided for in God's care.   **10 f.** John's acquaintance with the name of Malchus shows his familiarity with the scene. What John especially shows is that our Lord offers Himself freely.

18, 12-27: Peter's Denial. Cf. Matt. *26,* 57-75; Mark *14,* 53-72; Luke *22,* 54-62. John again adds to the Synoptic narrative those details which answer his purpose.   **12.** John alone records that Jesus was bound at the garden. *The Tribune,* in Greek *chiliarch,* was the commander of a thousand.   **13.** Cf. Luke *3, 2;* Acts *4, 6.* Though no longer High Priest, Annas wielded great authority, and was deeply involved in this effort against our Lord. *Who was high priest:* cf. *11,* 49 ff.

There has been question whether 15-23 describe a trial

before Annas or before Caiphas. (a) Most ancient interpreters take 24 to imply that all that precedes it took place before Annas, and that no action before Caiphas is noticed by John. The present examination but little resembles the trial before Caiphas in the Synoptics. (b) Several modern interpreters place 24 after 14, taking 15-23 entire as a trial before Caiphas. These hold that John would hardly omit altogether the trial at which Christ was condemned. (c) Still others, but with less probability, would read "had sent" for *sent* in 24. Whatever the correct explanation, John evidently aimed at no complete account of our Lord's judicial examination. The ground had been adequately covered by the Synoptics.

15. *Another disciple:* with St. Augustine, some commentators now doubt that this was John the Evangelist. *Courtyard:* the better houses had both a forecourt and an inner court; the latter is meant here. Annas and Caiphas may have occupied sections of the same residence. This would easily explain 18 and 25. **16.** At the door from the street to the forecourt, the cautious portress had probably refused Peter admission. **17.** Peter's first denial. The "maid" of the Synoptics is here called the portress. There is some precipitation in Peter's answer; he is thinking only of his personal danger. **18.** The night was well advanced, and, at this time of the year, could be very cold. **19-23.** If this *high priest* is Annas, John is relating a preliminary examination omitted by the Synoptics. **20 f.** Our Lord appeals to the law governing a Jewish trial: cf. Num. *35,* 30; Deut. *17,* 6; *19,* 15; *3* Kgs. *21,* 10-14. He exonerates His disciples. **22 f.** His answer to this unjust blow, given with either hand or rod, is again an appeal to the Law. **24.** If the above scene was before Annas, our Lord's appeal for justice may have resulted in His being turned over to Caiphas, the official High Priest. **25-27.** The second and third denials of Peter. Points of apparent divergence from the Synoptic account are probably due to the Evangelists' variance in the details they select, and also to their manner of stating the substance of a saying rather than its precise words. **25.** It is supposed that Peter is still in the place where he was at v.17; thus our Lord, in passing from Annas to Caiphas, had not left the building. **26 f.** The Synoptics suppose a lapse of time between the second and third denials.

18, **28-40: Jesus before Pilate.** Cf. Matt. *27,* 2.11-25; Mark *15,* 1-15; Luke *23,* 1-25. Throughout the scene John shows our

Lord preserving a majestic dignity. The Synoptics present more details of His sufferings. **28.** *Prætorium:* the residence of the Roman procurator in Jerusalem. Its location is disputed. *It was early:* probably soon after 6 A.M. *Defiled:* cf. Acts *10,* 28; *11, 3.* A gentile home was "unclean" because of the possibility of dead being buried there.

This verse is of great importance for the chronology of the Passion. (a) "To eat the passover" means to eat the paschal lamb; cf. *13,* 1.29; *19,* 14. The day would then be the 14th Nisan. (b) Such a trial and execution were hardly possible on the 15th Nisan, a day of sabbatical rest. (c) The method of computing the day of the month from the new moon admitted of variation. Hence it is quite possible that for Galilean pilgrims Thursday was the 14th, while for the Jews of the city Friday was the 14th Nisan. John would follow the official date in relating a motive based upon it.

29. Pilate, as procurator, was Rome's fiscal and judicial agent in Judea. **30.** *A criminal:* a vague reply, more out of anxiety to win their purpose than out of insolence. Their real motive would have no force. **31.** Pilate's reply reveals an impatience fully in keeping with what we know of his feelings toward the Jews. Under the Roman provincial system, the Sanhedrin had power to judge according to Jewish Law. Pilate may have suspected their real purpose, and therefore chided them; at least, he provoked an open declaration of their intention. **32.** Cf. *3,* 14; *8,* 28; *12,* 32; Matt. *20,* 19 and parallels. Crucifixion was a gentile method of execution. Stoning was a Jewish penalty for blasphemy; cf. Lev. *24,* 10-16. But our Lord had foretold that He would die on a cross. **33.** Seeing Pilate unwilling to do their bidding, the Jews advanced their political motive. Cf. Luke *23,* 2. **34.** Our Lord's answer implies a distinction. If the question originated with Pilate, it would involve kingship in the temporal sense; if it came from the Jews, it might have the Messianic meaning. **35.** *Am I a Jew?* therefore he knew nothing of the Messianic king. *What hast thou done?* shows that he had no suspicion that Christ pretended to temporal kingship, but was merely investigating an accusation. **36.** Our Lord cannot be a temporal king of the Jews, or they would not have surrendered Him. **37.** *Bear witness to the truth:* to reveal the truth and to have it accepted; cf. *1,* 9; *3,* 11; *9,* 5; etc. *Everyone who is of the truth:* those who have accepted His word; cf. *3,* 19-21; *8,* 47; etc. **38.** *What is truth?* Pilate shared the speculative skepticism of Rome, that uncertainty of mind which was born of the current philosophies, Stoicism, Epicureanism, etc. Convinced that there is no danger in his prisoner, Pilate again seeks to throw the case back upon the accusers. *Went outside again* supposes Mark *15*, 8; Luke *23*, 4 f. **39 f.** The designation *King of the Jews* may have been a recognition of their Messianic hope. In this pretext for a release Pilate shows his weakness; he could not rightly speak of releasing one not yet convicted.

19, 1-16: The Scourging and Crowning. Cf. Matt. *27*, 26-31; Mark *15*, 15-20; Luke *23*, 4 f. 13-16.20-25. Evidently supposing the Synoptic account, John is brief as usual, but displays more clearly Pilate's inclination to release our Lord. **1-3.** Scourging was a normal preliminary to crucifixion. Matthew and Mark report the incident only in this character, and therefore after the death sentence; whereas Luke, like John, shows that Pilate hoped the scourging might suffice for Christ's discharge. This extremely cruel practice was called "an intermediary death," and in fact, the victim often died under the lashes. The crowning with thorns was a spontaneous jest of the soldiers. The purple garment was probably a faded military cloak. In all of this Pilate was quite arbitrary. John tells of three admissions on his part that he found no guilt in Christ.

4 f. Pilate makes two appeals for the prisoner's release. First, *I find no guilt in him,* and yet, to satisfy you, I have scourged Him. Again, the demonstration of our Lord's pitiable condition: *Behold the man!* as if such a man could be a king! **6.** These appeals are rejected. The Synoptics suppose that the crowd joined in the cry, *Crucify him!* but John here reveals that the officials were behind it. *Take him yourselves,* etc.: if you dare crucify an innocent man. **7.** The law referred to is Lev. *24,* 16, against blasphemy. Clearly they understood the title "Son of God" in its literal force; otherwise it would not have been blasphemous: cf. *5,* 18; *7,* 20; *8,* 58; *10,* 33. The political charges having failed, the Jews now admit their real accusation.

8. This fear could have two causes, the rising fanaticism of the Jews, and the possibility of some divine power in our Lord. **9.** Christ's silence recalls Isa. *53,* 7. **10 f.** Pilate's pride merited the rebuke that, without permission from God, he would be powerless. Cf. *18,* 6. *He who delivered me:* i.e., Caiphas, as head of the Sanhedrin. *Has the greater sin:* to both Pilate and the Jews God had made available this power over His Son, but both sinned in using it. The sin of the Jews was the greater because they had initiated the process, they had evidence of His divinity, and yet their conduct was inspired by hatred. Pilate's sin was the less because he was an instrument, was acting under the moral stress of fear, and had far less light. **12.** *From then on*, or as the Greek text can also be rendered, "on that account." This gave Pilate a new incentive to dismiss the case. It also led the Jews to a final and a winning argument. To their charge of Christ's pretension to a throne Pilate could not be indifferent; to be so would endanger his title to be considered the *friend of Caesar*.

13. *The judgment seat* was a rostrum or curule chair from which sentence was formally pronounced. *Lithostrotos:* a Greek term meaning in general a stone pavement, here probably the court outside the prætorium. *Gabbatha:* a cleared space or court. Cf. Acts *12,* 4; *16,* 19; *18,* 12. **14.** *Preparation Day:* originally meaning preparation for the paschal supper, the expression later denoted the preparation also for the Great Sabbath. Cf. Mark *15,* 42. According to John's dating either sense might apply here. *The sixth hour:* cf. Mark *15,* 25, where from 9 to 12 may be meant, while John more precisely sets the scene nearer 12. Neither aims to give the exact hour. *Behold your king* is ironical. **15.** The retort of the chief priests is decisive. *We have no king but Caesar* is their final repudiation of Jesus as the Messias. **16a.** Pilate acceded to their wish, though the execution of the sentence was in the hands of the Roman soldiers. Cf. Luke *23,* 25; Acts *2,* 23; *3,* 15. **16b.** In the Greek text this is read with 17, omitting the words *and led him away*.

19, **17-24: The Crucifixion.** Cf. Matt. *27,* 31-56; Mark *15,* 20-41; Luke *23,* 26-49. **17.** *Bearing the cross:* the victim usually carried his own cross, probably his arms tied to the cross-beam only. The assistance of Simon, mentioned in the Synoptics, was found necessary after the procession left the praetorium. *The Skull:* as the Aramaic *Golgotha* is explained in Matt. *27, 33;* Mark *15,* 22. **18.** *Two others:* the fulfillment of Isa. *53,* 12, "He was reckoned among the wicked." Cf. Mark *15,* 28. **19 f.** John and Luke tell us that the inscription was written in three languages. It was usual thus to publish the crime of the offender. In this case Pilate was obviously deriding the Jews. Each of the Evangelists gives only the sense of the inscription, and thus their words can differ. **21 f.** Pilate had no mind to change what he had written. **23f.** The clothes of the condemned were the spoil of the soldiers who executed the sentence. Here the *garments* included all the under clothing, while the *tunic* was the outer cloak. A prediction of this scene is found in Ps. *21,* 19. The first verse of this same Psalm is quoted by Christ in the words *Eloi, Eloi, lama sabacthani (*Mark *15,* 34).

19, 25-30: The Death of Jesus. **25.** John mentions those particulars in which he was involved. The older and more general view is that there were three women in this group: Mary, wife or mother of Cleophas, who was *his mother's sister,* the Blessed Virgin, and Mary Magdalene. This last is probably to be distinguished, by her title "the Magdalene," from Mary the sister of Martha and Lazarus. **26 f.** The doctrine of Mary's maternity over all Christians does not rest on this text, although its words may be thus accommodated. **28.** Cf. Ps. *68,* 22, but also Ps. *21,* 16. **29.** *Common wine:* the common dry wine of the soldiers, often called "vinegar." *A stalk of hyssop:* this stalk is frail and only about eighteen inches long. Hence Lagrange explains: hyssop was wrapped around the end of a rod to serve as a sponge. **30.** *It is consummated:* Christ has accomplished His mission from the Father. Cf. Heb. *10,* 7. This was the price paid for our salvation. Cf. Rom. *3,* 25 f; *1* Cor. *6,* 20; 7, 23; *2* Cor. *5,* 15; Gal. *3,* 13; Eph. *2,* 13; Heb. *10,* 10-14; *1* Pet. *1,* 18 f.

19, 31-42: The Burial. **31-37:** The lance. The incident is recorded only by John. **31.** Cf. Deut. *21,* 23. If this Sabbath was the 15th Nisan, we can understand its especial solemnity. *That their legs might be broken:* the *crurifragium* was a means of hastening the death of one crucified. The femur was fractured with a rod, the additional pain compensating for a shortening of the agony of the cross. **32 f.** The thieves, still alive, died under this treatment. Jesus, however, was already dead. **34.** *One of the soldiers:* tradition, following the "Gospel of Nicodemus," names him Longinus, deriving the name from the Greek term for "lance." *Opened:* the Greek means more exactly "pierced." *There came out blood and water:* John seems to consider this a prodigy. But if John intended thus to prove a real death of Christ as against the Docetists, the phenomenon can readily be understood as natural. The commonest explanation is that the lance pierced the heart of Jesus, the *water* being the lymph escaping from the pericardial sac (cf. *1 John 5,* 6). **35.** John declares himself an eyewitness of this event. *And his witness is true:* note the force of this in *4,* 23.37; *6,* 32; *7,* 28; *15,* 1: it is a genuine testimony worthy of all credence. *And he knows:* generally taken of John himself; but some would refer it to Christ. Thus John would call upon Christ to witness the truth of his statement. *That you also may believe:* that the reader may see in this, as John does, the fulfillment of Scripture. **36.** *For* connects this verse with the foregoing as its explanation; cf. Ex. *12,* 46; Num. *9,* 12; Ps. *33,* 21. Christ fulfills the type of the Paschal Lamb. Cf. also *1,* 29.36. **37.** Cf. Zach. *12,* 10; Apoc. *1,* 7. The sense of the Hebrew text is expressed.

38-42: The burial. Cf. Matt. *27,* 57-61; Mark *15,* 42-47; Luke *23,* 50-56. **38.** *Arimathea:* probably Ramathaim, today er-Ram, to the north and west of Jerusalem, and the home of Samuel. Joseph probably gathered courage from seeing this death of the Master. Cf. Isa. *53,* 9. **39.** For Nicodemus cf. *3,* 1 ff; *7,* 50 f. *Myrrh* is an odorous resin; *aloes,* a scented wood. *Mixture:* a variant reading has "roll." Cf. Ps. *44,* 9; Prov. *7,* 17; Cant. *4,* 14. *A hundred pounds:* about 70 pounds avoirdupois. **40.** *In linen cloths:* possibly in the form of a shroud. Cf. *11,* 44. *Jewish manner of preparing for burial:* this involved three operations: the washing of the body (not mentioned in the Gospels), the anointing (the concern of the women), and wrapping in cloths over which the preservatives had been spread. The haste of this temporary internment admitted only of the shroud and the spices. **41.** Matthew tells us that the tomb belonged to Joseph of Arimathea.

3. THE RESURRECTION OF JESUS *20-21*

With the Synoptic account of the Resurrection already known, John has only to supply what pertains to his thesis. All the Gospels are manifestly concerned with the presentation of evidence of our Lord's actual rising from death. The brevity of each narrative and the difference in the details selected make it difficult to arrange the whole in the exact order of occurrence.

20, 1-18: Mary Magdalene. Cf. Matt. *28,* 1-10. 29 f; Mark *16,* 1-8; Luke *24,* 1-12. **1.** *The first day of the week:* our Sunday. It was later known to the Jews as "the day of the Christians." *While it was still dark* suggests the early hour and the zeal of the women. It can mean the dusk just before sunrise, and hence it does not conflict with Mark's note "the sun being now risen." The women (Mark and Luke) also saw that the stone had been removed from the entrance of the tomb.   2. Mary thought first of informing the Apostles. Peter here assumes his place as head of the group: John, under his peculiar title, is at his side. *We:* including the other women.

3-10: With the exception of Luke *24,* 13, this is peculiar to John.   3 f. The Apostles, though devoted to Christ, had no thought of the Resurrection.   5. John merely satisfied himself that the body was gone, and then waited for Peter. *Stooping down* because of the low entrance.   6 f. Peter, true to his character, entered at once and inspected the tomb. The cloths were evidence that the body had not been stolen, otherwise its wrappings would not have been removed.   8. When John saw this evidence he *believed,* was convinced that Christ had risen, while Peter only wondered: cf. Luke *24,* 12.   9. *They did not understand:* John's faith sprang from what he had seen, not from any anticipation of a resurrection; and this explains also why Peter only pondered what he had seen. *The Scripture:* the texts later used by the Apostles; cf. Acts *2,* 25-27. 31; *13, 33;* *1 Cor. 15,* 3 f. Our Lord Himself had foretold it (cf. *2,* 22; *16,* 16 f), and the Jews seem to have understood Him (Matt. *27,* 63). It is frankly admitted here that He was imperfectly understood by His Apostles.

11-18: Mary Magdalene, the first to see the risen Christ. **11.** She could but weep, still thinking the body had been carried off.   **12 f.** Rewarded by the sight and the words of the angels, she is not yet consoled like the other women. But a greater consolation is reserved for her.   **14 f.** He may have somewhat altered His appearance; cf. *21,* 4; Luke *24,* 16. She thought Him the gardener. *If thou hast removed him,* etc.: her preoccupation is still uppermost.   **16.** *She said to him:* the Greek text adds "in Hebrew," i.e., in Aramaic. Mary had again turned toward the tomb, when she joyfully recognized His voice pronouncing her name. *Master* (literally "My Master") expresses her deepest devotion.   **17.** *Do not touch me:* the present imperative implies that she had already done so (cf. Matt. *28,* 9 f). *I have not yet ascended:* the relation of these words to the foregoing is variously explained. (a) They are usually taken to express a reason why Mary should not touch Him; but here again opinion varies. (b) A more satisfactory explanation makes this statement parenthetical: "do not touch me (for I have not yet ascended; I shall soon), but go, tell this to the brethren," etc. He thus urges her not to delay. *My Father and your Father* is intentionally phrased so as to express the difference between His relationship and theirs.   **18.** *These things:* Christ's message.

20, 19-23: The Disciples. Cf. Mark *16,* 14; Luke *24,* 36-43; *1* Cor. *15,* 5.   **19.** John dates this on the day of the Resurrection. *It was late:* the time is not exact, but we may suppose it late enough to permit the return of the two disciples from Emmaus. *The doors ... closed:* the supernatural manner of our Lord's entrance showed the glorified state of His body. *For fear of the Jews* explains both the gathering of the Apostles and the fastening of the doors. *Peace be to you:* the usual Aramaic salutation, but pregnant now with deeper meaning.   **20.** *He showed them:* evidently, as in the Synoptics, they had to be convinced of His presence in the same body. The Greek text reads: "seeing the Lord ... rejoiced."   **21.** *As the Father,* etc.: this is the definitive mission of the Apostles.   **22 f.** The power to forgive sins was conferred in a special way, for it was one of the principal features of their mission. The Council of Trent (Sess. 14, cap. 5-6; Denzinger, 899-902) has defined that this verse proves a ministerial power to forgive sins. Cf. Luke *24,* 49, where the Apostles are told to await the power from on high; yet on this occasion the Holy Spirit was actually given them. The term *receive* implies "here and now." The Spirit is the principle of their new life, but here He confers a special power. *Breathed:* cf. Gen. *2,* 7; Wisd. *15,* 11; Ezech. *37,* 9; external sign of this power. Its nature, the forgiving or retaining of sins, is made very clear. Here that power is definitively conferred. The forgiveness is to be effected through an act of their judgment, not merely through the faith of the penitent.

20, 24-29: Thomas. The other disciples also had hesitated, and were reproved for doubting; cf. Mark *16,* 14. John's present aim is to show that even the Apostles were not credulous, but had to be convinced.   **24.** Cf. *11,* 16. Various reasons have been assigned for the absence of Thomas; we know only the fact.   **25.** *Put my hand into his side,* where the wound was larger than those of the nails.   **26.** *After eight days:* a week later, on the eighth day. This may be the first indication that the Sunday was even now hallowed. Cf. *1 Cor. 16,* 2; Apoc. *1,* 10. *Again* suggests that they did not thus gather every day. **27.** Using Thomas' own words, Christ condescends to his demands. The words may have been accompanied with gestures. **28.** The language of 29 makes it uncertain whether Thomas touched the wounds. Thomas recognized God in the fact of the resurrection of Christ. His faith is now complete, acknowledging our Lord's supremacy and divinity. This is the faith Christ required of them, the faith they were to create by witnessing to Him and to His Resurrection. **29.** Thomas is thus criticized only for the slowness of his faith, his need of compelling evidence. The Greek text, omitting *Thomas,* reads "dost thou believe?" *Blessed are they:* i.e., still more blessed, more pleasing to God.

20, 30-31: The Evangelist's Epilogue. To the strong profession of faith just recorded John deems it appropriate to add this statement of his thesis, propounded to engender that very faith. Some authors are of the opinion that the Gospel originally ended here. Still others would transfer these verses to *21,* 23, making them the conclusion of *21.* There is not sufficient reason for either view; see following introduction to *21.* **30.** *Many other signs:* as, e.g., in the Synoptics. John claims to have selected only those best suited to his purpose. *In the sight of his disciples,* who have since been His witnesses. **31.** Cf. *1,* 14-18. Faith and its reward run all through the Gospel. The content of faith is that Jesus, who lived a human life among them, is the Messias promised in the Old Testament; that He is the eternal Son of God, equal to the Father, who took on Himself this humanity in order to give us eternal life. Cf. *3,* 15 f; *5,* 24; *6,* 35; etc. *In his name:* cf. *15,* 4 f; Acts *4,* 12.

Chapter 21 is quite generally regarded as an appendix to the Gospel. Catholics who favor this view conclude that John wrote it later. Some rationalists deny its Johannine authorship. However, the explanations do not agree. One may admit that the argument of the chapter is supplementary without being therefore obliged to give it a later date. It is an appendix, in that it contains two thoughts not directly involved in the author's thesis: the primacy and the martydom of Peter, and the fate of John himself. This proves neither a later date nor another hand than John's; and against this latter the internal evidence is strong. The **miracle,** though pertinent to the general thesis, is here recorded as the occasion of the two further points just indicated as the author's present aim.

21, 1-14: The Manifestation in Galilee. This miracle is obviously distinct from that related in Luke *5*, 1-11.   **1.** *After these things:* the time is indefinite, as in *5*, 1; *6*, 1; *7*, 1. *Sea of Tiberias:* cf. *6*, 1. Only John uses this uncommon title for the Lake of Genesareth, or Sea of Galilee.   **2.** *Nathanael:* probably Bartholomew (cf. *1*, 45 ff), as accompanying these others, all of whom are Apostles. *Sons of Zebedee:* not elsewhere so called by John, whence some consider this a gloss originally intended to identify the *two others*.   **3.** The Apostles had repaired to Galilee, as instructed (Matt. *26*, 32; *28*, 7.10; Mark *14*, 28; *16*, 7), but this implies no permanent return to their trade of fishing. At all events, the initiative came from Peter. Night was considered the best time for fishing.   **4.** Their failure to recognize Him was probably due to alteration in His appearance. Cf. Mark *16*, 12.   **7.** Again Peter and John hold the center of the scene. John recognized our Lord in the miracle; Peter's zeal made him impatient of the slow-moving boat. *Tunic:* in this case an outer garment or a tradesman's apron. *Stripped:* he had laid aside this outer garment.   **9.** The fire with the fish and bread laid upon it is another surprise.   **10.** To satisfy all, more fish was needed.   **11.** Peter again assumes the lead, the others assisting him. In the miracle there is clearly some symbolism representing the relation between Christ and the Apostles in their work. The unbroken net is the Church in its indefectibility; in the work of Christ, Peter is the leader; the great number of fish is the great success of their mission; in all, they can do nothing without Christ.   **12.** The Greek text and some MSS of the Vulgate read, "none of his disciples dared ask him."   **14.** *Now a third time:* cf. *20*, 19. 26. This is the third time our Lord manifested Himself to the group; other appearances are not excluded.

21, 15-23: The Primacy of Peter. Cf. Matt. *16*, 17 f. The triple profession of love elicited from Peter recalls his three denials of Christ. When predicting these denials our Lord had hinted at Peter's death; cf. *13*, 37 f. Now he foretells this death as martyrdom. The primacy of Peter is abundantly confirmed in Acts. **15.** The meaning is: lovest thou me more than these do? In the three interrogations, the terms for *lovest thou* and for *feed my* *lambs* vary both in Latin and in Greek. Probably we should take them as synonyms, without substantial difference in meaning, or at most with a difference in degree of love. *Son of John* is the "Bar—Jona" of Matt. *16,* 17. The special questions put to Peter indicate that what is said applies to him alone, i.e., the particular charge of being shepherd to Christ's flock. The image of the flock as representing God's people is familiar from the Old Testament. Cf. *10.*

18-23: The martyrdom of Peter, and the fate of John. Recall in this connection the image of the good shepherd in *10:* he must be willing to lay down his life for the sheep.   **18.** *Thou didst gird thyself:* an image of freedom of action. *Stretch forth thy hands:* generally understood of extending the hands for crucifixion. *Another will gird thee:* reflecting the natural resistance to death.   **19.** John, writing long after this utterance, can affirm that it described Peter's crucifixion. *Follow me:* v.20 implies that our Lord had risen and moved away, thus calling Peter to follow Him in a physical sense.   **20 f.** Again we see the intimate association of Peter and John. The latter also had risen and was following. *And what of this man?* Peter had evidently understood the reference to his own martyrdom, and his interest in John led to this question about the latter's destiny. *Until I come,* as in *14,* 3, can mean Christ's coming for the individual soul in death. *What is it to thee:* the only thing that should concern Peter is his correspondence to the summons, *Follow me.*   **23.** There was a two-fold misunderstanding of our Lord's words. (a) They thought He had said, "John will remain . . ." (b) They took the words "till I come" to refer to the Second Coming of Christ. The false rumor was based on these mistakes. John denies it with some emphasis, and insists that Christ had affirmed nothing concerning his death.

21, 24-25: Second Epilogue. In *20,* 30-31, there was some pertinence in the Evangelist's advertence to his thesis. Having now added a supplementary event and a prediction about Peter and himself, he can bring his Gospel to a close. Again the epilogue grows out of what has just been said. Having placed himself in relation to Christ and to Peter, he identifies *the disciple whom Jesus loved* as the author of the Gospel. There is neither any conflict between the two epilogues nor any reason for questioning the authenticity of the second.   **24.** The author of this testimony to Christ's divinity is the disciple just referred to, whom we otherwise know to be John the Apostle. *And we know:* the plural is variously explained. John joins with himself the church at Ephesus; or these two verses were added by a group, probably the elders at Ephesus, who thus sanction the veracity of John's testimony. But there are other possibilities. *We* is not editorial. The plural might refer, as in *1,* 14 ff, to the others who were witnesses of Christ's life. Or, it can include with the author the divine source of his inspiration (cf. *he knows,* in *19,* 35).   **25.** *Many other things:* as in the first epilogue. The hyperbolic form of the statement is evident; it means that the author could continue to adduce other evidence for his thesis if necessary.

WILLIAM L. NEWTON

ACTS OF THE APOSTLES

Introduction

Nature and Purpose. This book is an account of the earliest days of the Church, covering about thirty-five years from the Ascension to the close of St. Paul's first Roman imprisonment in 63 A.D. Without aiming at a complete record of the events, St. Luke follows the plan outlined in *1*, 8: *you shall be witnesses for me in Jerusalem and in all Judea and Samaria and even to the very ends of the earth*. His chief interest is in beginnings, in the spread of the gospel into new territory rather than in the fortunes of churches once they have been established. After a description of the foundation of churches in various parts of Palestine and at Antioch in Syria he traces the development throughout the Roman Empire almost exclusively in the labors of St. Paul. For this reason his book is properly called, as in the Greek text, "Acts of Apostles," not "The Acts of the Apostles," as though it tried to cover at least the principal features in the careers of all the Apostles. Perhaps at first the title was simply "Acts," and the distinguishing term "of Apostles" was added later when apocryphal "Acts of Peter" and the like were being circulated. The prominent part played by the Holy Spirit in these narratives has led to the description of the book as the Gospel of the Holy Spirit.

Since St. Luke gives a good deal of attention to St. Paul's previous relations with the Roman authorities in the provinces, he may have had as a secondary purpose the vindication of the Apostle at the time of his trial in Rome by showing that he had always been judged innocent whenever the Jews brought him to trial before Roman officials.

Date and Place of Composition. The Acts was written about 63 A.D. and probably at Rome. The last verses say that St. Paul preached in Rome for two years while still under arrest. St. Luke had accompanied him on his eventful voyage from Palestine to Rome, since in chapters 27 and 28 he uses the pronoun "we," and he was still with him when the Apostle was looking forward confidently to being soon acquitted by the Roman court (Philem. 22-24). Since the history is carried no further, it is only reasonable to conclude that the book was written at the close of the two years of imprisonment, but before St. Paul's release. According to the decision of the Biblical Commission we hold that St. Luke finished the Acts at the end of St. Paul's first imprisonment at Rome.

The manner in which St. Luke speaks of St. Paul's contacts with the Roman authorities not only in the provinces but also at Rome shows that he still expects official Rome to take a tolerant attitude toward the Christian religion. If he had been writing after the outbreak of the persecution of the Church, started by Nero in A.D. 64, he might not have had this hopeful outlook.

Canonicity. The following testimonies prove that the Acts forms a part of the inspired word of God.

Eusebius, writing between the years 311 and 325, places the Acts among the books received by all as inspired. Since as historian he had carefully investigated the teachings of his predecessors, his testimony shows that the tradition in the Church regarding the inspiration of the Acts flows untroubled from apostolic times. Tertullian, who died about 240-250, blames the heretic Marcion for rejecting the Acts; hence the book must have been received in the Church at that time. Epiphanius says that the Ebionites, after rejecting the Acts, edited another book of Acts, and that Theodotus, the disciple of Valentinus, used the Acts to support his heresy. These heretics were of the second century. The apocryphal "Testimony of Benjamin," written in the second century, asserts that the words and actions of Paul are contained in the sacred books.

Allusions to the Acts or quotations from it are found in Clement of Rome (*1 Cor. 2;* cf. Acts *20,* 35); Ignatius Martyr (*Smyr.* 3 and *Magn.* *5,* 1); Polycarp (*Phil. 1,* 2 and 2); Justin (*Tryph. 36*); and the heretics of the second century. The Letter of the Churches of Lyons and Vienne, written about the year 177 to the churches of Asia and Phrygia on the martyrs of Gaul, says, "They prayed for those who tortured them, saying like Stephen, that perfect witness, 'Lord, do not lay this sin against them.' " (Eusebius, *H.E.* V, 2: cf. Acts *7,* 60).

Genuineness. The Acts was written by St. Luke, the author of the Third Gospel. Cf. the decision of the Biblical Commission, June 12, 1913, Denz. 2166 ff.

External historical evidence. Irenaeus (c. 140-202) proves from the Acts that Luke was the companion of St. Paul. The Muratorian Fragment (c. 170) says, "The Acts of all the Apostles were written in one book; Luke included for the excellent Theophilus all the events which had happened in his presence." Tertullian (150/160-240/250) defends the Acts against heretics and often cites it as Luke's. Clement of Alexandria (c. 150-215) writes, "Luke in the Acts of the Apostles relates that Paul said, 'Men of Athens, I see that in every respect you are extremely religious.' " (*Strom* V, 12; cf. Acts *17,* 22). Origen (185-244) says, "Luke who wrote the Gospel and the Acts."

Internal evidence. Compare Luke *1,* 1-4 with Acts *1,* 1.2. The opening sentence of the Acts refers to the Third Gospel as "the former book" and describes it as an account *of all that Jesus did and taught from the beginning until the day of His ascension.* The address is to Theophilus for whom the Gospel also was written. The style and diction in the Third Gospel and the Acts are the same. The Greek is less colored by Aramaic than in the other Synoptics; there is the same easy use of official, military, and medical terms, and the same careful consideration for the Gentiles.

The "we-sections," *(16,* 10-17; *20,* 5 — *21,* 18; *27,* 1 — *28,* 16) show that the writer was with St. Paul on some of his journeys. Of the Apostle's various companions only St. Luke could have been with him on all the occasions referred to. St. Mark is excluded because he was not on the second journey *(15,* 36-41; *16,* 10). Timothy was at Troas when the author was with St. Paul at Philippi *(20,* 4-6). Neither Titus nor Silas was with St. Paul toward the close of his first Roman imprisonment (Col. *4,* 10-14; Philem. 23. 24; Acts *28,* 16).

The Acts as History. St. Luke is our best source of information for the first days of the Church. Scattered notices in the Epistles tell us much about the different Christian communities, but the only orderly presentation of the early history of the Church must be sought in the Acts. Inspiration guarantees the absolute reliability of the narrative; independently of inspiration, the high historical character of the Acts is established on the same standards of judgment that are applied to other books claiming to be historical.

The author's serious purpose to write history, not fiction, is clear from the whole tenor of the Acts, and in the opening verses of his Gospel St. Luke expressly declares that his purpose is to prove the verity of the instruction received by Theophilus by a narrative of things which had taken place. He could not build such a demonstration on fiction. This evidently holds good for the Acts also.

His sources of information were abundant and trustworthy. He could draw on his own personal experience for many of the events of St. Paul's apostolate; for the other facts he could consult St. Paul, with whom he was closely associated for many years, and other Apostles and disciples whom he met at intervals.

His narrative touches secular history at many points: Claudius, the Roman Emperor, Herod Agrippa, the First and Second in Judea, Sergius Paulus, proconsul of Cyprus, Gallio, proconsul of Achaia, Felix and Festus, procurators of Judea, Drusilla, Bernice, the numerous departments of the Roman Empire with their various laws and customs. All these play their part in his record, and in every case they are presented with perfect fidelity to what is known of them from profane sources. Though it seems he did not have at hand copies of St. Paul's Epistles, he never contradicts them, but often sheds new light upon them. Avoiding a dry chronicle, St. Luke groups his material without obscuring for a careful reader the chronological sequence of events, while throughout he maintains a high standard of dramatic power and literary excellence.

It is not to be supposed that, in recording the speeches made by the actors in his history, St. Luke hands down the exact words spoken on the occasion. He had to translate some of them from Aramaic, and for most of them his purpose was merely to give a summary of the main ideas expressed. This in no way detracts from the historical accuracy of the speeches, but it explains how the record of them is at times colored by St. Luke's own style.

COMMENTARY
Prelude *1,* 1-26

1, **1-14: The Ascension.** **1-3.** Like the Third Gospel the Acts is dedicated to Theophilus, probably a noble Roman converted by St. Luke. The scope of the previous book is recalled; it embraced the deeds and teaching of Jesus from the beginning down to the Ascension. Our Lord taught chiefly by example during the many years of His Hidden Life and then for about three years by both word and example in His Public Life.

Connecting the two books, St. Luke gives a summary of the events after the Resurrection. Jesus communicated with the Apostles for forty days, not staying with them continuously, but appearing among them from time to time according to His own wise counsels, first in Jerusalem, then in Galilee, and finally in Jerusalem again. These frequent appearances are the *proofs* of v.3. Some of them are recorded in the Gospels.

In these interviews He explained the Scriptures to the Apostles, making their hearts burn at the glorious meanings hidden in the ancient prophecies. In a special way the Apostles were instructed about the Kingdom; though it is to be consummated only in heaven, it begins on earth in the Church to which Christ was to leave the carrying on of the work which He had begun. His part, as direct visible work, was to end with the instruction, strengthening, and sending forth of His chosen disciples; the application of the fruits of the redemption was to be their task, though He would be with them always. Since they still had only hazy notions on many points concerning the organization and administration of the Church, it was to the unfolding of these matters, we may conjecture, our Lord devoted Himself during the forty days, *speaking of the kingdom of God.* Convinced beyond all possibility of doubt, they were now to be His witnesses, fully qualified to testify to this crowning miracle of His Resurrection and to use it to prove the divinity of their Master, the conqueror of death, the author of life, the God-Man.

4. Our Lord's final interview with His disciples. They were then back in Jerusalem; as formerly, Jesus joined them and ate with them. It was His will that they should promulgate His law in the capital city of the old theocracy, the site of the temple of God, the religious center of the Chosen People; the prophets had foretold that from it would issue the glad tidings of the new dispensation. Our Lord intended, as we learn from the sequel, to manifest His power there in an extraordinary way at the very beginning of His universal dominion over the hearts of men. This intention He partly disclosed in the command He gave them; they were to remain in the city to receive the promise of the Father. They had frequently heard this promise from His own lips, for before His Passion He had told them of the Holy Spirit who was to be sent by the Father to help them establish the Kingdom. "And I will ask the Father and he will give you another Advocate to dwell with you forever, the Spirit of truth." "But the Advocate, the Holy Spirit, whom the Father will send in my name, he will teach you all things and bring to your mind whatever I have said to you" (John *14*, 16.26; cf. John *15*, 26; *16*, 7-15). Cf. also Luke *24*, 49.

5. Then Jesus linked this concluding scene of His earthly career with the prediction made by John the Baptist at the beginning of the Public Life, "I indeed baptize you with water . . . But he who is coming after me . . . will baptize you with the Holy Spirit and with fire" (Matt. *3*, 11). After a short time they would receive the copious outpouring of the Holy Spirit, their souls being imbued with His divine light and strength as the body in Baptism is immersed in water.

6. The second part of the scene is enacted at the place of the Ascension. Jesus had led them out of the city toward Bethany (Luke *24*, 50). Not having yet received the full enlightenment of the Holy Spirit, the disciples still clung to some of the popular errors about the Messias. The whole nation was longing for a military hero who should break the chains that held them in subjection to Rome. There were no limits to the disciples' confidence in our Lord's power; they knew Him as the supreme Lord of heaven and earth, and they took it for granted that His purpose included the establishment of the Kingdom in glory. The only question was about the precise time; they wished to know whether it was to come soon, whether perhaps it might not be included in the promise of the Father which, as He had just told them, was to be fulfilled in the near future.

7. Jesus answered with patient kindness. God had fixed the opportune time for manifesting His power; it need not concern them. 8. Still indirectly He supplied the answer to their

question. Yes, it was with the coming of the Holy Spirit that He would begin the establishment of His Kingdom, but it was to be a spiritual kingdom to lead men along the road of truth and virtue to eternal life. From the Holy Spirit the Apostles were to receive strength to preach the gospel, not to the Jews only, but to all nations throughout the world. It was only because of His divine power and foreknowledge that Our Lord could declare that they were to be His witnesses everywhere, for to the human eye they were poorly equipped for such a worldwide mission. But the spiritual forces of heaven were behind the chosen band, preparing them in a way the world could not understand, and with the passing of the centuries even the most skeptical would be forced to admit that in truth the Church had changed the face of the earth. It is the beginning of this gigantic work that St. Luke is recording in the Acts, and everywhere in the narrative the prominent feature is the Holy Spirit, the Spirit of Christ, instructing, guiding, strengthening, and consoling.

9-11. Having given His last command to the Apostles, our Lord was raised up and carried from their sight to take His place at the right hand of the Father in heaven. Lost in wonder they could only stand there, gazing longingly after Him till two angels in human form appeared and recalled them to themselves with the promise of Christ's second coming at the end of the world.

12-14. Then they went back to the city to await the descent of the Holy Spirit. The upper room where they stayed is referred to as a special and known place by the definite article; it was probably the one where the Last Supper had been eaten, and perhaps it was in the house of Mary, the mother of St. Mark (cf. *12,* 12). There the Apostles, the Brethren of the Lord, the pious women who had followed our Lord during His ministry, and the Blessed Virgin, assembled regularly for prayer.

1, **15-26: Matthias Chosen.** At the proposal of St. Peter they filled the place in the twelve left vacant by the defection of Judas. From the beginning St. Peter is shown exercising the primacy conferred on him by Christ. He indicated the course of action, the assembly selected the candidates, but the choice was left to God. Since the Apostles were to lead men to Christ, the first requisite was that the candidate should have shared in their experiences from the early days of the Public Life, and the first duty of an Apostle was to testify to the great historical fact of the Resurrection. The Resurrection is the central proof used by the Apostles to show the mission and divinity of our Lord; cf. *3, 15; 1* Cor. *15,* 12-22.

20. Both Psalms quoted by St. Peter are Messianic directly or indirectly. The Holy Spirit is said to have spoken by the mouth of David because God is the principal author of the Scriptures.

I: THE CHURCH IN PALESTINE AND SYRIA 2, 1–12, 25

1. Growth of the Church in Jerusalem 2, 1–8, 3

2, 1-13: Descent of the Holy Spirit. 1. *Pentecost* celebrated the conclusion of the harvest and commemorated the giving of the Old Law. It occurred fifty days after Passover, hence the expression, *when the days . . . were drawing to a close.* This becomes the birthday of the universal Church. *They were all together:* most likely the one hundred and twenty persons spoken of in *1,* 15, were all present. **2.** The sound came suddenly so as to attract more attention; though not caused by the wind, it was like the roaring of a mighty gale; it was so loud that it was heard throughout the city, or at least by those in the vicinity who then spread the news. This marvel recalls the thunder and lightning on Mt. Sinai at the promulgation of the Old Law, and God's breathing upon the first creation. **3.** Besides, over the heads of the disciples were seen tongues which had the appearance of fire; they all spread out from one central flame; they were distributed so that one flame in the shape of a tongue was above the head of each. The tongues were symbols of speech, of preaching, of the oral testimony which was to be the medium for announcing the Messianic message (Matt. *28,* 20; Rom. *10,* 14-16). They were like fire because the Messias was to baptize in the Spirit and fire (Matt. *3,* 11); He came to cast fire upon the earth (Luke *12,* 49); all things are to be purified by fire *(1* Pet. *1,* 7); and God is a consuming fire (Deut. *4,* 24).

4. There was a corresponding interior marvel, for all were filled with the Holy Spirit. Their souls, as appears later, were endowed with supernatural light, courage, and zeal. Only one effect, the gift of tongues, is here noted because of its striking nature. The Greek term used for *to speak* means utterance of a solemn and wondrous kind; the subjects they spoke of are hinted at in v.11, *the wonderful works of God;* they did not preach, but proclaimed God's praises. The miracle was in their use of languages unknown to them and not in the hearing of their audience (cf. *1*, Cor. *14*, 21; etc.).

5 ff. The crowd was composed of men from all parts of the civilized world. They were all *Jews* by birth or conversion and had come either to take part in the great feast or to dwell near the temple and be buried in the land of their fathers. The disciples went out, it seems, and mingled with the crowd and so each of the spectators could hear one or more of the disciples speaking in that person's own native language. Naturally there was general astonishment at this strange occurrence and no explanation could be found for it. But some of the crowd, not even taking the trouble to look for an explanation, tried to ridicule the whole affair by saying that the disciples were intoxicated.

2, 14-36: Peter's Discourse. As in the first chapter, St. Peter acted as the primate and leader of the Apostles. He began his address with confidence, prudence, and solemnity, and showed a mastery of the Scriptures; this revealed the effects of Christ's closing instructions and of the grace of the Holy Spirit. He may have spoken in Greek, since this language was known to very many in the crowd; but more likely he spoke in Aramaic, commonly used by the Jews in Palestine and by many of the Dispersion.

After refuting the calumny of drunkenness he explained from the prophet Joel the true cause of the marvels they had beheld; this was the outpouring of the promised Spirit. Then he developed his proposition: this divine Spirit, who according to Joel was to be given in the Messianic times, had been sent by Jesus whom they had crucified, but who had been raised from the dead in fulfillment of the prophecy of David and was then seated at the right hand of God, and who was in this way shown to be the Messias and true Lord.

14-16. St. Peter began kindly without denouncing the malice of the mockers. No one feasted early in the day; good Jews did not eat before the fourth hour (10 a.m.). Prudence demanded that he should say nothing in the beginning about Jesus and the Holy Spirit, so he showed first that the present event had been foretold by the prophet.   **17 f.** The *last days* are the times of the Messias; *all flesh* means all men; hence the universality of the gospel is announced on the very first day of the Church's existence. The effect of this outpouring of the Spirit was to be the speaking of salutary things, things hidden and future.   **19 f.** In the usual prophetic manner Joel saw the whole Messianic period as a unit and linked the beginning with the end and added what will happen at the end of the world. **21.** With this divine manifestation was united the promise of salvation.

22. The fulfillment of this prophecy offered the true explanation for the present marvels, for the Messias had already come and was no other than Jesus of Nazareth. St. Peter took up first the well-known miracles of Jesus; as *miracles,* they showed God's power; as *wonders,* they excited the astonishment of the beholders and so secured their attention; and as *signs,* they pointed out the worker as one approved by God.   **23.** But Jesus had been put to death; here was the great scandal. This death however, far from having been forced upon Him, was part of the plan of God. Foreseeing the malice of the Jews, God decreed to permit it to run its course.   **24.** But then came the miracle of the Resurrection. *Hell:* i.e., "sheol," the place where the just were detained after death; it was the unseen world.   **25-28.** Since the Resurrection had not been strikingly presented to the public, St. Peter first showed that it had been foretold and then proved its reality by the testimonies of the witnesses. David was not speaking of himself, for like other men he died and was buried and his tomb was before their eyes in Jerusalem.   **29-32.** But David was speaking of Christ who fulfilled the prophecy. For God raised Jesus from the dead, and the witnesses were St. Peter and the other Apostles.

33. With the fact of the Resurrection admitted, it was easy to believe in Christ's exaltation. What they had just beheld, this outpouring of the Holy Spirit, was simply one of the effects of this exaltation of Jesus. The people had heard the various languages spoken by the disciples, had seen their enthusiasm, and probably also the tongues of fire; they had at least some vague idea of the Holy Spirit from Joel (cf. Ezech. *11,* 19; *36,* 26; Isa. *44,* 3; Matt. *3,* 11). *Exalted* may have special reference to the Ascension.   **34 f.** The exaltation is further described in the words of David; the Jews themselves had acknowledged that David spoke of Christ, for Jesus had used this prophecy to prove the divinity of the Messias (Matt. *22,* 44 f.) Seated at the right hand of God, Christ is in complete triumph. As a sign of victory oriental conquerors at times placed their feet on the necks of the vanquished.

36. The address is brought to a strong and compelling conclusion. Jesus, whom they had crucified, had risen from the dead, was exalted to heaven, and was then sitting as Lord and Christ (Messias) at the right hand of God. The whole nation should realize this, for they were the chosen people of God to whom had been made the promises which were now fulfilled.

2, 37-41: The Result. **37.** The Holy Spirit added persuasive force to the words of the Apostle. Feeling the sharp sting of contrition, many asked for further directions. The entire nation had sinned at the crucifixion; perhaps some of those present had added their voices to the shouts demanding Barabbas and condemning Jesus; but now the prayer "Father, forgive them" is answered. **38.** Like Jesus and the Baptist, St. Peter called for repentance (cf. Luke *24,* 47), since their words showed that they believed. They must be baptized in the name of Jesus, i.e., as a profession of faith in Him; the formula to be used in Baptism is not indicated here, but in the last command of our Lord (Matt. *28,* 19). Then he revealed the purpose of Baptism; it was for the remission of sins and the receiving of the Holy Spirit. Justification is not something merely negative, the removal of that which makes us enemies of God; it is also positive, the infusing of sanctifying grace and all the wonderful gifts of the Holy Spirit coming and taking up His abode in the sanctified soul which then becomes the temple of God.

39. After this exhortation and promise St. Peter gave them a solid reason for hope and confidence: the promise made by Joel was to them and to their children. Still it was also universal. *To all who are far off:* these were not the Jews scattered in Gentile lands since they had already been called in God's special choice of the nation (Matt. *4,* 14; *22,* 3). St. Peter was well aware of the future calling of the Gentiles from our Lord's words (John *10,* 16; Matt. *28,* 19; Mark *16,* 15; Luke *24,* 47; Acts *1,* 8; Eph. *2,* 11-13.17), and Jesus had explained to the Apostles the meaning of the Messianic prophecies. **40.** Burning with zeal, he urged them to save themselves from the punishment in store for their sinful fellow citizens. It was an evil and adulterous generation (Matt. *12,* 39), obstinately holding to the wicked ways on which it had entered, just as had been often foretold (cf. Matt. *8,* 12; *21,* 41; Rom. *10,* 21; Isa. *65,* 2). St. Peter knew that a large part of the Jews would remain obdurate. Only the substance of his discourse is given by St. Luke, and so on this and similar occasions there need be no surprise at great results from a few words.   41. Three thousand were then added to the original group.

2, 42-47: Fervor of the Early Church. Brief sketches of the manner of life among the early Christians feature this part of the Acts (cf. *4,* 32-35; *5,* 12-16). From them and from other scattered notices it is clear that from the beginning the Church was a well defined and perfect society, independent of the old organization of Judaism. Externally they were a closely knit body to which unbelievers were not admitted; converts were added to this nucleus *(1,* 15; *2,* 41-47; *5,* 13). As an organized society they are called "the Church" *(5,* 11). They took care of their own poor *(2,* 45; *6,* 1). This moral unity is also expressed by the names used for the faithful; they are "brethren" (like the members of a family), "disciples" (like a school of thought), "saints" (like a society or group consecrated to God). Authority was vested in the Apostles under the leadership of St. Peter; discipline was enforced *(5,* 1 Ananias) and commands issued *(6,* 3). Their beliefs were those taught by the Apostles *(2,* 42) and they had distinct religious rites: Baptism *(2,* 38), the Holy Eucharist *(2,* 42), the common prayers *(2,* 42 where the definite article is used in Greek), Confirmation (perhaps *8,* 17; *2,* 38), the imposition of hands and Orders *(6,* 6).

Because of the miracles, which were blessings for the whole neighborhood, and because of the virtuous lives of the disciples, the people honored them and regarded them with reverential fear; in this way Providence gave the new seed time to take root *(2,* 43.47; *4,* 21; *5,* 13). The Jewish authorities were disturbed, but at a loss how to proceed against the new teaching. After the cure of the lame man *(3,* 1) they arrested St. Peter and St. John, but fear of the people made them cautious and they had to be content with vain efforts to intimidate the Apostles. When fresh miracles increased the popularity of St. Peter, they again threw the Apostles into prison; exasperated by their constancy, they even thought of putting them to death, but Gamaliel intervened, warning them against extreme measures, and so they had them scourged and then dismissed them with threats *(5,* 40).

The disciples frequented the temple, for there was to be no sudden break with the past, but they had their own regular assemblies for prayer, instruction, and the Holy Eucharist. Their life was marked by simplicity, zeal, fortitude, and thanksgiving. So unstinted were they in manifesting their brotherly love through works of charity that they could be said to have but one heart and one soul (*2,* 42 ff; *4,* 24-37). Many even sold their possessions for the common good, and all were ready to help the needy. That this is the true meaning of the passages referring to the selling of property comes out clearly in the case of Ananias (*5,* 4), whose punishment for lying preserved discipline and prevented disorders. There was no "communism," but the more fervent voluntarily disposed of their goods. When property was sold, the proceeds were entrusted to the Apostles to be used as a common fund; the later destitution of the community was not due to this practice, but to persecutions and a general famine (Acts *11,* 29; *1* Cor. *16, 3*).

3, **1-11: A Lame Beggar.** Having pictured the peaceful life of the first Christians, St. Luke passes on to report a miracle that led to persecution. The event is not dated; there may have been a long interval between the second and third chapters.

The scene is in the temple at the time of the evening sacrifice (about 3 p.m.). The *gate called Beautiful* was probably the one leading from the Court of the Gentiles to the Court of the Women, and *the porch called Solomon's* was the colonnade running along the eastern side of the Court of the Gentiles.

3, **12-26: Peter's Discourse.** St. Peter follows the lines of his first discourse, calling for repentance and faith in Jesus, but he dwells more on their sin in crucifying Jesus (13-17), on the prophecies about the sufferings of the Messias (18), and His prophetic office (22-24), and on His Second Coming at the end of the world (20-21). He indicates the divinity of Jesus by calling Him *the Holy and Just One* and *the author of life* (15). He again stresses the special blessings of the Jews (25), but by the words *to you first* (26) he shows that he realized the Gentiles would be converted later.

4, **1-22: Arrest and Release of Peter and John.** *The priests* were particularly concerned because the teaching was being given in the temple precincts by unauthorized persons, and the

Sadducees because it contradicted their denial of the resurrection of the dead (cf. Acts *23*, 6-9).

8-12. Answering the assembled Sanhedrin, St. Peter again touches on the chief points in his previous discourses, insisting on Christ's crucifixion and resurrection and His present power and glory; salvation is to be found only in Him, not in the Jewish Law.

4, 23-31: **Thanksgiving.** With the threats ringing in their ears, they had special need of courage to continue their preaching; miracles would lend weight to their words.

4, 32-37: **Manner of Life of Christians.** Cf. *2*, 42-47.

5, 1-11: **Ananias and Sapphira.** Ananias need not have sold the land nor given the whole price to the common fund; he sinned by lying about the price and pretending to hand over all he had received. **3.** *Tempted:* i.e., successfully; why did you yield to temptation? Cf. v.4. The Holy Spirit is clearly spoken of as a divine Person, guiding and protecting the Church; cf. *11*, 12; *13*, 2; *16*, 6 f. **11.** God's purpose in punishing this sin so directly is partly revealed; fear served to preserve discipline and respect for the authority of the Apostles. It is not necessary to conclude that Ananias and his wife were eternally lost; they may have received the grace of repentance. (Cf. Commentary on *1* Cor. *5*, 5.)

5, 17-42: **Arrest of the Apostles.** A full assembly of the Sanhedrin is called for the trial of the Apostles. After St. Peter's spirited defense, the council is ready to put them to death, but on the advice of Gamaliel they decide to let them off with a scourging and another warning. **28.** *Bring this man's blood:* to accuse us of killing Him and to stir up the people to seek our lives in revenge. **29.** St. Peter repeats the facts of our Lord's Passion and Glory and adds the testimony of the Holy Spirit as manifested on Pentecost and on other occasions. **34.** *Gamaliel:* grandson of Hillel and the leader of the more liberal party among the Pharisees. He was the most prominent rabbi of that time and the teacher of St. Paul (Acts *22*, 3). **36.** Rationalists think they find a contradiction here between St. Luke and Josephus in the dating of Theodas (Greek, *Theudas*), but without reason. Revolts were frequent at that period, and

Theudas was a common Jewish name. Josephus speaks of a man of this name who led a revolt in A.D. 45, while Gamaliel refers to the leader of an earlier revolt.   **37.** Judas rebelled against the Roman census of 6 A.D., and his party lived on in the "Zealots" and "Assassins" till the destruction of the nation in 70 A.D. Josephus speaks of his exploits.   **42.** *As the Christ:* i.e., the Messias. Acceptance of our Lord by faith involved acceptance of all His teachings which, of course, included His divinity.

6, 1-7: The Deacons. With the growth of the Church the number of widows needing help increased to such an extent that the original simple method of distributing alms to them seemed inadequate; at least the foreign-born Jews, called *Hellenists,* thought that their widows were being neglected. All those so far converted were Jews; but while some had been born in Palestine and used Aramaic in their synagogues, there were others who had been born in foreign lands (the Diaspora or Dispersion), or of parents born abroad and who spoke Greek in their synagogues. They were, of course, real Jews and so entirely distinct from the pagans; this distinction is clearly marked in Greek, the foreign-born Jews being called Hellenists *(6, 1),* while the pagans are called Hellenes *(11, 20; 14, 1).* The Vulgate does not bring out the distinction, calling both classes Greeks. To meet the new situation the Apostles ordained seven assistants; their primary task was the distribution of food (or of money for food) to those in want, but they also administered Baptism and gave catechetical instruction. Their usual name "deacons" *(1 Tim. 3, 8)* is derived from the Greek verb used here for *serve (6, 2).* There is nothing in the text obliging us to suppose that these seven were all Hellenists, i.e., Jews from the Diaspora.

6, 8 — 7, 1: Stephen's Arrest. Stephen, one of these seven deacons, was remarkable for his lively faith and unflinching courage. He preached with zeal and effect, confirming his words with miracles. His eloquence aroused the anger of the Jews and especially the active opposition of one (or perhaps several) of the Hellenistic synagogues in Jerusalem. His opponents were from various parts of the world, from Lybia, Egypt, Cilicia, and the Roman Province of Asia, together with the Libertines or freedmen, the descendants of the Jews whom Pompey had transported as slaves to Italy in 60 B.C. Unable to meet Stephen in argument they secured false witnesses to testify that he had blasphemed by speaking against Moses and the temple. Then they excited the crowd and some of the rulers, and with the help of this mob they dragged him off to be tried before the Sanhedrin.

7, 2-60: Stephen's Discourse and Martyrdom. Stephen easily refuted the direct charges; he praised Moses as the one who had given them the words of life (30), and he appealed to the prophets who had asserted that no temple could be an adequate dwelling for God (49). But the main line of his defense was an attack on their stubborn lack of faith and obedience to God; tracing their history from the early days, he showed that they had always resisted God and that they had lately displayed the same evil disposition in betraying and murdering the Just One, Jesus. In blind rage his accusers clamored for his condemnation, rushed upon him, dragged him outside the city, and stoned him to death.

In this trial the legal forms seem to have been carried out essentially, though the actual condemnation and execution were hurried through under pressure of the mob. The Sanhedrin had been deprived of the power of inflicting capital punishment, but if the Romans pressed the matter, they could have answered that the execution was really the work of the crowd which had become unruly. It is quite possible that at this time there was a change of Roman procurators, and so the administration of affairs was confused for the moment.

Some of Stephen's statements differ from the accounts in the Old Testament. Abraham's call (2) came in Haran according to Gen. *12,* 1; Joseph's kindred (14) numbered only seventy according to Gen. *46,* 27; the tomb bought by Abraham (16) was at Mambre (Hebron) according to Gen. *50,* 13. In these and similar instances Stephen was following the readings in the Septuagint, the Greek translation of the Hebrew Old Testament, or current traditions, or he was summarizing from memory. As an inspired writer St. Luke gives a faithful record of what was said without guaranteeing the accuracy of the statements themselves. Doctrinally Stephen follows closely in the footsteps of St. Peter who from the beginning, by proclaiming Jesus as the sole source of salvation for all men, had announced the inefficacy of the Old Law and the universality of the New. In his trial before the Sanhedrin the conditions were quite different from those at the trials of the Apostles. Fear of the people's resentment then prevented the rulers from resorting to extreme measures, but at the trial of Stephen the court was surrounded by a howling mob demanding his condemnation. Whatever they may have thought of the legal charge lodged against him, the real cause of their rage and of their stoning him was his proclaiming Jesus as the Just One, seated at the right hand of God (7, 52.55).

8, 1-3: Persecution. With the death of Stephen there arose the first general persecution of the Church. So far only the Apostles had been molested, but now the faithful and especially the deacons were involved. It was a general and devastating attack and *Saul* stood out conspicuously among the persecutors, going from house to house in search of victims and dragging them off to prison. In consequence large numbers of the faithful fled from Jerusalem; the Apostles however remained. Though probably not entirely exempt, they were treated with less rigor on account of their reputation and miracles; in persecuting them later *(12)* even King Herod moved cautiously to see how the people would react.

2. THE CHURCH IN JUDEA AND SAMARIA 8, 4 — 9, 43

The persecution furnished a great occasion for the promulgation of the gospel through the rest of the country. In His wisdom God had provided the necessary peace at the beginning till the Church was established and organized in Jerusalem; then out of the evil of persecution He drew the blessing of the spread of the glad tidings. The divine plan for the conversion of the Gentiles was well known to the Apostles, but the Jews were to be given the first chance and the work was to be confined to them until God revealed how and when He wished the Gentiles to be admitted. Throughout this section there is question of Jews only, with the doubtful exception of the eunuch in *8,* 26 ff. In the eighth chapter the chief regions of Palestine are expressly mentioned: Judea and Samaria (*8,* 1.4) and the cities along the coast (Gaza, *8,* 26; from Azotus to Cæsarea, *8,* 40). In the narrative of St. Paul's conversion there are disciples in distant Damascus (*9,* 2.10.19), and after the persecution has died down, St. Peter visits the churches and makes new converts in Lydda, Sharon, and Joppa (*9,* 31 ff), while Galilee is incidentally named among the regions where the Church is enjoying peace (*9,* 31).

8, 4-25: Samaria. It is quite in the style of St. Luke to select one or two events for detailed treatment and to omit or mention briefly the rest. The chief evangelist in Samaria was Philip, one of the seven deacons. Both the district and its capital city were called Samaria. He performed many miracles in the city and made numerous converts including the sorcerer Simon who had previously won great popularity there by his magical practices. When the news of his success reached Jerusalem, the Apostles sent St. Peter and St. John to consolidate the work; nothing is implied against the primacy of St. Peter since he was one of those who made the decision, and surely a superior can even yield to the wishes of his community to undertake some special work for the common good. The deacon Philip could baptize, but only the Apostles could confer the sacrament of Confirmation. In Baptism the Holy Spirit is received in the remission of sins and the infusion of sanctifying grace, but Confirmation brings a fuller infusion of the Holy Spirit with strength to profess the faith courageously.

Simon the magician, either never sincerely converted or lapsing quickly from his first fervor, had an eye chiefly for the miraculous powers granted to the faithful. He had kept close to Philip in constant wonder at his miracles, but when he saw the Apostles conferring the Holy Spirit, he could no longer conceal his worldly attitude and offered them money for a share in their power. His only reward was a scathing rebuke from St. Peter who exhorted him to pray for forgiveness. From this Simon is derived the word "simony" because he attempted to buy a power of a spiritual nature.

On their way back to Jerusalem the Apostles preached the gospel in the country of Samaria. This work marks a step towards the universal preaching of the gospel, since the Jews held the Samaritans in abhorrence and excluded them from their religion. But it did not open the door for the Gentiles because the Samaritans were the descendants of the Israelites who had been left in the country when the bulk of the population had been carried into exile. These had intermarried with the pagan colonists, but their religion, however contaminated, was founded on the books of Moses, which they cherished, and prescribed circumcision and the observance of the Mosaic Law.

8, 26-40: An Ethiopian. The nationality of this convert is disputed. If he was a Gentile, his conversion was a strictly private affair and did not change the practice of preaching to Jews only. No solid reason has been given to prove that he was not a Jew. He is called an *Ethiopian,* but the Jews in Jerusalem at the first Pentecost are also called by the names of the countries from which they came (Acts *2,* 9-11). He is styled a *eunuch* and such mutilation was forbidden to Jews, but their history shows that they were always lax in observing the Law, and the term "eunuch" was sometimes used simply to denote an office without implying mutilation. His bewilderment in trying to understand the Messianic passage from Isaias shows the need of the guidance of the Church for the interpretation of Scripture.

9, 1-30: The Vision of Saul; His Baptism and Zeal. The Acts contain three accounts of St. Paul's conversion. In *9,* 3-18 it is St. Luke's narrative; in *22,* 6-16 it is St. Paul's summary to the Jews at Jerusalem when they were seeking to kill him; in *26,* 12-18 there is another summary by St. Paul before King Agrippa and Festus, the Roman procurator. Critics have vainly tried to discredit these accounts by claiming that they involve contradictions. (a) It is claimed there is contradiction between *9,* 7 where St. Paul's companions are said to have heard the voice speaking to him, and *22,* 9 where it is said they did not hear it. The explanation is found in the different meanings of the word "hear"; they heard the sound of the voice, but did not understand what was said. (b) In *9,* 7 it is said his companions stood speechless, while in *26,*14 all are said to have fallen to the ground. What is emphasized in *9,*7 is that his companions remained speechless; no attention is given to the effect the light had on them. (c) Another supposed contradiction concerns the source from which St. Paul learned his future vocation; in *9,* 15 f and in *22,* 15 he learns it from Ananias, but in *26,* 16 he learns it directly from Christ. The solution is that in *26,* 16 St. Paul, speaking in summary form, refers to Christ calling him without saying whether He spoke to him directly or through the medium of Ananias, or both.

2. Being Jews, the disciples in Damascus could be denounced by the synagogues there as guilty of blasphemy, just as those in Jerusalem had been. The civil ruler would not interfere, since in localities where they were numerous the Jews usually secured the privilege of holding their own courts for Jewish offenders. The synagogues would either acknowledge the authority of the Sanhedrin or could be expected to yield to its wishes for the

extradition of the converts, especially if these had fled from Jerusalem to escape prosecution.   **3.** The light was brighter than the sun which was then shining in all the splendor of noon (*26,* 13).   **4.** Jesus identifies Himself with His disciples; so close is His union with them that to persecute them is to persecute Christ Himself. St. Paul is given his first glimpse of the doctrine of the Mystical Body which he develops in Eph.; Col.; *1* Cor. *12,* 12 ff.   **5.** *It is hard . . . Lord said to him:* not in the Greek, though the first part is in Acts *22,* 10 and *26,* 14. This proverb is drawn from the image of an ox goaded by its driver which by kicking merely increases the pain of the goad. Like many others St. Paul had previously been urged toward faith in Christ by the preaching of the Apostles and by the example of the disciples, especially in the persecution, and now there was the special extraordinary urging of this vision. The kicking might in itself indicate the pricks of conscience, but up to this time St. Paul had given no sign that his conscience was uneasy; as in the image of the goaded ox, the struggle is against something external. Francis Thompson's "Hound of Heaven" is built on this same idea that it is hard to resist Christ.

12. This vision, granted to St. Paul while our Lord was speaking to Ananias, prepares him for the visit of Ananias and the recovery of his sight.   **16.** The cross must be borne by all the followers of Jesus, but especially by those who distinguish themselves in His service.   **17.** This laying on of hands was not sacramental, but was for the purpose of restoring his sight and preparing him for Baptism.

22. After the preliminary preaching of v. 21, St. Paul retired for a while to Arabia (Gal. *1,* 17), probably to a deserted spot near Damascus, for quiet study and prayer. Returning to the city, he boldly confronted the Jews and demonstrated that Jesus is the Messias by carefully elaborated arguments drawn from the Scriptures and from the facts of His life.   **26-30.** This visit to Jerusalem lasted only fifteen days (Gal. *1,* 18). As in Damascus, he entered into argument with the Hellenistic Jews, but they refused to listen to one whom they regarded as an apostate. From v. 27 it is clear that at the time of his conversion St. Paul had not only heard Jesus speaking to him, but had also seen Him. After returning to his native Tarsus St. Paul occupied himself with preaching to the Jews in the neighboring districts of Syria and Cilicia (Gal. *1,* 21) till Barnabas told him of the great work to be done among the Gentiles at Antioch (Acts *11,* 25).

9, 31-43: Peter Visits the Churches. The violence of the persecution gradually spent itself and the Church throughout Palestine enjoyed a period of peace. Then St. Peter as supreme pastor made a tour of all the churches, administering Confirmation to those who had been baptized by the evangelists and by his preaching and miracles increasing the number of converts.

3. SPREAD OF THE CHURCH TO THE GENTILES 10, 1—11, 30

The importance of the conversion of Cornelius can hardly be exaggerated. Only by understanding that here by special divine intervention the door was opened for the admission of the Gentiles into the Church without circumcision or the Law can we rightly estimate the subsequent development of the Christian community. Beyond all doubt Cornelius and his household were Gentiles and without circumcision, and their conversion marked a radical departure from the previous policy of preaching only to the Jews. St. Peter explicity recalled what an abominable thing it was for a Jew to enter the house of a Gentile and explained his conduct by referring to the revelation given him in his recent vision (*10, 28*). He asserted that from God's favor to Cornelius he had learned that Gentiles who feared God and practised virtue were acceptable to God (*10, 34 f*). His companions were astonished that the grace of the Holy Spirit was poured out upon the Gentiles also (*10, 45*). When St. Peter returned to Jerusalem, he had to meet the criticism of whose who had heard of the baptism of Cornelius. He had to explain how clearly God had manifested His will regarding the Gentiles, and it was only then that they submitted and praised God, acknowledging that the way to eternal life had been thrown open by God Himself to the Gentiles as well as to the Jews (*11, 1-18*).

10, 1-33: Cornelius: The Visions. The Roman procurators for Judea and Samaria had their headquarters at Caesarea. The cohort was probably composed of a thousand volunteer Roman citizens, a hundred of whom, called a "century," were under the command of Cornelius. The cohort may have been stationed at Cæsarea, or Cornelius may have been detached for special work.

10, 34-43: Peter's Discourse. As in his earlier discourses, St. Peter dwells upon Christ's death and resurrection, the testimony of the Apostles, and the fulfillment of the prophecies, but he gives more space here to our Lord's ministry and to the experiences of the Apostles after the Resurrection.

10, 44-48: The Baptism. Like St. Paul (*1 Cor. 1,* 14-17), St. Peter seems to have been accustomed to leaving the actual baptism to his assistants. By staying with them St. Peter gave further practical proof that he considered the old Jewish restrictions as ended.

11, 1-18: Explanation at Jerusalem. The objectors "strain at a gnat and swallow a camel"; the great innovation was the baptism of the uncircumcised. Perhaps they tried to accept that fact as something exceptional or they expected the Gentile converts to be relegated to a position of inferiority like that of the devout pagans who were allowed to attend services in the synagogues. But in 18 they acknowledge that they are members of the Church in the full sense.

11, 19-30: The Converts at Antioch. Having disposed of this first series of events, St. Luke returns to pick up the story of other bands of evangelists who, like Philip, had left Jerusalem shortly after the martyrdom of St. Stephen. They preached throughout Phoenicia along the west coast and in the island of Cyprus, finally coming to Antioch, the capital of Syria. So far their preaching had been addressed to Jews only, but on entering Antioch some of them announced the gospel to the Gentiles also. This remarkable change of method can be explained only by the arrival of the news about St. Peter's action in receiving Cornelius into the Church. This set up a new rule for the preaching of the gospel; the Jews were still the privileged class and so the gospel was to be presented to them first, but the Gentiles were also to be converted. The first group of missionaries to reach Antioch had not heard of the conversion of Cornelius and preached to Jews only. They hesitated to change their method without more definite instructions from Jerusalem. But these later arrivals had heard the news and accepted it as practical guidance.

That the conversions at Antioch came after those at Cæsarea, and that the evangelists there were not acting on their own initiative, is strikingly confirmed by the attitude taken at Jerusalem. St. Peter had been called upon to give his reasons for admitting Gentiles, but when the news from Antioch reached the mother church in Jerusalem, no astonishment was expressed. They had learned God's will regarding the Gentiles, and the normal consequence could only be that the Gentiles would eagerly enter into eternal life through the door now thrown open to them. The next step would be to see that the new community was properly organized, and so the authorities at Jerusalem immediately sent Barnabas to Antioch for that purpose.

The church at Antioch stands out as the first Christian community made up chiefly of Gentiles. The blessing of God was upon the work among the pagans, while here as throughout the Diaspora the converts from Judaism seem to have been few. The community immediately showed its ardent spirit of charity by voting to send help to the churches in Palestine which were threatened by the famine foretold by the prophet Agabus, and it soon became a center for missions to other countries. Yet it was always in close association with Jerusalem, from which came its first evangelists, and Barnabas who gave it its regular organization and further instructions in the faith. As a consequence this city was well prepared to serve both as a starting point and as a support to St. Paul and his companions when they set out on their great missionary journeys. As a final distinction it was here that the faithful were first called Christians; the name was probably bestowed in derision by the city rabble, noted for its wit, but *1* Pet. *4,* 16 shows that it was soon commonly accepted as a title of honor.

CONCLUSION: PERSECUTION OF THE CHURCH BY HEROD AGRIPPA *12,* 1-25

12, 1-17: Peter in Prison: His Deliverance. After St. Peter's return to Jerusalem another persecution broke out under the leadership of King Herod Agrippa I, the grandson of Herod the Great, and the last king of the Jews. His early years were spent in Rome where he enjoyed the favor of Caligula who on his accession in A.D. 37 gave him the tetrarchies of Philip and Lysanias and the title of king. Two years later, when Herod Antipas was banished, he received Galilee and Perea also, and in A.D. 41 Claudius gave him Judea and Samaria, so that he ruled over all the land of his grandfather. His persecution of the Christians was in harmony with his policy of courting favor with the people by a display of zeal for the Law. He reversed the method of the earlier attack by striking now at the leaders.

The vividness of the narrative of St. Peter's delivery from prison shows that it came directly from the actors, probably from St. Mark the Evangelist, to the house of whose mother St. Peter went for a hasty visit. He sent a message to James and the brethren, *and went to another place.* As all Palestine was under the judisdiction of Agrippa, St. Peter must have left that country. There is a good tradition that around A.D. 42 he went to Rome, and the persecution of Agrippa seems to have occurred in 42.

12, 18-25: Herod Punished. Tyre and Sidon were commercial cities on the coast north of Palestine and depended on Galilee for their grain. Josephus gives a similar account of Herod's death and adds some details: the occasion was the celebration of games in honor of the Roman Emperor and Agrippa was wearing a robe of silver that shone brilliantly in the sun; hardly had he accepted the people's salutation implying that he was a god when he saw an owl seated above him and took it as an omen of death; severe pains immediately seized him and continued until his death five days later. The Romans then resumed direct rule over all of Palestine.

II: THE CHURCH IN ASIA MINOR AND EUROPE THE MISSIONARY JOURNEYS OF ST. PAUL 13, 1 – 28, 31

1. First Missionary Journey 13, 1 – 15, 35

13, 1-12: Antioch: Cyprus. St. Luke centers his attention almost exclusively on St. Paul for the rest of the Acts. Like the other prophets and teachers mentioned here, he probably had already been consecrated bishop. On reaching Cyprus he followed the established rule of preaching first to the Jews. The work among the Gentiles is illustrated in St. Luke's usual method by the selection of one significant event: the conversion of the proconsul was undoubtedly followed by that of many other Gentiles.

13, 13-52: Antioch in Pisidia: Paul's Discourse. St. Mark was probably dismayed at the decision to attempt the evangelization of the interior of the mainland. Antioch was about one hundred miles from the coast, and the journey meant stiff and dangerous mountain climbing. It was part of the Roman Province of Galatia, but St. Luke prefers the older local names. **15.** It was customary to invite visiting scribes to address the meeting.

16-41. This is the only example given in the Acts of a discourse by St. Paul in a synagogue, though his usual approach to the Jews was through speaking in their synagogues. He follows the same general lines as St. Peter in the earlier discourses.

16-22. The introduction won the good will of the audience by recalling the special blessings and protection of God in the course of their national history. **20.** Using round numbers, St. Paul allows four hundred years for the sojourn in Egypt, forty for the wandering in the desert, and ten for the conquest of Canaan.

23-25. Coming to more recent manifestations of providence in their behalf, he brought in the name of Jesus, calling Him *Savior,* and connecting Him with the most glorious period of their history by pointing to His descent from David. God had prepared the way for the coming of Jesus by sending John the Baptist. After preaching repentance, the recognized means for winning the divine favor, John had clearly testified that he was only the precursor of the Messias and immeasurably inferior to Him. Since the Jews everywhere regarded John as a prophet, this appeal to his testimony had strong argumentative force.

26-37. But there was the objection that Jesus had been rejected by their rulers and condemned to death on the cross. St. Paul replied that this death had been foretold by the prophets and so was an essential part of the designs of almighty God who used the Passion and Death of Jesus to bring about the glory and triumph of His Resurrection. That Jesus truly rose from the dead was proved by the testimony of the disciples who had seen Him, and they were then still living and boldly standing forth among the people as witnesses of this tremendous event. Scripture had prophesied the Resurrection no less clearly than the Passion of Christ. The words cited in 33 are used elsewhere by St. Paul to proclaim the eternal generation of the Son and His priesthood (Heb. *1,* 5; *5,* 5). All three ideas are included in the words of the Psalm; the foundation of Christ's glory is His divine Sonship. Becoming man, He redeemed the human race by the sacrifice of the cross; in His resurrection He is manifested in the glory due Him as the Son of God and the King of all men. The words from Isaias in v.34 are used to confirm the quotation to follow in the next verse from Psalm *15.* In Isaias God assured the people that He would faithfully fulfill the promises made to David. Psalm *15* contains the promise of the resurrection of the Messias, since the promise could not have referred to David personally who suffered the decay of the grave just as other men do. The promise however found its fulfillment in Jesus, for He is the *Holy One* whom God did not permit to *undergo decay. Did undergo decay:* return to the state of death in which decay naturally follows.

38-41. Justification can come only through faith in Christ. The efficacy of grace compared with the powerlessness of the Old Law, is a favorite topic in the Pauline Epistles. The final exhortation takes the form of a threat in the prophetic language of Habucuc. The Prophets are spoken of as a group because the Old Testament had three divisions, the Law, the Prophets, and the Writings. Men may find it hard to believe in the wondrous works of God, but rejection of Christ will bring down upon them the full force of the divine wrath.

42-52. **43.** This is the only place in the Acts where *worshipping* and *converts* are joined. Proselytes in the full sense took on all the obligations of the Mosaic Law. The persons meant here did not go so far; they remained uncircumcised, but professed belief in the one true God and frequented the synagogues and observed some of the Jewish ceremonial law. **51.** Shaking off the dust had the symbolic meaning that there was defilement in everything that had come into contact with the unbelievers. Cf. Matt. *10,* 14. *And went to Iconium:* they did so after spending some time preaching in this region.

14, **1-6: Iconium.** Iconium was about eighty miles east of Antioch. Like Lystra and Derbe it lay in the great plain at the foot of the Taurus mountains. The whole region belonged to the Roman Province of Galatia, but Iconium was on the frontier of the older districts known as Phrygia and Lycaonia and the inhabitants considered themselves Phrygians.   **4.** *Apostles:* used in a wide sense of missioners here and in v. 13.   **5.** *Rulers* of the Jews: had the civil rulers been hostile, they would have arrested the missioners instead of letting the mob rule.

14, **7-19: Lystra.** Lystra was about twenty miles south of Iconium.   **12.** Mythology had a legend of Jupiter and Mercury being entertained in Phrygia by the virtuous peasants, Baucis and Philemon (cf. Ovid, *Metamorphoses, 8,* 611 ff). Mercury was the eloquent messenger of the gods, and St. Paul was here the chief speaker. **13.** The preparations for sacrifice were the first intimation the missioners had of what was afoot; they had not understood the acclamations made in the Lycaonian dialect. **14-16.** Speaking to a pagan audience, St. Paul begins with arguments drawn from pure reason. The Apostles were men like themselves with human needs and cares, but their purpose in coming to them was to show them the vanity of worshipping imaginary deities and lifeless idols and to lead them to a knowledge of the true God, who was the creator of all things in heaven and on earth. In the past God had permitted them to go their own dark way with only the light of reason for their guide. Even that light should have been enough to reveal His existence and His good will toward men, for He was the source of all the natural blessings which brought them joy and satisfaction. St. Luke gives us only this opening address, but it is easy to see how the reference to God's natural manifestations of Himself opened the way for speaking of His supernatural revelation which came to a climax in the sending of His own Son as the teacher and redeemer of mankind.

14, 20-27: Derbe; the Return. Derbe was near the eastern frontier of the province of Galatia, about thirty miles southeast of Lystra. **21.** *Kingdom:* as consummated in the next life. **22.** *Appointed:* the Greek verb means literally "to choose by a show of hands," and is not restricted to ordination; but it has the force here of ordaining priests, as priests would be needed to carry on the religious services. **24.** They had been in Perge before, but perhaps had pushed on to the interior without stopping to preach. It was the capital of Pamphylia and about eight miles from the coast. Attalia was on the coast about fifteen miles from Perge. **27.** *No little time:* perhaps the winter of 48/49 A.D.

15, 1-5: Dissension at Antioch. Some of the Jews were shocked by the fact that Gentile converts were admitted to the Church without circumcision. Pride of race prevented them from seeing that the Mosaic Law was no longer binding. They were perhaps converted Pharisees from Jerusalem (cf. 5). The question was too fundamental to be settled at Antioch, so they determined to appeal to the Apostles and presbyters at Jerusalem.

15, 6-35: Peter's Decision: Advice from James: The Decision. God's will in the matter had been clearly shown in the conversion of Cornelius; those who tried to force the unbearable yoke of the Old Law on the Gentile converts would be acting against God. The only source of justice and salvation was faith in Jesus Christ. In this way St. Peter implicitly excluded the need of Judaism for any one, whether he were Jew or Gentile. St. Paul and Barnabas recounted the miracles by which God had sanctioned their preaching.

Then St. James the Less, seconding St. Peter, added the testimony of Scripture and the proposal that, to make it easier for Jews to associate freely with the Gentile converts, certain restrictions should be placed on the Gentiles. They were to abstain from bloody meats, from anything strangled, from food previously offered to idols, and from immorality. His reason for offering these restrictions was that the Jews readily took offense at these practices since the weekly reading of the Law of Moses in the synagogues continually reminded them that such things were unlawful. A decree embodying these restrictions, and the express declaration that the Mosaic Law was not necessary for the Gentiles, was then drawn up and entrusted to a delegation, including St. Paul and Barnabas, to be presented to the churches of Syria and Cilicia.

This discussion is described by St. Paul also in Gal. 2, but he confines himself to the private consultation with the Apostles, while St. Luke records the public deliberations. The differences in the two accounts are easily reconciled by keeping in mind that as a general historian St. Luke lays less stress on what is personal to St. Paul. When he says that St. Paul went to Jerusalem as a delegate of the church at Antioch, he does not contradict St. Paul's statement that he went up because of a revelation, since such a revelation might have induced St. Paul to suggest the appeal to Jerusalem or to accept a place on the delegation. The omission of Titus in the Acts is explained by supposing that Titus was not a delegate or that he is included in the number of those who went along with Paul and Barnabas as delegates; the Greek text in *15, 2* reads "some others of them" and does not necessarily mean that all the other delegates belonged to the party opposed to St. Paul.

Some try to identify this meeting at Jerusalem, mentioned in Galatians, not with Acts *15*, but with the "alms-visit" of Acts *11*, 30; *12*, 25. Their main argument is that in Gal. *1* and *2*, St. Paul is proving that he did not learn his gospel from men, but from God, and to prove this he had to enumerate all his visits to Jerusalem and so he could not have omitted this "alms-visit." But this argument is founded on a misunderstanding of the relation between Gal. *1* and Gal. *2*. The first chapter disposes of the charge that St. Paul was a mere disciple of the other Apostles, and it closes with a picture of St. Paul as already launched on his apostolic career and as preaching the gospel in Syria and Cilicia with such effect that his reputation was known even in distant Judea. In Gal. *2* the theme is quite different. St. Paul had also been accused of teaching doctrines contrary to that of the other Apostles, and here he refutes the accusation by showing that his doctrine had been approved by the other Apostles at Jerusalem. Besides, it is most unlikely that so important a question as the admission of the Gentiles without circumcision would have had to be decided twice by meetings of the Apostles. Again Acts *11, 30* implies that at the "alms-visit" the Apostles had already been scattered by the persecution of Herod Agrippa, since the alms were to be delivered to the elders of the church at Jerusalem.

Gal. *2* makes no mention of the letter drawn up at the council, but there was no need for St. Paul to refer to it, since its main point, the freedom of the Gentiles, was perfectly clear from the welcome given him and his doctrine by the other Apostles. The negative part of the letter, the restrictions placed on the Gentiles, was only disciplinary and was intended to be enforced only at times and places where the difficulties were similar to those in Syria and Cilicia. The small number of Jews in Galatia would hardly have created such difficulties.

2. Second Missionary Journey *15, 36 — 18, 22*

16, 6-10: Departure for Macedonia. Lengthy disputes have been carried on over the meaning of *through Phrygia and the Galatian country*. Galatia had a restricted and a wide meaning; its restricted meaning refers to the region originally occupied by the conquering Gauls and called Galatia after them, while the wide meaning refers to the Roman Province of Galatia which embraced also the districts to the south and west of that region. Special importance attaches to the question because St. Paul wrote an Epistle to the Galatians. The "North Galatian theory" holds that the churches addressed in that letter were in the original Galatian territory; the "South Galatian theory" places them in the southern part of the Roman Province which was evangelized by St. Paul on his first journey. A third theory, claiming that both sections were addressed, has found few advocates, since the crisis which called forth the letter seems to have affected a rather restricted area. There are two questions here: did St. Paul ever preach in North Galatia? and did he send his letter to North or to South Galatia? There seems to be no great difficulty in separating these two questions.

St. Luke's narrative seems to call clearly for preaching in the northern region. St. Paul had already visited Derbe and Lystra (*16*, 1) and, it would seem (from *cities* in *16*, 4), also Iconium and Antioch of Pisidia. The Greek text, accepted by nearly all critics, reads: "They passed through Phrygia and the Galatian country, after having been forbidden by the Holy Spirit to preach the word in Asia." Asia is the Roman Province of Asia to the west of the regions through which St. Paul had been traveling. Excluded from Asia, he went through Phrygia and the Galatian country. The Province of Galatia cannot be meant, for he had been in that province all the way from Derbe. The reference must then be to Galatia proper, the northern part of the province. This need not imply a long stay there nor an extended journey to the more remote parts; the population was scattered and St. Paul may have pushed on as well as he could. On reaching the northern part of this Galatian territory he was east of Mysia, a part of the Province of Asia, and he thought of continuing northward into Bithynia. Being forbidden by the Holy Spirit once more he turned west and went through Mysia to Troas on the coast. This gives an intelligible itinerary. On his third journey St. Paul visited these regions in reverse order; starting as usual from Antioch in Syria, he "went through the Galatian country and Phrygia in order, strengthening all the disciples" (*18*, 23). This shows that converts had been made on the earlier journey. No difficulty is caused by the omission in *18*, 23 of Derbe and the other cities of the south, because St. Luke tries to be brief and usually calls attention only to new features in the apostolic ministry. The new feature in *18*, 23 is the strengthening of those lately converted, the older churches having already had a second visit and a strengthening of their faith in *16*, 5.

16, 35-40: Freedom. **37.** It was unlawful to scourge a Roman citizen.

17, 1-9: Thessalonica. The Apostle traveled southwest along the great Egnatian highway which led from Dyrrachium on the Adriatic through Thessalonica, Apollonia, Amphipolis, and Philippi to Byzantium. Thessalonica was about a hundred miles southwest of Philippi.   2. After the three weeks in the synagogue the work seems to have been continued for a considerable time among the Gentiles, perhaps with Jason's house as a center (5). In his letters to this church St. Paul writes as though most of the converts had been won over from paganism.   6. *Magistrates of the city:* called "politarchs" in the Greek text, an example of St. Luke's accuracy in the use of local titles.   7. As in the trial of our Lord, the charge is that of treason to the Emperor by trying to set up a new kingdom.

17, 10-15: Berœa. Since "we" is not used after *16,* 16 until *20,* 5 f, St. Luke seems to have remained at Philippi to care for the new community, as Silas and Timothy later remained at Berœa (15). Berœa was about fifty miles southwest of Thessalonica.

17, 16-21: Athens. In this center of culture St. Paul encountered entrenched intellectual paganism and met with little success. Pride of intellect scorns the humility demanded by faith. Of the two most popular schools of philosophy, the Epicureans were materialists and set up pleasure as the highest good, while the Stoics were pantheistic and placed man's highest good in devotion to duty with a balanced contempt as armor against the ills of life.

17, 22-34: Paul's Discourse. St. Paul tried to bring them to recognize the one true God by dwelling on His presence in creation, His conservation, and provident care for the human race. He must be superior to their idols, since even man is above such material objects. God was then calling on them to repent of their past folly, for He had appointed a day of judgment and a Judge whom He had sanctioned by raising Him from the dead.   23. *Unknown god:* in superstitious fear lest they should omit any god from among their idols.   28. The quotations are from Epimenides, Aratus, and Cleanthes.   32. Neither Epicureans nor Stoics believed in the resurrection of the body or in the survival of the soul after death.   34. According to a tradition dating from the second century Denis (Dionysius) was the first bishop of Athens.

18, 1-11: Corinth. Corinth was the capital of Achaia and the headquarters of the proconsul who governed the province; it was noted for commerce, wealth, culture, and the worship of Aphrodite. **2.** The edict of Claudius, published about 50 A.D., was enforced for only a short time; Acts *28,* 17 shows that the Jews were then once more settled in Rome. In his life of Claudius *(25)* Suetonius seems to refer to this edict in the words "he banished from Rome all the Jews, who were continually making disturbances at the instigation of one Chrestus"; the fact probably was that the Jews had been trying to persecute the Christians, and the Roman authorities classed both factions as Jews. **3.** At Tarsus, the birthplace of St. Paul, one of the chief industries was the making of tent-cloth from Cilician goat hair.

18, 12-17: Gallio. Gallio became proconsul in the year 51/52 A.D. The charge was that St. Paul violated the Mosaic Law by teaching a religion condemned by the Jews, but Gallio dismissed the case since it was a mere question of their Law, all of them admitting a Messias and St. Paul simply recognizing him in Jesus while the other denied this.

17. *Sosthenes:* perhaps there were several presidents or several synagogues, or he was the successor of Crispus (8).

18, 18-22: Return to Antioch. After a year and a half spent in establishing and organizing the Corinthian church St. Paul left for Ephesus where he stayed only long enough to find out what promise it held for the gospel.

3. Third Missionary Journey *18,* 23 — *21,* 16

18, 23 — 19, 40: Return to Ephesus. **25.** This shows how strong and widespread was the influence of John the Baptist. **28.** Apollos became an innocent cause of the factions at Corinth which are denounced in *1* Cor. *1-4.*

19, 3. These men had not heard John the Baptist's designation of Jesus as the One for whom he was merely preparing the way. Apollos was better instructed, though he knew nothing of Christian Baptism *(18,* 25). **6.** On first meeting them St. Paul thought that perhaps they had not been confirmed. Now after having them baptized he administered the sacrament of Confirmation, according to a common opinion.

9. To *in the school of one Tyrannus* Codex D adds "from the fifth to the tenth hours." This may be founded on good tradition and would then imply that during the morning and evening St. Paul worked at his trade and did his preaching during the customary hours of rest, from about noon till four o'clock.
10. From Ephesus as a center St. Paul evangelized the whole region, probably by sending out missioners. On his second journey he had stayed at Corinth for eighteen months, and now he spends over two years at Ephesus (cf. *20,* 31). This modifies the common idea that he was continually hurrying from place to place. Though an extensive traveler, he was also a great organizer and took care that everything should proceed regularly in the churches he founded.

13. Not only pagans, but Jews also were deeply interested in trying to control the spirit world, and so magic spells and formulas were highly esteemed. It was considered of the first importance to call the spirit by his proper name, and these names were among the high priced secrets of the sorcerers.
14. *High priest:* in a wide sense this designated a member of one of the priestly families from which came the high-priest in the strict sense (cf. *4,* 6).

18. Old superstitions die hard, and many converts had not realized the evil of magic or had lapsed into it again. **19.** The estimated price would amount to about $10,000 but money had far greater purchasing value then. **21.** Feeling that he had already established churches in the most promising fields of the Orient, St. Paul hoped soon to visit Rome on the way to new regions farther west.

24. The temple of Diana at Ephesus was one of the wonders of the ancient world. The *shrines* were statues of Diana standing under a canopy, and were widely used as charms. **25-28.** Loss of trade was what was most feared, but zeal for religion was put forward to support it. **33.** By denouncing the Christians, Alexander was to keep the crowd from blaming the Jews.

***20,* 1-6: Macedonia and Greece.** As the first person plural "we" is again introduced here, it is natural to suppose that St. Luke had remained at Philippi after the events of *16,* 17 ff and had been employed there in guiding the church.

20, 7-16: Troas. St. Luke had just asserted that the boy was dead. In embracing him St. Paul recalled him to life and after the miracle assured them that he was alive. **11.** *Ate:* of the Eucharist. Had it been ordinary food, it would have been eaten before the celebration of the Eucharist according to the custom of that time. **13.** Perhaps the twenty-mile journey by land was quicker and permitted him to stay longer with the disciples at Troas.

20, 17-38: Discourse at Miletus. St. Paul speaks with personal feeling like a father to his children; this is the only example of a discourse to his disciples. He recalls how strenuously he had worked among them and foresees danger for himself at Jerusalem (17-24); he warns them to have great care of their flock, especially guarding it against false teachers (25-31); and in farewell exhorts them to labor unselfishly as he had done, not only in teaching, but also in manual work so as to have the means for helping the poor and sick (32-35).

21, 1-16: Tyre: Ptolemais and Cæsarea. At this time and during St. Paul's two-year imprisonment at Cæsarea *(24-26)* St. Luke doubtless heard from Philip many details about the ministry both of our Lord and of the early evangelists. **11.** In the symbolic manner of the Old Testament prophets Agabus bound himself with St. Paul's girdle to signify his arrest at Jerusalem.

4. Imprisonment in Palestine 21, 17 — 26, 32

21, 17-26: Jerusalem. **20.** *Thousands:* in Greek "tens of thousands," but used as a round number to indicate a great multitude. As at the beginning (*2,* 46), the converts continued to attend the services in the temple. **21.** The charge was false, if it meant directly persuading Jewish converts to abandon their customs; but it was true in the sense that he taught that there was no longer any strict obligation for Jews to observe these customs. **23.** *Vow:* the Nazirite vow. Their heads would be shaved on the seventh day, and the necessary sacrifice offered on the eighth. **24.** *Sanctify:* since he had just arrived from the Diaspora St. Paul would need levitical purification, and besides he had still to offer the sacrifice for his own vow *(18,* 18). Since the hair was to be burnt in the fire of the sacrifice (Num. *6,* 18), St. Paul perhaps again shaved his head. **25.** *We ourselves have written:* at the Council of Jerusalem (c. *15).* This is not an assertion of the authority of the church at Jerusalem over the other churches; it simply states the fact that there the assembled Apostles and presbyters, including St. Paul himself, had passed this decree.   **26.** *Days of purification:* in accord with Num. *6, 5.*

21, 27-39: Paul's Arrest.   **28.** *Into the temple:* inside the barrier separating the Court of the Gentiles from the inner court into which Gentiles were forbidden to enter under pain of death. **31.** *Tribune of the cohort:* the commander of the garrison in Jerusalem. The procurator had his headquarters at Cæsarea and usually left to the tribune the task of keeping order in Jerusalem. A cohort, the tenth part of a Roman legion, numbered five or six hundred heavy armed soldiers with about an equal number of light or auxiliary troops. This would give the tribune about a thousand men, and the Greek term here is "chiliarch," the commander of a thousand. In *23, 23* the escort given St. Paul amounts to four hundred seventy. This seems a large proportion of the garrison, but with the exception of the seventy horsemen they would all return the next day (*23, 32*).   **38.** Josephus gives contradictory accounts of this affair of the Egyptian, and the tribune may be stating only the popular version. If he erred in the number or place, the mistake was his; St. Luke is responsible only for accurately reporting what the tribune said.

22, 22-29: Paul's Citizenship. Their fury was aroused again by the reference to the Gentiles because in their eyes St. Paul's great offense was that he had been putting the Gentiles on a level with the Jews.   **24.** *Barracks:* the Antonia at the northwest corner of the temple area with stairs leading down to the Court of the Gentiles.   **28.** In the corruption of the Empire the rights of Roman citizenship were for sale. St. Paul inherited his citizenship, but it is not stated how his father obtained the privilege.   **29.** The binding is that in preparation for scourging (25). He would still bear whatever fetters they thought necessary to keep a prisoner safe.

22, 30 — 23, 11: The Sanhedrin.   **30.** Inquiry by torture being now out of the question, the tribune decided on an investigation before the Sanhedrin.   **23, 2.** *Ananias,* high priest about 47-59 A.D., was called by Josephus "a great hoarder of money" and was slain during the internal strife that accompanied the final revolt against Rome.   **5.** In this investigation conducted by the Roman tribune the High Priest may not have been conspicuous, and St. Paul may not have known him personally.

23, 12-21: A Conspiracy. Such a murderous plot fits in well with the violence that marked this period during which the way was being prepared for the revolt against Rome and for the war that followed.

23, 22-35: To Cæsarea. **27.** Claudius dresses up the story to put himself in the best light. **31.** Antipatris was about thirty miles northwest of Jerusalem. The rest of the journey to Cæsarea would be along the plain of Sharon and less exposed to attack.

24, 1-9: The Accusation. **2.** Tertullus, a professional orator, began his presentation of the case in the accepted style of the schools with a bit of flattery. In reality Felix was a rascal and, as Tacitus puts it, "he wielded royal power with the soul of a slave," and "thought he could commit all crimes with impunity" (*Hist. 5*, 9; *Annals 12*, 54). He had previously been accused at Rome, but had secured his acquittal through the influence of his brother Pallas, a favorite freedman of the Emporer Claudius. His wife was Drusilla, the sister of Herod Agrippa II and Bernice, and he had won her from her husband, the king of Emesa, by the help of a Jewish magician. On his recall to Rome about 58 A.D. he had to face accusations laid against him by the Jews of Cæsarea. **5-8.** Tertullus advanced three charges: St. Paul had stirred up seditions among the Jews throughout the world, he was the soul of the Nazarenes (a Jewish name for the Christians), and he had profaned the temple.

24, 10-21: The Defense. St. Paul skillfully began by winning the good will of the judge, but he confined himself to facts; Felix had been in office for a long time and was acquainted with local conditions. Then with calm dignity he met each of the charges. He was not a disturber of the peace, for he had been in Jerusalem for only a short time and had comported himself quietly. Nor was he a heretic, since he believed in the Law and the Prophets, and Christianity was the flowering of all that was essential in Judaism. Finally he had not profaned the temple for, when attacked, he was performing customary religious rites there. If he had done anything amiss, his first assailants, the Asiatic Jews, should have been on hand to accuse him. In his trial before the Sanhedrin the only fault laid to his charge had been his belief in the resurrection of the dead, a doctrine held by most of the Jews.

24, 22-27: The Prisoner. Living at Cæsarea where there were many Christians, Felix was in a position to judge how unfounded were the accusations made against the Apostle. Besides, his wife Drusilla must have known of the hostility between Jews and Christians. He postponed his decision, however, on the pretext of wishing to consult the tribune, but in reality, after the manner of the worst type of Roman official, merely to play fast and loose with the Jews and to receive bribes from St. Paul. After two years, with a trial at Rome confronting him, he was chiefly concerned with placating the Jews.

25, 1-12: Festus. The two years' inactivity had not lessened the Jews' animosity against St. Paul. Backed by a shouting crowd, they had at first tried to overawe the new procurator and force him to condemn St. Paul at once *(25, 15 f)*. This failing, they wanted him to order the trial held at Jerusalem. **9.** The injustice of this request is indicated by his motive which was to begin his term of office with the good will of the Jews. **11.** St. Paul knew the hopelessness of justice in such a trial and also perhaps the plots against his life. Once he exercised his right as a Roman citizen to be tried before the Emperor at Rome, Festus had to admit the appeal.

25, 13-27: Agrippa. **13.** *Agrippa* was the son of Herod Agrippa I, who had persecuted the Church (Acts *12*), and the great-grandson of Herod the Great. He had received Chalchis, to the north of Palestine, from the Emperor Claudius and later was given Batanea and the vicinity. He had authority over the Jews in religious matters. During the Jewish war he sided with the Romans. His sister Bernice lived with him after the death of her first husband, Herod of Chalcis, and rumor accused them of incestuous intimacy. Then she married Polemon, king of Pontus, but later king merely of a small section of northwestern Cilicia; but she soon left him and returned to her brother. She found favor with both Vespasian and Titus during the Jewish War, and Titus would have married her in Rome, had it not been for the popular outcry. **20.** In reality he was afraid of the Jews. **21.** He seems aware that he has put himself in a false position by forcing St. Paul to appeal instead of releasing him.

26, 1-23: Paul's Discourse. St. Paul reviewed the main facts of his life, conversion, and mission to the Gentiles. He showed that his doctrine was in conformity with the prophets and Moses who had announced that Christ was to die, rise again, and bring light to Jews and Gentiles. With no understanding of such matters Festus soon broke in with the exclamation that the Apostle must be insane from too much study. But St. Paul was speaking mainly to Agrippa and he then appealed to him to support his statements by asking him directly whether he did not believe in the prophets. The king evaded the question with the ironical remark that St. Paul had almost persuaded him to become a Christian. His words might also have the force of "Dost thou think thou canst so easily make a Christian of me?" Then Agrippa closed the interview, no doubt fearing other embarrassing questions, since he was a renegade to the Jewish religion. As Festus had no jurisdiction after the appeal to Cæsar, it was easy for the king to suggest politely that St. Paul might have been released if he had not appealed.

4 f. Till about his fifteenth year St. Paul lived at Tarsus and then went to Jerusalem to study the Law in the School of Gamaliel (22, 3). His earlier years are perhaps covered in the phrase *among my own nation,* referring to the Jews of the Diaspora as distinguished from those of Jerusalem. *Sect:* as in 5, 17, a party having its own proper tenets, but still in union with the main body.   **8.** Agrippa was deeply imbued with pagan ideas and probably had given up all belief in a resurrection of the dead. As always, the fact of Christ's resurrection is the central argument. **9-11.** His zeal in persecuting showed how hard it would be to convert him.   **16-18.** For the sake of brevity he attributed directly to our Lord the message he had received from Him through Ananias, or having received the message at the time of the vision, he heard it again from Ananias.   **21.** Particularly because he put the Gentiles on the same plane as the Jews, teaching that for all salvation came through faith in Jesus.

5. IMPRISONMENT IN ROME 27, 1 — 28, 31

27, 1-13: Departure for Rome. They sailed about the middle of August on a ship bound for Adrumythium, a port on the west coast of Asia Minor not far from Troy. From there they could have crossed to Macedonia and proceeded along the Egnatian Way to Dyrrachium and then across the Adriatic to Brundisium. The prevalent summer winds being from the west, they sailed close to the east coast of Cyprus and then coasted along Asia Minor till they reached Myra, a quiet port used by Egyptian grain ships when westerly gales prevented them from sailing directly across the Mediterranean to Italy. From Myra they could creep along the coast, from island to island, and make Italy in about ten days. A ship from Alexandria happened to be about to sail for Italy, and the centurion decided to take it and so save the long journey by land across Macedonia.

27, 14-26: A Storm. After passing Gnidus they were beyond the shelter of the land and were caught by winds from the north which drove them in a southwesterly direction toward Crete. It was only with difficulty that they managed to get past the headlands of Crete, but, this danger over, they found shelter in the port of Good Havens on the southern side of the island. Progress had been slow, they were already off their course, and the season for navigation was almost closed. St. Paul's suggestion that they be content to winter there was overruled, and with a favorable breeze they set out for Phœnix, a more commodious harbor about a day's sail along the Cretan coast. But soon they were caught by a gale from the east-northeast. After a fearful run of over twenty miles they got a little respite under the lee of Cauda and took advantage of it to haul in the long boat and to undergird the ship by passing cables under the hull and drawing them tight with a windlass, a custom said to be still in use among sailors caught in a storm that threatens to tear the ship asunder. With all sails lowered, the ship simply drifted before the storm and they feared they would be carried toward the Syrtes of Africa, the terror of ancient seamen, consisting of shifting sandbanks with strong currents that made destruction almost certain.

27, 27-44: Shipwreck. Breakers ahead gave warning of land, and they cast anchors from the stern for fear of running on the rocks. They were off the northeastern shore of Malta and directly before them they saw a small bay with a stretch of sandy beach on the far end on which they hoped to run the ship. But at the northeast corner of the bay there was a strip of land separated from the mainland by a channel about a hundred yards wide, and the current of this channel caught them on the starboard as they tried to sail into the bay. The ship struck a shoal of clay, and the prow stuck fast while the stern was buffeted by the waves. Those who could swim were ordered to leap off and, once ashore, to help rescue the others who were to save themselves as best they could by clinging to the wreckage of the ship.

28, 1-10: Malta. **2.** *Natives:* in Greek "barbarians," in the sense of those who did not speak Greek, but with no implication of savagery; in the essentials of civilization many isolated peoples of that time surpassed the degenerate Greeks and Romans. Malta is about sixty miles south of Sicily and was a part of the Province of Sicily. **4.** *Justice:* perhaps, as among the pagans, personified as a goddess presiding over justice. **6.** They were well acquainted with the natural effects of the bite of this poisonous viper. As at Lystra (*14,* 7 ff), the wonder-worker is proclaimed a god. **7.** *Head man:* in Greek "the first," the official title of the one who ruled the island as representative of the praetor of Sicily; the title is found in inscriptions referring to Malta and is another proof of St. Luke's carefulness in details. **9.** St. Luke "the physician" shows marked interest in the cure of physical ailments in this section (3-9). Miraculous power is implied in the specific cases, but the general term in v.9 may well mean that at least some were cured after being given medical treatment.

28, 11-16: To Rome. **12 f.** *Syracuse:* the capital of Sicily. *Rhegium:* on the "toe" of Italy. *Puteoli:* in the bay of Naples, about one hundred forty miles from Rome. **15.** Two towns on the Via Appia, *Market of Appius* being about forty miles from Rome, and *Three Taverns* some ten miles nearer the city. St. Paul was always highly appreciative of human kindness. **16.** He would have to stay in this lodging, and his wrist was chained to that of his guard who would naturally be changed frequently. This freedom was granted to distinguished prisoners who were not considered dangerous. Tiberius treated Herod Agrippa I in the same way.

28, 17-31: At Rome. **17.** Not being guilty, he would have preferred a trial by the Jews; but as there was no hope of justice from them, he had been forced to appeal. Before the Emperor he intends merely to prove his own innocence without bringing any charges against his nation.   **20.** *Hope of Israel:* the Messias whom the Jews awaited and whom St. Paul proclaimed in Jesus.   **21.** Distracted by national troubles and aware of the weakness of their case, the authorities at Jerusalem had sent no instructions to Rome regarding Paul, but the local Jews must have heard of his work in the Diaspora.

26 f. These terrible words of Isaias had been used by our Lord in denouncing the disbelief of the Jews (Matt. *13,* 14 ff).   **31.** St. Luke closes on a note of triumph; St. Paul is free to preach to all who came to him. These final words were evidently written at the end of the two years' imprisonment, but before the acquittal of St. Paul. The Jews seem to have let the case go by default. Many details of these two years in Rome are found in the "Epistles of the Captivity" (Ephesians, Philippians, Colossians, and Philemon) which were written at this time and show that the Apostle expected to be released soon.

WILLIAM A. DOWD, S.J.

THE LIFE AND EPISTLES OF ST. PAUL

Origin. St. Paul, the Apostle of the Gentiles and Chosen Vessel, was born in the prosperous city of Tarsus of the Roman province of Cilicia (Acts *9*, 11; *21*, 39; *22*, 3) of Jewish parents (Acts *21*, 39; 2 Cor. *11*, 22), who were descended from the tribe of Benjamin (Rom. *11*, 1; Phil. *3*, 5). As he was a "young man" at the stoning of St. Stephen (Acts *7*, 58) and calls himself "an old man" when writing to Philemon (v.9) about the year 62/63, we may conclude that he was born around the beginning of the Christian era.

He was a citizen not only of Tarsus (Acts *21*, 39) but also of Rome from birth (Acts *22*, 27 f). From this we may infer that his father must have been a man of means and a prominent as well as influential member of the community.

The Apostle had two names: Saul, a Hebrew name—in memory of the first king of Israel, who was also a member of the same tribe (Acts *13*, 21)—and Paul, a Roman name. In his Epistles he always makes use of the name "Paul." In the Acts "Saul" appears up to the time when he meets Sergius Paulus, the proconsul of Cyprus (Acts *13*, 9) and thereafter only three times in reported speech (Acts *22*, 7.13; *26*, 14).

Education. In his youth Paul had acquired a threefold education. First, he learned the Greek language in his Tarsian environment, as is evident from his later skill in writing his Epistles. The Bible of the Jews of the Dispersion was also the Greek Septuagint version. Secondly, his father probably initiated him into his own trade, which was that of tent-making, and thus Paul later during his apostolic labors was able to defray the cost of his food and lodging by the work of his own hands (Acts *18*, 3; *20*, 34; *1* Cor. *4*, 12; *1* Thess. *2*, 9; *2* Thess. *3*, 8). Such an occupation was in accordance with a Rabbinical maxim: "Whoever does not teach his son a trade, teaches him to be a brigand." Thirdly, in his father's house at Tarsus his religious education was strongly Pharisaic (Acts *23*, 6). To complete his schooling Paul was sent to Jerusalem, where he sat at the feet of the learned Gamaliel and was taught according to the strict acceptation of the ancestral Law (Acts *22*, 3). Here he also acquired a good knowledge of exegesis and was trained in the practice of disputation. He apparently returned to Tarsus before the public ministry of Christ opened in Palestine, for he never refers to any personal acquaintance with Christ during the Savior's mortal life.

Some time after the death of our Lord, Paul returned to Palestine. His profound conviction and emotional character made his zeal develop into a religious fanaticism against the infant Christian Church. He took part in the stoning of the first martyr, St. Stephen, and in the general and fierce persecution of the Christians that followed (Acts 7, 58; 8, 3; 26, 9-11; 1 Cor. 15, 9; Gal. 1, 13; 1 Tim. 1, 13). With the proper authorization from the high priest, he departed for Damascus to arrest the Christians there and bring them bound to Jerusalem (Acts 9, 1 f).

Conversion. As he was nearing Damascus about noon, a light from heaven suddenly blazed around him. Jesus with His glorified body appeared to him and addressed him, turning him away from his apparently successful career. By omnipotent grace an immediate transformation was wrought in the soul of Paul. In a miraculous manner he was suddenly converted to the Christian faith and arose as an Apostle (Acts 9, 3-19; 22, 6-16; 26, 12-18).

He remained some days in Damascus after his baptism and confirmation (Acts 9, 10-19) and then went to Arabia (Gal. 1, 17), the kingdom of the Nabatæans, possibly for a year or two, to prepare himself, according to a common opinion, in seclusion and solitude by prayer and meditation for his future missionary activity. Having returned to Damascus, he remained there for some time and preached in the synagogues that Jesus is the Son of God. For this he incurred the hatred of the Jews and had to flee from the city (Acts 9, 23-25; 2 Cor. 11, 32 f). He then went to Jerusalem to see Peter (Gal. 1, 18) to pay his homage to the head of the Church. Later he went back to his native Tarsus (Acts 9, 30) and began to evangelize his own province (Gal. 1, 21) until called by Barnabas to Antioch (Acts 11, 25). After one year, on the occasion of a famine, both Barnabas and Paul were sent with alms to the poor Christian community at Jerusalem (Acts 11, 27-30). Having fulfilled their mission, they, together with John Mark, returned to Antioch (Acts 12, 25).

Journeys. Soon after this both Paul and Barnabas were selected by the Holy Spirit for a special task (Acts 13, 1-3). With John Mark as an assistant they made a first missionary journey (44/45—49/50 A.D.) visiting the island of Cyprus, then the regions of Pamphylia—where Mark left them to return to Jerusalem—Pisidia and Lycaonia, all in Asia Minor, and establishing churches at Pisidian Antioch, Iconium, Lystra and Derbe (Acts 13-14).

After the Apostolic Council of Jerusalem, Paul, accompanied by Silas and later also by Timothy and Luke, made his second missionary journey (50—52/53 A.D.), first revisiting the four churches previously established by him in Asia Minor and then passing through Galatia (Acts 16, 6). At Troas a vision of a Macedonian was had by Paul, which impressed him as a call from God to evangelize Macedonia. He accordingly sailed for Europe, and preached the gospel in Philippi, Thessalonica, Berœa, Athens and Corinth. Thence he returned to Antioch by way of Ephesus and Jerusalem (Acts *15, 36-18,* 22).

On his third missionary journey (53/54—58 A.D.) Paul visited nearly the same regions as in the second, but made Ephesus, where he remained nearly three years, the center of his missionary activity He laid plans also for another missionary journey intending to leave Jerusalem for Rome and Spain. But persecutions by the Jews hindered him from accomplishing his purpose. After two years of imprisonment at Cæsarea he finally reached Rome, where he was kept another two years in chains (Acts *18, 23—28,* 31).

The Acts of the Apostles gives us no further information on the later life of the Apostle. We gather, however, from the Pastoral Epistles and from tradition that at the end of the two years St. Paul was released from his Roman imprisonment, and then traveled to Spain (Rom. *15,* 24.28), later to the East again, and then back to Rome, where he was imprisoned a second time, and in the year 67 was put to death. Being a Roman citizen he was beheaded.

Epistles. St. Paul's untiring interest in and paternal affection for the various churches established by him have given us fourteen canonical Epistles. It is, however, quite certain that he wrote other letters, which are no longer extant. Mention is made in *1* Cor. *5,* 9 of a previous letter, in Col. *4,* 16 of a letter to the Laodiceans, in Phil. *3,* 1 of still another letter.

The place assigned to the Epistles in the Canon of the Tridentine Council and in our Bible is the same as that occurring in the Council of Laodicea about 360 A.D. They are not arranged according to chronological order. In the first place are given the Epistles addressed to communities, according to the relative dignity of the church receiving the Epistle and the length of the subject matter; and in the second place we have those addressed to individuals; and finally, the Epistle to the Hebrews, which was the last Epistle received by the entire Church into the Canon.

The Epistles are replete with doctrine. Some of the Fathers call them a storehouse of Theology. St. John Chrysostom, the greatest interpreter of the Apostle, compares them to inexhaustible mines of precious metals and to unfailing springs that flow the richer the more we draw from them. St. Thomas declares that they contain practically all of Theology. The central and pivotal thought contained in nearly all the Epistles is the universality of salvation through Christ, which can only be attained through a living faith in Jesus Christ and His gospel.

Language. All of the Epistles, even those to the Romans and Hebrews, were written in Greek. Though St. Paul on occasions could speak that language with grace and elegance, he did not strive after literary elegance in his compositions. His language stands midway between the classics and the ordinary language of the papyri. Because of the pressure of his work and care he usually dictated his Epistles and wrote the final salutation with his own hand (Rom. *16*, 22; *1* Cor. *16*, 21; Gal. *6*, 11; *2* Thess. *3*, 17). At times it is evident that the scribe did not write fast enough for him, with the result that new ideas and pictures arose in his active mind. So overflowing were his thoughts that the rules of grammar and style were often neglected. As a consequence a mode of expression or an entire sentence is now and then difficult and obscure for us (*2* Pet. *3*, 16).

And yet, in spite of these grammatical faults and irregularities of style, no one can read the Epistles of St. Paul without being amazed at his natural eloquence. He delights in picturesque expressions and metaphors, questions and exclamations, climaxes, puns and antitheses, and many other figures of speech (*1* Cor. *9*, 1-13; *13*, 1-3; *2* Cor. *4*, 8-12, *6*, 4-10). His style is also strong and vigorous. St. Jerome remarks that the words of the Apostle Paul seem to him like peals of thunder. His mental acumen and depth of feeling impart to his language loftiness, amazing power and beauty.

Form. The external form of the Epistles is the same as that found in ancient secular letters. At the beginning are given the name of the writer, the name of the person or persons addressed and the greeting. Then there are usually added an act of thanksgiving and some eulogistic words for the addressees. In the body of the Epistle we generally find first dogmatic truths developed and proven, and then practical exhortations for religious life. The Epistle is concluded with personal messages, greetings and blessings.

JOHN E. STEINMUELLER

THE EPISTLE TO THE ROMANS

Introduction

Genuineness. The Pauline authorship of the Epistle to the Romans, attested by the Epistle itself (*1*, 1), and accepted without hesitation by the early Church, has, in modern times, never been called into question except by the most extreme critics.

The Epistle was unanimously recognized by the ancient Church as the genuine work of St. Paul. The earliest explicit testimony to this effect is found in the Muratorian Canon (160-170 A.D.), and in the writings of St. Irenaeus (about 185 A.D.). Subsequent witness to the Pauline authorship of the Epistle is frequent and unanimous.

Evidence of the existence of the Epistle long before the explicit testimony to its Pauline authorship is found, many think, in *1* Peter, in which there are resemblances of ideas and expression which can scarcely be explained except on the hypothesis that St. Peter was familiar with the Epistle to the Romans. It is generally admitted that there is literary dependence of the Epistle of St. James on Romans. Be that as it may, there are passages in the Epistle of St. Clement of Rome to the Corinthians (98 A.D.), of St. Ignatius Martyr (110-117 A.D.), and of St. Polycarp (about 120 A.D.), which are undoubtedly inspired by St. Paul's Epistle to the Romans. This testimony is ample to establish the authenticity of the Epistle.

Integrity. Serious doubts have arisen as to whether we have the text of the Epistle today as it originally came from the pen of St. Paul's secretary, Tertius. The passages on which doubt has been cast are chapters fifteen and sixteen and in particular the doxology, *16*, 25-27. In regard to the doxology, it is omitted in some few later MSS, while some of the early MSS put it at the end of chapter fourteen and others at the end of chapter sixteen, while Codex Alexandrinus puts it in both places. The text used by St. John Chrysostom, by St. John Damascene, and by Theodoret, had the doxology at the end of chapter fourteen. It is argued that this uncertainty as to the place of the doxology is an indication that it is a subsequent interpolation. It is further argued that the style of this doxology is not Pauline, that it was not his custom to terminate his Epistles with a doxology, and that it does not fit in with what goes before it. These arguments, however, are far from conclusive in view of the fact that all the oldest and best MSS such as Sinaiticus, Vaticanus, Bezae, Claromontanus, and the recently discovered Chester Beatty Papyri, contain it and thus witness to its genuineness. It sums up all the great thoughts of the Epistle, and although St. Paul does not thus end his other Epistles, there is no reason why he should have to conform all his letters to a rigidly fixed and stereotyped rule. Therefore, there is no serious reason for denying or doubting the genuineness of *16*, 25-27.

In regard to *15* and *16*, although all existing Greek MSS contain them, it is generally admitted that there was extant in the early Church a recension which omitted these chapters. The doxology, placed, as we stated above, in certain manuscripts at the end of *14*, seems to indicate that the Epistle ended there, for such a solemn doxology would hardly have been put elsewhere than at the end of an Epistle. Various explanations of the omission have been suggested. The following two are highly probable. (a) An edition of the Epistle was prepared for use in the liturgical services of the Church. Since the last two chapters were largely personal and hardly suited for public reading, they were omitted from the lectionaries, i.e., copies of the Scriptures prepared for church use. The doxology, however, which terminated the entire Epistle, was appended at the end of *14*. (b) Some heretic, possibly Marcion, may have mutilated the Epistle by eliminating the last two chapters for doctrinal reasons. The first of these hypotheses satisfactorily accounts for the different positions of the doxology in various MSS.

There is, however, no reason, drawn from the character of these two chapters, to doubt their Pauline authorship. Their style and diction are certainly Pauline, and they have several touches which are strikingly characteristic of the Apostle of the Gentiles. But, while the genuineness of these chapters is generally accepted, it has been supposed by some critics that *16* belonged originally to the Epistle to the Ephesians, and by the mistake of a copyist, was appended at the end of Romans. But there is no strong argument in favor of this hypothesis, and against it we have the unanimous testimony of all the MSS. All of them conclude the Epistle to the Romans with *16*.

Date and Place of Composition. It is practically certain that the Epistle was written at Corinth. It seems to have been entrusted to Phœbe to bring to Rome (Rom. *16*, 1-2). Now Phœbe was a member of the church at Cenchræ, a small city, a port of Corinth, on the Ægean Sea. Moreover, St. Paul sends to the faithful at Rome the greetings of his host, Gaius (*16*, 23), who is probably the Gaius whom St. Paul baptized at Corinth (*1* Cor. *1*, 14). Finally St. Paul sends the greetings of Erastus whose name elsewhere (*2* Tim. *4*, 20) is connected with Corinth. It seems then, that the Epistle was written from Corinth.

But St. Paul made two visits to Corinth. His first visit (Acts *18*, 1 ff) lasted a year and a half. His second visit lasted three months (Acts *20*, 1-3). When he wrote to the Romans he was on the point of going to Jerusalem to present to the faithful of the mother-church the collection which he had made for them in Macedonia and Achaia (Rom. *15*, 25 f). As a matter of fact, St. Paul journeyed to Jerusalem for this purpose after his third missionary voyage (Acts *19*, 21; *20*, 1-3; *24*, 17). At this time among his companions were Timothy and Sopater or Sosipater (Acts *20*, 4) who send their greetings to the Romans in the Epistle (*16*, 21). There is no doubt, then, that St. Paul wrote to the Romans during this three months' stay in Achaia at the close of this third missionary journey, just prior to setting sail for Jerusalem. This was during the winter of either the year 57 or 58 A.D.

The Church at Rome. There were many Jews at Rome. Many of them, brought there originally as slaves during the Roman wars in Palestine, in the course of time won their freedom, and were granted many concessions by the Roman emperors.

Their first contact with Christianity came on the first Pentecost, when, as Acts tells us, there were among those who heard St. Peter's inaugural sermon "visitors from Rome." It is not improbable that some of these were among the three thousand converted and that they carried back with them to Italy the first seeds of the Christian faith. Jews, moreover, were constantly migrating to Rome, and among them, undoubtedly, were some who had embraced Christianity. Thus the nucleus of a Christian church was formed at Rome. At the time St. Paul wrote his Epistle to the Romans, he had many friends among them.

The official foundation of the Church at Rome is attributed by early tradition to St. Peter. We read in Acts *12*, 17 that, after the persecution of the Church by Herod Agrippa, St. Peter went to "another place." Certain statements of early ecclesiastical writers make it highly probable that this "other place" was Rome, which accordingly was visited by St. Peter about the year 42 or 43 A.D. But be that as it may, it is historically certain that the Prince of the Apostles went to Rome some time during his life, and that he suffered martyrdom there. The flourishing condition of their faith when St. Paul wrote to the Romans seems to demand that some one of the Apostles taught and inaugurated the church there. But early writers mention no other name than that of Peter. Among other witnesses to the presence of St. Peter in Rome are Clement of Rome, Ignatius Martyr, Irenaeus, and Tertullian. No other city has ever claimed the honor of possessing the tomb of the Prince of the Apostles.

Although the first Christians at Rome were in all probability converts from Judaism, at the time St. Paul wrote to the Romans the faithful were, according to the opinion of the great majority of critics, largely converts from Paganism.

Occasion and Purpose. In St. Paul's day the only safe way in which letters could be transmitted to friends in other cities was to entrust them to travelers who were journeying to the city where the addressees of the letter resided. At this time Phoebe was setting out for Rome, and St. Paul seized the opportunity to send by her to the Roman Christians the Epistle to the Romans. St. Paul, too, was planning soon to journey to Spain by way of Rome in quest of new fields in which to exercise his apostolic zeal (*15, 24*). This journey would give him the long desired opportunity to see the Christians at Rome, whose faith "was proclaimed all over the world," and to satisfy his longing to impart "some spiritual grace" to them and to "be comforted" by the faith which was common both to himself and to them. (*1, 8-12*). In order that his visit might be more profitable and enjoyable, St. Paul wrote to them to prepare them for his coming, and to introduce himself to them.

St. Paul recognized that the Roman church was destined to play a conspicuous and most influential part in the future destiny of Christianity. He wanted their assistance as he was about to set out for the mysterious and distant west to conquer new fields for Christ. And while writing to them he took occasion to set forth the fundamental doctrine that Christianity is the only way of salvation, and the corollary of this thesis, that the Mosaic Law is powerless to sanctify or save. St. Paul had briefly set forth this thesis in the Epistle to the Galatians, but since he had penned that document, he had much time to reflect and further to evolve and develop his argument. He wished to present this doctrine to a wide circle of readers, but until the present time, the opportunity to do so had not offered itself to him. At length, finding himself with leisure during the winter at Corinth while he was waiting for the sea to become propitious for his voyage to Jerusalem, and intending to introduce himself to the Romans by letter prior to his visit with them, he decided also to propose his doctrine with a more or less elaborate proof and explanation. In directing this treatise to the church at Rome, Paul assured it a wide circulation, for Rome was the hub of the world; people flocked to it from all quarters. Thus he could give to the Romans and through them to the Church Universal of that day and of subsequent ages the fruits of his long meditations on a problem which was then sorely vexing the Christian Church, and threatening to throw it into schism—the problem of the relationship of Christianity and Judaism, of the Church and the Synagogue.

COMMENTARY

Introduction 1, 1-17

1, 1-7: Greeting. **1.** *Servant of Jesus Christ:* as the agents of God in the Old Testament were termed "servants of God" (Ps. *17,* 1; Amos *3,* 7). *Called to be an apostle:* the details of St. Paul's vocation may be read in Acts *9,* 1 ff. *The gospel of God:* either the content of the revealed message, or the work of propagating it by preaching. In v.16 it means the whole economy of Redemption, the body of revealed truth. God is its author. **2.** *The prophets:* not only the works of the writing prophets, but all the Old Testament Scriptures.

The mention of the *Son* occasions a digression, not uncommon in Paul's letter's, which reaches to v.6. **3.** As regards His human nature, the Son was of the *offspring of David.* **4.** Cf. note to text. *The holiness of his spirit:* more probably designates the divine nature; nowhere in Scripture is the expression used of the Holy Spirit. Cf. *1* Tim. *3,* 16; *1* Pet. *3,* 18. *Resurrection from the dead:* either the moment from which His glorification began, or the chief proof of His divinity. *Jesus Christ:* in apposition to *His Son* in v.3. **5.** *To bring about obedience, etc.:* the Gentile world was in a special way the field of Paul's mission. Cf. Acts *9,* 15; Gal. *2,* 7. The object of his mission was *his name's sake,* i.e., that He might be acknowledged by the Gentiles as Son of God and the Savior of mankind. **6.** *To be Jesus Christ's:* the call is from the Father, and is to membership in Christ's Mystical Body; cf. Rom. *8,* 30; *1* Cor. *1,* 9; Gal. *1,* 15. **7.** *God's beloved:* the reason of their call. *Saints:* through grace they become partakers of the divine life; they are made holy by incorporation in Christ's Mystical Body.

1, **8-12: Commendation.** Paul does not say that the Romans' faith is above that of the other churches, but that it is noteworthy. **9.** *In my spirit:* with his whole heart and soul. **11.** *Spiritual grace:* any gift proceeding from the Holy Spirit. **12.** A modest clarification of his meaning.

1, **13-15: Why He Wishes to Visit Them.** **13.** The hindrance to his visiting Rome is explained in *15,* 20-25 as preoccupation in other missionary fields. *Results:* the making of new converts. **14.** *To Greeks and to foreigners:* the distinction is between those imbued with Graeco-Roman culture and those devoid of it.

***1*, 16-17: Theme of the Epistle.** **16.** In the eyes of the Gentiles the gospel was considered foolishness (*1* Cor. *1, 23*); hence Paul declares he is not ashamed of it, asserting its true worth. *Power of God:* divine in origin, it cannot be impeded by men. *Unto salvation:* its aim is the liberation of mankind from sin and eternal death, and their introduction to grace and eternal life. *To everyone who believes:* it is available to all on the condition of faith. *To the Jew first:* in virtue of the Messianic promises, the Jews were given the first opportunity of hearing the gospel message. This subject is enlarged upon in *9-11*. It was St. Paul's practice in each city to go first to the Jews. Cf. Acts *13*, 15.46; *28*, 17. **17.** *The justice of God:* that holiness which has God for its author, and is imparted to all who believe in Jesus Christ as their redeemer; that vital principle by which man lives the supernatural life of grace. It is in contrast with the justice which is from the Mosaic Law (Phil. *3,* 7-9; Rom. *10, 3*). This is the main thesis of *1-8*. *From faith to faith:* i.e., faith increasing in intensity. This faith is an internal assent to the truths revealed; it is made active by love (Gal. *5, 6*); it results in union with Christ (Gal. *2, 20*; 2 Cor. *13, 5*; Eph. *3, 17*).

I: DOCTRINAL *1*, 18 — *11*, 36
THE GOSPEL THE POWER OF GOD FOR THE SALVATION OF ALL WHO BELIEVE

1. HUMANITY WITHOUT CHRIST *1*, 18 — *3*, 20

***1*, 18-23: The Pagans Adore Idols.** **18.** *The wrath of God:* God's vindictive justice in punishing sin. While this is ultimately reserved for the Last Judgment, it is visited upon men even in this life, hence it is *revealed*. *Hold back the truth of God:* wickedness is a force opposed to truth which, if unrestrained, would expand in justice. **19.** *What may be known of God:* God's existence and something of His nature may be known by the light of natural reason. **20.** The evidence which created things offer in proof of the existence of God, and in manifestation of His attributes, is such as to leave the man of normal understanding *without excuse.* Cf. Wisd. *13,* 1 ff.

21. *Did not glorify him as God:* by rendering Him a service commensurate with His nature and with His claim upon them. *Vain . . . darkened:* persistence in ungodliness led to failure in mind and heart in the things pertaining to God. **22 f.** They thus reached the supreme degree of folly manifest in their idolatry.

1, **24-32: Punishment of Idolaters.** *God has given them up:* the terrible consequence of their error. This is repeated three times, each time introducing a new degree of their fall and a new penalty. **24.** *They exchanged the truth of God for a lie* (as in v.23); and God gave them up to *uncleanness.* **25-27.** They *served the creature rather than the creator;* and God gave them up to *shameful,* unnatural *lusts.* **28.** They *resolved against possessing the knowledge of God;* and God gave them up to a *reprobate sense.* Their moral sense, the judgment directing their moral acts, no longer discharged its function. **29-31.** The result was moral chaos, social disruption. **32.** A further definition of the *reprobate sense:* they lost view of the terrible sanction, and even applauded iniquity.

2, 1-11: All Will Be Rewarded or Punished. In this Paul declares that the Jews likewise, in spite of their claims to a privileged position, are far from salvation and are inexcusable.

1. *Wherefore* connects this passage with *1, 32.* The argument is: if the pagan is inexcusable, how much more responsible is the Jew who, in spite of the revelations he enjoys, is guilty of the same sin. **2.** *According to truth:* according to man's responsibility, and not misguided by racial privilege. **4.** *Riches of his goodness,* etc.: God has shown patience with the Jews in not inflicting on them a like punishment; but this indulgence is only to offer time for repentance. **5.** In contrast with this leniency is their *hardness and unrepentant heart. The day of wrath:* the day of judgment. **6.** *According to his works:* this clearly indicates that, while Paul denies that the works of the Law give salvation, he does not disparage good works in themselves. Cf. *1* Cor. *3,* 8.13-15; Gal. *6,* 7-9; Eph. *6,* 8; Heb. *6,* 10.

7 f. The divine judgment will divide men into two classes. *Life eternal* will be the reward of the good; *wrath and indignation* (cf. *1,* 18) that of the *contentious, who do not submit to the truth,* by refusing the revelation of Christ. **9.** *The soul of every man:* the soul as responsible for sin, but the punishment of the body is not excluded. *Jew first:* as first to receive the opportunity

of salvation. *Greek:* the Gentile as distinguished from the Jew.
11. The divine judgment unswayed by external conditions, as in v.2.

2, 12-16: Gentiles to Be Judged by the Natural Law. Here Paul digresses somewhat from his main line of thought to explain how the Gentiles can be judged though they have no written law. They have not had the advantages of the Mosaic Law, but they have known the "ordinance of God" (*1,* 32). This knowledge came to them through the natural law, an objective norm of conduct.
12. There is, therefore, one norm for the Gentiles, another for the Jews.   **14 f.** *Law unto themselves:* without the revealed Law, the Gentiles depended on the moral dictates of their natural reason. *Conflicting thoughts:* the conflict of pros and cons in their natural judgment.   **16.** *Will take place:* these words are not in the original, but are supplied to make clear the connection with what precedes. Some would attach v.16 to v.13, considering vv.14 and 15 a parenthesis. *My gospel:* Paul had more occasion than the other Apostles to emphasize the fact that the Gentiles were free from the Mosaic Law, and yet subject to divine judgment.

2, 17-24: The Jews Transgress the Law. The Jews had many blessings not enjoyed by the Gentiles, but they failed to take advantage of them.   **17 f.** *If thou art called Jew:* a name of honor in virtue of divine favors granted the race. These favors are enumerated: possession of the safe guidance of the revealed Law, the glory of being the chosen people of God, knowledge of God's will through the prophets, and thus enabled to discern what is good and pleasing to God.   **19 f.** The Greek text connects this immediately with v.18 by "and." *A light:* Wisd. *18,* 4 speaks of the pure light of the Law given to the world through the Jews. In Jewish thought all those without the Law are unwise.   **21 f.** Through these questions the Apostle shows how the Jews had failed to profit by their advantages. Cf. note to text.   **23 f.** The full indictment against the Jews.

2, 25-29: True Circumcision. St. Paul's point of view here is that of the old dispensation prior to the promulgation of the new. It does not conflict with what he says in Gal. 5, 2-6. Christianity abrogated the Mosaic Law and nullified its rites.   **25 f.** *Has become uncircumcision:* circumcision without the observance of the entire Law availed nothing; one might as well have been without the circumcision. On the other hand, the uncircumcised Gentile, observing to the best of his knowledge the natural law, could be pleasing to God.   **27.** *Will judge:* as the men of Nineve (Matt. *12,* 41).   **28 f.** The conclusion, then, is that a sincere internal disposition and not an external circumstance merits *praise from God.* This is often demanded by Paul even of Christians.

3, 1-8: Objections Answered. As in *2,* 1, the answer is made to an individual, an imaginary objector to his argument.   **1 f.** The first objection: the requirement of circumcision was then of no advantage to the Jew, there was no advantage in being a Jew. Answer: their great advantage was that they enjoyed the *oracles of God.* By this expression is meant the revealed Law, or perhaps the Scriptures as a whole.   **3 f.** The second objection: some Jews have not believed. Answer: the infidelity of man in no way weakens the absolute fidelity of God. The quotation from Ps. *50,* 6 refers to the manifestation of God's justice in punishing sin, and His rising victorious thereby over the charges of the insolent.   **5 f.** The third objection: then wickedness has the advantage of manifesting God's justice. Before answering, Paul makes an apology for even expressing the thought. The answer: God is to judge and condemn sinners. This He could not do if sinners rendered Him a service by their sins.   **7 f.** The fourth objection: what redounds to the glory of God does not merit punishment. The answer reduces the objection to an absurdity: *do evil that good may come from it. 9-11* will return more formally to these Jewish objections.

3, 9-18: The Scriptures Attest Universal Sin. The Jews indeed (*3,* 2) have had special advantages, but their failure to avail themselves of their privileges has made them equally sinners with the Gentiles. This draws the conclusion of the thesis begun at *1,* 18. Proof is found in the words of their Scriptures. The citation in 10-18 is found in the Vulgate in Ps. *13,* 1-3; but in the Hebrew it is derived from various other Pss.   **10-12.** Ps. *13,* 1-3;   **13.** Ps. *5,* 11; Ps. *139,* 4;   **14.** Ps. *9,* 28;   **15-17.** Isa. *59,* 7 f;   **18.** Ps. *35,* 2.

3, 19-20: This Concerns the Jews.   19. *Subject:* the Greek term means "liable."   **20.** *By the works of the Law,* etc.: in

2, 13 Paul has admitted some efficacy in the Law, but here he denies that it reached to the justification of a man. The most the Law could do was to indicate a line of action; the inner disposition, or the strength to follow that line, could not be supplied by the Law.

2. Salvation Through Faith in Christ *3, 21 — 4, 25*

This is the main thesis of the Epistle (cf. *1,* 16 f). It is first developed didactically (*3,* 21-30), and then illustrated from the Old Testament (*4*).

3, 21-26: Justice Comes through Faith in Christ. **21.** *But now* is to be taken in a temporal sense. *Justice of God:* the holiness which comes not through the Law, but by another means arranged by God and foretold in *the Law and the Prophets,* i.e., in the Old Testament Scriptures.   **22.** This means is *faith in Jesus Christ.* The justice is available to all men, *without distinction,* who fulfill the conditions of this faith.   **23.** *The glory of God:* the grace of God which has its consummation in eternal glory.   **24.** *Freely:* justice is a free gift of God, conditioned on faith, but in no way merited. *By his grace:* the quality imparted to the soul, transforming it from the state of sin to that of holiness. Even the act of faith cannot be elicited without the aid of God's actual grace. *Redemption which is in Christ Jesus:* this is a frequent thought in St. Paul: cf. *1* Cor. *6,* 20; *7,* 23; Gal. *3,* 13. It is found in other New Testament writings: cf. *1* Pet. *1,* 18; *2* Pet. *2,* 1; Apoc. *5,* 9. Man is bought back, or ransomed, by Christ from his slavery to sin; being a slave, he was unable to secure his own liberation.   **25.** The price paid by Christ was His death on a cross. *Set forth:* publicly exhibited; or, according to another interpretation, foreordained. *Propitiation:* a victim who, by His blood, expiated the offense of man's sin. Cf. Heb. *9,* 22 ff. *Through faith:* the disposition required for sharing in the benefits of this sacrifice. *To manifest his justice:* the purpose of God in thus setting forth His Son. In this sacrifice God exacted adequate atonement for the sin of the world, and thus His justice was made manifest.   **26.** Former times witnessed God's patience in dealing with sinful man; now His justice stands out in the sacrifice of His Son. His goodness and mercy are manifest in that He allows men who have *faith in Jesus* to share in His merit.

3, 27-31: Justification Excludes Boasting. **27.** *Law:* here a system of justification. *Of works:* the Mosaic Law, which prescribed merely a course of conduct. *Law of faith:* which imparted the aid necessary for the accomplishment of God's will. **28.** *Justified by faith:* i.e., "first justification," the introduction of sanctifying grace into the soul. This cannot be merited. The *faith* which is the condition of this gift includes sorrow for sin and the disposition to do whatever God requires. St. Paul does not intend to deprecate the works of the Christian life. **29.** Since there is but one God, the means of attaining this holiness are the same for all men. **31.** This verse forms a transition to the next passage.

4, 1-8: Abraham Justified by Faith. **1.** The question amounts to this: what kind of justice did Abraham acquire, that of works or that of faith? **2.** From the fact that Abraham has no reason to boast it follows that he was not justified by his works. **3.** But there remains the question: how then was he justified? The answer is provided in the quotation from Gen. *15,* 6: it was his faith that led to justification. The object of Abraham's faith was the divine promises of a numerous posterity, the possession of the land of promise, and that through his offspring all nations should be blessed. Cf. Gen. *12,* 2 f; *15,* 5; etc. St. James in his Epistle (*2,* 21 ff) uses this same text (and also Gen. *22,* 9 f) to establish the need of good works, but he is speaking of a speculative faith and of works which follow from sound Christian life. St. Paul's argument has to do with works which precede justifying faith. St. James is dealing with the preservation and increase of grace; St. Paul with the winning of initial justification. **6-8.** In the citation from Ps. *31,* 1 f, *forgiven, covered, not credit* are emphatic of the idea that sin no longer exists in the one whom God has justified. The text shows that forgiveness is due to God, and not to the sinner himself.

4, 9-12: Justified before Circumcision. **9.** The citation in 7-8 is from David who lived long after circumcision was prescribed. Do his words also apply to those who lived before it was prescribed? The answer is again found in the case of Abraham. **10.** The fact of Abraham's justification is announced in Gen. *15,* and therefore prior to the prescription of circumcision, which is told in Gen. *17,* 10 f. **11 f.** It was by providential design that Abraham was justified before his circumcision. He thus became the spiritual father of Gentiles and of Jews who would imitate his faith.

4, 13-17: Not Justified by the Works of the Law. But now the question rises: if faith centered in the promise which was to be realized through the Law, is it not in some way dependent on the Law? Paul shows in answer that the promise was not contingent upon the Law.

13. *Heir of the world:* this promise is not expressed in Genesis in these terms; it refers to the Messianic blessing which would bring the whole world into the spiritual Israel.   **14.** When this promise was made the Law did not exist. If the promise is conditioned on the Law, then the faith of Abraham would be rendered useless. It was faith, not the observance of the Law, that brought about his justification.   **15.** Law merely indicates what must be done, what avoided. It does not of itself supply the inner strength required for its observance. The inevitable transgressions of law provoke the lawgiver to wrath and call for punishment. But where there is no law on the observance of which the promise is contingent, there is no transgression to impede the fulfillment of the promise.   **16.** Therefore all comes by faith. *A favor:* a pure gratuity on the part of God. *That it might be secure:* independent of the uncertain observance of the Law, and firmly assured to those who share the faith of Abraham, both Jews and Gentiles.   **17.** Two instances of divine omnipotence on which rested Abraham's faith: (a) God's power to give life to the dead; (b) His power to call nonexistent things into being.

4, 18-22: The Strength of His Faith.   **18.** *Hoping against hope:* when according to human calculations there was no hope, Abraham believed the promise of offspring made him by God. **20-22.** There are three stages to Abraham's faith: (a) the difficulties involved did not weaken his faith; (b) he did not waver; (c) he was strengthened in faith, thus giving glory to God.

4, 23-25: The Model of Our Faith.   **23 f.** A similar unwavering faith on our part will be credited to us unto justification which is conditioned on such a faith.   **25.** *Who was delivered up:* by His death Jesus atoned for sin and merited for us justification. *Rose again,* etc.: the meritorious cause of justification is the death of Christ; His resurrection supplied the evidence which leads to faith in Him, and hence to our sharing in His merit.

3. THE SUPERABUNDANCE OF THIS JUSTIFICATION 5, 1-21

Before going on to discuss the nature of Christian life, St. Paul pauses here to contemplate the extent of this gift of God. The passage includes two considerations: (a) the hope that springs up in the justified soul (1-11); (b) the havoc and wretchedness from which justification rescues us.

5, 1-9a: Christ's Death Assures Us Hope and Peace. **1.** *Let us have peace:* some MSS read "we have peace." The sense is: let us enjoy the peace we have. The peace is that resulting from the destruction of the enmity between God and sinful man (Col. *1*, 21). **2.** Through Christ also we have *hope of the glory,* the final consummation of our state of grace. It is called *glory of the sons of God* not only because prepared for us by God, but because it is a sharing in His own divine glory. *Of the sons* is not in the Greek text. **3 f.** In this state of grace even the trials of life contribute to our hope, and thus become a cause of exultation. Cf. Jas. *1*, 3. **5.** Further, our hope is supported by the love of God *poured forth into our hearts by the Holy Spirit.* The Holy Spirit dwells in us, so that we have become His temples (*1* Cor. *3,* 16). **6-8.** The Greek text reads 6 as a declarative sentence. The divine love evident in Christ's death for us sinners is further support to our hope.

5, 9b-11: Christ's Death Assures Our Salvation. This states more positively in the form of an *a fortiori* argument the completeness with which our hope rests on our Lord Jesus Christ.

5, 12-14: In Adam All Have Sinned. **12.** *Therefore* is not to indicate this passage as a conclusion following upon 9-11. It is rather a transitional particle. The comparison begun with the word *as* is not completed. It would be: so through one man justice entered the world, etc. (cf. *1* Cor. *15,* 21 f). The Greek text reads "the sin," i.e., the original sin, the cause of humanity's ills. *Entered into the world:* i.e., into the consciences of men. *Death:* the reference is to the penalty attached to eating the forbidden fruit (Gen. *2,* 16 f; *3,* 19). *All have sinned:* the thought here is not of actual sin, but of the sin committed by Adam and transmitted to all his posterity. This is explained in the following verses. The doctrine of original sin, for which this is the classical scriptural evidence, has been defined by the Council of Trent.

13 f. These verses form a parenthesis inserted to prove that all mankind has sinned in Adam. *Since death reigned,* i.e., came upon all men, *from Adam until Moses,* i.e., prior to the positive law, it must have been due to their sharing in Adam's sin. St. Paul supposes Adam the moral as well as the physical head of the human race. Cf. notes to text. *A figure:* how Adam is a type of Christ is explained in the passage following.

5, 15-19: Grace and Life Superabound through Christ. 15.
Not like the offense, etc.: Adam is a type of Christ by a double contrast. First contrast: his sin led to death, through Christ grace abounds. *The many:* all humanity. *Much more:* the redemption by Christ is more efficacious. *Grace of God:* His infinite mercy and benevolence. *The gift:* the work of Redemption. *Grace . . . of Jesus Christ:* justification, sanctification, in their plenitude were in Christ, and from Him passed in varying degrees to mankind.

16. Second contrast: in the case of Adam it was one sin that led to a universal condemnation; in the case of Christ the redemption was from countless sins.   **17.** *Grace:* divine benevolence, the source of the gift. *Gift of justice:* all that is implied in our redemption by Christ. The reign of death is spoken of in the past, *death reigned,* because its dominion was broken by the death and resurrection of Christ. The reign of the faithful in life is still in the future, because the life of grace here will be perfected only in heaven.   **18.** *Justice* here means a just act, an act of obedience. *To all men:* St. Augustine explains: "Not that all men attain to the grace of Christ's justification, but that all who are reborn unto justification, are reborn only through Christ." The grace of Christ is sufficient for all men, it is efficient only to the faithful (St. Thomas).   **19.** *Constituted:* in neither case is this a mere extrinsic imputation; the term does not allow that sense.

5, 20-21: Purpose of the Law.
That the offense might abound: St. John Chrysostom: "After the Law had been given and became known, many sins followed because of the innate malice

of men and the stirring up of concupiscence." If interpreted as a purpose, this can be explained as a means of humiliating man by making him conscious of his weakness and of the need of Christ (St. Augustine).

4. JUSTIFICATION AND THE CHRISTIAN LIFE 6, 1 – 8, 39

St. Paul now turns to the exercise of justice in the Christian life. A man, though justified, must yet struggle against sin. In this struggle the Law is of no service. Success can be had only through grace. He concludes with a description of the eminent effects of grace in the soul.

6, 1-11: Christians Dead to Sin. **1.** An objection. Cf. *3, 1-8*. *Continue in sin:* probably the inclination to evil remaining even after the remission of original sin. *Grace:* God's generous goodness and mercy which lead Him to pardon sin. Some Christians may have thought that since grace was so abundant they need no longer worry about sinning. **2.** Its negative answer. *Dead to sin:* as the soul is separated from the body in physical death.

3. Here begins a more positive answer. *Baptized into Christ Jesus:* incorporated in Christ, so intimately united to Him that we die mystically to sin as He died really on the cross to satisfy for sin. Through Baptism this incorporation in Christ's Mystical Body is effected. Cf. note to text. **4.** *Through the glory of the Father:* through His power. *So we may also walk:* to be clothed with Christ (Gal. *3*, 27); it is no longer we that live, but Christ lives in us (Gal. *2*, 20). **5.** *United with him:* the term is used of things that grow together, as of plants. The Latin can mean, "planted together with him." The term *united* can be retained if understood of a very close union. By Baptism we are made one with Christ in His death; we should be one with Him also in the new life we enjoy. Many have understood *we shall be* as referring to the resurrection to glory.

6. *Our old self:* man with his evil inclinations and unruly passions. *The body of sin:* the body subject to sin, serving sin by yielding to evil inclinations. *Slaves to sin:* personifies sin. **7.** As a slave by death is removed from the dominion of his master, so the body, through mystical death in Baptism is released from servitude to sin. **8.** *Live together with Christ:* in the present context this means more probably a virtuous

Christian life in this world; though the future resurrection to glory, as the complement of the life of grace, is not excluded. **10.** *He died to sin:* He voluntarily undertook to pay the penalty of sin—death. It would never again be necessary for Him to undergo death. *He lives unto God:* for the praise and glory of God.

6, 12-14: The Reign of Sin. This is a warning against the false security that might result in the minds of some from the above references to the death of sin. **12.** *Mortal body:* the evil inclinations remain in the flesh until it puts on immortality. **13.** *Weapons:* a military metaphor. **14.** *Not under the Law,* etc.: unlike the Law, grace gives the strength needed to accomplish God's will.

6, 15-23: Slavery to Sin. A further precaution against a misunderstanding of his words, urged by means of a simile drawn from the familiar customs of slavery.

15. The difficulty: freedom from the Law might be interpreted by some as freedom from all restraint. **16.** Cf. our Lord's words: "No man can serve two masters" (Matt. 6, 24). The two masters here are *sin* and *obedience* to God. This is the major premise of Paul's argument. **17 f.** The minor premise. *Form of doctrine:* a norm of life to which Christians must conform. *You have been delivered:* i.e., in which you have been instructed. The passive implies they owe their conversion to God. This is also implied in *set free from sin*. The conclusion of the argument is too obvious to require expression. As slaves of justice they must render faithful service.

19. *I speak in a human way:* as imposing a precept; or, as apologizing for speaking of their service to God as slavery. *Slaves of uncleanness:* the vices of impurity, so prevalent among the Roman pagans. *Iniquity:* or rather "lawlessness," the absence of all moral law. *Unto iniquity:* service to these vices (personified) leads to their having complete mastery. **20.** *Free as regards justice:* not subject to, i.e., removed from justice. **21.** Some would punctuate this verse to read: What fruit had you then from those things? You are now ashamed of them, for their end is death. **23.** The metaphor is somewhat altered. Two pay-masters are paying their servants. Note the difference between *wage* and *gift*. Sin merits *death;* grace which leads to *life everlasting*, is due to God's gratuitous generosity.

7, 1-6: Christians Freed from the Law. Now St. Paul explains and proves that the faithful are not subject to the Law, a proposition already stated (*3*, 21; *6*, 14).   **1.** Jurisprudence was the peculiar genius of the Romans, and they would readily understand this principle. *Law* in this and the following verses may refer to law in general; though Paul has ultimately in view the Mosaic Law.   **2.** *Law of the husband:* the law which binds husband and wife.   **3.** Although both the Roman and Mosaic laws recognized divorce, Paul is now interested only in the general principle.   **4.** The illustration applied to the Mosaic Law. Cf. note to text. *Bring forth fruit:* under the Law sin brought forth evil works; joined now to Christ, Christians are expected to bring forth good works. It will be noticed that this application does not hold strictly to the illustration. St. Paul's main purpose is to establish the liberation of the faithful from the Law. For this the illustration is apt.   **5.** *When we were in the flesh:* still under the sway of disorderly, fallen human nature. *Aroused by the Law:* stimulated by legal prohibitions. This will be explained further in the passage following immediately. *Fruit unto death:* this has been pointed out in the description of human nature without Christ (*1-3*).   **6.** *By which we were held down:* this refers not so much to the Law as to man's evil inclinations and passions which dominated him under the Law. *In newness of spirit:* by our incorporation in Christ through Baptism a new principle of life is given us, the indwelling in us of the Holy Spirit. *Oldness of letter:* the written Law now old and discarded.

7, 7-12: The Law the Occasion of Sin. St. Paul faces another objection. His language may have aroused the suspicion that he regarded the Law as evil. In answering the objection he explains the relation of the Law to sin.

7. *I did not know sin,* etc.: probably the innate tendency to sin inherited from Adam. *I had not known lust,* etc.: probably any illicit desire. Good tendencies can with difficulty be distinguished from evil by a man unless a positive law teaches him. Others explain: while a man knows in theory right from wrong, he experiences no attraction to evil until a prohibition is placed upon it.   **9.** *Once upon a time:* i.e., in his childhood, when he had no knowledge of the Law. Others think Paul assumes the person of Adam and refers to the divine prohibition relative to the tree in Paradise. In this latter interpretation *revive* would mean "come to life."   **10.** *Died:* lost his innocence. That which in itself was contemplated to promote life, working in fallen nature actually brought about death. **11.** *Taken occasion:* the command became a stimulus to concupiscence, a goad to evil. **12.** It was sin and not the Law which led to the commission of evil. Therefore the Law in itself is good and holy.

7, 13-23: Sin the Cause of Death. A further objection is faced: if the Law is not the cause of sin, it is at least the cause of death. Paul explains that it is rather sin that causes death.

13. The divine purpose in the Law was that sin might be recognized as sinful. The malice of sin is known in all its perversity when it employs the Law to lead man into rebellion against his Creator. *Sin* is here again personified; it is the evil inclination in man resulting from original and leading to actual sin. **14.** *Spiritual:* opposed to man's evil instincts, appealing to his rational faculties. *Carnal:* man as ruled by his disorderly passions. *Sold into the power of sin:* given over, as a slave, to do the bidding of sin. **15.** St. John Chrysostom: "I am in darkness, I am carried away, I submit to spiteful abuse, I do not know how I am overturned." **16.** This struggle, between what is known as good and an inclination to evil, is an approval of the Law in spite of the actual transgression of its precepts. **17.** *I do what I do not wish:* this does not relieve man of responsibility for his sin; it rather makes emphatic the struggle within man against his inclinations to evil. **18.** *In my flesh no good dwells:* the bodily inclination is to moral evil. *To wish is within my power:* the rational faculties recognize moral good. **21-23.** *This law:* a constant experience, not a positive law. *Law of God:* the revealed Law. *Law of mind:* the Law of God recognized and approved by the mind. *Law in my members:* disordered human nature.

7, 24-25: Deliverance Due to the Grace of God. **24.** *Body of this death:* a slave to sin, destined for death as the penalty of sin. **25.** The Greek text reads "Thanks be to God through our Lord Jesus Christ." The sense is the same as that of the Vulgate: the only way out of the hopeless struggle is afforded us by God through our Lord Jesus Christ. *Therefore,* etc.: man left to himself is unable to escape the evil.

8, 1-4: The Faithful Need Fear No Condemnation. The security which Christianity gives to its adherents is based on two conditions: (a) a man must be incorporated into Christ; (b) he must live according to the spirit and not according to the flesh. Both of these are made possible through Christ.

1. *Therefore* introduces the conclusion following upon 6-7. *There is now no condemnation,* etc.: the threats and penalties of the Law are nullified, since the Law itself was abrogated by the death of Christ. Through union with Him the faithful are enabled to live the life of the spirit. **2.** This provides the reason. For the expressions involving the term *law* cf. 7, 21-23. *Law of spirit of life:* a rule of life apprehended by the mind illumined by grace which comes from union with Christ through Baptism. It is a *law* because it is a permanent state; and it is in opposition to the "law of sin." **3 f.** *It was weak,* etc.: good in itself, the Law could not overcome the rebellion of the flesh against the inner man (cf. *8,* 12.16.22). *His Son:* the Greek reads: "His own Son." This emphasizes the relation of the Father to Jesus Christ and their community of nature. *In the likeness,* etc.: like to us in everything except sin (Heb. 2, 17; *4,* 15; 2 Cor. 5, 21). *Justification of the Law:* the Law was given as a means of securing the promised justification. *According to the flesh:* guided by one's evil instincts. *According to the spirit:* guided by reason enlightened by the grace of the Holy Spirit.

8, 5-11: The Flesh and the Spirit. This develops the thought expressed in *8, 4.* **5.** *Mind the things:* desire and seek after. *The things of the flesh:* cf. Gal. *5,* 19. *Things of the spirit:* cf. Gal. *5,* 22. **6.** *Inclination:* in Greek from the same root as *mind* in 5. It includes both disposition of mind and inclination of will. The flesh and the spirit lead in opposite directions. **7.** *Hostile:* the Greek is more expressive: "is enmity against God." *Nor can it be:* this refers to *inclination.* **8.** *Carnal:* is the same as "according to the flesh."

9. *If indeed:* this does not question the fact that the Romans are spiritual, but indicates the reason why they are so. It is, however, possible to fall from this state. The *Spirit of God* and the *Spirit of Christ* is the Holy Spirit who proceeds from the Father and the Son. *Belong to Christ:* as a member of His Mystical Body. **10.** *If Christ is in you:* another way of describing the intimate union with Him in the Mystical Body. *The body is dead:* destined to die as the result of original sin which once infected it. *Life:* justification is the root of spiritual life.   **11.** *Bring to life,* etc.: in the resurrection of the body to glory.

8, 12-13: A Solemn Warning. The faithful, as already shown, owe nothing to the flesh. They owe everything to the Spirit. If they would live again according to the flesh they must die. To avoid that death, they must put to death the yearnings and inclinations of the flesh. This supposes that grace, powerful though it be, can be lost.

8, 14-17: The Faithful Sons of God.   14. *For* connects this passage with the preceding. *Are led:* man is led by the Holy Spirit in accord with his nature, which is endowed with intellect and free will. It is possible for man to resist this guidance.   **15 f.** These verses form a parenthesis. *Spirit of bondage:* the attitude of mind of a slave who serves out of a fear of punishment. *Spirit of adoption as sons:* by which they regard God as their father and address Him as such. *Abba:* an Aramaic word to which its equivalent, *Father,* is added.   **17.** As suffering was a prerequisite to the glorification of Christ (Phil. *2, 9*), so it is, in God's design, required that we suffer in union with Him. Cf. Matt. *16, 24* f.

8, 18-22: Yearning of All Creation.   18. *The sufferings:* those referred to in v.17. *Are not worthy to be compared:* our sufferings in this life are momentary and passing; future glory will be eternal.   **19.** *Creation:* all created things, men and angels excluded. *Awaits:* impatiently expects, looks for. *Revelation of the sons of God:* the day of general judgment when the faithful will be manifested the sons of God and enter into their glorious inheritance.   **20.** *For* indicates the reason for this "eager longing." *Vanity:* the law of change and corruption, the disorders in the world. *Not by its own will,* etc.: through no fault of its own, but by God, to make the world share in the punishment of man's sin. Cf. Gen. *3, 17* f. *In hope:* as creation shared in the penalty of man's sin, so it shares in his hope of justification.   **21.** The doctrine of a new heaven and a new earth. Cf. Isa. *65, 17;* 2 Pet. *3, 13;* Apoc. *21,* 1.   **22.** Paul's readers would know from Gen. *3, 17* that the state of created things was not what it was intended to be. They also would know that creation's groaning would not go unheeded by God.

8, 23-25: Yearning of Human Beings. 23. *First-fruits:* the life of grace in this world. *Adoption as sons . . . redemption of our body:* in this glorification in the general resurrection at the end of time. 24. *In hope we are saved:* i.e., in the hope of eternal glory. Cf. Eph. *1,* 14; *4,* 30. Hope necessarily has as its object an absent blessing.

8, 26-27: The Holy Spirit Aids Us. 26. *In like manner:* as all creation and we also groan for liberation. *Our weakness:* probably, as in the following sentence, our not knowing *what we should pray for.* Others explain: our inability to formulate prayers acceptable to God. *Unutterable groanings:* there may be reference here to the gift of tongues (*1 Cor. 14*). The Greek can be translated, "unspoken (or silent) groanings." 27. What we cannot express is known to God.

8, 28-30: God's Designs. 28. *Those who love God:* i.e., all true Christians, the "saints" of 27. The Greek reads: "he (God) makes all things work together unto good. . . ." All the circumstances of life whether favorable or unfavorable, God employs as a means of further sanctification of the faithful. *Through his call:* His efficacious call. 29. *Foreknown:* the first step in "His call," implying favor, choice. *Conformed to the image of his Son:* to be like Him in the resurrection and glorification of their bodies. *That he should be,* etc.: in His human nature He was "obedient unto death" and thus merited glory for all mankind. We share in that glory as His brothers, He being the eldest, the *firstborn.* 30. The other steps towards glorification. *He has glorified:* the past tense used of a future event sure of realization. The certitude in this case is on the part of God. He cannot fail on His part; man on his part must co-operate with the grace.

8, 31-34: Unshakable Hope in God. 31. *What shall we say:* i.e., what conclusion shall we draw from what has been said? *Who is against us:* what does it matter how powerful the enemies lined up against us? 32. There is emphasis on *His own Son,* the Son of God by His nature, while we are sons by adoption. 33. No accusation against us is valid, since He declares us just. 34. No one can condemn us, since Jesus Christ, the supreme judge of all men is our advocate.

8, 35-39: Indomitable Love of Christ. **35 f.** *The love of Christ:* i.e., the love Christ has for us; though the common opinion among ancient interpreters considered it the love we have for Christ. *Shall tribulation,* etc.: none of these adversities, all of which Paul himself had experienced, could weaken Christ's love for us, nor should they make us doubt His love. The citation from Ps. *43* refers to the persecutions suffered by God's people of old for their faith **37.** *We overcome:* the Greek term is stronger: "we more than emerge victorious." And this is owing to Christ who, in His love for us, supplies the needed courage and strength. **38 f.** An enumeration of our most formidable opponents: *death,* the most terrifying; *life,* the most alluring; *angels,* the most powerful; *present* or *to come,* all-embracing in time; *height* or *depth,* all-embracing in space. No *creature* can separate us from that profound love that God has for us in Christ Jesus.

5. The Problem of the Rejection of Israel 9, 1 — 11, 36

St. Paul has completed his exposition of the gospel of God active in the world for the sanctification and glorification of all who accept it in faith. He now passes, without formal transition, to the sad condition of his own unbelieving people, the Jews. He considers the problem of Jewish reprobation, first from the side of God *(9, 6-29),* and then according to its causes in the Jews *(9, 30 — 10, 21).* He concludes with the consoling prediction that Jewish unbelief is only temporary, and that in the end the Jews will be converted *(11, 1-36).*

9, 1-5: Paul Grieves for the Jews. An exordium to this section. **1 f.** *In Christ:* in union with Him, and as His minister. *I do not lie:* a negative repetition of the thought for emphasis. *In the Holy Spirit:* St. Thomas: "Because conscience sometimes errs unless set right by the Holy Spirit, St. Paul adds 'in the Spirit'." **3.** *Anathema ... from Christ:* cf. note to text. The statement may have been inspired by Ex. *32,* 31 f. **4 f.** Privileges that make the Jews' unbelief a cause of deeper sadness. *Adoption as sons:* in their election as God's chosen people. *The glory:* the visible evidence of God's presence among His people, more especially the luminous cloud that enveloped the tabernacle (Ex. *40,* 32). *The covenants:* made with Abraham and Moses. *The legislation:* the Mosaic Law. *The worship:* of the true

God by a divinely given ritual. *The promises:* of the Messias. *The Fathers:* cf. Heb. *11. From whom is Christ:* the greatest favor bestowed on them by God. *According to the flesh:* in His human nature, an insinuation there is another nature in Christ. *Who is over all things,* etc.: perhaps one of the most direct and forceful statements of the divinity of Christ in the New Testament. It asserts the reality of His human nature, and the union of the human and divine natures in the one divine personality.

9, 6-13: God's Election Depends on His Free Choice. Paul first considers why it is that the Jews, already so highly privileged, have not recognized Jesus as the Messias.

6 f. *The word:* the Messianic promises made by God to Israel. *Not all Israelites:* there is a spiritual Israel in which these promises have been realized; and not all who are carnal descendants of Abraham belong to it. Abraham had other children, but the promise descended only through Israel (Gen. *21,* 1-12). **8-13.** Further proof is adduced from Genesis that the true posterity is that which inherited the promise by God's design. **9.** Isaac's birth was the result of divine intervention (Gen. *18,* 10.14). **10-13.** In the case of Esau and Jacob the matter is even clearer; for the choice of Jacob was made before the birth of the children (Gen. *25,* 21 f). God's choice was not based on merit; it was a free choice. The citation from Malachias (*1,* 2) shows the ultimate reason for God's choice was His love.

9, 14-18: God Is Not Unjust. **14.** An objection emphatically rejected. **15.** No solution is offered for the proposed difficulty. Rather a text (Ex. *33,* 19) is cited as further evidence of the freedom of God in bestowing His favors. **16.** Faith is God's gift and not the product of man's will or effort. *Who runs:* suggestive of exertion, striving. *God's showing mercy:* the generous love of God extended to whom He wills. This does not set aside the need of man's co-operating with the grace; here Paul is not dealing with that aspect of the doctrine. **17 f.** Cf. notes to text.

9, 19-24: His Power and Glory. **19.** Another difficulty is raised. **20.** There is some insolence in the statement, and Paul answers it by pointing to man's need of modesty in dealing with God: his position is that of the clay in the hands of the potter. *Reply:* The Greek term implies opposition, contradiction. The one point of comparison in the illustration is the potter's liberty to make of the clay any kind of vessel he wishes. **22 f.** Here is applied the principle declared in 18. The sentence is not completed, Paul leaving the apodosis to be supplied from the context. *Wishing:* either "although He wishes," or "because He wishes." In the first case God gives time for repentance to sinners though His justice impels Him to punish; in the second He restrains His hand for the present in order to vindicate His justice more strikingly on the day of judgment. *Ready for destruction:* having prepared themselves by their wickedness for punishment. *Vessels of wrath . . . of mercy:* the sinners and the just.

9, 25-29: Witness of the Old Testament. **25 f.** These words from Osee refer to the promise of Israel's conversion, which Paul regards as a type of the Gentiles' call to faith. Cf. notes to text. **27 f.** The text from Isaias *(10,* 22 f) deals literally with the few to be saved from the judgment God would visit upon Israel. The few foreshadowed the small number of Jews who embraced Christianity. *The word . . . in justice* was that which brought the Assyrian invasion on Juda in punishment for unbelief. **29.** Isa. *1,* 9 is another proof that the situation in the time of Isaias foreshadowed the incredulity of the Jews towards Christianity.

9, 30-33: Jews' Refusal to Believe. The first human reason for the reprobation of the Jews: they refused to believe and hence bear the responsibility and guilt. **30.** *What then,* etc.: what conclusion follows from the above premise? *The Gentiles:* in Greek without the article, hence better "some Gentiles." As a class the Gentiles had not pursued justice. **31.** On the other hand the Jews had pursued justice through the Mosaic Law. But misunderstanding that Law they did not reach the *law of justice,* i.e., justice through the merits of Christ. **32.** Their misunderstanding consisted in thinking the *works* of the Law could merit this justice, whereas it is a gift of God independent of works and conditioned on faith. **33.** This had been foretold by Isaias *(8,* 14; *28,* 16). *Rock:* a term applied to God in the Old Testament: a solid foundation and source of confidence to those who believe; a cause of ruin to unbelievers. Cf. *1 Pet. 2,* 6. It was Christ's death which scandalized the Jews *(1 Cor. 1,* 23; cf. Luke *2,* 34).

10, 1-4: Ignorance of the Justice of God. St. Paul explains more in detail why the Jews failed to recognize Jesus as the Savior. **1.** *Brethren:* The Christian brotherhood, not the Jews. **2.** *Zeal:* the Jews had in fact an ardent desire to vindicate the honor of God. *Not according to knowledge:* the misunderstanding of 9, 31 f. **3.** *Seeking to establish their own:* to acquire justice by their own merits. **4.** *Consummation:* the end to which the Mosaic Law led. Cf. Gal. *3,* 24. The function of the Law was to guide its followers to Christ. When it reached that end, when Christ came, the Law ceased.

10, 5-13: This Justice Comes through Faith. **5.** The allusion is to Lev. *18,* 5. Cf. note to text. *Shall live by it:* in the difficult and limited way explained in 7. **6-8.** The words of Moses (Deut. *30,* 11-14) now spoken by justice personified. It is uncertain whether the text is interpreted in a mystical sense or merely accommodated to the present argument. *The word is near:* the word of faith preached by the Apostles is accessible. *In thy heart:* the soul, the source of thoughts, desires, affections. **9.** True faith must precede the confession. St. Paul reverses this order. **12.** *No distinction,* etc.: Jesus is Lord of all, as (*3,* 29) God is the God of all, not merely of the Jews. This is aimed at Jewish prejudices against the Gentiles. **13.** These words of Joel (*2,* 32), spoken of God, are here applied to Christ.

10, 14-21: Refusal to Believe the Gospel. **14.** The question is rather a rhetorical figure than an objection. **15.** *Unless they are sent:* alludes to the apostolic office. This is clearer in the Greek, for apostle means "one who is sent." *How beautiful,* etc.: the coming of the gospel missionaries is a glad occasion because of their message. **16.** *But all did not obey:* implying the objection: therefore it must not have been sufficiently authenticated. Paul answers by showing from Isaias (*53,* 1) that this unbelief had been foretold. The citation refers to the rejection of the Servant of Yahweh, who is Christ. **17.** *The word:* i.e., the command of Christ to His Apostles; others understand: the word concerning Christ. **18.** They cannot plead that the message was not preached to them. The words from Ps. *18,* 5, describing how creation universally proclaims the glory of God, are used here by accommodation. **19.** Another objection: Israel may not have understood (*known*) the message.

The response is: numerous Gentiles without the previous advantages of Israel understood it; hence again the Jews are without excuse. This was foretold by Moses (Deut. *32*, 21): God, provoked by the worship which Israel gave to idols, would punish them by making them jealous of those *not a nation, a senseless nation,* i.e., of an inferior people not the elect of God. **20.** *Dares to say:* because the declaration was offensive to his contemporaries. Yet the saying (Isa. *65*, 1) is realized in the conversion of the Gentiles. **21.** Isaias also (*65*, 2) in this same connection predicted the unbelief of the Jews.

11, **1-36:** St. Paul has shown that God is completely free in the distribution of His gifts. He has demonstrated that Israel's reprobation was due to their own wilful obstinacy and blindness. But now, lest this contention irritate and repel the Jews, he shows that in their rejection God's gracious purpose and providence were at work.

11, **1-6: A Remnant of the Jews Will Be Saved.** **1.** The question is rhetorical. Three times in the Old Testament God had promised "not to reject His people" (*1* Kgs. *12,* 22; Pss. *93,* 14; *94,* 4 [Greek]). Paul's own call is proof that the promise has been kept. **2.** The same answer, negatively for emphasis. *Whom he foreknew:* the practical knowledge which implies His favor and loving kindness. *The account of Elias:* the story of Elias is found in *3* Kgs. *17 — 4* Kgs. *2;* this event is in *3* Kgs. *19.* **3 f.** The text is cited more as an illustration than a proof of the fact that God reserves a faithful few for Himself. **5 f.** The word *left* is not in the Greek. If there are a few out of Israel faithful to Christ, it is owing to the call of grace and not to the merit of works.

11, **7-10: Witness of the Scriptures.** This is the conclusion to vv.1-6. **7.** *Was seeking:* the Greek reads in the present, "seeks," for Israel is still seeking the justice of God through the Law. **8.** The citation is probably a fusion of Deut. *29,* 4; Isa. *6,* 9; *29,* 10. It describes the way in which God deals with persistent sinners. So it has happened to the Jews. **9 f.** The citation from Ps. *68,* 23 f, is to the same effect. In punishment for their pride and hypocrisy, the light of grace was not granted the Jews; and the very Law which was their privilege and joy became their bondage.

11, 11-12: Israel's Fall the Gentiles' Salvation. **11.** *Stumbled as to fall:* fall in such a way as to be unable to rise. *Offense:* in Greek their "false step" or "failure." How their fall brought about the conversion of the Gentiles is recorded in the Acts of the Apostles, especially in the account of St. Paul's missionary journeys. *May be jealous:* through seeing the divine blessings enjoyed by the Gentiles be made to desire them, and thus be brought to Christ. *Riches:* in this way the world and the Gentiles were enriched with divine blessings. *Their full number:* the final conversion of the Jews. If the evil which befell the Jews was of such utility to the world, of how much greater utility will the good of their conversion be!

11, 13-22: The Gentiles Must Be Humble. **13 f.** Since the major part of his readers would be Gentiles, this seems to be an apology for his having devoted so much attention to the Jews. *As long as:* better, "inasmuch as." *Apostle of the Gentiles:* cf. Acts 9, 15; 22, 21. *I will honor,* etc.: the Greek reads, "I honor." The sense is: I devote myself to it with zeal. The more success he has the sooner will come the *jealousy* spoken of in 11 f. **15.** Cf. 11. *Reconciliation:* before their conversion the Gentiles were enemies of God. *The reception:* receiving them back into the Messianic kingdom. *Life from the dead:* either great spiritual blessings, or more probably, the resurrection to glory at the end of the world. We cannot conclude from this that the end of the world will follow immediately upon the conversion of the Jews.

16. By these two analogies it is shown that in spite of its temporary lapse, Israel continues to be the chosen people of God. *The first handful of dough:* offered to God as a token from each new batch according to Num. *15,* 19-21. *Root . . . branches:* the patriarchs were holy, and their posterity also. **17 f.** The latter analogy suggests another, that of grafting a branch from a wild olive tree onto a cultivated tree. The Jews for centuries were nurtured by God; the Gentiles in a sense left to themselves. The figure is a warning to Gentile converts, inclined to despise all Jews, not to forget the source of their Christian blessings: they do not support the stem, the stem rather supports them. **19 f.** The objection is true, but the warning in 17 f still holds. In any case their election was an unmerited gift, that of faith. **21.** A further and more serious warning. **22.** Two attributes of God, evident in the case of the Jews and Gentiles, make the warning more pointed. *If thou abidest in goodness:* continue in the faith, doing what is worthy of God's goodness to man (Chrysostom).

11, 23-24: Israel Can Yet Be Saved. Continuing the analogy, Paul asserts a truth implied in the above warning, and in line with his purpose in this chapter. The analogy is not perfect, nor strictly according to nature, but the lesson is made clear by means of it.

11, 25-29: Israel's Final Conversion. Vv.23 f. dealt with the possibility of conversion. Paul now makes the formal prediction that after the conversion of the Gentiles all Israel will be saved. **25.** *I would not,* etc.: what he is about to say is of no slight importance. *This mystery:* this being in the future and known only to God and to those to whom He reveals it. *Wise . . . in conceits:* in their pride condemning the Jews. The mystery consists in the termination of Israel's *partial blindness* when the *full number,* i.e., the Gentiles in general, have entered the Messianic kingdom. **26 f.** *All Israel:* i.e., Israel in general, not necessarily each individual Israelite. Proof of this is adduced from Isaias (*59, 20; 27, 9*). *From Jacob:* i.e., from the Israelites. Neither text indicates the time of fulfillment. **28.** *In view of the gospel:* i.e., to facilitate its spread. *They are enemies:* they resist the Son of God. *In view of the divine choice:* the election of Israel as God's chosen people. *The fathers:* the patriarchs. *The gifts:* enumerated in *9, 4*.

11, 30-32: Ultimate Triumph of God's Mercy. **30.** *Did not believe:* the Greek reads, "Did not obey." St. Paul has in mind their condition as described in *1,* 18-32. Their emergence from that state is described as "obtaining mercy." **31.** *Have not now believed:* in the Greek, "have not now obeyed." *That they too may obtain mercy:* God's purpose in permitting both Jews and Gentiles to lapse into a state of rebellion and sinfulness was to manifest His mercy. *Mercy:* having as object the helpless, the miserable. **32.** *God has shut up,* etc.: has imprisoned all in their own disobedience (as the Greek reads). In this state they are helpless.

11, 33-36: God's Ways Unsearchable. **33.** The Greek reads: "and of the wisdom." *Depth:* immensity, the unfathomable profundity of these three divine attributes. *Riches:* the divine treasures of goodness and grace. *Wisdom:* the infinite prudence of God in disposing and directing all things. *Knowledge:* the knowledge of all things. *Judgments:* His merciful decrees dispensing grace, and His condemnation and punishment of men for their disobedience. *Ways:* His manner of dealing with His creatures.   **34.** The citation from Isaias *(40, 13)* repeats the thought of 33.   **35.** Job *41, 2:* God being infinitely rich cannot be enriched by anyone.   **36.** All things derive *from him,* they continue in existence *through him,* they are destined and ordered *unto him.*

II: MORAL

THE DUTIES OF CHRISTIANS *12, 1 — 15, 13*

12, 1-2: Conclusion of the Foregoing.   1. *Therefore* connects what St. Paul is about to say with the entire foregoing theme. *I exhort:* a gracious term, but implying authority. *By the mercy:* as abundantly proved in the Epistle so far. *Your bodies . . . a sacrifice:* the body is the victim, in this figure, the soul the priest who immolates it. This is accomplished by putting to death the sinful inclinations of the body, and by directing it towards good works. This is *spiritual service,* spiritual because deriving its value from the mental disposition accompanying it. Cf. Pss. *39,* 7-9; *50,* 18 f; Prov. *16,* 5; Ecclus. *35,* 2-5.   **2.** *Be transformed:* by an inner and permanent change. This is effected by a renewing of the mind, by rejecting the principles of the world and by adopting those of Christ. *What is the good,* etc.: in the Greek these three adjectives are nouns: what is morally good, what is acceptable, what is perfect. There is an intentional gradation in the enumeration. All these make up the *will of God.*

12, 3-8: Humility and Concord.   3. *By the grace:* the authority of his apostolic office. Cf. note to text.   **4 f.** On the analogy of the human body St. Paul deals here with the unity and variety of function of the members in Christ's Mystical Body, the Church. Cf. *1* Cor. *12,* 12-31.   **6-8.** First, the variety of functions. These are all gifts or graces bestowed by God on individuals more for the interests of the Church than for the advantage of those receiving them. *Prophecy:* an insight into truths otherwise hidden from human reason; its purpose is edification *(1* Cor. *14,* 1 ff). *According to the proportion of faith:* that the faith may be confirmed by it, not against the faith (St. Thomas). *Ministry:* including various works. *In ministering:* do so assiduously and disinterestedly. *He who teaches:* one gifted as a catechist. *In teaching:* with zeal. *He who exhorts:* one who brings consolation, encouragement, guidance. *In exhorting:* to be content and faithful. *He who gives:* one prompted to almsgiving. *In simplicity:* out of a pure motive. *He who presides:* one with the gift of ruling. *In carefulness:* in diligence. *He who shows mercy:* one gifted to help the afflicted. *In cheerfulness:* with joy of heart and affability.

12, 9-21: Fraternal Charity. The principle of unity in the Mystical Body.   **9.** *Without pretense;* sincere, and yet reasonable, for it must *hate what is evil* but *hold to what is good* in a person.   **10.** *Fraternal charity:* a strong and tender family affection. *Anticipating,* etc.: each regarding the other as more deserving (cf. Phil. 2, 3).   **11 f.** The true Christian must be *fervent,* and from this springs the joy of *hope,* which in turn begets patience. This is all sustained by *persevering in prayer.*   **13.** *The saints:* fellow Christians. *Share the needs:* moved to sympathy with those in need and disposed to assist them. *Hospitality:* a prominent form of charity among Christians.   **14.** *Bless:* to wish another well.   **16.** *Of one mind:* the consciousness of oneness with all Christians, the lowly as well as the rich and powerful.   **17.** This precept includes all men, not only their fellow Christians.   **18.** Peace involves more than one party. The Christian, for his part, is instructed to maintain peace.   **20.** By showing kindness to an enemy he is made ashamed and induced to repent of his attitude.

13, 1-7: Obedience. St. Paul here takes up the obedience of Christians to civil authority.   **1.** *Higher:* said of any civil official in relation to private individuals. *No authority except from God:* i.e., legitimate authority exercised for the common good in accordance with the laws of God and man's inalienable rights. Society derives from the nature of man, and hence from God the author of that nature. Paul is not here concerned with illegitimate civil rulers.   **2.** *Condemnation:* pronounced by civil authority but ratified by God.   **3.** *A terror:* a cause of fear.   **4.** *The sword:* the symbol of power to punish. Again in this statement Paul has in view legitimate government.   **5.** *For conscience' sake:* because the authority is derived from God.

6. *This is also why:* i.e., for conscience' sake. *Unto this very end:* the gathering of taxes for the maintenance of the government. **7.** *To all men:* in the context, to all who are in a position of authority. *Tribute:* levies on real and personal property. *Taxes:* import and export duties.

13, 8-10: Charity a Social Duty. **8 f.** The obligation of fraternal love is a debt that can never be fully paid. *The Law:* the ethical element of the Mosaic Law, as regards our neighbor. The order in which Paul enumerates the commandments is not that of Ex. *20,* 13 f; Deut. *5,* 17 f. But ancient manuscripts also vary in this regard. The Clementine Vulgate, with some support in the codices, adds: "Thou shalt not bear false witness." **10.** *Love* excludes any desire to injure another and impels one to fulfill his obligations to another. Thus it reaches the negative and positive purpose of the Law.

13, 11-14: The Spirit of Christ. **11.** *And this do:* i.e., the content of the exhortation beginning at *12,* 1. *The time:* the present life, that in which man works out his salvation. *Sleep:* the state of sin and indifference to the concerns of salvation. *Salvation is nearer,* etc.: we are nearer to eternal salvation than at the time of our conversion. **12.** *The night:* the present life; *the day:* eternal life. *Works of darkness:* sinful deeds; *armor of light:* virtue, especially faith, hope and charity (cf. Eph. *6,* 13-17; *1* Thess. *5, 8*). **13.** *As in the day:* when one's actions are observable. **14.** *Put on the Lord,* etc.: to imitate Christ, to live the life of Christ. *As for the flesh,* etc.: this does not forbid reasonable care of the body, but rather such a care of it that leads to sinful desires. This passage led to the conversion of St. Augustine *(Confessions,* viii, 12).

14, 1 — *15,* 13: This concluding exhortation is a particular application of the precept of charity. It deals with the "weak" and the "strong." The "weak" were a small group that abstained from meat, apparently drank no wine, and regarded certain days to be kept holy by not working, by fasting, and by religious services. They considered this their conscientious duty, yet were so weak that they might be led by the example of others to violate their convictions. Their "weakness in faith" was an unenlightened faith and a poorly informed conscience. The "strong" were those who had no scruples in regard to food and drink, no prejudices about certain days being more sacred than others. The "weak" were inclined to condemn the "strong"; the "strong" were inclined to despise the "weak."

14, 1-11: Mutual Forbearance. **1.** *Receive:* to take to oneself with brotherly affection. *Without disputes,* etc.: more probably "without passing judgment on their opinions." **2.** *One believes:* i.e., has the firm conviction. *Let him eat:* the Greek reads simply "eats." **3.** *God has received him:* i.e., has admitted him to the Church without imposing on him these restrictions. **4.** Judgment belongs to our Lord. *But he will stand:* implies an approval of the more liberal of these parties. **5.** *Let everyone be convinced,* etc.: what is important in this difference of opinion is that each should act with a certain conscience. **6.** Each party acts in good faith, thinking by his actions to please the Lord. **7 f.** The life of the true Christian, in all its activities, is dedicated to the Lord; the last act of his life, his death, is no exception to the rule. **9.** This is but following the purpose of the Redemption. Cf. Phil. *2, 9* f. **10.** Since we all belong to God, it is His and not our place to exercise judgment. To Christ the Father has committed all judgment (John *5, 22*). *Judgment seat of God:* the Clementine Vulgate has "judgment seat of Christ." But there is no difference, since Christ is God. Cf. "tribunal of Christ" in *2* Cor. *5,* 10. **11.** The citation made freely from Isa. *45,* 23 f deals rather with the universal sovereignty of God, but the sovereign is the supreme judge.

14, 12-23: Charity and Peace. **12.** *For himself:* and not for the actions of others. **13.** *Judge:* used in two senses: first, to condemn another's action; second, to decide upon a course of action. *Stumbling-block . . . hindrance:* an occasion of another's falling into sin, here by making him do what is against his conscience. **14.** This settles the matter in favor of the "strong." *But to him who regards,* etc.: this is not subjectivism, but allows conscience its true place in the moral responsibility for an act. **15.** *Grieved:* perhaps the remorse of conscience for having done what he considers wrong. *With your food:* one of the points on which the two parties differed. Charity requires that we sacrifice our freedom if its use leads to the spiritual fall of another. **16.** *Our good:* our freedom to eat. **17 f.** There are more essential things to give us concern in the *Kingdom of God,* the Mystical Body, pursuit of which both pleases God and wins the approval of men. **20.** *The work of God:* either the Christian community which is threatened with disruption by this controversy; or, more probably, the "weak" party, "for whom Christ died" (v.15). **21.** *Is offended,* etc.: perhaps not mere synonyms, but progressive stages in scandal that leads to sin. **22.** *Keep it,* etc.: act according to your convictions unless that would give scandal; in that case do not act in the presence of the "weak." *Blessed,* etc.: who does not condemn himself in action by what he approves in conscience. **23.** *Hesitates:* acts with a doubtful conscience. *Not from faith:* not from a conscience enlightened by the faith.

15, 1-6: Self-Denial and Patience. **1.** *Infirmities:* those referred to in the above but inclusive of any imperfection that might require forbearance. *Not to please ourselves:* the prerequisite of forbearance. **2.** The pleasing of others (cf. *1* Cor. *9,* 20-23) is not an end, but a means to their spiritual good. **3.** Ps. *68,* 10 is the plea of the just man that his zeal for the house of God has brought on him the maledictions directed by the wicked against God. The verse is prophetic, at least typically, of Christ, and illustrates His selflessness. **4.** Such lessons are preserved for us in Scripture that, learning patience from them, we may attain that hope of eternal life which is our great consolation and support in trial. **5.** *May . . . God . . . grant:* these are all gifts which God alone can grant. *According to Jesus Christ:* as He taught by word and example. **6.** *You may . . . glorify:* the ultimate reason for the preservation of this unanimity. *The God,* etc.: the preferable construction is: "the God of our Lord and the Father of our Lord"; His God as the author of His humanity, His Father as having begotten Him from eternity.

15, 7-13: Mercy. Paul exhorts that, since all have been the objects of God's mercy, they should in turn show mercy to one another. **7.** *Receive one another:* as in *14,* 1, to treat one another as brothers. *As Christ:* the standard of all Christian conduct. **8 f.** *Minister of the circumcision:* to fulfill the covenant made with the patriarchs, the sign of which was circumcision. *To show God's fidelity:* this fulfillment was a confirmation of God's fidelity to His promise of the Messianic salvation. *Because of his mercy:* the call of the Gentiles was an act of pure mercy on God's part. The texts cited in vv.9-12 predict this relation of the Gentiles to God. In Ps. *17,* 50 David is the speaker. He is a type of Christ, and his prediction is fulfilled in the conversion of the Gentiles.   **10.** *He says:* i.e., God, the author of the Scriptures. The text is found in Deut. *32,* 43 in the Septuagint. Moses is inviting the Gentiles to join with Israel in praising God. The invitation was accepted when Christ came. **11.** The same invitation is found in Ps. *116.* The invitation is prophetic in that it implies acceptance at some future time. **12.** The text from Isaias *(11,* 10) is from the Septuagint. *Jesse:* the father of David, from whom the Messias was descended. **13.** A prayer suggested by the citation in v.12. This brief prayer embodies the main ideas of the entire Epistle. *Joy and peace in believing:* faith gives rise to hope of eternal life in which the Christian finds peace and joy. *Abound in hope:* from the *God of hope,* who is the source and the object of supernatural hope. *And in the power,* etc.: we should read rather: ". . . abound in hope by the power of the Holy Spirit." He, dwelling in the Christians, is the immediate source of hope.

Conclusion *15,* 14 — *16,* 27

Personal Explanations and Greetings

15, **14-21: Apostle of the Gentiles.**   **14.** This repeats the praise of *1,* 8. *Full of love:* in the Greek, "full of goodness," i.e., of true Christian disposition. *Knowledge:* fully acquainted with the teachings of the faith. These are characteristics of the Roman Christians as a group. Hence there is still need that they *admonish one another.*   **15 f.** *Here and there:* by some referred to *boldly* in the sense of "boldly to a certain degree." *Because of the grace:* his divine commission, as immediately described. *Sanctifying:* cf. note to text. The phrase might be rendered: "dealing in priestly fashion with the gospel of God" (Boylan). Though another may have established the church at Rome, in virtue of his mission to the Gentiles, Paul has the right to address them.   **17.** *This boast:* of having prepared the Gentiles as an oblation acceptable to God. *In Christ Jesus:* by whose grace he has accomplished this *work of God.*   **18 f.** Paul further explains his boast. *What Christ has wrought:* he is careful to attribute his entire success to Christ. Cf. 2 Cor. *3,* 5; *10,* 17. *Signs and wonders:* miracles, wonderful to the beholder, and signs of supernatural origin. *Round about:* more probably his circuitous missionary journeys from Jerusalem as far west as Illyricum, a territory he may have visited on his second journey, though there is no other record of this. *Completed,* etc.: in the sense that he had preached in the principal cities of this vast region. **20 f.** Does he not violate this rule set for himself in planning to go to Rome (*1,* 15)? It was his purpose to do no more than visit Rome (*15,* 24), and impart grace (*1,* 11 f) in doing so. The Scripture justification for his policy is found in Isa. 52, 15.

15, 22-29: **St. Paul's Plans.** **23.** *Having no more work:* i.e., having no more scope or opportunity, according to his rule (vv.20 f). *A great desire:* based probably on news of the church at Rome heard from such as Aquila and Priscilla (Acts *18,* 2; *1* Cor. *16,* 19). **24.** *When I set out for Spain:* it was not his intention to remain at Rome. *Sped on my way:* according to the custom, to be accompanied some distance by them. These escorts probably also supplied the needs of the missionaries. *Enjoyed being with you:* or, with the Greek, "after having satisfied my longing for your company." **25-27.** *To minister,* etc.: to deliver the alms gathered in Macedonia and Achaia (cf. *2* Cor. *8,* 1 — *9,* 15). This voyage is recounted in Acts *21.* This helps date the Ep. to the Romans at the close of his third missionary journey. *Their debtors,* etc.: this may be a suggestion that the Romans treat their Jewish-Christian neighbors with kindness. Cf. *11.* The debt of the Gentiles to the Jews was that from the latter came Christ and His missionaries. **28.** *Delivered:* the Greek term means "put the seal upon." The meaning probably is the completion of this project on behalf of the poor in Jerusalem. *Proceeds:* more literally "fruit." Cf. *1* Cor. *9,* 11. It is very probable that St. Paul did eventually reach Spain, but the circumstances of the journey are unknown. **29.** *The fullness,* etc.: the generous outpouring of Christ's grace.

15, 30-33: **Request for Prayers.** St. Paul knows that trouble awaits him in Jerusalem (Acts *20,* 22 f). **31.** *The unbelievers:* the Jews who not only had refused to believe, but had made every effort to impede his apostolic labors. *May be acceptable:* even many of the Christian Jews were unfriendly to him because of his teachings relative to the Mosaic Law; hence he is not sure they will receive his alms in the right spirit. **32.** *That I may come,* etc.: he even foresees the possibility of being prevented from going to Rome. **33.** *The God of peace:* a title occasioned by the trouble he anticipated in Jerusalem.

16, 1-2: Commendation of Phœbe. A Gentile convert, from her name and the designation *sister,* a deaconess, since she was *in the ministry of the church* (*1* Tim. *5,* 1 ff), Phœbe probably was the bearer of this letter to the Romans. Cenchræ was some eight miles from Corinth, and its seaport on the Ægean sea. It is not known how she assisted Paul and others; probably in a financial way.

16, 3-16: Greetings to Individuals. **3 f.** Prisca (Priscilla) and Aquila had helped Paul and the young churches at Corinth and Ephesus (Acts *18,* 2 ff; *1* Cor. *16,* 19; *2* Tim. *4,* 19). We do not know how they risked their lives for Paul. **5.** *The church . . . in their house:* the early Christians met in the larger houses for religious services. **6-16.** Nothing is known with certainty of those whose names are mentioned here. Some of the names are Greek, others Latin, and that of *Mary* is Hebrew. *Rufus* may be the one mentioned in Mark *15,* 21, since the Second Gospel was written at Rome.

16, 17-20: Warning to Trouble-makers. **17.** *Watch:* to observe closely, to mark. **18.** *Do not serve Christ,* etc.: probably in the sense that they lived more for riches. *Smooth words and flattery* were their means. **19.** *For* connects with "I exhort" of v.17; or we must supply "A word to you is sufficient, for your faith . . ." *Wise:* discreet; *guileless:* uncontaminated by evil. **20.** *Satan:* who inspires these trouble-makers to disturb the peace of the Church. *The grace,* etc.: only this help will enable them to triumph over Satan.

16, 21-24: Greetings from Corinth. **21.** *Timothy:* he was also with St. Paul when he wrote 2 Cor., Phil., *1* and 2 Thess., and Philem.; he was himself the recipient of two letters from St. Paul. *My fellow-laborer:* this may be needed since Timothy was probably not known to the Romans. *Lucius:* probably the one from Cyrene mentioned in Acts *13,* 1. *Jason:* may be Paul's host named in Acts *17,* 5 ff. *Sosipater:* may be the one introduced in Acts *20,* 4. He and Lucius were Macedonians. **22.** *Tertius:* this is the only instance in which the scribe or amanuensis interjects his own greetings in Paul's Epistles. He was probably acquainted with the Romans. **23.** *Gaius:* probably the same as the one given in *1* Cor. *1,* 14. According to the Greek he was host of the whole church, i.e., the Christians probably met in his home. *Erastus:* hardly the one mentioned in Acts *19,* 22; but probably the one of *2* Tim. *4,* 20. *Quartus* is otherwise unknown. **24.** This verse is not in the Greek; it is found in a few Latin codices which recopied it from 20, with the addition of *all.*

16, **25-27: Doxology.** **25.** *My gospel:* this passage makes it evident that St. Paul's teaching in no way differed from that of the other Apostles, for he had as yet not preached to the Romans. What this gospel was is explained by the expression *and the preaching of Jesus Christ:* i.e., concerned with Jesus. *According to the revelation,* etc.: the preaching and gospel dealt with the manifestation of the *mystery* explained in Eph. Cf. Eph. *3,* 6. *Kept in silence:* from all eternity the divine plan, known only to God, was to redeem mankind through Jesus Christ. **26.** *Now:* since the advent of Christ. The mystery had been revealed, or rather foretold, in an obscure way by the Prophets. *Made known,* etc.: by the preaching of the Apostles. *According to the precept,* etc.: all this was according to the decree of God.

This doxology sums up the teaching of the Epistle. *God who is able to strengthen:* the gospel is the power of God *(1,* 16). *My gospel:* Paul's commission to preach to the Gentiles *(15,* 14 ff). *The preaching of Jesus Christ:* cf. *10,* 5-14. *The revelation of the mystery:* the subject of *9-11. Through the writings of the prophets:* cf. *1,* 2; also his constant appeal to the Old Testament in this Epistle. *According to the precept,* etc.: cf. *10,* 15, and *1,* 1. *Obedience to faith:* cf. *1,* 5; the aim of all his apostolic effort.

JOSEPH L. LILLY, C.M.

THE FIRST EPISTLE TO THE CORINTHIANS

Introduction

Corinth was already an old city some thousand years before Christ. The embellished Greek city became the leader of opposition against Rome and was destroyed in 146 B.C. In 46-44 B.C. it was rebuilt as Colonia Julia Corinthus by Julius Caesar who peopled it with Italian freedmen and dispossessed Greeks. Augustus added his own name to its already complex title and made the city the capital of Achaia.

Corinth is situated on the narrow isthmus which joins the Peloponnesus with the mainland of Greece and separates the Gulf of Corinth from the Gulf of Saron. This location made the city a natural trade center. The population of the city and its ports at the time of St. Paul has been estimated at six hundred thousand of whom two-thirds were slaves. There were many of the laboring class, but still enough men of leisure to maintain a certain excellence in rhetoric and philosophy. This city of mixed stock retained much of Greek culture, but it was one of the few in Greece to import the Roman games. The numerous temples to Greek, Roman, and Oriental gods did little to check the rampant immorality. Vice formed a part of the popular worship shown to Aphrodite, the Greek Venus. "To live as a Corinthian" was to live in debauchery.

The Church. On his second missionary journey Paul was forced by Jewish opposition to leave Thessalonica and Berœa (Acts *17,* 1-15). He went by boat to Athens. After his poor success there with idle speculators (Acts *17,* 16-34), he passed at once to Corinth (Acts *18,* 1). There he was joined again by Silas and Timothy who had stayed in Macedonia (Acts *17,* 14; *18,* 5). Encouraged by a vision of the Lord who told him "I have many people in this city," Paul labored at Corinth for about two years (Acts *18,* 9-11.18).

Authenticity. There is scarcely any other ancient book the authenticity of which is so firmly established. About 95 A.D. St. Clement of Rome wrote to the Corinthians, "Take up the Epistle of the Blessed Paul the Apostle . . . he wrote in the Spirit to you concerning himself, and Cephas, and Apollos, because even then you formed factions" (*1* Clem. *47*; cf. *1* Cor. *1,* 12). In the first years of the second century St. Ignatius of Antioch refers so frequently to the contents of this letter without directly quoting it that he seems to know the letter by heart. St. Polycarp, a disciple of St. John the Apostle, writes, "Or do we not know that the saints shall judge the world as Paul teaches" (Polycarp to the Philippians *11*, 2; cf. *1* Cor. *6*, 2). St. Irenaeus in Gaul, a disciple of this Polycarp, quotes the Epistle about sixty times, frequently giving Paul's name or his accepted title, "the Apostle." Also in the second century Clement of Alexandria refers to its contents about 150 times and Tertullian of Carthage about four hundred times, both mentioning Paul expressly as the author on more than one occasion.

Occasion and Purpose. On his third missionary journey Paul settled down for three years at Ephesus (Acts *20*, 31). Excepting from November 10 to March 10 when the Mediterranean was generally considered closed because of stormy weather, commercial relations between Corinth and Ephesus were frequent and passengers were often aboard the ships. Members of the house of a lady named Chloe brought word to Ephesus of divisions in Corinth (*1*, 11). Apollos who had continued Paul's preaching at Corinth (Acts *19*, 1) was now back with Paul at Ephesus (*16*, 12). Stephanas, Fortunatus and Achaicus (*16*, 17) had recently brought a letter in which the Corinthians submitted some of their problems to Paul (*7*, 1; *8*, 1). Paul now writes to correct abuses and to answer questions.

Place and Date. It is clear that St. Paul writes from Ephesus for in *16*, 8 we read, "I shall stay on at Ephesus until Pentecost." This also shows that it is towards the end of a stay at Ephesus that he is writing. This would scarcely have been during the brief stay at Ephesus immediately after leaving Corinth (Acts *18*, 19-21) and therefore must refer to the three years' stay on his third missionary journey (Acts *18*, 23 — *19*, 40; *20*, 31). Now Paul left Corinth shortly after Gallio became proconsul of Achaia, which according to profane chronology was probably in the year 51/52 A.D. (Acts *18*, 12-18). Allowing about five years for the "some time longer" that Paul continued at Corinth (Acts *18*, 18), the journey back to Antioch (Acts *18*, 18-22), the "some time" spent at Antioch (Acts *18*, 23), the mission through Galatia and Phrygia (Acts *18*, 23), and the three years at Ephesus, it is about 57 A.D. when Paul now writes. The same approximate date is reached if one calculates forwards from the time when Festus replaced Felix as procurator at Cæsarea, which was between 59 and 60 A.D., after Paul had been captive two years (Acts *24*, 27).

COMMENTARY

INTRODUCTION 1, 1-9

1, 1-3: Greeting. **1.** Paul is a real *apostle,* like the Twelve, called and sent immediately by Christ to witness to His Resurrection. God took the initiative when Paul was minded rather to persecute the Church (Acts *9,* 3 ff). *Sosthenes* was probably the president of the synagogue at Corinth mentioned in Acts *18,* 17. If so, he was converted so as to become *our brother* and is now associated with St. Paul at Ephesus.   **2.** They are sanctified, consecrated, set apart at Baptism by incorporation *in Christ Jesus.* All Christians are called saints by St. Paul (*6,* 1 f; *7,* 14; *14,* 33; Rom. *1,* 7; *8,* 28; etc.), an external distinction which calls for internal holiness. *Their Lord and ours:* the word "Lord" is added here to complete the sense. Paul would emphasize the fact that Jesus Christ is the Lord of all, possibly because some at Corinth consider Christ as in a particular way their own (cf. *1,* 12-13; 2 Cor. *10,* 7). It is possible that "theirs and ours" refers to "every place": *their* place would then be Achaia of which Corinth was the capital, and *ours* the places evangelized by Paul and Sosthenes; or Gentile places as opposed to Jewish places. **3.** Salutation; cp. Rom. *1,* 7. God the Father and Jesus Christ are equally the source of grace.

1, 4-9: The Gifts of God.   **5.** *In all utterance and in all knowledge:* the spiritual gift of knowledge for instruction. The Corinthians received all that they were prepared to receive. St. Paul later speaks of the limitations of their knowledge (*3,* 1 ff). **7.** *The appearance of our Lord Jesus Christ:* this refers to the Second Coming of Christ at the end of the world but can also be applied to His coming to the individual at death in the particular judgment. Christ is pointed out as the center of all things by the fact that "Christ" appears ten times in the first nine verses of this Epistle.

I: PARTY SPIRIT 1, 10 – 4, 21

From the first four chapters of this Epistle it appears that there were party divisions at Corinth. (a) The number of these parties is not certain. All admit that there were parties of Paul, Cephas (Peter) and Apollos. Some claim that these are only instances of an indefinite number of parties. Some authors deny that there was a party of Christ. In *3,* 22 f mention is made only of the parties of Paul, Cephas and Apollos. Clement of Rome later in an epistle to the Corinthians mentions only these three parties (*1 Clem. 4,* 7). Yet the words "I am of Christ" in v.12 appear exactly the same as the slogans of the parties which precede. The words "Has Christ been divided up?" refer to the Mystical Body: there is only one head and there should be only one body. (b) Concerning the nature of these parties, let it be said first that Paul, Peter, and Apollos were not the leaders of these factions. St. Paul disclaims any such responsibility (*1,* 13-16); Peter and Apollos are in no way reproached in this letter. Neither were the factions based on doctrinal differences: no indication of doctrinal differences is found in the letter; Paul wishes Apollos to return to Corinth to continue his preaching (*16,* 12). There may have been some error in Corinth concerning the resurrection (cf. *15*), but the error is not linked up with any of the factions. The divisions seem to be based on personal preference for one Apostle or preacher over another because of his manner and method, or because of personal acquaintances, or because of some association in the reception of the sacraments, etc. These preferences were harmless in themselves, but were the occasion of hurt to Christian unity and were probably leading to a real danger in the so-called party of Christ. *The party of Paul* may have arisen among the defenders of Paul against his critics alluded to in this Epistle. The defenders probably tended to exaggerate his viewpoint on the principle of Christian liberty and the abrogation of the Law. *The party of Apollos* was probably formed among those who preferred his rhetorical style, allegorical interpretation, and metaphysical mind. Apollos followed St. Paul in evangelizing the church at Corinth (Acts *18,* 24-28). *The party of Cephas* probably consisted of a few converts from Judaism who boasted of their acquaintance in Palestine with the head of the Church, and were inclined to attach undue importance to Jewish observances. There is no evidence that St. Peter ever preached in Corinth. The lack of refutation of this party may be due to its insignificance or to respect for the person of Peter. *The party of Christ* may have arisen as a reaction to the other parties. It is believed that some proud men, considering themselves superior, refused to pledge fealty to mere man, and professed themselves to be subordinate to Christ alone. These would easily consider their own fancies as divine inspirations. There is evidence throughout the Epistle of such a self-sufficient group (cf. 2 Cor. *10, 7*).

1, 10-16: Nature of the Division. **10.** *Mind* refers to speculative understanding; *judgment* to practical decision, to disposition of mind. **11.** *The house of Chloe:* probably slaves of a Corinthian woman otherwise unknown. **12.** *Each of you* need not be taken as absolutely universal. *I am of Paul,* etc.: slogans of the various parties. Those who deny the existence of a party of Christ interpret *I am of Christ* as the cry of St. Paul giving a slogan that should be used by all in one Christian party. *Cephas:* St. Paul always uses the Aramaic name of Peter, reminiscent of Peter's call to be the foundation rock of the church. **13.** *Has Christ been divided up?* Division in the Church, the body of Christ, is like a division of Christ Himself. A party of Christ, appropriating Christ to themselves, would also be an attempt at dividing Him. *Baptized in the name of:* literally "immersed into." Baptism makes us Christ's. **14-16.** Paul is glad that he did not personally baptize many Corinthians, since he might thus unwittingly have become the occasion of these coteries. *Crispus* had been president of the synagogue before Sosthenes (Acts *18,* 8.17); *Gaius* or Caius was Paul's host on a later visit to Corinth (Rom. *16,* 23), *Stephanas* was one of Achaia's first converts, now visiting Paul (*16,* 15.17).

1, 17-25: Salvation Not by Wisdom of Words. Some at Corinth found that Paul possessed less wisdom than Apollos. Paul defends his position to keep his authority. Wisdom for the Greeks was applied to the skill of the philosopher, of the artist, and even of the artisan. Some Corinthians wanted more eloquence. Paul found it best to give the Corinthians the doctrine of Christ's salvation without embellishments of oratory for fear that in their case the form would detract from the matter. The conviction of Christian faith rests not on the power of words but on the power of the cross (*1, 17; 2,* 1-5).

True wisdom, as inherited by the Christians from the Old Testament, consists in the knowledge and love of God. Man by abusing his natural intelligence through voluntary ignorance of God and His law loses true wisdom and gives the name to a fraud (cf. Rom. *1,* 18-32). This false wisdom is put in quotation marks in the text. Such "wisdom" could not grasp the wisdom of elementary Christian doctrine, and those Christians who were not yet fully spiritualized were unable to receive the loftier truths of faith.   17. Paul's mission was *not* so much *to baptize* as *to preach*. Christ (John *4*, 2) and Peter (Acts *10*, 48) also usually allowed disciples to perform the baptismal rite.   19. God promised to deliver the Jews from the Assyrians by His own power without Egyptian aid which the leaders were contriving to bring in.   20. Apollos was a wise man with the Greeks, Paul a scribe among the Jews. There were others, but not many worldly wise in the early Church. *Disputant of this world:* one whose intelligence is totally absorbed by things of this world without consideration of the supernatural (cf. Isa. *19*, 12; *33*, 18). 21. The wisdom of God, although manifest in the world, was often not found by the intellect of fallen man (cf. Rom. *1*, 18-23). 22. The Jews looked for a Messias conquering by miraculous power, the Greeks looked for a leader conquering by reason. 25. The least wisdom and power in God is superior to man's best.

1, 26-31: **Their Case an Example.** When the Corinthians were called there were not among them many possessed of what the world esteems, a show of wisdom, noble birth, influence, wealth. These are from God though not beyond natural attainment. The Corinthians can boast only of the supernatural which **does not** come in any way from themselves.

2, 1-5: **Paul's Method of Preaching.** In Christ God bears witness of His love for man.   2. Paul preached Christ, His Messiasship and divinity, His power and virtue, His ministry and Church, but at the center of all His cross. The Corinthians needed this approach for humility and sobriety.   3. Paul had been weakened and disquieted by persecution in Macedonia, indifference at Athens, and the reputation of the Corinthians. 4 f. Faith is not based on external oratorical argumentation but on conviction given through the internal power of the Spirit.

2, 6-16: **True Wisdom.**   6. The wisdom of the spiritually mature touches on more difficult doctrines, as the priesthood of Christ (Heb. *5,* 11 — *6, 3*), and embraces more profound knowledge of truths already learned in the catechism. Religious knowledge can and should keep pace with sound mental development. *The* *rulers of this world:* "world" or "era" implies present time and visible universe. The rulers may be the leaders of thought, such as philosophers, but Paul probably also refers to the devil. The devil has not dominion, but "this world" as opposed to the "world to come" is largely evil. The leader of the forces of evil is the devil who has gained a large following among men of the world. **7.** This wisdom was God's secret, but is now revealed and can be known with His help. **8.** Cf. 6. Here probably political leaders and the devil. Pilate and the Jewish leaders would not have crucified Christ if they had not blinded themselves to His wisdom. The devil would not have instigated the crucifixion had he realized that the death on the cross meant the eventual ruin of all diabolical power. **10-16.** The *spirit* is the soul, the principle of life. It is used of man's intellectual soul. But in the Old Testament it is used particularly of God who is pure intelligence. The Holy Spirit who dwells in the soul of the just gives him a new sharing of the divine life, of the divine mind, a superior intellect and will. Paul usually calls this elevated soul *the spirit*. The one who possesses it is spiritual. The spiritual man, i.e., every Christian, receives at baptism, (a) the Spirit, i.e., the divine indwelling appropriated in the Holy Trinity to the Third Person, and (b) the spirit, i.e., the elevated intellect which knows now by faith and is to see later by the beatific vision.

By contrast the merely natural man is called sensual (14) or carnal (*3*, 1) because his fallen nature so often manifests itself as merely animal life, intent upon things of the body. He has intelligence but does not show it. So also the Christian who does not live according to the spirit, is called sensual and carnal. **14-16.** The eye is not naturally able to perceive infra-red rays. God could give it this power. So also the spiritual man in addition to his natural knowledge is able to know things beyond the powers of a merely natural man. He has more fully the mind of Christ, which is the supreme mind of God ("the Lord").

3, 1-4: They Cannot Receive Full Doctrine. **1.** *Little ones:* spiritually immature. **3.** *Jealousy and strife* are mentioned among "the works of darkness" in Rom. *13*, 13; among the works of the flesh in Gal. *5*, 20.

3, 5-9: The Office of God's Ministers. **5.** *Of him whom:* the Greek text has "servants through whom you have believed." **6.** *Planted . . . watered:* Paul first preached the gospel, sowing the seed of faith (Acts *18,* 1 ff); Apollos later by his preaching nourished the seed (Acts *18,* 27); but God gave faith, grace, and growth in the spirit.   **7.** In comparison to God the preachers are of no account.   **8.** *Are one:* either have the one office of servant, or are united; therefore there should be no division by reason of them (cf. *16,* 12). *Reward:* wages, hence, by good works with grace we can merit. *Labor:* refers to the personal effort; "work" in the following verses refers to the external product.   **9.** *Tillage:* cultivated field for which God and His helpers work. This leads up to 22-23. Paul now changes to the metaphor of a building, developed in the following verses.

3, 10-17: Their Responsibility and Reward.   **10 f.** *Builder:* literally a master-builder. Paul laid the foundation, the doctrine of Christ and Him crucified (2, 2); others built by further doctrinal instruction.   **12.** *Gold, silver, precious stones* represent solid and fruitful doctrines; *wood, hay, straw* help to build and hence are not errors or heresies, but are flimsy materials, such as frivolous teachings, useless stories or light ramblings that have little influence for good.   **13.** *The day of the Lord:* i.e., the last judgment. *The day is to be revealed in fire:* this probably refers to the fire of the final general conflagration (2 Pet. *3,* 7). This suggests the metaphor of final judgment by fire which is developed in the following verses.   **14 f.** *Reward:* since this reward is distinct from salvation it must refer to the special reward for preaching. *Yet so as through fire:* when fire breaks out in the building those who are constructing with perishable materials will be scorched by the burning of their work. Purgatory is not here referred to directly, since this is the last day when purgatory will cease. But by analogy this implies that those who die before the last day will have to suffer for their venial sins before gaining salvation. It is possible that the comparison is suggested because of fire in purgatory, but the fire cannot be proved from this text.   **16 f.** *God dwells* in the Corinthians individually and collectively: God's dwelling is the *temple.* Those who attempt to destroy God's temple (third class of workers) by teaching error will be deprived of eternal salvation.

3, 18-23: Pride Not to Be Taken in Man.   **18 f:** Cf. *1,* 25.   **19.** The worldly wise who are ready for anything (Greek text) for their own gain are often trapped in their devices.   **20.** Ps. *93,* 11 has "thoughts of men." Paul adapts the wording while keeping the sense. **21 f.** All persons, things, and occasions are given us to reach God through Christ. **23.** Christ's human nature is subject to God (*15, 28*).

4, 1-5: Ministers of Gospel Judged by Christ. 4. Paul refers to the work of his apostolate, in which he is not conscious of fault; *not thereby justified* does not therefore directly refer to the justification of the sinner through grace. His apostolate was probably not perfect; only God is an infallible judge. **5.** So also the Corinthians should avoid judging the preachers, which leads to division. They have not sufficient evidence now, but they will have it on the last day when they judge the world with Christ (cf. *6,* 2 f). *Everyone will have his praise:* Paul takes it that all the preachers will merit praise; if there be any blameworthy they will be blamed.

4, 6-21: Corinthians Contrasted with Apostles. Perfect contentment will be found in the Messianic kingdom only in its final stage when Christians will reign with Christ. The Corinthians are so self-satisfied that they seem to expect nothing better than their present state while the Apostles, who would be expected to be admitted first to glory, are still suffering with Christ. The first part of this is ironical. **6.** *Transgressing what is written:* in the Greek this comes immediately after *learn;* literally, that in us you may learn the "not beyond what is written." It is probably proverbial. *What is written* almost invariably refers to the sacred writings. In this case Paul would be referring to Old Testament teaching on humility in such passages as those quoted in *1,* 19.31; *3,* 19 f. **7.** *Who singles thee out?* probably implies the answer "no one"; there is no foundation for the personal advantage claimed in party division. Even if there were it would be, as all else, from God and not an occasion for boasting. **9.** An allusion to the Roman "circus" where men condemned to death had to fight men and beasts for the spectators, some admiring and sympathizing, others scorning and gloating. **13.** *Offscouring:* what is removed in cleaning, similar to refuse. **16.** *As I am of Christ:* probably added from *11,* 1. **17.** Timothy's mission is mentioned in Acts *19,* 21 f. *My ways:* teaching and example. **18 f.** Paul had promised a return visit (Acts *18,* 21), and intends to keep his promise (*16,* 5-8). Some adversaries seem to have boasted that they had so undermined his authority that he would be afraid to return (cf.

2 Cor. *10-13*). Paul will look for the fruits of preaching not vain eloquence.

II: MORAL DISORDERS 5, 1 — 6, 20

1. THE INCESTUOUS MAN 5, 1-13

5, 1-8: Action to Be Taken. **1.** *His father's wife* is used instead of stepmother because of the wording of the prohibition in Lev. *18*, 8; Deut. *22*, 30. If *2* Cor. *7*, 12 refers to this case the father is still living. The wording implies concubinage or attempted marriage. Such marriage was forbidden by Roman law, but the law was not always observed; the impediment was possibly concealed from the authorities. Andocides, Cicero and Caius are indignant at similar marriages which were rare even among the pagans. At the time of Paul most Jews felt that conversion to Judaism broke all former family ties. Possibly some converts to Christianity made the same claim. **5.** Although some authors think that Paul pronounces here a real sentence, it seems more probable that Paul is comparing the action he has already taken in mind with the slowness to act at Corinth. He urges at least excommunication (v.13). *To deliver such a one over to Satan for the destruction of the flesh:* (a) All authors take *flesh* at least in a moral sense. In contrast to spirit, flesh is used of man's evil tendencies (cf. *2*, 10-16) which are to be destroyed by this punishment. Some say that flesh is to be taken also in a physical sense so that the body suffers if this sentence is carried out (cf. Ananias and Sapphira, Acts *5;* Elymas, Acts *13*). (b) *Deliver to Satan* (cf. *1* Tim. *1,* 20) refers principally to excommunication. The offender deprived of the sacraments and other helps of the Church will find the struggle against Satan more difficult and be brought to repentance. Many think that the sentence implies some degree of diabolical possession. Such possession does not directly affect the soul, but can harass the body (Mark *9,* 17.25). The Apostles may have had the power of loosing devils (cf. Mark *5,* 11-13), as they had that of driving them out.

5, 9-13: Punishment by Excommunication. V.9 refers to a lost letter. **10.** *Of this world:* i.e., non-Christians, those "outside" (12). **12.** Bad Christians hurt the Church in pagan eyes because they saw its teaching only in the lives of its members. Avoidance of those leading un-Christian lives showed disapproval and prevented corrupt influence.

2. Lawsuits before Pagans 6, 1-11

6, 1-11: Public Litigation. The Jews had courts of their own, tolerated and sometimes recognized by the Romans. Early Christians may have shared this privilege with them. In any case if the litigants accepted the authority of a Christian court, most cases could be looked upon by the Romans as settlements out of court. **1.** *Unjust:* i.e., "unbelievers" (6), not having the faith that brings supernatural justice; as a group in comparison to the Christians they were unjust. *Saints:* Christians (cf. *1, 2*). **2.** *The saints will judge the world* in union with Christ their Head, the supreme Judge of men and angels. The judgment will be that of Christ, but also of His body, the Church, which judges with Him just as it rises with Him and reigns with Him. **4.** The order here given is clearly ironical, *to shame* them. **V.5** shows that they should select a "wise man" to judge. But even the least Christian has an advantage over the wisest pagan (cf. *2,* 15).

7 f. A lawsuit implies at least a defect of judgment in one of the parties. Since each is apt to be prejudiced in his own case, it would be better to expose oneself to suffer wrong rather than run the risk of doing injustice. **11.** Most of the early Christians were adult converts; some had led a life of sin in paganism. *Washed . . . sanctified . . . justified* are equivalent terms for Baptism which cleanses us from sin, restores justice, and gives us the life of sanctifying grace.

3. The Evil of Immorality 6, 12-20

6, 12-20: Sacredness of the Body. **12.** *All things are lawful for me:* more commonly these words, twice repeated in the verse, are taken as a quoted principle of the Corinthians, and the words following "but" in each case, as Paul's comment on it. In any case it is clear that Paul agrees with the principle, but not with all applications of it. Christian liberty was open to abuse. (1) Things lawful in themselves may not be expedient in certain circumstances, as in the case of eating food sacrificed to idols (*10,* 23 ff). (2) The exercise of freedom may lead to slavery if a good thing, as drink, is abused. (3) God's law in nature, in revelation and in His representatives must be obeyed. St. Augustine expresses Christian freedom: "Love God and do what you will." Paul here discusses only the keeping of freedom of use within the bounds of nature. **13.** The belly was made to receive food, and man may always satisfy with moderation his natural appetite for food for the preservation of his individual life. The belly and food will disappear, in contrast to the body which remains in the resurrection.

But *the body is not for immorality:* literally, "fornication." Paul does not deny that the organs have their proper function and that the appetite is good. The appetite is ultimately for the preservation and propagation of the race. (1) For the preservation of the race it is not necessary for each individual to satisfy his appetite (cf. 7); (2) the propagation of the race calls for education of the children, therefore, the appetite may be satisfied only in marriage. Paul implies this by referring to the abuse by the name of a crime, "fornication." That name is sufficient to indicate the natural malice. Paul develops the malice particular to Christians in this act. Our bodies which will rise and remain after the resurrection are one with Christ because it is only by union with Him that we have the power to rise (cf. *15*). With Him we are one spirit (contrast to flesh), *15,* 45-49. Husband and wife are, in the words of Adam, "in one flesh." This is a permanent moral union, not dependent merely on the use of marriage (Matt. *19,* 6). By the union with a harlot, the Christian, who is one with the body of Christ, is one with a professional sinner. The Greek text has *"is* one body" not *"becomes* one body." **18.** *Every sin . . . is outside the body,* etc.: no sin is entirely outside the body; the brain is used in a sin of pride. Some sins are directly against the body, as drunkenness and suicide. (1) Some say that Paul means that every sin tends to something outside the body, to drink, wealth, honor, etc. But fornication just uses and abuses the body. If it is objected that the fornicator tends to the body of another, they answer that the two are one body. Suicide is ruled out of consideration since it is an unusual sin. (2) St. Thomas holds that in other sins the body is only an instrument, but in fornication it is the object and end. (3) Taken in the context where Paul deals with the special malice of sin in a Christian, the sense seems to be that no sin goes so directly against the dignity of the Christian body, which is one with Christ, as fornication by which man is one with a professional sinner. **19 f.** The Greek reads "Your body is a temple of the Holy Spirit." Any sin against the body is a profanation of a sacred temple. To this is added injustice since we have been bought by Christ to become His slaves, *bought at a great price,* the price of His blood (Acts *20,* 28; *1* Pet. *1,* 18-19).

III: ANSWERS TO QUESTIONS 7, 1 — 11, 1

1. Marriage and Celibacy 7, 1-40

The Corinthians evidently wrote to ask whether it is good for man not to touch woman. Paul fully endorses the principle, but shows great care in the application of it. Some at Corinth seem to have questioned the lawfulness of marriage, considering all sexual intercourse as sinful. Lax morals sometimes lead to such exaggerated reactions. Paul makes application of the principle to the married, to virgins, and to widows.

7, 1-7: Advice to the Married. **2.** *Let each man have his own wife, etc.* That Paul does not make marriage obligatory is clear from 7, 8 and his own example. He is here giving the general rule in opposition to what some Corinthians wanted to make general on the principle that it is good not to touch woman. **3 f.** One has not the right to apply the principle in question independently of the partner, because there is question of a debt in justice. Paul does not deal with possible conflict of rights. **5.** Abstention is permitted on the following conditions: (1) mutual consent, (2) a good motive, such as prayer (cf. 7, 32-35), (3) a time limit (to be determined by the self-control of the weaker party). **6.** *This* (1) is referred by some to v.2, meaning there is no command to marry; (2) others connect it with what immediately precedes, saying that there is no command to *return together* if they have self-control; (3) more probably in view of Paul's dealing with an exaggeration of the good of abstinence, he hastens to add that he is not commanding abstinence, in marriage, even for the best motives. **7.** St. Paul's celibacy is universally recognized in tradition. Celibacy is a higher state when properly motivated (see below). Similarly abstinence in marriage is praiseworthy. But such continence is a special gift of God, and therefore is not to be general, as some of the Corinthians would have had it. *Each one has his own gift from God, etc.:* implies that one who is married may have a special grace superior to spiritually motivated virginity.

7, 8-11: Advice to the Unmarried. **9.** *It is better to marry than to burn:* i.e., with concupiscence that is not controlled.

Paul does not deal here with the higher motives for marriage but with the danger to which some expose themselves by remaining single. Paul supposes that they are free to marry; those separated from their spouse are "to remain unmarried" (v.11); those bound by vow must keep their promise (*1* Tim. *5*, 11 f). **10 f.** Paul reminds those who follow the advice given in v.9 that marriage once completed is a permanent state by the command of Christ (Matt. *5, 32;* etc.). If there is just cause for separation, remarriage is forbidden during the life of the spouse (Rom. *7, 2*).

7, 12-16: Obligation of the Believing Spouse. The interpretation of this *Pauline privilege* is well given in the present practice of the Church. In general the marriages of infidels are held to be indissoluble. For Paul's exception the following conditions are required. (1) Neither party is a Christian (baptized) at the time of the marriage. (2) One party is later baptized, the other remaining unbaptized. (3) The unbaptized party has already "departed" or now "departs," either (a) by actually leaving, for any reason provided it is not unjustly caused by the convert; or (b) although remaining, by refusing to continue the union in peace for religious or moral reasons. (4) The unbaptized party, when possible, is given due opportunity either to become converted or to decide whether or not to continue the union in "peace." When these conditions are realized the baptized party may break the bond and remarry.

12. *For to the rest I say, not the Lord:* some connect these words with what precedes and interpret Paul as saying that the Lord commands concerning the stability of marriage, but Paul speaks concerning *the rest*, i.e., vv.8-9. More probably the sense of *to the rest* is determined by what follows, i.e., to converts married before baptism to one who remains unbaptized. That Paul refers to marriage before baptism is clear from (a) *if any brother has*, not "if any brother takes"; (b) the reference to consent, shows that there is an occasion, viz. baptism, for a renewal of consent. **14.** The sanctity is an external sanctity (cf. *1*, 2). The Christians are the new Chosen People, set apart especially from pagan unbelievers who because of idolatry and sin were considered *unclean* as a class. The Christian party, one body with the unbeliever, effects by the union that external sanctity of the spouse and their children. Infant baptism was at this time exceptional. **15.** *Departs:* the Greek literally means "separates himself." The reference to *peace*, disturbed by the baptism of one, indicates that there is question not only of physical separation, but also of moral separation. *Is not under bondage in such cases* shows that the marriage was real, that the bond may now be broken, that the parties then are free to remarry. **16.** An answer to a possible objection that the newly baptized Christian should remain with an unbaptized spouse in any circumstances to bring about his or her conversion.

7, 17-24: No Change to Be Sought. Baptism is a rebirth; the beginning of a new life. Dissatisfied with their past, some took occasion of the spiritual renewal to change their whole mode of life. The change permitted in the previous section is exceptional. In general, Paul now insists, Christians should continue after Baptism in their former state of life, since these circumstances are relatively unimportant and adaptable to Christian living.

18. Some Jews in the Greek world, ashamed of their origin, used artificial means to remove its physical traces (cf. *1* Macc. *1,* 16). Some Christians may have considered this a necessary break with the past. Other Jewish Christians held that all were obliged to observe the Old Testament initiation (Acts *15,* 5). **21.** *Make use of it rather: it* could refer either to freedom or to slavery. Hence, according to some, as an exception to the advice against change, Paul urges the slaves to become free if possible. But the context rather implies that Paul even in this case would have converts slow to change. The Greek could be translated "but even when thou canst become free," etc. Some masters after the baptism of the household offered liberty to the slaves. The change was not always beneficial. There is question here only of general advice, requiring an examination of circumstances in each case. **22 f.** All Christians are freed by Christ from the yoke of sin, only by becoming completely subject to Christ who bought us with His blood. The comparison taken from slave trade led to our use of the word "redemption." Even in slavery they obey God rather than man (cf. Eph. *6,* 5-9; Col. *3,* 22 – *4,* 1; Philem.).

7, 25-35: The State of Virginity. **25.** *Virgin* is used of both sexes (27 f; Apoc. *14,* 4). Christ recommended perfect chastity (Matt. *19,* 12); Paul, faithfully breathing the spirit of Christ, repeats this advice for those who can take it (Matt. *19,* 12; v.7). **26.** *Present distress:* many non-Catholic authors translate this as "imminent distress" and refer it to Paul's thought of the

7, 28-36 FIRST CORINTHIANS

possible nearness of the end of the world (cf. Luke *21,* 23; Matt. *24,* 6 f. 9. 21). The shortness of time (29) would make change useless. But the word for *present* less commonly means imminent (in *3,* 22 it is opposed to "to come"); the practice recommended should stand on better foundation than possibility; the *distress* at the end of the world shows no advantage to virginity now. The Fathers find in the *present distress* the worries and troubles of this world, which particularly in a divided life tend to impede prayer (32-34).   **28.** *Tribulation of the flesh:* the same as "present distress," v.26.   **29-31.** All things in this world are passing away, its goods, its joys, and its sorrows. Therefore, in using its goods one must keep detached so as not to consider them his last end; and in bearing its sorrows one must keep hope for a better life with God. Even the longest life in this passing world is *short* compared to eternity.   **32-34.** In the married state one cannot plan even spiritual practices independently of one's spouse; and the greater dependence on the world's goods makes detachment more difficult and distractions more frequent. Paul is comparing state to state, not individual to individual. **34.** *Holy in body and in spirit; holy* is to be taken in its original meaning of "set apart." Even the body of the virgin is set apart for the Lord. Such consecration should be accompanied by moral holiness.   **35.** Paul insists that there is no obligation to follow this advice (28); that the vocation is not ordinary is clear from 7, 2.7.

7, 36-38: Duty of Father to Virgin Daughter. This section supposes a custom wherein a father or guardian had an obligation to seek a suitable husband for his daughter or charge. Similarly among Christians an obligation of preservation was felt, when virginity had been agreed upon by father and daughter. The advice to the daughter and her freedom from obligation is treated above; Paul now adds a few words for fathers and guardians. Like any other method of determining a state of life, this one was open to abuse, but in the end probably led to less hardship than arises from immature independence. That tyrannical despotism is not thought of is clear from Paul's insistence, *if so it ought to be done, being under no constraint, but is free to carry out his own will.* The daughter's consent is implied in the Greek text of v.36 "Let them marry."   **36.** *Incurs disgrace:* probably from the danger of seduction when one who had planned permanent virginity is being courted. *Over age:* at that time one over twenty. *Let him do what he will:* i.e., let him do what he has in mind. According to the context it means let him give her in marriage to her suitor. The Greek text implies a period after *he does not sin,* i.e., in giving her in marriage. Then instead of *if she should marry* the Greek reads "let them marry." 37. *He who stands firm in his heart:* i.e., has no cause for fear concerning his charge.

7, 39-40: Widows. 39. *In the Lord:* i.e., in the body of Christ, the Church. This can be taken as a general counsel to marriage within the Church. 40. The reasons for the greater blessedness of those who remain widows are clear from Paul's previous advice to the unmarried.

2. Idol Offerings 8, 1-13

This section really extends from *8, 1 — 11, 1.* Notice that the discussion is resumed in *10, 14.* The apparent digressions form really a part of this central theme: although one may feel free to eat things sacrificed to idols, it is not always well to use one's rights, as illustrated by Paul's foregoing his rights (*9, 1-27*). Eating these foods may be connected with idol worship, and the Christians are no more secure from relapse into idolatry than were the Jews (*10, 1-13*). The background of this problem was complex. A large number of animals were sacrificed daily on the pagan altars in the busy temples at Corinth. Some of the meat was burned for the god, some was taken by the priest for the service, and some was eaten by the votary as communion; the rest was sold for general use. The meat was served on the temple grounds, which became the most frequented restaurants. The markets obtained this meat for a low price so that the poor could have the meat cheap. Even if the Christians would avoid the tables around the temple and refuse to patronize the butchers who handled these meats, they would still be embarrassed in the houses of pagan friends who might be serving food that had been offered to idols. They presented these cases to Paul and sought his advice. At Jerusalem when the Apostles had decided that in general the Gentile Christians were not to be held to the observance of the Mosaic law, a restrictive clause was added forbidding the use of food sacrificed to idols (Acts *15*). But this was a local and temporary measure to spare the feelings of Christians, newly converted from Judaism, in places where these were numerous. The restriction did not seem to apply in Corinth at this time.

8, 1-6: General Principles. **1-3.** Even if some had taken a logical and correct attitude towards the present problem they could not put it into practice without consideration for less favored brethren. Charity, which is one with the love of God, commends us to God more than knowledge. **4-6.** Even though idols are known as gods and lords, there is only one God and one Lord. The gods and lords then are nothing. Something offered to nothing is nothing changed. Therefore food sacrificed to idols is the same as ordinary food.

8, 7-13: Practical Rules. *Food does not commend us to God:* eating in itself is an indifferent act. Apart from the motives for eating it is not a source of merit. But circumstances may make it wrong. A brother, who, through lack of proper knowledge, is still *idol-conscious* through habit of considering an idol some reality, thinks it wrong to eat of food offered to them. When he sees a fellow Christian, who has proper knowledge in this matter, eating without qualms where such food is served *(an idol place)*, the weak brother may be led to do what he thinks is wrong. The *knowledge* of one has become the occasion of sin to the other. **13.** Paul would rather not eat at all than be the occasion of the loss of a soul to the crucified Christ. *Scandalize* is not to be taken merely of a shock in innocent matters, which is often temporary and sometimes hypocritical, but of a real occasion of sin even to an erroneous conscience.

3. PAUL'S RIGHTS AS AN APOSTLE 9, 1-27

9, 1-14: His Claim of Rights. **1.** *Am I not free?:* since Paul is an Apostle, the highest rank in the Christian hierarchy, it would be expected that he too has rights in Christian liberty; yet he does not always use these rights. *Am I not an apostle?:* Paul claims this title in the restricted sense (cf. Gal. *1,* 11 — 2, 10). *Have I not seen Jesus our Lord?:* an Apostle in the strictest sense was one who had received his instructions and commission directly from Christ and was a witness of the Resurrection (cf. Acts *1,* 21 f). Although Paul had not been instructed and sent by Christ during His earthly ministry, he had received his doctrine and mission in visions of the risen Lord (*1* Cor. *15,* 8;

Gal. *1,* 12.15 f). **2.** His work among the Corinthians marks him (as with a *seal*) as a genuine Apostle. **3-14. Paul has a right** to support of the body for his spiritual ministry (4 and 11). His arguments are taken from common practice among men (soldiers, v.7; vine-growers, v.7; shepherds, v.7; farmers, v.10), from the Mosaic Law (vv.9 and 13 referring to Deut. *25,* 4; Num. *18,* 8. 31; Deut. *18,* 1 ff), and from the law of Christ (v.14, referring to Christ's teaching now recorded in Matt. *10,* 10; Luke *10,* 7).

5. *A woman, a sister:* the Greek text has "a sister woman." The sense is much controverted. This much is generally agreed upon. (1) Some of the Apostles (even those of them who were related to Christ, and even Peter) used their right of obtaining support for themselves and also for a woman (whether wife or not) who looked after their material needs and possibly gave some secondary help in the ministry; (2) Paul, in spite of his right, supported himself and saved the Church further expense by not having the help of a woman. Some interpreters think that the "sister woman" was a wife, but the more common opinion is that St. Paul was writing of a Christian woman who ministered as an apostle, but that he did not think of her as a wife.

9, **15-18: Reason for Not Using Rights.** In 12 Paul had said that he had not used his right to support from the Corinthians; cf. *4,* 12: "we toil working with our own hands." At Corinth he had worked at tent-making (Acts *18,* 3). Some punctuate 17 f to read "If I do this willingly, I have a reward; but if, without willing it, I have a stewardship entrusted to me, what then is my reward?" The comparison is clear: if one takes up work of his own accord he gets a salary, but if service is imposed upon him, as upon a slave, he gets no pay but fulfills his trust faithfully in self-interest since even his life may be at stake. Paul thinks of his days as persecutor of the Church, and of his miraculous call. In his humility he forgets his promptness to obey the call and his generous sacrifices in a wide and fruitful ministry. All this for him is just the work of God, he has no title to glory nor to the special reward of a preacher (cf. *3,* 14 f). He seeks the extra glory and reward by working without charge.

9, **19-23: Paul Is All to All.** **19.** *Slave:* the figure of *9,* 17 is continued—full service without compensation. *The more:* from the Greek and the Latin masculine, it is clear that this refers to individuals, not to greater rewards.   **20-22.** Paul makes as many concessions as possible to customs, and prejudices at the sacrifice of his own preferences, though without sacrificing Christian freedom from the Mosaic Law (21a), nor Christian obligation to the law of Christ and God (21b).

9, 24-27: He Makes Sure His Reward. This serves as a link with what follows: Paul is not self-reliant for the obtaining of even the essential reward, whereas some of the Corinthians appear over-confident.   **24.** *So run:* there are many crowns in the course to heaven, but we should give all our effort as if there were but one.   **25.** *From all things:* i.e., from all that would make them less fit for victory. *A perishable crown:* i.e., a garland of leaves and passing fame.   **26.** *Not as without a purpose:* one's course should always be directed towards the goal or finishing line, God and His heavenly home. By mortal sin we run in the opposite direction; by venial sin we divert to the side.   **26 f.** Not as boxers swinging wildly, we should strike well what is to be subjected—our lower selves. *Chastise:* some Greek MSS read "subject." *Lest perhaps after preaching to others I myself should be rejected:* allusion is probably made to the herald who announced the rules for combat and would naturally be expected himself to abide by the requirements.

4. AGAINST OVERCONFIDENCE 10, 1-13

10, 1-11: Warning from Old Testament. St. Paul shows that although the Jews were the objects of God's special favor through miracles which were types of the supernatural gifts given to Christians, some fell away. Christians should take heed lest they fall. Note that in comparisons the illustration is sometimes described in words that apply strictly only to what is being illustrated; so here *baptized* and *spiritual.*
1. *Under the cloud:* the Jews were guided by a cloud (Ex. *13,* 21; *14,* 19). *Passed through the sea:* through the waters of the Red Sea the Jews passed safely from the bondage of Pharaoh to the promised land, as Christians pass through the waters of Baptism from the bondage of sin to the kingdom of God.   **2.** *In Moses:* these miracles caused the Jews to identify themselves with Moses as their leader, as Baptism identifies us with Christ who is God.   **3.** The manna (Ex. *16,* 15) and the water from the rock (Ex. *17,* 6; Num. *20,* 11) fed the bodies of the Jews in

the journey to the home God was giving them, as the Eucharist nourishes our souls on the way to our heavenly home. **4.** *The spiritual rock:* the real source of these favors was not the material rock (for instance) but the spiritual rock, the Christ pre-existent as the Second Person of the Blessed Trinity. God is called Rock in the Hebrew text of Deut. *32,* 4.15-18; Isa. *17,* 10; etc., as the solid support of His people. **5.** Of those over twenty at the time of leaving Egypt only Josue and Caleb lived to enter the Promised Land (Num. *14,* 29 f; *26,* 63-65). **7.** *Sat down to eat and drink and rose up to play:* refers to Ex. *32,* 6 when some of the Israelites ate food that had been sacrificed to the golden calf and engaged in religious dances before it. **8.** *Fornication:* probably refers to the sins with the daughters of Moab when twenty-four thousand were killed (Num. *25,* 1.9). The difference in number may arise from a fault in copying, common in numbers. Some think Paul is still speaking of the idolatry (often called fornication) referred to in the previous verse. The Vulgate gives the number slain on that occasion as twenty-three thousand; but the more probable reading in Ex. *32,* 28 is three thousand. **9.** *Tempt Christ:* they provoked Christ (cf. *10,* 4) by lack of confidence, and were bitten by the "fiery" (poisonous) serpents (Num. *21,* 5 f). **10.** Of the numerous occasions when the Jews *murmured* (cf. Heb. *3,* 10.17) this probably refers to Num. *16* when Core, Dathan, and Abiron rebelled and were swallowed up by the earth. According to Wisd. *18,* 25 this punishment was effected by a *destroyer,* i.e., by an angel, as in the slaying of the firstborn (Ex. *12,* 29). **11.** *Final age of the world:* i.e., the Messianic period, inaugurated here now and consummated eternally in heaven.

10, **12-13: Application.** Temptation is never beyond our strength of resistance; God always gives sufficient grace.

5. DISCUSSION OF IDOL OFFERINGS RESUMED *10,* 14 — *11,* 1

10, **14-22a: The Table of the Lord.** Paul refers here to those cases where the eating of food offered to idols is directly connected with idolatrous worship. **14-16.** This shows clearly that the Christians have a sacrificial communion, by comparison with the sacrifice of the Jews (v.18) and with that of the pagans (v.19); and that this communion is the body and blood of Christ. *We bless:* Christians consecrate through and in union with the priest. **17 f.** Eating a common food which becomes our body and blood gives us to some extent one common body and blood. So also the sacrifice of the Jews eaten in part by the votary and consumed in part at least symbolically (as through burning) by God brought union with God. *Israel according to the flesh* refers to the unconverted Jews as opposed to the Christians who have inherited the things promised to the Jews. **19 f.** Cf. *8,* 4-6. **20.** *They sacrifice to devils and not to God:* (Deut. *32,* 17) worship of man-made gods was itself evil and accompanied with many evil practices, and the devil is the leader of the forces of evil. Thus the pagans, who tried in sacrificial banquets to unite themselves with their gods, became one with the devil. **21.** *Table:* i.e., altar (Council of Trent; Ez. *34,* 16; etc.). **22.** *Provoking to jealousy:* by diverting devotion. *Are we stronger than he,* to oppose Him?

10, **22b-30: Practical Directions.** **22b.** Cf. *6,* 12. **23 f.** No man can *seek his own interests* without considering *those of his neighbor.* **25-27.** It is not necessary to inquire at the market or at the home of a pagan whether or not the food has been offered to idols, because God has made everything good. **28 f.** If someone mentions that the food has been sacrificed he implies either that the meal is being considered as sacrificial or that he believes it wrong for a Christian to eat it. **29b.** A conscience which judges rightly will not be condemned, just because someone may erroneously consider the action wrong. **30.** A sincere grace before and after meals implies that the one who eats does not consider his eating as displeasing to God.

10, **31 — *11,* 1: Give No Offense.**   **31.** An indifferent action is meritorious when done with at least a virtual intention of glorifying God.

IV: RELIGIOUS GATHERINGS *11,* 2-34

1. THE HEADDRESS OF WOMEN *11,* 2-16

11, **2-16: Rules for Men and Women.** From these rules it is evident that it was the custom at this time to keep the head uncovered as a sign of authority, and to use a veil as a sign of subjection. Among the Greeks women had their heads and foreheads veiled in public. Some seemed to wish to discard this usage on the plea of Christian liberty. A certain amount of conservativeness in manners is usually a help to right living. Paul sets first the right order of authority: God, Christ, man, woman (3). "Man" and "woman" are taken in a collective sense; and the order is based upon creation (8 f). Individual exceptions in certain circumstances are not barred. Paul shows the fitness of the existing customs. There are two principal reasons, one taken from nature, the other from creation. (a) Nature: woman's hair grows more abundantly and is more permanent than man's hair. The woman is ashamed to lose this natural feminine adornment by having her head shaved as many men did. Nature provides a covering for her head which gives feminine attraction, so it is appropriate that a veil should symbolize feminine subjection. If she removes the veil to be like man, let her be shaved like men (5-6.15). Men on the other hand are ashamed if their hair makes them look effeminate (14). (b) Creation: man was created to the image and likeness of God (Gen. *1,* 27) and woman was made from man. Therefore man directly reflects God. The intellect and will which make man like to God are reflected most in the head. Therefore it is appropriate that man keep his head uncovered so as to reflect God's glory (7-10; cf. *2 Cor. 3,* 18). Lest it be thought that Paul is slighting woman, he adds 11 f. Eve came from God through Adam, man comes from God through his mother.

Paul seems to suspect that some might find his reasoning too subtle and *be disposed to be contentious,* so he asserts the authority of custom in the various Christian communities (16). **4 f.** *Prophesying:* cf. 14. The current practice of woman prophesying in public is there forbidden entirely. **10.** *Because of the angels:* who witnessed the order established at creation and now assist at the Christian religious service since Christ is also the head of angels. The order of subjection and service should be "on earth as it is in heaven."

2. The Eucharist *11,* 17-34

***11,* 17-22: An Abuse.** In the early Church the Eucharistic service or Mass was connected with the agape or love-feast. This was done in imitation of Christ's last supper and in anticipation of the heavenly banquet. Each brought food according to his means—an occasion for almsgiving. The free mingling of all classes developed humility and the union of charity. The climax of the meal was the Holy Communion, Christ's gift of loving union with Him and with one another (cf. *10,* 16 f). This agape was open to abuse and had to be abandoned early. In spite of its sacred purpose (a) distinct groups formed (18 f); (b) instead of having the food in common some ate their own even before the assembly, thus embarrassing the poor (21 f.33 f); (c) some even drank to excess (21). **19.** Factions and defections are permitted by Providence to prove the virtue of the sincere.

11, 23-34: Institution of the Eucharist. **23.** The Eucharistic teaching here, as now held in the Church, is sufficiently clear. See note to text. *Betrayed:* the sins of the Corinthians in connection with the Eucharist recall Judas who ate with Christ while planning to help in the death of Christ (27). **26.** *Proclaim the death of the Lord until He comes:* the Eucharist commemorates and continues the sacrifice of the cross until the coming of Christ at the end of the world. **28.** *Prove himself:* each should ask himself, as the Apostles inquired among themselves, whether he is the traitor. **29.** *Without distinguishing the body:* i.e., from ordinary food such as was taken previously at the agape. **30-32.** Physical punishments of sickness and death for sin were to be expected in an age when miracles also restored life and health. Thus Ananias and Sapphira (Acts *5,* 5.10); the incestuous man (*1* Cor. *5,* 5). Since those who *are judged* guilty are *chastised by the Lord,* so that they *may not be condemned,* it is implied that even those who die as a temporal punishment are given opportunity to repent and avoid eternal punishment.

V: THE SPIRITUAL GIFTS *12,* 1 – *14,* 40

The essential gift of the Holy Spirit is what we call grace, i.e., a sharing in the divine life. God's life is the knowledge and love of Himself. Now we know and love God indirectly in His creatures, especially in man who is created to His image and likeness; but this grace gives us now the power of knowing and loving God directly. This gift which can in some way be identified with the love of our neighbor is described in *13.* But besides this internal gift of the Spirit, and as a proof of its bestowal, the Holy Spirit also gave external miraculous favors which form the central theme of this section. These were promised by Christ (Mark *16,* 17 f) and first distributed on Pentecost day (Acts *2*). These are unmerited, supernatural and passing gifts, granted for the general good of the Church. (a) They are unmerited, given by the *Spirit, who divides to everyone according as he will* (*12,* 11), and found possibly even in one who *has not charity* which is necessary for merit (*13*). (b) They are passing inasmuch as God may remove the special favor through no fault of ours whereas sanctifying grace is habitual. It seems from what follows that one or more of these gifts was possessed with a certain stability so that one would be considered a prophet, another possessed of various tongues, etc. Yet there could be change (*14,* 1.13). The gifts are passing also inasmuch as they have been more or less common according to the needs of the Church; most extensive in the first days, rare today. Origen notices their decline in his day (early 3rd cent.). (c) They are granted for the general good of the Church, the Body of Christ (*12,* 4-31; Eph. *4,* 12). Paul, therefore, values them in proportion to their utility to the Christian community (*14*). Although they do not directly bring special benefit to the recipient spiritually, nevertheless they can be a source of spiritual profit to the possessor (*14,* 4). (d) These gifts are supernatural in their cause. The effect in some cases is clearly beyond human powers. In other cases while the effect may not be above the ability of man, it does elevate the particular individual above his natural limitations. As in other cases of the supernatural, so also here the gift is frequently given to one who already has an extraordinary natural aptitude.

It is not possible or necessary for us now to make a clear distinction between all of the various gifts. Some names may be practically synonymous. One person may possess simultaneously several gifts. An attempt is made here to group the gifts in order of importance.

(1) Gifts of knowledge for Christian instruction. (a) *Apostle:* (*12,* 28; Eph. *4,* 11; Rom. *16,* 7); a missionary; the title is not limited to the Twelve and Paul who have it in a restricted sense. (b) *Prophet:* (*12,* 28) one who speaks for God and with His authority. Among the revelations they received were sometimes future events (Acts *11,* 27 f; *21,* 10 ff). They exhorted and strengthened (Acts *15,* 32), edified, encouraged and consoled (*14,* 3), and read men's hearts (*14,* 24 f). This gift was given to some extent to women (*11,* 5; Acts *21,* 9). (c) *Evangelist:* (distinct from the sacred writers) probably one appointed to strengthen new churches, but not to found them (Acts *21,* 8; 2 Tim. *4,* 5). (d) *Teacher:* probably an inspired catechist (Rom. *12,* 7; Eph.

4, 11; *1* Tim. *4*, 13.16). (e) *Exhorter:* a preacher with special powers of appeal for good living (Rom. *12*, 8; *1* Tim. *4*, 13; Acts *4*, 36). (f) *Utterance of wisdom:* for explaining high religious speculation (*12*, 8). (g) *Utterance of knowledge:* for explaining elementary truths by human analogy (*12*, 8). These last two gifts were probably possessed in varying degrees by those mentioned before, e.g., the prophet probably had the utterance of wisdom, and the teacher the utterance of knowledge.

(2) Gifts of organization. (a) *Pastor:* grace for the office of presbyter-bishop "to rule the Church" (Eph. *4*, 11; Acts *20*, 28). This gift is not to be confused with the power conferred only by Sacred Orders. (b) *President:* a head of a group of presbyter-bishops, or an organizer of charities or other works (Rom. *12*, 8; *1* Thess. *5*, 12; *1* Tim. *5*, 17). (c) *Ministry:* helper in Church organization (*16*, 15; Rom. *12*, 7).

(3) Gifts of miracles. (a) *Faith:* a supernatural instinct, founded on the theological virtue of faith, to know that God in a given case will miraculously manifest His power (*12*, 9; *13*, 2; Matt. *17*, 19). (b) *The working of miracles:* the general gift of miracles (*12*, 10.28). (c) *The gift of healing:* limited to cures (*12*, 9.30).

(4) Gifts of service. (a) *Almsgiver:* one who received supernatural tact in service of the poor with simplicity of intention (Rom. *12*, 8). (b) *Showers of mercy:* who assisted the unfortunate, sick, and prisoners "with cheerfulness" (Rom. *12*, 8). (c) *Services of help* (*12*, 28): similar to the two previous gifts. (d) *Administrator:* probably an organizer of charities (*12*, 28).

(5) Gifts connected with prayer. To show the universality of the Christian faith and its ultimate purpose of praising God, the power to pray in tongues unknown naturally was frequently given as a sign of the reception of grace at baptism. Some had the gift of *kinds of tongues,* others the *interpretation* of these tongues, others both (*14;* Acts *2*, 5-11; *10*, 46; etc.).

1. Their Distribution *12*, 1-31

12, 1-26: A Principle of Discrimination. **1 f.** Service of dumb idols is irrational, but Christian service must be intelligent. **3.** Not all that is marvelous is supernatural. There are various tests of the supernatural but a basic one is that of faith. If a man directly curses Jesus, or does it indirectly by denying His doctrine as false, he has not the gift of God; but one who sincerely confesses the divinity of Jesus and therefore holds all

His teachings as true may have his gift by union with the Holy Spirit. *Anathema* is a Greek word meaning "set up," and was applied to the offerings hung up in the temples (cf. Luke *21*, 5). In the Septuagint it was used of persons and things given over to destruction by God and for God. In the New Testament it is used of the objects of divine malediction.

12-26: While St. Paul develops this doctrine of the organic nature of Christian society in connection with the supernatural gifts, it applies with equal force to the various natural gifts which God has given for the benefit of all. As the hand does not seek to harm the eye in the body, so in society there is no place for unethical competition. But as the hand protects the eye and the eye warns the hand of harm, so in society, which is the body of Christ, there should be a holy rivalry to make the best use of one's talents for the benefit of all. Although the talents are different, all are necessary; and each should be satisfied to use whatever God has given him, be it small or great, to the best advantage of all. **22.** The heart and lungs are weak in themselves, yet most necessary; so also a member of society who appears weak may be filling an important role. **23 f.** If a part of the body is deformed or ugly it is not needlessly exposed but receives greater attention so as to remove or cover the unsightliness, so in the body of Christ charitable care should be given to the less honorable members. **25 f.** Any repression of talents hurts society, development of talents makes it prosper.

***12*, 27-31: Christ's Mystical Body.** **27.** Cf. Eph. *4*, 1-16. **28-31.** This is the application of 12-26 to the role of the supernatural gifts in the Church, the body of Christ. The order of the gifts mentioned here and elsewhere in the New Testament is given in the introduction to this section.

2. A Digression on Charity *13*, 1-13

Charity above all other spiritual gifts, and charity in the use of the spiritual gifts, is the theme of this beautiful hymn. The charity directly spoken of is the love of one's neighbor, but that this is one with knowledge and love of God, our supernatural life, is clear from 12 f. Cf. Matt. *22*, 38 f.

***13*, 1-7: Its Excellence.** **1.** Charity is a *more excellent* gift than the *greater* gifts of which Paul has been speaking (cf. introduction to this section). *Tongues of men and angels:* the gift of tongues was given primarily for the praise of God; hence this deals with the external praise of God in the languages of all men and in the manner of the angels. *As sounding brass:* the Greek omits *as*. Confused noise from a dead instrument is contrasted with the sweet music of the spirit.   2. *Prophecy . . . faith* are the spiritual gifts (cf. introduction to this section). *To remove mountains* is proverbial of doing the impossible.   3. *Distribute goods:* this is the gift of the almsgiver (see above). *Deliver my body to be burned:* instead of *to be burned* many MSS read "that I may glory." If *to be burned* is retained, as it probably should be, it refers to heroic self-sacrifice for some good natural motive. The two acts in this v. are philanthropy but may not be charity, which is one with the love of God. They may help the beneficiaries but they *profit me nothing*.   6. Justice and *truth* are interchangeable: truth corresponds with what is in our mind, or with reality; justice corresponds with what is in God's mind, and with His law. So also injustice (*wickedness*) and falsehood correspond.   7. *Bears with:* the Greek means to keep or to cover. Here it may mean to bear patiently and with self-control; or to cover over with silence, to excuse. Charity *believes* the best until evil is evident, then it *hopes* for reform; but even if the evil remains beyond hope of change, charity endures it with patience.

13, 8-13: Contrast with Other Gifts.   8. The spiritual gifts may be removed in this world through no fault of the possessor, and they will certainly not be found in heaven; but charity remains here as long as there is no grievous sin, and remains everlastingly in heaven.   9-12. Charity enables us to see and love God in our neighbor. This charity will remain but will be changed in heaven. Now we are, with grace, in our spiritual childhood. Even with the highest gifts of knowledge, such as in prophecy, we know God only in part, reflected in His creatures as *through a mirror*. The metal mirrors of those days reflected poorly, *in an obscure manner*. But when grace turns to glory and we are spiritually mature, our charity shall be based on the direct sight of God, *face to face;* we shall know Him directly as He now knows us.

Note that throughout this passage charity is not a blind sentimental attraction but a love of esteem and appreciation that is found in the intellect and will. The Greek word for charity means just this. Grace is a sharing of the life of God, which consists in knowing and loving Himself.

3. THE GIFTS OF TONGUES AND PROPHECY *14*, 1-40

One of the most remarkable of the spiritual gifts was the gift of tongues. It was given to all at the first coming of the Holy Spirit (Acts *2*, 4) and on numerous occasions thereafter (Acts *10*, 46; *19*, 6). (1) The *nature* of the miracle is not that those gifted with tongues spoke their own language and were understood in the various languages of their hearers, but that they spoke one new language or several new languages successively. This is clear from the need of an interpreter (*14*, 6-19). (2) The *purpose* of the miracle was not for preaching but for praising God. (a) At Jerusalem (Acts *2*) they all speak at once, one in one language, another in another, and they speak "of the wonderful works of God" (Acts *2*, 4-13). (b) When Peter speaks there is no evidence of the gift of tongues (Acts *2*, 14 ff). (c) St. Paul here contrasts it with prophecy which is for preaching (cf. *14*, 22-25). (d) It is said to be for speaking to God (*14*, 2.16.28).

The gift of tongues was fascinating and, therefore, much desired. Without depreciating this gift St. Paul urges rather the use of prophecy as being more beneficial for building up the Church.

14, 1-5: Superiority of Prophecy. **2.** *No one understands:* i.e., excepting the few, if any be present, whose language he speaks. *Mysteries* generally means truths once hidden by God but now revealed; here it means that what is spoken is hidden by reason of the language used. **4.** *Edifies himself:* i.e., builds up his spiritual life; prayer always does this even if the language is not understood, provided only the soul is united with God. **5.** If he interprets he leads others also to his prayerful inspiration.

14, 6-19: Tongues Require Interpretation. The point made in the last verse is here developed by comparisons. **13-15.** When the spirit, man's higher self, consisting principally of good will, is united to God there is the essence of prayer; but still greater advantage is derived if to this is added also the natural understanding of what is said. **16.** *Amen* means "so be it." How can one approve of your prayer and make it his own if he does not understand it?

14, 20-25: **Functions of These Gifts.** **21.** Paul gives the sense of Isa. *28,* 11 where the prophet threatens that God will punish the Israelites by permitting them to be conquered by people of a foreign tongue, the Assyrians. The last part of this verse is a summary of Isa. *28,* 13. **22.** This punishment at the hands of foreigners was a sign to the unbelieving Jews, so now also the gift of tongues is a sign to unbelieving heathens. This miraculous sign of the divine character of the Christian faith was no longer needed by the faithful who profited more by the instruction given by the prophets. **23-25.** On the other hand the disorderly use of the gift of tongues might make one who does not know the Church have the impression that it is irrational. The unbeliever as well as the believer will benefit more by the inspired preaching of the prophets.

14, 26-33a: **Practical Directions.** **28.** *Church* refers not to a building but to an assembly. *Speak to himself and God:* to the sole hearers, not that he addresses himself. **29.** *Let the rest act as judges:* to see whether the gift is genuine (*1* Thess. *5,* 20 f). **30-33.** Religious enthusiasm among the pagans carried one temporarily out of his mind so that he had no control over his religious emotions, but the gifts of the Holy Spirit are always under one's control. If they are not under control, they are not genuine.

14, 33b-40: **Order Necessary.** **34.** It is only the public exercise of the gift that is forbidden to women. **35.** They are also forbidden to raise questions and engage in discussion in the public assembly; they can do this through their husbands. **36.** The church at Corinth cannot set up customs and traditions independently of the other members of the one Church of God. **37 f.** No one may appeal to a gift of the Spirit to go contrary to the external authority set up by the Lord in the Apostles and their successors.

VI: THE RESURRECTION *15,* 1-58

Christ's resurrection was difficult for pagans to believe. When the Athenians heard it some began to sneer (Acts *17,* 32); Festus thought Paul mad when he spoke of it (Acts *26,* 24). At Corinth some questioned the fact of our resurrection, others also the resurrection of Christ (12).

15, 1-11: Christ's Resurrection. **1.** *Being saved:* the process is continual from the first grace through perseverance to glory. **4.** *According to the Scriptures:* Christ's death is foretold in Isa. *53,* 4-9; His burial and resurrection in Isa. *53, 9;* Pss. *6, 3; 15,* 10; Jonas *2,* 1 f (cf. Matt. *12,* 40). **5.** *He appeared to Cephas:* i.e., Simon Peter (Luke *24,* 34). **6 f.** In the Gospels there is no specific mention of the apparition to the *five hundred brethren* nor of the one to *James.* The accounts are nowhere complete. The appeal to many living witnesses is forceful. **8.** *Born out of due time* as an apostle, a witness to the Resurrection (cf. *9,* 1). This may refer either to the fact that he was chosen before he was properly prepared, or to the fact that he was selected after Christ's earthly life during which the others received their mission.

15, 12-19: The False Doctrine. Paul sees an inseparable connection between the resurrection of Christ, our resurrection, faith, and the remission of sins. (1) *If the dead do not rise neither has Christ risen:* this is clear if resurrection is denied as a possibility; but Paul probably means that our life is one with Christ's, so that if we do not rise, it is because He has not risen. (2) If Christ did not rise, our *faith is vain,* because the Resurrection is at once an object of faith and the strongest proof for the truth of faith, and the habit of faith is the beginning of life. If the preaching of the Apostles was false in this, their claim to infallibility in the truths of faith is wrong. (3) *If Christ has not risen* . . . you are still in your sins, because the Resurrection is the necessary triumphal complement of the death on the cross; if the resurrection of Christ does not give us eternal life, neither does the death of Christ save us from the death of sin. Then Christians who sacrifice pleasure for future hopes are most *to be pitied.* Cf. Rom. *4,* 25.

15, 20-28: Christ the First-fruits. **20-23.** Just as by being one with Adam, the physical and moral head of the human race, sin began in us and its principal punishment, death, so Christ, our new head, brings grace and life. He as first-fruits has risen, bringing promise of the full harvest—our resurrection. **24-28.** Although Christ has merited the subjection of all evil, not all things will be subjected to Him till the end of the world (Heb. *2,* 8), when the last enemy, death, will be destroyed and proper order restored (cf. *11, 3*).

15, 29-34: Practical Faith. Christians show by their actions that they believe in a future life. (a) **29.** *Baptism for the dead,* received by substitution without sacramental value for a dead catechumen who had been baptized only in desire. (b) **30 f.** Paul's sacrifices show hope of a future life. **32.** *Fought with beasts:* probably not to be taken literally of being thrown to the beasts, because it was forbidden to inflict this punishment on a Roman citizen. Paul was now suffering strong opposition at Ephesus (cf. *16, 9*; Acts *19,* 13-40). **33.** This proverb counteracts the slogan of the worldly given in the previous verse.

15, 35-44a. The Mode of the Resurrection. One reason for objecting to the resurrection of the body was the difficulty of understanding how a corrupted or corruptible body would rise to an incorruptible life (35). Paul does not pretend to explain exactly the manner of the resurrection, but he shows by comparison how this objection does not disprove the fact. **37 f.** The seed in the ground dies and yet produces a body which, though different, is one with the seed: wheat seed produces only the wheat plant, rye seed only the rye plant, etc. The same applies to various kinds of animal flesh (39). This variety of plants and animals and also of the heavenly bodies (40 f), each with its individual characteristics reflecting the divine glory, points to the omnipotence of God who can surely change our bodies from corruption to incorruption, from an instrument of sin to a sharing in the divine glory (42 f). God can do this in such a way that the *spiritual body* is in some way identical with the *natural body* (44a).

15, 44b-49: The Natural and the Spiritual Body. **45.** Paul again contrasts the first head of the human race, Adam, with the final head, Christ. Using the words of Gen. *2, 7,* Adam became *a soul,* but Christ became *a spirit.* In Paul's terminology the "soul" (like the flesh) represents the natural man in his weakness, "spirit" the man elevated by grace which is the seed of glory (cf. *2,* 10-16). While Adam is *living,* Christ is life-giving. At the Resurrection the spirit of Christ, i.e., His grace, became life-giving, continuously generating His life of grace in us. **47-49.** As we naturally resemble our father who came ultimately from the earth, so also we must resemble our heavenly Father whose life is mirrored by Christ who came from heaven (Matt. *5, 48;*

John *14*, 8 f). **49.** Some MSS give *let us bear* as a statement of fact: "we shall bear."

15, 50-58: Final Glory of the Body. **51.** As this reads it would mean that although all shall rise, not all shall be changed to glory, for the wicked shall rise only for continued punishment. This is scarcely a mystery (a secret of God). The reading in most Greek MSS better fits the context: "We shall not all sleep (die), but we shall all be changed." **52.** This is connected with "we shall all be changed." It shall be done quickly. *Trumpet* signifies metaphorically a signal, or sign, (the voice of Christ, John *5*, 28; the voice of an archangel, *1* Thess. *4*, 16). **55.** *O death, where is thy sting:* some understand this as the sting that causes death, others take it better of the sting by which death hurts us. **56.** Sin and death go together (Rom. *5*, 12) but it is only sin that can hurt us permanently. The Law by pointing out sin without giving strength to resist it, tended to increase sin (Rom. *7*, 13 ff).

Conclusion *16*, 1-24

16, 1-4: The Collection. **1.** The Apostles requested a remembrance of the poor in Jerusalem (Gal. *2*, 10). No particulars of the collection in Galatia are known. The collection is referred to in 2 Cor. *8-9;* Rom. *15*, 26; Acts *24*, 17. **2.** *On the first day of the week:* Sunday was already the Christian sabbath (cf. Acts *20*, 7). Every Sunday, besides contributing to local current needs at the liturgical meeting, the Christians will set aside a certain sum for this special collection. Instead of *whatever he has a mind to* the Greek reads "to whatever extent he prospers" or "whatever he can afford." **4.** *If it is important enough:* Paul urges them to generosity by promising to bear their offering himself if it is sufficiently large. Note that travel in those days was more of a burden than a pleasure, and that Paul was accustomed to meet his own expenses (cf. *9*, 15-18). The offerings were abundant (2 Cor. *8-9*) and Paul did take them up to Jerusalem himself (Acts *20-21; 24*, 17).

16, 5-12: St. Paul's Plans. **5 f.** Paul did *pass through Macedonia*, staying in Greece (Corinth) for "three months" (Acts *20*, 1-3). *Wherever I may be going:* he was not sure at this time that he would be going to Jerusalem (4). **7.** Paul could have gone directly from Ephesus to Corinth, but he chose rather to wait so as to have more time with them.   **8.** Paul is writing from Ephesus.   **9.** Instead of *evident* the Greek has "effective." Paul is speaking both of his opportunities for good and the Ephesians' need of him in the present opposition. The opposition eventually led to a violent uprising (Acts *19*, 23 ff).   **11.** *Let no one despise him:* ten years later Paul wrote to Timothy, "Let no man despise thy youth" (*1* Tim. *4*, 12). Timothy was now scarcely more than twenty-five.   **12.** The Corinthians had requested the return of Apollos. He was a favorite at Corinth (*1*, 12; *3*, 4-6; Acts *18*, 24-28). There is no jealousy or rivalry between him and Paul. Apollos' insistence on delaying his return to Corinth indicates that he was quite vexed that his name should be used in the Corinthian factions (*1*, 12). *With the brethren:* i.e., with those who had brought the message from Corinth and were now returning with Paul's letter.

16, 13-24: Final Directions and Greetings.   **13 f.** The allusion to the party division in the previous verse suggests the exhortation to set aside childish strife and to be united in charity. **15.** Some MSS add the name of *Achaicus* here as in 17; but the best Greek MSS omit both *Fortunatus* and *Achaicus* in 15. The first converts of Achaia were *the household of Stephanas* (*1*, 16). They had volunteered special service to their fellow-Christians.   **16.** St. Paul asks for their deference and recognition and service.   **17 f.** *Fortunatus* and *Achaicus* are not mentioned elsewhere. Paul feels that these three, who probably are visiting with him now, make up the void created in his heart by the absence of the Corinthians.   **19.** Proconsular Asia is the western extremity of Asia Minor. Ephesus is its capital. At Corinth on his first visit Paul stayed with Aquila, a Jewish tent-maker, and his wife Priscilla (Acts *18,* 2 f). They accompanied Paul to Ephesus (Acts *18*, 18). They are later found at Rome (Rom. *16,* 3-5). Their house at Ephesus was used for Christian worship. **20.** *A holy kiss:* our liturgy still keeps this in the "kiss of peace."   **21.** The last verses are written by Paul personally; for the rest he used a scribe. This was a sign of the genuineness of his letter (2 Thess. *3*, 17; cp. 2 Thess. *2*, 2).   **22.** *Anathema:* cf. *12*, 3. *Maranatha:* an Aramaic expression current in the early Church, like *alleluia*. It may be derived from *Maran atha,* "the Lord comes," or from *Marana tha,* "our Lord, come." It expressed a warning of the coming of Christ the Judge, whose coming is the hope and wish of the just. **24.** Concerning the last words, St. John Chrysostom writes, "The Apostle shows that he has written not from anger or indignation, but from the care he has for them, since after so great an accusation he does not turn away from them, but loves them and esteems them." *Amen* is probably a liturgical addition.

EDWARD H. DONZÉ, S.M.

THE SECOND EPISTLE TO THE CORINTHIANS

Introduction

Occasion of this Letter. This second canonical letter of St. Paul to the Christians at Corinth was proximately called forth by a report of Corinthian conditions made to the Apostle by his legate Titus. All authors are agreed on this point. But what was behind the news brought by Titus? Why was Titus making a report? And what matters did his account cover? Again authorities are in agreement that Titus was reporting results and effects on the faithful at Corinth of a censuring letter which St. Paul had previously written them during his stay at Ephesus. While in Ephesus the Apostle had heard on good authority that all was not going so well at Corinth. Factions and moral disorders were already in existence there which needed immediate and vigorous correction, and which only St. Paul himself could properly handle.

Now we know that faults and abuses of this kind are referred to and denounced in the first Corinthian letter that has come down to us. Was it then the effects and results of this letter that Titus reported to his master and that occasioned our *2 Corinthians*? All authorities until recent times have answered in the affirmative. Most modern scholars, however, looking more carefully into the contents of the two Corinthian letters we have are disinclined to accept this ancient opinion; they do not believe that our *1 Corinthians* furnished the remote occasion of our *2 Corinthians*. And in support of their view these recent scholars say, first, that it is very difficult to reconcile the language of *2 Cor. 2*, 3-11 and *7*, 8-12 with what St. Paul wrote in *1 Cor. 5*, 1 ff about the incestuous man. There seems to be question of a different kind of offender in *2 Corinthians*, and his offense would seem to be of a personal nature against St. Paul himself or some other individual, rather than against the whole community, as in *1 Cor. 5*. Secondly, these authorities hold that the description in *2 Cor. 2*, 3-5; *7*, 8, of a letter written the Corinthians, and of his feelings while writing it, does not easily fit or harmonize with anything in our *1 Cor.* Finally, the defenders of this new opinion point to *2 Cor. 2*, 1; *12*, 14; *13*, 1, as referring to a visit to Corinth subsequent to the time of founding the Church there. St. Paul could not speak of his first visit to Corinth, which met with unexpected success, as one of sorrow; nor could he call his forthcoming visit the "third" one, unless he had made a journey thither after his first long sojourn in that city when establishing the Corinthian church.

For these reasons it is the modern and, we think, the more probable view that St. Paul, in order to set matters right at Corinth, went there in person on a second visit some time during his stay at Ephesus; and, failing in his mission, he returned to Ephesus and thence wrote the severe letter to which he refers in 2 Cor. *2, 3-5; 7, 8,* on the effects of which Titus later made a report to him, and which therefore was the remote occasion of this present letter, our *2* Corinthians. In this opinion, then, St. Paul wrote four letters to the Corinthians: the one spoken of in *1* Cor. *5, 9,* which is lost; our *1* Cor.; this intermediate letter written while he was at Ephesus; and our *2* Cor.

Integrity. But where is that third and intermediate letter? Has it also not come down to us? On this point scholars are not in agreement. Some say that it is wholly lost. Others believe that we have it, at least in part, in *2* Cor. *10-13,* which consequently was written before *2* Cor. *1-9,* and which at some later date was joined by a copyist to our present *2* Cor. A less probable view regards *2* Cor. *10-13* as written shortly after *2* Cor. *1-9,* in response to fresh bad news from Corinth.

The reasons for believing that *2* Cor. *1-9* and *10-13* do not belong to one and the same letter are the marked differences in tone and content between the two parts. It is observed that the first seven chapters of this letter are a joyful acceptance of Titus' report and a mild defense of the work and life of St. Paul and his companions, ending on a note of complete harmony; while *8* and *9* constitute a tactful and restrained handling of the collection to be finished at Corinth for the poor in Jerusalem. Then suddenly, contrary to his practice in all his other letters and contrary to all psychological rules, he manifests an extraordinary anger, which is unequaled elsewhere in his writings, against his enemies at Corinth who were trying to destroy his authority and his work. And what the Apostle says in these closing chapters is possibly aimed at the Corinthian church as a whole, and not at a disloyal minority. How, we are asked, could St. Paul have so sharply terminated a letter, which in the first part is so mild and conciliating?

On the other hand, those who defend the unity of our *2* Cor. point to the evidence of all our extant MS copies of the Epistle, which give no signs of dislocation or separation, and to unbroken tradition down to recent times. They also feel that the differences between these two parts of the letter have been unduly stressed.

Date and Place of Writing. This letter was written towards the close of St. Paul's third great missionary journey, and therefore very probably some time in the autumn of the year 57 of our era. He had expected to remain in Ephesus until Pentecost of that year, and then go to Corinth by way of Macedonia, meeting Titus at Troas and there receive from him a report on Corinthian conditions. But opposition and disturbances at Ephesus compelled him to leave that Asian capital sooner than he had intended, and so it was in Macedonia, very probably at Philippi, that he and Titus met, that the report from Corinth was made, and that our *2* Corinthians was written.

Character of the Letter. This is doubtless the most personal of all St. Paul's writings, not excepting his letters to Timothy, to Philemon and to the Philippians. It is also the most vehement and polemical, especially in the four closing chapters, though the Epistle to the Galatians is similarly strongly apologetic and self-defensive in its tone and contents. In no other letter of St. Paul have we so complete and intimate a portrait of the great Apostle. As our *1* Cor. gives us a detailed picture of the early Church and its teaching, beliefs and practices, so this letter affords us the most minute description of the person and character of St. Paul that we have.

Purpose. St. Paul wrote this letter in defense of his own person, authority and work, as well as of his associates in the ministry; to urge the collection that was being made at Corinth for the poor in Jerusalem; and to reply to the accusations of his bitter enemies.

COMMENTARY

Introduction 1, 1-14

These verses introduce the present Epistle. They contain the salutations of St. Paul and Timothy to the church at Corinth, acts of thanksgiving for favors received, a request for prayers, and the reason for the Apostle's confidence in future prayerful help from the Corinthians. Some authorities prefer to close the Introduction with v.11 rather than with v.14, but this is a matter more of choice than of importance.

1, 1. *Timothy:* the fellow-worker and beloved disciple of St. Paul, who was well known to the Corinthians, is to be a witness of all the Apostle says in this letter. *Our brother:* literally, "the brother." *Saints:* St. Paul frequently speaks of Christians in this manner, because their very vocation was a call to holiness and sanctity. *The whole of Achaia:* Achaia was a Roman province including the Peloponnesus and north Greece as far as Macedonia. Corinth was its capital. Not only Corinth, but the outlying churches of this entire province, evangelized by St. Paul or his associates, were addressed in this letter.   **5.** The sufferings our Lord endured for our salvation are repeated to a greater or less extent in His followers, as members of His Mystical Body, for the salvation of souls and the spread of the gospel. This verse should be compared with Col. *1,* 24.

8. *Affliction* very likely refers to some bad news from Corinth about conditions there. Or the reference may be to the riot caused by Demetrius, the silversmith. See Acts *19,* 23 ff.   **9.** *Death sentence* means the sentence, the judgment, the expectation of death. The Apostle was ready to die, so bad did he feel. He felt as if the sentence of death had already been pronounced upon him, and he was ready to go.   **11.** St. Paul often speaks of the power of intercessory prayer and the necessity of thanking God for benefits received.   **13.** *We write:* refers to his other letters to the Corinthians.   **14.** *The day,* etc.: i.e., the day of judgment.

I. PERSONAL DEFENSE 1, 15 — 7, 16

1, 15 — 7, 16. Here we have the first main division of this letter, in which we find a general apology for the life and ministry of St. Paul and his companions. In this section the Apostle defends (a) himself and his actions, *1, 15 — 2, 17*; (b) his liberty as a minister of the New Law, *3, 1 — 5, 10*; (c) his sincerity as an ambassador of Christ, *5, 11 — 7, 1*; (d) his previous letter, *7, 2-16*.

1. THE APOSTLE EXPLAINS HIS DELAY 1, 15 — 2, 17

The Judaizers at Corinth who were causing all the trouble accused St. Paul of fickleness and lightmindedness because he had changed his plan about visiting them. The Apostle now replies to them, telling them that it was their own disorders and sins that made him stay away. He pardons the one who caused so much sorrow and relates his reaction to the good news brought him by Titus.

1, 15-22: **He Is Not Fickle.** **15.** St. Paul begins here to refute the accusation of his enemies that he was fickle. His original plan was to pay two visits to Corinth, one on his way to Macedonia, and the other on his return. This was the *double grace*, or joy or favor, that he wanted to give them; but something prevented the carrying out of this arrangement, and they accused him of being fickle and lightminded. **18 f.** The character and habitual manners of the Apostles were not vacillating and changeful, but uniform and reliable, like their preaching, and like the Christ whom they preached. **20.** Christ fulfilled all the promises made by God through the prophets, and thus He made possible the "Amen" by which we acknowledge that fulfillment. *Amen* is a Hebrew word meaning "so be it"; an expression of approval. **21 f.** *Anointed us* doubtless refers to the call and commission of the Apostles to preach the gospel and do the work of the ministry. *With his seal* means with His authority. God has prepared the Apostles for their work; He has *sealed* them, marked them, as His own by gifts of the Holy Spirit which made their mission manifest. This sealing, or giving of gifts, is a pledge for the future, when the Spirit will be given more fully. Reference also to the effects of Baptism is implied here.

1, 23 — 2, 4: **His Wish to Spare Them.** **23.** This verse and 2, 1 strongly suggest a visit to Corinth subsequent to the founding of the Church there. St. Paul's first coming to that great city was not *in sorrow*.   **2, 1-4:** These verses seem clearly to refer to a visit and to a letter other than our *1* Corinthians intervening between the Apostle's first visit to Corinth and the writing of this present letter.   **2.** *The very one* refers to the community taken as an individual.   **3.** *As I did* refers to what he told them in that intermediate letter—the severe rebuke which he was obliged to give them; or to his decision not to come to them again in sorrow.

2, 5-11: He Pardons the Offender. It is not certain who this offender at Corinth was or what his crime was. See Introduction. *But in a measure:* i.e., the offender has not so much pained the Apostle, as the whole Corinthian church.   **7.** St. Thomas and many theologians find here and in 10 an argument for the power of the Church to grant indulgences that avail before God.   **9.** Now that the Corinthians have shown obedience to St. Paul's demands, he forgives the offender in question, as he has asked them to do.   **10.** *In the person of Christ* means in the presence, or with the authority of Christ.

On vv.5-11 we may observe, with most modern interpreters, that they do not seem to refer to *1* Cor. *5,* 1-8, though all the Fathers, Tertullian alone excepted, identified the offender here with the incestuous man of *1* Cor. *5,* 1 ff. V.6 here does not agree with St. Paul's pronouncement in *1* Cor. *5, 5.* And if we compare this present passage with *7, 8-16* below, we shall see that there is question here of something quite other than sex enormity.

2, 12-17: Thanksgiving for Good News. *Troas* was a Mysian sea-port on the northeastern shore of the Ægean Sea. See map. **14.** *Leads us in triumph:* as in Col. *2,* 15. The Apostle is overjoyed at the good news from Corinth which Titus has brought. His sense of dependence on God makes him represent God as always leading His Apostles in a victorious, triumphal procession; but *in Christ,* which shows the intimate relationship between the Leader and the led.   **16.** The preaching of the Apostles meant eternal life to those who obeyed it, but to those who refused it, eternal death (Mark *16,* 16). *That leads to death,* etc.: the best Greek reading here has "from death to death . . . from life to life," apparently implying a progress from bad to worse, and from good to better.   **17.** *Adulterating:* literally the term means "huckstering," i.e., corrupting for sordid gain. *In Christ:* i.e., as members of Christ and ministers of His word.

2. THE APOSTLE DEFENDS HIS ASSURANCE 3, 1 — 5, 10

The adversaries at Corinth also accused St. Paul of self-commendation and boasting. They said he had to resort to such language and such practices in order to get a hearing. But the Apostle in this section of his letter tells them that he needed not to commend himself, since the Corinthians themselves were his recommendation and testimonial. And as to boasting, the very grandeur and excellence of the ministry committed to him and his companions made them speak with boldness and assurance; they could not do otherwise. And yet they are at all times aware of the frailty and weakness of their own lives and personal efforts, as contrasted with the sublimity and perfection of the work God has entrusted to them. In their labors and sufferings they are sustained by the hope of a glorious hereafter. Hence they seek only to please Christ, their future judge.

The conception of Christian theology and of the apostolic ministry, in its preaching, its labors, its sufferings, and its hopes given in *3, 1 — 5, 10* constitutes one of the finest passages in all the writings of St. Paul.

3, 1-3: They Are His Commendation. *Again to commend,* etc.: implies that the Apostle had already been accused of self-recommendation, perhaps in *1* Cor., or in a lost letter written after our *1* Cor. This matter of self-recommendation reappears in *10,* 12-18 and *12,* 11. If those chapters were written before this one, they throw light on the present verse. See Introduction. 2 f. The Corinthian church was St. Paul's letter of commendation. It was a letter inscribed on his heart and open to the inspection of all the world. It was a letter dictated by Christ and written down by the Apostle by means of the Spirit. Finally, it was a letter written on the hearts of the Corinthians themselves.

3, 4-11: Excellence of the New Law. 5. While St. Paul refers here directly to the work of the Corinthian missionaries, he lays down the general principle that of our natural strength and ability we are not able even to think, much less to wish or to do, anything supernaturally good and meritorious of life eternal. The beginning, as well as the completion, of each and every salutary act requires the grace of God. Such is the doctrine of the Church against the Pelagians, who denied all need of grace, and against the Semi-Pelagians, who denied the necessity of grace for the beginning of a salutary act.

6. *The letter kills:* i.e., the outward observance of law without an inner spirit is ruinous. Some authorities think "letter" here means the Mosaic Law, and "spirit" the gift of the Holy Spirit; the two are thus contrasted in Rom. 7, 6.   **7.** *Ministration of death:* i.e., the Mosaic Law, which had no power, apart from faith and grace, to save from spiritual death; and yet, being divine in its origin, it was accompanied, when given on Mt. Sinai, by a glorious manifestation which shone on Moses' face (Ex. *34,* 29-35).   **8.** *Ministration of the spirit* means the New Law, the gospel.   **9 f.** The glory of the New Covenant so completely transcends that of the Old that the latter has by comparison no glory at all.   **11.** The splendor which accompanied the giving of the Old Law, like the Law itself, was passing.

3, 12 — *4,* 6: **The Veil Is Taken Away.**   **12.** Such hope, of one day enjoying the fullness of the glory spoken of in 8.   **13.** The sense of the Greek is "that the children of Israel might not gape upon the end of that which was being destroyed."   **14.** *Their minds:* i.e., the minds of the Jews who would not accept the gospel.   **17.** *The spirit* is the Holy Spirit mentioned above, in 6.8, the life and principle of the New Law, inasmuch as the Holy Spirit is the Spirit of Christ, or inasmuch as *the Lord* here does not mean Christ, but God, the life-giving Spirit of the New Covenant (6; *1* Cor. *15,* 54), in contradistinction to the letter of the Old (Cornely). But it is difficult to see how *Lord* here can mean Yahweh, to whom the Jews as a people had always turned. There seems rather to be question of Christ to whom they refused to turn. When, therefore, the Jews shall have turned from the letter of the Law which kills to the Spirit of the gospel which gives life, the blindness of their minds shall disappear and they shall be freed from the servitude which now enslaves them (Callan). *There is freedom:* i.e., freedom from the ceremonial precepts and bondage of Mosaic legislation but not from its moral teaching.

4, **1.** *Mercy* means the Apostle's supernatural vocation to preach the gospel.   **2.** *On the contrary:* literally, "But"—a strong adversative. So it is used often elsewhere. *We avoid:* there is no implication here that St. Paul had once lived in moral laxity. **4.** *The god of this world:* i.e., of this "age," as it is in the Greek. The reference is to Satan whom our Lord called "the prince of this world" (John *12,* 31; *14,* 30; *16,* 11), and whom St. Paul elsewhere designates as "the prince of the power of the air" (Eph. *2,* 2). Satan is called the god of this wicked age, in so far as it lives according to his maxims, obeys and serves him; and he, in turn blinds the minds of his unbelieving followers, leading them away from the faith by evil suggestions, so that the light of the gospel, whose object is *the glory of Christ,* does not shine unto them. *The image of God:* Christ is the image of God, (a) as having the same nature as the Father; (b) as being the Son of the Father; (c) as being equal to the Father (St. Thomas).

6. *For God, who commanded,* etc.: the best Greek reading here is, "For God who said, 'Out of the darkness light shall shine.'" The radical reason why the Apostles preach Jesus Christ and not themselves, is that such is the will of God, who in the beginning of the world made light shine out of the darkness, and who through Christ had made the light of faith shine in the hearts of the Apostles in order that, through their preaching, they might enlighten the world with a knowledge of the glory of God, as it was revealed in the person of Christ, i.e., in His divinity, His actions, His doctrines, etc.

4, 7-18: **Frailty and Support.** **7.** *This treasure,* etc.: the gospel message and the Christian ministry are discharged by frail human beings. **10.** The Christocentric mysticism of St. Paul is said to reach its height here, where it involves bodily, as well as spiritual fellowship with our Lord. **13.** The preachers and believers in the gospel are sustained by the same faith which saved the just of the Old Testament; they believed that Christ would come, we believe that He has come (St. Augustine). **14.** *Who raised up Jesus:* the best Greek reads, "Who raised up the Lord Jesus." *And will place us with you,* in heaven. The Apostle here, as in *5,* 1-8, speaks as if he did not expect to be alive at the Second Coming of Christ; whereas in *1* Cor. *15,* 51 f, he speaks as though he might live to see that event. This shows that he had no revelation in the matter: he knew "not the day nor the hour" (Matt. *25,* 13).

17. The greatness of their future rewards sustains the Apostles and the faithful in the trials and sufferings of the present life. This verse proves that the good works of the just on earth are meritorious of eternal life, as the Council of Trent teaches.

18. The passing afflictions of the present life are building up an immeasurable and eternal weight of glory for those whose eyes are fixed, not on the things of the visible universe, but on the lasting realities of the invisible spirit world. The same thought is found in Rom. *8, 17* f.

5, 1-10: Reward after Death. **1.** *Earthly house:* literally, "Tent-dwelling," i.e., a dwelling that has only a transitory existence. "The camp-life of the Israelites in the wilderness, as commemorated by the annual feast of Tabernacles, was a ready and appropriate symbol of man's transitory life on earth." **2.** *To be clothed over,* etc.: the Apostle wishes that he might attain immortality without passing through death, that he might take on the resurrection body over his natural body (*1 Cor. 15, 51*). **3.** *Clothed, and not naked:* i.e., living, not dead, when Christ comes. The dead who shall have lost their bodies at the Second Coming shall be clothed anew. This is the most probable explanation of a very difficult verse. **5.** *The Spirit as its pledge:* the Holy Spirit received in Baptism is the earnest, the warrant, a foretaste of eternal life.

8. The Apostle means to say, if death is necessary to be with Christ, then welcome death. This verse affords a clear proof that purified souls immediately after death are admitted to the vision of God (St. Thomas). **10.** *All of us must be made,* etc.: all men, even children who die before the use of reason, must appear in the General Judgment. Sinless children will be present then, "not to be judged, but to see the glory of the judge, in order that both the mercy and the justice of God may be manifested in their case" (St. Thomas).

3. THE APOSTLE DEFENDS HIS SINCERITY 5, 11 — 7, 1

What the Apostle has just been saying about the lofty motives that guide his life and actions was related not in self-praise, but for the sake of the Corinthians. He and his fellow-workers are directed by love of Christ. The Corinthians ought to imitate the character of the Apostles and avoid the sins and vices of the pagans around them.

5, 11-19: His Labor for God and Souls. **11.** *The fear of the Lord:* i.e., the fear inspired by the thought of the judgment to come. *To God we are manifest:* the Apostles have to resort to persuasion to convince men of their sincerity, but not so with regard to God who reads their hearts and souls.   **13.** When the Apostle spoke of the graces and privileges he had received from God, his adversaries accused him of madness; but he spoke thus for the glory of God and the welfare of the faithful.   **14.** *For the love of Christ,* etc.: i.e., the love Christ has for us restricts the energies and activities of the Apostles to the things that pertain to the service of God and the salvation of human souls, to the exclusion of all selfish interests. And the reason for this is that since Christ died for all men, for the salvation of all, therefore all have died in Him, i.e., have participated in His death, sharing in its merits, so far as Christ is concerned.   **16.** *Have known Christ according to the flesh:* St. Paul before his conversion had the ordinary Jewish idea of "the Christ," i.e., the Jewish Messias; but now, enlightened by the Spirit, he knows Him to be the Son of God, the Redeemer, the Sanctifier, the Reconciler.

17. The change in the Apostles, which the preceding verse describes, is now extended to all Christians. *If then any man is in Christ:* through Baptism he has become a new creature, morally and spiritually.   **19.** God was reconciling the world to Himself *in Christ,* i.e., through Christ, in virtue of Christ's merits, (a) by wiping out men's sins, for which Christ atoned, and (b) by confiding to the Apostles the office of preaching the gospel, of administering the Sacraments, etc.

5, 20 – 6, 13: Ambassadors of Christ.   **21.** For our salvation the sinless Son of God, in the person of Christ our Redeemer, became identified with our sins; so that God "suffered Him to be condemned as a sinner, and to die as one accursed" (St. Chrysostom).

6, 1. *Working together:* i.e., the Apostles, working together with God, entreat that the faithful remember their obligation to the grace which God has given them in converting them to Christianity.   **2.** The Apostle now cites the prophet Isaias (*49,* 8) to show why the Corinthians should heed his exhortation without delay. The prophet represents God as addressing His Servant, the Messias, and through Him His people, assuring Him that His prayers and labors for the salvation of mankind have been heard. Now St. Paul here means that the Messianic time spoken of by the prophet has come, and that therefore everyone should profit by the graces now given, because, if they are abused, there will

be no hope of salvation, since another Messias shall not come.
3. *To anyone:* The Greek has, "In nothing." After the parenthesis in 2 the thought goes back to 1. The meaning is that the apostles, St. Paul and his companions, give offense in nothing, i.e., they avoid everything in the exercise of their ministry, and in their dealings with men, that might bring any blemish on their profession and thus keep people from the gospel. If a preacher or a Christian leads a life that is out of harmony with his profession, he gives occasion to men of despising his faith.
4. *Let us conduct ourselves,* etc.: in the Greek it is, "But in everything commending ourselves," etc. *In much patience,* etc.: the Apostle here begins an enumeration of nine classes of things that tried his patience and the patience of his companions.

6 f. St. Paul now mentions nine other practices by which he and his fellow-workers commended themselves in their ministry. **8-10.** In a series of antitheses St. Paul now shows how, under all conditions of life, he and his companions conducted themselves as became their high office and ministry. *Dishonor, evil report, deceivers,* etc.: referring to things done and said by enemies of the apostles. **11-13.** St. Paul explains why he has spoken so freely to the Corinthians; it is because he loves them.

6, 14 — 7, 1: Avoid Marriage with Unbelievers. **14.** Some critics find a break in the thought from this verse to 7, 1, which they feel warrants the opinion that we have here an interpolation belonging to some other letter of St. Paul, perhaps to the lost letter spoken of in *1 Cor. 5, 9*. But the interruption in the line of thought here is not too much for a writer like St. Paul, and hence the argument against its integrity is without sufficient grounds. The Apostle is warning in 14 against sharing pagan ideas and practices in general and especially in marriage. **15.** *Belial:* or Beliar—a Hebrew word meaning "nothingness," "uselessness"; in a secondary sense it means extreme wickedness, and it was commonly understood by the Fathers as a designation for the devil, or Satan. **16-18.** These verses from the Old Testament are cited to show God's paternal care for all His faithful children. **7, 1.** Christians cannot avoid sins of the flesh and of the spirit, neither can they attain perfect holiness of life, unless they have a salutary fear of God. "Love begets security, but sometimes causes negligence, but he who fears is always solicitous" (St. Thomas).

4. THE APOSTLE DEFENDS HIS PREVIOUS LETTER 7, 2-16

The Apostle now returns to the appeal of *6, 13*, and asks the Corinthians to show towards him the charity which he has manifested towards them.

7, 2-7: Love for the Corinthians. 2. The Apostle is here refuting the accusation of his enemies. 6 f. The news brought from Corinth by Titus greatly consoled St. Paul. There had been joy in the arrival of Titus, but more joy in his news.

7, 8-16: Their Repentance. 8. Here again the Apostle seems to be referring to a lost letter, or to a letter much more severe than our *1 Cor.* 10. *Surely tends,* etc.: or "a repentance that is steadfast unto salvation," i.e., penance that "is not repented of" (as it is in the Greek), but endures unto salvation. 12. *Who suffered the wrong:* would seem to refer to St. Paul himself. The person *who did the wrong* can hardly mean the incestuous man of *1 Cor. 5,* 1 ff. *The zeal we have for you:* the best Greek reading here has, "The zeal you have for us."

II: THE COLLECTION FOR THE POOR CHRISTIANS IN JERUSALEM 8, 1 – 9, 15

In this second main section of *2 Cor.*, St. Paul is dealing with the collection for the poor Christians in Jerusalem. The Apostle is deeply concerned over this collection. The need was pressing. But there were also other considerations in this matter. A generous collection at Corinth would be a special sign of the unity between that Gentile church and their Jewish brethren so far away, and at the same time it would be a telling proof of the Apostle's own authority which had been recently questioned. Again, there were the adversaries at Corinth and opponents at Jerusalem. How would these look upon this collection?

8, 1-6: Example of the Macedonians. 2. *Testing of tribulation* means the many afflictions experienced by the Macedonians, which made them sympathetic towards the sufferings of others and willing to help those in distress. Here we should read the narrative in Acts *16-17.* 4. *Saints:* the poor Christians of Jerusalem. The Apostle had previously spoken of a collection for those needy brethren in *1 Cor. 16,* 1-4, and had sent Titus to start it, perhaps soon after the arrival of our *1 Cor.*   5. *Gave themselves, first,* etc.: i.e., they put their lives and persons at the disposition, first of Christ, and then of the Apostles for Christ's sake.

8, 7-15: Exhortation.   8. *Zeal of others:* i.e., the earnestness of the Macedonians.   10. *A year ago,* etc.: this can hardly mean that twelve months had intervened since the writing of *1 Cor. 16,* 2, because that Epistle was written in the spring, and *2 Cor.* appears to have followed very probably in the succeeding autumn. Perhaps the collection had been decided on some time before *1 Cor. 16,* 2 was written; or St. Paul might have been reckoning according to the Macedonian year which, like the Jewish civil year, may have begun in autumn. In this latter supposition *a year ago* would mean "last year."   13. *Equality:* it is not necessary that this be thought of as realized in the present.   14. *Your abundance:* Corinth was a center of great wealth. *Their abundance* may refer to the spiritual plenty of the Christians in Jerusalem, or to the material goods which might be theirs at some future time.   15. The Apostle now cites a passage from *Ex. 16,* 18, to illustrate how there should be equality in temporal goods among the Christians, just as of old God so distributed the manna in the desert that all had what was necessary, superfluities being made to supply needs. Those who gathered more manna than the others had not in the end more than they needed, while the others had all that they required.

8, 16-24: The Mission of Titus.   18. *The brother:* some hold this refers to St. Luke. *And we have sent:* this is an example of the epistolary aorist in Greek, referring to a present action which will be past when the letter is received.   20. St. Paul took care that the collection should be handled by several, so as not to give his enemies any chance to accuse him of misappropriating the money.   22. *Our brother:* i.e., some trusted fellow-Christian, who was to be the third delegate to go to Corinth.

9, 1-15: The Collection.   4. *For having been so sure:* more literally, "in this matter," i.e., according to the meaning of the Greek here, in the confidence or hope he had placed in the Corinthians.   9. The Apostle confirms what he has just said by citing Ps. *111,* 9. The just man *scatters* his gifts as the sower his grain, and *his justice remains,* etc.: i.e., the remembrance of his good deeds will never be forgotten; his reward will await him hereafter. This is the most probable meaning of *justice* here. **10.** The Greek of this verse runs as follows: "Now he that ministers seed to the sower and bread to eat, will also provide and multiply your seed," etc. These words are a quotation from Isa. 55, 10. What the prophet says of the rain from heaven St. Paul applies to God's ordinary providence, which not only will enable a charitable man to give, but will also increase his temporal possessions, *the fruits of* his *justice,* i.e., the reward of his virtue.   **11 f.** The recipients of the collection would render abundant thanks to God. The value of the service the Apostles will render will be seen in such thankfulness as well as in its material benefits.   **15.** *Unspeakable gift:* i.e., the gift of faith and charity spoken of in the preceding verse.

III: THE APOSTLE DEFENDS HIS APOSTOLATE
10, 1 — 13, 10

This is the last main division of the letter. See remarks on this part in Introduction. St. Paul is here vindicating his person and apostolic authority against the attacks of vicious enemies.

10, 1-18: His Authority. St. Paul first urges his readers to amend their lives, so that when he arrives in Corinth he may not be forced to use his powers against any of them. Let no one underestimate him and his authority. When he comes, he will be prepared to take action, if necessary. His boasting, unlike that of his critics, is not in the labors of other men, but only in the realms assigned to him by Christ. Let him who would glory, do so in the Lord.

1. *Am mean,* etc.: thus did the Apostle's adversaries speak about him.   **2.** The Apostle hopes that it will not be necessary when he arrives in Corinth to give proof and demonstration of the boldness of which he is accused. *Walking according to the flesh:* regulating their life and conduct according to human and worldly standards.   **4 f.** Faith is a gift of God and is above reason, though not contrary to it. Everything that opposes faith is wrong and must be set aside or destroyed. *Bringing every mind,* etc.: i.e., all the designs and workings of the natural reason that are opposed to the gospel must be brought into subjection, making all obedient to the faith of Christ. True faith consists not only in the assent of the intellect, but also in the submission of the will to God's revelation. The evidence for faith is not sufficient to force the intellect, but the will freely determines to move the intellect to accept revelation and give its assent. **6.** When once the faithful of Corinth have submitted to the gospel and apostolic teaching, St. Paul will deal with the rest as they deserve.

7. *Look* may also be interrogative. *That he is Christ's:* the Apostle cautions his adversaries to reflect that, if anyone considers himself a minister of Christ, he must not overlook the fact that Paul and Timothy are also equally ministers of Christ and preachers of the gospel. **8.** The Apostle is confident of his divine authority and will not hesitate to use it if necessary.

12-18: The work entrusted to St. Paul was designated and defined by God Himself, and the Corinthians were included within those divinely appointed limits. **12.** *We, on the contrary, measure,* etc.: our English versions, like the Vulgate, have perhaps missed the meaning here, because they have failed to take account of the words, "they do not understand," which occur at the close of this verse in nearly all the MSS and in the citations of many of the Fathers. Hence the clause should read: "They measure themselves by themselves, and compare themselves with themselves, and (so) they do not understand." The general sense is: "They make fools of themselves, measuring themselves by their own standards" (Rickaby). **15 f.** The Apostle is referring to his opponents at Corinth who have obtruded themselves into the field of his own labors and commission.

11, 1-33. In the preceding chapter St. Paul vindicated his apostolic authority against his adversaries in Corinth and warned them to reform their ways before he would arrive. In this chapter he makes a comparison between himself and them in order to show his readers how far superior to his opponents he really is. If this seems like self-praise, it is done only for the sake of the faithful. The Apostle is glad that he preached the gospel at Corinth without pay; he has all the reasons for boasting which his critics have; he rejoices at the thought of his apostolic labors and sufferings.

11, 1-15: He Preached Gratuitously. 1. *Foolishness* is the folly of self-praise which the Corinthians have forced the Apostle to indulge in for the moment. Had they remained loyal to him and to his preaching, such folly would have been unnecessary. *Bear with me,* may also be translated, "You do bear with me." **2.** *Divine jealousy:* St. Paul's zeal for the Corinthians was like God's for the people of Israel. At the time of their conversion the Corinthians were *betrothed* to Christ, through faith and Baptism, and he hoped to present them on the day of judgment *a chaste virgin,* i.e., as free from corruption in faith, to their heavenly spouse. **4.** The supposition here is impossible since there is only one Christ, one Holy Spirit, one heavenly gospel. This verse may also be understood as ironical praise of an actual situation, thus: It is a fine thing you are doing in tolerating newcomers with their preaching of another Jesus, and their offering of a different Spirit and gospel. **5.** *Great apostles* may refer ironically to the false leaders at Corinth, or sincerely to the Twelve. St. Paul was certainly far superior to the former, and not inferior to the latter. **12.** The pseudo-apostles at Corinth took pay for their ministry, and if St. Paul did also, their practice would have justification before the people. That was what they wanted.

11, 16-33: His Ministry of Labor and Suffering. **17.** *Not according to the Lord:* i.e., not in keeping with the general rule given by Christ to His Apostles (Matt. *6,* 1-6; Luke *17,* 10; *18,* 11-14). But St. Paul was justified in deviating from this rule in order to counteract the bad influence of his enemies. **18-21.** The Corinthians ought to be able to put up with a little foolishness from St. Paul since they themselves are so wise, and since they willingly endure all kinds of outrages from their false leaders. **22 f.** Over against the sham boasting of his opponents, St. Paul cites his own origin and some of his many labors and sufferings for Chirst and the gospel. **28.** *Those outer things:* the sufferings so far cited were incidental. What really mattered was his anxious concern for all his churches; this was a daily responsibility. **32 f.** *Damascus* was the capital of Syria. It goes back to the days before Abraham (Gen. *14,* 15) and was founded by Uz, the grandson of Sem. It is situated at the eastern foot of the Anti-Libanus on the high road of commerce between Egypt and Upper Syria and between Tyre and the East.

There is some difficulty in explaining the isolation of the incident about Damascus here. Perhaps the adversaries had attributed St. Paul's departure from Damascus to cowardice, and denied that he had been in real danger.

12, 1-21. In addition to his superior labors and greater suffering for the gospel, St. Paul can prove his divine commission by the extraordinary graces and gifts with which God has favored him. He therefore cites some of his visions and revelations. Could his adversaries boast of anything of the kind? And yet he prefers to glory in his infirmities. If his converts had been true to their Apostle, all this apparent boasting, so distasteful to him, would have been unnecessary. After all, he is not on trial before the Corinthians. God is his judge.

12, 1-6: His Revelations. 2. *A man in Christ:* St. Paul humbly speaks of himself in the third person. *Whether in the body,* etc.: the Apostle at the time was totally abstracted from the senses, as in ecstasy. *The third heaven:* i.e., paradise, the abode of the blessed. The *fourteen years ago,* spoken of in this verse, would be shortly before the beginning of the first missionary journey, that is, around 43-44 A.D., if this Epistle was written around 57-58 A.D.

12, 7-10: His Infirmities. 7. *Thorn for the flesh:* or "thorn in the flesh," seems to refer to some physical defect or chronic illness. 9. *And he has said to me:* the use of the perfect tense here implies that the force of the reply continues. *Strength is made perfect,* etc.: the power of God is most perfectly realized and appreciated when human strength is wanting, that is, when weak human agents are made use of to accomplish great results.

12, 14-21: His Third Visit. 14. *This is the third time,* etc.: in view of *13,* 1 this can only mean that the forthcoming visit to Corinth would be his third. See Introduction. 15. Some critics make the second clause of this verse independent, and read it interrogatively: "If I love you more abundantly, am I to be loved the less?"

16. *But be it so:* it would seem from the preceding verse that less love from the Corinthians was to be St. Paul's reward for his more abundant love of them. But the accusation against him could take another form. His enemies perhaps said, or would say, that he was cunning enough to get money from the Corinthians through his legates. 18. There is some confusion about the mission of Titus referred to in this verse. Perhaps we shall

encounter fewest difficulties if we suppose three visits of Titus to Corinth. Thus, (a) an earlier one in which he and a *brother* started the collection for the poor in Jerusalem, to which the present passage and *8,* 6 seem to allude; (b) the visit following the painful letter (*2,* 13; *7,* 6.13); (c) the visit on which he and two brethren were to complete the collection (*8,* 6.17.18.22). **19.** *We speak before God in Christ:* i.e., before God as our judge, and as ministers of Christ.   **21.** *Who sinned before,* refers to some former period in the Christian life of the Corinthians, perhaps before St. Paul's second visit, the painful visit.

13, 1-10: Warnings. Returning to the thought of v.20 of the preceding chapter, St. Paul again warns, and this time more explicitly, that he will be ready to exercise his authority, if necessary, when he arrives at Corinth. He hopes there will be no need for severity. It is the spiritual good of the faithful, not a display of his power, that he seeks.

1. *The third time:* this surely implies that he had visited Corinth twice before. *On the word,* etc.: he means to say that he will take action in a strictly legal manner, such as that spoken of in Deut. *19,* 15.   **2.** According to the Greek reading of this verse, the words, "the second time," should be inserted after *when present.* The sense is: I have warned before, when present the second time, and now, being absent, I warn again those who sinned before, and all similar sinners, that if I come again, etc. **3.** *Do you seek,* etc.: the best Greek here is not interrogative, but begins with a conjunction meaning "because" or "seeing that." The sense is elliptical.   **4.** *We also are weak:* i.e., we share in the sufferings of Christ. In Christ weakness and death were followed by life and strength, and so it is with His Apostle.

5. Christ dwells in the intellect by faith, in the heart and affections by charity. St. Paul asks the Corinthians to test their own standing in the faith. This they can do since Christ is among them.   **7.** St. Paul hopes there will be no reason at Corinth which will deserve the exercise of his divine authority. He is not eager to draw attention to his God-given powers.   **8.** The Apostle can work only according to the demands of truth, and he implies that what truth now demands of the Corinthians is good conduct and obedience.   **10.** St. Paul is conscious of the authority which he has from the Lord, but its purpose is primarily constructive, and not destructive.

Conclusion 13, 11-13

These verses form the conclusion of the Epistle, and they fittingly close *10 — 13, 10*, whether or not we regard those chapters as written before *1-9*. These final verses contain a kindly appeal for the practice of the good conduct of which the Apostle has been speaking just above.

13. Of the three genitives of this verse, the first two are subjective, and perhaps the third also, meaning the source and cause of the *grace, charity, fellowship,* spoken of. It is probable, however, that the third genitive is objective, signifying participation of the Holy Spirit.

Charles J. Callan, O.P.

THE EPISTLE TO THE GALATIANS

Introduction

The Galatians Addressed. It is not certain whether this Epistle was written by St. Paul to the churches in the southern part of the Roman Province of Galatia (Antioch, Iconium, etc.), which he established on his first missionary journey (Acts *13*, 13 ff; *14*, 1 ff), or to the churches of Galatia proper, the northern part of the Roman Province of Galatia. These last were probably visited by the Apostle on his second and third missionary journeys (Acts *16*, 6; *18*, 23). Some authorities infer from Acts *16*, 6 that the Epistle was addressed to both regions. Cf. Commentary on these passages in Acts.

The solution of this question makes little difference as regards the teaching of the Epistle, that is, that justification and salvation are not to be had through the Mosaic Law, but through faith in Christ.

The Purpose of the Epistle. Shortly after the foundation of the church at Galatia, the new converts were disturbed by the arrival of certain Judaizers, who, contrary to the teachings of St. Paul, claimed that for salvation it was necessary to be circumcised and to conform to the Mosaic observances (Gal. *3*, 1-4). More boldly they maintained that Paul was not a real Apostle, and that his teaching concerning the Mosaic Law was not authorized by the true Apostles, Peter, James and John. They sought therefore to undermine his authority. Hearing also from these Judaizers, who had come probably from Jerusalem or Antioch in Syria, that the preaching of Paul had been subjected to examination at the Council of Jerusalem (Gal. *2*, 1-10), and that Peter had openly disagreed with Paul at Antioch (Gal. *2*, 11-15), the Galatians were perplexed, and they were about to accept this new doctrine (Gal. *1*, 6).

When St. Paul learned of this situation, he hastened to correct the errors, and to prevent the defection of the Galatians from the true faith, by sending the present Epistle. He saw that the situation was serious, but not entirely desperate (Gal. *5*, 10). Were it possible, he would have gone to them in person (Gal. *4*, 20); but he had to be content with defending his person and his teachings in this letter. His first concern was to defend the divine origin of his teaching and authority; and secondly he proved that justification is not through the Mosaic Law, but through faith in Jesus Christ.

Date of the Epistle. The Epistle was sent to the Galatian church after Paul had paid a second visit to the community (Gal. *4*, 13). Furthermore, it must have been written soon after this second visit, otherwise the complaint of Paul about the sudden change of mind amongst the Galatians (Gal. *1, 6*) could not clearly be explained. Therefore, the most probable place of its composition is Ephesus, where Paul went immediately after his stay in Galatia (Acts *18,* 23 ff); and there is probability that it was written at the very beginning of his activity in Ephesus, about the year 54 A.D., for Paul was stirred with the deepest anxiety for his converts, and certainly would not have delayed in admonishing them.

The Authenticity of the Epistle. The authenticity of the Epistle is generally acknowledged today because of both internal and external evidence. Its language is typically Pauline, being a vivid picture of severity tempered by tenderness; while the account of the Galatian situation is such that a later forger would have been unable to invent it. The Apostolic Fathers, such as Ignatius and Polycarp, quote from the Epistle, and Irenaeus and Clement of Alexandria among others directly testify to St. Paul's authorship.

The Style of the Epistle. The Epistle is vigorous in its style and character. Its theme is similar to that of the Romans, but in style the latter Epistle is more expository. Galatians lacks the digressions found in the other Pauline Epistles, nor does it have the usual closing salutations, a sign of the haste with which it was sent.

COMMENTARY

Introduction 1, 1-10

1, 1-5: Greeting. The abruptness with which he plunges into his theme, and the absence of all commendation of the Galatian Christians are signs of the Apostle's intense feeling of displeasure and grief at the readiness of the Galatians to desert the principles he had given them.

1. *An Apostle:* as being a witness of the Resurrection (Acts *1,* 22), and immediately appointed by Christ (Gal. *2,* 8), and consequently equal to the Twelve. *Not from men:* (as the source) his opponents were self-appointed teachers, his commission as a teacher came from God. *Nor by man:* (as the medium) he was not authorized by men, as were the deacons (Acts *6,* 5), and Timothy (*1* Tim. *4,* 14). *But by Jesus Christ:* i.e., St. Paul was designated apostle by Christ Himself. *Who raised him from the dead:* it was the special function of the Apostles to preach Jesus and the Resurrection (Acts *17,* 18). **2.** St. Paul, out of custom, includes his companions such as Timothy and Erastus in his salutation. The mention of churches indicates that several, if not all, were guilty of the same error. **3.** In his charity St. Paul wishes the Galatians the grace of God, which they have not esteemed as they should. **4.** In this verse the Galatians are reminded that they have been freed from sin, not through the Mosaic Law, as the Judaizers were claiming, but through Jesus Christ, who died for them.

1, 6-10: Surprise and Rebuke. St. Paul immediately begins his theme. **6.** This verse contains a rebuke sterner than Paul is wont to give. He marvels that they are so *quickly deserting,* i.e., so quickly after the temptation came upon them, and not so quickly after their conversion. **7.** *Which is not another gospel:* there is only one gospel of Christ, while the Judaizers were preaching among the Galatians serious doctrinal errors as if they were the gospel. **8.** *Anathema:* cursed, excluded from the kingdom of God. **9.** St. Paul reminds the Galatians of the warning he and his companions had given them before, possibly on his second visit, against false teachers. **10.** Here St. Paul observes that he did not make it his primary object to please men. He is evidently repeating charges made against him. The Apostle asks the Galatians to judge for themselves whose favor he is seeking; if he were trying to have the favor of the Judaizers he would have continued defending Judaism, but then he would not be the servant of Christ.

I. PERSONAL DEFENSE 1, 11 — 2, 21

1. A Defense of His Apostolate 1, 11-24

St. Paul asserts, in refutation of his opponent's accusations, that his gospel is from God. As a proof he recalls his conduct before his conversion on the road to Damascus. A conclusive proof that his work among the Gentiles was regarded as a mission from God was the praise given his missionary labors by the churches of Judea, who knew the former persecutor of the Christians only by reputation.

1, 11-24: Not of Human Origin. **11.** *I give you to understand:* expressing Paul's earnestness *(1 Cor. 12, 3; 2 Cor. 8, 1). Not of man:* i.e., not according to a human standard. **12.** This is a formal declaration that the doctrine Paul taught was revealed directly to him by Christ, and not learned from the Apostles or members of the Church. This doctrine embraced the whole preaching of Christianity, the mysteries of the life, passion and resurrection of Christ. **13.** *You have heard:* i.e., from different sources, but notably from his enemies, the Judaizers. **14.** *The traditions of my fathers:* the explanatory additions to the Mosaic Law (Mark 7, 3-13). Paul, like the other Pharisees, had regarded these traditions as sacred as the Law itself. **15.** Before Paul was born, it pleased God to choose and predestine him to be the Apostle of the Gentiles. **16.** *To reveal his Son in me:* i.e., to grant him knowledge of Christ as the Messias and Savior, compared with which Paul counted all other things as worthless (Phil. 3, 8). *Flesh and blood:* human counsel. Paul is contrasting Christ, who spoke to him, and mortal men, whom he did not consult. *Immediately:* the Apostle is not insisting so much on the prompt obedience he showed to his call, as upon the divine origin of his apostolate. **17.** So clear and certain were the call and revelation, he did not even go to Jerusalem to confer with the other Apostles. *Arabia:* the country east and southeast of Palestine, stretching from the Euphrates to the Red Sea. Paul in mentioning his retirement into Arabia, where there was no one who could instruct him, wishes to prove that he did not receive his gospel from men. In that country he devoted himself to meditation and prayer; and according to some Fathers he also engaged in preaching.

18. This verse shows St. Paul's respect for St. Peter. He went to make the latter's acquaintance, although he had nothing to learn from Peter. It was not necessary that he go up to Jerusalem to learn the gospel, and so he remained with Peter only *fifteen days*. **19.** On this occasion Paul saw a good deal of Barnabas (Acts *9*, 26 f). He does not mention this, since he is speaking only of the Apostles. James is called *the brother of the Lord*, meaning His relative. His father was Cleophas or Alpheus, and his mother was the sister of the Blessed Virgin. Therefore he was Christ's cousin. **20.** St. Paul considered it very important to insist that what he had just said about his independence of the twelve Apostles was undeniably true. **21.** From Acts *9*, 29 ff we know that Paul was persecuted at Jerusalem by the Hellenistic Jews, and was obliged to flee. He went to Tarsus in Cilicia, where he was later found by Barnabas (Acts *11*, 25 ff), who brought him to Antioch. In this verse Paul is probably referring to his ministry with Barnabas in and about Antioch, the metropolis of Syria (Acts *11*, 25-30).

2. A Defense of His Gospel 2, 1-21

2, 1-10: Approved by the Apostles. Having shown that his call and his knowledge of the gospel came directly from God, St. Paul now proceeds to refute another argument of the Judaizers, namely, that he did not have the approval of the Apostles.

1. Fourteen years after his first visit to Jerusalem (*1*, 18), St. Paul went to the Holy City again, accompanied by Barnabas and Titus. It is true that he had been in Jerusalem in the meantime to bring alms as narrated in Acts *11*, 29 f; but he omits any mention of that visit, because it had nothing to do with his present concern, i.e., showing his approval by the other Apostles. **2.** St. Paul was prompted to undertake the journey by divine revelation, and also because the Christians of Antioch sent him (Acts *15*, 2), just as St. Peter went to Cæsarea, Cornelius sending for him and the Holy Spirit bidding him go (Acts *11*, 11 f). He explained his preaching to all the faithful, but more particularly to the Apostles, the *men of authority*, for their approval. He needed no assurance for himself, but he wished to guard his future labors, as well as his past against his enemies. *Run:* a metaphor based on the games of those days.   3. This is a proof that Paul's preaching was not in vain. Not only did the Apostles not reject his gospel, but they did not even require the circumcision of Titus, his companion, notwithstanding strong pressure from the Judaizers.   4. In spite of the Judaizers, Titus was not circumcised. If it was a mere matter of expediency, as in the case of Timothy (Acts *16, 3*), St. Paul himself could practise circumcision, for his doctrine was that such circumcision was no bar to salvation. In the case of Titus, however, the Judaizers held that salvation was dependent upon the rite. *Brought in secretly, who slipped in:* i.e., they entered the Christian Church stealthily. *Spy upon:* to find some flaws in *our liberty,* our freedom from the Mosaic Law.   5. *An hour:* i.e., a moment.

6. In 3-5 St. Paul indulged in a digression about the case of Titus; he now returns to the thought of 2. He states that the Apostles added nothing new to his gospel. *What they once were,* etc.: i.e., although they were older in the apostolate than Paul, and had lived with the Lord, that was not of present importance. As regards his gospel there was a complete accord between them.   7-9. The Apostles did not correct Paul's teaching. On the contrary, they saw that his mission too was divinely ordained, and that they and he had been assigned by Providence to different spheres of a common work. These verses do not mean that Paul's mission was to be exclusively among the Gentiles, or that of James, Cephas, and John exclusively among the Jews. We know from the Acts that St. Paul always began with the Jews, although his converts were principally Gentiles. Peter and John later worked among the Gentiles, but Paul was the first Apostle to labor among them on a large scale. *The right hand of fellowship:* solemn approval. *James:* the Lord's cousin, and the first bishop of Jerusalem; *Cephas:* Peter's Aramaic name, by which he was known among the Jews.

2, 11-21: Paul Reproves Peter. Immediately after the Council at Jerusalem, Paul and Barnabas, accompanied by Judas and Silas, conveyed the decrees to Antioch, as related in Acts *15, 30.* It appears that St. Peter very shortly followed them there, and the occurrence here referred to took place at that time. What Peter was blamed for was that while he associated and took food with the Gentiles, without regard to what was allowed or forbidden by the Mosaic Law, he varied this custom on the arrival of certain legates from St. James, the bishop of Jerusalem. He then withdrew from his association with the Gentiles, fearing that he would scandalize the circumcised.. This showed a lack of consideration, and the Gentiles were seriously offended. It was not Peter's purpose to compel the Gentile converts to adopt Jewish rites; but he over-indulged the prejudice of his countrymen against eating in common with Christians who would not add to their Christianity the extra observance of the Mosaic Law.

11. *To the face:* i.e., directly to him, and perhaps openly and publicly.   **12.** St. James possibly sent these legates to collect alms for the poor in Jerusalem.   **13.** The authority of Peter influenced the other Jewish Christians in Antioch, and even Barnabas, to follow his example.   **14.** *Walking uprightly:* i.e., they were not showing the courage of their convictions. St. Paul spoke to St. Peter in a public assembly of both the Jewish and Gentile Christians, perhaps during the agape or love-feast. *Compel:* the moral constraint which the example of Peter put on the Gentile Christians. The address to Peter probably ends here.

15. In this verse Paul showed that, notwithstanding what he had said, he was keenly sensible of the privileges of the Jews; and as a pious Jew shared in the horror felt at the idolatry and immorality of the Gentiles.   **16.** The precepts of the Mosaic Law were ceremonial, such as circumcision, and moral, such as the Commandments. The Judaizers insisted on the observance of the ceremonial precepts or works. Such prescriptions of themselves had no power to save, as salvation depends on faith in Christ.   **17.** By thus denying the efficacy of the Law, we class ourselves with those outside the Law. If the Law were really of force, we should thus make ourselves sinners.   **18.** In other words, if after abandoning the Mosaic Law to seek justification through Christ, one returns to its practice, he sins, since he knows justification is had only through faith in Christ.   **19.** Cf. note to text.   **20.** Christ, the source of grace and justice, who died for me, is now the source of my spiritual life, a life which I have not attained by the works of the Law.   **21.** These words were directed against the Judaizers, endeavoring to lead the Galatians away from the true gospel. For these Christ died in vain, since seeking salvation through the Law they nullified His death.

II: DOCTRINAL 3, 1 — 4, 31

1. Justification From Faith Not From the Law 3, 1-29

St. Paul has just stated (2, 21) that to seek salvation through the Mosaic Law is to render null the death of Christ; and reflecting now on the situation in Galatia, he undertakes to prove the doctrine that justification is rather through faith in Christ.

3, 1-6: Proved From the Galatians' Experience. 1. *Who has bewitched you:* as if by an evil eye. Through Paul's teaching Jesus Christ crucified was made to appear before their eyes as if actually existing in the flesh. Some Vulgate codices and the Clementine edition after *bewitched you* add "that you should not obey the truth." 2. They had received the Holy Spirit with His sanctifying grace and His special gifts through Paul's preaching, and their believing in Christ crucified. 3. Are you so foolish, that beginning your salvation by the reception of the *Holy Spirit* you seek the completion of it in the carnal ceremonies of the Mosaic Law (*the flesh*)? 4. The Galatians suffered many persecutions for their faith. These will not be in vain. 5. By *Spirit* the Apostle is referring to internal gifts, such as knowledge and wisdom (*1 Cor. 12,* 4 ff); by *miracles* he means the outward manifestations of the Holy Spirit received after Baptism, e.g., the gift of tongues (Acts *10,* 46), or that of prophecy (Acts *19,* 6). There could be no doubt that these gifts were not given because of the observance of the Mosaic Law, since the majority of the Galatians were converts from paganism, and had not been circumcised (*4,* 8). 6. For this text (Gen. *15,* 6) cf. the Commentary on Rom. *4,* 3 ff.

3, 7-9: The Example of Abraham. 7. *The men of faith:* those who make faith, and not the ceremonial works of the Law, the principle of their religious life. 8. *The Scripture* is here personified. Paul refers to the Holy Spirit speaking in the Scripture. The meaning is that the Holy Spirit, the Author of Scripture, foresaw before the giving of the Law that God the Father had determined to justify the Gentiles by faith. The citation is a fusion of two passages, Gen. *12,* 3 and *18,* 18. *All the nations:* i.e., all those, whether Gentile or Hebrew, who shall imitate the faith of Abraham.

3, 10-14: The Nature of the Law. 10. Those who trust in an inherent sanctifying power in the works of the Law are under a curse. The Law gave no help towards keeping its mandates; and thereby multiplied sins. **11.** No one is made just before God by virtue of the Law; true justice comes only through faith. **12.** The Law does not rest on faith, because faith is concerned primarily with internal dispositions, while the Law regards only external acts. The citation is from Lev. *18, 5,* meaning that he who keeps the Law shall live; but St. Paul points out that keeping the Law is impossible without some help, which the Law itself cannot give. **13.** *Curse:* i.e., an execration, an expression or sentence of reprobation. The citation illustrates the way Christ redeemed us. He took upon Himself all the maledictions of the Law in order to liberate those who were under the Law. Crucifixion was not a Jewish method of execution, except under circumstances quite unusual, e.g., Num. *25, 4.* **14.** This verse indicates the purpose for which Christ died on the cross.

3, 15-18: The Promise of God. **15.** *Brethren:* the Apostle now speaks with the affection of a master for his disciples; he returns to his usual kindliness. *I speak after the manner of men:* I borrow an example from human life. St. Paul refers to the alteration of a will after the death of the testator. **16.** *The promises:* the plural is used because the promise was several times repeated to Abraham, and renewed to Isaac and Jacob. *Offspring:* i.e., the promise was made to Abraham and to his offspring, Christ, hence no one can have a part in this inheritance except in Christ, united to Christ by faith and love. **17.** St. Paul is either counting the time between the last renewal of the promise made to Jacob in the land of Canaan (Gen. *46,* 3-4) and the giving of the Law, i.e., he is reckoning the period during which the Jews were in Egypt, which according to the Hebrew text of Ex. *12,* 40 was 430 years; or he is following the Septuagint version of Ex. *12,* 40, which gives 430 years as the length of the period from the entrance of Abraham into Canaan until the departure of the Hebrews from Egypt. **18.** The conclusion now arrived at is that the inheritance of Abraham, Canaan, was not of the nature of wages for the observance of the Mosaic Law. God gave it to Abraham by promise, i.e., as a gratuitous gift. We must note that Paul uses inheritance here in a spiritual sense, as embracing all the blessings of which Christ is the source. Canaan was a type of these.

3, 19-29: The Purpose of the Law. St. Paul anticipates the objection that if the Law does not in any way affect the promise made to Abraham, why was it given? It was a divine institution, in no way opposed to the promise, a protection to the Hebrews, a moral guide to lead them to Christ.

19. The Law was a temporary measure. It revealed to man his sins and infirmities, without giving him the grace to overcome them, and so indirectly it multiplied man's sins; but it made him long for the help of a redeemer. That the angels had a part in the making of the Mosaic Law was a Jewish tradition, based on Deut. *33,* 2.   **20.** The employment of Moses as an intermediary showed that at the giving of the Law a contract was made, God binding Himself to bless the Hebrews as long as they observed His Law (Ex. *19,* 5-8). But in the giving of the promise there was but one, i.e., God; and so, no subsequent contract could in any way violate or impair this promise.   **21.** This is another part of the objection. Does it not follow that the Law was opposed to the promise? No. Such would have been the case if the Law was intended to accomplish what had been promised. Then justice, or supernatural holiness would have been by the Law, not by faith, and then the Law would have been against the promise (Rom. *4,* 13 f).   **22.** Contrary to the supposition of the preceding verse, the entire Old Testament shows that all men were enslaved by sin, in order that the inheritance promised to Abraham might be given to all who believe, not for their observance of the Mosaic Law, but by the faith of Jesus Christ (*1 Cor. 1,* 30). Cf. Rom. *3, 9* ff.

23. *Faith:* i.e., the object of our faith, Christ. *Shut up:* as captives and prisoners under the bonds of the Law. *For the faith that was to be revealed:* i.e., waiting for the coming of Christ. **24.** The *tutor* in Greek and Roman households was a faithful slave charged with the protection of the younger children. He took the children to school, and guarded them from danger on the way. In this sense the term is here applied to the Law, which led to the school of Christ, the acceptance of whose teaching is called faith.   **26.** The emphasis is on the word *all.*   **27.** *Put on Christ:* being clothed with Christ is another way of stating the incorporation in Christ which is the result of faith and Baptism. Cf. Rom. *13,* 14.   **28.** It now follows that religiously no differences, national or social, exist because of the identity with and in Christ.   **29.** This verse contains the conclusion that if all are united with Christ, then all are heirs to the inheritance promised to Abraham, and to his offspring, Christ.

2. Christians Live in a State of Freedom 4, 1-31

4, 1-7: Slavery and Freedom. **1.** *He differs in no way from a slave:* i.e., in the eyes of the law. The father of this child is supposed to be dead. St. Paul is only making a comparison, and every comparison is imperfect. Our Father in heaven never dies. **2.** *Guardians* were controllers of his person, and *stewards* of his property.   **3.** The application is now made. The meaning is that the ritual of Judaism was an elementary instruction preparing the way for the perfect doctrine of Christianity. Cf. Col. 2, 20.   **4.** *Fullness of time:* i.e., the time appointed by God the Father for the fulfillment of the promise made to Abraham. *Born under the law:* born of a Hebrew mother, and therefore subject to the obligation of the Law.   **5.** He was born under the Law, that He might redeem the Jews, who were under the Law. He was born of a woman, and became our brother, that He might elevate us all, Jews and Gentiles, to the dignity of adopted sons of God.   **6.** Cf. note to text.   **7.** The conclusion is that the Galatians are now the adopted sons of God and heirs to the inheritance. A better reading is: "You are no longer."

4, 8-20: No Return to Slavery. The Galatians had been slaves to material things. Now that they know God since their conversion, they should not return to religious slavery, which they would do by accepting the Law.

9. *Known by God:* i.e., they were the recipients of his graces. *Weak and beggarly elements:* the various heathen ceremonies. They were *weak,* because unable to lead men to salvation; and *beggarly,* since at their best they were only shadows of good to come.   **10.** Here Paul speaks of some of the Jewish practices which the Galatians were following, namely Sabbaths, the observance of the new moon, and other recurring festivals of the Jewish calendar.

12. *I have also become like you:* after his conversion he became a Gentile, he was free from the Law. *You have done me no wrong:* i.e., since you have not injured me personally, I feel free to plead with you.   **13.** *Infirmity* according to the majority of the Fathers means the trials Paul had in founding the

Galatian churches; according to most modern authors it means an illness of some kind, e.g., malaria, or a disease of the eyes. **14.** Instead of despising him, they received him as if he were an angel, or even Christ Himself. **15.** They congratulated themselves that they had Paul with them as their teacher. **16.** The *truth* refers to some preaching by Paul on his second visit which caused offense to the Galatians. **17.** *They would estrange you:* they would separate you from your true friends, Paul and his companions, or the Christian community. **18.** Cf. note to text. **19.** Paul regards the Galatians as through the eyes of a mother, who suffered in giving them Christian birth, and who would suffer the same pangs again to preserve them from perversion. **20.** *I wish,* etc.: to know more accurately their condition of mind; then he could change his voice according to the circumstances.

4, 21-31: Ismael and Isaac. **23.** *According to the flesh:* according to the ordinary laws of nature. *In virtue of the promise:* cf. Gen. *17*, 16.19; *18*, 10. **24.** *Allegory:* i.e., these characters have a higher meaning. *These:* namely, the two women. *Two covenants:* the Old and the New Testaments. The first was from Mt. Sinai, where it was contracted between God and the Hebrews. It brought them under the yoke of the Law, serving God as slaves and for fear of punishment. **25.** Paul shows the relation between Agar and Sinai, emphasizing the fact that she represents the Old Testament, which supposes subjection. The Mount, situated in Arabia, is appropriately connected with the allegory of Agar, the mother of the Arabs, for from Ismael sprang the principal Arab tribe. Many MSS have "For Agar is Mount Sinai in Arabia"; but this reading is less probable, where Agar is a gloss that has crept into the text. *Which corresponds with the present Jerusalem:* i.e., it is in the same class with the city which is now the center of Judaism, which holds her children in slavery. **26.** By *above* Paul does not mean exclusively the Church Triumphant, for he says that she is our mother. *Free:* i.e., not subject to the Law. **27.** The *barren* and the *desolate* spiritually represent the Gentiles, while the one *that has a husband* is the Synagogue; although this prophecy of Isaias (*54*, 1) literally refers to Jerusalem during and before the captivity. It implies that the children of the Catholic Church would be beyond all comparison more numerous than the Jewish nation. **29.** All that we read in Gen. *21*, 9 is that the son of Agar played with Isaac, but from

Sara's indignation as well as from Jewish tradition we may conclude that there was something offensive, something of a mockery in that playing, which Paul here calls a persecution. Likewise, the Judaizers now persecute Paul and the faithful Christians. **30.** Sara told Abraham to cast out Agar and her son. In like manner, the Galatians should expel the Judaizers. Cf. Gen. *21,* 10. **31.** We are not bound to the obligations of the Mosaic Law, from which Christ has emancipated us.

III: MORAL 5, 1 — 6, 10

1. GENERAL COUNSELS 5, 1-26

5, 1-6: Circumcision Now Voidance of Christ. If the Galatians regarded the Law as necessary for salvation, then they must renounce Christ.

1. This may be said to be the key-note of the Epistle. **2.** Paul speaks with the authority of an Apostle against those who think of circumcision as a complement to Baptism. Christ will be of no advantage to them, since they do not regard Him as the source of salvation. **3.** If circumcision is obligatory, so must the entire Mosaic Law be. **4.** You have put yourselves outside the pale of Christianity; you are incapable of receiving the grace of Christ. **5.** *We in the spirit:* i.e., in the grace of the New Law, wait for, or "look forward with intense longing" to the hope of justice. The latter is the hoped for crown of justice (*2* Tim. *4,* 8). *In virtue of faith:* not by the works of the Old Law. **6.** The Greek Fathers consider "working through" to be passive, i.e., "actuated by" charity.

5, 7-12: Judgment on Seducers. **7.** *You were running well:* i.e., you were doing well as Christians. The metaphor is taken from the races among the Greeks. The *truth* means the gospel. **8.** To believe that circumcision is necessary for salvation is not from the eternal Father. **9.** *Leaven:* here signifies a bad influence, generally in Scripture it is a symbol of evil. *A little:* i.e., though the Judaizers be few, they can do great harm. **10.** *I have confidence,* that when you have fully considered the question, you will retain no opinion which is not part of the doctrine I taught you. The author of all your confusion will pay the penalty for his guilt by the justice of God. **11.** This falsehood regarding the practice of Paul must have been circulated by his enemies, probably because he induced

Timothy to receive circumcision (Acts *16, 3*). As he observes, his persecution by the Jews was a sufficient proof that his doctrine on this point did not suit their views. *Then is the stumblingblock of the cross removed:* i.e., if Paul still preaches the necessity of circumcision, then the cross of Christ as a means of salvation, so scandalous to the Jews, is removed. Such a supposition is clearly absurd.   12. St. Paul ironically wishes that the Judaizers would go further, and insist even upon castration.

5, 13-26: How Christians Should Live.   13. *To liberty:* freedom from the ceremonial observances of the Mosaic Law. They are not, however, free of the moral precepts, so as to indulge the lower tendencies of the flesh.   14. *The whole law:* as it regards our relation to others. Cf. Lev. *19,* 18.   15. If they act like wild beasts, they will ruin their Christian community entirely.   16. Paul now explains what is meant by the liberty spoken of in 13. *The Spirit:* the life of grace in man. *Lusts:* the evil inclinations of our lower nature.   17. The reference is to the struggle of man against the weakness of his nature. The second part of the verse recalls that the flesh desires what is carnal, pleasant, and temporal, but the Spirit, what is spiritual, holy, and eternal. These are so opposed that frequently we cannot without the Spirit's help do all we would.   **19-21.** Paul does not enumerate all the works of the flesh but some specimens which he reduces to four classes: sins of luxury, of false worship, against charity and against temperance.   **22 f.** The Apostle now opposes the fruit of the Spirit to the works of the flesh. He uses the singular *fruit* to show that they are all united in charity. They are the effect of the grace of the Holy Spirit acting on the powers and faculties of human nature. *Patience, modesty,* and *continency* are wanting in the Greek. *Faith* does not mean the theological virtue, but fidelity. Those who practise the above-mentioned virtues are not under the Mosaic Law. **25.** It is not enough to have received grace, but we must lead a holy life.   **26.** Paul saw in *vainglory* a special danger to the Galatians, leading to challenges (*provoking one another*), and rivalries (*envying one another*).

2. Specific Counsels 6, 1-10

6, 1-5: Fraternal Correction.   1. With kindness and humility they should instruct those who err. *Caught:* i.e., led away by passion or surprise into a fault.   2. They should *bear one* *another's burdens,* or moral defects; and thus they will *fulfill the law of Christ,* which is reduced to charity.   **3.** Anyone who thinks himself anything is guilty of self-deception, because in the presence of God we are all nothing.   **4.** Let everyone examine his own life and actions. If he finds something good, he may rejoice, but not by passing judgment upon others.   **5.** Each one must answer for his own sins at the judgment-seat of Christ. Cf. *1* Cor. *4,* 4 f.

6, 6-10: Good Works.   **6.** *Share all good things:* i.e., give some of his temporal possessions.   **7.** God, who sees and knows all, is not mocked by false excuses that you may offer for the neglect of this duty.   **8.** The harvest depends chiefly upon the kind of seed that is sown, and upon the soil in which it is sown. This life is the time of sowing, eternity the time of reaping. *Sows in the flesh:* i.e., in the pleasures and ambitions of this world. *Reap corruption:* i.e., punishment, eternal death. *Sows in the spirit:* i.e., if he performs good works, which proceed from the grace of God in his soul, he will *reap* as his harvest *life everlasting.* This verse proves that good works done through grace can merit heaven as their recompense.   **9.** There is question here of perseverance in doing good.   **10.** During this life we should do good to everyone, but especially to Christians.

Conclusion 6, 11-18

6, 11-18: Summary.   **11.** St. Paul concludes the Epistle with his autograph in large characters to impress on the Galatians the importance of what he wrote. The remainder of the Epistle was probably written with his own hand and not by his secretary.   **12.** Here we have the real motive of those who were endeavoring to lead the Christians of Galatia astray. They wished to avoid being persecuted as believers in a crucified Messias. *In the flesh:* in a worldly way.   **13.** *Boast of your* (circumcised) *flesh:* they would boast of having induced you to submit to circumcision.   **14.** Paul will glory only in the cross of his crucified Savior, the one true source of justification. The Jews considered the cross a sign of ignominy, but to the Christians it was the cause of salvation, and the chief point in the preaching of Paul and of the other Apostles (*1* Cor. *2,* 2).   **15.** *A new creation:* the supernatural state of grace, which transforms one's mind, heart, and actions.   **16.** *This rule* refers to the glorying in the cross of Christ, and to a new creation. *Israel of God:* as opposed to Israel according to the flesh. It embraces all true Christians, whether of Jewish or Gentile origin.   **17.** The reference is to a practice according to which slaves bore marks made by burning or incision, which were indications of proprietorship. Paul was the property of his divine Master, and his marks were, e.g., the scars of wounds he received from the scourging at Antioch, and the stoning at Lystra.   **18.** The same benediction concludes the Epistle to Philemon. The word *brethren* put last, testifies that, notwithstanding their mistakes, Paul tenderly loves the Galatians.

JOHN F. ROWAN

THE EPISTLE TO THE COLOSSIANS

Introduction

The Churches of the Lycus Valley. The Lycus River rises in the mountains of Southern Phrygia and flows west and northwest for more than twenty miles before joining the Mæander. In ancient times Colossæ was situated on the left bank of the Lycus at the northern roots of towering Mount Cadmos. Laodicea, a larger and a wealthier city, lay about ten miles further down the valley. The great trade route from the Euphrates to Ephesus passed through both towns. Several miles directly north of Laodicea was Hierapolis, a famed shrine and a much-frequented health resort.

The Christian community at Colossæ was not founded by St. Paul. He had been very close to the Lycus Valley when on his third missionary journey he traveled from Antioch through Galatia and Phrygia to Ephesus (Acts *18*, 23; *19*, 1); but several passages of his letter to the Colossians (*1*, 4.9; *2*, 1) find their most natural explanation in the supposition that St. Paul had never preached there. During his long stay at Ephesus (55-57 A.D.), however, he must have converted many visitors from other Asian cities, who in their turn preached the gospel in their native towns (Acts *19*, 10.26; *1* Cor. *16*, 19). Among them were Philemon and Epaphras, both natives of Colossæ, perhaps, too, Nymphas of Laodicea (Col. *4*, 12.15; Philem. 19). Under St. Paul's direction (*1*, 7) Epaphras spread the good news of the gospel in the three cities of the Lycus Valley (*4*, 13).

Occasion of the Epistle. At first the converts of Epaphras, especially those in Colossæ, made great progress in their new life of grace and faith (*1*, 4-8; *2*, 5; Philem. 5-7). But before many years had passed, a tendency to superstition inspired by the unusual natural phenomena observable in the valley (hot springs, underground rivers, and earthquakes), and an inclination to syncretism based on the mixed character of the population, gave rise to errors in belief and practice with which Epaphras felt he could no longer cope unaided. He went to Rome where St. Paul had for two years (61-63 A.D.) been awaiting trial before Cæsar. From the report of Epaphras the Apostle saw that a new danger, more subtle than the heresy he had condemned in his letter to the Galatians, now threatened the Asian Christians. To warn them of their peril and to furnish them with weapons for defense, he wrote the Colossians a letter embodying an energetic protest against the heretical opinions and a more complete instruction on Christian doctrine and practice than they had as yet received.

Place and Time of Composition. A comparison of several passages in the four letters of the first captivity of St. Paul in Rome makes it likely that they followed each other in this order: Colossians, Philemon, Ephesians, and Philippians, and that they were written in the spring of 63 A.D. Though the Apostle knew the outcome of his appeal to Cæsar's tribunal (Acts 25, 11 f) might yet be fatal, he nevertheless hoped to be liberated and he planned to send Timothy on to Philippi before him (Phil. 2, 19-24). Philippians, then, can be placed at the very end of St. Paul's imprisonment. When the Apostle wrote to Philemon, his hope of a release was only a little less strong (Philem. 22). Since Tychicus, accompanied by Onesimus who had with him the letter of recommendation to his master Philemon, was the bearer of both Colossians (Col. 4, 7-9) and Ephesians (Eph. 6, 21-22), these three letters will have been written a few months before Philippians; and Colossians before Ephesians, which is more general in its theme and in detail more developed.

The Colossian Errors. To understand fully St. Paul's Epistle to the Colossians it is important that one have a clear idea of the nature of the errors which he sought to refute. While the Apostle's first readers knew quite well the persons and the teachings he condemned, for us many of his expressions referring to them lack clearness. However, by grouping his direct and indirect references to the sources, the essence, and the objective of the heresy, we can form a picture of it definite enough to serve as a foil for the revealed truths which St. Paul opposed to the errors.

The false teachings were drawn not from the gospel, the true message of salvation through Christ preached by divinely commissioned apostles, but from human sources. The contribution of the pagan environment is marked by terms that were current in the mystery-cults (2, 8 f), such as "elements" and "fullness"; perhaps, too, by the words philosophy, mystery, knowledge, life, member, perfect. What had been borrowed from the Old Testament was wrongly interpreted and applied (2, 16-23). Based ultimately, then, on the fallible mind of man, the "philosophy" was a system void of real content, but capable of leading weak souls astray (2, 8.18). Yet the false teachers took a superior attitude toward their fellow-Christians (2, 4.19.23). Censorious of the morals of others, they followed practices of devotion and asceticism calculated to give the impression of a more perfect religion.

The central point in the heretical teaching seems to have been the worship of angels (2, 18). It was thought that they, the Elements of the World, spirits animating the natural forces and governing the stars and planets, controlled human destiny. In some measure the evil in the world was due to their influence. They were a series of spiritual beings, more or less divine, interposed between God and man. Christ, then, was not the only mediator; these angelic powers, too, must be appeased by man in his striving for reconciliation and union with God.

To this purpose the Colossian heresiarchs proposed a religious system in which ideas borrowed from the mystery-cults were joined to observances taken from Judaism. The body, a conglomerate of matter controlled by the elemental spirits, must be purified by ascetical practices; to win the favor of the angels as "Lords of Time," enthroned in the heavenly bodies, feasts were to be kept each week and month and year (2, 16-23). By meticulous adherence to such rules might one hope to be freed from evil.

St. Paul's effort to offset the danger is a model of constructive criticism. He attacks the false teaching rather than the heretics, who remain unnamed. His method is positive. The true doctrine must be derived from authentic sources: men whom God has sent, either directly, as the Apostle himself (1, 26); or indirectly, as Epaphras, his disciple (1, 7; 4, 12 f). In this original gospel message is clearly contained the doctrine of Christ's divinity and His pre-eminence over all creatures, even the angels (1, 15-20). His redeeming death is all-sufficient (1, 14.22); its effects have, indeed, to be applied to individuals (1, 24), but cannot be supplemented or supplanted by the action of beings subordinate to Christ (2, 9 f).

Away, then, with the system of belief and practice (2, 8-23) that would prolong the subjection of man to the powers of evil vanquished in the death of our Savior! Lift up your hearts (3, 1-4)! Otherworldliness, not serving the Elements of the World, must be the mark of Christian asceticism. Virtues, good habits that bring peace and joy to the soul, must take the place of pagan sins (3, 5-17). Christians, in fact, are other Christs (3, 3 f).

COMMENTARY

Introduction 1, 1-14

1, 1-2: Greeting. **1.** In five other letters (*1.2* Thess., *2* Cor., Phil., Philem.) St. Paul associates with himself his disciple Timothy, who, as some commentators suggest, may have written this letter at the Apostle's dictation. But see *4, 14* and 18. *Apostle:* writing to Christians whom he does not know personally, St. Paul appeals to his infallible authority as an Apostle chosen by God. **2.** *Holy:* separated off from the world and consecrated to God in Baptism (*2, 12; 3,* 1-3). *Faithful:* either having the faith, believing; or steadfast and true. *Grace, peace:* the friendship of God, bringing with it tranquillity of soul.

1, 3-8: Thanksgiving. **4.** *Love* must be joined to *faith,* else faith is dead (Gal. *5,* 6; Jas. *2,* 23-26). Epaphras had reported (8) an instance of their love shown especially in works of charity towards fellow-Christians. **5.** *Hope* means here the things hoped for, as in Titus *2, 13.* Believing in God and loving Him, the Christian strives by leading a life of virtue to attain to salvation and an eternal reward, the full possession of God in the beatific vision. In Christian hope, true love of self is selfless, because it looks to God's greater glory. St. Paul himself labored with the eternal reward in mind (*1* Cor. *9,* 23-25; *2* Tim. *4,* 8). **6.** *Whole world:* a hyperbole, as in 23; similarly *1* Thess. *1,* 8; *2* Cor. *2, 14;* Rom. *1, 8.* The gospel bears fruit and grows through the good works and spiritual progress of the faithful and through new converts being won to the faith. *The grace of God:* the free gift from God of redemption and salvation through Christ preached in the gospel. **7.** *Epaphras,* who had preached the gospel in Colossæ and the adjacent towns (*2,* 1; *4,* 12-13), shares with Tychicus (*4, 7*) the honorable title of a companion of St. Paul in the service of Christ. **9.** The Apostle prays constantly that the knowledge they received (6) may grow until they are filled with it. *Spiritual:* coming from the Holy Spirit (Jas. *3,* 17). Of the gifts of the Holy Spirit, *wisdom* directs the mind to refer all things to the supreme cause, while *understanding* enlightens us concerning their true nature.

1, 9-14: Prayer for Their Progress. **10.** The gifts for which St. Paul prays will spur the Colossians on to greater charity and give them a deeper insight into God's will. **11.** Besides *knowledge,* they will need God's help to bear their trials patiently. **12.** Christian *patience* is always tinged with joy, enabling the faithful to accept hardships with rejoicing and even to render thanks to God for them. **13.** To the metaphor of *light* for the life of grace corresponds *darkness* as the realm of sin and Satan (John *3,* 19 f; Acts *26,* 18; Eph. *5,* 8; *1* Thess. *5, 5*; *1* Pet. *2, 9*). **14.** The transfer occurs at Baptism, when we put on Christ (Gal. *3,* 27; Rom. *6,* 4 f). We are reborn in Christ. In that rebirth, the grace by which we are justified and our sins are forgiven is granted us through the merits of His Passion. (The Council of Trent, Session vi, chapter 3). The *kingdom* (of the Son) is the Church Militant.

I: THE PRE-EMINENCE OF CHRIST 1, 15 – 2, 3

1, 15-20: God, Creator, Head. **15.** *Image:* the same expression is used of Christ in *2* Cor. *4,* 4 and of man in *1* Cor. *11,* 7. An image is both derived from an archetype and resembles it. Man, created to God's image (*3,* 10; Gen. *1,* 26 f), is like God in his being endowed with spiritual faculties and supernatural grace. Christ is God's image in that He possesses the divine nature (*1,* 19; *2, 9*). Through the image we come to know the invisible God, the Father, whom we cannot see directly (John *1,* 18; *14,* 9; *17,* 21-26; Matt. *11,* 27; Luke *10,* 22). *The Firstborn of every creature:* to the firstborn belong both priority and pre-eminence. Here, then, St. Paul combines two ideas: the Second Person of the Blessed Trinity, incarnate in Christ, is "born, not made," and for that reason is above all creatures; besides, existing from all eternity, He is prior to creatures made in time.

16. The sense of *in him* is explained by the formulas *through him* (John *1,* 3) and *unto him* (Rom. *11,* 36): the Word is the efficient and final cause, the beginning and the end (Apoc. *22,* 13) of all created things, corporeal and incorporeal; in Him all things are centered. *Thrones,* etc.: the terms can refer to good and bad angels indiscriminately. See *2, 15* where demons, and Eph. *3,* 10 where good angels are meant. St. Paul names four of the angelic orders (Eph. *1,* 21; Rom. *8,* 38). The Thrones are not mentioned elsewhere. The Apostle began here to attack the false teachers

(*2*, 10.15.18) who raised angels to a place of mediatorship between God (or Christ) and men.

17. The Son is supreme because all created things came into being through His agency, have permanence in Him, and exist for Him who is God (Acts *17*, 28; Heb. *1*, 3). **18.** The same person, the Word made flesh (John *1*, 14), Jesus Christ, is also *the head,* that is, the principle or source, *the beginning,* of the supernatural life of grace in *the Church,* which is *His* Mystical *Body* (see Eph. *4*, 15 f; *5*, 23. 29 f). Since He first conquered death in His Resurrection (*1* Cor. *15*, 20; Apoc. *1*, 5), He is *first-born* (in a sense different from that of 15) *from the dead.* As Creator and Redeemer, in the created world and in the spiritual order represented by the Church, He must be held to be chief and leader. The false claims that St. Paul has in mind in *2*, 10.19 are thus exploded.

19. The ultimate reason of Christ's pre-eminence is now given. *God the Father,* not present in either the Greek or the Latin, is supplied as subject, for here the thought of 12-14 is again taken up. It was God's free choice that in Christ there should reside permanently a *fullness* of graces deriving from the fullness of the divinity (see *2*, 9 below). **20.** As sinners we were God's enemies, standing in fear of His wrath and punishment; reconciled and in peace with Him, we have the hope of union with Him in heaven (Rom. *5*, 1-11; *2* Cor. *5*, 18-21). *Earth, heavens:* the benefits of the Redemption are shared by all creatures; not only by men but also by inanimate nature (Rom. *8*, 20 f) and by angels (Eph. *3*, 10), who rejoice in the deeper knowledge they have gained of God's plan of salvation.

1, 21-23: **Conciliator.** **21.** What the Apostle has affirmed of all mankind he now applies in particular to the Colossians. Before their conversion they were, as pagans and sinners, far from God, in thought and deeds His enemies. **22.** Christ's atoning death, His sacrifice of Himself on the cross, has made them here and now, even before the final judgment, worthy of God's friendship. *His body of flesh:* Christ's human body, subject to sufferings and death, to distinguish it from His Mystical Body, the Church (18). **23.** Steadfast in faith, the foundation of their spiritual structure, the Colossians must hold firmly to the hope, based on the promise of an eternal reward, contained in the gospel (5). To give up this hope for the false teachings of a sect would be to cast aside what the whole world (a hyperbole as in 6) has embraced; for to them Epaphras had preached the same doctrine his master St. Paul had taught in other cities.

1, 24 — 2, 3: Center of Preaching. 24. A consoling passage, showing how intimately we can be united with our Savior. The tribulations St. Paul has borne, such as the labors enumerated in 2 Cor. *11,* 23-28, and his imprisonment in Rome, far from making him lose courage, are a source of true joy, for the infinite merits of Christ's Passion and Death (Heb. *9*) are thus applied to all the members of the Mystical Body. One with Jesus in his spirit of sacrifice, the Apostle too will be put to the supreme test (2 Tim. *4,* 6), that of giving his life (John *15,* 13).

25-27. *Minister:* servant. St. Paul had received a special mission to the Gentiles (Acts *9,* 15; *13,* 2-3; *22,* 21; Gal. *1-2*). *Fulfill:* to aid in bringing about a full realization of God's plan of salvation by preaching the gospel everywhere (Rom. *15,* 19). *Mystery:* see *1* Cor. *2,* 7; Eph. *3,* 1-13. Not a teaching reserved for a small number of initiates, as in the mystery cults; but a truth which cannot be fully known to reason unaided by revelation. *Hidden:* that the Gentiles were to enter the New Covenant had been foretold by the prophets of the Old, as the Apostle himself points out in Rom. *1,* 2; but in the New Testament this design of divine Providence has been made so much more clear that the ancient promise can be spoken of as obscure or kept in silence (Rom. *16,* 25). As a short, comprehensive formula of the mystery, St. Paul chooses the words: *Christ in (or among) you, your hope of glory.* Before their conversion the Colossians were, like other pagans, without hope (*1* Thess. *4,* 13); now their hope of a reward in heaven rests with Jesus (*1* Tim. *1,* 1).

28. His work as Apostle to the Gentiles entails *admonishing,* turning men from their sins to lives of virtue; and *teaching,* imparting full knowledge concerning God's revelation. To emphasize the all-embracing character of his mission, St. Paul thrice repeats *every man.* **1.** Consequently he has at heart the spiritual welfare even of communities not founded directly by himself. **2.** When they have learned to love one another for God's sake, their Christian life will become richer in their perceiving fully God's purpose in revealing Himself in the person of Christ (John *17,* 21-23. 26). **3.** *Hidden:* in Christ all spiritual wealth lies, not concealed, but buried, awaiting discovery by those who ask for it in seeking Him.

II: WARNINGS AGAINST FALSE TEACHERS 2, 4 – 3, 4

2, 4-7: A General Admonition. **4.** *Persuasive words:* the false teachers boasted of their views as embodying a higher form of religion. See 8, 18, 23. **5.** *Orderly array, steadfastness:* the Greek words are military terms, which may have been suggested to St. Paul by the soldiers under whose guard he was placed. The condition of the Colossians reported by Epaphras was good (*1,* 3-8); though there was danger, the cancer of heresy had not yet developed beyond the first warning stages. **7.** *Rooted, built up:* like a tree, drawing nourishment from the ground through its far-reaching roots, the Colossians derive their life of faith from Christ; He is the solid foundation of their spiritual edifice. For the same metaphors see Eph. *2,* 20-22; *3,* 17.

2, 8-15: Speculative Errors. **8 f.** *Philosophy:* it is likely that here St. Paul takes up, in order to refute them, the very terms under which the Colossian heretics proposed their errors. They called their system of religious teaching "philosophy"; but it was mere sophistry, inefficacious and deceptive, based on the teaching of men, not on a revelation from God. *Elements of the world:* see Gal. *4,* 3.9. The *world* is this earth as the kingdom ruled by Satan (John *12,* 31); the *Elements* are rebellious angels, thought of as controlling not only the simple substances of earth, air, fire, and water; but also the sun, moon, and stars, and through them the destiny of man. The Greek word for *elements* had, in the first century, been turned from its original meaning to designate these demons. St. Paul, too, speaks of them as rulers of this world (*1* Cor. *2,* 6.8; Eph. *6,* 12). In the oriental mystery cults it was pretended that the initiated could by various ceremonies and renunciations placate these spiritual powers, free themselves from cosmic influences, and rise to a higher sphere of life, thus achieving "fullness" of being. With beliefs of this nature some Phrygians, perhaps of Jewish origin, had combined practices borrowed from the Old Testament into a system that neglected the fundamental revealed truths of Christ's divinity and His sole mediatorship. There is a true sense of "fullness," the sum-total of divine being in Christ. *Bodily:* in Him the divine nature resides through union with a human body in the person of the Word (John *1,* 1.14).

10. *Head:* used here in a sense different from that in 19 and *1,* 18. Christ as God is the chief, the leader and ruler of the angels (they are not members of His Mystical Body). In the next four verses St. Paul reminds the Colossians of various aspects of our participation (John *1*, 16; 2 Pet. *1*, 4) in the fullness of the Godhead in Christ.

11 f. The physical circumcision, of which the false teachers perhaps boasted and to which some Colossians may have been attracted, St. Paul's readers do not need; for they too were circumcised, in a spiritual sense, when they threw off the tyranny of the body and its sinful desires. A similar contrast between an outward and an inward circumcision, an external rite and a moral change, is made in Rom. *2*, 25-29. Renouncing sin is the Christian circumcision. Received at Baptism (for the symbolism of death and resurrection see Rom. *6*, 3-6), its effect must be sustained through mortification (*3*, 5. Rom. *8*, 13). **13.** *The uncircumcision of your flesh:* as pagans they were enemies of God (*1*, 21), spiritually dead, not only because of their evil deeds, but also because lacking the circumcision, both physical and spiritual. See the more detailed statement of the thought of this verse in Eph. *2*, 1-12.

14. St. Paul speaks of the obligation men owed to God as of a decree or bond that stood against them, a note of indebtedness which they had countersigned. The Jews were bound in a special manner by the Law of Moses, signed when they agreed to observe the conditions of the Sinaitic pact (Ex. *19*, 8; *24*, 7); the Gentiles were indebted to God, in a more general sense, by the law written in their hearts (Rom. *2*, 12-15). In cancelling our bond, God not only forgave us our sins (debts, Matt. *6*, 12), but did away with the instrument of our indebtedness. This benefit God conferred on us when He suffered His Son, who took our sins upon Himself, to be nailed to the cross (*1*, 14. 20 f; John *3*, 16). Thus was the curse of the Law annulled, the ancient Covenant abolished (Gal. *3*, 13; Eph. *2*, 15 f). **15.** But Christ's death on the cross effected even more: the evil angels were deprived of their power over men (John *12*, 31). To bring home this truth St. Paul here uses two metaphors: that of a victorious leader setting up or otherwise displaying the weapons of his conquered enemy as a trophy, and of a triumphal procession in which the vanquished are led chained to the victor's chariot.

2, 16-23: Erroneous Practices. **16 f.** From the Mosaic religion, it would seem, the Colossian heresiarchs had taken over the observance of weekly, monthly, and annual holydays, and restrictions in eating and drinking; they condemned faithful Christians for not following them in these practices. The distinction in the Law between clean and unclean animals was binding on all Jews; restrictions in matters of drink were imposed on priests when serving at the sanctuary or incurred by the vow of Nazirites (Num. *6, 3*). St. Paul warns his readers that since they possess the reality, the Christian life, foreshadowed by the feasts and fasts of the Old Testament, they must not submit to the false teachers. **18.** *Worship of Angels:* a cult paid not to good, but to fallen angels (Matt. *25,* 41; 2 Pet. *2,* 4; Jude 6; Apoc. *12,* 7), evil spirits, by whom it was imagined the heavenly bodies and the earth's constituent elements were animated. See 8 f. By their self-imposed asceticism and the keeping of seasonal feasts regulated by the planets, the Colossian heretics sought to win the favor of these hostile spirits. They pretended thus to have a higher knowledge of spiritual things than was preached in the gospel, an attitude quite foreign to the true humility that leads to the acceptance of God's revelation. **19.** See Eph. *4,* 15 f. The false teaching, finally, is a denial of the truth of Christ's Mystical Body (*1,* 18; *2,* 10): from our head, who is God, we, the members, derive our life of grace and our spiritual growth. We do not become divine; but with God's help, we can perfect our nature through the practice of virtue, thus becoming more like to God.

20 f. *Lay down the rules:* in Greek the verb is passive, and can be translated: "Why do you submit to ordinances?" The Latin suggests that the Colossians had begun to apply such rules to one another. Since by dying with Christ (*2* Tim. *2,* 11) the Christians are freed from the angelic powers that dominated the world through sin, the asceticism by which men denied themselves food and drink to placate them serves no purpose. **22.** A parenthesis: food and drink are destined for the use of man. The quotation is from Isa. *29,* 13; our Lord, too, uses it in repudiating a similar false teaching of the Pharisees (Mark *7, 7*). **23.** The religion of the heretics is *superstition* (in the Greek: a self-chosen form of worship), their asceticism self-imposed, their attitude toward the body unnatural. In the end their practices, centering the mind as they do on material things rather than on moral principles, lead to giving way to the sensual appetite.

3, 1-4: **Mystical Death and Resurrection.** To those who believe in the gospel, who live a new life of faith, other realities than the material things of earth are all-important. Like the martyr Stephen (Acts 7, 55), they bear witness by their attitude to their risen Lord, occupying a position of honor and authority in heaven. Hidden now from men, their spiritual life will one day be made manifest to all.

III: THE IDEAL CHRISTIAN LIFE IN THE WORLD.
3, 5 — 4, 6.

3, 5-11: Renounce Vices. **5.** *Members:* it may be that here there is another use of the language of the false teachers. They recommended feasts and fasts as a means "to mortify earthly members," men who clung to worldly things instead of the vaunted higher Christianity of the heretics. St. Paul takes up their words: "Indeed, 'put to death your earthly members,' i.e., in the light of Rom. 6, 19-21, the vices to which you are inclined." He then names several typical sins by which the pagan world was bound to Satan, adding covetousness to carnal sins, for the voluptuary needs money to obtain satisfaction of his evil desires. For more complete lists see Gal. 5, 19-21; Rom. 1, 29-31.   **7.** *Life* for them had at one time (see *1,* 21; *2,* 13) consisted in these sinful pleasures.   **8.** *Put them all away:* the sins of their pagan years, both what has been listed and is yet to be mentioned. *Anger:* vehement outbursts of temper. *Wrath:* a feeling of anger nursed in the soul. *Malice:* a disposition to harm others. To these the Apostle adds sins of the tongue.   **9 f.** The precepts given in 5-8 are repeated under the figure that lies behind *2,* 11. See Rom. *6,* 6. Through the working of the Holy Spirit (Titus *3,* 5) the new man within us (2 Cor. *4,* 16) is ever renewed, recast in a perfection modeled on God's holiness.   **11.** Here, in the new man, distinctions based on religion, culture, and social standing fall away. In Christ we receive everything (Rom. *8, 32*).

3, 12-17: Practise Virtues.   **12.** The virtues here recommended were sadly lacking in pagan life (Rom. *1,* 31).   **13.** This verse is connected by Eph. *4,* 32 with *2, 13,* where forgiving is attributed to God (the Father).   **14.** *The bond of perfection:* achieving a permanent union among the members of the Mystical Body, a condition in which the virtues recommended in the previous verses will be continuously practised.   **16.** The teaching of Christ is to be embodied in songs, and inculcated and spread through singing them in religious gatherings.   **17.** *The* *name,* in scriptural usage, is the person as known. All the speech and all the actions of a Christian must proceed from a principle in union with Christ. A habitual reference of our words and deeds to God suffices. See also Eph. *5,* 20 and *1* Cor. *10,* 31.

3, 18-21: The Christian Family. **18.** *In the Lord:* as disciples of Christ. The manner and degree of subjection is determined by Christian teaching. See *1* Cor. *11,* 3-16; Eph. *5,* 22-23. **19.** The recommendations made in 12-13 are to be applied by husbands to their wives. If they find that their spouses do not come up to their expectation in all things, they may not be harsh, irritable, or sullen towards them; but must practise kindness, meekness, and longsuffering. **20 f.** The dutiful subjection of children will be easier if the commands of fathers are not given in a haughty manner or with too great severity.

3, 22 — 4, 1: Slaves and Masters. **22-25.** To slaves a high ideal of service is here proposed. "Consider yourselves slaves of our Lord Christ." In their relation to God the slaves stand on an equal footing with their masters (*4,* 1). But the latter have claims on them, based on custom and civil law, which are not to be set aside. Even at laborious tasks the slaves are to work with a will, without complaint, acknowledging thus the wise designs of God's providence, who requires such obedience of them. From their earthly masters they receive neither salary nor inheritance, but our Lord will repay them in full. Wrong done to masters will also be requited; even the Christian slave who wrongs a pagan master may expect no benefit of faith. **4, 1.** Your Master in heaven will deal with you as you deal with your slave, a special application of Matt. *8,* 2. Job (*31,* 13-15) was a master who kept in mind this rule of final retribution.

4, 2-6: Prayer and Prudence. **2.** Our Lord's precept of watching, of being at all times spiritually prepared for His Second Coming (Matt. *24,* 42-44), can be observed by maintaining a spirit of prayer, keeping our souls turned toward God, especially by thanking Him for His bounty. **3.** *Us:* as in *2* Thess. *3,* 1, St. Paul here uses the plural pronoun in referring to himself. His only concern is the spread of the gospel; he does not request prayers for his liberation from prison **4.** *Ought:* see *1* Cor. *9,* 16. **5 f.** All Christians share in the apostolate at least in this that they must give to outsiders the example of a holy life. Their conversation, flavored with spiritual reflections, must be pleasing and discreet, especially when questions concerning Christian practices are put to them. See *1* Pet. *3,* 15.

Conclusion *4,* 7-18

4, 7-9: Tychicus and Onesimus. 7. Tychicus was from Asia (Acts *20,* 4) and like Trophimus (Acts *21,* 29), with whom he is associated, he was probably a native of Ephesus. 9. Notice that a slave is being recommended on equal terms with a free fellow-Christian.

4, 10-14: From Paul's Co-workers. 10. *Fellow-prisoner:* the same epithet is given to Epaphras in Philem. 23 (and to Andronicus and Junias in Rom. *16,* 7). It can mean that without bearing chains these disciples shared the Apostle's imprisonment. *Mark:* this recommendation of the relative of St. Paul's sponsor shows that the difficulties spoken of in Acts *15,* 39 and Gal. *2,* 13 had been amicably settled. 11. *Jesus . . . Justus:* Jesus here and in Luke *3,* 29 (like Jason in Acts *17,* 6; Rom. *16,* 21) is a Greek form of Josue (Acts *7,* 45; Heb. *4,* 8), a common name among Jews; the surname Justus, the Latin equivalent of Sadoc (Matt. *1,* 14), was also given to Joseph Barsabbas (Acts *1,* 23) and Titus, St. Paul's host at Corinth (Acts *18,* 7). Some of the Jewish converts at Rome did not support St. Paul. See Phil. *1,* 15-17. 12 f. *Epaphras:* see Introduction and *1,* 7. From this passage we learn that he was a Colossian, and can gauge his zeal for the welfare of the churches in the Lycus Valley. The pastoral solicitude of the disciple was modeled on that of his master (*1,* 28; *2,* 1 f). 14. *Physician:* added perhaps to distinguish the Evangelist from other Christians who bore the same or a similar name (Lucius, Acts *13,* 1; Rom. *16,* 21). St. Paul was, no doubt, indebted to him for medical care. See Gal. *4,* 13; *2* Cor. *12,* 7-9. *Demas:* the absence of a note of praise may indicate that he was the scribe who wrote this letter at the Apostle's dictation. But cf. *2* Tim. *4,* 9.

4, 15-18: A Message for the Laodiceans. 15 f. Concerning the view that the "Letter of the Laodiceans" is the same as Ephesians see the Introduction to that Epistle. The exchange of the Apostles' letters between Christian communities and their public reading in the churches (*1* Thess. *5,* 27) was the first step in their canonization. **17.** These words need not imply that Archippus was slack in the performance of his duty; they may, like *2 Tim. 4, 5*, be an encouragement to face difficulties bravely. **18.** Other examples of the Apostle's signature at the end of letters: *2 Thess. 3,* 17; *1 Cor. 16,* 21; Gal. *6,* 11. The hampering shackles move the prisoner of Christ to repeat his request (*4, 3*) for the intercession of his readers. *Grace:* St. Paul ends, as he began, with a prayer for God's favor.

MAURICE A. HOFER

THE EPISTLE TO PHILEMON

INTRODUCTION

Occasion and Purpose. Philemon, a well-to-do Colossian Christian, was a convert and close friend of the Apostle, St. Paul. That he was a citizen of Colossæ is inferred from Col. *4, 9*, where his slave Onesimus is said to be "one of you." Since Philemon owned slaves and was able to succor his fellow-Christians in their need (5-7), we may safely judge that he possessed wealth. The Apostle gives clear marks throughout this letter of his affection for Philemon; that he had converted him to Christianity is the meaning of 19. Now, one of Philemon's slaves had, perhaps after robbing his master (18), fled to Rome. There he met St. Paul, and was won for the new life of faith (10 f.) The Apostle found him useful, either as a personal attendant or in his work of evangelizing; but however much he wished to keep him, he must practise the doctrine he preached and restore the slave to his owner. He gave him this letter of recommendation, in which he requests, both as a personal favor (9.20) and because Onesimus now is a fellow-Christian (16), that his master receive him kindly. More, he almost asks in 21 for the slave's manumission.

Time and Place of Composition. The note to Philemon, a pendant to Colossians, was written at Rome in the spring of 63 A.D. See Colossians, Introduction.

COMMENTARY

1-3: Address and Greeting. **1.** *Prisoner:* because this letter is more a personal appeal than an authoritative command St. Paul here substitutes an allusion to his bonds for the more usual title of Apostle. See also Eph. *3,* 1; *4,* 1. **2.** *Church:* the Christians met at his home for religious services.

4-7: Philemon's Faith and Charity. **4.** *Always* can also be drawn to the main clause. **5.** Faith in Jesus shown in charitable deeds toward the saints (see Col. *1,* 4). **6.** May thy sharing of material possessions from motives of faith result in deeper appreciation among thy fellows of the graces our Savior has granted them as Christians.

8-20: Plea for Onesimus. **9.** *Since,* etc.: in Greek: "Paul an old man—for such I am." **10.** St. Paul is the spiritual father of his converts (see *1* Cor. *4,* 15). **11.** The Apostle is alluding to the slave's name—Onesimus, in Greek, means "profitable, useful." **16.** The stress lies on *as.* The runaway slave Onesimus will, on his return, be treated as a fellow Christian.

21-25: Hopes, Greetings, Blessings. The Apostle is confident not only that his request will be more than fully granted, but that he will soon have the happiness of reunion with his friends. He repeats the greetings already expressed in the longer letter (Col. *4,* 10-17), and closes with his benediction. **21.** St. Paul does not ask directly that Philemon grant Onesimus his freedom (see *1* Tim. *6,* 1 f). **22.** The Apostle expected soon to be freed at Rome (see Phil. *2,* 24).

Maurice A. Hofer

THE EPISTLE TO THE EPHESIANS

INTRODUCTION

The Recipients. The Epistle to the Ephesians is addressed "to all the saints who are at Ephesus and faithful in Christ Jesus." The address is thus in the very great majority of ancient authorities; but there is a minority of considerable weight which leaves out the words "at Ephesus." These words are omitted in the codex *Vaticanus* and the codex *Sinaiticus*, our oldest manuscripts (but supplied in both by later hands). They are expunged in a later corrector of codex 67. St. Basil tells us in the fourth century that they were not in the ancient manuscripts. They were not accepted by Origen, who tried to give an acceptable meaning to 2 when they are left out: "To those who really are." St. Basil followed him. St. Jerome, who had read Origen, does not think of omitting the words "at Ephesus" but refers to the speculation about "those who are." Tertullian, who combatted Marcion's ascription of the Epistle to the Laodiceans (those referred to in Col. *4, 15* f) appeals to the tradition of the Church, which ascribed our Epistle to the Ephesians, but without mentioning the "at Ephesus" of our text. A mistake of Origen, followed by St. Basil, regarding the traditional text, is hard to explain; but it would be much harder to explain that a text which supposes the Epistle to have been addressed to the Ephesians came to be so well established if St. Paul did not put it there.

Some able scholars have recently revived the view of Marcion that our Epistle is really the letter of which St. Paul speaks at the end of Colossians. There was a letter to come *from* the Laodiceans; but it is not spoken of as a letter written for them. If Ephesians was a circular letter it could have been the letter from Laodicea. The bearer, Tychicus, might have left it at Laodicea when he brought to the Colossians the one addressed to them specially.

That Ephesians was a circular letter is the view of a great many scholars of the present day. This view is suggested chiefly by its contents. It is felt that if the Epistle were addressed exclusively to the Ephesians it would be much more personal. There is no reference in it to the three years the Apostle spent at Ephesus, where he must have made many friends. One has but to think of his parting from the presbyters of Ephesus at Miletus (Acts *20,* 17-38) to find it strange that he has no special greetings for anyone, and that he seems to know about the faith of the Ephesians only by hearsay.

We may regard it as a letter written by St. Paul after he had finished his Epistle to the Colossians. He had vigorously condemned errors and teachers of errors; now in a calmer mood he sets forth objectively the truths on which he has been meditating and utters a hymn of praise to God who is so infinitely good to man. The dogmatic portion of the Epistle, as well as the moral teaching, sets forth universal principles. Nothing seems restricted to the use of any one church. On the other hand, it is confined to a definite group of churches; Tychicus, who is to tell the recipients everything concerning Paul's circumstances and to comfort their heart, must not be thought of as visiting all the churches of the world (Eph. 6, 21 f). Most likely St. Paul had in mind the Christians of the province of Asia or of some part of it. He had particularly in mind those who did not know him personally and on account of them he would avoid personal references to the past. We have no right to bind the versatile St. Paul to one kind of letter.

Authenticity. The external evidence in favor of the Pauline authorship is very strong. Ephesians belongs to that group of thirteen Epistles about whose inspiration and authorship the early Church was always sure; there is not the slightest doubt about the matter expressed in ancient documents. This, as in the case of other books so well guaranteed by ancient tradition, is decisive for any one who can and will weigh the evidence. Going back into the obscure period before St. Irenaeus and the other writers of about 200, one finds such indications as one might expect of its existence. About 140 A.D. the heretic Marcion regarded it as the work of St. Paul though he thought it was an Epistle written to the Laodiceans. Hippolytus tells us it had been used by the Gnostics Valentinus, Basilides, Ptolemy. This writing, then, went back at least to the earlier part of the second century (before 130 A.D.) when the Church and these heretics used the same Scriptures. There are citations or reminiscences in the works of still earlier writers. It was probably used by the very early Epistle of Barnabas; it is morally certain that it was used by the Apostolic Fathers, St. Polycarp, St. Ignatius and Clement of Rome (before 100 A.D.). No forger could have got his book accepted so early by all the leaders of the Church.

The opponents of the authenticity of the Epistle do not deny that there is very strong external evidence in favor of the tradition, but they maintain that there is something in the work itself which does not allow an enlightened critic to declare that it comes from Paul, or at least to be sure of it. The only way to answer their objections completely is to go into minute details about the vocabulary, style, and the doctrine of the Epistle. Catholic scholars, and many who are Protestants and even rationalists, have done this in learned commentaries and special studies. While we may keep our opponents in mind in our short commentary here, we prefer to set forth positively our reasons for regarding our Epistle as a genuine expression of St. Paul's mind and heart. He has a manner of his own which is inimitable. Father Prat says that anyone who has read even a few paragraphs of the Apostle will recognize his work anywhere; one may recognize him even in Ephesians, which seems to be now an object of special attack, provided he does justice to the Apostle's command of words, his tendency to overload his sentences when setting forth his dogmatic teaching, and his ability to develop doctrine without altering it.

COMMENTARY

Introduction 1, 1-14

1, 1 f: Greeting. 1. *At Ephesus.* While not writing exclusively for the church at Ephesus (see Introduction) St. Paul had it specially in mind. Tychicus, the bearer of this Epistle as well as that to Colossæ, would on his journey from Rome to Colossæ land at Ephesus and the Ephesians would be the first to read what was in all probability a circular letter.

1, 3-6: The Eternal Plan of the Father. The passage in 3-14 forms one long sentence in the original—a rhapsody on God's blessings to Christians.   3. *Blessing on high:* "on high" in this text means literally "in heavenly places" and might be translated "in heaven." Heaven is always imagined as above the earth. The Father has poured down upon us His heavenly blessings. In *4, 8* Christ is pictured ascending to heaven and sending down gifts for men. The blessings here spoken of are holiness and the adoption of sons. On this adoption see Gal. *4, 6* f, where St. Paul distinguishes those who are sons by adoption from Him who is God's real Son, such by His essence. It is *in* the real Son of God that we are called. The expressions "in Christ" and "in the Spirit" are brief summaries of the doctrine of the Mystical Body of Christ which is the principal topic of this Epistle. We are in Christ as we are in the atmosphere which surrounds and sustains us; His action upon us is like that of the soul upon our body: "It is now no longer I that live, but Christ lives in me" (Gal. *2,* 20). The Father having in mind this intimate union, predestines us from all eternity to become His sons.   6. *Unto the praise of the glory:* God who is infinite perfection cannot act for any other end than His own glory. His goodness to us, as manifested by His grace, contributes to His glory.

1, 7-10: Realized in the Son.   7. *We have redemption through his blood.* Redemption means buying back. When one buys back a slave and gives him liberty one is said to redeem him. Now men were in slavery to sin, which in Rom. *6* is pictured as a tyrant. By His loving sacrifice of Himself on the cross, Christ expiated for our sins as a victim of propitiation. For this loving sacrifice we should be grateful not only to the Son but also to the Father "who so loved the world as to send His only begotten Son into the world," not to condemn the world but to save it by the Cross: cf. John *3, 14-17*. In this supreme manifestation of divine love, justice and charity meet. **8 f.** To enable us to appreciate God's blessings we have the gift of *wisdom*. It makes us understand the *mystery of God's will according to his good pleasure*. St. Paul here as elsewhere emphasizes that grace is given us through the free choice of the Father. The *mystery* is a truth hitherto hidden but now revealed: the call of all men to salvation, both Jews and Gentiles. He says this first in the rather obscure words of 10, about the Father's plan *to re-establish all things in Christ*. The Greek seems to mean "to reunite all things in Christ." Men are reconciled to their heavenly Father in Christ; angels are reconciled with men. Even inanimate beings share in this reconciliation. St. Paul probably thinks, too, of the union of man with man, of church with church.

1, 11-14: Fulfilled through the Holy Spirit.

12. *We who before hoped in Christ* refers to Jewish Christians, such as St. Paul himself, who were prepared for the coming of Christ, who hoped in Him. *You too:* Jews who have believed in the gospel. **13.** When they believed, Christians were *sealed with the Holy Spirit*. The present possession of the Holy Spirit marks those who will belong to God in the future; it is a *pledge* of their inheritance of the blessedness of heaven. A pledge or earnest is the first payment, guaranteeing full possession of our heritance. It will be ours on the last day, the day of full redemption, when our soul and body shall be redeemed: cf. Rom. *8, 23*. The gift of grace is followed according to God's will by that of heavenly glory; and although those who have received grace are not confirmed therein and may, through their own fault, fall away from Him who is the source of our life, our present grace gives us well-grounded assurance. We are saved in hope. We have only to let ourselves live. That the Apostle considers us free under the action of grace would be evident from the very fact that he constantly urges us to walk worthy of the vocation unto which we are called.

I: DOCTRINAL *1, 15 – 3, 21*

1. The Church Is One with Christ *1, 15 – 2, 22*

1, 15-23: Thanksgiving and Prayer. St. Paul thanks God for the faith and love of the recipients of his Epistle, of which he *has heard;* and prays that they may come to a greater realization of God's benefits, the glories He has in reserve for Christians and the greatness of His power exerted for the benefit of those who believe, as it was exerted in the resurrection of Christ, whom He has made to sit on His right hand. He has given him as head, above all, to the Church, which is His body.   **23.** This body is *the fullness of him who is wholly fulfilled in all.* The expression is obscure. Father Huby, S.J., (in *Verbum Salutis ad 1.*) translates: "The Church which is his body, the fullness of Him who fills everything in every way." Father Prat, S.J. (*Theology of St. Paul,* 8th ed., p. 363) renders the passage: "God has given Christ as an incomparable leader to the Church, which is His body, the complement of Him who completes himself entirely in all (His members)." This is clearer but it does not take sufficient account of the meaning which the word rendered *fullness* has elsewhere; and moreover it sounds strange that Christ should be said to need a complement, however the idea be explained. Father Lemonnyer, O.P., (in his translation and commentary of St. Paul's Epistles, *ad 1.*) is a little clearer when he translates, "He has placed everything under His feet and made Him head of the whole Church which is His body, the complement of Him who perfects all in all."

Head of the Church, Christ is the source of all the graces which develop life in her and she assimilates with herself the members given her by faith and Baptism.

2, 1-10: All Brought into Christ's Life. The recipients of this Epistle to whom St. Paul adverts principally are converted pagans. He draws of the pagan society from which they came a picture something like that of Rom. *1.*   **2 f.** They were before their conversion subject to *the prince of the power of the air about us,* i.e., the devil. They had been subject not only to him, but to the desires of their own unregenerate nature. Hence they were *children of wrath,* i.e., the wrath of God was upon them. But not only the Gentiles, but the Jews also had incurred the wrath of a just God.   **4 ff.** But *God who is rich in mercy, by reason of the great love wherewith he has loved* us, even when we were dead (spiritually) *by reason of our sins, brought us to life together with Christ*. Those who are thus quickened share in the privileges of their divine Head. They sit with Him in heavenly places: already their glorification has begun. The Apostle, however, has said that we are called to holiness (*1, 4*). Here he tells us that we must bring forth good fruits: God has made good works ready beforehand *that we might walk* therein.   **8.** By grace we are saved and not through works. We are not justified by the observance of the Mosaic Law or of any other law. Nothing done before we became Christians could save us from wrath. A soul in mortal sin cannot merit. It cannot of itself even believe: the very beginnings of faith require grace. Once justified, however, men are able to perform meritorious works and St. Paul constantly urges us to live worthy of the vocation to which we are called (*4, 1*). If our natural works had been enough to transfer us from the state of sin to the state of grace, we might be proud of ourselves. But St. Paul insists that we should not boast (*1 Cor. 1, 29*).

2, 11-22: Gentile and Jew United. St. Paul recalls that the Gentiles for whom he writes were formerly sinners, devoid of the privileges granted Israel. They were not cheered by the hope of a Messias as the Jews were; and although they had worshipped many gods they were as a matter of fact without a God in this world. From this condition they were saved by the blood of Christ; through Him, the wall of separation between Jews and Gentiles had been broken down. Jews and Gentiles were made fellow-citizens. Jesus had caused His gospel to be proclaimed throughout the world; salvation is accessible to all without distinction of race. They are all *members of God's household*. **20.** This verse refers more likely to the prophets of the New Testament, so closely connected with the Apostles in *4, 11*.

2. Paul's Commission to Preach the Mystery 3, 1-13

3, 1-6: Paul Instructed.   1. St. Paul speaks of himself as Apostle of the Gentiles, and deals with the subject of the *mystery* confined to him. This mystery, hitherto veiled, was that Jews and Gentiles should be united in the Church.

3, 7-13: Assigned to Preach to the Gentiles. St. Paul had been made a herald of the gospel, though he was the weakest of men, by the power of Almighty God. He had been empowered to announce among the Gentiles the mysterious plans of God. This economy, or dispensation, fills even the angels with admiration. He had helped open to men confident access to God. Remembering this, his readers will not be down-hearted when they hear that he is in prison. His sufferings, arising from his endeavors to make Christ known to the Gentiles, were their glory.

It may be remarked here that a forger would hardly have thought of keeping people from being scandalized by his imprisonment; neither would he have called St. Paul "the very least among the saints" (8).

3. A Prayer for His Readers 3, 14-21

This prayer is addressed to the Father, from whom all the choirs of angels as well as all the families of men have their name and their existence. The word "father" in Greek and Latin suggested the thought that God is the head and the source of all families in heaven and on earth. Recalling that this God is rich in grace and in every other glorious perfection, St. Paul begs that He may powerfully strengthen his readers in *the inner man.* This "inner man" may be the man spoken of in Rom. 7, 22, where we read that the inner man delights in the law of God though his nature is too weak to live up to it. The "inner man" is not then the "new man," regenerated by the Holy Spirit, but the man endowed with reason and conscience who approves the better things. The next petition, that Christ may dwell in our hearts by faith, indicates how one may be strengthened; the Apostle prays that we may be rooted and grounded in love and enabled to understand the love of Christ for us—its *breadth and length and height and depth.*   **18 f.** Some commentators think St. Paul is speaking of the greatness of the mystery as he has set it forth—its *height,* etc. But though the construction is not smooth, the immediate context indicates rather love as the object of knowledge. *In order that you may be filled unto the fullness of God,* i.e., (according to Newman) the fullness "of which God is the fountain-head."   **20 f.** Doxology: blessed be God who is capable of granting us even more than we ask for or think of!

II: MORAL 4, 1 – 6, 20

1. For Christians in General 4, 1 – 5, 20

4, 1-6: Unity in the Mystical Body. **1.** *Therefore:* the moral part of the Epistle is closely connected with the previous teaching. Christians must behave themselves as befits their calling. First of all there must be unity in the one body of the Church. Cf. *1 Cor. 1, 13*: Christ is not divided. As whenever he urges unity, the Apostle urges humility as a preliminary condition: cf. Phil. *2*, 3-11. Unity also requires meekness and loving patience in bearing with one another (longsuffering). **3.** *Careful to preserve the unity of the Spirit in the bond of peace.* Unity exists already; we must give diligence to preserve it. **4.** The essential unity of the Church arises from the fact that it is one body, animated by one Spirit (the soul of the Mystical Body). There is *one Lord,* Christ; *one faith,* which is fixed on Christ; *one baptism,* in which men confess their faith; *one hope,* by which all are cheered. There is one God and Father of all who is *above all,* and *throughout all, and in us all.*

4, 7-16: Diversity of Graces. The gifts chiefly in view are those which Christ communicates to us after the Ascension; cf. *1 Cor. 12-14.* The Spirit of Christ gives to each Christian his allotted measure of grace according to the function he has to fulfill. **7 f.** To illustrate the idea of Christ's sending us spiritual gifts, St. Paul cites freely Ps. *67,* 19, which depicts a great triumph of God in His temple. Where the Psalm says that the Lord receives gifts from men, the Apostle says that Jesus "gave gifts to men." **9.** Our Lord comes down to this earth by the Incarnation and then ascends into heaven, whence He sends the Holy Spirit. *He led away captives:* these captives, says St. John Chrysostom, are the enemies of Christ. The figure is that of a triumphant conqueror leading his captives in his train. There is no reason to dwell on the nature of the captives. **11.** *That he might fill all things:* with His power and His gifts. In particular (11 ff) He gives gifts to the Church. He has established for her some as *apostles* (like the Twelve, St. Paul, St. Barnabas and others of inferior rank sent forth to establish churches); *prophets* who speak God's word not only concerning the future but the present and the past; *evangelists* who complete the work of the Apostles in the building up of the churches; *pastors and teachers,* doubtless established with some permanence in particular places.

St. Paul does not aim at completeness.   **12 f.** *For the work of the ministry* appears to refer to the use of the gifts rather than to pastoral duties. All so endowed *build up the body of Christ,* contribute to the growth of the Church. All together as one body, we have come to the same faith and a true knowledge of Christ the Son of God; and we have reached the perfection of the grown man, that maturity of our Christianity which will make us fit to receive the fullness of the gifts of Christ, to be the perfect receptacle of His graces. One who pursues this perfection is no longer a child, *tossed to and fro and carried away by every wind of doctrine.*

4, 17-24: Change of Self. See Col. *3,* 5-11 on taking off the *old man* and putting on the *new.*   **18.** The *life of God* is that which God lives and which He communicates to men. On the corruption of paganism see Rom. *1,* 18-32.   **21.** Paul is evidently not writing exclusively to the Ephesians he had himself evangelized.   **22.** The *old man* is the unregenerate man, not under the influence of the Holy Spirit; the *new man* is he who is under the influence of the Spirit.   **24.** *In justice and holiness of truth* seems to mean simply, in true justice and true holiness.

4, 25 — 5, 20: Vices to Be Avoided. Six are specified: lying, anger, stealing, bad language, bad temper, lust. Although lying is the only one pointed out as unbecoming fellow-members of the body of Christ, the same can be said in regard to all vices and sins.   **5, 1.** The thought of imitating God leads to other practical warnings.   **8.** *Light in the Lord:* a metaphor familiar to St. Paul. See Rom. *13,* 11 ff; *1* Thess. *5,* 4 ff. To be a child of the light we must avoid sin and error. Doubtless 14 gives us a formula known to Paul and his readers, borrowed from the liturgy.   **20.** *Giving thanks always:* the Apostle realizes very keenly the greatness of the spiritual benefits granted him and his converts. Thanksgiving occupies an important part of his Epistles. The songs of the Christians to which he refers are very likely songs of thanksgiving.

2. The Christian Home 5, 21 — 6, 9

5, 21-33: The Wife and the Husband. The union of Christ and His Church is symbolized by the union of man and wife.

The great mystery St. Paul refers to is the union of Christ and the Church: it is a hidden truth now made known. It is certain that the *mystery* here cannot mean "sacrament."

6, 1-4: Children and Parents. Children must obey their parents *in the Lord,* i.e., in a Christian manner, in accordance with the directions of Christ. *First commandment with a promise:* this promise was at first a temporal one, it promised temporal happiness in Palestine. It came later to signify chiefly spiritual blessings. Some, however, see in St. Paul's words an indication that God will reward filial devotion with blessings even in this world; but the promise must not be taken too absolutely. God may have other plans for us, and blessings greater than long life.

6, 5-9: Slaves and Masters. Without discussing the legitimacy of slavery, here or elsewhere, the Apostle urges slaves to do well what their condition requires, serving Christ by the elevation of their motives. **9.** Mindfulness of the equality of master and slave would in time do away with slavery.

3. The Christian Warfare 6, 10-20

6, 10-17: The Armor of God. Strengthened in the Lord, made powerful by our fellowship with Him, we share already in what is His. **10.** Armor represents divine help in the warfare of the Christian. **11.** *Armor of God:* i.e., supplied by God. **12.** Our warfare is not only against our own nature and other men, but against the hierarchies of the evil spirits and their wickedness. The *sword of the Spirit* is the word of God. In Isa. *11,* 4 the Messias overcomes His enemies by the "breath of his lips," i.e., by His word. The sword is provided by the Spirit.

6, 18-20: Assiduous Prayer. The thought of the Spirit given utterance leads to the thought of another kind of utterance, viz., intercessory prayer. The Apostle asks that we pray at all times under the influence of the Spirit, for all Christians and especially for St. Paul himself, who needs supernatural strength to keep on preaching the gospel. Of that gospel he is an ambassador in chains.

Conclusion 6, 21-24

May St. Paul's readers have peace, love and faith! He is about to send them Tychicus, who will acquaint them with the circumstances in which the Apostle finds himself at Rome. He hopes that the faithful may be comforted by this fellow-worker of St. Paul.

WENDELL S. REILLY, S.S.

THE EPISTLE TO THE PHILIPPIANS

Introduction

Philippi was famous in history. When Philip of Macedon, father of Alexander the Great, refounded the city about 360 B.C., he gave it his name. Three centuries later the place was the scene of the battle in which Augustus and Antony defeated Brutus and Cassius (42 B.C.). In St. Paul's day the city was the capital of a district of Macedonia, a trade center due to its site on the Egnatian Way linking Rome and the East, a military colony whose citizens enjoyed the special privilege of *jus italicum*, the full rights of those living in Italy itself. The Acts describe St. Paul's ministry there (*16*, 9-40).

Occasion. At least twice afterwards St. Paul visited Philippi (Acts *20*, 1-6). But whether present or absent, he was in the heart of its people. They were so dear to him, that the church at Philippi can be called his favorite. The immediate purpose of the letter is to express gratitude for the gift his converts had sent him. Learning that St. Paul had been imprisoned at Rome, the Philippians had sent Epaphroditus, one of their number, to the Apostle with money for his needs and with instructions to assist him by any service possible. While devoting himself unsparingly to the saint, Epaphroditus fell sick and nearly died. On his recovery he seems to have had a touch of homesickness, and the Apostle determined to send him back to Philippi. This present letter praises his work highly and serves to forestall any suspicion that Epaphroditus was deserting his post.

At the same time Paul warns the congregation against two dangers. One comes from the Judaizers, Jewish converts who wished to maintain as obligatory certain practices of the Old Law. He reminds the Philippians that all grace comes from Christ alone and, once the law of Grace began, the Mosaic rites ceased to have any efficacy. The second danger consists in the dissensions which had sprung up among the Philippians. These are indicated in his appeal to the two women, Syntyche and Evodia, that they be reconciled. It has been said that the church at Philippi offered to the Apostle a bright picture marred by only one shadow, the darkness of disunion and uncharitableness. However, Paul feels certain his readers will correct this defect, and joy breaks forth so constantly in his letter that with good reason its theme has been found in the words, *Rejoice in the Lord always; again I say, rejoice* (*4*, 4).

The Time and Place of Composition can be gathered from the Epistle itself. The Apostle writes as a prisoner expecting an early release. This imprisonment would seem to be that which early tradition places at Rome during the years 61-63 A.D. The mention of the *prætorium* (*1*, 13) and *Cæsar's household* (*4*, 22) lends support to this interpretation. The hope of a speedy release (*2*, 24) would indicate a period near the end of the captivity so that we may conclude the Epistle came from Paul's hand in the spring or summer of 63 A.D.

COMMENTARY

INTRODUCTION *1*, 1-11

1, 1-2: **Greeting.** 1. Timothy had helped found the church at Philippi (Acts *16*), and during St. Paul's third missionary journey had twice visited it, once when sent there from Ephesus (Acts *19*, 22), a second time when returning with Paul from Corinth to Jerusalem (Acts *20*, 3-6). Notice that in the greeting St. Paul omits his ordinary title of Apostle, as if hinting he has no need to vindicate his authority over them. He writes as a friend to a friend. *Bishops and deacons* designate two grades of church ministers, as appears from the meaning of the terms: bishop (overseer), and deacon (assistant). At the time of the letter's composition these words did not possess their present technical ecclesiastical meaning.

1, 3-11: **Thanksgiving and Prayer.** A thanksgiving to God and a prayer for special graces begin every letter of St. Paul, except those to the Hebrews and Galatians. Here the Apostle thanks God for their generous assistance in the diffusion of the gospel and prays for charity and unity, thus foreshadowing his later earnest plea for oneness of heart and mind. 6. God *has begun a good work* in them because not even the first step in the way of salvation came from them. An important text for the doctrine of grace. *The day of Christ Jesus:* the day of the Last Judgment when they will stand before the tribunal of Christ. 7. *As sharers in my joy:* the Greek text reads "of my grace." The Greek text refers to the grace or favor Paul enjoys by preaching to the Gentiles. The faithful by their prayers unite themselves with St. Paul's missionary work and sufferings for Christ. A beautiful exemplification of the communion of saints. 8. Instead of his own heart, he puts the heart of Christ. 9. He prays that they may abound ... *in discernment,* for charity, like zeal, must be according to knowledge. Misguided charity should not make them welcome false teachers, such as the Judaizers, nor cause them, as in the case of Evodia and Syntyche, to tolerate petty squabbles which disrupt church unity.

I: PERSONAL NEWS 1, 12-26

1, 12-20: Propagation of the Gospel. **12.** The Philippians had inquired eagerly as to St. Paul's condition. Wholly immersed in the things of Christ, he speaks of his present situation only as it concerns the spread of the gospel. Not a word about his own health or the conditions of his prison life. **13.** The *praetorium* designated properly the space in the camp reserved for the praetor, then it came to mean any residence of a governor (the place in which Pilate judged our Lord is called a praetorium), finally it could mean the praetorian regiments. Here the term refers to the praetorian guard. The different soldiers who had charge of Paul, discussing with their comrades this unusual prisoner, would make known the cause of Christ for which he suffers. **15-17.** Some preachers, jealous of St. Paul's success and prestige, wish to show they can accomplish as much as he, and think their success will make him bitter. Evidently there is nothing wrong with the faith they preach for St. Paul passes no censure upon their doctrine. **19.** The present situation will not necessarily bring freedom to the Apostle, as is clear from 20-21, but will aid his eternal salvation. Two forces contribute to that end, the prayer of his faithful and the spirit of Jesus Christ, Christ Himself being the giver and the gift bestowed.

1, 21-26: Sentiments of St. Paul. **21.** *To die is gain:* because death brings immediate union with Christ in glory. **23.** *To depart and to be with Christ:* death will immediately bring him into the life with Christ in glory. Commentators observe that the verse presupposes that after the death of Paul some time will elapse before Christ comes in glory, an indication that the Apostle does not think the Second Coming of Christ is imminent. **25.** Paul knows he will abide with them. This knowledge comes not through a revelation, but from acquaintance with the attitude of the Roman officials toward his case. Fr. Knabenbauer proposes another interpretation. The Apostle has expressed doubt whether he will live or die (20 and 2, 17). Therefore the present sentence means: I am convinced that, if I remain, it will be for your benefit.

II: EXHORTATION 1, 27 – 2, 18

1, 27-30: Firmness. **28.** *This is . . . a reason:* not that the adversaries are now lost, but they are now on the path which leads to destruction while the persecuted Christians are in the way which leads to salvation.   **30.** The Philippians have seen in their city the persecution Paul endured, and they have heard of his trials at Rome. Their suffering for the faith makes them imitators of Paul and thus of Christ.

2, 1-11: Unity and Humility.   **1.** If there are *any feelings of mercy* in them, they should spare him the sorrow caused by their differences. If you would relieve my sufferings, he says, heal your factions, forget your squabbles—so you will bring me happiness.   **3.** *Contentiousness:* in Greek, "party spirit," "factiousness"; this should not motivate their actions. Selfishness can masquerade under the guise of group loyalty and thus become an indirect source of vainglory. *Each one regard the others as his superiors:* St. Thomas explains that one does this by realizing his own sins and defects and considering the effects of God's grace in others.   **5.** As an example of perfect humility and complete surrender of all honor due to Him Christ is set forth for their imitation. Three stages are mentioned in the existence of the God-Man: (a) His eternal life before the Incarnation, 6; (b) His earthly life, 7-8; (c) His exaltation, 9-11.

6. *By nature God:* literally "in the form of God." "Form" signifies that which underlies the essential attributes; and in this context means *nature,* just as the "form" of a slave (i.e., man, 7) stands for human nature. The early Fathers used this text against the Arians who wished to show that the Son was a god of a lower order than the Father. *A thing to be clung to:* literally "a thing to be snatched at eagerly," the Greek word occurring only here in the New Testament. The Son of God did not think He must selfishly cling to all His glory and enjoy to the full His dignity, but He laid them aside to become man. Others interpret: although He fully realized His equality with God was no usurpation, yet He put it aside.

7 f. *He emptied himself:* by becoming man. This emptying or annihilation of self consisted in His putting aside the glory of God and forsaking all honor due Him as God-Man to become like us in lowliness. *Taking the nature of a slave:* assuming human nature. *Appearing in the form of man:* outwardly those who saw Him would take Him for an ordinary mortal, not recognizing His divinity. The humiliation is delineated in three steps, His birth, His life and His death. Becoming man Christ could have taken to Himself a glorious human body free from suffering and need, such as He had after the Resurrection. He might have appeared suddenly upon this earth as a full-grown man, as Adam was created or as He Himself appeared after the Resurrection. Instead of dying, He could have been carried up to heaven alive as Elias in the fiery chariot, or He could have ascended into heaven directly from the Last Supper. Setting aside all these glorious aspects of human life, He chose to be born a babe in Bethlehem, to live poor and unhonored, and to die as a criminal upon the cross.

9-11. After degradation came the state of glory when Christ's majesty became manifest at the Resurrection. *The name* can mean a particular title, or stand simply for honor, rank, dignity. If Paul means the Father conferred a title upon the Son, it could be the "Son of God" (St. Thomas), or "Lord" (Huby). More probably the meaning is: God gave Him honor above the honor of all creatures. **10.** *At the name of Jesus:* or, before the majesty of Jesus. *Every knee should bend:* acknowledging His supreme rule. St. Paul applies to Christ a text in which God speaks of the submission all beings owe to their maker (Isa. *45, 23*). The angels in heaven, men on earth and the demons under the earth adored Christ glorified. **11.** *The Lord Jesus Christ:* each one of the titles brings out a special feature of Christ's dignity: *Jesus,* or savior, the man; *Christ,* the anointed king; *Lord,* or God.

2, 12-18: Fear and Joy in Serving. **12.** Greater caution against falling is required during Paul's absence. He counsels *fear and trembling,* not servile cringing before a dread God, but filial reverence and self-distrust. Elsewhere Paul speaks of chastising his body lest when he had preached to others he himself be rejected (*1 Cor. 9, 27*). **13.** Your salvation demands your best efforts, for it is God's work and He gives you strength to accomplish it. Grace precedes and accompanies every act that leads to salvation. **15.** You are *children of God;* behave as such without reproach before men, and with your soul given wholly and sincerely to God. You are the light of the world. **18.** Paul urges them to rejoice in his sufferings as he endures them for their sake. Death would be a gain, for then he would win Christ. It would even be a joy, for so he could suffer for his beloved Philippians.

III: TIMOTHY AND EPAPHRODITUS 2, 19-30

2, 19-24: Timothy. **20.** *Like-minded:* i.e., to St. Paul; or the Greek could mean equal or like to Timothy. This disciple is Paul's beloved son in the Lord. Probably other comrades then present with the Apostle did not show eagerness for the hardships of a journey to Philippi. **22.** They recognize his worth and tested virtue, for Timothy from the first days of the church has been a familiar figure to them.

2, 25-30: Epaphroditus. **25 f.** Epaphroditus, sent by the Philippians to assist St. Paul, has recovered from his dangerous illness but seems to be homesick and upset at the thought that his friends may be worried about his condition. **28.** The realization of his helper's sorrow and the Philippians' anxiety bring grief to St. Paul. The return of Epaphroditus to Philippi will restore joy to his fellow townsmen, thus lifting one burden from the heart of Paul. The trials of the prison, however, still remain for the Apostle so that he says (in the Greek text), "I may be less sorrowful."

IV: WARNING AGAINST FALSE TEACHERS 3, 1 – 4, 1

Some Jewish converts to the faith wished to retain the Mosaic Law in the Church and taught that circumcision was necessary to salvation, or at least that its observance made one a better Christian. Such adversaries appear in other letters, especially that to the Galatians.

3, 1-6: The Christian Spirit. V.2 shows an abrupt change. Probably news had just reached Paul that danger from Judaizing Christians threatened his beloved church. He characterizes the adversaries as *dogs,* either because he casts into their teeth the reproach they uttered against all Gentiles, or because, like mad dogs, they rage shamelessly against the converted Saul. They are *evil workers* who claim to lead men to salvation, but actually harm them. They would advocate *mutilation*—here the Apostle uses a play on words, as if in English one substituted incision for circumcision—because they insist upon circumcision after its religious meaning has been abrogated by the New Law. Cf. 2 Cor. 5, 12.

3. *We are the circumcision:* of the heart and not the flesh, the true Israel or Chosen People of God. *Glory in Christ Jesus:*

as the sole and sufficient source of salvation. *Have no confidence in the flesh:* i.e., on our descent from Abraham or any privileges of birth or natural advantages.   **5.** *Circumcised on the eighth day:* as the Law prescribed. *Of the race of Israel:* i.e., not a Gentile proselyte. *Of the tribe of Benjamin:* the only one of the twelve patriarchs born in the Promised Land, from whom descended Saul the first king, the tribe which with Juda alone remained faithful after the schism. *Hebrew:* in race and language, brought up in Jerusalem at the feet of Gamaliel. To these advantages of birth and training he adds his own accomplishments. He was a strict Pharisee, zealous even to persecuting the Christians, a blameless observer of the minutiae of the Law. His brow wore a halo of legal sanctity (Huby).

3, 7-16: Renunciation for the Sake of Christ.   7. On becoming a Christian, St. Paul realized these former advantages were useless for salvation, in fact they were a real obstacle because they hindered him from seeing that all justice and grace come only from Christ.   **9.** *I may . . . be found in him:* as the branch in the vine, not with any justice I could acquire by my own merits, but only with the justice which comes from God through faith in Christ.   **10.** To come to know Him, experience His power and the might of the risen Christ. For we are risen again with Christ, living the life of the Spirit, although this life has not yet unfolded itself in its full beauty. *Fellowship of his sufferings:* one with Christ in Gethsemani and on Calvary, in order to live with Him the life of Easter and Pentecost.   **12.** *Christ Jesus has laid hold of me:* Saul had sought to do many things against the name of Jesus. Overpowered by the vision on the road to Damascus, Paul now pursues Christ that he may be like Him.

3, 17 — 4, 1: Followers and Opponents of the Cross.   18. *Enemies of the cross of Christ:* either Judaizers (see notes to text), or bad Christians who refuse to accept the obligations of their state.   **19.** *Their God is their belly:* sensual sins such as gluttony and lust.   **20.** *Our citizenship:* our true country is heaven. The Greek word for citizenship may designate a commonwealth and have the sense, "we are a colony of heaven, our true country." The Philippians enjoyed the rights of *jus italicum,* citizenship rights equal to the rights of those dwelling in Italy itself. *Await a Savior:* at the end of the world, who will perfect the salvation of His chosen ones, freeing them from the infirmities and limitations of corruptible and mortal flesh. **21.** *Who will refashion the body of our lowliness:* as members of Christ's Mystical Body we are made like Him in suffering and shall also be similar to Him in glory, even in body.

Conclusion 4, 2-23

4, 2-3: Concord. **2.** Evodia and Syntyche, two women who had helped Paul in the work of the gospel, have caused dissensions in the church. He exhorts them to have one mind and one heart as members of Christ, engaged in the work of God. Some conjecture they were deaconesses. **3.** *My loyal comrade:* some one well known to them. It may have been a Christian of Philippi named Synzyge (comrade) or Epaphroditus, the bearer of the letter. *Clement:* so common a name that most commentators do not attempt to identify him. Jerome considers him to be Pope Clement who wrote a letter to the Corinthians toward the end of the first century. *Whose names are in the book of life:* as belonging to God, not that they can be certain of their salvation. Sin can blot out their names from the book.

4, 4-9: Peace and Joy in the Lord. **4.** *Rejoice in the Lord:* the keynote of the letter. This delight, spiritual and independent of material conditions, shows itself in St. Paul writing from prison, an ambassador in chains. **5.** *The Lord is near:* the thought of Christ's coming to reward them gives strength and adds substance to their joy. Paul often speaks of the General Judgment as if it were at hand, in order that they may always be ready. **6 f.** A peace which the world cannot give will stand sentinel over the fortress of their hearts. **9.** By word and much more by example St. Paul has taught them.

4, 10-20: Their Gift. **10.** *Has revived:* they had helped him with alms, then for a time no assistance came. At last through Epaphroditus a generous gift reached him. *Lacked opportunity:* St. Paul hastens to excuse the absence of gifts. There is no tone of reprehension. The cause which prevented their sending gifts is unknown. **11 f.** Grateful for their charity, he still maintains his apostolic independence. As preaching Christ, he welcomes all situations, becomes all things to all men. **13.** Not through Stoic self-sufficiency, but from Christ comes his power.

15. *No church . . . but you:* allowing them to assist him by money gifts constituted an exception to St. Paul's custom of supporting himself by manual labor. *Partnership in . . . giving and receiving:* or an account of payments and receipts. The Philippians have paid by their alms, St. Paul acknowledges the receipt of the money. Chrysostom understood the phrase of an exchange: the faithful give material goods for the support of their Apostle who repays them with spiritual gifts, preaching, etc.

17. *Accumulating to your account:* in reality they are giving to God who rewards them in good measure at high interest. **18 f.** God will repay them from His boundless treasure. **20.** God inspired the Philippians to make this gift, thus showing His love for them and for the Apostle, and strengthening the love between Paul and his converts. Realizing this the saint breaks forth into praise of Him who orders all things sweetly.

4, 21-23: Farewell. **22.** *Cæsar's household:* the term, although it can extend to people not living at Rome, would primarily designate the Roman palace and as such adds support to the tradition that this letter was written at Rome.

JOHN J. COLLINS, S.J.

THE FIRST AND SECOND EPISTLES TO THE THESSALONIANS

INTRODUCTION

Thessalonica. The city of Thessalonica, known at the present time as Saloniki, or Salonica, was in St. Paul's day the capital of Macedonia. It was situated on a fine harbor in the Thermaic Gulf (Gulf of Salonica). First known as Thermæ, because of the hot springs in its vicinity, it was in the fourth century B.C. named Thessalonica after the half-sister of Alexander the Great. In 168 B.C. Thessalonica along with the rest of Macedonia was conquered by the Romans and made a part of the Empire. The Romans connected Thessalonica with Italy and the East by constructing the great military road, called the Egnatian Way. The natural and political advantages of Thessalonica made it a prosperous and important city. In the time of St. Paul its population was composed chiefly of native Greeks, although Jews and Romans were there in large numbers.

The Church at Thessalonica. Paul founded the church at Thessalonica during the early part of his second missionary journey, assisted by Silvanus (Silas) and perhaps Timothy. At least Timothy acted as Paul's representative there shortly after Paul and Silvanus were compelled to flee to Berœa.

After his arrival in Thessalonica Paul, following his usual procedure, entered the synagogue in that place and preached to the Jews, who willingly listened to him on three successive Sabbaths. He succeeded in making some converts among the Jews and proselytes. Apparently after three weeks he was no longer allowed to speak in the synagogue, and he turned his attention to the Gentile population of the city. Here he met with great success, converting, as St. Luke says, "a large number" (Acts *17*, 4). The numerous conversions stirred up the envy of the unbelieving Jews who induced the rougher elements among the Gentile population to create a riot against Paul and Silvanus, and to charge them before the city magistrates with treason against Cæsar. As a result Paul and Silvanus had to leave the city hurriedly under cover of darkness. Protected by an escort of converts they proceeded to Berœa, where again Jews from Thessalonica compelled Paul to leave. This time he went to Athens, where after a brief stay he proceeded to Corinth (Acts *17*, 5-15; *18*, 1-5).

It seems likely that Paul had been in Thessalonica at least two or three months before his forced departure. Such a length of time would be needed for him to bring about the organized and flourishing state of the church which the two Epistles to the Thessalonians reflect. Then too, the report of the good example of the Thessalonians, which had been spread through the other Christian communities, presupposes that the church at Thessalonica had an existence of more than a few weeks before Paul's departure from it.

Occasion of the Epistles to the Thessalonians. 1 Thessalonians. While at Athens St. Paul learned that his departure had not put an end to the persecution of the converts in Thessalonica. Consequently he sent Timothy to them to instruct and strengthen them in their sufferings (*1 Thess. 3,* 1 f). Shortly after Paul's arrival at Corinth Timothy brought a report of conditions in the church at Thessalonica, and his report occasioned Paul's First Epistle to the Thessalonians. In general the condition of the church was most favorable. The Thessalonians had remained firm in the faith in spite of persecution, and their conduct served as an example to other Christians in Macedonia and Achaia. But evidently they had misunderstood Paul's preaching on the Second Coming of Christ. Consequently it became necessary to give them fuller instruction on this topic. The Thessalonians labored under the impression that the departed brethren would not share as fully in the triumphal return of Christ as those who would be alive on that day. The thought that the dead would be at a disadvantage gave them great grief. Paul informs them that such grief is unnecessary. Those alive at the *parousia* will have no advantage over those already departed.

Certain faults in conduct were also making their appearance in the church, and Paul feels it necessary to deal with them in this Epistle.

2 Thessalonians. St. Paul's instruction in *1* Thessalonians about the Second Coming of Christ did not have the desired effect. Some were convinced more than ever that the Day of the Lord was at hand, and consequently spent their time in idleness. Upon learning these facts Paul hastened to write a second letter in order to remove all misunderstanding on the subject of the *parousia,* or Second Coming of Christ.

Time and Place of Composition. It was during his long stay of more than eighteen months at Corinth that Paul wrote *1* and *2* Thessalonians. (Cf. Acts *18,* 5.11; *1* Thess. *3,* 6.) Both Epistles were probably written in 51/52 A.D. and they are generally thought to be the earliest of all Paul's Epistles.

Authenticity. *1* Thessalonians. The authenticity of this Epistle is well supported by external and internal evidence.

(a) *External evidence.* From the earliest time this Epistle is ascribed to Paul. It is listed as his in the Muratorian Canon, and by Marcion in his canon. It is quoted as Paul's by Irenaeus, Clement of Alexandria,

Justin Martyr, Tertullian, and Eusebius. It bears Paul's name in the earliest Greek codices and Versions, such as the Old Latin and Syriac.

(b) *Internal evidence.* The ideas and style are thoroughly Pauline. In *1* Thessalonians are found such common Pauline teachings as the death and resurrection of Jesus Christ (*1*, 10; *4*, 14; *5*, 10); the resurrection of the human body (*4*, 14-17); the divinity of Christ (*1*, 9-10); the mediatorship of Christ (*5*, 10); the call of the Gentiles to the faith (*2*, 12). Written to a church composed mostly of Gentile converts it contains, as would be expected, no quotations from the Old Testament.

2 **Thessalonians.** (a) *External evidence.* This Epistle was also acknowledged as Paul's from the earliest times. It is attributed to Paul in ancient canons of the Scriptures and in the MSS and Versions. The volume of testimony for its Pauline authorship is even greater than that for his authorship of the First Epistle.

(b) *Internal evidence.* The style and thought of this Epistle are also characteristically Pauline. There is a close similarity in word and expression between it and the First Epistle. The objection that the Second Epistle contradicts the teaching of the First Epistle is not a valid one. Paul made no statement in the First Epistle as to the time of the *parousia,* or Second Coming of Christ. He merely said it would come suddenly and unexpectedly. In the Second Epistle he gives fuller information. The *parousia* is not imminent because as yet the great apostasy has not occurred and Antichrist has not appeared. The Second Epistle is only more complete than the first on the subject of the *parousia.*

COMMENTARY

THE FIRST EPISTLE TO THE THESSALONIANS

INTRODUCTION 1, 1-10

1, 1: Greeting. Paul associates himself with Silvanus (Silas) and Timothy in the opening salutation, because they were with him in Corinth at the time this letter was written, and because both were known to the Thessalonians. Silvanus assisted Paul in the foundation of the church at Thessalonica (Acts *17*, 4.10). Timothy, too, may have been one of its founders. If not, he at least represented Paul there shortly after the latter's expulsion from the city (*1* Thess. *3*, 2). *The church . . . in God the Father and in the Lord Jesus Christ:* through the grace of Christ the members of the Church live in union with Jesus Christ and God the Father. Cf. John *17*, 22; *1* John *2*, 24.

1, 2-10: Thanksgiving for Their Faith.   **2 f.** The theological virtues of faith, hope, and charity, which are the foundation of all holy living, are conspicuous in their lives.   **4-10.** The virtuous conduct of the Thessalonians edifies the Christian world and commands its praise.   **4.** Paul proceeds to describe the manner in which their call to the faith took place.   **5.** His preaching was accompanied by the performance of miracles *(in power)*, and by an extraordinary display of spiritual gifts *(in the Holy Spirit)*, such as the gifts of tongues and prophecy. *In much fullness:* refers to Paul's preaching, rather than to the miracles and gifts of the Holy Spirit. It describes the full conviction, sincerity, and energy with which he preached.   **6.** *Imitators of us:* cf. *1* Cor. *4*, 16; Phil. *3*, 17. *In great tribulation:* refers to the persecution and suffering recorded in Acts *17*, 5-13. They experienced interior joy, a gift of the Holy Spirit, under external suffering. "But the fruit of the Spirit is . . . joy" (Gal. *5*, 22). Cf. Acts *5*, 41; *2* Cor. *7*, 4.   **9.** *They themselves:* the Christians of the other churches in Macedonia and Achaia, and all the other places which Paul had visited or heard from. *From idols:* the members of the church at Thessalonica were drawn mostly from the ranks of the Gentiles.   **10.** *Wrath to come:* eternal punishment, as is clear from *5*, 9, and Rom. *5*, 9; cf. *1* Thess. *2*, 16; *2* Thess. *1*, 9.

I: PAUL'S PAST RELATIONS AND PRESENT INTEREST
2, 1–4, 12

2, 1-12: His Mission among Them. **1.** *Not in vain:* refers to the fruitful effects of his preaching and not to the effort which his preaching cost him. Cf. 2, 15. **2.** *Shameful treatment at Philippi:* cf. Acts *16,* 22-24. *And amid much anxiety:* the tumult stirred up against Paul and his converts by the Jews. Cf. Acts *17,* 5-10. **3.** *Nor from impure motives:* the impurity referred to is that of intention, and not of doctrine. No defective motives prompted his ministry of preaching, his motives were pure and disinterested. **4-6.** Paul replies to specific charges made against him. In his preaching, a task imposed on him by God, he seeks only God's approval. **7.** *Nurse:* his conduct towards them was as gentle and kind as that of a mother for her infants. He acted towards them both as a father (11) and a mother. **8.** *Our own souls:* he was prepared to lay down his life for them. **9.** Cf. *2* Thess. *3,* 8. Paul's trade was that of tentmaking. Cf. Acts *20,* 34; *1* Cor. *4,* 12. **12.** *His kingdom and glory:* the call to the faith has its culmination in the life of glory in heaven.

2, 13-20: Thanksgiving for Their Constancy. **13-16.** Paul thanks God for the constancy of the Thessalonians under the persecution which they suffered from their own countrymen. **13.** *Who works in you:* the Vulgate "who" refers to God. The Greek relative pronoun can be rendered either "who" or "which," accordingly as one refers it to "God" or "word." In either case God's grace in a man is rendered fruitful by co-operation. **14.** The pagan hostility stirred up by the Jews against the Christians of Thessalonica did not cease with the departure of Paul (cf. Acts *17,* 5-10), but became more widespread. **16.** *Always filling up the measure of their sins:* they continue to increase their guilt, and, therefore, render themselves even more deserving of eternal punishment. *The wrath of God:* cf. *1,* 10 and *5,* 9. The severest kind of punishment God can inflict on the unbelieving Jews is eternal damnation. Possibly Paul also has in mind two other manifestations of God's anger towards the Jews, namely, the destruction of Jerusalem, which was to occur about twenty years later, and the exclusion of the Jews as a nation from the Church. The latter is not to be final, inasmuch as Israel will enter the Church after the fullness of the Gentiles. Cf. Rom. *11,* 25-27.

17. *More than ordinary efforts to hasten:* literally, have hastened more abundantly. **18.** *Satan hindered us:* "us" refers not to Silvanus and Timothy, but to Paul. Satan using evil men, such as Paul's Jewish and Gentile opponents, prevented him from returning. Or it may be that Satan by causing illness prevented Paul's return. Cf. 2 Cor. *12*, 7. **19.** Paul regards his converts as trophies of victory. Cf. 2 Cor. *1*, 14; Phil *4*, 1. It will give him intense joy to present them to Christ on the last day. *Coming:* i.e., "parousia," a word used four times in *1* Thess. (*3,* 13; *4,* 15; *5,* 23), occurs here for the first time. Literally, it means "presence," used in profane literature of the official visit of a ruler, but in the New Testament it is a technical term for the Second Advent or Coming of Christ, which forms the principal subject of the two Epistles to the Thessalonians.

3, 1-13: **The Mission of Timothy.** **1-5.** Paul being unable to go to Thessalonica sends Timothy in his stead. **1.** Cf. Acts *17,* 15 f. Paul received Timothy's report at Corinth. Cf. Acts *18,* 5. **3.** *We are appointed thereto:* cf. Acts *14,* 21; 2 Tim. *3,* 12. **5.** *The tempter:* i.e., the devil. He is also given this name in Matt. *4, 3.*

6-10. Timothy's favorable report of the faith and constancy of the Thessalonians was a source of great joy to Paul. **8.** Their constancy in the faith gave him new life and energy. **10.** *More and more:* literally, "more abundantly." *Those things lacking to your faith:* he wished to give them fuller instruction in the teachings of Christianity. He had been forced to leave Thessalonica before completing their instruction. Cf. Acts *17,* 1-10.

11-13. Paul's prayer is that God may direct him to them, and that they may live in perfect holiness. **11.** Under the term *God* which is the subject of the verb *direct,* St. Paul groups two persons who share the Godhead, namely, the Father and the Lord Jesus Christ.

4, 1-12: **Exhortation to Chastity and Charity.** God's will concerning their conduct is that they should become holy by observing the precepts of Christ. All sins of the flesh are to be avoided. **1.** *Make even greater progress:* literally, "to abound more." **3.** *Immorality:* the common pagan approbation and practice of immorality made it necessary for Paul to stress its malice and warn his converts to avoid it. Cf. *1* Cor. *5,* 11; *6,* 9.13.15-20; *1* Tim. *1,* 10.

4. *His vessel:* it is uncertain whether "vessel" refers to a man's body or to his wife. The first seems preferable. St. Chrysostom, Theodoret, Tertullian, Cornelius A Lapide, as well as many moderns understand "vessel" in the sense of "body." Hence Paul is saying: "Let a man control the unruly passions of his body." Actually "vessel" stands for body in *2 Cor. 4, 7*. Paul's own use of the term, therefore, favors the meaning "body." Profane authors, e.g., Cicero and Lucretius, refer to the body as a vessel. The meaning "body" fits in well with the context. Having said: "You should abstain from immorality," he at once gives the reason, namely, because the body is to be sanctified by holy use, and is not to be dishonored by sexual sin. Cf. *1 Cor. 6,* 13, "The body is not for immorality."

Others, e.g., St. Augustine, St. Thomas, Estius, as well as many moderns, take "vessel" in the sense of "wife." St. Augustine, following the reading "possess a vessel," interprets as "make a proper and faithful use of marriage." Cf., however, *1 Cor. 7, 2,* where Paul says: "have his own wife," and not "have his own vessel." Others believe that the Greek verb has the meaning "acquire" rather than "possess." Hence they translate: "Acquire a wife, take a wife." Wives are to be acquired in a holy and honorable manner.

6. *In the matter:* the Vulgate reading may also mean "in business." But the context, which is concerned with avoidance of sins of impurity, favors the rendering "in the matter." Cf. *4,* 11; *1 Cor. 6,* 1; *2 Cor. 7,* 11; *2 Tim. 2,* 4. The offense to be avoided is the sin of adultery. **8.** *These things:* i.e., this teaching. **10.** That is, towards all the Christian communities in Macedonia. **11.** *Minding your own affairs:* literally, do your own work, i.e., tend to your daily duties and occupations. He is referring especially to those who indulged in idleness because they believed the Second Coming of Christ was at hand.

II: THE SECOND COMING OF OUR LORD *4,* 13–*5,* 11

4, 13-18: Witnessed by the Dead. At the Second Advent of Christ the just shall rise first, and then both the living and the risen just shall be taken up in the air to meet Christ and share alike in everlasting life. The time of Christ's Second Coming is unknown and uncertain. Vigilance, therefore, is necessary. **13.** *Those who are asleep:* i.e., the dead. **14.** After *so* understand the words: "we believe that." *Fallen asleep through Jesus:* those who die united to Jesus by grace shall share in His glorious resurrection. There is a necessary bond between the resurrection of Christ and that of the just. Cf. *1 Cor. 15,* 13-16.20-21.50-55. *Through Jesus* can also be construed with *bring.* **15.** The evident meaning of Paul's words here is that the just who are alive at the last day will pass into the glorified state without experiencing death. This meaning is supported by the Vatican Codex reading of *1* Cor. *15,* 51: "We shall not all sleep (die), but we shall all be changed."

In the word of the Lord: either the teaching of Christ, such as that recorded in Matt. *24,* 31.36.44; or a special revelation given to Paul. He is insisting that his teaching on the *parousia* is that of Christ Himself. This point is helpful for an interpretation of the words which follow, because Christ did not teach that the *parousia* was imminent.

We who live, who survive (also 17): the traditional interpretation, namely, that Paul is expressing ignorance as to the time of the *parousia,* and is not asserting, nor implying that he and his readers will survive until the *parousia,* is supported by the true notion of inspiration, the rules of logic, the usage of Greek grammar, and a decision of the Biblical Commission. (1) The Catholic notion of inspiration requires that no erroneous statement be made by the sacred writer. Paul, therefore, writing under inspiration cannot erroneously assert that he and his readers will be alive at the *parousia.* (2) The rules of logic do not permit one to say that although Paul teaches no error, he nevertheless, by his form of expression, indicates that his personal opinion was that the *parousia* was imminent. No such indication or connotation can be admitted. The same words of Paul cannot be interpreted to mean lack of knowledge as to the time of the *parousia,* and also conviction, conjecture, personal opinion, or implication that the *parousia* is imminent. (3) Greek syntax permits one to translate: "we, if we be alive, if we survive," and such a translation removes all difficulty. Or, Greek grammar through the figure of *enallage,* whereby one person of the verb is substituted for another, permits one to regard the pronoun "we" as equivalent to "they." We Christians, i.e., those who shall be alive, who shall survive. This figure also occurs in *1* Cor. *11,* 31-32, where "we" is used in place of "you." (4) In its decision of June 18, 1915, the Biblical Commission declared that Catholic exegetes are not permitted to say that on the question of the *parousia* Paul is expressing his own human views into which error and deception can enter. Catholic interpreters, therefore, maintain that on the subject of the Second Coming of Christ Paul is setting forth Christian doctrine and not personal opinion or conjecture.

Paul teaches the Thessalonians that those living at the Second Coming of Christ will have no advantage over those who have died, because the living will not go before them to glory nor receive glory without them. The Thessalonians, therefore, need not grieve over the lot of the dead.

16. The *cry of command, voice of archangel,* and *trumpet of God,* probably signify the same thing, namely, God's command to the dead to arise. Cf. Matt *24,* 30 f; Luke *21,* 27; Acts *1,* 11; 2 Thess. *1,* 7.

5, 1-3: Time Unknown. The day of the Lord will come unexpectedly as a thief in the night. **1.** Paul in 2 gives the reason why it is unnecessary for him to write to them about the time of Christ's Second Coming. When among them he had already informed them that it will occur suddenly and unexpectedly at a date which is uncertain. The time is uncertain, because no man knows it. Cf. Matt. *24,* 36; Mark *13,* 32.

5, 4-11: Be Always Prepared. **4.** Since the day of the Lord is uncertain as to time, there remains the possibility that it may occur during the lifetime of the Thessalonians. Paul neither asserts nor implies that it will. He merely points out the practical conclusions the Thessalonians should draw from such a possibility. It should not terrify them, because their virtues render them ready to meet Christ. It should prompt them always to live in a state of readiness. Paul's admonition is the same as that of Christ: "Therefore you must also be ready, because at an hour you do not expect, the Son of Man will come" (Matt. *24,* 44). **8.** Cf. Eph. *6,* 14-17. Paul again stresses the theological virtues of faith, hope and charity. **9.** *Wrath:* cf. *1,* 10; *2,* 16. **10.** *Whether we wake or sleep:* Paul says that it does not matter whether they be found among the living or the dead at Christ's Second Coming. For both groups will enjoy immortal life with Christ.

Conclusion 5, 12-28

5, 12-22: Obedience, Patience and Charity. Paul exhorts them to practise obedience to lawful superiors, to be patient, to correct the idle, and to be charitable. **12.** The existence of a hierarchy is indicated. **13.** *Be at peace with them:* a variant reading has: "Have peace among yourselves." **14.** *Irregular:* i.e., the idle who refused to work because they thought the *parousia* near at hand. These were a source of trouble, because their views alarmed others, and their failure to work made it necessary for others to support them.

16-22. Various other recommendations. **19.** Make use of the charismatic gifts such as tongues and prophecy. Cf. *1 Cor. 12-14.* **20.** *Prophecies:* inspired instructions of those who had the gift of prophecy. **21.** They are to make certain that the gifts are genuine (cf. *1 Cor. 12,* 1-3).

5, 23-24. Final Blessing. *Spirit* is the human soul as the principle of intelligence and will. *Soul* is the human soul as the principle of animal and sensitive life, hence as the seat of the affections and feelings. *Spirit, soul,* and *body,* are the terms in which Paul sums up man and his activities. Cf. *1 Cor. 15,* 44.46; Heb. *4,* 12. Consequently he prays that under God's grace they may be wholly sanctified.

5, 25-28. Greeting. 26. The kiss was a liturgical practice.

THE SECOND EPISTLE TO THE THESSALONIANS
INTRODUCTION *1*, 1-12

1, **1-2: Greeting.** See *1* Thess. *1*, 1.

1, **3-10: Their Faith and Constancy.** Thanksgiving for their progress in faith and virtue, and for their constancy under persecution. Paul regards the marvelous operations of grace in the Thessalonians as gifts for which God is to be thanked. **4.** In praising God he likewise praises them. **5.** *Proof:* i.e., token or indication. Their persecutions and sufferings show that God will have to punish and reward in another life. Cf. 6 f. **7.** *At the revelation of the Lord Jesus:* i.e., at the *parousia,* or Second Coming of Christ. Cf. Matt. *24,* 30-31; John *1,* 51. **8.** *In flaming fire:* this is descriptive of the appearance of Christ. *Those who do not know God:* i.e., the pagans. *Those who do not obey the gospel:* perhaps his reference is principally to the Jews, though the words are applicable to all who deliberately reject the teachings of the faith. **9.** *Eternal ruin:* a proof of the eternity of the sufferings of the damned. *Away from the face of the Lord:* in their punishment the damned are forever separated from the sight of God. Cf. Matt. *7,* 23; *25,* 41. *From the glory of his power:* the damned will not share in the glory of the just, which is communicated to them through the power of Christ.

10. *To be glorified in his saints:* the glory imparted to the elect redounds to Christ Himself who obtained it for them. *To be marvelled at in all those who have believed:* an instance of Hebrew parallelism in which the same thought is expressed in slightly different form. The blessings to be enjoyed by those who have believed are due to Christ. Men will praise and admire Christ for what He has done. The Christians of Thessalonica are entitled to heavenly glory for they are numbered among the believers. They became such by accepting Paul's testimony, i.e., his preaching.

1, **11-12: Prayer for Their Glorification.** Paul prays that the Thessalonians may obtain the glory of heaven. **11.** Grace and glory complete God's call to the faith. "And those whom he has called, them he has also justified, and those whom he has justified, them he has also glorified" (Rom. *8,* 30). **12.** All sanctification and glorification are achieved through the grace of Christ.

I: THE SECOND COMING OF OUR LORD 2, 1-17

2, 1-12: Preludes to the Second Coming. The Second Coming of Christ will be preceded by a great apostasy and the advent of Antichrist. **2.** *Spirit . . . utterance . . . letter:* indicate three possible sources of their belief that the *parousia* is imminent. *Spirit* refers to some falsely claimed revelation; *utterance* may be a statement of Paul's which was misunderstood, or wrongly attributed to him; the *letter* seems to be one forged in Paul's name (cf. *3, 17*).

3 f. As written by St. Paul this is an unfinished sentence. The words *the day of the Lord will not come* have been inserted to complete the sentence, which in the original is elliptical. Paul warns them not to be misled in any manner into believing that the *parousia* is close at hand. Those who say it is about to happen are deceiving them. A great apostasy must occur and Antichrist must appear before the advent of the last day. Paul had given them this information before, and he now repeats it in order that they may banish the fears which have arisen at the thought of an immediate return of Christ. The *apostasy* will be some great religious defection in which men will revolt against God and His representatives. Perhaps it signifies that after the gospel has been preached to all nations, there will be a general apostasy from the faith. Cf. Luke *18, 8*: "Yet when the Son of Man comes, will he find, do you think, faith on the earth?" *The man of sin:* some Greek MSS read, "the man of lawlessness." The reference is to Antichrist. As Christ was the best of men, so Antichrist will be the worst of men (St. Thomas). He shall be a "son of perdition," i.e., entirely deserving of eternal punishment. Antichrist will be *revealed* when he appears publicly before the world claiming the allegiance of all men. *Who opposes . . . God:* two characteristics of Antichrist will be his impiety, whereby he will exhibit himself as the enemy of God, and his pride, whereby he will attempt to set himself above God, the saints, and whatever men, whether truly or falsely, regard as sacred and holy. *So that he sits in the temple of God,* etc.: what Paul means by sitting in the temple of God can only be conjectured. It may perhaps signify that Antichrist will succeed in establishing the worship of himself in Christian churches by inducing men to accept him as the true Christ.

5. Paul reminds them that while at Thessalonica he had told them all this. That he had given much more information to them on the same subject is evident from the following verses. But what this information was is unknown to us today. Had the Thessalonians transmitted it we should know what and who it is that prevents the appearance of Antichrist. **6.** *What restrains him:* Paul refers to some obstacle hindering the advent of Antichrist. We do not know what the obstacle is. St. Augustine in his day confessed complete ignorance of its nature. **7.** *The mystery of iniquity* is already at work. The evil power of which Antichrist is to be the public exponent and champion is now operating secretly in the world. *Only that he who,* etc.: here Paul speaks of the obstacle as a person. Above he had spoken of it as a thing. Someone is now restraining the evil force, but the time will come when he shall no longer obstruct it. When this happens Antichrist will appear (v.8). Some make the conjecture that Michael, the Archangel, and his heavenly army are the obstacles.

8. *Whom the Lord Jesus shall slay,* etc.: i.e., after some time has elapsed Jesus by a word of command will destroy Antichrist. *Will destroy with the brightness of his coming:* when Christ appears in glory at the end of the world He will quickly inflict death on Antichrist. **9-11.** The operations of Antichrist are described. **9.** By the aid of Satan Antichrist will perform prodigies, which men will falsely regard as miracles. **10.** The false miracles will succeed in deceiving all those who have preferred to follow erroneous opinions rather than accept the doctrines of Christ. As a consequence they will be led by Antichrist to adopt sinful practices. Cf. John *3,* 19; Rom. *1,* 25. **11.** Their love of error shall have its natural effect, namely, blindness to the truth. God will abandon them to their intellectual blindness. **12.** The consequence of willful unbelief is eternal condemnation.

2, 13-17: Thanksgiving for Their Election. Having spoken of those who have wilfully rejected the gospel, Paul calls to mind some who have readily accepted the teachings of the gospel, namely, the Thessalonians. He thanks God for calling them to the faith, and prays that they may be granted final glorification in heaven. **13.** *First-fruits:* i.e., they are numbered among the earliest believers in the gospel, and as such are destined for salvation. Some Greek and Latin MSS read: "From the beginning"; i.e., God has called them to the faith and to salvation from all eternity. **15.** *Teachings:* all his instructions, whether given orally or in writing. All the doctrines of Christianity, therefore, are not contained in the New Testament Scriptures, and consequently oral tradition has equal weight with the written Word.

II: EXHORTATION 3, 1-15

3, 1-5: Request for Mutual Prayer. Paul requests the Thessalonians to pray for him and for the spread of the faith.   **1.** *May run and be glorified:* may the teachings of Christ be rapidly diffused throughout the world and be held in honor.   **2.** Paul had previously experienced the enmity and opposition of the Jews in Thessalonica and Berœa (Acts *17,* 5-13), and now at Corinth he was meeting hostility from the same quarter (Acts *18,* 5-6). It is very likely that he is here referring to the Jewish opponents of the gospel.   **3-5.** Paul expresses the confidence he has in them, and prays God to give them continued patience and charity.

3, 6-15: Against Idleness. Once more the idleness of certain ones is condemned. Corrective measures must be taken to stop a further spread of this abuse. Paul urges all to continue in the practice of virtue.   **6.** *Irregularly:* cf. *1* Thess. *5,* 14. In *1* Thess. *4,* 11, and in his preaching (*infra* 10), he expressly instructed them to work and perform their daily duties. In addition he himself had set them a good example in this matter.   **8 f.** Although Paul had the right to demand support from them he did not exercise it. Instead he supported himself by the labor of his own hands. Cf. *1* Thess. *2,* 9; Acts *18,* 3; *20,* 34; *1* Cor. *9,* 14-15; *2* Cor. *11,* 7-13.   **10.** Paul indicates one of the means to be used in repressing culpable idleness. Cf. 12 *infra* for another.   **11.** The meddling of the idle probably consisted in the attempt to persuade others that the *parousia* was at hand, and that it was useless to work.   **14 f.** Formal excommunication is not spoken of, but rather a social and religious ostracism having for its purpose the correction of the offending Christian.

3, 16-18: Final Blessing and Greeting.   **17 f.** At this point Paul takes the pen from his secretary and inscribes the final greeting and blessing in his own handwriting. *Thus I write:* i.e., this is my handwriting. His personal signature in all future letters will indicate that he is the author of them. By this personal signature his readers will be protected against forged letters.

CHARLES J. COSTELLO, O.M.I.

THE PASTORAL EPISTLES
Introduction

Three Epistles of St. Paul (*1* and *2* to Timothy and the Epistle to Titus) are called "pastoral" because they are addressed, not to any church as a group, but rather to its head or pastor for his guidance in the rule of the church. All three are closely connected in form and content. Their genuineness is attested by the universal and persistent tradition of the Church from the outset as shown in early ecclesiastical records; and their style in some places is so characteristically Pauline as to preclude a forgery. As well on historical grounds as by reason of ecclesiastical tradition we must admit that St. Paul was twice a prisoner at Rome. These Epistles were written during the interval between his liberation from his first captivity, 63 A.D. (Acts *28*, 30-31), and his death, probably in the year 67.

THE FIRST EPISTLE TO TIMOTHY

St. Timothy was of Lystra in Lycaonia. Born of a Gentile father and a Jewish mother, he may have been converted on St. Paul's first visit to Lystra (Acts *14*, 7 ff). On the Apostle's second visit, Timothy enjoyed general esteem and was recommended as a suitable co-laborer of St. Paul despite his youth (Acts *16*, 2 ff). Thereafter Timothy was Paul's most beloved disciple; and in the Acts *16-20* we read of his missionary labors with Paul during the Apostle's journeys up to his final return to Jerusalem. The Acts make no further mention of Timothy, but from the Epistles of the Captivity we learn that Timothy was at Rome during the Apostle's first captivity, and in Phil. *2*, 19 there is the indication that Timothy will be sent shortly to Macedonia.

To explain the occasion and purpose of this Epistle, we assume that when Paul was released from the first Roman imprisonment, he probably visited Spain and soon returned to the Orient. The Epistle to Titus (*1*, 5) notes a brief visit to Crete. Proceeding to Ephesus, Paul found serious disorders due to certain teachers of the same mentality as the agitators of the Lycus valley (see Colossians). They dealt with myths and genealogies of a Jewish nature and urged abstinence from certain foods and other ascetical practices based on a dualistic mysticism. They would later, St. Paul says, prohibit marriage; and they were covetous and troublesome. Paul's stay in Ephesus was short, for he was anxious to reach Macedonia, so he left Timothy to handle the situation. Later he wrote him precise instructions in this Epistle. While not remarkable for its good order, a twofold thought is dominant. Timothy must energetically combat these false teachers and actively engage in the work of organizing the community. The thought of the Apostle moves restlessly back and forth on these two points, fully aware of the danger that threatened from his own experiences of the insolence of the false teachers.

THE SECOND EPISTLE TO TIMOTHY

When St. Paul wrote to Titus, he was somewhere in Macedonia and declared his intention of spending the winter at Nicopolis in Epirus (Titus 3, 12). This Epistle to Timothy finds him at Rome, and already a prisoner for some time. We conjecture from the Epistle that he had visited Ephesus and was arrested there but had been remanded to Rome for trial. The Apostle has conducted a fairly successful defense but expects an unfavorable sentence. His imprisonment is strict, nearly all his companions except Luke have left Rome and in his isolation he feels the need of seeing Timothy. This letter is an urgent invitation to join him, yet the Apostle is concerned to strengthen the soul of his beloved disciple to act vigorously against the separatist teachers during the time he will continue at Ephesus. Written 66 A.D., it is the last extant letter of St. Paul and in its moving urgency may be justly considered his spiritual will and testament.

THE EPISTLE TO TITUS

St. Titus was a Gentile and was probably converted to the faith by St. Paul, who styles him "his son." Paul refused to circumcise him even though at the Council of Jerusalem the Judaizers insisted that Titus submit to the rite. Some years later Titus conducted a successful commission for Paul at Corinth, reconciling this church with the Apostle (2 Cor. 7, 13-15; 8, 16-24). We lose sight of him after this until this Epistle which finds him in the island of Crete. While Titus is mentioned in the Epistles to the Galatians and 2 Cor., his name does not appear in the Acts of the Apostles. In explanation of this remarkable omission, it has been suggested that Titus was a relative of St. Luke, the author of Acts, who hid his own identity as well as that of his relatives, somewhat as St. John did in the Fourth Gospel. 2 Cor. 8, 18 f lends likelihood to his view.

The examination of this Epistle shows that the situation in Crete and the mission of Titus correspond exactly to Timothy's in Ephesus. St. Paul on his way from Rome had stopped on the great island and found the Christian community without due organization and further deeply disturbed by false teachers. On his departure after a short stay, he charged Titus with the work of remedying the situation. From Macedonia probably, some time later, Paul wrote this Epistle with exact directions to his disciple, urging him to zeal in the work. Since the Epistle is not so anxious in tone as those addressed to Timothy, we gather that the recipient was more mature and not so timid, which too is in keeping with their former conduct, particularly in their relations with the church of Corinth (2 Cor.).

COMMENTARY

THE FIRST EPISTLE TO TIMOTHY

This Epistle, like 2 Timothy contains instructions for Paul's beloved disciple, Timothy, whom the Apostle had left at Ephesus to handle a very serious situation. Certain teachers had gained authority in the Christian community and were working to introduce a false asceticism, in which appeared Jewish practices concerning distinctions of legally clean or unclean food. There also appeared certain Phrygian influences which would lead to the prohibition of marriage. Underlying their practices was a dualistic mysticism, holding the existence of a supreme Good and a supreme Evil, the latter being identified with matter. Pride and avarice were joined in the teachers, who disturbed families, particularly by their influence over women.

To his anxiety for the Church is joined in the Apostle the fear (in great part subjective) that Timothy because of his youth, natural timidity and delicate health will not correct the evils and exert his authority against these teachers with the full vigor that the situation demands. This attitude towards Timothy accounts for the urgent tone of the letter, as also for the lack of order in developing the various points.

INTRODUCTION *1*, 1 f

1, 1-2: Greeting. Though writing to his own disciple, **Paul,** because of the seriousness of the situation, feels that he must insist on his divine appointment to the apostleship. *God our Savior:* a title applied to God Himself only in the Pastoral Epistles of St. Paul, in St. Luke *1,* 47, and in St. Jude 25; elsewhere in the New Testament it is reserved to Jesus. In the Old Testament the phrase is frequently applied to the Lord. Christ is not so much the object of our hope as its author by the Redemption.   **2.** *Grace, mercy and peace:* a formula peculiar in St. Paul's writings to Timothy. The *mercy* is manifestly the mercy of God, which is the principle of salvation. This formula is found also in *2* John 1.3.

I: AGAINST FALSE TEACHERS 1, 3-20

1, 3-7: **Timothy's Mission at Ephesus.** This section has one of the characteristics of the Pauline writings. He here begins a sentence, which is left unfinished grammatically, and does not resume his thought until *2, 1.* All of *1* is a digression or lengthy parenthesis. The purpose of St. Timothy's mission is to repress the lying teachers and triumphantly establish the true gospel ideal, which is charity in a pure heart and a good conscience and an unfeigned faith. This is the theme of the entire letter.

3. We suppose that St. Paul was released from prison at the period indicated at the close of the Acts of the Apostle and that he visited Ephesus on his way to Macedonia.   **4.** The *genealogies* are legends built up around biblical characters.   **5.** The *charge* is Timothy's commission to handle the situation at Ephesus. *Unfeigned faith:* an expression deliberately chosen to set off the unmixed faith, which is in Christ alone, from the mixed faith of the false teachers adulterated by Jewish and other practices.

1, 8-11: **Rôle of the Law.** A digression to clear up the point insisted on by the false teachers in their false conception of the Mosaic Law and its part in the Christian life, for they posed as doctors of the Law. We should expect such an error in Ephesus where the Jewish colony was fairly large as Acts *17-20* shows. **9.** The sense is that the Law is not framed primarily for good people but for sinners; the list here is not complete.   **11.** *Gospel of glory:* the gospel has as its subject matter as well as its purpose the glory of God. *Blessed* or "happy" as an epithet of God, while current in Homer and Hesiod, is found nowhere else in the Bible.

1, 12-17: **The Apostle's Own Life.** A new digression connected with the mention of the gospel by association of ideas or even of words. The thoughts he presents show him to be an attentive and reflective observer, and the passage is so personal that it guarantees the authenticity of the Epistle.   **16.** *Mercy* recalls the greeting in *1, 2.* God's mercy was manifest in His calling Paul to the office of Apostle, though formerly Paul had been a most bitter adversary of Christianity.

1, 18-20: **Fidelity to Vocation.** Timothy's mission is difficult, but to inspire confidence in the young disciple, the Apostle recalls to him certain prophecies about him, because of which the presbyterium, or college of priests, imposed hands on him. Timothy appears to have been designated in the name of the Holy Spirit as co-laborer of St. Paul by inspired discourses pronounced by Christian prophets. It reminds one of the scene recorded in the Acts of the Apostles *13*, 1-3, where St. Paul himself, with Barnabas, was pointed out by the Holy Spirit to preach the gospel to the Gentiles.   **20.** Hymeneus is mentioned again in *2 Tim. 2*, 17, in connection with Philetus, as among those in error about the resurrection of the dead. It is not certain that this Alexander is the coppersmith of whom Paul complains in *2 Tim. 4*, 14. The chastisement, inflicted doubtless on the occasion of his visit to Ephesus, is excommunication, as is suggested by the parallel found in *1 Cor. 5*, 4 f.

II: PASTORAL CHARGE 2, 1 — 3, 13

2, 1-7: Directions on Prayer. V.1 resumes the thought of *1, 3*, from which Paul digressed in 5. The Apostle described the duties incumbent on Timothy by reason of his commission.   **2.** Prayers are to be offered for princes and those in authority. A certain amount of anxiety underlies this recommendation; the general situation of Christians, and particularly their relation with civil authority, no longer appears to be that of the time when St. Paul wrote the optimistic passage found in Rom. *13*, 1-7.   **6.** This witness is the gospel itself, whose content is described in 5 in an entirely Pauline fashion. Note the Apostle's insistence that Jesus is a man, hence His mediation extends to all men. From these recommendations one feels that there existed in the church at Ephesus certain exclusivist tendencies, which would restrict the Christian idea of salvation and practically reserve its benefits to a privileged caste.

2, 8-15: Women in Public Assemblies. St. Paul speaks to the leaders of the Ephesian community.   **8.** We are here reminded of the *graffiti* and frescoes of the catacombs, in which are pictured the orants or prayers, standing erect, arms raised, with palms turned upwards.   **15.** Motherhood is sanctifying, though not the only means by which women may be sanctified and saved.

3, 1-7: Qualities of a Bishop. One of the principal objects of Timothy's mission is evidently to provide for the good organization of the Ephesian church. It is not a question of the first organization (Acts *20,* 17, ff), but of a reorganization, with leaders capable of nullifying the disastrous activity of the false teachers and remedying the disorders already consequent on it. The danger of these false teachers is predominant in Paul's thought and inspired his instructions.

1 f. In the Commentary on Titus *1,* 7 it is shown that presbyter and bishop are convertible terms in the Pastoral Epistles. *Married but once:* there is here a question of forbidding not merely simultaneous polygamy, but even successive legitimate marriages. The command is a prohibition to select as bishop one who had married more than once. Priestly celibacy as a law is of later ecclesiastical institution. From *1* Cor. we know it was a matter of counsel even in apostolic times.

3, 8-13: Qualities of a Deacon. V.11 deals more likely not with women in general, nor even wives of deacons, but with those who are to be promoted to the office of deaconess. These were women assigned to certain activities favoring the gospel, which are not precisely defined, but were mainly of the material order. Thus deaconesses did not preach. They may have assisted at the baptism of women, cared for the sick, etc. The case of Phœbe (Rom. *16,* 1-2) as well as the present passage seems to imply the existence of deaconesses in St. Paul's lifetime. The Apostle does not designate precisely the functions of deacons and deaconesses.

III: AGAINST FALSE DOCTRINE *3,* 14 — *4,* 16

3, 14-16: Pillar and Mainstay of the Truth. Abandoning the pastoral theme, St. Paul returns to polemics, with this passage serving as transition.   **16.** The gospel itself is the mystery of godliness, because godliness is its peculiar benefit. The content of this mystery is briefly described in a rhythmic composition generally believed to be a fragment of a liturgical hymn. It is the Christ, who has been manifested in the flesh in His appearance on earth in the Incarnation (Phil. *2,* 5). His justification in the Spirit is His Resurrection (Rom. *1,* 3 f). He appeared to the Angels in His Ascension, and His elevation in glory includes His glorification on earth in the preaching and faith, as well as the glory He enjoys in heaven seated at the right hand of His Father.

4, 1-5: Lying Teachers. The multiplying of these teachers and their influence was manifestly a source of scandal to sincere believers, which St. Paul undertakes to remove by recalling that the appearance of false teachers had been foretold by the Holy Spirit.

1. Seemingly some early Christian prophet, whom we cannot identify, spoke by the Holy Spirit the express words that St. Paul cites here. **2.** Fugitive slaves and criminals were often branded with a hot iron on the forehead to identify them as such. The false teachers bear a like mark of infamy in their conscience. **5.** The *word of God* may be His creative word, but more probably it refers to the words of inspired Scripture pronounced as prayers over food before meals. The same forbidding of food is referred to in Col. *2, 21* f. The prohibition of marriage is rather Phrygian and Asiatic than Jewish. Mutilation was held in honor by the priests of Cybele in Phrygia and by those of Diana-Artemis even at Ephesus. Behind all these prohibitions and all this mystical asceticism revealed in the Pastorals and in the Epistles of the Captivity, there are tendencies, if not a fixed doctrine, that are decidedly dualistic. Already apparent in the Essenes, dualism is found in Philo and a little later is prominent in Neo-Platonism. Hence it is not surprising to meet the error in Asia Minor, even at the period when the Pastoral Epistles were written.

4, 6-10: Piety and False Asceticism. **8.** Corporal discipline or asceticism is the ensemble of the prohibitions imposed by the Ephesian teachers. Piety or the worship of God is the Christian life, as St. Paul conceives and imposes it in the moral part of his Epistles. **9.** The *saying* is to be found in 8 where are set forth the opposite conceptions of Christian salvation, in terms reminiscent of the Epistles of the Captivity.

4, 11-16: Zeal in His Office. Timothy is to give himself entirely to his commission. This passage gives the impression, already referred to, that in the Apostle's opinion Timothy's attitude lacks definiteness and decision.

IV. DUTIES TOWARDS THE FLOCK 5, 1 – 6, 19

5, 3-8: Widows. The Apostle defines a widow in the Christian sense of the word. There is a play on the Greek word for *widow* which means "desolate" or "isolated."

5, 17-25: Presbyters. **18.** To the citation of Deut. *25, 4*, St. Paul adds a saying of the Lord mentioned in St. Luke, which Gospel was probably composed before this Epistle.

22. The rite of laying on of hands appears in the New Testament in connection with (a) healing of the sick; (b) imparting of the Holy Spirit; (c) imparting of the spiritual gifts fitting one to exercise a special office in the community. This last is the case in Acts *6, 6*, where the Twelve laid hands on the first seven deacons; in *13, 3*, where the prophets and teachers of Antioch laid hands on Sts. Paul and Barnabas; in *1* Tim. *4,* 14, telling us that a college of presbyters, perhaps that of Lystra, laid hands on Timothy; and finally in the present verse where Timothy is warned not to lay hands hastily on anyone. It is clear in all the cases mentioned that this rite of laying on hands can be carried out only by certain persons.

Presbyters (Priests). This term does not appear in St. Paul's writings outside the Pastoral Epistles. In his first Epistle to Timothy he mentions a presbyterate, seemingly a college of presbyters, and this is connected with Acts *14,* 22, where Paul and Barnabas are said to have appointed presbyters in every church. In 17 the presbyters are described as presiding officers, and some of them at least are given to preaching and teaching. Their functions are sufficiently important to require that they be chosen with great discretion. Their rank is so high that no accusation is to be heard against them unless supported by two or three witnesses, that is, in juridical form. The public admonition to be inflicted on them is of a nature that will affect the community deeply. In the Epistle to Titus (*1,* 5), Titus is ordered to institute presbyters in each city of Crete. In these passages then "Presbyter" is an official title. Yet in *1* Tim. *5,* 1 the same word appears to be understood in its original sense of "old man." See Titus *1, 7* for the relation of priest and bishop.

23. It is hard to say what had led Timothy to total abstinence from wine. St. Paul urges that he use some on account of his health. Possibly the Apostle wished him to react against a false asceticism.

6, 1-2: Slaves. This passage shows that the gospel had introduced enough of practical fraternity to create a seeming danger of too great familiarity between different social classes.

6, 3-10: Lying Teachers. St. Paul returns for the last time to the false teachers. Empty discussions and exploiting of piety for the sake of gain describe their whole conduct, with the result that many wander far from the faith, inasmuch as their teachers are not given to the sound speech that is of Jesus Christ in the teaching that is conformed to godliness, but to discussions of the Law and Jewish traditions.

6, 11-16: Final Plea. Here is an extraordinarily urgent return to the point, betraying a certain anxiety. In terms particularly solemn the Apostle urges Timothy to be faithful. No other writer at the close of the first century would have so written of the preferred disciple of St. Paul.

12. Since the confession of Timothy is compared to the confession of Christ before Pilate, its content must be the same, namely that Jesus was the Christ. This was the great Christian confession (Rom. *10,* 9); it was made by Timothy either at his baptism, or when the priests of Lystra laid hands on him, or before the civil magistrates. Perhaps the last is the most probable hypothesis. **14.** The Second Coming of Christ will be accomplished in due time. The perspective appears to be rather distant.

6, 17-19: The Rich. These verses are the counterpart of *6,* 1-2, to slaves.

Conclusion *6,* 20 f

6, 20-21: Exhortation and Greeting. **20.** The *trust* is the sound teaching, the true gospel. The expression is Pauline, the Apostle often referring to the gospel as a deposit entrusted to him.

THE SECOND EPISTLE TO TIMOTHY

INTRODUCTION *1*, 1-5

1, 1-2: **Greeting.** *In accordance with the promise of life:* of which Christ is the source. See Commentary on Titus *1*, 1-2.

1, 3-5: **Thanksgiving and Prayer.** **3.** Timothy's attachment to God is inherited. The addition of the words, *with a clear conscience,* excludes the worship of God for one's selfish interest. The idea recurs constantly in the Pastoral Epistles. The prominence that it has in the mind of St. Paul is due without doubt to the conduct of the false teachers. It is however found before in *1* Thess. *2*, 3 f; *2* Cor. *1*, 12. **4.** Timothy's tears make us think of some touching scene of farewell like that of Acts *20*, 17 ff. We cannot say with certainty what the occasion was. Perhaps it was at Paul's departure for Macedonia (*1* Tim. *1*, 3); or more probably at the time when the Apostle, again a prisoner, left for Rome. **5.** This *unfeigned faith* is a part of the ideal St. Paul opposes to the mixed concepts of the Ephesian teachers (*1* Tim. *1*, 5). The Apostle refers to a faith in Christ alone, and free from attachment to Jewish fables and religious tendencies of pagans. *Eunice:* see Acts *16*, 1.

I: PASTORAL CHARGE *1*, 6 – *2*, 13

1, 6-14: **Paul's Example.** The passage, *1*, 6-11.13, is intended to encourage Timothy by strengthening his soul, which appears timid and hesitant. It is supposed that he is acquainted with Paul's imprisonment and the incidents that led up to it. In fact the impression is given that he witnessed them.

6. This gift which Timothy, according to *1* Tim. *4*, 14, owes to the presbyterate of Lystra, is by the intermediary of St. Paul himself, as this verse shows. It is reasonable to connect the two conferrings. Paul and the priests of Lystra together laid hands upon Timothy. The spiritual gift, not otherwise specified, had rendered Timothy fit to become the co-laborer of St. Paul. In *4*, 5 he is qualified as an evangelist, but we need not insist too strictly on the meaning of the title. **8.** The new imprisonment of St. Paul and the reaction it doubtless had on the affairs of Ephesus and the personal standing of Timothy could not fail to affect his state of mind profoundly. **12.** St. Paul cites to Timothy his own example. In spite of his trials he experiences no confusion. The gospel still continues to be what it is, and he himself is its herald, its apostle and teacher. The deposit which God can keep until the coming of Christ is the gospel, the sound teaching. See *1* Tim. *6, 20*, and *2* Tim. *1, 14*.

1, 15-18: Loyalty and Defections. **15.** *All in . . . Asia:* the Christians of Asia, particularly those personally attached to Paul. All have abandoned him. The context suggests that the Apostle had expected them to send delegates to Rome to visit him, and perhaps serve as witnesses in the course of his trial (*2* Tim. *4, 16*). They did not do so. Phigelus and Hermogenes must have been outstanding persons on whose attachment St. Paul had counted, for their desertion affected him painfully. We need not understand their fault as that of defection from the faith, nor should we extend their disloyalty beyond this single instance. From the references to Asia, we are led to consider western Asia Minor, or even Ephesus itself, as the setting of the troubles in the course of which St. Paul was again thrown into prison. **16.** Onesiphorus is unknown. He must have been a member of the church at Ephesus where his family still resided. The expression of St. Paul conveys the impression that this generous person was dead at the time he wrote.   **17.** The position of the Apostle is no longer that of his first imprisonment. His sequestration is stricter and his relations with the Roman community are restricted and perhaps secret. One had to seek him out diligently to find him. The journey of Onesiphorus to Rome implies that the second imprisonment of Paul had already lasted some time.

2, 1-7: Devotion to His Office.   **2.** The reference is to the preaching of Paul, of which Timothy was an habitual hearer. In confirmation of the facts which he relates, the Apostle cites the words of numerous witnesses. In *1* Cor. *15*, 3-11, where he deals with the resurrection of Christ, we can form an idea of his manner of preaching. Now a prisoner the second time, he has no hope of deliverance. He sees his death as near, and is concerned with assuring the perpetuity of the gospel and its transmission. Hence the command given Timothy, to entrust it to capable men with the charge to transmit the teaching received from him. **3-7.** It seems that Timothy, desirous of imitating his master in all things and also because of his timidity, took care of his own needs by his personal work and refrained from asking any support from the community, even refusing what they offered. The result was that his ministry suffered from this care for the necessaries of life. *Worldly affairs:* the generic name for all occupations which tend to assure subsistence for those who are given to them. One might even gather from the context of the exhortations that Timothy had created by this very conduct a plausible reason for escaping to a certain extent from the difficult duties of his office. In any event, that was the result, and St. Paul in truly paternal fashion calls it to his attention. Let him enjoy the rights that belong to the cultivation of the field in which he works (*1* Tim. *5,* 17 f; *1* Cor. *9,* 1-12).

2, 8-13: The Thought of Christ. St. Paul tries to set before Timothy considerations that are most suited to encourage him. He bids him recall the Christ, asserts that he himself in spite of his chains stands firm, and puts before his eyes the great Christian hope.   **8.** The emphasis is on *rose from the dead.* This doctrine Timothy is to remember above all.   **11.** The expression here, so peculiar to the Pastoral Epistles, refers less to the word of God itself or the teaching of Christ, than to the apostolic teaching promulgating the word or teaching and commenting on it. The preaching of the Christian apostles is true. This solemn attestation is notable inasmuch as it always refers to the reality of the salvation which is in Christ.

II: FIDELITY TO HIS OFFICE 2, 14 — 4, 8

2, 14-18: False Teachers. Having comforted Timothy, St. Paul resumes the principal question of the new teachers. He fixes the conduct of Timothy towards them.   **16.** Hymeneus and Philetus must have understood the resurrection in a purely spiritual fashion. We have here doubtless the extreme consequence of the ascetical and mystical ideas of Ephesian and Cretan teachers. And yet this excess seems to have been peculiar to these two persons. The exceptionally grave sentence pronounced by St. Paul against Hymeneus (*1* Tim. *1,* 20) confirms this impression. See however *2* Tim. *2,* 8-13, where the Apostle sees fit to affirm solemnly the reality of the Christian hope.

2, 19-26: The Faithful Servant.   19. To outward appearance all is upset. But in reality the solid foundation, established by God, on which the Church rests, stands firm. This foundation must be Jesus Christ, but the Apostle does not stress the point. That which interests him is to assert the permanence and cohesion of the building set on this foundation. Outwardly its lines appear to totter, its very existence is threatened by the multiplying of false teachers and their followers.   **20.** An abrupt change of metaphor. The Christians are now compared to different vessels which are part of the furniture of the house. Materially it is the same image as in Rom. *9,* 19-24, but in Timothy all the vessels are figures of Christians.

3, **1-9: Against New Teachers.**   **1.** Compare with *1* **Tim.** *4,* 1. It is in virtue of prophecies recalled in the first Epistle that St. Paul announces the appearance of difficult times.   **5.** *Avoid these* refers to the false teachers among whom Timothy lives. They belong already to that class of Christians who the Spirit had foretold would appear in later days. St. Paul in speaking of the later days is thinking of the general crisis that is to precede the reappearance of Christ. The forerunners are already visible. On this final crisis, see *2* Thess. *2,* 5-12.

8. According to Jewish traditions, Jamnes and Mambres are the names of the Egyptian magicians who opposed Moses before Pharaoh.   **9.** St. Paul believes that the folly of the false teachers will be exposed. The thought differs from *2,* 16 f. In the first passage he stressed the dangerous nature of these discussions, which destroy all piety, and he does not pronounce on their future fate. See also *3, 13.*

3, **10-17: Paul's Example and Doctrine.** Once more Paul exhorts Timothy to stand firm and fulfill his office. The letter grows more and more intimate and urgent. It sounds like the spiritual last will and testament of the Apostle.   **10 f.** All these persecutions Paul had suffered in the very country of his disciple. This is his reason for giving them prominence. During these persecutions Timothy's mother and Timothy himself were converted and joined St. Paul (Acts *13,* 49 — *14,* 19).   **14.** The teachers referred to are the mother and grandmother of Timothy, as well as Paul himself. Timothy knew the Scriptures, thanks to his mother and grandmother, both fervent Jewesses. The intention is to move Timothy and touch his heart, as is evident. The soul of St. Paul has unlimited resourcefulness.   **16.** *All Scripture:* a technical term, designating the books which Jews and

Christians held as sacred. In 15 it refers in the concrete to the Scriptures of the Old Testament. In declaring that all Scripture is inspired by God, St. Paul does no more than recall an idea admitted by all. **17.** *Man of God:* see *1* Tim. *6,* 11. This title appears to be reserved to workers in the gospel alone and it belongs to them in a special fashion. It marks them out as persons who belong to God and are attached to His service.

4, **1-5: Preach Sound Doctrine.** Solemn in tone, the exhortation supposes that Timothy was overtimid in exercising his office. **3.** A new allusion to the prophecies concerning aftertimes. See *3,* 1 and *1* Tim. *4,* 1. All these disorders had already begun to appear.

4, **6-8: Reward.** The Apostle sees that death is near. In the long leisure of his imprisonment, separated from his most faithful disciples, very much restricted in his relations with the Roman church, and then under surveillance, he takes occasion to reflect on his career as a whole. It has been long, with many trials. But he has done his duty and the work entrusted to him. He awaits from the just Judge the crown that he has earned by the faithful service of more than thirty years. **8.** His *crown* is the sign and reward of justice. Then for Timothy's comfort he adds that the same crown awaits all those that love the manifestation of the Lord.

Conclusion *4,* 9-22

4, **9-15: Paul's Loneliness.** Timothy seems to be fully acquainted with St. Paul's situation. He knows in particular the persons who had accompanied Paul to Rome. This is explained if Paul departed from Ephesus. **9.** St. Paul finds himself deserted. Luke only is with him of his personal company. See 20. He has sent Tychicus to Ephesus, to replace Timothy, who will soon be free, and Paul invites him to come to him. **10-12.** St. Paul's expression appears to indicate that all these persons had followed him to Rome. In Col. *4,* 14, Demas was mentioned in a very detached fashion. Crescens is otherwise unknown, but his departure does not seem to be a defection; nor does that of Titus. Mark and Tychicus are known from Col., Eph., Titus. **13.** Some think that *the cloak* is rather a wrapper or container for books, but the meaning is far from being certainly estab-

lished. It may well be a heavy winter cloak that Paul desires.
14. This *Alexander, the coppersmith* was evidently known to Timothy. He is certainly an Ephesian and may be the Alexander mentioned in *1* Tim. *1,* 20. Had he followed St. Paul to Rome as an accuser? Or had he simply been involved in the events that led up to Paul's imprisonment, and did that occur at Ephesus? The first supposition is more probable in the context. In any event, at the time of writing Alexander is back in Ephesus, since Timothy is warned against him.

4, **16-18: His Trial.** **16 f.** It appears quite doubtful that the allusion is to his first Roman imprisonment and the final defense that gave him his liberty. We have here rather a reference to a recent incident in his late trial. Even though none who should have supported him did so, yet he was successful at least provisionally in his first appearance. It had been for him further an occasion to preach the gospel to his judges and those assisting, as he emphatically states in 17. Impressed as he was by the Roman majesty, he considered as the consummation of his career as a preacher this privilege of bearing witness to the gospel before a tribunal seated in the capital of the world, before a cosmopolitan crowd, such as would be found in imperial Rome. The *lion's mouth* is the danger from which he escaped. **18.** The end of the verse explains the deliverance he expects, namely, death.

4, **19-22: Greetings.** Erastus, mentioned in Rom. *16,* 23, is a city treasurer of Corinth. Trophimus was an Ephesian of pagan origin (Acts *21,* 29). **21.** These are members of the Roman church known to Timothy from his stay at Rome during Paul's first imprisonment. Linus, according to St. Irenaeus, is St. Peter's successor.

THE EPISTLE TO TITUS

INTRODUCTION 1, 1-4

1, 1-4: Greeting. The greeting is long and solemn and one is surprised to find St. Paul stating his titles in a letter to one of his most faithful disciples. But this letter, unlike the note to Philemon, has an official character. For while it is addressed directly to Titus, it is evident that the Apostle composed and dictated it in view of Cretan disturbers. It is to them and against them that he cites his titles. Finally, and above all, the emphasis is not on the titles themselves, but upon the clauses added to them, *in accordance with the faith of God's elect,* etc.

1. *Servant of God:* elsewhere St. Paul styles himself "servant of Christ" (Rom. *1,* 1). This is in the style of the ancient chanceries of the east, "servants of the king." Likewise priests used this title in reference to the god they served. *In accordance with the faith,* etc.: phrases that determine the sphere of Paul's apostleship as well as the norm of its exercise. The faith of the elect of God is the common faith (4), the faith of the universal body of the saints, in opposition to the special faith of the Cretan teachers. The *knowledge of the truth which is according to piety* is in turn opposed to the new doctrines, which substitute for piety, or the true Christian life, an asceticism which is dubious and compatible with immorality.

I: PASTORAL CHARGE

1, 5-16: Titus' Mission, and Special Needs in Crete. His mission is twofold, to organize the communities, in placing leaders over them, and to teach the members of the community the true Christian life. This first section, *1,* 5-16, completes the directions already given on the organization of the churches.

6. Paul demands in the Pastoral Epistles that the presbyter-bishops and deacons be married but once; but this is no absolute condemnation of second marriages, since the Apostle positively recommends that young widows remarry (*1* Tim. *5,* 14). **7.** The term "bishop" is substituted quite abruptly here for that of "priest," as in the Apostle's address to the priests of Ephesus (Acts *20,* 17. 28). It characterizes better than the term "priest" the office of the person in question. "Bishop" signifies inspector, overseer, superintendent. St. Paul uses the title in this sense, and drops it to substitute at the close of the verse the still more expressive term "steward of God."

Bishops. The title "bishop" like that of "priest" is not, in Greek, of Christian origin. Used first by St. Paul in addressing the priests of Ephesus (Acts *20,* 28), it reappears in the greeting of the letter to the Philippians (*1,* 1), then twice in the Pastorals (*1* Tim. *3,* 1-7 and Titus *1,* 7). From these passages it follows, (a) that the title fits a number of personages in each church; (b) that it is in fact convertible with "priest." The fixed elements of an organized community are clearly enumerated in the greeting of the Epistle to the Philippians. These are the saints, or simple members, the deacons or lower officials, the priest-bishops, the highest local authorities. Did they form a college, "the presbyterate" (*1* Tim. *4,* 14)? And was there a president as its head? The answer to both questions is possibly, even probably, affirmative. But to judge by the language of St. Paul, the title of bishop at this date was not reserved to such a president.

10-16: In enumerating the qualifications demanded of priests, and in the special urgency of establishing them in every city, St. Paul is making provision against the large number of teachers in Crete, who, like those at Ephesus, are empty talkers and disturbers that must be silenced.

12. This hexameter is from the Minos of Epimenides, a Cretan poet who may have lived in the sixth century B.C. Here and in the verse following St. Paul speaks no longer of the false teachers themselves, but rather of the body of Christians who accepted their teaching.   **14.** The disturbers' teaching has two elements: the one speculative and made up of Jewish fables; the other practical, containing precepts that forbid certain foods.   **15.** The reference is to food, with the distinction of clean and unclean under the Mosaic Law.

II. CHARGE TO TEACH THE CHRISTIAN LIFE, 2, 1–3, 11

2, 1-10: Different Classes. Morality had suffered much because of these disorders. Titus is commanded to make Christian morality and true piety prevail. Certain special moral lessons are to be inculcated.   **2.** Here there is question of elderly men not priests.   **4.** The false teachers are accused in *1,* 11 of disrupting families. As at Ephesus, they urged women, even married, to the perfect life as they conceived it in pseudo-asceticism. The

husband, the children, the care of the home were unimportant.
8. Those of the opposite camp may be pagans or Satan. It is not a question of false teachers who are rather responsible for these disorders (*1* Tim. *5,* 14; *3,* 6 f).

2, 11 — 3, 7: Changed Life. The kind of life described briefly by St. Paul in the foregoing verses, and of which he now sums up the principal traits, is what he calls elsewhere "piety." This and not the pseudo-asceticism of the false teachers represents the true Christian life, that the grace of God, manifesting itself in Jesus Christ, proposed to establish.

14. This indifference to good works doubtless gained support for the movement combated by St. Paul. Yet it was attached to ascetical and mystical theories.   **15.** The Apostle fears that Titus himself does not show the desired vigor and does not assert his authority with sufficient emphasis. The impression is gathered that the Cretans, left to themselves a long time, have assumed the habit of independence and are little disposed to be guided even by a delegate of an Apostle. Christianity must have been introduced early into Crete, perhaps without the intervention of any notable person. There were Cretan Jews or Proselytes in the audience of St. Peter on Pentecost day (Acts *2,* 11).

3, 1. Submission, obedience, readiness for every good work— these make up the proper attitude of the Christian towards those in authority, especially civil authority (Rom *13,* 1-7; *1* Tim. *2,* 1 f).   **2.** St. Paul now passes on to their dealings with private individuals.   **3-7.** These verses are to be compared with Eph. *2,* 1-9.   **5.** The Holy Spirit is the agent in the bath of rebirth and renewal. The reference is to Baptism.

3, 8-11: Good Works and Truth. These verses take up and develop the thought of *2,* 15, and the theme of *1,* 10-16.   **8.** The *saying* is the content of 4-7 preceding.   **9.** There was question before in *1* Tim. *1,* 4 of genealogies, where they are treated as fables that simply give rise to disputes. Here they are considered as empty researches and are manifestly the same as those referred to in *1,* 14 of this Epistle, where they are said to be Jewish fables, legends apparently, grafted especially on the stories of Genesis. The Jewish apocrypha written in the century before and after the birth of Christ are full of this type of speculation.

Conclusion 3, 12-15

3, 12-14: Closing Messages. **12.** The mission of Titus at Crete is only temporary. Nicopolis, where Paul proposes to see him, is Nicopolis in Epirus, one of the most important cities on the west coast of Greece, having been made a Roman colony by Augustus after his naval victory at Actium close by. In *1* Tim. *3,* 14, St. Paul hopes to rejoin Timothy at Ephesus shortly. Here he says he will pass the winter at Nicopolis, indicating a recent decision and a change in former plans. We are led to believe then that the letter to Titus is later than *1* Timothy.

13. Zenas is otherwise unknown. He would seem to be a Roman lawyer rather than a Jewish doctor of the Law. Apollos is mentioned in *1* Cor. and in Acts *18,* 24 ff. Zenas and Apollos had agreed to carry this letter to Titus, but would pass beyond Crete. **14.** *Our people:* the general run of Christians, in contrast with officials like Titus himself, should take care of material affairs.

3, 15: Greeting. *In the faith* replaces the customary formula "In the Lord." This change and the relief given to the faith, conceived of as adherence to sound doctrine, can be explained by the dangers to which the gospel was exposed at Crete.

Leo P. Foley, C.M.

THE EPISTLE TO THE HEBREWS

Introduction

Title. In the earliest Greek MSS, the title is simply "To the Hebrews." The name of St. Paul is generally added to the title in Latin MSS and in Greek MSS of late origin. It must be remembered, however, that the titles were not a part of the original Scriptures, but were added when the Scriptures began to be copied for general circulation.

Canonicity. This Epistle was always accepted in the East as canonical, that is, as Divine Scripture. It was also accepted in the West at first, for St. Clement of Rome, in his letter to the Corinthians, which was written before the end of the first century, quotes it extensively as Scripture. Whatever may have been the reasons, doubts did arise about the Epistle during the three following centuries. Some would attribute the difficulty to the use made of it by heretics; others, to uncertainty regarding its Pauline authorship. The Epistle does not appear on some private lists such as the Muratorian Fragment, and was even explicitly denied to be Scripture by private individuals such as Gaius, a priest of Rome. In spite of the doubts raised against it, every official pronouncement in the West was true to the early tradition. Thus the Council of Hippo (393), the two Councils of Carthage (397, 419), and the letter of Pope Innocent I to Bishop Exuperius of Toulouse (405), list the Epistle as Scripture. After the fourth century, it was again accepted everywhere as canonical.

Authorship. The question of the Epistle's authorship is complicated and vexatious. In the East, St. Paul was generally considered to be the author, but authorship was viewed in the broad sense as embracing whatever had been planned by an author even though actually put together by another. In the West, however, from the second to the latter half of the fourth century, there was not only considerable doubt, but even actual denial of Pauline authorship. Then for more than a thousand years, East and West agreed in ascribing the Epistle to St. Paul. With the rise of the new learning and the advent of Protestantism, the question was again taken up by both Protestants and Catholics.

At present non-Catholics generally deny the Pauline authorship of the Epistle, although many will admit readily that the Epistle emanates from the Pauline circle. They ascribe the Epistle for various reasons to Barnabas, Luke, Clement, Apollos, Priscilla, or other companions of St. Paul. Catholics, since the decree of the Biblical Commission (June 24, 1914), maintain the Pauline authorship of the Epistle at least in the sense that it was conceived by him and written under his direction. Its thought is thoroughly Pauline, and much of its phraseology is also distinctly Pauline. Such ideas as are peculiar to the Epistle are in perfect harmony with the rest of Pauline thought. While St. Paul's name and customary greetings are missing at the beginning of the Epistle, his usual salutations and good wishes are to be found at its close. Then too there is the intimate reference to Timothy his beloved disciple. The excellent literary style, however, is generally superior to that found in the other Epistles of St. Paul.

Time and Place of Composition. Various dates, from 60 to 96 A.D., have been assigned to the Epistle. The several warnings against the seductiveness of the Jewish sacrificial system indicate that the temple of Jerusalem was still standing, and consequently a date must be sought prior to its destruction in 70. The persecution of 62, in which St. James the Less, Bishop of Jerusalem, was slain, would explain very well the occasion of the Epistle, and would indicate a date sometime in 63. The last verses also point in the same direction. The writer is expecting Timothy, with whom he hopes to be able to see his readers soon. According to Phil. *2,* 19-30, St. Paul had promised to send Timothy to the Philippians on a mission as soon as the issue of his own trial was certain. It was probably Timothy's return from the Philippian mission, which St. Paul alludes to in Heb. *13,* 23. St. Paul was released in 63.

The place of composition is not mentioned, but Rome seems to be the most probable. Codex Alexandrinus has "From Rome" as a part of the inscription, and the Peshitto Syriac gives Italy as the place of writing. The greetings of the "brethren from Italy" (*13,* 24) would likewise indicate some city in Italy.

Destination. The Epistle was written apparently to encourage a community of Jewish Christians, who were in danger of relapsing into Judaism. Although the community is not named, most of the indications point to Jerusalem. Alexandria, Cæsarea in Palestine, Ephesus, Antioch of Syria and Rome have also been suggested.

COMMENTARY

I: SUPERIORITY OF THE NEW DISPENSATION OVER THE OLD *1, 1–10,* 18

1. A Superior Mediator *1,* 1–*4,* 13

1, 1-14: **Christ Superior to the Angels.** Christ, the Mediator of the New Covenant, is as Son of God superior to the angels and to Moses, the mediators of the Old Covenant. **1.** *Sundry times and divers manners:* the revelation through the prophets while fragmentary, was nevertheless extensive in content and varied in form. The same God who spoke through the prophets, now speaks finally through His Son. The continuity and progressive character of revelation are implied. **2.** *Last of all in these days:* (Greek: "At the end of these days.") The phrase indicates the Messianic age, and implies that the revelation through the Son is superior and final. *Appointed heir of all things:* Christ was appointed heir of all things according to His human nature; according to His divine nature He already possessed all things. *By whom also He made the world:* cf. John *1, 3;* the Word is the Creator of the world.

3. *The brightness of His glory:* As the rays of light stream forth from the sun, so the Son is "Light of Light." *The image of His substance:* i.e., as the imprint made by a seal. Both figures attempt to describe the Son of God. *Upholding all things:* the Son keeps all things in existence by His divine power. *Effected man's purgation from sins:* Christ through His human nature redeemed mankind. *Taken his seat:* Christ finished satisfactorily the work undertaken in His human nature, and returned with His human nature to heaven. *At the right hand* denotes dignity and not place. It had been foretold that the Messias would sit at God's right hand (Ps. *109,* 1). **4.** *Having become:* i.e., in His human nature at the moment of His Incarnation. Some Greek Fathers interpret *having become* as "having been shown to be," and refer the expression to the glorification of Christ through His Resurrection and Ascension. *Has inherited,* as man. *A more excellent name:* i.e., that of Son of God. According to His divine nature, He possessed this name from eternity, and as man, He received it from the first moment of His Incarnation. It was due Him by reason of His origin. The name of Son of God is far superior to that of angels (messengers).

5-14: Five arguments based on seven texts from the Old Testament (Septuagint) follow, to prove the superiority of the Son over the angels. The Old Law was given to Moses through the angels (Deut. *33,* 2; Acts *7,* 53; Gal. *3,* 19); the New Law was given through Christ, who was God's own Son.

5. The first argument is based on the name "Son." To no angel had God ever given the name of son; but of the Messias to come He had said *"Thou art my Son, I this day have begotten thee* (Ps. *2,* 7). Some of the Fathers interpret these words of the eternal generation, while others take them as referring to the Incarnation. St. Paul (Acts *13,* 33) takes them apparently of the Resurrection. A second proof based on the name of son is taken from 2 Kgs. *7,* 14: *I will be to him a father, and he shall be to me a son,* words which in their literal sense refer to Solomon, and in their typical sense to the Messias of whom Solomon was a type.

6. The second argument that Christ is superior to the angels is based on God's command that the angels worship the Messias (Ps. *96,* 7; cf. also Deut. *32,* 43 in Septuagint). *And again:* if the "again" is taken with *he says,* it indicates simply another quotation, and the reference would be to the first introduction of the Son at the time of the Incarnation or Nativity. If, however, the "again" is taken with *he brings,* the reference is to a second introduction of the Son at the Last Judgment. *Firstborn:* some take the term here in its natural sense as referring to the Messianic office of the Word Incarnate in the sense of Ps. *88,* 28, where God says of the Messias to come, "I will make him my firstborn, high above the kings of the earth." In this latter case "firstborn" is equivalent to "heir."

7-9. The third argument, showing Christ's superiority over the angels, is the fact that the angels are only servants, likened to the winds and lightning (Ps. *103,* 4) which God uses as His agents; whereas of the Son it is said that He is an anointed ruler, who Himself is called God (Ps. *44,* 7 f).

10-12. The fourth argument is that Christ is the Creator, who is beyond all time and change. St. Paul here applies words to Christ which in Ps. *101,* 26-28, are applied to God, indicating thereby that Christ is God.

13 f. The fifth argument, drawn from Ps. *109,* 1, is that Christ shares the throne of God. The angels are but ministering spirits.

2, 1-4: Warning and Exhortation. Practical exhortation and warning based on the preceding chapter. It is the first indication of the dangers which threaten the readers. See *4, 1-10; 6, 1-8; 10, 26-31.*   **2.** *Words spoken by angels:* the Old Law had been given to Moses on Sinai through the ministrations of angels (Acts *7, 53;* Gal. *3, 19).*   **3.** *Confirmed unto us by those who heard him:* i.e., by the Apostles. The Jerusalem church had been organized on the first Pentecost by Peter and the other Apostles. **4.** *By signs and wonders, and by manifold powers:* i.e., by various kinds of miracles. *By impartings of the Holy Spirit:* i.e., by an outpouring of many charismatic gifts such as prophecies, speaking in tongues, etc. (Rom. *12,* 6; *1* Cor. *7, 17; 12, 4-11).*

2, 5-18: Christ Suffered for His Brethren.   **5-9:** The argument broken off at *1,* 14 is here taken up, and additional proof is given to show that Christ is superior to the angels. He is the Lord of the Messianic kingdom.

5. *He has not subjected:* not to the angels, but to the Son as man, has God the Father subjected *the world to come,* i.e., the Messianic kingdom. The Messianic kingdom inaugurated by Christ will not have its consummation until the Last Judgment, when He will be acknowledged as Lord by all.   **6-8.** A quotation from Ps. *8,* 5-8 describing the perfect man, which is here applied to Christ.   **6.** *Man* and *Son of man* by Hebrew parallelism have the same meaning.   **7.** *A little lower:* i.e., in degree. Some translate "for a little while lower."   **8.** *For in subjecting all things to man, he* (God) *left nothing that is not subject to him* (man). All things will be subject to this ideal man, leaving nothing for the angels to rule over. *But now we do not see:* i.e., all things do not as yet acknowledge the sovereignty of Christ, the ideal man to whom the Psalm is here applied.   **9.** *Crowned with glory and honor,* etc.: cf. Luke *24,* 26; Phil. *2,* 8-11. *By the grace of God:* probably, God's mercy towards us. *Taste death:* i.e., to experience death, to die. *For all:* Christ died for all mankind.

10-18: A digression to explain Christ's humiliation in being for a time lower than the angels. In God's providence, the way to glory for mankind in its fallen state is through suffering. Since this is God's plan, it was fitting that the Savior of men should share man's nature and tread the path of suffering to glory.

10. *For whom are all things,* etc.: namely, God as the final and efficient cause of creation. *Who had brought,* etc.: the time of the verb in the Greek text is rather present, meaning "in bringing many sons into glory." This fits the context better. The past tense, as in the Vulgate, would refer rather to God's eternal decree, to His predestinating many to glory. For the same idea cf. Eph. *2, 5-7. To perfect:* to lead to a goal, to complete a process. It refers here to the "crowned with glory and honor" of 2, 9.   **11.** *All from one:* from one family, with God as the father of all. *Not ashamed to call them brethren:* Christ and men have the same human nature, but there is here an implication that Christ is more.

12 f. Three quotations from the Old Testament are given to prove that the Messias, since He was to have a human nature, would consider men as His brethren. The first is from Ps. *21, 23* where the Messias is represented as giving thanks to God for His exaltation after the humiliation of the Passion. The quotation is introduced here to show that the Messias called other men His brothers. Christ from the cross applied the first verse of this Psalm to Himself ("My God, my God, why hast thou forsaken me"; Matt. *27,* 46). The other quotations are from Isa. *8,* 17 f where they are continuous. Here they are introduced separately. In Isaias they refer literally to the prophet himself, and typically to the Messias. Our author takes them as Messianic. In the first of these, the Messias is represented as trusting in God, and hence as being human. In the second He is shown as closely associated with God's children (men), really as constituting one family with them.

14 f. Christ is to suffer and die like other men in order to free men from the fear of death. *Blood and flesh in common:* having a human nature. *Destroy:* "render impotent," according to the Greek. By dying Christ proved that death was the entrance to eternal life.   **16.** Christ took a human and not an angelic nature, since He came to redeem mankind and not angels.   **17 f.** These verses give another reason for Christ's humiliation and sufferings, namely, that He might have a sympathetic understanding of our needs, and be our mediator with God. The Jewish High Priest on the Day of Atonement offered a sacrifice for the sins of the people. Christ offered up His sacrifice of the cross for the sins of all mankind.

3, **1-7: Christ Superior to Moses.** The author now proceeds to prove that Christ is superior to Moses. The Jews believed that **through** Moses they had received God's final and complete revelation. It was necessary to prove to them that Christ had greater authority than Moses.

1. Christ is both Apostle and High Priest, for as Apostle He was sent to establish our religion, and as High Priest He offered its one sacrifice.   **2.** *Who is faithful to Him who made him:* i.e., loyal to God who invested him with his office. *In all his house:* Num. *12,* 7. God's house is the organized society in which God dwells, the whole family of God. It includes both the Jewish and Christian dispensations.   **3 f.** The argument here is that Moses is a part of the "House of God," but that Christ was the creator and builder of that House. In *1,* Christ was represented as the creator of all things; here He is represented as the architect and builder of the House or Religious Family of God. The new dispensation is a continuation of the old, and both were planned and established by Christ Himself, hence Christ must be superior to Moses.   **5.** Moses was but a servant in that House; Christ is the Son over the House.   **6.** Perseverance in confidence and hope is necessary, if we are to be considered as belonging to that House of God.

3, **8-18: Exhortation.** Another warning, which continues to *4,* 13. The Promised Land of Palestine was a type of the joy and rest of the heavenly home to which Christians have been called. Just as the Israelites in the desert, because of their unbelief, were excluded from their destined repose in the Promised Land, Christians through lack of faith may never reach the joy and repose of heaven. Ps. *94,* 8-11 is quoted, commented upon, and applied to the readers, who were in some danger of losing their faith.   **16.** The Greek here may be translated as two questions: "Who were they, who having heard did provoke? Were they not all those who were led out of Egypt by Moses?"

4, **1-13: Our Promised Land.** The Rest promised by God to His people was more than a mere repose in the Land of Canaan. It was to be a participation in His own Sabbath Rest entered upon after His work of creation. That Rest had been offered to the Israelites of old, but they failed to obtain it because of their unbelief. It was promised again in the time of David, centuries after the occupation of the Land by Josue. That promise of God regarding His Rest still remains, and is accessible to the Christians. As of old, unbelief may prevent one from entering. Faith and labor are necessary.

2. A Superior High Priest *4, 14 — 7, 28*

4, 14-16: Confidence in Christ. The author now passes on to another argument proving the superiority of the new dispensation over the old, namely, the superiority of Christ's priesthood to the priesthood of the Old Law. This argument with some degressions will continue to *10, 18.*

14. *Passed into the heavens:* in contrast with the High Priest of the old dispensation, who entered into the Holy of Holies, where God dwelt in a special manner, only on the Day of Atonement, Christ the High Priest of the new dispensation has entered into the very presence of the Father Himself, in His heavenly home.

5, 1-10: Christ the High Priest. Christ has all the qualifications of a high priest, namely, a human nature, sympathy for human infirmity, and appointment by God.
1. *Gifts and sacrifices:* the unbloody and bloody offerings.
5. Christ was called to the high priesthood by the Father. The words of Ps. *2, 7* are understood here as being addressed by the eternal Father to Christ at the moment of the Incarnation. It was then that Christ became the mediator between God and man. **6.** *According to the order of Melchisedech:* according to the type of Melchisedech, or according to the manner of Melchisedech (Ps. *109,* 4; cf. 7, 1-10). **7.** Christ as man surrendered himself absolutely to God. **8.** *Learned obedience:* the earthly life of Christ was a process of doing God's will. **9.** *When perfected:* having brought to completion His mission.

5, 11-14: Importance of the Doctrine. As the readers by their neglect of the Faith have grown dull of understanding in spiritual things, it is difficult to explain the high priesthood of Christ to them. *Rudiments of the words of God:* the fundamentals of Christian doctrine.

6, 1-3: An Appeal for Progress. **1.** *Dead works:* i.e., sinful works. **2.** *Doctrine of baptisms,* may refer to the distinction in effects between the baptism instituted by Christ, and other ritual ablutions. *Laying on of hands:* the sacrament of Confirmation.

6, 4-8: Danger of Apostasy. **4.** *Impossible:* i.e., morally impossible, or very difficult. Those who have apostasized refuse to repent. *Enlightened:* baptized. One baptized passed from the darkness of sin to the light of divine grace. Baptism was sometimes called "Illumination" by the early Fathers. *Heavenly gift:* redemption. *Partakers of the Holy Spirit:* sharers in the charismatic gifts, such as speaking in tongues, prophesying, etc. **5.** *Tasted the good word of God:* experienced the consoling effects of the gospel. *Powers of the world to come:* the miracles of the Messianic age. **6.** *And then have fallen away:* have rejected completely the Christian religion. *To be renewed again to repentance:* or better, according to the Greek, "To renew (them) to repentance," thus making the impossibility on the part of the preacher. It would be practically impossible for a preacher to awaken repentance in such perverts, for whom Christ has become an object of hate and mockery. They are likened to barren soil (8).

6, 13-20: Certainty of God's Promise. **13.** *When God made his promise to Abraham:* see Gen. *12,* 2 f; *13,* 14-17; *15,* 5 ff; *17,* 5 ff; *22,* 16 f. **15.** *After patient waiting:* many years intervened between the promise made to Abraham and the birth of Isaac.

16-18. The promise made to Abraham finds its complete fulfillment in the Christians, who by faith are the true descendants of Abraham and heirs of the promise (Gal. *3,* 29). *Two unchangeable things:* namely, God's promise to Abraham, and the solemn oath with which He confirmed it. **19.** *Anchor . . . reaching even behind the veil:* our hope is anchored to Christ in heaven. The allusion is to the veil which hung before the Holy of Holies in the temple, and behind which the High Priest entered on the Day of Atonement. Christ, our High Priest, has entered behind the veil which separates God's presence from this world. **20.** *Forerunner:* Christ has entered into heaven first, to prepare a place for us (John *14,* 2 f).

7, 1-28: The argument for the superiority of Christ's over the Levitical priesthood, which had been broken off at *5,* 10, is now resumed. Melchisedech, a priest of the true God, received tithes from Abraham, the father of the Jewish people, and blessed him (Gen. *14,* 18-20), hence Melchisedech must be the greater. The Levitical priesthood, although originating several centuries later, is represented likewise as paying tithes to Melchisedech in the person of their ancestor Abraham, thus indicating inferiority to Melchisedech's priesthood. It had been foretold that the Messias would be a priest according to the order of Melchisedech (Ps. *109*, 4), hence the priesthood of Christ, who was the expected Messias, must be greater than the Levitical priesthood.

7, 1-3: Melchisedech More Than Abraham. **1.** *The most high God:* the true God. **2.** Justice and peace had been predicted of the Messias. **3.** *Without genealogy:* not that Melchisedech had no father and mother but that Scripture mentions none. His priesthood was not based on family descent as was that of the Sons of Levi, but was a personal prerogative. In this it resembles the priesthood of Christ. As nothing is said in Scripture of his death, his priesthood is looked upon as continuing forever, and hence as typical of the Son of God. Ps. *109*, 4 speaks of the Messias as a priest forever according to the order of Melchisedech.

7, 4-10: Melchisedech More Than Levi. **8.** The Levitical priests died and had successors. The priesthood of each ceased with death. Neither the death of Melchisedech, nor any successor to him is ever mentioned in Scripture, hence his priesthood is considered as living on. **9.** The tribe of Levi was set aside for priestly functions under the Mosaic Law. This priestly tribe is represented here as paying tithes to Melchisedech through their forefather Abraham.

7, 11-17: Levitical Priesthood Imperfect. **11.** *Order of Aaron:* Aaron, the first High Priest, was of the tribe of Levi. **12.** The Levitical priesthood had failed, and with it the Old Law. A new priesthood had been foretold, and that meant also a change of law, a new system. **16.** *Law of carnal commandment:* by family descent, or birth. *According to a life that cannot end:* by the power of life inherent in His nature as Son of God, which death was not able to dissolve; hence His priesthood continues forever. **17.** Christ conquered death by His resurrection, and hence His priesthood continues forever as foretold in the Psalm (*109*, 4) quoted.

7, 18-19: Superseded by Priesthood of Christ. **18.** The Old Law had failed, and was abrogated. **19.** *Better hope:* through the gospel.

7, 20-25: A Priest by Divine Oath. 20. God made a solemn oath that Christ was to be a priest forever. The priests of the Old Law were established in their office without an oath. Christ's priesthood, therefore, must be superior. 24. Since He always lives, He has no successor.

7, 26-28: Sinless and Perfect. The sinless Christ is the ideal High Priest. He offered up the one supreme sacrifice of Himself for the sins of the people, and needs not to offer up daily sacrifices for sins as did the Jewish priests. 28. *The Law:* the Mosaic Law. *The word of the oath which came after the Law:* Ps. *109,* in which the oath regarding the priesthood of the Messias is contained, was written centuries after the Mosaic Law had been given. The Mosaic Law made mere sinful men to be priests; this oath set up the sinless Son of God as a priest forever.

3. A Superior Covenant 8, 1-13

8, 1-5: Christ in the Heavenly Sanctuary. The Jewish priests exercised their functions in an earthly sanctuary (tabernacle), which was but a copy and shadow of the true sanctuary of heaven, where Christ, the ideal High Priest, exercises His functions seated at the right hand of God.

8, 6-13: Mediator of a Superior Covenant. Christ is the mediator of a new and better covenant based on better promises. This is proved by a quotation from Jer. *31,* 31-34, where God revealed to Jeremias that the Old Covenant was unsatisfactory, and would be superseded by a new and a better one. The New Covenant would effect the forgiveness of sins, which the Old Covenant was unable to do. The knowledge of God will ultimately be a common possession of all.

4. A Superior Sacrifice 9, 1 — 10, 18

9, 1-28: A comparison is made between the earthly sanctuary where the Jewish priests functioned, and the heavenly sanctuary into which Christ has entered to make intercession for us (1-14). The sacrificial value of Christ's death is then explained (15-28).

9, 1-5: The Earthly Sanctuary. **2.** *Tabernacle:* i.e., tent. The large tent set up by Moses for religious worship consisted of two compartments, each of which is here referred to as a tabernacle. The outer tabernacle was known as the Holy Place; the inner one, as the Holy of Holies. There was a veil before each compartment. **4.** *Golden censer:* the Holy of Holies was never entered without incense (Lev. *16*, 12). Some believe that the golden altar of incense is meant, but that stood in the Holy Place, immediately in front of the veil, but not in the Holy of Holies. *Ark of the covenant:* the sacred box in which originally were kept the two tablets of stone on which the ten commandments were engraved (Ex. *25*, 10-22). *Golden pot containing the manna:* see Ex. *16*, 32-34. *Rod of Aaron:* see Num. *17*, 1-10. **5.** *Cherubim:* see Ex. *25*, 17-22; *37*, 6-9. *Mercy-seat:* the golden lid of the ark, which was sprinkled with sacrificial blood by the High Priest on the Day of Atonement (Lev. *16*, 14 f). The presence of God rested above the mercy-seat (Ex. *25*, 22; Num. *7*, 89).

9, 6-10: A Type of the Heavenly Sanctuary. The Holy of Holies is a type of heaven; the Holy Place (first tabernacle) of the Old Covenant. Heaven was not opened until after the death of Christ when the veil of the temple was rent (Matt. *27*, 51).

9, 11-14: Christ the High Priest and Victim. **11.** *Tabernacle not made by hands:* namely, heaven. Some interpret it of Christ's body. *Entered once for all . . . into the Holies:* into the sanctuary of heaven. Since His blood was of infinite value, there was no need to repeat the sacrifice. **13.** *Cleansing of the flesh:* the Jewish sacrifices removed ceremonial uncleanness, but could not remove sin. **14.** *Dead works:* sins.

9, 15-17: Redemption through Christ. The Greek word for "covenant," also means "testament" in the sense of a last will. The covenant which Christ came to establish was also His testament or will, and hence His death was necessary before the will could be effective. The death of Christ was an essential condition for the establishment of the New Covenant which brought about man's salvation.

9, 23-28: The Blood of Christ. **23.** *Copies of the heavenly realities:* the Mosaic tabernacle had been made after a pattern shown to Moses on the Mount (Ex. *25,* 9). It was but a type of the heavenly sanctuary. The heavenly sanctuary could not be cleansed like the earthly, but sins could be removed through the sacrificial blood of Christ, and man enabled to enter.   **25.** The sacrifice of Christ was complete, and did not need repeating.   **26.** *At the end of the ages:* in the Messianic times. **27.** *The judgment:* most probably the Last Judgment is meant, rather than the particular judgment immediately after death.   **28.** *The sins of many:* see Isa. *53,* 12. Christ died for all, but all do not avail themselves of salvation. *With no part in sin:* our Lord's first coming was in connection with sin; His Second Coming will be in glory. The allusion is to the re-appearance of the High Priest from the Holy of Holies.

10, 1-10: One Sacrifice Supplants Many.   **1.** *Shadow . . . image:* as a sketch is to the picture.   **5-7.** A prophecy from Ps. *39,* 7-9, picturing the dispositions of the Messias in submitting Himself absolutely to the will of God. Obedience is more than mere sacrifice of animals. A body would be prepared for the Messias, through which the sacrifice of perfect obedience to God's will would be accomplished. *A body thou hast fitted to me:* the Hebrew has, "Ears thou hast opened for me." The meaning is similar, as through the ears the will of God would be known. *Head of the book:* the Hebrew has "roll of the book." Books were written on strips and wound on rolls.   **10.** *Will:* the will of the Father.

10, 11-18: Its Eternal Efficacy.   **16 f.** Jeremias is quoted again to prove that there will be no further need of any sacrifices offered for sin. Christ's sacrifice is all-sufficing; the Jewish sacrifices are obsolete.

II: EXHORTATIONS *10,* 19 — *13,* 17

1. To Perseverance in Faith *10,* 19 — *11,* 40

10, 19-25: First Motive: The Judgment.   **20.** Jesus led the way to heaven through His Passion and Death. Heaven is now open for all.   **22.** *Hearts cleansed:* an allusion to the sprinkling with blood in the Jewish rites. Our hearts are cleansed with the blood of Christ. *Washed with clean water:* an allusion to Baptism.   **24.** Faith and hope must issue in charity and good works. **25.** The chief purpose of the Christian assembly was to celebrate the Holy Eucharist (*1 Cor. 11,* 20). *Day drawing near:* the destruction of Jerusalem, or more probably the last day.

10, 26-31: Guilt of Apostasy. **26.** As in *6,* 6, it is the sin of apostasy which is meant. They have rejected Christ's sacrifice, and there is no other. **27.** Apostasy was punished with death under the Mosaic Law; here eternal punishment is threatened. The allusion is to Isa. *26,* 11 according to the Septuagint. **28.** See Deut. *17,* 6. **29.** See Ex. *24,* 8. **30.** See Deut. *32,* 35-36; also Rom. *12,* 19. **31.** See Deut. *5,* 26.

10, 32-39: Second Motive: Trials Well Borne. Words of encouragement and hope. **32.** *Enlightened:* by faith received in Baptism. **37.** See Ag. *2,* 7; also Isa. *26,* 20 (Sept.). *He who is to come:* Christ as Judge. **38.** See Hab. *2,* 4; also Rom. *1,* 17; Gal. 3, 11. The quotation from Habacuc is rendered rather freely from the Greek, which again differs somewhat from the Hebrew. The underlying thought, however, is the same. As it was faith in God which brought the just Hebrew through the terrible trial of the Chaldean invasion, so now it would be faith in Christ which would bring the Christians through the difficulties which they were facing (probably the persecution before the destruction of Jerusalem in 70 A.D.).

11, 1-40: Third Motive: Old Testament Examples. **1.** Faith is here described in some of its effects, and not really defined regarding its own nature. In the last verse of the preceding chapter, faith was spoken of as essential to salvation; here it is stated that faith is the very foundation (*substance*) of salvation (*things to be hoped for*). Faith also furnishes clear proof (*evidence*) that what God has testified to exists even though not seen (*things that are not seen*). Although the Greek word for "substance" has also the subjective meaning of "assurance," the word for "evidence" never has the subjective meaning of "conviction," which would be necessary for the translation adopted by some that "Faith is the assurance of things to be hoped for, a conviction of things not seen." **2.** *Had testimony borne to them:* Scripture testifies to their faith. **3.** That the visible world came through creation is known by faith.

4. *Abel's* sacrifice was acceptable to God because Abel had faith, which Cain lacked (without faith it is impossible to please

God, 6). *Through which:* namely, faith. *Though he is dead he yet speaks:* see Gen. *4*, 10 regarding the voice of Abel's blood crying to heaven for vengeance.

5. *Henoch:* see Gen. *5*, 21-24. Henoch's faith is inferred from the fact that he pleased God. *Pleased God* is found in the Septuagint; the Hebrew has "walked with God." **6.** Faith in God's existence and in His providence are the two most fundamental acts required for salvation.

7. *Noe:* see Gen. *6*, 1 — *9*, 29. *Condemned the world:* for its unbelief regarding the coming flood. *Was made heir:* was given a right to the inheritance which would come through Jesus Christ. See 39 f below. *Justice which is through faith:* and hence not a merely legal justice.

8. *Abraham:* see Gen. *12*, 1 ff. **10.** *City that has the foundations:* the heavenly Jerusalem. See also *12*, 22 and *13*, 14; Gal. *4*, 26; Apoc. *21*, 2. This *city* is in contrast to the "tents" in which Abraham and his family lived.

11. *Sara:* see Gen. *17*, 15; *21*, 1-3; Rom. *4*, 19. *He who had given the promise was faithful:* see also *10*, 23. God is faithful to His promises. **12.** See Gen. *22*, 17; *32*, 12.

13. *Without receiving the promises:* without having the Messianic promises realized during their lives. Minor promises, however, had met their fulfillment.

17. *Offered Isaac:* see Gen. *22*, 1-18. The great promises were to be realized through Isaac, yet Abraham was ready to sacrifice him when God requested it.

19. *Received him back as a type:* as a type of the Resurrection. Isaac, who had been given up to death, was restored to Abraham.

20. *Isaac blessed Jacob and Esau:* see Gen. *27*. **21.** *Bowed in worship towards the top of his staff:* the Hebrew of Gen. *47*, 31 has, "bowed in worship towards the head of his bed." Jacob had exacted a promise from Joseph to bury him in the Promised Land, convinced in faith that his posterity would possess it as God had promised. He thanks God, either leaning upon his staff, or prostrate on his bed. "Bed" and "staff" are spelled with the same Hebrew consonants. **22.** *Joseph:* see Gen. *50*, 24.

23. See Ex. *2*, 2.

26. *Reproach of Christ:* Moses suffered disgrace (reproach) because of his belief in divine promises which had their fulfillment in Christ. The whole Old Testament dispensation was given as a preparation for Christ, and the sufferings endured

under that dispensation by God's faithful servants may be looked upon also as having been endured for Christ. See Pss. *88,* 50 f; *68,* 9 f; Rom. *15,* 3.   **27-29.** See Ex. *12,* 12 ff.
30 f. See Jos. *2-6.*
33. *Obtained promises:* various minor promises and not the great Messianic promise mentioned in 39 below. *Stopped the mouths of lions:* see Dan. *6,* 22.
34. *Quenched the violence of fire:* see Dan. *3,* 17-25; *1* Mach. *2,* 59.   **35.** See 2 Mach. *7,* 29; also *3* Kgs. *17,* 23; *4* Kgs. *4,* 36.
38. *Caves:* see 2 Mach. *6,* 11. The sufferings alluded to in 36-38 are of the Machabean times.
39. These heroes of the faith did not live to see the Messias, but they have been admitted through their faith into the blessings brought by Him; and they share in them equally with us. They have been admitted into heaven which was opened by Christ's redemptive work.

2. OTHER VIRTUES *12,* 1 — *13,* 17

12, **1-13: Constancy.**   **1 f.** The allusion is to the contests in the Greek amphitheater, with crowds of spectators watching. Jesus is our goal.   **5.** See Prov. *3,* 11 f. God punishes as a loving father.   **12.** See Isa. 35, 3.   **13.** Prov. 4, 26.

12, **14-17: Peace and Holiness.** This includes various exhortations.   **15.** *Root of bitterness:* see Deut. *29,* 18.   **16.** *Profane:* because he bartered away his spiritual privileges.   **17.** *No opportunity for repentance:* his act was irrevocable. He had bartered away his inheritance, and forfeited the blessing (Gen. *27,* 34.38). Even though Esau had repented sincerely, the lost blessing could not be regained.

12, **18-29: Sinai and the New Sion.** See Ex. *19-20;* Deut. *4,* 11 ff. Sinai as representing the Old Covenant.   **22-24.** Sion, the heavenly Jerusalem, representing the New Covenant.   **23.** *Firstborn:* in the sense of heirs. We are co-heirs with Christ, and our names are written in heaven.   **24.** *Sprinkling of blood:* the blood of Abel called for vengeance; that of Jesus calls for mercy.   **26-28.** God's voice shook the earth at Sinai (Ex. *19,* 18). At the end of time, God will again shake the whole world (Ag. *2,* 6.7.22), only the spiritual realities of the new dispensation *(a kingdom that cannot be shaken)* will remain.   **29.** *Our God* *is a consuming fire:* He will ultimately destroy all His enemies. (See Deut. *4,* 24; 2 Thess. *1,* 8).

13, 1-4: Brotherly Love and Purity.   **2.** *Entertained angels:* as did Abraham and Sara (Gen. *18*); Lot (Gen. *19*); Manue (Judg. *13*); Tobias (Tob. *12,* 15).   **4.** *Let marriage be held in honor with all:* let all honor marriage. It does not mean that all should marry. See *1* Cor. 7.

13, 5-6: God Will Never Fail You.   **5.** Deut. *31,* 6; Jos. *1,* 5.   **6.** Ps. *117,* 6; Matt. *6,* 31. 34.

13, 7-17: Loyalty to Christ and Superiors.   **8.** Jesus remains ever the same.   **9.** His doctrine is unchangeable. The Jewish distinctions of foods, which did not profit even the Jews who practised them, have been abrogated. It is the grace of God which counts.   **10.** *Altar* in this verse has been interpreted variously of the altar of Calvary, the Eucharistic altar, or the altar of the heavenly sanctuary. All three are but different aspects of the one supreme sacrifice of Christ. *From which they have no right to eat who serve the tabernacle,* indicates that Christians "eat" of their altar, and the reference seems to be to the Holy Eucharist. See also *1* Cor. *10,* 20 f; *11,* 23-28.   **12.** *Outside the gate:* Calvary was outside Jerusalem.   **13.** *Outside the camp:* namely, of Judaism. *Bearing his reproach:* the shame of His cross.   **15.** *Through him:* namely, Christ, through whom we have obtained access to God. *Sacrifice of praise:* see Lev. *7,* 12. Under the Old Law, it was a thank-offering for a favor graciously bestowed by God. Jewish teachers had a saying that one day all offerings but the thank-offering would cease. *Fruit of lips:* in Isa. *57,* 19 the fruit of lips is called peace. The sacrifice of praise was the highest form of peace offering. In Ps. *49,* 14, the sacrifice of praise is placed above all animal offerings.

Conclusion *13,* 18-25

23. *Set free:* either from prison, or from some task.   **24.** *The brethren from Italy:* Christians living in Italy. Those who hold that the Epistle was not written in Italy interpret *the brethren* as the Italian companions of the author, who were sending their greetings.

Edward A. Cerny, S.S.

THE EPISTLE OF ST. JAMES

Introduction

St. James. The author of this Epistle calls himself James (*1*, 1). This name occurs several times throughout the New Testament. First, there is James the Apostle, or James the Greater, son of Zebedee and brother of St. John the Apostle (cf. Matt. *4*, 21; *27*, 56; Mark *1*, 19; Luke *5*, 10; etc.). But St. James the Greater is not the author of this Epistle, for he was martyred by Herod Agrippa I, probably in the year 42 (Acts *12*, 1 f); whereas, it is evident from the contents of the Epistle that it was written at a later date and only after Christianity had been widely propagated.

The Synoptics often mention another Apostle, James, the son of Alphæus (cf. Matt. *10*, 3; Mark *3*, 18; Luke *6*, 15; Acts *1*, 13). This Apostle is usually identified with James the Less (Mark *15*, 40), and James, "the brother of the Lord" (Matt. *13*, 55; Mark *6*, 3). This identification is confirmed by tradition found in the writings of Clement of Alexandria, Origen, St. Jerome, St. Augustine and others. Consequently, James "the son of Alpheus," or "the Less," or "the brother of the Lord," must be the James mentioned in the inscription of our Epistle: "James the servant of God and of our Lord Jesus Christ" (*1*, 1).

But in what sense was St. James the Apostle, "the brother of the Lord"? Certainly the expression is not to be understood in the strict sense of the term: for this would contradict Sacred Scripture (Luke *1*, 34) and tradition. Neither does the term imply that James (or others) was the son of St. Joseph by a former marriage. Nothing in Sacred Scripture indicates this and tradition is strongly against it. A careful analysis of certain texts confirms the tradition that St. James and our Savior were first cousins. St. Matthew (*13*, 55) and St. Mark (*6*, 3) call James, Joseph, Simon, and Jude "the brethren of the Lord." Evidently, these are not real brothers of our Lord; for elsewhere both St. Matthew (*27*, 56) and St. Mark (*15*, 40) refer to a certain Mary, standing at the foot of the cross, "the mother of James and Joseph." This Mary is probably identified with Mary of Cleophas, the sister of the Blessed Virgin (John *19*, 25). If this is true, then, James and Joseph were first cousins of our Lord.

At Jerusalem St. James was greatly respected by all classes, so that even his fellow-countrymen gave him the surname "The Just." After our Lord's resurrection he was favored with a special apparition (*1* Cor. *15*, 7). His authority and prudence were greatly respected by the early Christian Church in Jerusalem: St. Luke clearly shows this in the Acts of the Apostles. First, on the occasion of St. Peter's arrest by Herod, and his liberation; "Tell this to James and to the brethren" (*12*, 17).

Secondly, at the Council of Jerusalem (Acts *15*) all those present accepted the disciplinary points suggested by St. James for the conduct of the Gentile converts. And finally, when St. Paul visited Jerusalem, at the end of his third missionary journey, St. James, acting as the head of the church there gladly received him and suggested a plan to him whereby the opposition of his enemies might be overcome (Acts *21*, 18-25). St. Paul calls him a "Pillar" of the Church (Gal. *2*, 9).

Aside from these facts little more is known about the person and life of St. James. According to the historian Hegesippus (Eusebius, *Hist. Eccl.* II, 23) he led an austere and holy life, and suffered martyrdom at the hands of the Jews by being thrown from the pinnacle of the temple. Josephus Flavius (*Antiq.* XX, 9, 1) relates that he was martyred about the year 62.

Authenticity. Although Eusebius (*Hist. Eccl.* II, 23; III, 25) enumerates the Epistle of St. James among the disputed writings, he himself recognized it. In the sixteenth century Erasmus and Cajetan raised some doubts about its authenticity. Luther called it "an Epistle of straw, and unworthy of an Apostle." The reason for his antipathy is evident; for the teaching of the Epistle undoubtedly condemns Luther's favorite doctrine, that faith alone without good works suffices for salvation. But Catholic tradition, from the latter part of the first century and the beginning of the second century, has maintained unbrokenly that St. James is the author of this Epistle. Here we can but briefly consider the evidence upon which this tradition rests.

First, a casual examination of early ecclesiastical literature reveals that this Epistle exercised a great influence on the writings of such men as St. Clement of Rome, St. Ignatius Martyr and St. Polycarp; traces of it are also found in the Didache and the Letter to Diognetus. It is shown that there are at least fifteen places in the Shepherd of Hermas which evidently allude to the Epistle of St. James.

Secondly, this Epistle was not only known throughout the early Christian Church, but it is often explicitly attributed to St. James. For example, Clement of Alexandria and Origen cite it as Scripture and as the work of St. James. Other ecclesiastical writers, such as St. Athanasius, St. Cyril of Jerusalem, and St. Epiphanius, ascribe the work to St. James. St. Jerome says explicitly that St. James was the author of this Epistle. Finally, this Epistle is mentioned in all the Councils in which a list of sacred and inspired books was drawn up, for example, the Third Council of Carthage, and those of Florence and Trent.

Thirdly, a careful study of the Epistle itself reveals that it does not in any way contradict the external evidence in its favor, but confirms it. The author betrays his Jewish origin: he is quite familiar with the Old Testament and quotes it frequently (cf., e.g., *5*, 10-18; cp. also *2*, 8 with Lev. *19*, 18). Again, the Epistle is thoroughly Christian: cf., e.g., *1*, 16.19; *2*, 4 where St. James employs the strictly Christian terminology, "my beloved brethren." Finally, his exhortations recall vividly the teaching and preaching of Christ (cp. *5*, 12; *5*, 2.3; *2*, 13; *3*, 12; *1*, 12; *3*, 1.2; *1*, 9.10 with Matt. *5*, 34-37; *6*, 19; *7*, 2.16; *10*, 22; *12*, 36; *18*, 4). St. James was not in this Epistle concerned so much about dogmatic questions, as he was about moral perfection and the necessity of performing good works. He drew upon the words of Christ, observed them literally and exhorted others to follow him.

Purpose and Occasion. In the opening chapter, St. James addresses his Epistle "to the twelve tribes that are in the Dispersion" (*1*, 1). Those who understand these words in their more obvious sense think St. James intended his Epistle primarily for the Christian converts from Judaism who lived in the Jewish colonies outside of Palestine. But the Epistle does not seem to have been restricted to this group of Christians. Some exegetes understand the opening verse to mean "the true Israel of God," that is, "all Christians throughout the entire world, without reference to their Jewish or Gentile origin, thus taking the dedication in a metaphorical sense" (Eaton, *The Catholic Epistles*, p. 92).

While no mention is made of local churches, in all probability this Epistle was first sent to some particular church. According to some critics, its immediate destination was the church at Antioch in Syria, at that time the nearest center of Christianity to Jerusalem. If this is true, the actual writing of the Epistle was probably occasioned by the particular needs of the church at Antioch and the vicinity; at least parts of the Epistle especially applied there. Moreover, in the opinion of some authors, St. Paul's Epistle to the Romans and his exposition of the doctrine of justification were also the occasion of this Epistle. There are no real contradictions between the teachings of these two Apostles. If St. James had in view St. Paul's teaching on justification, the most that can be said is that he rebuked those Christians who misinterpreted and distorted it.

The present Epistle is written in the form of a sermon or moral exhortation. Its teaching is practical and to the point; for St. James knew the needs and deficiencies of the people concerned. His purpose in writing, therefore, was threefold: (1) to encourage Christians in their trials; (2) to refute erroneous doctrines which were beginning to appear; and (3) to promote virtue and holiness in Christians.

Time and Place of Composition. Nothing definite is known about the time St. James wrote this Epistle. But its date of composition can be calculated approximately. St. James suffered martyrdom in the spring of the year 62; if we suppose that St. James knew of St. Paul's Epistle to the Romans we must place his Epistle after the year 58. As regards the place of composition, it is commonly believed that St. James wrote the Epistle at Jerusalem, to which his ministry seems to be confined.

COMMENTARY

Introduction 1, 1

In his salutation St. James, like Sts. Peter, Paul, and Jude, calls himself a *servant,* that is, a slave, out of humility, signifying thereby his absolute dependence upon God and Jesus Christ. *Greeting:* i.e., "joy be with you." This was an ordinary form of address used in public and private letters (cf. Acts *15,* 23).

I: EXHORTATION TO PATIENCE IN TRIALS 1, 2-18

1, 2-12: **Wisdom in Trials.** St. James exhorts the faithful to bear joyfully and patiently their trials and afflictions. **2.** *Trials* are understood here in the sense of misfortunes and persecutions due to external causes, rather than temptations in the moral sense of the term. **4.** *Patience* is the guide to perfection, and its perfect exercise together with other virtues, especially charity, helps men to lead a more virtuous life and obtain eternal salvation (*5,* 11; cf. Rom. *5,* 3 ff). **9-11.** The poor should *glory,* that is, rejoice in the dignity which faith and grace have bestowed upon them. The rich Christian should be happy to lower himself before God and recognize how perishable are his goods.

1, 13-18: **Sources of Evil and Good.** In contrast to the preceding section, St. James found it necessary to speak of temptation in the stricter sense of the term. **13.** *He is tempted by God:* the Greek reads, "I am tempted by God." *God is no tempter to evil:* according to the Greek text the meaning probably is that God is not subject to temptation, neither does He tempt any man. **14 f.** St. James therefore explains the true origin of temptation. **17.** In sharp contrast to his explanation regarding the origin of sin and its effect, death, the Apostle emphatically declares that every good gift and every perfect gift comes from God. *The Father of Lights:* literally, a reference to the celestial bodies, but here it is understood metaphorically as is seen in the following (cf. *1* John *1,* 5; Eph. *5,* 8). **18.** *Begotten us:* i.e., in our regeneration through Baptism; note the contrast between the generation of evil and our creation. *By the word of truth:* St. James probably had the gospel in mind, but undoubtedly he is also referring to the Word of God (John *1,* 1-3). This is further illustrated when he reminds us that we are *the first-fruits of his creatures:* the figure is taken from the Jewish ritual of offering the harvest to God, by presenting the first sheaves at the temple. As Christians, we are His offspring, not merely because He is our creator, but because by Baptism, without any merits of our own, we have been spiritually regenerated (cf. *1* Pet. *1,* 3.23; Gal. *3,* 27).

II: LIVING AND ACTIVE FAITH *1,* 19 — 2, 26

1, 19-27: Hearers and Doers of the Word of God. **19.** They should be *swift to hear* the word of God, and *slow to speak* in controversy. **21.** The Greek expression usually refers to the lack of physical cleanliness. In the present context it is used metaphorically of what defiles the soul. *Abundance of malice:* i.e., not merely excessive, but manifold wickedness which must be discarded before the word of God can be received (Rom. *1,* 16; Col. *3,* 8; *1* Pet. *2,* 1). *Ingrafted word:* i.e., a word which is not ours by nature. **22-26.** *But be doers of the word and not hearers only:* this forms one of the main themes of the Epistle, which is more fully explained in the following chapter. Throughout the Epistle we notice a gradual development of the doctrine. Hitherto St. James has spoken about faith which is proved chiefly by enduring trials, and he points out that this is a necessary condition of efficacious prayer. Then he goes on to show the necessity of practising charity without which faith is unprofitable (cf. *1* Cor. *13,* 2). Here he reminds his readers of the need of practising as well as hearing the word of God, that is, the word of God must be made the rule of their lives (Matt. *7,* 21). This in turn suggests that real charity is complemented by fruitful works (*1* John *3,* 17.18), and it is exemplified by an illustration: a mere hearer of the gospel is like a person who looks at himself in a mirror, and going away immediately forgets what he looks like; the *mirror* is the gospel which shows us what our lives should be. On the other hand, the Christian who not only hears, but practises what he hears, examines closely *the perfect law of liberty,* i.e., the gospel which frees him from sin and the bondage of the Old Law. In it he sees not only his imperfections, but the ideal of Christian perfection. *Blessed in his deed:* i.e., shall merit eternal happiness as the reward of following Christ's teachings. **26.** *Religion* means religious observance.

27. Consequently, doers of the word of God or religious-minded persons must be charitable in speech, pure of heart, and ready to help the poor and afflicted.

2, 1-13: Impartiality. **1-3.** The connection with the preceding is not immediately evident, but the warning against partiality very likely follows the reminder to be doers of the gospel and not hearers only, because partiality manifested among Christians was a sign that they had not fully followed the principles laid down in the gospel (Acts *10*, 34; Rom. *2*, 11; etc.). **4-6.** St. James condemns particularly those who showed undue preference to the rich, and contempt for the poor in the synagogue (church), and courts of justice. *Judges with evil thoughts:* i.e., evil-minded. But God the just judge acts otherwise and will reward the pious poor man, for he is an heir of the kingdom of heaven (cf. Isa. *3*, 15; Amos *4*, 1). **7.** *Good name:* perhaps a reference to the name of God invoked in Baptism, or the name Christian which was fast becoming known throughout the early Church (Acts *11*, 26). **8.** The *royal law* is that of fraternal charity which forbids hatred of one's neighbor (Lev. *19*, 18; Matt. *5*, 43-47; John *13*, 34 f.). **10-13.** The Apostle concludes by saying that the whole law and each article of it must be observed; he who despises the supreme law-giver or offends against one precept of the law, manifests his insubordination, and in a sense is deficient in regard to the whole law (cf. Matt. *5*, 18 ff). **12.** The *law of liberty* is a law of love which man obeys freely and gladly, because to serve Christ means freedom. **13.** *Mercy triumphs,* etc.: i.e., God's mercy is not opposed to, but above, his justice.

2, 14-26: Practical Faith. Here the Apostle develops the Pauline teaching "faith working through charity" (cf. Gal. *5*, 6; *1* Cor. *13*, 2; *2* Thess. *2*, 17). **17.** Faith devoid of the vivifying principle of charity and good works is like the body without the soul. **19.** The Greek reads: "Dost thou believe that there is one God?" This is perhaps a continuation of an imaginary discussion or a new argument. Faith is intellectual assent given to revealed truths, and once we admit the existence of God we should obey his commandments and love all men for His sake. Otherwise, we should be no better than evil spirits whose faith begets profound fear (cf. Mark *1*, 24; Luke *8*, 28).

St. James enforces his arguments by presenting a few examples of practical faith from Sacred Scripture. **21.** Abraham's willingness to comply with God's commands was the crowning act of a life of faith. His readiness to sacrifice his son was a proof of his faith and love of God (Gen. *22*; Isa. *41*, 8; Rom. *4*, 2 f; Heb.

11, 7).   **25.** Rahab also, by the courage of her convictions, manifested her practical faith in God (Jos. *2*, 1-24; Heb. *11*, 31). As a result she was rewarded by becoming an ancestress of our divine Savior (Matt. *1*, 5).

III: THE HAZARD OF TEACHING   *3*, 1-18

3, 1-12: **Abuses of the Tongue.** The relation between the preceding and this section is seen in the grave dangers in false teaching and abuse of speech to faith and morals. The Apostle warns his readers against undue eagerness to teach (cf. Matt. *12*, 37; Rom. *2*, 19 ff; *1* Cor. *12*, 28; *14*, 26-40; Eph. *4*, 11).   **1.** *Not many teachers:* i.e., let not many of you strive to become teachers, seeking the honor but with insufficient appreciation of the responsibility; this thought leads St. James to speak of the sins of the tongue. The Greek text reads for the latter part of the verse "we shall receive the greater judgment," that is, a stricter judgment (Luke *12*, 48).   **6.** *The very world of iniquity:* the tongue resembles the wicked world in so far as it can inspire and cause all kinds of sin. *Course of our life:* literally, "the wheel of our nativity" or "nature," that is, man's life from beginning to end. *Set on fire by hell:* i.e., the source of its evil activity (Matt. *5*, 22; Mark *9*, 43).   **7 f.** All kinds of animals can be tamed, but no human skill or industry can tame the tongue; for *and the rest* the Greek text reads "and sea creatures."   **9 f.** St. James goes on to point out the evil and incompatible uses to which the tongue is put.   **11 f.** The justice of this rebuke is illustrated by several examples taken from nature: the purposes of nature are clear, and what is true of the natural order, is true also of the moral order (cf. Matt. *7*, 16 ff).

3, 13-18: **True Wisdom.** After the digression of 3-12, regarding the vices of the tongue, St. James returns to the subject mentioned in the beginning of the chapter, namely, the inordinate desire of teaching.   **13.** The qualities of a true teacher are knowledge, good example, and meekness. True wisdom is at once practical and gentle.   **14.** On the other hand, false wisdom is characterized by strife and party spirit.   **15-17.** A further description of the qualities of earthly and heavenly wisdom.   **18.** The effect of true and heavenly wisdom is peace; peace and righteousness go hand in hand. *The fruit of justice* may mean justice itself reaped by the peacemaker, or the fruit which the seed of justice produces.

IV: SPECIAL ADMONITIONS *4, 1 — 5, 6*

4, 1-10: Sources of Discord. In a series of introductory questions St. James denounces the Christians for their contentions, envy, and moral depravity. **2.** *You do not have because you do not ask:* i.e., do not pray to God properly. **3.** Some Christians perhaps objected that they did present their petitions to God, but they went unanswered. The Apostle, however, points out that they asked *amiss*, i.e., without the proper dispositions and motives. **4.** *Adulterers:* the Greek reads "adulteresses." The context shows that St. James has spiritual infidelity particularly in mind. The figure is quite common in the Old Testament and is used of the Jews who often fell away from the worship of the one true God: God and the world are rivals, and it is impossible to serve both at the same time (cf. Isa. *57,* 3-10; Ezech. *16,* 15-24; Os. *2,* 2-6; Matt. *6,* 24; Luke *16,* 13; 2 Cor. *11,* 2; *1* John *2,* 15). **5.** The words of this verse are not actually found in Sacred Scripture, but are the expression of several scriptural passages showing the Holy Spirit's concern for the true Christian (cf. Wisd. *11,* 25; Isa. *49,* 15; *66,* 13; Jer. *31,* 3; Zach. 2, 8; John *3,* 16 f). The context clarifies an otherwise obscure verse: in Sacred Scripture God is often described as a jealous God who will not give His glory to another (Ex. *20,* 5; *34,* 14). The idea is here applied to the Holy Spirit dwelling in the soul of the Christian; the Holy Spirit loves that soul so intensely that He entertains feelings analogous to envy if another supplants Him there. **6.** Moreover, God has a greater claim because He gives *a greater grace,* i.e., more valuable gifts than the world or its votaries can offer. In order to receive these more abundant graces humility is the most necessary disposition required. **7-10.** St. James concludes the section by laying down certain definite practical norms of conduct for Christians. In this respect his admonitions especially resemble our Lord's Sermon on the Mount (cf. Matt. *5,* 1-12).

4, 11-17: Presumption. **11.** An admonition against the sin of detraction; he who *speaks against the law and judges the law,* offends against the new commandment of fraternal charity (*1,* 19.20.26). **12.** The person who rashly judges his neighbor presumptuously usurps a prerogative of God (Matt. *7,* 1; Rom. *2,* 1; *14,* 4). **15.** The Greek reads: "you are mist"; and, "if the Lord wills, we shall live and do this or that." These two conditions should always be expressed or implied in the accomplishment of any good work. Life is uncertain and it is foolish to make far-reaching plans (cf. Job *7, 7*; Ps. *101, 4*; Prov. *27, 1*; Wisd. *9, 9-14*; Matt. *6, 34*; Luke *12, 16-21*; Acts *18, 21*). St. James does not condemn the prudent foresightedness of Christians, but those practices which are contrary to Christian virtue. Such conduct is presumptuous in so far as it disregards God's providence.   **17.** The instruction is complete: hereafter if they continue to offend against any of the points in which they have been instructed, their sin will be more grievous.

5, 1-6: The Unjust Rich. According to the more common view St. James in these verses denounces the rich pagans who cruelly and unjustly mistreated the Christians; but his condemnation of riches can apply also to the rich Christians guilty of similar crimes (cf. 2).   **2.** *Weep and howl,* etc.: i.e., in anticipation of their future punishments.   **3.** *Last days* probably refer to the time of the destruction of Jerusalem; many, however, see in them a reference to the Final Judgment.   **5.** *In the day of slaughter:* according to the context the words are to be understood in the sense that, just as animals are fattened and prepared for slaughter, so the rich by over-indulgence and dissipation are simply preparing themselves for the day of reckoning when they shall become the victims of God's justice.   **6.** *The just:* many understand "the Just One," and see here some allusion to the passion and death of Christ. More probably the expression refers to the class of the just who offered no resistance.

Conclusion 5, 7-20

5, 7-12: Patience in Affliction. Here St. James returns to the theme of *1, 3-4*.   **7.** *Be patient:* i.e., be long-suffering till the final judgment.   **8.** The *Coming of the Lord is at hand:* (cf. Matt. *24;* 1 Cor. *16,* 22), i.e., the day of the Second Coming. Many Christians hoped it would be soon, but its time was not revealed even to the Apostles.   **12.** A caution against the frequent vice of imprudent and unnecessary oaths (cf. Ex. *20, 7*; Matt. *5, 34* ff).

5, 13-18: Last Anointing, Confession and Prayer.   **13.** God is the source of consolation, and Christians should commune with Him in prayer (*1* Cor. *14,* 15; Eph. *5,* 19).   **14.** According to the teaching of the Council of Trent (Sess. 14, c. 1), St. James promulgated here the sacrament of Extreme Unction. *Is any one among you sick:* by "sick" is understood those laboring under a severe bodily infirmity, or in danger of death from sickness. *Let him bring in the presbyters of the Church:* i.e., one of the priests. *In the name of the Lord:* i.e., in the person of the Lord, or by the command and authority of Christ (cf. Mark *6,* 13). **15.** *The prayer of faith:* i.e., a prayer which proceeds from faith and is grounded on the faith of the Church. *Will save:* i.e., restore the sick man to health if it is expedient, or save his soul in the life to come (cf. Mark *16,* 18). *And if he be in sins:* understanding all kinds of sins some of which for some reason may not have been remitted by penance. **16.** There are three possible interpretations for the words *confess, therefore, your sins to one another:* (1) confess your offences against one another, and mutually ask pardon of each other; (2) confess your faults to one another for the purpose of seeking counsel and assistance by mutual prayer; (3) confession of sins in the sacrament of Penance. Most Catholic authors believe that St. James here admonishes Christians to confess their sins to the priests in the sacrament of Penance. Consequently, *to one another* is taken in the restrictive sense (cf. Rom. *15,* 7; *1* Thess. *5,* 11; Eph. *5,* 21), i.e., "in accommodation to the subject matter of the precept, of such as are empowered to hear confession and bestow absolution" (MacEvilly). *Pray for one another:* especially for sinners. *That you may be saved:* the sense of the Greek is "that you may be healed." Consequently, "pray for one another" for the purpose of obtaining forgiveness from God and the gift of final perseverance.

5, 19-20: Conversion of a Sinner.   **20.** *Will save his soul from death:* i.e., the soul of the sinner from spiritual death in this life, and eternal death in the life to come (cf. Matt. *18,* 15 ff). *Will cover a multitude of sins:* (cf. *1* Pet. *4,* 8; Prov. *10,* 12), i.e., the sins of the converted persons; "but indirectly reference is made to the sins of the man who exercises the good work of converting his neighbor; for, by this act of charity, he will obtain from God the remission of his own sins, or an increase of grace to persevere in justice, and the remission of the temporal penalties due to his sins already remitted" (MacEvilly).

CHARLES H. PICKAR, O.S.A.

THE FIRST EPISTLE OF ST. PETER

Introduction

St. Peter. St. Peter, also called Simon, was the son of a certain John of Bethsaida (John *1*, 40-42). The preaching of John the Baptist had drawn him to the shores of the Jordan where Andrew, his brother, introduced him to Jesus. Both brothers were fishermen (Matt. *4*, 18); Peter's residence was situated at Capharnaum, a city on the Lake of Genesareth. The Evangelists, indeed, speak of Peter's mother-in-law; still, *1* Cor. *9*, 5 is not conclusive evidence that Peter's wife was alive even at the time of his first acquaintance with Christ.

On the occasion of their first meeting Christ promised to confer on Simon the surname of Cephas; its Greek equivalent, Petros, was later Latinized and has since become the name by which the Prince of the Apostles is best known. Peter did not immediately leave all things and follow Christ. The miraculous catch of fishes mentioned by St. Luke (*5*, 1-11) marks the beginning of Peter's permanent association with our Lord. He was one of the few to witness the restoration of Jairus' daughter; on the Mount of Transfiguration and during the Agony in the Garden Peter was a privileged witness; he had even been invited by Christ to walk upon the waters (Matt. *14*, 22-32). On the occasion of his profession of faith near Cæsarea Philippi, Christ promised to confer on him the power of the keys, the symbol of supreme authority (Matt. *16*, 13-20).

St. Peter is spoken of frequently in the Gospel account of the events of Holy Week. He and John were commissioned by Christ to prepare the Cenacle where Christ instituted the Holy Eucharist. Informed that Christ's betrayer was in their midst, Peter asked John to question Jesus concerning the identity of the betrayer. Christ's prophecy of Peter's denial, of His special prayer for Peter's faith, and Peter's vehement protestation, "Even if I should have to die with thee, I will not deny thee" (Mark *14*, 31) were uppermost in Peter's mind when he drew the sword in defense of Christ (John *18*, 10 f). Peter's repentance after his denial of Christ was quick and sincere. He and John were first among the Apostles to explore the empty tomb; to Peter and the disciples were the women commissioned to announce the Resurrection (Mark *16*, 7). St. John, the disciple whom Jesus loved, saw fit to record for posterity Simon Peter's threefold attestation of love for Christ (John *21*, 13-17).

During the days preceding the first Pentecost, Peter's was the directing hand in the events which resulted in the election of Matthias to the apostolate (Acts *1*, 15-26). On Pentecost, Peter's discourse in the defense of the Apostles and the cause of Christ resulted in the conversion of about three thousand souls (Acts 2, 14-41). St. Peter was the guiding genius of the infant Church; he defended Christ before the Sanhedrin, suffered imprisonment and disgrace, but was always buoyed up with the one thought, "We must obey God rather than men" (Acts 5, 29). Peter continued his activity by visiting and comforting the churches in Judea and Samaria until about the year 42; at this date according to one tradition he journeyed to Rome. He returned to Jerusalem and presided at the first council in the history of the Church about the year 49. He then retired to Antioch (Gal. 2, 11-14); according to a tradition handed down by Eusebius, the Church historian, Peter continued there as Bishop for some time. His missionary endeavors led him most likely through the provinces of Pontus, Galatia, Cappadocia, Asia and Bithynia, and then back to Rome where he established his episcopal see. He remained at Rome more or less continuously until the year 67 when he was martyred as foretold by Christ (John 21, 18 f).

The Author. Catholic tradition has ever maintained that St. Peter, the Prince of the Apostles, is the author of 1 Peter.

External Evidence. As is to be expected, traces of the Epistle appear at the end of the first century. The epistle of Clement of Rome addressed to the Corinthians reproduces in 49, 5 the saying of 1 Pet. 4, 8, "Charity covers a multitude of sins." Indirect references to the Epistle occur frequently in the Shepherd of Hermas written about the year 135. On the authority of Eusebius, writers of the second century, such as Papias of Hierapolis and Polycarp of Smyrna, definitely cited the Epistle. These citations are not ascribed to St. Peter; but with the ecclesiastical writers of the third century, the Petrine authorship of the Epistle stands as an established fact. This explicit statement of authorship is attested by Irenaeus in Gaul, Clement of Alexandria, Origen, Tertullian and Cyprian.

Internal evidence. 1 Pet. 1, 1 describes the author as "an Apostle of Jesus Christ"; 5, 1 declares him to have been "a witness to the sufferings of Christ." These statements are confirmed by other references, particularly 1, 8 where the author indirectly speaks of himself as a disciple of Christ. The similarity of content and development between the Epistle and the preaching of Peter as recorded in the Acts of the Apostles—that the suffering and resurrection of Christ constitute the foundation of our Christian hope—confirms its Petrine authorship. At the close of the Epistle the author sends greetings from "my son, Mark," the spiritual child and co-laborer of St. Peter. A final reference to the Petrine authorship, "The church which is at Babylon," points to Rome as the place of its origin.

Date. The author's familiarity with the Epistle to the Ephesians places the earliest date of composition about the year 63. But the burning of Rome on July 18, 64 A.D., with its subsequent persecution of Christianity, demands that the Epistle be written prior to this date.

COMMENTARY

Introduction 1, 1-12

1, 1-2: Greeting. The literary form of the salutation resembles that of the Pauline Epistles. **1.** *Sojourners of the Dispersion:* cf. note to text. *Pontus, Galatia, Cappadocia, Asia and Bithynia* comprise the greater part of Asia Minor. **2.** Man's election stated in relation to the most Blessed Trinity: sanctification is effected in time by the Holy Spirit (*1* Thess. *4,* 7 f) according to the eternal foreknowledge of God the Father (Rom. *8,* 28 ff; **Eph.** *1,* 4); in virtue of this man becomes the servant of Christ; he participates in the new covenant of grace through the blood of the God-man, Jesus Christ (cf. Ex. *24,* 8).

1, 3-12: Thanksgiving. A doxology comparable to the great rhapsody upon Christian salvation (Eph. *1,* 3-14). **3.** The living hope of Christians is eternal and divine because: (a) it is founded on divine mercy (Rom. *5,* 8-11); (b) its source is the God-man who "dies now no more" (Rom. *6,* 9). Furthermore, this *hope begets confidence.* **4.** The essential character of the reward promised by this living hope is its *incorruptible inheritance;* from the viewpoint of duration it is everlasting (Heb. *9,* 15); in view of its character, one *undefiled and unfading.* **5.** This reward depends neither on fortune nor human causes, for the power of God assures it (Matt. *6,* 19 f), and the virtue of faith will maintain it through trial and temptation (Phil. *4,* 13) *ready to be revealed in the last time* (7.13). **6.** *Over this you rejoice:* i.e., the prospect of a reward so great (Matt. *25,* 34). **7.** The Greek text heightens the comparison between the value of *gold* and *faith:* gold, though refined by fire, perishes; a tempered faith endures forever. But even as gold must be purified by fire so must faith likewise be tried in "the furnace of humilation" (cf. Ecclus. *2,* 1-5; Prov. *17,* 3). *The revelation of Jesus Christ* refers to the glorious Second Coming of Christ. **8.** The readers had not, as St. Peter, seen Christ. **9.** *The salvation of your souls:* this constitutes the reason for their *joy unspeakable and triumphant* (cp. *1* Cor. *2,* 9). **10.** The activity of the Old Testament *prophets* demonstrates the importance of this salvation which is within reach of the readers of the Epistle. **11.** Dan. *9,* 22-27 foretold the time of its advent; Isa. *9,* 1-7; *11,* 1-5 and Mich. *5,* 2 prophesied its circumstances. The prophecy of salvation disclosed to the people of Israel came by the *Spirit of Christ,* the eternal Logos (*1 Cor. 10,* 4.9). The fact that Christ, the Son of God, trod this path of sorrow is consoling; His example is inspiring and encouraging. **12.** *Those who preached the gospel,* etc., refers to the events of the first Pentecost (Acts *2). Into these things:* i.e., the fruition of those graces foretold by the prophets.

I: GENERAL COUNSELS OF CHRISTIAN HOLINESS
1, 13 — 2, 10

1, 13-21: Filial Obedience and Fear. **13.** *Therefore:* i.e., in view of this living hope (3). To gird up one's loins: a metaphor based on oriental manner of dress. A cincture or girdle was an important article of men and women's apparel. The loose robes were fastened closer to the body as a preparation for work. This passage calls for preparedness in fighting the battles of life according to a Christian's knowledge of spiritual and material values. *Be sober:* the exercise of cool, dispassionate reason. **14.** The Apostle counsels the readers against the immoral habits of former days; these contradict their present status in the household of God (Eph. *5,* 8; Rom. *8,* 12 f). **15.** An exhortation in keeping with the standards of their present character (Rom. *8,* 14). **16.** The superiority of the New Covenant over the Old compels one to strive for greater holiness (Heb. *3,* 4 ff; *10,* 1; Gal. *4,* 22-31). **17.** *Father:* with reference to the Lord's Prayer (Matt. *6,* 9 ff; Luke *11,* 2 ff). *Behave yourselves with fear in the time of your sojourning:* this cautions the readers against the error of the Israelites who believed that mere carnal descent from Abraham admitted one to the promises made to the children of Abraham (Luke *3,* 8). God's children must prove themselves such by God-fearing works (Matt. *5,* 44 ff). **18.** *The vain manner of life handed down from your fathers* refers to an ethic devoid of moral and religious principles in which birth and circumstances had formerly placed the readers (Eph. *4,* 17 ff). *Redeemed . . . not with perishable things:* Christ alone was able to purchase man's redemption (Rom. *5,* 18 f; Gal. *3,* 22). **19.** *A lamb without blemish,* etc.: a reference to the Messianic prophecy of Isa. *53,* 7, and to the preaching of John the Baptist

1, 20 — 2, 7 FIRST ST. PETER

(John *1,* 29). The animals used in sacrifice had to be without physical blemish (Lev. *1,* 10; *3,* 6); Christ is without moral blemish. **20.** *Foreknown,* etc.: the decree of man's redemption through the blood of Christ is from eternity; its fulfillment has been wrought and *manifested in the last times* (Luke *24,* 25-27) *for your sakes* (John *17,* 1-2). **21.** *Through him you are believers in God:* the resurrection of Christ and His subsequent glory have firmly established their belief in God; thus, faith becomes the guiding principle in life and the source of hope.

1, **22-25: Brotherly Love.** **22.** Once a Christian has purified his heart from selfish and worldly motives because of *obedience to charity* he must sincerely love his neighbor (*1* John *3,* 18). **23.** The duty to love one's neighbor is founded on the supernatural rebirth of a Christian *from . . . seed incorruptible . . . through the word of God* (i.e., the message of the gospel) sown in one's heart. **24.** The Apostle here invokes the testimony of Sacred Scripture to prove that God's word alone remains forever. Vv.24 and 25 are a free rendering of Isa. *40,* 6-8. **25.** Since *the word of the Lord endures forever* the Apostle insinuates that the obligation to love one's neighbor is forever incumbent upon Christians.

2, 1-10: Growth in Holiness. **1.** The greatest and first commandment in the gospel is the love of God; the second is like it, the love of one's neighbor (Matt *22,* 37 ff). Hence, *all malice, and all deceit,* etc., must be put aside. **2.** The positive obligations imposed upon the neophytes in order to attain the end of their spiritual rebirth is *salvation. Pure spiritual milk:* the pure and unadultered word of God announced to them in the gospel of Christ. *To salvation:* omitted in some Greek MSS but undoubtedly genuine. **3.** *If, indeed:* a Greek expression equivalent to "since." **4.** Christ is the cornerstone (Isa. *28,* 16), rejected indeed by the chief priests and the Pharisees (Matt. *21,* 42-45), which now sustains the Mystical Body of Christ. **5.** *Living stones; a spiritual house:* metaphors frequently found in the Epistles of St. Paul; cf. *1* Cor. *3,* 16 ff; *2* Cor. *6,* 16; Eph. *2,* 19-22; *1* Tim. *3,* 15; Heb. *3,* 6; *10,* 21. **6.** A free citation of Isa. *28,* 16. The Apostle cites it as a scriptural guarantee that *he who believes in it shall not be put to shame.* **7 f.** Ps. *117,* 22 and Isa. *8,* 14 are given a Messianic interpretation; Christ and His message is *a stone of stumbling and a rock of scandal to unbelievers. For* *this also they are destined:* the punishment preordained by God because of their unbelief (Luke *20,* 17 f). **9.** On the contrary believers in Christ become: (a) *a chosen race,* more so than the Hebrews who were selected by God from among the nations (Isa. *43,*20-21); (b) *a royal priesthood,* because of the spiritual sacrifices they offer to God through Jesus Christ (*2,* 5); (c) *a holy nation,* a title bestowed on the Mosaic theocracy (Ex. *19,* 6) and more fully realized in the New Testament; (d) *a purchased people,* i.e., a people redeemed by the blood of Christ. **10.** This reference to the prophet Osee (*2,* 23-25) proclaims the striking contrast between the present and former state of the readers. The Apostle cites the prophecy in favor of the Christians—Jew and Gentile—redeemed by Christ, and who now constitute God's people.

II: PARTICULAR COUNSELS OF CHRISTIAN CONDUCT 2, 11 – 4, 6

2, 11-12: Good Example. **11.** The principles of the world strive against the Christian economy; hence, the followers of Christ must conduct themselves as *strangers and pilgrims* toward a world through which they journey to their heavenly home (Rom. *8,* 5). *Carnal desires* are utterly opposed to the welfare of the soul (Rom. *8,* 7 f). **12.** *Whereas they slander you:* the Christians had experienced the stinging scourge of slander; but truth must conquer. Thus hope remains that the Gentiles *may through observing you by reason of your good works glorify God.* Cf. Matt. *5,* 16. In this manner a Christian's present conduct will contribute additional glory to God *in the day of visitation,* i.e., when their present calumniators receive the light of God's grace.

2, 13-17: For the Citizen. **13.** A Christian's civic duties. The Apostle demands obedience to civil authority in virtue of its divine origin (Rom. *13,* 1-7). *To every human creature:* i.e., to every ordinance created by man. *To the king as supreme:* i.e., the foremost authority in the state. **15.** Compliance with one's civic duties, furthermore, silences *the ignorance of foolish men,* i.e., the slanderers of Christianity. **16.** *Freemen:* Christian freedom liberates one from the observance of the Mosaic Law and the law of sin but not from civil ordinances; cf. Rom. *7,* 4.6; *8,* 2; 2 Cor. *3,* 17. A false conception of Christian freedom gave rise to the antinomian tendencies which are denounced in 2 Pet.

2, 13 f. 17-19. **17.** In keeping with the command of Christ: "Render, therefore, to Caesar the things that are Caesar's; and to God the things that are God's" (Matt. 22, 21).

2, 18-25: For the Slave. **18.** Slaves were especially liable to misunderstand the Christian concepts of freedom and equality (*1 Cor. 12,* 13; Gal. *3,* 28); hence the opportuneness of this counsel. The welfare of the Church and of the slave demanded it (cf. *1* Cor. *7,* 20-24). *Servants:* the Greek text reads "domestic slaves." **19.** An obedience which because of religious motives sees in the command of a superior the command of God; hence, an obedience dependent neither on one's feeling nor on the person of the superior. **20.** Maltreatment patiently borne is a service acceptable to God and meritorious for the servant. **21.** The truth of this statement (20) is supplemented by the example of Christ. **22.** Christ, too, suffered unjustly. **23.** *But yielded to him,* etc.: the Greek reads, "but committed himself (i.e., his cause) to Him that judges justly," i.e., to God, His Father. He suffered patiently. **24.** He suffered for us. *By his stripes you were healed:* the scourging, so well known to slaves. **25.** The first part of the verse was probably suggested by Isa. *53,* 6 f: Christ was led as a sheep to the slaughter.

3, 1-7: For the Wife and the Husband. **1 f.** The example of a religious and God-fearing wife can gain the soul of a non-Christian husband whom the word of God has not reached. Today *braiding of hair, or wearing of gold, or putting on robes* (alluring apparel) may appear rather common; but one cannot judge by modern standards the impression these feminine artifices created nineteen centuries ago. **4.** As this verse points out, attractiveness of heart is preferable to that of the body. Meekness and a retiring spirit are incorruptible riches in the sight of God. **6.** Probably this verse was directed to women converts from paganism. St. Paul had reminded the Galatians (*3,* 7-9) of their spiritual relationship with Abraham; St. Peter now exhorts Christian wives to adopt the new standards of their spiritual mother, Sara, the wife of Abraham. Hence, even the threat of a non-Christian husband should not cause them who fear *no disturbance* to waver in their Christian obligations. **7.** The law of Christ binds husband and wife alike. A husband's rule of life may be summed up in one word, *considerately;* thus in keeping with the teaching of Christ and not according to the will of the flesh. Woman as well as man lives in the Mystical Body of Christ; any action on the part of man which destroys this harmony or which sins against marriage—a type of the union existing between Christ and His Church—tends to nullify that man's prayer.

3, 8-12: In Christian Charity. **8.** An exhortation to mutual charity among Christians. **9.** An emphatic assertion regarding the necessity of charity toward non-Christians (cf. 2, 12 f). **10.** A free citation of Ps. *33,* 13-17 which confirms the exhortation of 8-9.

3, 13-22: In Christian Suffering. **14.** Cf. Matt. *5,* 10. *Have no fear of their fear:* do not be intimidated by the prospect of persecution. **15.** *Hallow the Lord Christ in your hearts:* this reference to Isa. *8,* 13 in the context emphasizes the contrast between the brief moments of pain here below and the eternal bliss of heaven. For, as the prophet continues, "let Him be your fear and let Him be your dread." *A reason for hope:* i.e., the glorious Second Advent of Christ, the King and Ruler and Judge of mankind; cf. Matt. *24,* 30; Mark *14,* 62; Acts *17,* 7. **16.** *Fear:* i.e., with deference. **17.** Cp. with *3,* 14. **18.** The blessing begot of persecution and endured for the sake of justice is demonstrated by the life of Christ: He suffered for another's guilt; He was the innocent victim of suffering. His sufferings, however, gave birth to our redemption and to a new activity of Christ's soul in limbo. Note the reference to this thought in the Apostles' Creed: "He descended into hell," etc. **19.** *In which:* i.e., His soul. Christ's body, indeed, lay in the sepulchre; His soul, however, abode *in prison,* i.e., in limbo where the souls of the just awaited the opening of the gates of heaven. *Preached:* the Greek verb insinuates the announcement of good news. **20.** Jewish theology had refused to the victims of the Deluge participation in the Messianic kingdom; however, the blessings of Christ's suffering extended even to these souls who had put off conversion until the last moment. *Eight souls:* Noe and his wife; his three sons and their wives. *Saved through water:* freed from the persecution of a recalcitrant world by the waters which lifted the ark above the earth. **21.** *Its counterpart, Baptism, now saves you also:* the Greek reads, "Baptism which now in antitype also saves." *Not the putting off,* etc.: Baptism is not a mere ceremony, as circumcision: it directs man to walk in the newness of life, "as dead to sin, but alive to God in Christ Jesus" (Rom. *6,* 11).

Through the resurrection of Jesus Christ: the source whence Baptism has its efficacy. **22.** *Swallowing up death that we might be made heirs of eternal life:* missing in the Greek text. The final and supreme blessing accruing from the sufferings of the God-man is His elevation to the right hand of God in heaven where He continues His supplication for us.

4, 1-6: In Christian Faithfulness. **1.** In view of the blessings consequent upon Christ's suffering the Apostle exhorts his readers to *arm* themselves *with the same intent. He who has suffered:* the Christian who endures contempt or even violence rather than associate himself with former companions in sin; such have definitely severed relations with their former sinful ways. *From sins:* not in the Greek text. **3.** This verse suggests that the Epistle was addressed, in part at least, to converts from paganism. *And unlawful worship of idols:* the Greek reads, "unlawful idolatries." **4.** The sinful character of many pagan gatherings compelled the Christian converts to remain aloof from them; this silent rebuke stirred pagan resentment which vented itself by calumniating the Christians. **5.** The hope and encouragement of Christians: their calumniators will find a just judge in Christ who is ready to pass sentence upon them. **6.** *The gospel preached even to the dead:* a reference to Christ's activity in limbo (cf. *3,* 19). The fruits of redemption were applied to the souls in limbo; these were awarded a merciful judgment before the tribunal of Christ, an ordeal which all *men in the flesh* must undergo.

III: CHRISTIAN SERVICE AND THE COMING JUDGMENT 4, 7 – 5, 11

4, 7-11: Mutual Charity. **7.** *But the end of all things is at hand:* "but" introduces a new trend of thought suggested by 6. The truth of Christ's Second Coming and the judgment must ever be present to the Christian (cf. Luke *12,* 35-38. 42-48; Matt. *24,* 43; *25,* 1-13); hence, the increased need of prudence and watchfulness in prayer (cf. Jas. *5,* 8; Phil. *4,* 5; Apoc. *22,* 12). **8.** *Charity covers a multitude of sins:* charity is a visible sign of that invisible bond of unity and brotherhood in Christ. St. Peter's exhortation is based on Prov. *10,* 12; it has no reference to charity as atonement for sin (cf. Matt. *6,* 14; Luke *7,* 47), nor as an instrument of conversion (cf. Jas. *5,* 20). The abuse that

Christians must endure at the hands of the pagan should not be increased by strife and hatred among the Christians themselves (cf. *4, 4*). **9.** *Hospitality:* the reception, entertainment, and relief of travelers. St. Paul recommends this virtue to all Christians (Rom. *12,* 13; Heb. *13,* 2) and particularly to bishops and priests (*1* Tim. *3,* 2; Titus *1,* 8). The thought of the coming judgment may explain St. Peter's pressing injunction for hospitality, a virtue which Christ, the Judge, specially enjoins upon all Christians (cf. Matt. *25,* 35.43). **10.** A reference to the spiritual gifts enjoyed by some of the early Christians (Rom. *12,* 6-8; *1* Cor. *12,* 4-10); let the recipients of such gifts humbly dispense them in view of the common good. **11.** *If anyone speaks:* the two following examples clarify the injunction of 10; (1) when an individual dispenses his gifts of prophecy and teaching to the brethren, let him clearly indicate that it is God's message, not his; (2) let the individual who *ministers* spiritual or material comforts disclose that these gifts come from God.

4, 12-19: Blessings of Persecution. **12.** In *1,* 6; *2,* 18 ff; *3,* 13 ff; St. Peter dealt with the problem of suffering and persecution in the life of the individual Christian; he now discusses the same problem but in relation to the Christian community. That trials should come to the community with the acceptance of Christ is to be expected. **13.** Persecution makes the Christian community conformable to the suffering Christ—a condition for participation *in the revelation of his glory.* **16.** *Christian:* probably a term of contempt conferred on the brethren by gentile scorn. **17.** The merciful hand of God permits persecution and trial in order to purify Christians on earth prior to their glorious life in eternity. But what, by inference, must be *the end of those who do not believe the gospel of God?* **18.** A free citation of Prov. *11,* 31 (Septuagint) which suggests the terrible punishment in store for *the impious and the sinner.*

5, 1-4: For the Ministry. **1.** An exhortation to the ecclesiastical heads of the community. A *witness to the sufferings of Christ:* St. Peter witnessed the physical sufferings of Christ and bore witness to the doctrine of Christ by his preaching and suffering. *Partaker also of the glory,* etc.: St. Peter advances his personal hope of future reward as a safeguard to the bishops against enticements of worldly ambitions suggested in 2-3.

5, 5-11: Counsels to the Laity. **5.** *You who are younger:* a counsel addressed to the inferior ministers to respect the office of the presbyters. **6.** *Under the mighty hand of God:* a phrase frequently met with in the Old Testament in connection with the deliverance from Egypt, Ex. *3,* 19; Deut. *3,* 24; *4,* 34; hence a plea for resignation in the vicissitudes of life coupled with complete trust in the omnipotence of God. **8.** God's loving care of man is not to encourage heedless and reckless living because *the devil, as a roaring lion, goes about seeking someone to devour.* **9.** Forewarned, forearmed; such visitations are the inheritance of Christians (cf. *4,* 12 ff; 2 Tim. *3,* 12). **10.** The foundation of a Christian's hope is the assurance that God will *perfect, strengthen and establish us* in our call unto *His eternal glory.*

Conclusion 5, 12-14

12. *Silvanus:* the bearer of the Epistle. Many scholars consider him identical with Silas mentioned in Acts *15,* 22; and Silvanus of 2 Cor. *1,* 19; *1* Thess. *1,* 1; 2 Thess. *1,* 1. Literary considerations suggest the hypothesis that this disciple had a share in editing the Epistle. The second part of the verse impresses upon the readers that the grace of faith *is the true grace of God* which leads to eternal life. **13.** *Babylon:* Rome, then the center of paganism. *Mark:* the Evangelist who wrote the Second Gospel, a disciple of St. Peter and a companion of St. Paul. **14.** A Christian greeting found also in the Epistles of St. Paul (*1* Cor. *16,* 20. 23; 2 Cor. *13,* 12; *1* Thess. *5,* 26. 28; Eph. *6,* 23 f). *Grace:* the Greek text reads, "peace." A prayer for peace—peace of heart and eternal peace with God—concludes the Epistle. *Amen:* missing in the Greek text.

Charles G. Heupler, O.F.M. Cap.

THE SECOND EPISTLE OF ST. PETER

Introduction

The Readers Addressed. The Epistle expressly states that it is a second communication written to this circle of readers by "Peter, a servant and apostle of Jesus Christ" (*1*, 1; *3*, 1). These Christians were in all probability the same as in *1* Pet. *1*, 1; i.e., inhabitants of Asia Minor. The progress of antinomian tendencies, the abuses subsequent upon a false interpretation of Christian freedom, and the detailed explanation of God's punishment visited upon sin in the past are factors that point to a community recently converted from paganism.

Authorship. *External evidence.* The external evidence in favor of the Petrine authorship is not so early as that of the other Epistles in the New Testament. At the end of the fourth century both the eastern and western churches acknowledged the authenticity of the Epistle. Prior to this time the attitude of the eastern church found more favorable expression among ecclesiastical writers than in the West. Individuals who quote the Epistle as the work of St. Peter are: Cyril of Jerusalem, Cyprian, and Origen. According to Photius (*Bibl. cod.* 109), Clement of Alexandria wrote commentaries on the Catholic Epistles; Eusebius informs us that Clement definitely included the antilegomena in his writings. Possible echoes or traces of the Epistle appear in the writings of Irenaeus of Gaul, Tertullian, and Theophilus of Antioch.

Internal evidence. The evidence contained in the Epistle pointedly indicates its Petrine authorship. 2 Pet. *1*, 1 declares the author to be "Simon Peter, a servant and apostle of Jesus Christ." This statement is further confirmed by *1*, 16 where the author speaks of himself as a witness of the Transfiguration; then, too, *1*, 14 has direct reference to Christ's prophecy concerning the death of St. Peter (John *21*, 18 f). The author's knowledge of early human history through Jewish sources and the promises made in the Old Covenant marks him a Jew by education. Finally, the insistence in 2 Pet. that a Christian's hope will find its complete realization with the advent of Christ's Second Coming closely unites the thought of this Epistle with that of *1* Pet. The stylistic differences between *1* and *2* Pet.—a fact which occasioned the reserve of early ecclesiastical writers toward 2 Pet.—may be explained in the light of St. Jerome's statement that as the occasion demanded St. Peter employed different secretaries. Thus, in spite of all difficulties of criticism—external and internal—sufficient reason does not exist to deny this indirect Petrine authorship. From a Catholic point of view, the authentic declaration of the Church has definitely settled the question in favor of the inspiration of this Epistle.

Date. The composition of 2 Pet. is subsequent to *1* Pet. which was written about 63 or 64. The year of St. Peter's martyrdom is 67. St. Peter's reference to the nearness of his death (*1*, 13-14) places the composition of 2 Pet. about the year 67.

COMMENTARY

Introduction 1, 1-2

1, 1-2: Greeting. The literary form of the salutation is similar to that found in *1* Pet. and many of the Pauline Epistles. **1.** *Simon Peter:* Jesus and His disciples seem to have continued the use of the name Simon after the events narrated in the Gospel of St. John *1,* 42. *A servant and apostle:* a combination found also in Rom. *1,* 1 and Titus *1,* 1. This fact gave to St. Peter the privilege and right to address his fellow Christians. *An equal privilege of faith:* a statement which emphasizes the absolutely gratuitous gift of faith common to both the readers and St. Peter. *Through the justice,* etc.: the effective cause of salvation for all men, Jew and Gentile, in Jesus Christ. **2.** *May grace and peace:* cf. *1* Pet. *1,* 2. Grace is the foundation of Christian peace; it perfects a Christian's knowledge concerning the goodness and will of Christ, our God, which leads to everlasting life (John *17,* 3).

I: CHRISTIAN VIRTUE
ITS NECESSITY AND MOTIVES *1,* 3-21

1, 3-7: Life of a Christian. **3.** *All things pertaining to life and piety:* the Apostle reminds his readers that all graces conducive to supernatural life come through Christ. The *knowledge* of Christ through faith brings within reach of all participation in His *glory and power* as once witnessed at the Transfiguration and now manifested in heaven. **4.** *Through which:* i.e., Christ's glory and power. These assure to men the realization of *the very great and precious promises* whereby they may overcome their former selves and *become partakers of the divine nature* through sanctifying grace. Participation, however, in the divine nature demands immunity *from the corruption of that lust which is in the world.* **5.** *Do you accordingly on your part,* etc.: the Greek text reads: "For this very reason, employing all care, supply . . ." An exhortation to a virtuous life which secures the gift of faith imparted to the readers. *Supply your faith with virtue:* faith to be active, to be pleasing to God, must be supplemented by a virtuous life; for, "just as the body without the spirit is dead, so faith also without works is dead" (Jas. 2, 19-26). Next follows a list of virtues which a Christian must acquire and cultivate. **6.** *Your knowledge with self-control:* a virtuous life begets a true knowledge of God as well as self-knowledge. Self-knowledge teaches one to differentiate between love of oneself and selfishness; it shows the necessity of moderation and self-discipline. *Patience* is the fruit of self-control and thence comes *piety*. Patient resignation alone can behold the finger of God in the trials and temptations that beset a Christian life. **7.** *Your piety with fraternal love:* piety will always express itself in charity, i.e., the love of God and of one's neighbor. And so faith, the beginning of supernatural life, finds its perfect realization in charity which is the bond of perfection (Col. 3, 14).

1, 8-15: **Necessity of Virtue.** **8.** When virtue and knowledge are rooted in and blossom through faith Christians are *neither inactive nor unfruitful.* **9.** Contrariwise, the Christian who does not walk worthily of God, he who does not bear spiritual fruit, *has forgotten that he was cleansed from his former sins* (Col. 1, 10); he has renounced his baptismal vows; he has foresworn his allegiance to God. **10.** *Make your calling and election sure:* in Baptism our old self has been crucified with Christ that we may no longer be the slaves of sin (Rom. 6, 6); the Spirit, too, testifies that we are the sons of God, heirs indeed of God and joint heirs with Christ (Rom. 8, 16 f) provided we strive to secure our position *even more by good works.* Many Greek MSS lack the phrase, "by good works." **11.** *This way:* a life in accord with one's baptismal vows, in keeping with the spirit whereby the deeds of the flesh are put to death (Rom. 8, 12 f). **12.** The intrinsic value of the great and precious promises (4) conferred on the Christians through Christ impels the Apostle further to instruct them and to take such precautions as will insure their calling and election (10). **13.** *Tabernacle:* literally, a tent, a transient shelter, a hut. The metaphor reminds the readers of the soul's temporary abode in the human body here on earth. **14.** The obligation to rouse his readers becomes more imperative in view of his approaching death foretold by Christ (cf. John 21, 18 f). Ancient commentators were of the opinion that St. Peter had received a special revelation of his imminent death. This tradition has been immortalized by Henryk Sienkiewicz in his *Quo Vadis?* **15.** *You may have occasion:* in all probability St. Peter refers to the instructions contained in this Epistle which will serve his readers as a guide and reminder of their Christian obligations.

1, 16-21: Sovereignty of Christ. **16.** *Fictitious tales:* the doctrine of false teachers centered about the Second Coming of Christ. The millennium of happiness which Christ's Second Coming will inaugurate was misinterpreted by carnal men. *"Where is the promise or his coming?"* they asked (3, 4), forgetful of the fact that *"one day with the Lord is as a thousand years"* and that *"the Lord does not delay in his promises"* (3, 8 f). St. Peter now reaffirms *the power and the coming of our Lord Jesus Christ.* The proof: his own experience of Christ's power and glory at the Transfiguration. **17 f.** St. Peter not only saw the glory that was Christ's; *on the holy mount* he himself had heard Christ proclaimed as the Son of God by that *majestic voice* of God the Father. Cf. Matt. *17,* 1-8. **19.** Furthermore, the Messianic prophecies of the Old Testament reassure us of the truth of Christ's Second Coming. The Scriptures are *surer still* because infallible. Many of these prophecies were then accomplished facts; these guarantee the fulfillment of the prophecies yet to come to pass. These unfulfilled prophecies St. Peter now compares *to a lamp shining in a dark place* directing Christians until the dawn of their accomplishment. **20 f.** Finally, prophecy comes *not by will of man,* i.e., by some natural medium whereby man can know the future. Prophecies, then, which are inspired by God, as Scripture in general, are vouched for by the truth of God Himself; hence, absolutely true.

II: FALSE TEACHERS 2, 1 — 3, 13

2, 1-3: Punishment of Lying Teachers. **1.** *False prophets among the people:* as the pseudo-prophets were a curse in Israel; so false teachers now plague St. Peter's readers. These charlatans of religion deny by their immoral life the doctrine and the truth of Christ, a crime that spells its own doom. **2.** *The way of truth will be maligned:* the lofty concept of Christian morality will be identified, in the mind of the Gentiles, with wanton conduct once Christ's moral law is replaced by a standard dependent upon popular and sensual appeal. **3.** Then as now poor dupes will sacrifice hard-earned savings to enrich the purse of these religious quacks. But their sentence of condemnation is passed; their doom is sealed.

2, 4-10: Warning from the Past. The terrifying punishments for similar sins in the past should confirm the readers in their Christian manner of life.   4. First instance: the fallen angels. The sin of the angels brought swift destruction; from the heights of heaven they were cast into the depths of hell. *Tartarus:* in Greek mythology Tartarus was a place of punishment for crimes committed against the gods. St. Peter here accommodates his language to the mind of his readers; such usage would readily be understood by converts from paganism. *Kept in custody for judgment:* the final and irrevocable judgment will be passed upon the fallen angels at Christ's Second Coming.   5. The second instance: the Deluge. *The ancient world:* the world previous to the Deluge. *Noe, a herald of justice:* while the ark was being built, Noe preached penance without avail to a recalcitrant world. *With seven others:* Noe's wife, his three sons and their wives.   6. The third instance: the destruction of Sodom and Gomorrah. This catastrophe stands as a warning, an object lesson for all times; the stench of unnatural lusts hastens the inevitable punishment of God.   7. A God-fearing life stands forth as a reproach to wanton conduct; Lot's faithfulness to God made him the object of derision and contempt among his fellow citizens.   8. To have consented to *their wicked deeds* would have freed his soul from that particular torment for a while; God, however, in His own good time delivered Lot (7); but these cities He destroyed from the face of the earth.   9. These examples prove that God knows how and when to deliver the just; that the destruction of the wicked does not slumber (3).   10. *Follow the flesh,* etc.: these vices are particularly subject to the avenging hand of God because they undermine the faith and morality of men.

2, 11-22: The Vices of Heresy. *Majesty* is predicated of personages of high estate, whether divine or human. St. Peter understands this (1) of Christ; (2) of Church, State, and the angels. The connotation is: self-willed and lustful men acknowledge no authority. Just how these false teachers derided the majesty of the angels remains a difficult problem; possibly their doctrinal beliefs or ritualistic practices detracted from the angelic nature. 11. Angels by nature are superior to men; their conduct worthy of imitation by men. Still in the presence of God the angels of heaven refrain from abusing the fallen angels.   12. *Irrational animals,* etc.: animals are created for man's use; hence born for *capture and destruction*. Now, false teachers and their victims are comparable to irrational animals when they defame things of which they are ignorant. For, (1) human nature is inferior to the angelic; (2) these teachers lack spiritual insight, theirs is a physical not a spiritual life; their principles and manner of life are sensual and mundane. They *deride what they do not understand,* they *will perish in their own corruption.* **13.** *Daylight revelry:* sensual indulgence and drunken orgies. Their licentious conduct *while banqueting with you* are *spots and blemishes* on Christian gatherings. **14.** A further description of the conduct of earthly-minded men. *Children of a curse:* i.e., doomed to eternal destruction (*2, 3;* cp. Eph. *2, 3*). **15.** These false teachers had known the *right way* but avarice turned them away from God; they capitalize on wrongdoing for the sake of gain as Balaam, tempted by the gold and silver of King Balac, had a mind to do. *Bosor:* the father of Balaam. Num. *22,* 5 calls him Beor; Bosor may be a corrupt spelling of the name or possibly a play on words in the Hebrew language signifying, "son of flesh." **16.** This incident is related in Num. *22-24.* **17.** These teachers of men are as *springs without water* which confuse the weary traveler, as a mirage which taunts parched lips. Their doctrines resemble *mists driven by storms* full of false promise but with never a quickening drop of living water. And because they have robbed men of the true light, *the blackness of darkness* (recesses in the depths of hell) *is reserved for them.* **18.** An explanation of the metaphors in 17. *High sounding, empty words:* religious teaching devoid of all spirituality. *Sensual allurements,* etc.: an artifice that flatters the senses while it restrains in moral darkness *those who are just escaping from error.* **19.** Christian freedom from the Mosaic Law, "the freedom wherewith Christ has made us free" (Gal. *4, 31*), had occasioned disturbances among the Galatians (*5,* 13) and also at Corinth (*1* Cor. *6,* 12-14). Here, too, the mask of freedom served but to enslave its devotees; *for by whatever a man is overcome, of this also he is the slave.* **20.** Through Baptism Christians have escaped *the defilements of the world;* such a travesty, however, of Christian freedom renders one's latter state *worse . . . than the former.* Cf. Matt. *12,* 45; Luke *11,* 26 f. **21.** An individual's accountability before God is greater in proportion to the liberality with which grace is conferred (cf. Heb. *6,* 4-6; *10,* 26). **22.** St. Peter's characterization of false teachers and their followers who have turned their backs on God.

3, 1-13: The Second Coming.   **1-3.** In his first Epistle St. Peter endeavored to set forth the basis—the living hope—on which Christians must evaluate the issues of life (*1 Pet. 1, 3-12*). The sword that St. Peter held over the head of the recalcitrant was the Second Coming of Christ with its glorious vindication of His faithful followers (*1 Pet. 4, 7 — 5, 11*). Thus this doctrine of Christ's glorious Second Advent forms the foundation of Christian hope.   **4.** *Where is the promise,* etc.: this scornful denial of Christ's prophecy (*Matt. 25, 31-46*) motivated the moral extravagances of the false teachers. *The fathers:* the first generation of Christians. The false teachers interpreted the death of so many Christians who had not lived to see the Second Coming of Christ as a proof for their denial of the doctrine itself; hence, St. Peter's warning that they "abuse what they do not understand" (*2, 12*). Next follows a rebuttal of the objections against the doctrine of Christ's Second Coming. First objection: *All things continue as they were:* the fathers have died; the world continues as it was; and so all things will continue as from the beginning.   **6.** Rebuttal: *By these means:* i.e., the word of God and water; hence, the Deluge brought about the destruction of the first creation.   **7.** This same word of God that created the world and decreed the Deluge reserves *the heavens that now are, and the earth* for destruction by fire. Second objection: Many of the fathers have fallen asleep and Christ's coming has still not materialized (*4*). Rebuttal: A thousand years compared to eternity is but a pittance of time; how then can a man, whose span of life is but three score and ten years judge in these matters?   **10.** The sudden and unexpected advent of the *day of our Lord* will leave no time for repentance. Judgment will be pronounced upon all men; the earth itself will be purified by fire. Jesus, too, had spoken of the destruction of heaven and earth (*Matt. 24, 35; Luke 21, 33*); St. Paul, of a judgment by fire (*1 Cor. 3, 13; 2 Thess. 1, 8*); and St. John, "of a new heaven and a new earth" (*Apoc. 21, 1*). *The heavens . . . and the elements:* the sun, moon and stars which St. Peter distinguishes from *the earth, and the works that are in it.*   **11 f.** The final dissolution of all things will put an end to false doctrine with its sensual way of life; but a God-fearing life will continue to enjoy the happiness of salvation. *You who await and hasten towards,* etc.: such who pray for and await the perfect realization of God's kingdom.   **13.** The ordeal by fire will not result in a complete destruction of the universe but will rather purify and cleanse it of all dross (Isa. *65, 17; 66, 22*). God's curse on the earth by reason of Adam's fall will no longer mar the purified heavens and earth.

Conclusion 3, 14-18

14. An Exhortation to prepare for the coming of Christ so as *to be found by Him without spot and blameless, in peace.* **15.** Account Christ's delay in coming as salvation, i.e., as a time of grace in which men may repent and be saved. Peter's condemnation of license and disorder (*2,* 11-12) and the necessity of a Christian way of life find ample support in the writings of *our most dear brother Paul.* Probably Peter had in mind the Epistles addressed to the churches at Rome, Ephesus and Colossæ. *All his epistles:* St. Peter need not have had information of every letter written by Paul; still in view of Peter's residence and detention in Rome, his intimate acquaintance with Mark and Silvanus, he certainly was in a position to know practically all of Paul's Epistles.   **16.** *Things . . . which the unlearned,* etc.: St. Paul's insistence on justification by faith in Jesus Christ and the helplessness of the Mosaic Law gave rise to a movement which despised all the injunctions found in the Law. False teachers capitalized on this antinomian tendency; they taught that the moral obligations mentioned by the Law no longer bound Christians. St. Paul's emphasis on Christian freedom (Gal. *4, 31*), justification by faith independently of the Law (Rom. *3, 28*), and the curse of the Law (Gal. *3, 10*), made difficult reading for recent and poorly instructed converts to Christianity.   **17.** A final exhortation to resist the lawless, i.e., *the foolish* who may seek to ensnare you.   **18.** The Epistle closes as it began (*1, 2*) with a hope and a prayer for the spiritual advancement of his readers. *To him be glory,* etc.: the doxology.

Charles G. Heupler, O.F.M. Cap.

THE FIRST EPISTLE OF ST. JOHN

Introduction

Authenticity and Canonicity. This Epistle was written by the author of the Fourth Gospel. This is evident from the similarities in language and vocabulary, in style, in dogmatic content and expression, all of which cannot be explained adequately by the hypothesis of imitation. Elsewhere it is demonstrated that St. John the Apostle is the author of the Fourth Gospel; hence he is also the author of this first Epistle. This is corroborated by the author's claim to be a witness of that to which he testifies (*1*, 1-3; *4*, 14).

St. Irenaeus is the first extant author to ascribe this Epistle explicitly to St. John the Evangelist. The author of the Muratorian Fragment calls attention to the identity of authorship of this Epistle and the Gospel as evidenced in the prologues of both. Eusebius places this Epistle among the "undisputed books." Other external evidence can be cited in abundance.

Time and Place of Composition. This Epistle was written either shortly before the Fourth Gospel to serve as its introduction, or, as seems more probable, shortly after, to serve as a kind of postscript. This helps explain the absence from the Epistle of the usual epistolary salutations. It was written most probably towards the end of the first century, when the christological heresies alluded to in the Epistle began to make their appearance; hence between the years 85 and 95 A.D.

There is no exact information regarding the place of its composition. From its close connection with the Gospel, however, we may conjecture it was written at Ephesus, where, according to Irenaeus (*Adv. Haer.* III, 1, 1) St. John published his Gospel.

Destination and Purpose. It is highly probable that this was a circular letter directed to the churches of Asia Minor which came under the jurisdiction of St. John. Despite the absence of the usual salutations, the epistolary character is evident from the general tone, and from certain phrases (*2*, 1.7.12-14).

The author writes mainly to preserve his readers in possession of eternal life, in communion with God and Christ. The polemical motive, less evident than in the Gospel, is present in the emphasis placed on the necessity of faith in the Incarnation, and in the condemnation of the false teachers who deny this fundamental truth. The author's purpose will be seen more fully in the Commentary.

COMMENTARY

Introduction 1, 1-4

1, 1-4: The Witness to the Word of Life. The author omits the formal greetings usual to epistles, his own name and that of those to whom he writes, possibly because these were no way in doubt, and because of his preoccupation with his message. It is to be noted that 2 constitutes a parenthesis, and that the relative clauses of 1 form the object of *we announce* of 3.

1. *The Word of life:* i.e., as in the Prologue of the Fourth Gospel, the personal Word of God, the Logos, the Second Person of the Blessed Trinity. *From the beginning:* this refers not to the beginning of the Christian dispensation, but to eternity. Hence the Word is eternal and divine. He is called the *Word of life* because life constitutes the very essence of Divinity (John *1,* 1.4; *6,* 35; *8,* 12; etc.), and because He shares it with those who come to Him. *Handled:* referring probably to the events that took place after the Resurrection (Luke *24,* 39; John *20,* 27 f), though John's experience at the Last Supper may also be in view (John *13,* 25). **2.** *And the Life was made known:* i.e., the divine Word, who is life itself and the source of all life, manifested Himself, or became visible to man through the Incarnation. **3 f.** The preaching and witness of the Apostles has a definite purpose. The term *fellowship* is in a sense the key word of the entire Epistle. It means intimacy, intercourse, communion, a joint sharing. Christians are partakers of the good things which they possess in common from God and Jesus Christ. The term is also characteristic of St. Paul's writings. The Christians to whom St. John addresses this letter already possess this divine fellowship; their increase in faith and in this sharing in the divine life is the Apostle's purpose in writing.

I: GOD IS LIGHT 1, 5 — 2, 27

In this section the Apostle warns and encourages his readers to walk in the light, i.e., to conduct themselves as befits children of God. To walk thus is effective of that fellowship or communion with God already referred to. The Apostle lays down certain precepts which his readers are to follow if they would walk in the light. There are, first, a series of positive precepts (*1,* 5 — 2, 11), and then some negative precepts (*2,* 12-27), in which he also indicates his reasons for writing.

1, 5 — 2, 2: Walk in Light. **5.** *From him* may refer either to God or to Christ; probably the latter, indicating His general teaching. *God is light:* a definition of God by one of His attributes. The term *light* is frequent in Sacred Scripture, and is a favorite of St. John in his Gospel. It is indicative of the infinite truth and holiness of God, and here especially of the latter. The Christian religion is light both to the intellect (as truth to be believed) and to the will (as truth to be put into practice). *In him is no darkness:* the negative expression of the same thought, a semitism. *Darkness* is used to symbolize error and evil, perhaps because error blinds the intellect, and evil strives to hide itself. Darkness is the domain of Satan (Eph. *6*, 12; Col. *1*, 13; *1* Pet. *2*, 9); sins are the works of darkness (Rom. *13*, 12; Eph. *5*, 11).

6 f. The term *walk* in biblical language frequently indicates a moral and not a physical action, as is the case here. Fellowship with God and a life of sin are incompatible concepts (cf. 2 Cor. *6*, 14). *Are not practising the truth:* Christian truth is a rule of life to be lived and practised. This is a condemnation of all those heretics who looked upon the truth merely as something to be believed speculatively. **7.** The advantages of walking in the light are two: fellowship with one another, and purification from sin. Fraternal charity is both a sign and an effect of communion with God. *Cleanses us from all sin:* the *sin* from which we are purified has been variously understood: concupiscence, attachment to earthly things, sins already remitted, human frailty.

8-10. Some of those to whom John was writing considered themselves free from sin, and hence in no need of this purification. This attitude is self-deception, and it makes God a liar. *If we acknowledge our sins:* very probably a reference to sacramental confession; it is in John's Gospel we find the words instituting the sacrament of Penance (*20*, 22; cf. also Jas. *5*, 16). *He is faithful and just:* the hope of complete pardon. *We make him a liar:* God has expressly stated in Scripture that all men are sinners, and that God sent His Son to save men from their sins (Mark *2*, 17; Rom. *3*, 23-26).

2, 1. *My dear children:* indicating the close personal relations between St. John and his readers. *That you may not sin:* the Christian ideal is to avoid all sin. The term *advocate* means a defender, consoler, mediator, intercessor. Cf. John *14*, 16.26; *15*, 26; *16*, 7. The Holy Spirit continues the work of Christ here on earth. Christ, however, remains our advocate in heaven. Cf. Rom. *8,* 34; Heb. *7,* 24-25; *9,* 24; *1* Tim. *2,* 5. **2.** *He is our propitiation:* i.e., the essential office of Jesus consists in His act of expiation.

2, 3-11: Observe the Commandments, Especially Charity.
3. *We know him:* only he who keeps God's commandments can be said to know Him (cf. John *14,* 15. 21; *15,* 10). The knowledge is practical. **4.** *Is a liar:* the lip-service of those who, like certain gnostics of John's time, claim a superior knowledge of God but consider themselves exempt from the observance of the commandments. **5 f.** *Abides in him:* a frequent term in John's writings.

7-11. The commandment of brotherly love is singled out for special mention because of its practical importance. It is called both an *old* and a *new commandment.* That it is old, cf. Lev. *19,* 18; Tob. *1,* 19 f; *4,* 16. It is new because its full meaning in theory and in practice was revealed to us in the teachings and example of Christ. **11.** For St. John there is no middle way: not to love is to hate; and he who refuses to love is still in darkness.

2, 12-17: Reasons for Writing. **12-14.** The different categories addressed, *children, fathers, young men,* refer to physical and not to spiritual stages. **15-17.** The term *world,* frequent in the writings of St. John, usually designates the men who are in the world, sometimes without any moral qualification (John *3,* 16 f); more often, however, it designates the unbelieving enemies of God, and the kingdom of darkness, of Satan, which is opposed to the kingdom of light and of God (cf. Matt. *13,* 22; John *1,* 10; *12,* 31; Jas. *1,* 27; *4,* 4). St. John makes the disordered love of the world and its goods incompatible with true love of God. He illustrates this inordinate love with three examples. The *lust of the flesh,* the demands of the flesh replacing obedience to God; the *lust of the eyes,* unbridled curiosity to see and try all, even what is contrary to the will of God; the *pride of life,* the inordinate pomp and display of earthly honors, dignities, possessions. *Is passing away:* the goods of this world are doomed to perish. Worldly pleasures cannot satisfy the human heart; God alone remains eternally, and he who is of God, and not of the world, also remains forever, because he obtains eternal life.

2, 18-27: Against False Teachers. **18.** In St. John the term *hour* is used to designate an extended, indefinite period (John *4,* 21. 23; *5,* 25; *16,* 2), or a period determined by God for some definite action (*2, 4;* 7, 30; etc.). The expression, *last hour,* recalls the Old Testament terms, "the last days," "the end of days" (Isa. *2,* 2; Jer. *23,* 20; etc.), which usually refer to the time of the Messianic era. Hence the expression here probably signifies this final period of the world, the Christian era, during which Christ will establish His reign in triumph over Satan (Dan. 7, 14). The Apostle does not say how long this period will continue. St. Augustine: "The last hour is a long hour, nevertheless it is the last." *As you have heard:* from the predictions of Christ (Matt. *24,* 5. 24), and from the teaching of the Apostles (Acts *20,* 30; *1* Tim. *4,* 1; *2* Tim. *4,* 3; *2* Pet. *2,* 1-3). The coming of Antichrist is a proof that the last hour is at hand.

Antichrist is coming: "Antichrist" means both adversary, and one who is taken for Christ, who usurps the honor due to Christ. This double characteristic is also found in the "man of sin" of whom St. Paul speaks (*2* Thess. *2,* 3-12), and in the beasts of the Apocalypse (*13; 16,* 13; *19,* 19; *20,* 1-3. 7-10). Commentators are divided in their identification of Antichrist. Some prefer to see in him an individual, the great adversary of Christ, who at the end of time will appear and embody in himself all the world's hatred of Christ. This seems to be the meaning of the term as used by St. John. Others, however, prefer to see in Antichrist a collective personage, a personification of the enemies of Christ, of the many antichrists who have been, and are continually at work to destroy the kingdom of Christ.

19. *But they were not of us:* they belonged to the Christian community, but were Christians in appearance only. Christ had predicted that His kingdom would have a mixture of wheat and weeds (Matt. *13,* 24-30). **20.** *An anointing:* probably the sacraments of Baptism and Confirmation. The *Holy One* can mean anyone of the three Divine Persons, or the three Persons in the one God, although the Holy Spirit is perhaps meant primarily. Enlightened by the Holy Spirit, the Christians possess a knowledge which enables them to distinguish between truth and error, between false and true Christians. **21.** *No lie is of the truth:* hence these false teachers never really belonged to the Church. **22.** Their principal error is the denial that Jesus is the Christ, that is, they denied the Incarnation, separating Jesus from the Christ, as did the followers of Cerinthus (cf. *4,* 2 f). This is ultimately a denial of the fundamental doctrine of the divinity of Christ.   **23.** There can be no communion with the Father if one denies the Son. The Father can be known only through the Son (*1*, 2; John *5*, 23; *8*, 19. 42; *14*, 6; *15*, 23).   **24.** Note the importance given here to the traditional teachings concerning Christ which are preserved by the Church.   **25.** The life of grace, lived here in union with Christ, is the prelude to life eternal (John *3*, 15; *4*, 14; *6*, 40; etc.).   **26 f.** The Apostle again summarizes his purpose in writing, which is to warn his children, and to urge them to persevere in Christian truth. It is to be noted that the interior teaching of the Holy Spirit, which is here referred to, in no wise excludes but rather includes the testimony of the Scriptures, or the infallible teaching authority of the Church established by Christ.

II: GOD IS JUSTICE 2, 28 — 4, 6

God is just, therefore we must do justice; and everyone who does justice is born of God. Fellowship with God, which is the main thought of the Epistle, is here shown to consist in a real, singular sonship, effected not by our own faith or love, but rather by the love God has for us. Justice is not the cause of our spiritual generation, by which we call God our Father, but rather its effect. God's love and grace have made us His children, and as children of God we must practise justice.

2, 28 — 3, 6: Children of God.   **28.** *When he appears:* this probably refers to the Second Coming of Christ at the end of time. Those who walk in the light of the commandments need not be disturbed over this.   **29.** *Has been born of him:* the practice of justice is made possible through a supernatural rebirth, by which man is said to be born of God. *Of him* refers to God the Father, and not to Christ, since the faithful are never called children of Christ; they are rather called the brethren of Christ, and coheirs with Him (Rom. *8*, 17).   **3, 1.** The principal effect of God's love for us is that it makes us children of God, permitting us to live in the closest union with Him. Our sonship remains adoptive, but vastly superior to human adoption. *Because it did not know him:* the world rejects Christ, therefore also the knowledge of the Father, therefore also this divine sonship of the followers of Christ.   **2.** This divine sonship which is really a fact here in this life, is but the germ and root of a more glorious condition. *When he appears:* may apply either to God or the final coming of Christ. In either case it refers to our condition after the life of this world, a condition that is described as *being like to him,* i.e., to God. *For we shall see him,* etc.: cf. *2* Cor. *3,* 18; Col. *3,* 3 ff; *1* Cor. *13,* 12 f. This is the beatific vision in which the human mind, aided by the light of glory, will behold the entire Godhead, though not wholly or comprehensively. All this *we know:* i.e., from the teachings of Christ (John *17,* 24). **3.** *Makes himself holy:* by works of penance and justice. *As he also is holy:* i.e., God, or Jesus Christ, the sinless, just and holy One (*2,* 6; *3,* 5. 7. 16).

4. *Commits sin:* the Greek term implies sin with full knowledge and deliberation. *Iniquity also:* sin is lawlessness, impiety, a revolt against the will of God as expressed in His commandments. It is always and entirely incompatible with the justice demanded of the Christian. **5.** The gravity of all sin is shown by two considerations: the Son of God became man in order to *take our sins away;* and *sin is not in him,* i.e., in Jesus Christ. **6.** The highest Christian ideal. The more fully we know Christ's life, and dwell in Christ, the closer we come to the ideal of sinlessness after the pattern of Christ.

3, 7-15: Children of the Devil. **7.** *Let no one lead you astray:* a warning against the errors of the gnostics. **8.** *Is of the devil:* the sinner is said to be the son of the devil because every sin is in some way dependent on, and inspired by him. *From the beginning:* it was through him that sin entered the world (Wisd. *2,* 24 f). All moral evil in the last analysis is reducible to Satan, the leader of the forces of evil. *The works of the devil:* i.e., sin. **9.** This does not mean that a Christian is impeccable; rather the impossibility is relative. Grace can be lost through serious sin. **10.** *In this:* i.e., in the avoidance of sin.

10 b. The term *brother* refers primarily to the members of the Christian community; virtually its import is universal. **11.** Cf. *2, 7-11.* **12.** *Not like Cain:* the spirit opposed to brotherly love is hatred, which finds its supreme expression in murder according to the example of Cain (Gen. *4,* 5 ff). Cain's *works were wicked,* and those of Abel *just* because of their dispositions (cf. Heb. *11,* 4). **13.** *The world,* i.e., men not regenerated by grace, is filled with the spirit of Cain. Our Lord had predicted this hatred of the world (John *15,* 18-25). Persecution was undoubtedly raging at the time these lines were written. **14.** *We have* *passed* from the *death* of sin to *life* in God. True fraternal charity has its origin in the love of God, and hence it is a sign and proof that we possess divine fellowship.   15. Hatred is put in the same class with murder. Our Lord speaks similarly of anger (Matt. *5,* 22).

3, 16-18: True Charity.   16. The example of Christ's love is the extreme opposite of Cain's hatred. His example is set before us for our imitation.   17. The aid envisaged here is primarily material, but it does not exclude spiritual aid which is governed by the same principle.   18. This is the old proverb: actions speak louder than words (Jas. *1,* 22; *2,* 15 ff).

3, 19-24: A Good Conscience.   19. *In this:* i.e., in the practice of brotherly love. *We know,* etc.: i.e., that we really enjoy the friendship of God. The consciousness of practising charity both in word and in deed gives us this assurance.   20. This verse is obscure because *blames* is without an object. It may refer either to our past sins or to lack of brotherly love. Further, its construction is not clear. Going with the preceding verse, it means that we can have confidence notwithstanding our past sins. Construed with the following it means that God will not allow us to go unpunished for such want of brotherly love. 21. The testimony of a good conscience engenders confidence and hope. Cf. Acts *24,* 16; *1* Cor. *4,* 4; *9,* 27.   24. *By the Spirit:* i.e., the Holy Spirit who gives to the Christian assurance of this divine fellowship.

4, 1-6: True and False Spirits.   1. *The Spirits:* i.e., the teachers, many of whom were false, the antichrists referred to in *2,* 18-28. *False prophets:* here one who speaks falsely in the name of the true God. The spirit of evil and darkness inspires these false prophets.   2 f. The criterion which the Apostle here offers his readers is very simple: the confession of the doctrine of the Incarnation. This remains the sense even with the reading of the Greek. Cf. note to text. *Now is already in the world:* the heresies rampant at the time were especially christological and soteriological.

   4. *Greater is he,* etc.: God's superiority over Satan is obvious; John rather insists that the source of victory is God, and not themselves.   5 f. This emphasizes the antithesis between the world and Christ, between those who are of the world and those

who are of God. This distinction between acceptance and rejection of the teaching authority of the Apostles is always a criterion of true and false teachers.

III: GOD IS LOVE 4, 7 – 5, 17

Again the Apostle returns to the theme of love as the basis of our knowledge of and communion with God. Again he insists, more emphatically, on the inseparability of love of God and love of neighbor, and on the necessity of faith in Jesus Christ. Without this faith there can be no real motive for fraternal charity. We are asked to love simply because God is love, and this love of God is shown in the Incarnation. Hence if we have no faith in the Incarnation, we have no real motive for love of our brother.

4, 7-21: Love Unites Us with God. The Apostle here continues the thought developed in 3, 11-24.   **8.** *For God is love:* a definition of God by one of His attributes—by one that sums up, it seems, the very essence of the Godhead. It is the basis of fraternal charity, as it is the test of our knowledge of God.   **10.** The love that unites us to God does not spring from ourselves; it began in God, even before we knew Him, in the plan of the Incarnation and the Redemption. Christ is the *propitiation* (cf. 2, 2).

12 f. *No one has ever seen God:* i.e., with the eyes of the body (John *1*, 1; *1*, 18); nor has anyone with his merely natural powers ever seen God as He is in Himself. We have, however, practical assurance of communion with Him *if we love ane another,* for then *God abides in us.* Fraternal charity thus attests the fact of the divine indwelling, but does not produce it. *He has given us of his Spirit:* further confirmation of our communion with God. The Holy Spirit is by appropriation spoken of as the author of this divine indwelling.

14-16. God's love in sending His Son is an historical fact that has its witnesses, among them John himself. The Apostle makes his own profession of faith in this fundamental truth. *To know:* used with reference to the truths which the light of faith brings; *to believe:* used with reference to an act of adherence to the person of Jesus.

17 f. The fruits of love. Cf. *5*, 14-17. *He* obviously refers to Jesus. *In this world:* best understood as a parenthetical statement. The comparison, therefore, is simply between Christ and ourselves. Our condition is like that of Christ, if we have faith and charity; hence there is no need to fear Christ as our judge. *Fear:* i.e., servile fear. Love does not banish that fear which springs from love, or filial fear.   **19 f.** A recapitulation in forceful language of the great commandment of love.   **21.** The unity of the commandment of love is based not only on the very nature of things, but also on a positive command of God, revealed through Jesus Christ (Matt. 22, 37-40; Luke *10,* 27; John *13,* 34).

5, 1-5: The Basis of Love.   1. *The Christ:* i.e., the promised Messias, the legate of God. Cf. *2,* 22; *5,* 5. *Him who begot:* i.e., God. *The one begotten:* primarily Jesus, and then all those who are born of God to a supernatural life through grace, and by extension all men.   **2.** This is the converse of *4,* 20 f. It confirms the unity and the inseparability of the two great loves. **3.** Cf. *2,* 5; 2 John 6; John *14,* 15; *15,* 10. The observance of the commandments is not difficult for the children of God. It is grace that makes them easy of fulfillment. Cf. Rom. 7, 14 ff. **4 f.** The victory comes ultimately from our faith, which is the root and foundation of love *(3,* 9; Rom. 7, 25). This faith centers in the belief *that Jesus is the Son of God.*

5, 6-13: Witnesses to Christ.   6. *This is he:* i.e., Jesus. *Who came:* into the world in the Incarnation. *In water and in blood:* both prepositions here indicate instrumentality. *It is the Spirit,* etc.: a more probable reading has, "It is the Spirit that bears witness because the Spirit is the truth."

Tertullian, and those who follow him, interpret this verse as a polemic against Cerinthus who denied the divinity of Christ. St. Augustine, and his followers, refer it rather to the water and blood issuing from the side of Christ when pierced with the lance. The former interpretation takes the *water* as that of Christ's baptism, and the *blood* as the symbol of His death. Thus the divinity of Christ was manifested both at the beginning and at the end of His ministry. Cf. Matt. *3,* 13 ff; John *1,* 32-42; Matt. 27, 54. Hence the error of the teaching that the Son descended upon the man Jesus at His baptism, but deserted Him before His death. The other opinion holds that John is rather answering the Docetae, who maintained that Christ suffered only apparently. The water and blood issuing from Christ's side prove His death to have been real. By His real death on the cross He brought about the remission of sins. The Spirit is witness of these facts especially at the baptism of Christ, and on Pentecost. He now bears witness through the infallible teachings of the Church. He is indeed the Spirit of truth (John *14,* 17; *15,* 26; *16,* 13).

7. Cf. note to text. According to the reading of the Vulgate, there are three heavenly witnesses corresponding to the three witnesses on earth, the three divine Persons. Their witness is indicated in the Scriptures. *These three are one:* i.e., in nature. 9. *The testimony of men:* this seems to refer to the witness of the water, blood and Spirit, either as visible witness in the order of creation, or as fulfilling the conditions of reliable human witness. 10. Refusal to believe means the rejection of this reliable witness; hence it amounts to blasphemy. Cf. John *8,* 39-44. To accept the testimony through faith requires the grace of God which is offered to all.

12. *He who has the Son:* i.e., through faith and love. He already possesses eternal life because grace is the seed of glory.

5, 14-17: Confidence in Prayer. 14. *Towards him:* i.e., toward Christ. The main quality our prayers must have is to be *according to his will.* Such prayer is answered. Cf. John *14,* 13; *15,* 7. 16; *16,* 26 f. 15. Cf. John *11,* 41 f; *16,* 26 f.

16. Cf. note to text. 17. The Greek probably reads, "All lawlessness is sin, and there is a sin not unto death." The Apostle wishes to keep his readers from sin, and yet would not have them become discouraged should they fall into a sin not unto death. Cf. *3,* 4.

Conclusion 5, 18-21

These verses are a summary of the chief thoughts of the Epistle. 18. Cf. *3,* 6-10. *Preserves him:* i.e., as long as he co-operates with grace. *The evil one,* etc.: although the devil will tempt him, he will be preserved. Cf. John *10,* 28; *14,* 30; *17,* 11. 12. 16; Eph. *6,* 11 ff. 19. This again emphasizes the division of men into two general categories: the children of God and the children of the world. Cf. *3,* 1-10. 20. *We know,* etc.: with a knowledge clarified by faith. This knowledge unites us most intimately with the true Son of God. *He is the true God:* an explicit declaration of the divinity of Jesus Christ. The Son is called *life eternal* also in *1,* 2; *5,* 11 f; John *1,* 4; *3,* 36; *5,* 26; etc. 21. Idolatry in the strict sense was prevalent in many forms in those days.

THE SECOND AND THIRD EPISTLES OF ST. JOHN

Introduction

Authenticity and Canonicity. It is quite commonly admitted that the second and third Epistles of St. John were written by the one author. That this author was John the Apostle may be established on the following evidence.

External evidence. The Muratorian Canon, St. Irenaeus, Clement of Alexandria, Origen, Denis of Alexandria, all witness to this opinion. The writers of the church of North Africa, Tertullian and St. Cyprian, do not mention these Epistles, but in a council held at Carthage in 256 A.D. the Bishop Aurelius quotes 2 John 10 as words of John the Apostle. Eusebius held these Epistles authentic, though he enumerates them among the "antilegomena." St. Jerome also testifies that while many ascribed them to the presbyter John, he himself held them to be written by John the Apostle. In the fourth century both Epistles were commonly recognized as authentic, and are found in the canons of the Synods of Hippo (393) and Carthage (397). They are not found in the Peschitto, and possibly not in the writings of St. Ephrem. Their very brevity and their private character are sufficient to explain this tardiness in the general recognition of their canonicity.

Internal evidence. It is very probable that the "Presbyter" in the writings of Papias is John the Apostle, known by that title throughout Asia Minor. The tone of the letters, one of gentle and affectionate authority, agrees with this. The vocabulary of both letters contains characteristic Johannine words and expressions.

Time and Place of Composition. It is impossible to determine anything with certitude touching these points for either Epistle. It is probable that they were written in Asia Minor, some time after the composition of the first Epistle.

Destination. The second Epistle is addressed to "the Elect Lady." It seems probable that this title refers to a church rather than to an individual. The references within the letter are plural (6.8.10.12), and the use of the same title in 13 seems to imply that one "elect" group is addressing another. The third Epistle is addressed to a certain Gaius, a man of character and authority. His further identification is uncertain. Cf. Acts. *19,* 29; *20,* 4; Rom. *16,* 23; *1* Cor. *1,* 14.

Purpose. The second Epistle has practically the same purpose as the first: to confirm the recipients in the truth, i.e., in the faith and in love. It cannot be determined if a special occasion provoked the letter. The third Epistle envisages the conditions in a particular church, in which the authority of St. John was being rejected by a certain Diotrephes.

COMMENTARY

THE SECOND EPISTLE OF ST. JOHN

1. *The Presbyter:* i.e., the elder. Cf. Commentary on *1* Tim. *5, 17* ff. The title is also used for all the Apostles (*1* Pet. *5, 1*), and indiscriminately for bishops and priests (Acts *11, 30; 15, 2; 16, 4; 20, 17; 21, 18*; etc.). It is indicative of authority. *Whom I love in truth:* i.e., with a sincere love, or for the sake of the truth which is common to them both.   **4.** *Walking in truth:* i.e., according to the teachings of the faith. Cf. John *10, 18; 1* John *3, 23; 4, 21.*

7. Cf. *1* John *2, 19; 4,* 1-6. These deny the reality of the Incarnation in one form or another.   **8.** The just through their good works truly merit their reward; sin, however, entails the complete loss of previously acquired title to reward.   **9.** It is only through faith and love that the divine communion is had and preserved. Heretics who deny the Son destroy this relationship; for he who denies the Son denies also the Father. Cf. *1* John *2,* 22-24; *5,* 12. 20.

10. This injunction is best understood in the light of oriental custom, in which such hospitality is a religious act. Such a greeting to a known heretic or sinner is equivalent to co-operation in his sin.

12 f. The *paper* was papyrus; the *ink* was made of pine-soot and glue dissolved in water and applied with a reed-pen. *The children of thy sister,* etc.: this seems to confirm the opinion that the recipients of the letter were a community. Otherwise the sisters would have the same name.

THE THIRD EPISTLE OF ST. JOHN

1. *Gaius:* this is a common Roman name, and is found often in the New Testament. Cf. Introduction. St. John is said to have appointed one of this name as bishop of Pergamon (*Const. Apost.* VII, 46). We gather from the Epistle that he was a fervent Christian, but it does not seem that he enjoyed an ecclesiastical office. He is known for his good deeds, and for his hospitality to the preachers of the gospel. He enjoys the confidence of St.

John, and probably for this reason is looked upon with suspicion by the Apostle's opponents.

2. *Even as thy soul prospers:* subtle but very high praise. **3.** *Walkest in the truth:* by his faith and works of charity. **5 f.** *Strangers:* the traveling missionaries. *See them off,* etc.: to provide them with means for their work, and see them along part of their journey. Cf. Acts *15,* 3; Rom. *15,* 24; *1* Cor. *16,* 6. 11; Col. *4,* 10. *Worthy of God:* i.e., with generosity.

7. *Of the name:* i.e., of Christ. *They have gone forth:* from their own homes to preach the gospel. *Taking nothing:* accepting no pay or material reward for their labors. Hence their need of help from such as Gaius.

9. Cf. note to text. Nothing certain is known of this Diotrephes. He may have been the bishop of the place, or a priest who, in the absence of the bishop, *loved to have the first place among them. Does not receive us:* either by refusing to admit St. John's authority, or by rejecting a former letter which St. John had written this church. **10.** *Does not receive the brethren:* the missionaries sent out by St. John. *Casts them out of the church:* this seems to imply that Diotrephes is bishop of the place, and the action is equivalent to excommunication. However, it is possible to interpret the phrase as a simple exclusion from the meetings of the faithful, and in that case Diotrephes could have been a simple priest. **11.** To do good is a sign of communion with God, and to do evil is a sign that one has not seen God. Cf. *1* John *2,* 3; *3,* 6. 10; *5,* 19. **12.** *Demetrius:* otherwise unknown. He may have been one of the itinerant missionaries recommended by the Apostle. *By the truth itself:* either by the very upright conduct of his life, or by the Holy Spirit speaking through inspired men, or through signs of the Holy Spirit working in and through Demetrius himself.

13. Cf. *2* John 12 f. **15.** The Apostle closes with a familiar Jewish greeting, which now has added Christian meaning. Cf. John *14,* 27; *20,* 19. 21. 26; *1* Pet. *5,* 14. There is no greeting from church to church, since the letter is private. Instead there is an exchange of greetings between their friends.

ALBERT G. MEYER

THE EPISTLE OF ST. JUDE

Introduction

Author. Ancient tradition attests that the author of this Epistle is St. Jude the Apostle. This tradition rests on a strong scriptural basis. The author designates himself "Jude the brother of James," and St. Luke twice mentions "Jude the brother of James" among the Apostles (Luke *6,* 16; Acts *1,* 13). The evident purpose of both writers is to identify Jude by his relationship to James. When St. Luke and St. Jude wrote (60-67), St. James the Less, son of Alpheus and Bishop of Jerusalem, was still prominent, if only in memory. St. James the Less was one of the "brethren of the Lord" (Gal. *1,* 19) and had a brother named Jude (Matt. *13,* 55; Mark *6,* 3). In Matt. *10, 3* and Mark *3,* 18, Jude is called Thaddeus and is joined with James, the son of Alpheus, in the list of the Apostles. Hardly more than his name is known from the New Testament. From the manner in which he identifies himself in this Epistle, one may rightly conclude that he preached the gospel to the Jews, as did his brother James. Tradition assigns Palestine and the neighboring districts of Arabia, Syria and Mesopotamia as the scene of his apostolic labors, and Beirut as the place where he suffered martyrdom.

Destination. Though not mentioned by name, those to whom the Epistle was written are indicated by its contents. They were well acquainted with the Old Testament (5 f.11) and with Jewish apocrypha and tradition (9.14). This marks them as converts from Judaism. More specific indication is the fact that St. Jude identifies himself as the brother of St. James, the well known and beloved Bishop of Jerusalem and author of an Epistle written to Jewish converts. A further indication is seen in v. 17, which implies that the gospel was preached to the readers of this Epistle by several Apostles. It is most probable that the Epistle was addressed to Jewish Christians of those districts in which tradition says St. Jude exercised the apostolate.

Date of Composition. It seems certain that the Epistle was written after the death of St. James (62 A.D.), for Hegesippus declares that there were no heresies or dissensions among the Christians of Palestine as long as St. James lived (Eusebius, *Hist. Eccl.,* IV, 22). This date is suggested as the earliest by the admonition to recall the teaching they had received from the Apostles, and by the fact that the errors mentioned in the Epistle appear more developed than when St. Paul wrote against the same teachings in the Pastoral Epistles. The date of writing cannot be placed after the destruction of Jerusalem (70 A.D.), for the wars that preceded this event caused a dissolution of the Christian communities in Palestine, and it is not likely that St. Jude would have omitted mention of this example of divine judgment. Finally, a comparison between this Epistle and 2 Peter shows that St. Peter knew and used the Epistle of St. Jude

Some modern authors outside the church claim that the Epistle was written in post-apostolic times. They base this claim on the assumption that the Epistle attacks gnostic errors of the second century. The assumption is proved false by the fact that the errors mentioned in Jude made their appearance in the days of the Apostles, as evidenced by the Epistles of St. Paul (Phil. *3,* 17 f; *1* Tim. *4,* 1 f; *2* Tim. *3,* 1 f).

Canonicity. The Epistle of St. Jude has been received by the Church as part of the inspired Scriptures from the earliest times. Witnesses from the fourth century are Sts. Jerome, Ambrose and Augustine, the synod of Laodicea and the Council of Carthage. Witnesses from the second and third centuries are the Muratorian Canon, Origen, Clement of Alexandria, Tertullian, St. Cyprian, the "Martyrdom of Polycarp"—witnesses to the faith of the Church from all parts of Christendom.

Some doubts were raised against the inspiration of the Epistle, especially among the Syrians. The cause of their doubts was probably Jude's quotation from the Book of Henoch. But not everything contained in an apocryphal book is false, and an inspired author may quote an uninspired book. Such a quotation does not imply approval of the book itself, but approval only of the part quoted, and is not contrary to inspiration.

COMMENTARY

Introduction 1-4

1-4: Purpose of Address. **1 f.** The writer identifies himself as the brother of St. James the Less (see Introduction). He calls himself a *servant of Jesus Christ,* a designation not only of every Christian, but especially of the Apostles who were in the service of the gospel (Rom. *1,* 1; Jas. *1,* 1). The Christians are *called* from sin to grace and salvation. The source of their call is the love of God the Father, and its purpose is to preserve them for Jesus Christ. St. Jude writes in order to warn them against teachers who were endeavoring to make them untrue to their call.

3. This verse affords no evidence that St. Jude wrote, or intended to write, a longer letter. It means that he had long planned to write, and the report of heretics in the Church moved him to write the present Epistle. In the face of danger, he exhorts the Christians to contend for the purity of the faith *once for all delivered,* i.e., preached by the Apostles as an unchangeable doctrine for all times. **4.** The heretics are characterized as ungodly men who have *stealthily entered in,* men who were not sincere in embracing the faith. They pervert the doctrine of grace and liberty by teaching the lawfulness of sensual indulgence (2 Pet. *2,* 19; Gal. *5,* 13). They *disown,* i.e., refuse subjection to Jesus, by which a practical denial of faith by sinful life is most probably meant. They *were marked out:* their fate was foreshadowed in the judgments of God upon sinners in times past.

I: WARNING AGAINST FALSE TEACHERS 5-19

5-7: Divine Judgments. **5.** All the adult Israelites who came out of Egypt, except Josue and Caleb, perished in the desert (Num. *14,* 35-38). The Lord first saved them from the Egyptians, but later destroyed them because of their unbelief—a warning that those who are untrue to faith will perish.

6. Many of the angels were not satisfied with the dignity conferred upon them in their creation and rebelled against God. They lost their dignity and were cast into hell. There they are kept in everlasting bonds, awaiting *the great day,* the day of general judgment, when final sentence will be passed on them and the fullness of penalty exacted (2 Pet. 2, 4; Job *4*, 18). The example warns that Christians lose their dignity by rebellion against God.

7. *Just as:* does not make this verse parallel with the preceding, as if St. Jude meant that the angels committed the same sin as the people of Sodom, for such crimes are impossible in the angels. According to Greek construction, this connecting particle can be understood as introducing another example. If any comparison is intended, it is between the punishment of the angels and the punishment of the cities mentioned. The people of Sodom and Gomorrah, and of the neighboring cities of Adama and Seboim (Deut. *29,* 23), practised unnatural vice against the sixth commandment (Rom. *1,* 27; Gen. *19,* 1 f). *Eternal fire:* the fire that destroyed these was eternal in its effect, for they were not rebuilt; the punishment of the inhabitants was also eternal like that of the angels.

8-13: Evil Life of Heretics. **8.** A general statement that the heretics are guilty of the same sins as those mentioned in the preceding examples. They defile the flesh with impurities and refuse to submit to authority (cf. 4). They deride *majesty,* i.e., the angels, who are called "majesties" (Greek text) because of their superiority to man. Cf. 2 Pet. *2,* 10.

9. In deriding the angels, the heretics do what even the Archangel Michael did not venture to do when disputing with the devil about the body of Moses which God caused to be buried in a secret place (Deut. *34,* 6). This dispute is nowhere else recorded in the Scriptures. Its cause is not given, but it is certain that the devil wished to make some evil use of the body of Moses. Clement of Alexandria and Origen say that St. Jude here quotes from the "Ascension of Moses," an apocryphal book. Though the fragments of this book now extant contain no mention of the dispute, it is possible that it originally contained the incident. Views on the origin of the book differ widely, some believing that it was not written until the second century, and its use by St. Jude is uncertain. He may be referring to an oral tradition, as St. Paul does on several occasions (Gal. *3,* 19; 2 Tim. *3,* 8; Heb. *11,* 37).

10. The heretics, having no understanding of the spiritual, deride whatever is above the natural. Their thoughts and actions are concerned with satisfying the natural powers and instincts they have in common with brute beasts. Given free play and indulgence, these natural powers become a source of destruction. **11.** The heretics are similar to certain sinners of the past: to Cain, in neglecting the warnings of God and following his own desires (Gen. *4,* 7); to Balaam, in endeavoring to seduce men for the sake of gain (Num. *31,* 16); to Core, in rebelling against God and the authority established by Him (Num. *16,* 1 ff).

12 f. A series of concrete examples describe the character and life of these men. They are *stains* on the congregation by their immoderate feasting. In emptiness of spirit they are like *clouds without water* carried about by the winds. Like *trees in the fall,* they are unfruitful of good. Uprooted from grace, they are *twice dead*—having died to paganism by their conversion and having died to grace by their immorality. Like *the sea* casting up uncleanness on the shore, their life brings to light the evil of their hearts. Separated from God, they wander like *stars* gone astray and disappear into eternal *darkness.*

14-19: Judgment of Heretics. **14 f.** St. Jude quotes a prophecy of Henoch, the seventh patriarch after Adam (Gen. *4,* 17). The heretics who fall under this condemnation are *murmurers,* lustful, and seekers after popularity for the sake of gain. The prophecy cited in these verses is also found in the apocryphal "Book of Henoch." Many Fathers believed that St. Jude quotes this book, a view that has found widespread acceptance. Others believe that St. Jude quotes oral tradition. If he quotes the apocryphal book—a possibility which must be allowed—he does not thereby approve the book or imply that it was inspired. The formula, "Henoch prophesied," introduces the prophet himself as speaking, and approval and prophetical character are given only to the words quoted.

17 f. The Apostles had warned against teachers who would scoff at the supernatural and live immoral lives. This appeal to the teaching of the Apostles does not imply that the Epistle was written in post-apostolic times or that its author was not an Apostle. St. Jude recalls to the minds of his readers what the Apostles had told them, without excluding himself from their number.

19. The heretics are *sensual men:* follow the lower dictates and instincts of nature; and *have not the Spirit:* are without the grace of the Holy Spirit. Some see in this verse an indication that the Epistle was written against second century gnostics who divided men into two classes: the sensual and the spiritual.

This classification, however, is current in the New Testament (see *1* Cor. *2,* 13f; *1* Thess. *5,* 23). In the Epistles of St. Paul, "sensual men" are those governed by the senses and lower faculties, and "spiritual men" are those governed by the higher faculties of the soul influenced by the grace of the Holy Spirit. St. Jude uses the term "sensual men" in the same sense, as seen from his description of the heretics in the preceding verses.

II: ADMONITIONS FOR CHRISTIANS 20-23

20-21: Perseverance. An admonition to remain steadfast in the faith, upon which alone true Christianity is founded. The grace for this blessing is obtained through prayer in the Holy Spirit (Rom. *8,* 26). They are exhorted to persevere in the love of God, by which they possess the firm hope of mercy when Jesus Christ comes in judgment. As elsewhere in the New Testament, the essence of Christianity is expressed in the formulas: faith-hope-charity, God-Christ-Holy Spirit.

22-23: Charity. The Christians should practise charity towards those who are led astray (Jas. *5,* 19f). The reading of these verses is uncertain in the Greek, but the better translation recognizes three classes. *Judged*, in the light of the Greek text, may be understood as "who waver." The Christians should endeavor to bring back all who have strayed from the faith. If their efforts fail, they are to show mercy and sympathy, but at the same time carefully avoid the danger of contamination by sinners.

Conclusion 24-25

The Epistle closes with a doxology similar to Rom. *16,* 25. The Apostle exhorts his readers to praise the eternal, almighty God who has become their Savior through Jesus Christ. His grace is sufficient to preserve them from sin. Guarded and preserved by grace, they will be able to stand unspotted in His presence.

RAYMOND F. STOLL

THE APOCALYPSE

Introduction

Title and Meaning. The word "apocalypse" means a revelation. God is the ultimate author of this revelation. He gave it to Christ, and Christ through His angel gave it to His servant John. The Apocalypse is a prophetic revelation in which the principal object is Christ manifesting Himself as Lord and Judge.

Purpose. The Christians, by their refusal to observe the official cult, had exposed themselves to persecution, and John, who had himself been exiled to the island of Patmos, writes to the seven churches of the Roman province of Asia, to stimulate their faith and fortitude. Persecution has come, and greater persecution will come, but the victory of the Church is certain, for Christ controls and executes the decrees of God during the course of history. The only danger is that Christians may permit themselves to be seduced, and become morally lax. The message of John is, therefore, a warning, a basis for hope, and an exhortation to reform.

Canonicity. The canonicity of this Book was never called into question in the Western Church. It stands in the Muratorian Canon without suspicion. In the Eastern Church, however, its place in the canon was long uncertain. The church at Alexandria accepted it. At Antioch it was rejected for a time. It was not originally in the Syrian Peschitto. The Eastern Church yielded, after a time, to the influence of Alexandria and the West, and admitted the Apocalypse to its canon. At the time of the Renaissance the old objections were renewed. Erasmus denied that John the Evangelist was its author. The early Protestants were divided, but the Book is not excluded from any Protestant Bible.

Author. The Apocalypse announces itself as written by one whose name was John. Tradition tells us firmly and almost unanimously that the author is John the Apostle, the author of the Fourth Gospel. The language, the doctrine, the characteristics of the Book confirm this. They manifest a Johannine hand.

Date and Place of Composition. On the authority of St. Irenaeus, we may hold as very probable that the Book was written towards the end of the reign of Domitian, and therefore about the year 96. The author himself tells us that it was written on the island of Patmos (*1,* 9-11).

COMMENTARY

Prologue *1,* 1-8

The Prologue indicates the source of the Apocalypse, its contents, and the happiness of those who receive and observe it. John salutes the seven churches of the Roman province of Asia which he addresses and calls down upon them grace and peace from God. The salutation is trinitarian in form; for the *seven spirits* refer to the person of the Holy Spirit. Of the three Persons he speaks more definitely of Christ in 5-7, and of God the Father in 8. Christ is the *faithful witness,* etc. *To him belong glory and dominion:* i.e., to Christ; because in His love He has redeemed us, He has endowed us with the offices of kingship and priesthood, and He will come again in the clouds. Of God the Father John does not speak in the third person, but introduces Him as declaring: *I am the Alpha and the Omega,* i.e., *the beginning and the end.* These explanatory words are probably interpolated from *22,* 13 where Christ speaks. Through the triumph of Christ, all things shall be consummated in Him, the beginning and end of all things, the all-powerful Lord of the past, present and future.

I: THE SEVEN LETTERS *1,* 9 — *3,* 22

1. Preparatory Vision *1,* 9-20

1, **9-11: John Told to Write His Visions.** John first sets forth his call and commission. *Because of the word of God:* i.e., because of his loyalty to the gospel. *In the spirit:* wrapt in ecstasy. *The Lord's day:* i.e., Sunday.

1, **12-20: Vision of the Son of Man.** Nearly every phrase of this description (13-16) and of His words (17-20) occurs again in *2* and *3.* See Dan. *7,* 9-14; *10,* 5 f. The description is symbolical. The long *garment* signifies a priest; the *girdle,* a king; *the whiteness of wool,* His purity and maturity and solid wisdom; the *eyes as a flame,* the power to penetrate and behold everything; the *feet like brass,* stability; the *voice of many waters,* a voice grave and majestic amid the sounds of the earth. The *two-edged sword* is a symbol of judicial authority; the Son of Man is the supreme judge. In 17 f the Son of Man declares who He is: the *First and the Last,* etc. In 19 f He bids John to write and explains the mystery of the seven *lamp-stands* which symbolize the churches, and the seven *stars* which symbolize their angels. These angels are the guardian angels of the churches, or the chief pastors or bishops, or the churches themselves.

2. The Letters 2, 1 — 3, 22

2, 1-7: To the Church at Ephesus. Ephesus at this time was the chief city of Ionia, the most important city in the province of Asia. It was a trading center, and also a center of Greek culture. Its magnificent temple was one of the wonders of the world. St. Paul had labored there; Apollos had preached there; Tychicus had met Paul there; Timothy had been its bishop.

The titles given Christ here, as at the opening of each of the seven letters, are drawn mainly from the vision in *1*. Christ is present personally in Ephesus; He is the supreme bishop, and holds in His hands not only the angel of Ephesus, but the angels of all the churches. The angel and the community merit praise because of their work against false doctrines and pseudo-prophets. But, *thou hast left thy first love:* they had become weary and were in danger of falling into a mechanical faith. *I will move thy lamp-stand:* i.e., remove or degrade or destroy their church. Nothing now remains of Ephesus but a few huts. *The tree of life:* an image of that heavenly bread by which souls participate in the life of God. It is not exclusively the Holy Eucharist.

2, 8-11: To the Church at Smyrna. Smyrna is now a large city with important commercial connections, located in the ancient province of Ionia, a little north of Ephesus. In olden times it commanded the trade of the Levant. In Roman times it was perhaps the most brilliant and splendid of the cities of Asia. It was famed for its worship of Dionysos and noted for its pagan temples. In one of these an inscription called Nero "the savior of the whole human race." Homer is said to have been born there; and Polycarp was martyred there during a popular outburst in which the Jews played a prominent part.

The church at Smyrna has had an honorable history. The letter sent to it is more uniformly laudatory than those sent to the other churches. It had kept the faith apparently through continual suffering. It was poor and oppressed, and not exposed to the dangers of riches. The Jews were hostile, and few of them adopted Christianity; but they were not really Jews, but a *syna-*

gogue of Satan. The title, *the First and the Last,* is most fitting in view of this persecution; for in all persecution, Christ, who had tasted and overcome death, is their Savior and King. He praises their fortitude. They are rich spiritually. *Ten days:* i.e., the persecution will be short, and they that are faithful until death will receive the crown of victory. *Shall not be hurt by the second death:* they shall not lose their souls, the death which admits of no resurrection.

2, 12-17: To the Church at Pergamum. Pergamum, or Pergamos, was a famous city of Mysia, about fifteen miles up the Caicus valley, and three miles north of the river. It was a little north of Smyrna. It lay apart from the great lines of commerce and was noted for its pagan ritual and was known as the chief center of the imperial worship. It was the birthplace of the physician Galen. It was celebrated for its invention and manufacture of parchment. Behind the city rose a high conical hill covered with pagan temples. In contrast to the "throne of God" this appeared to John as the *throne of Satan.* There especially was the cult of the Emperor, which menaced the very existence of the Church, for refusal to take part in it was treason.

The sharp two-edged sword: the irresistible power of the divine word. Cf. Heb. *4, 12.* This is the weapon with which Christ will subdue His enemies; with it He cuts off the diseased members of His Church. Christ praises the community for its steadfastness; their persecution has claimed at least one martyr, Antipas. Balaamites and Nicolaites taught the people to eat things sacrificed to idols and to commit fornication. An end must be put to these scandals or Christ will intervene quickly with the sharp sword of His mouth. *The hidden manna:* cf. John *6, 35.48.57.* He who eats the meat sacrificed to idols shall die. *A new name:* this expresses the change in the soul of the victor, a change depending on grace which only the soul of the victor can know.

2, 18-29: To the Church at Thyatira. This city was located between Pergamum and Sardis, and a little off the main road which connected these two cities. It was both important and wealthy. It was the holy city of the god Apollo Tyrimnaios. During the Roman period there seem to have been various mercantile guilds in the city, membership in which was a most important matter to every tradesman. The Christians objected to these guilds for two reasons: they were placed under the patronage of a pagan god; and their common banquets were celebrated with a revelry that was not conducive to morality. The dye trade was most prominent, and Lydia, the seller of purple at Philippi, was perhaps connected with it (Acts *16,* 14).

Son of God: this title is noteworthy, since "Son of Man" is used persistently throughout the Book. It fits the message, which breathes the language of sovereignty and omniscience. Christ knows that their works and charity, their faith, patience and ministrations are good, and He praises them for their progress in good works. *A Jezebel:* one who seduces the servants of God to the crime of the Nicolaites. Christ gives her time to repent but she will not, and a great tribulation will come upon her and upon her disciples. *Depths of Satan:* the pagan mysteries. *The morning star:* Christ Himself.

3, 1-6: To the Church at Sardis. Sardis, the modern Sart, now a village of squalid huts, was once the capital of the Lydian monarchy and was associated with the names of Crœsus, Cyrus and Alexander. It is south-east of Thyatira and east of Smyrna. The art of dyeing is said to have been invented here, and the city was noted for its dyed woolen fabrics. The special religion of Sardis was the worship of Cybele. While the three cities, Pergamum, Smyrna and Ephesus vied for the title of the First City of Asia, Sardis was a town of the past.

The seven spirits: the Holy Spirit who is here the Spirit of the Son. Cf. Gal. *4,* 6. Jesus distributes the powers of the Holy Spirit, upon which the life of the churches depends. If those who minister are without gifts it is because they have not asked for them. The angel of this church has not asked, and so his works and the works of the church are dead. But all is not lost. Repentance is necessary, otherwise Christ will come as a thief in the night and judge the church. *Have not defiled,* etc.: those whose works are good. *Shall walk . . . in white:* shall be victors over the beasts and their armies in *19.*

3, 7-13: To the Church at Philadelphia. This city was twenty-eight miles south-east of Sardis. In ancient times it was a rich and powerful city, connected by trade with a large district towards the east and north. It was thus well suited to be one of the central churches of Asia.

A door . . . opened: an opening for preaching the gospel,

referring also to the rapidity with which Christianity spread through the cities connected with Philadelphia. Cf. *1 Cor. 16,* 8 f; *2 Cor. 2,* 12; Col. *4,* 3. The whole region, however, was volcanic and the city suffered severely from earthquakes. While this reduced the population, its favorable location preserved the city from complete desertion. Of all the seven churches it had the longest life as a Christian city. From the words of John it is clear that its religion was of a high character, second only to Smyrna among the churches.

The true one: "truth" is a Johannine word, occurring twenty-three times in his writings and only five times in the rest of the New Testament. *The key of David:* supreme power, or the power of God. The expression has a Messianic significance (cf. *5,* 5; *22,* 16). To Christ belongs absolute authority to admit or exclude anyone from the city of David, the New Jerusalem. He has the same authority in regard to death (*1,* 18). He has supreme authority in heaven and on earth (Matt. *28,* 18). He is as a son over His own house (Heb. *3,* 6). Because Philadelphia has kept and advanced the faith, Jesus promises that the Jews will escape from Satan and come and worship at her feet. This conversion has nothing to do with the conversion of the Jews predicted by St. Paul (Rom. *11*). Because of her fidelity, the church here also will be spared the trials of persecution, a persecution that is to come upon the whole world. *A pillar:* i.e., a support in the Church of God.

3, 14-22: To the Church at Laodicea. This city was located on the south bank of the Lycus, six miles south of Hierapolis and ten miles west of Colossæ. It was founded by Antiochus II, and was named in honor of his wife Laodice. It was a small city until the Roman period, then it rapidly became rich and great. It was a banking center; Cicero proposed to cash his bills of credit there. It was noted for the manufacture of clothing and carpets from the native black wool. It was the seat of a flourishing medical school; Aristotle spoke of Phrygian powder, and later Galen of collyrium, both of which refer to an eye salve. Hence the allusion in *3,* 18. The site of Laodicea is now utterly deserted.

The Amen: this title is not found in the vision of *1.* It means truth, fidelity, perfection personified. It is here used as a personal name, and perhaps refers to the "God of truth" of Isa. *65,* 16. It contrasts with the inconstancy and mendacity of the Laodiceans. The angel of this church receives no praise. Rather the church is threatened with rejection, but the possibility of repentance is admitted in 18-20. Their material wealth is a deception, for they are spiritually *poor and blind and naked.* They need the *gold,* etc., which only Christ can give. Christ chastises them because He loves them; He is ready to give them what they need; whoever opens to the Lord will share his table with the Lord.

II: THE SEVEN SEALS 4, 1 — 8, 1

1. PREPARATORY VISION *4,* 1 — *5,* 14

4, 1-11: The Court of Heaven. Hitherto the scene of John's visions was the earth; now it is heaven, where there are no more fears and alarms, and no more wrongdoing. This vision is here interposed to remind us that all the decrees respecting the future rest with God and are executed in time through Jesus Christ. It is, therefore, a vision of the Master of human destiny and His court. The symbols of this vision, and of those to follow, were largely found by John in the Apocrypha, and some reach back to Babylonian astronomy. But John transformed what he borrowed; he clarified the meaning and gave the symbols a religious signification they did not originally possess.

This vision is logically connected with that of *1. The former voice:* the voice of Christ *(1,* 10). *In the spirit:* in ecstacy. He is spiritually transported to heaven, and there remains till the close of *9.* He does not name the One who sits on the throne in order to add to the majesty and mystery of the scene. He avoids anthropomorphic details, no form is seen but only lights of various hues flashing through the clouds around the throne. The *rainbow* in Genesis symbolizes mercy (Gen. *9,* 13).

4-7. *Twenty-four elders:* the saints of the Old and New Testament, or, according to some, angels who preside over the destiny of the world. *Lightning . . . thunder:* tokens of God's power, perhaps drawn from the scene of Mt. Sinai (Ex. *19,* 16). *Seven lamps:* a symbol of the Holy Spirit, in His sevenfold power. *A sea of glass:* the firmament above which is the throne of God. *In the midst of the throne:* under the throne and supporting it. *Four living creatures:* like those called Cherubim in Ezechiel *(9,* 20), an order of angels and apparently one of the highest orders. *Full of eyes:* indicating vigilance. *Like a lion,* etc.: the order in Ezechiel is man, lion, ox, eagle. The characteristics of these animals unite in the four creatures to give a picture of their action on the world or creation under God.

8-11. The imposing liturgy of the heavenly court. *Holy, holy, holy:* foreshadowing the blessing, *glory, honor, benediction* of 9. *Will worship him:* the kings of the earth are the enemies of God, but the kings of heaven pay Him homage. Nature, the angelic host, faithful humanity, represented by the figure of the twenty-four elders, unite to praise and glorify the majesty and holiness of God. The future tense, *will,* implies the eternal repetition of this worship.

5, 1-8: The Scroll and the Lamb. John's attention is now directed to the scroll in the hand of the throned One. As the vision in *4, 1* ff centers around God the Creator, so this vision attends to God the Redeemer who executes the divine decree.

A scroll: such heavenly books are mentioned by the prophets. Cf. Isa. *29,* 11; Ezech. *2, 9*; Mal. *3, 16*. *Written within and without:* with writing on either side. *Seven seals:* symbolizing that the divine judgments and counsels are a profound secret. What were the contents of the scroll? Probably all the future history described in the Apocalypse to its close. *Who is worthy?* no one in all creation was found worthy.

5-8. *The lion, etc.:* Christ. He, by reason of His divine Sonship and His victory is worthy to open the book, to reveal the secrets of God to mankind. Cf. Mark *12, 35*; Matt. *20, 41-45*. *A Lamb:* the lion of victory is now a lamb suffering for the sins of mankind. *Seven horns,* etc.: the horns denote fullness of power, the eyes denote omniscience, the spirits are the Holy Spirit. *And he came,* etc.: in the Greek the tenses here indicate the permanent results of the action.

5, 9-14: The Three Songs of Praise. **9 f.** The adoration of the Lamb by the living creatures and the elders. Their doxology has three members because the Holy Trinity resides in heaven. **11 f.** The adoration of the myriads of angels. Their praise is sevenfold, because Jesus completes the work of creation and dispenses the seven gifts of the Holy Spirit. **13.** The adoration of all creation. This doxology is fourfold, because of the four regions of the earth. **14.** Praise echoed in heaven.

2. The Breaking of the First Six Seals *6, 1-17*

The seven seals are but a prelude to the seven trumpets of *8-11.* The visions of the seals reveal things that will take place at the time of the sounding of these trumpets. They are, therefore, preparatory visions, revealing summarily the divine plan. Christ will make use of these natural plagues to punish His enemies. But when or how, or under what circumstances, will not be known until the angels sound the trumpets. Therefore, when John beholds these plagues, he beholds the preparations of heaven to chastise the world at a time and in a manner already decreed by God.

6, 1-8: The First Four Seals. The rupture of each seal is accompanied by a vision and by a command "Come!" This cry is answered by the appearance of a horseman whose rôle is signified by the color of his horse. The vision is based on the vision of Zach. *1,* 8; *6, 1-8.* The functions of the riders, however, are changed.

1 f. War. The horses used in Roman triumphs were white. Who is the horseman? Tradition answers that he is Christ, or the Word of God (*19,* 13).

3 f. Strife. Red naturally corresponds to the sword. This is a declaration to the Church that it must look for wars and rumors of wars, although the Prince of Peace has come. Cf. Matt. *24, 7;* Mark *13,* 8; Luke *21,* 10.

5 f. Famine. Black symbolizes famine. The extent and seriousness of the famine is described in 6. The whole of a man's pay goes for food which is held at a very high price. *Do not harm,* etc.: probably, men will have oil and wine in abundance, but suffer from lack of bread. Such a condition existed in the time of Domitian.

7 f. Pestilence. There is only one horse and its rider, yet it denotes both death and hell. This is the darkest and most terrible of the plagues. The pale-green, pallid and livid, is the color of death. *Over the four parts,* etc.: the Greek reads, "over the fourth part of the earth."

6, 9-11: The Martyrs. The fifth seal reveals the Christian martyrs. Persecution and martyrdom are foretold in the Gospels (cf. Mark *13,* 9-13 and parallels); here the predictions are partially accomplished. The persecution referred to here is perhaps that of Nero, but other persecutions are to follow. *Under the altar:* the type of an earthly altar, the altar of sacrifice or burnt offering. The martyrs are holocausts offered to God. Cf. Phil. *2,* 17; 2 Tim. *4,* 6. They are under the altar because, according to a Jewish tradition, God does not leave the souls of the martyrs in *sheol*.

6, 12-17: Signs on Earth and in Heaven. The effects that follow upon the breaking of the sixth seal are, in part, an answer to the prayers of the martyrs. It manifests the great conflict between the two "cities" which are the foundation of this Book. It has reference to the Second Coming of the Son of Man, and will find its ultimate and perfect fulfillment on the Day of Judgment. On that day the elect need have no fear, but the others, even the great of the earth, will seek protection from the rocks and in caves, but to no avail.

3. An Intermediate Vision, and the Opening of the Seventh Seal 7, 1 — 8, 1

The last vision closed with the question, "Who is able to stand?" The present vision answers that question. It is the second part of the tableau belonging to the sixth seal.

7, 1-8: Sealing of the Spiritual Israel. *The four winds:* denote days of trouble, as the winds sweep away the chaff. Cf. Jer. *49,* 36; Dan. *7,* 2. The angels appear as carrying out the designs of God. *From the rising of the sun:* the fifth angel comes from the region of light and salvation. The *seal of the living God* is an emblem of security. This symbolism is taken from Ezech. *9,* 4. *Servants of our God:* the faithful, who are to be marked as slaves were branded, probably with the name of God and the Lamb. As in Exodus (*12,* 7) the houses of the Israelites were marked with the blood of the lamb and their occupants preserved from the exterminating angel, so the servants of God, both Jew and Gentile, are preserved by this mark from the cataclysms of the last day.

The number is manifestly symbolical and indicates all those who will be saved. Cf. note to the text. The number twelve is maintained in the list of the tribes to show that in all changes the purposes of God stand firm.

7, 9-12: Blessedness of the Sealed. *After this:* introduces a new scene. The passage is closely linked to *4-6. White robes ... palms:* this symbolism may be drawn from the Feast of Tabernacles, or from the triumphal entry of Jesus into Jerusalem. In any case it represents victory. According to Tertullian these victors are those who overcome Antichrist; others hold they are the martyrs, or the Gentiles as opposed to the Jews in 7, 5 ff. It is more likely that they are identical with the hundred and forty-four thousand, that number now being indicated as meaning a countless throng. The sealing is an assurance that all God's servants will be safe in time of trouble; this vision tells us they have actually come safely out of it. All bear the emblems of purity, of victory, of felicity and triumph; and all worship and praise God with a loud voice. Their cry is an acknowledgment that their salvation is due not to themselves, but to their God and to the Lamb.

7, 13 — 8, 1: The Seventh Seal. The constancy of the victors is attributed to the blood of the Lamb, because the death of Jesus is the cause of this as of every other supernatural act. They are before the throne of God night and day and serve Him in His temple. The language is figurative, for in the heavenly Jerusalem there is no temple (*21, 22*). In 15 the verbs change from the present to the future; the vision becomes a prediction. None of the privations they endured for Christ's sake will trouble them; God will become their comforter. Finally, the Lamb opens the seventh seal. *Silence:* even the doxologies of heaven are hushed. This symbolizes the anxious expectancy of all creatures, and all the events which follow in the Apocalypse are in some way attached to the opening of this seal.

III: THE SEVEN TRUMPETS 8, 2 — 11, 19

1. Preparatory Vision 8, 2-6

Seven Angels with Trumpets. *Seven angels:* cf. Tob. *12,* 15; Luke *1,* 19. *Trumpets* have a traditional eschatological sense; they announce the Day of Judgment. Cf. Isa. *27,* 13; Joel *2,* 1; Matt. *24,* 31. But they also announce good tidings. Cf. Zach. *9,* 14. Here, in announcing the destruction of the profane world, they proclaim at the same time the triumph of the elect.

But before the trumpets are sounded, a liturgical tableau is unfolded. *Another angel:* some think this is Michael, others that he is the "angel of peace." *The altar:* corresponding with the altar of incense in the Jewish temple. The mingling of the smoke of the incense with the prayers of the people is an allusion to the intercession of the angels in behalf of men before the throne of

God. The *fire of the altar* thrown upon the earth is a token of the coming judgment. The *peals of thunder,* etc., herald the approach of the judgment.

2. THE FIRST SIX TRUMPETS 8, 7 — 9, 21

8, 7-12: The First Four Trumpets. The first trumpet's sound brings fire upon the earth, consuming a third part of it. The prohibition of 7, 1 is therefore abrogated. The allusion to the seventh plague of Egypt is obvious (Ex. *9,* 23-25), but here there is the addition of blood.

The sound of the second trumpet changes a third part of the sea into blood, bringing destruction to a third part of its creatures and the ships that sail upon it. This reminds us of the first Egyptian plague (Ex. *7,* 20 f) in which the waters of the Nile became blood, but here is introduced the great mountain.

At the sound of the third trumpet a third part of the waters of the earth were polluted, leading to the death of many. *Wormwood:* a bitter, nauseous plant, here representing trouble and calamities. Cf. Jer. *9,* 15.

When the fourth trumpet sounded the light of the world was reduced by one-third its intensity. This recalls the Egyptian plague of darkness (Ex. *10,* 21-23); but with this difference, there the Children of Israel had lights in their dwellings, while the rest of the land was dark, here the darkness is that which results from a withdrawal of a third part of the light of the sun by day and of the moon and stars by night.

8, 13: The Three Woes. The judgments invoked by these trumpets fall upon nature, the earth, the sea, the rivers and heavenly bodies; men are affected only indirectly. Now three special maledictions are to fall directly on the inhabitants of the earth.

9, 1-12: The Fifth Trumpet. This is the first "Woe" proclaimed by the eagle. *A star . . . fallen:* the emblem of a fallen star is used elsewhere in the Bible. Isaias (*14,* 12) speaks of Lucifer as fallen from heaven. The star here represents a spirit, and can be identified with Satan in 11. *Bottomless pit:* the dwelling of the evil spirits.   **2.** The first result of the opening of the pit: smoke issued in such volume as to obscure the atmosphere.   **3.** The second result: locusts with the power of scorpions. This is a reference to the eighth Egyptian plague (Ex. *10*, 12-15), but the locusts here are able to sting and cause torture. 4. Whatever the plague, it cannot hurt God's children (Luke *10*, 18 f).

There follows in 7-12 a picture of the locusts. From the description it is evident that these locusts are not natural. Their characteristics are partly drawn from the prophets (Joel *1*, 4 ff) and partly original with John. They symbolize the infernal spirits and Satan their leader.

9, 13-21: The Sixth Trumpet. The second "Woe" announced by the eagle. The plague under this trumpet resembles the last, though it is more aggravated in nature. Again there is a vast host, with the powers of the horse, the lion and the viper, but the destructive elements are increased. They have power not only to torment but also to kill a great part of the human race.

The golden altar: the same as in *8, 3*. *Who had been kept ready:* the time of the plague had been definitely fixed. Whatever the nature of the plague, it was an affliction that was both vindictive and medicinal. It was designed to bring about repentance. Two-thirds of the profane world had time to repent, but we are told that they did not. What about the Christians? We are not told; but the silence of the prophet leads us to suppose that they turned from blasphemy. Cf. the letter to Philadelphia.

3. An Intermediate Vision and the Seventh Trumpet
10, 1 — 11, 19

A twofold intermediate vision is interposed here as was the case between the opening of the sixth and seventh seals. These visions are linked with the sixth trumpet and with all the plagues that precede. They amplify the general notions already given of the struggle between the Church and Satan.

10, 1-11: The Angel with the Little Scroll. *Another angel:* an angel and not Christ. But he bears witness to Christ's power for he is clothed with a cloud, the token of the Divine Presence (Ex. *13*, 21). He also bears witness to God's mercy: the rainbow is on his head, a token of mercy and love. *His feet . . . like fire:* strong in the power of purification and judgment. *The sea . . . the earth:* this indicates that John has returned to earth. *Seven thunders:* agreeing with the seven seals, the seven trumpets, the seven bowls. *Seal up,* etc.: we do not know what the thunders revealed, perhaps they were secrets like those in 2 Cor. *12,* 4.

Lifted up his hand, etc.: an ancient sign giving emphasis to an oath (Gen. *14,* 22). The oath of the angel is to the effect there will be no longer any delay. This is in answer to the martyrs' cry (*6,* 10), "How long?" The period of their waiting is at an end. The image of eating the scroll is derived from Ezech. *3,* 3. It is sweet because of the divine revelation given to John, but bitter because of the events foretold.

11, 1-3: **The Measuring of the Temple.** It is not said by whom the reed was given, nor are we told who spoke the command. It is probably Christ, as in 3. *A rod:* i.e., a measuring rod. The measuring implies the protection of the temple from desecration. This symbolism has its basis in the Old Testament. Cf. Ezech. *40,* 3; 2 Kgs. *8,* 2; Zach. *2,* 2. The temple is the Church of Christ, and the sense of the measurement is the preservation of the Church and everything necessary to its worship—temple, altar and worshippers. The temple at Jerusalem, with its various parts, is the image. All was to be measured except the Court of the Gentiles. Why? Because that belongs to the profane, who, by their idolatry, are to trample on the Church of Christ for forty-two months. This refers to the conditions under which the Church must live. The number of months is symbolical of a limited time.

11, 4-14: **The Two Witnesses.** The symbolism of this section is in general sufficiently evident with the exception, perhaps, of the two witnesses. Some have thought them to be Elias and Henoch; others Elias and Moses; others Josue and Zorobabel. They have many of the characteristics of all the great prophets of Israel. But from the allusion in 4 to Josue, son of Josedec, and Zorobabel, John seems to imply that the two witnesses are the representatives of the civil and religious power among the children of God. Cf. Zach. *3,* 4. The two witnesses, then, represent the universality of Christian preachers and teachers; their mission is to combat the enemies of Christ and of His Church. In this office they shall have considerable power; it will resemble the power of the ancient prophets. But one day the world-power will be great enough to put them to death at Jerusalem, which signifies the whole of the profane world. It is *Sodom,* for it is a place of immorality; it is *Egypt,* for the world is in bondage to sin; it is Jerusalem, for it is the apostate place where Christ is hated. But there shall be a resurrection and Christ shall triumph; after three days and a half, symbolically, Antichrist shall be chained.

11, 15-19: The Seventh Trumpet. In the thought of John Christ is envisaged as already come; the *parousia* is accomplished. The profane world is destroyed and the children of God enjoy an eternal reward.

IV: THE SEVEN SIGNS *12, 1 — 15, 4*

We now enter upon the execution of the decrees of the "little open scroll." These decrees extend to all human history from the birth of Christ to the general judgment, and they present a recapitulation of *6-11* with the exception that all is now viewed from the side of the Church. John insists upon the lot of Rome, because his readers were more interested in that, but his vision looks beyond Rome. *12* is the preface of these final visions.

12, 1-6: The Woman and the Dragon. Cf. note to the text. The woman in pain is the bride of Christ, i.e., the Church. This is not merely the Christian Church, which did not give birth to Christ, but rather the Jewish Church which, through Mary, gave Him birth, and which, since Pentecost, has become the Christian Church. *Clothed with the sun,* etc.: images of unchangeable radiance. *Twelve stars:* the illustrious members of the Church, the twelve tribes or the twelve Apostles.

3. *A great red dragon:* some terrible and hostile power which, in 9, is identified with Satan. It is a spiritual power because it is placed in the category of the angels. *Seven diadems:* the royalty he arrogates to himself (Matt. *4,* 8 f); *ten horns:* the variety of his powers. *Red:* the color of fire, destruction.   4. He shuns the light, and hence he tears a third part of the stars out of heaven.   5. *A male child:* i.e., Christ. His title to rule is taken from Ps. *2,* 7-9. *Caught up to God:* Christ's ascension. Though the person of Christ is in John's mind, the evil one's hatred extends also to His Church. *A thousand two hundred and sixty days:* i.e., the forty-two months of *11,* 2.

12, 7-9: Michael Overcomes the Dragon. The battle in heaven is due to the dragon's effort to reach the Child. He is driven out by the angels. The adherents of the Mythical School hold that the symbols in this vision are taken from oriental legend. But the serpent in Genesis, many female personifications in the Prophets, and various metaphors in the Psalms, Prophets and the Apocrypha, are sufficient to provide the elements involved.

12, 10-12: The Song of Triumph. The song celebrates Christ's victory over Satan, but it allows for the continuation of the struggle between Satan and the Church.

12, 13-18: The Dragon and the Woman. *The woman:* i.e., the Church. *Two wings:* the power God gives His Church to escape the threat of the evil one. *Time and times,* etc.: an indefinite period agreeing with 6. Disappointed at not being able to trap the woman, the dragon turns his attack upon her *offspring,* that is, the faithful *who keep the commandments of God.*

13, 1-10: The Beast of the Sea. **1.** *A beast:* the image is based on Dan. 7, 4 ff. It combines in itself all the power and defects of the first three beasts mentioned by Daniel. **3 f.** *All the earth:* earthly men. Their wonder at the beast leads to their worshipping it.
5. The beast received this power either from the dragon or from God. In the latter case it was His permission. *Forty-two months:* cf. *11, 2; 12, 6.* This establishes a close relation between all these visions. The indefinite time in all extends to the Second Coming of Christ. During this time the beast will torment the saints, but though it will conquer their bodies it will not overcome the constancy of their souls (Matt. *10,* 28). **9.** The Apostle closes the section with a solemn warning: the Church must not oppose violence with violence, but suffer with the faith and patience of the saints.

13, 11-18: The Beast of the Earth. *Another beast:* resembling the first in having horns, in being wild, and in being inspired by the dragon; but the second beast is less monstrous, has only two horns, and apparently only one head. Its power lies in deception as well as in violence, and it seems to have more preternatural power. The whole of its work is directed to magnifying the first beast.

Many interpreters, both ancient and modern, have seen in this first beast the figure of Antichrist, his person, his kingdom and his power. The figure is borrowed from Dan. 7. According to the context, and on the analogy of the beast in Daniel, this beast is not an individual but a collective person, an empire or empires, i.e., a political power. The key to its identification is given in *17, 10*: the pagan and persecuting Roman Empire—but not exclusively, as will be seen. When he speaks of the "head" of the beast he seems to identify it with one of the Roman Emperors. But in *13, 12-14* the head is identified with the beast. There is thus a fluctuation in the symbolism between a determined individual and the empire he represents. Nero best fits the details, when an individual is in view. He concentrated within himself all the wickedness of the beast; and the image may have been suggested to John by the legend of *Nero redivivus*. But it is probable that John's intention is to show that the evil of Nero will return in one of his successors. This continual renaissance of the beast-Antichrist has been witnessed in every historical epoch, and will be witnessed until the forty-two months have run their course.

18. The name of the beast is a number: *six hundred and sixty-six*. The meaning of this number has perplexed interpreters. Perhaps the Christians of Patmos and Asia understood it, for John would hardly have presented them with a conundrum. If a traditional understanding of it existed it was lost before the time of Irenaeus. John hints that a special wisdom is required to know the meaning of the name. Some have thought that Nero is referred to, since the numerical value of the letters of his name, in Hebrew characters, is six hundred and sixty-six. Others, on the same basis, but using Greek letters, think of Titus. Still others hold it is Caligula. There are other opinions. But what is certain is that John warned the Christian Church that the beast, incarnated at first in the Roman Empire, would always act as Nero had acted, that its power would be very great, and that it would continue under various forms until the Second Coming of Christ.

The Second beast is a supporter of the first beast, but its power is of the moral and intellectual order. In *16, 13*; *19,* 20; *20,* 10, John gives it the name of False Prophet. It represents the Roman power in the provinces, and especially in Asia Minor. This power was both civil and religious. The pagan priesthood flattered the civil authorities. Its marvels are worked before the beast, i.e., before the Emperors; incense is burned before the beast, i.e., before the statues of the Emperors. The wrath of the priesthood of the official cult was directed against the Church, and Christians could neither buy nor sell because they did not bear the mark of the beast, i.e., did not worship the Emperor. Though the times have changed, the Church is still persecuted because it does not carry the mark of the beast and does not bow down before the world power or its false prophets. This will continue until the Day of the Lord.

14, 1-5: The Lamb and the Virgins. This vision is in contrast to the foregoing. The beast, followed by those who bear its mark, is opposed here by the image of the Lamb, Christ, with His countless faithful, who are marked with His name and that of His Father. **2 f.** The *new song* is one of sweetness and majesty. **4 f.** The reason only the faithful can understand the song is that they are *virgins*. Some would understand this term of spiritual virginity, i.e., as opposed to "fornication" which is so often used of apostasy. Others accept the term in its strict sense. This is the common Catholic opinion, and the one best supported in tradition. They follow the Lamb here on earth by the practice of the counsels, being faithful even to martyrdom, if that be necessary.

14, 6-11: The Three Angels. This vision corresponds closely to that of the seven seals and serves as a prelude to the judgments which God will pronounce over the beast. There are three phases: (1) preliminary proclamations of the angels (*14,* 6-11), corresponding to the four horsemen (*6,* 1 ff); (2) prediction of the happiness of the saints (*14,* 12 f), corresponding to the consolation of the martyrs (*6,* 11); (3) anticipated vision of the divine judgment, corresponding to the sixth seal (*6,* 12 ff).

The first angel promulgates the gospel and warns of impending judgment. *His judgment:* to be taken in a wide sense, especially of the judgment to be passed on the beast and the world-city Babylon. This judgment begins with the victory of the gospel.

The second angel follows the first, as the doom of the world-city follows the proclamation of the gospel. Pagan Rome was *Babylon* to St. John. Cf. *16-18.* Rome had not fallen when John wrote, but he uses the prophetic past, indicative of the certainty of accomplishment. The fall is more fully described in *16-18.*

The third angel foretells the certain punishment that will be imposed on all who *worship the beast,* i.e., the world power, and *its image,* i.e., the statues of the rulers. This punishment is described as eternal damnation, the eternity of hell is clearly taught in this verse.

14, 12-13: Blessedness of the Saints. This happiness is opposed to the condemnation pronounced on the followers of the beast. *Who die in the Lord:* i.e., not only martyrs, but all who die in union and fellowship with Him. This happiness begins at the moment of death, since *henceforth* refers to present time.

14, 14-20: Vision of the Judgment. This divine judgment is presented under two symbols: that of the harvest, and that of the vintage. The harvest is restricted to the good, while the vintage refers only to the evil. The symbols may be borrowed from Joel *3, 13.*

The harvest (14-16). Christ is the reaper, as is clear from the *white cloud* (Matt. *24,* 30), the *son of man* (John *5,* 27; Dan. 7, 13), and *the crown. In his hand a sharp sickle:* a sign that the harvest is ready for the reaper. *Out of the temple:* the inner shrine measured off in *11,* 1. It is the angel who, on the part of God the Father, calls to Christ to begin the reaping.

The vintage (17-20). *From the altar:* cf. *6,* 9 f; *8,* 5. *Clusters of the vine:* i.e., sinners. John sees them cast into the wine press of God's wrath. The treading of the wine press is a figure of divine vengeance. The red juice of the grapes suggests the shedding of blood. As the wine press usually stood outside the city, it is here a symbol of those outside the City of God. *A thousand and six hundred stadia:* the square of four multiplied by a hundred, a symbol of completeness, reaching to the four corners of the earth. The horses, perhaps, belong to the army of heaven (*19,* 14).

15, 1: The Angels and the Plagues. This is the seventh sign, appropriately having seven angels and seven plagues. It is a token of something to follow and not a mere empty wonder. This introduces a new set of scenes and serves as a transition to what follows.

15, 2-4: The Sea of Glass. This sea has already been referred to in *4, 6. Mingled with fire:* reflecting the lightnings that proceed from the throne of God. *Standing on the sea:* i.e., standing on its shore. *Harps of God:* belonging to the service of God. These victors are not only those who have reached heaven, but also the faithful in Christ who are yet on earth, since the life of grace is the prelude to that Glory (Phil. *3, 20*). *The song of Moses:* the song of victory over Pharaoh and his army (Ex. *14,* 26-31). This song was sung by all the faithful; that of the Lamb probably by the one hundred forty-four thousand, who alone could understand it.

V: THE SEVEN BOWLS 15, 5 — 16, 21

1. Preparatory Vision 15, 5-8

The Angels and the Bowls. As the seven signs correspond to the seven seals, so this group corresponds to the seven trumpets.

The temple of the tabernacle, etc.: the figure is from the tent built by Moses to house the ark of the covenant, or testimony. The ark was the perpetual witness of God's presence in Israel. Its heavenly counterpart is here the inner shrine of heaven which is now opened.   **7.** *Golden bowls:* the shallow cups which were used for incense. The plagues poured forth from the bowls are not entirely natural. They are not like the judgments of the trumpets, i.e., calls to repentance; they are rather plagues on those who have refused to repent and have rejected the Church of Christ.   **8.** *Smoke* is the symbol of God's presence. The vision resembles that of Isa. *6. No one could enter,* etc.: this may mean that the judgments of God are impenetrable and will not be understood until they are fulfilled at the time of Christ's Second Coming; or it may mean that God is inaccessible, i.e., no intercession will be able to arrest these judgments.

2. The First Six Bowls 16, 1-12

16, 1-7: The First Three Bowls. *A loud voice:* in 17 the voice is heard to come from the throne, the voice here seems to be the same; therefore God Himself gives the signal to pour the bowls of divine wrath upon the earth.

The result of the first bowl: *a sore and grievous wound.* This resembles the sixth Egyptian plague (Ex. *9,* 8-12). The second bowl: *the sea . . . became blood.* This resembles the first Egyptian plague in which the Nile was turned to blood. Cf. the second trumpet (*8,* 8). The third bowl: the effect is the same as that of the second bowl. They poured out the blood of the saints, now God gives them blood to drink. *The altar:* cf. *8,* 8 f; *9,* 1 f.

16, 8-12: The Second Three Bowls. The result of the fourth bowl: *mankind were scorched with a great heat.* The fifth bowl: the *kingdom* of the beast *became dark.* This resembles the ninth plague of Egypt (Ex. *10,* 22). But the failure of their light does not lead to repentance. The sixth bowl: the *waters* of the *Euphrates* were dried up. *That a way might be made ready:* in preparation for war.

3. AN INTERMEDIATE VISION AND THE SEVENTH BOWL *16,* 13-21

16, 13-16: The Unclean Spirits. In this vision the dragon, the beast of the sea, and the false prophet (the beast of the earth) reappear. *Armagedon:* i.e., the mountain of Megiddo. The neighboring plain of Esdraelon was the great battle-ground of the Holy Land. The name is used here symbolically: the war of the kings against the Lamb will terminate here. The name is a prophecy of disaster.

16, 17-21: The Seventh Bowl. The result of the seventh bowl is a great cataclysm. *Babylon the great:* i.e., Rome. And yet not even this worked repentance; men still blasphemed God. This, as the effects of the other six bowls, can be taken in the physical sense, of disease, death, war, etc., and in the moral sense of moral corruption and dissolution.

VI: BABYLON THE GREAT *17,* 1 – *19,* 10

17, 1-6: The Woman on the Scarlet Beast. *The great harlot:* the impure city where idolatry and false worship flourish. *Sits on many waters:* her domination extends over many peoples. Many kings have been seduced by her, they have fallen under the spell of her obscene idolatries. Why did John wonder? Was it because of the splendor of the woman, or because of the blasphemous names?

17, 7-18: The Angel's Explanation. *The mystery:* i.e., the hidden meaning of the woman and the beast. In this explanation the beast is identified with the beast of *13.* The wound of *13* is expressed in the words, *was and is not. Seven kings:* the thought is that the beast united itself with seven great kings; these all eventually fell; thus the beast itself is left as an eighth, then he goes to destruction. *Ten horns:* they are kings who have not yet risen to power. *For one hour:* i.e., a short time.

The angel's explanation of the vision is rather a figurative interpretation of the symbols. Babylon, as a symbol, was the great city whose splendor and power dazzled and then destroyed Jerusalem, the Holy City. Against Babylon the voices of the prophets had been lifted. Cf. Isa. *39,* 7 f; *13,* 19; *21,* 9. In her hostility to Jerusalem she became a type of later corrupt powers. In this vision Babylon stands for Rome, the center of idolatry and immorality, the seat of a world power that dominated kings and peoples. Rome in turn serves as a type of all those later powers that war against Christ and His Church.

The beast, as in *13,* symbolizes pagan Rome either in her successive kingdoms or in her succession of Emperors who embody the imperial power. It is on the beast, i.e., the Empire, that the woman, Rome, is seated. The seven mountains are the seven hills of Rome. The beast can further be identified as Nero. The kings are the Roman Emperors, but it is not easy to say which individuals are meant. The common opinion holds the five to be Augustus, Tiberius, Caligula, Claudius, Nero. The one who *is* would thus be Vespasian. The one yet to come is Titus. The eighth then is Domitian. The popular rumor had it that Nero would return again, typically; thus Domitian is a second Nero. A case can be made for this view, and yet it is open to serious objection. The first four Emperors were not serious adversaries of the Church. Moreover, Domitian did not drag down the Empire to ruin. It was at the time of Nero that the Roman Empire began its endeavor to swallow up the Church. It would seem, therefore, that the series of kings should begin with him. The five fallen kings would thus be Nero, Galba or Otho, Otho or Vitellius, Vespasian and Titus; the sixth would be Domitian; the seventh, who is to reign a short time, is an unnamed king, a successor of Domitian. He is not determined, but alluded to in order to justify the framework of the figure seven. The eighth is likewise not clearly determined. He is one in the series of innumerable successors of the seventh, and includes within himself all their ferocious vitality. He reproduces the persecuting power of Nero, the head which was wounded and healed, Nero revived, with which the beast of the sea is identified.

Some think the ten kings, represented by the ten horns, are the barbarians who will attack and destroy Rome. At first they will persecute the Church, but later they will turn upon and destroy Rome. This does not mean that they will cease to persecute the Church. These kingdoms are given to the beast and therefore war against the Lamb. But the Lamb will overcome them, and the kingdoms of the earth will finally disappear.

18, 1-3: The Fall of Babylon. In his explanation the angel proclaimed the overthrow of Babylon, the harlot. In this vision the dramatic action is resumed, but we do not see the actual overthrow of the city; we learn of it through the four agencies that participated in her destruction. The first agency is the angel who proclaims it (1-3).

Great authority: i.e., power given him for his work against Babylon. *The earth was lighted,* etc.: in contrast to the tinseled splendor of Babylon. Perhaps he executes his proclamation even as he pronounces it. Babylon falls because she is the seat of idols and licentiousness, the great enemy of mankind, sharing her luxury with kings and nations and merchants.

18, 4-8: Her Sins and Punishment. *Another voice,* the expression *my people,* and the mention of God in 5, lead to the supposition that this is the voice of the Lamb. *Go out from her:* in the sense of having no share in her works of darkness. *Render to her:* in this reading of the Vulgate the order is given to the faithful, who are to join in the retribution; in most Greek and some Latin texts the order is given to the angels of *17,* the ministers of divine justice.

18, 9-10: Dirge of the Kings. *Standing afar off:* fearing to be involved in her ruin. As the dirge in 16 and 19, their cry begins with "Woe" and ends with "in one hour."

18, 11-17: Dirge of the Merchants. This lamentation more clearly rises out of fear and selfishness. Cf. Isa. *23;* Ezech. *26,* 16.

18, 17b-20: Dirge of the Mariners. 20. The section closes with an invitation to the saints to rejoice, not because of the evil that has come upon the wicked, but because of the triumph of God over evil.

18, 21-24: The Angel's Promise. The taking up of the stone and casting it into the sea is a symbol drawn from Jeremias *(28,* 63), where the casting of a stone into the Euphrates signifies the sinking and destruction of Babylon. The symbol here has the same meaning: she will be found no more. The words "any more" are repeated six times in these verses as a funeral dirge over her departed greatness. The power of her wealth is gone, her industrial life has ceased, the grinding of corn is at an end, her domestic life is over, nuptials shall no more be seen there. These poetic verses are taken from Jeremias (*25,* 10). *In her was found blood:* to the other causes of her destruction this adds her guilt for slaying the martyrs, and other victims. With the destruction of Rome there disappears the first great world power hostile to Christianity.

19, 1-5: The Angelic Song. In *18,* 20 the saints were bidden to make merry over the destruction of Babylon; their voices are now heard. The first voices heard are those of the angels and the blessed in heaven. *Alleluia:* i.e., praise the Lord. This is the object of the song, praise of the justice of God which is manifest in the punishment of the harlot. On earth this judgment appears as future, but in heaven it is viewed as already made. *The twenty-four elders,* etc.: the point of view is that of *4* and *5. A voice came forth from the throne:* this is not the voice of God, for it speaks of "our God." Some have thought it the voice of Christ speaking as a man, since He occupies the throne of His Father (*7,* 17). *You his servants:* the voice exhorts the faithful on earth to join in the heavenly song.

19, 6-10: The Song of Triumph. The universal Church praises God. The fall of Babylon is assured, although it has not yet taken place. Its fall is a token of the final triumph of the Lamb. *Marriage of the Lamb:* the glorious marriage of the Church Triumphant and the Lamb in heaven. The spouse of the Lamb is the heavenly Jerusalem. This statement foreshadows the final chapters of the book (*21-22*). *Fine linen:* good works. We must note that these works are a gift on the part of God (cf. Eph. *2,* 10). *And he said to me:* the voice is probably that of the angel in *17,* 1.

VII: THE CONSUMMATION *19,* 11 — *22,* 5

19, 11-14: The Divine Warrior. This is the conquering Christ. The description corresponds essentially with the vision of *6,* but it is enriched by several traits taken from *1. Faithful and True:* a true king with power to fulfill all His promises and accomplish all justice. *Many diadems:* as opposed to the beast with his many crowned heads. *A name written,* etc.: expressing His essence, this name cannot be known without revelation, and even with revelation it cannot be comprehended. *The Word of God:*

the divine Word of the first chapter in John's Gospel. Only the Word with the Father and the Holy Spirit understand perfectly the meaning of the name. *The armies of heaven,* etc.: sharing in His triumph and glory.

19, 15-16: King of Kings and Lord of Lords. *A sharp sword:* the symbol of power. This name can be understood by men.

19, 17-21: Defeat of the Beast and the False Prophet. In 17 and 18 John is inspired by the symbolism of Ezech. *39, 17-20. The great supper:* in contrast to the marriage of the Lamb in 9. *Kings . . . tribunes . . .* etc.: the enemies of Christ belong to all classes. The war of justice is not between class and class in the world, but between the justice of God and the world power which is made up of many classes. The battle itself is not described; only the result is given. This pictures the destruction of the forces symbolized by the two beasts. We must remember that, although the details are realistic, the victory is essentially in the spiritual order and the "sword" is the power of divine persuasion through Christian preaching (Eph. *6,* 17). This vision will be realized perfectly at the time of Christ's Second Coming, but it is partially accomplished each time there is a great movement towards Christianity.

20, 1-3: Satan Chained. In *12,* 9 John has already envisioned the destruction of the dragon; now his chaining takes place. *The ancient serpent:* cf. Gen. *3. A thousand years:* a long but indefinite period. The imprisonment of Satan prevents him from seducing the nations and extending his kingdom; it does not deprive him of his army which he holds in readiness for another assault.

20, 4-6: Reign of the Saints. This vision is obviously in contrast to the last. *Men sat:* probably the martyrs and all who have not received the mark of the beast. *The rest of the dead:* those who have not died in Christ will not live or reign, spiritually or corporally. *The first resurrection:* the life after death which assures man escape from the "second death." *The second death:* eternal damnation. The *first resurrection* prepares for the second resurrection, i.e., the resurrection of the body at the general judgment. Those who die in sin can participate in neither.

20, 7-10: Satan Loosed. *The nations . . . Gog and Magog:* these names are derived from Ezech. *38* and *39;* they are also found in many Apocrypha and rabbinical writings. In Jewish literature they are used to designate the nations who would rise against the Messias. They are to be taken figuratively for the great army and its leaders who are to rise in battle against the people of God. The members of this army are innumerable: they are *as the sand of the sea.* They are ready for battle when their time comes; hence during the *thousand years* the rule of Christ is not universally and sincerely accepted. *The camp of the saints:* the Church. But Satan is defeated (cf. 2 Thess. *2,* 8). *Forever and ever:* this concludes the struggle between the powers of evil and those of good. All the enemies of Christ are overthrown, their judgment pronounced for eternity.

The thousand years. The question of the Millennium of *20* is one of the most discussed subjects in the Apocalypse. The outline is clear: an angel descends from heaven, binds the dragon and imprisons him in the bottomless pit for a thousand years, after this he is loosed for a short time. During this period the faithful live and reign with Christ. At the end of the period the dragon is loosed, the nations are led astray, and the camp of the saints is once more besieged. But then the dragon and his army are destroyed forever. The question is: what are we to understand as the duration of the thousand years? and what manner of kingdom is the reign of Christ and His elect?

The most common opinion is that the thousand years refer to a period of considerable length, much longer than the periods mentioned in *11,* 2-4; *12,* 2-6. But the term is not to be taken literally as some take it. The "chiliasts" maintain that Christ will reign on earth visibly and personally for the space of a thousand years, and they place this reign between the first and the second resurrections. This interpretation falls before 7-10. The number is rather symbolical, as are the other numbers in the book. The thousand years cannot begin with the Second Coming of Christ for that ushers in the end of time. They must, therefore, precede His coming. But does this Millennium follow chronologically the other prophecies in the Apocalypse, or does it coexist with them? It seems that it coexists with them, and is but another side of the future history of the Church. It is, therefore, the symbol of the spiritual domination of the Church Militant joined to the Church Triumphant for the glorification of Christ to the end of the world.

The character of this reign is purely spiritual. The martyrs, who are dead, and the confessors, some of whom are dead and some living, form part of the kingdom. Christ reigns over all these souls. Christian teaching is propagated in the world, Christian influence is exercised on human society, and thus is restrained the world power. In this sense Christians reign with Christ.

20, 11-15: The Last Judgment. *A great white throne:* the symbol of mercy that surpasses justice. The one who sat upon the throne is God the Father, though the Son shares the throne. *Earth and heaven fled away:* not that they were annihilated; the language is poetry: they fled before the majesty of God. *Hell:* probably to be taken as *sheol,* merely the residence of the dead, and not the place of the damned. *Scrolls were opened:* in these were recorded the works of men. *Book of Life:* that holding the names of the predestined. *The second death:* eternal death, condemnation to hell forever.

21, 1-4: New Heaven and New Earth. A renewal of all things had long been the hope of Israel. The prophets sanctioned it (Isa. *65,* 17), Christ spoke of it (Matt. *19,* 28), Paul mentioned it (Rom. *8,* 19). The fulfillment of this hope may seem to tarry (*1* Thess. *5,* 2; *2 Pet. 3,* 4) but a new heaven and a new earth will surely come (*2 Pet. 3,* 13). *The sea:* the beast's place of origin. All old and corrupt elements will be swept away. The heaven and earth will not be annihilated, but renovated. *New Jerusalem:* the Church of Christ; the Church in heaven is the same as the Church on earth, and we are citizens of heaven. *As a bride:* she is united by a sacred and indissoluble bond. She is the dwelling place of the elect, and there God dwells with them. The words "as their God" at the end of 3 are not found in the best Greek MSS.

21, 5-8: The Promise. *He who was sitting:* it is God who speaks and affirms His omnipotence, but it is the voice of God the Son. *These words:* the promises which at once follow. *To him who thirsts:* with an effective desire. *He who overcomes:* who is victorious in the struggle with evil. *The cowardly:* who by reason of their cowardice go down in the struggle and become *murderers,* etc. The sins here enumerated were those of the Roman Empire.

21, 9-21: The Heavenly Jerusalem. **9 f.** Its origin. *One of the seven angels:* probably he of *19,* 9. *The bride, the spouse:* the union of these terms expresses the virginal and perpetual purity of the mystic Spouse. She is the antithesis of the harlot, Babylon. *A mountain:* the elevation of spirit necessary for viewing the Church. *Coming down,* etc.: the Church comes from God and draws its glory from Him.

11-21. The description of the city. *The glory of God:* it possesses the presence of God and shows forth His brightness. The description, which follows to 23, relies on Ezech. *48, 30-35.* There are gates in the wall opening to the four quarters of the earth so that people from every direction may enter it. The angels guarding the gates are, perhaps, an allusion to Isa. *62,* 6. If by the names written on the gates St. John intends the twelve tribes of historical Israel, then by the twelve Apostles of 14 he points out the union of the Old and New Covenants. The names of the Apostles are written on the foundation stones. Perhaps the twelve gates rest on these stones. Thus even though the Church of the Old and New Covenants be represented, the work of the Apostles is especially prominent: they are the foundation on which the spiritual Jerusalem rests. The catholicity and apostolicity of the Church are indicated in these verses.

The measurement of the city. *A golden reed:* in keeping with the dignity of the angel and the splendor of the city. Cf. Ezech. *40-42. Twelve thousand stadia:* about nine hundred miles. It is uncertain whether this refers to one side or to the sum of its sides. In any case the city is of tremendous size, indicating the figures are to be taken as symbolical. *A hundred and forty-four cubits:* about two hundred and thirty feet. Hence the wall is out of proportion to the dimensions of the city; but the wall is not important in such a city of peace. *Angel's measure:* implying its accuracy.

The material of the wall. Each stone, perhaps, had for St. John a symbolical meaning, but we have lost the key to these symbols. Perhaps they refer to the diversity of gifts and graces received by the elect who dwell in the city. The general intention of the description is to convey an impression of the great beauty of the Spouse, who is without "spot or wrinkle or blemish" (Eph. *5,* 27).

21, 22-27: God and the Lamb Give It Light. *No temple:* the entire city has become the temple of God and the Lamb. There are no altars, for the time of intercession has passed. Note that God and the Lamb are equal, a new affirmation of the divinity of Christ. *Its gates shall not be shut:* an allusion to the custom of closing the gates of a city at night. This implies the universality of the Church. Nothing unclean will enter her gates: implying her sanctity. There is here a mingling of references to the Church Militant and Triumphant; but the scene embraces the entire future of the Church, Militant and Triumphant.

22, 1-2: The River and Tree of Life. These verses continue the description of the new heaven and new earth. They reflect the story of the first Paradise from Genesis (2): the river, the tree of life, the willing subjection to God. Yet the sense is more profound and spiritual. The river here proceeds from the throne of God and of the Lamb. All the older commentators apply this living water either to the person of Christ or to the heavenly beatitude which flows from Baptism. Others think of Baptism in this life, and of the Holy Spirit in the new Jerusalem; and this appears more exact. The work of the Holy Spirit is the work of sanctification which begins with Baptism. Again, the fruit of the tree of life is more abundant and the leaves more useful. The tree is related to the Word and the fruits are the symbol of the Eucharist. The leaves which heal the nations represent the sacrament of Penance. This pertains to the present life.

22, 3-5: The Throne of God and of the Lamb. This concludes the description of the new heaven and earth. The scene reaches beyond the resurrection. God will give to the elect the light of His glory with which they will be able to contemplate His face. The curse is lifted from man; he regains Paradise, sees God face to face and reigns with Him. Here ends the last prophetic vision of the Bible.

EPILOGUE 22, 6-21

22, 6-9: Confirmation. *He said:* i.e., the angel of *21, 9*. *These words:* all that we have read in the book. *The God ... of the prophets:* the gift of prophecy has its source in God. **7.** *I come quickly:* the speaker now is Christ, and His words confirm the declaration of the angel. The prophecies of the book will occur shortly, but in the eyes of God "one day is as a thousand years" (2 Pet. *3,* 8).   **8 f.** These verses repeat *19,* 10.

22, 10-15: Words of Christ. *The time is at hand:* the events foretold by John began to be accomplished in the first days of the Church. Daniel (*12,* 9-13) had been told to seal up his prophecies because the time was far distant when they would be realized. *He who does wrong,* etc.: this is a warning: the conduct of men will not change the divine plan; Christ will come quickly and reward each one according to his works. *I am the Alpha,* etc.: indicating that He has the power to reward or punish. *Who wash their robes:* freedom from sin gives the right to enter the City. *Dogs:* the pagans, or the baptized who return to their sin (*2 Pet. 2,* 22).

22, 16-21: Final Attestation. Jesus once again affirms that He is the author of the revelation transmitted by the angel to the seven churches. *I am the root,* etc.: i.e., the true Messias, the only source of hope. *"Come":* the Church and her children, moved by the Holy Spirit, thus call to the Messias. *I testify to everyone:* the speaker may be Christ, though John could so speak in his authority as prophet. John then adds a solemn warning against tampering with the words of the book. *I come quickly:* this is the final witness of the Savior, and the seventh time the expression occurs. This evokes from John the prayer, *Amen! Come, Lord Jesus! The grace,* etc.: the Apocalypse, being a letter, closes with a salutation.

JOSEPH S. CONSIDINE, O.P.

INDEX OF SCRIPTURE TEXTS

ABRAHAM: *Patriarch,* Matt. *1,* 2; *22,* 32; Luke *19,* 9; John *8,* 33; Acts *3,* 25; *7,* 2; Rom. *4,* 1-3; *9,* 7; Gal. *3,* 6-9; *4,* 22; Heb. *7,* 1-10; *Abraham's faith,* Rom. *4,* 1-25; Gal. *3,* 6; Heb. *11,* 8; Jas. *2,* 21-23; *spiritual children of,* Rom. *4,* 11 ff; Gal. *3,* 7. 29.

ABSOLUTION: *power of promised,* Matt. *16,* 19; *18,* 18; John *20,* 23.

ACCESS TO GOD: John *14,* 6; Rom. *5,* 2; Eph. *2,* 18; *3,* 12; Heb. *10,* 19 ff; Jas. *4,* 8.

ACCOUNTABILITY TO GOD: Matt. *5,* 12; Acts *5,* 41; *20,* 24; Rom. *5,* 2.

ACTIONS OF THE MIND: *good thoughts,* Matt. *5,* 8; Luke *1,* 66; *2,* 19.51; *6,* 45; Acts *8,* 30; *1* Cor. *7,* 34; *13,* 5; *2* Cor. *7,* 1; Eph. *1,* 4; Phil. *4,* 7; *1* Tim. *4,* 13; Heb. *12,* 2 f; *1* Pet. *4,* 1; Jude 17; Apoc. *7,* 14; *22,* 14; *evil thoughts,* Matt. *9,* 4; *15,* 19; Rom. *1,* 21; Jas. *2,* 4.

ACTIONS OF THE WILL: *good intentions,* Matt. *6,* 1-23; *10,* 42; *1* Cor. *10,* 31; Col. *3,* 17; *effort to improve,* Matt. *5,* 48; *19,* 21; Rom. *12,* 2; *1* Cor. *15,* 58; *2* Cor. *7,* 1; *13,* 11; Eph. *4,* 15.23 ff; Phil. *1,* 9-11; *3,* 12; *1* Thess. *4,* 1-3.7; *5,* 23; *2* Tim. *3,* 16; Heb. *12,* 14; *1* Pet. *1,* 14-16; *evil intentions,* Matt. *6,* 2; John *6,* 26.

ACTIONS IN WORDS: *good words,* Matt. *12,* 34 f; Eph. *4,* 29; Col. *4,* 6; Titus *2,* 1.8; Jas. *1,* 19; *3,* 2.13; *evil speech, 1* Cor. *15,* 33; Eph. *4,* 29; *5,* 3 f; Jas. *1,* 26; *3,* 5-12; *1* Pet. *3,* 10; *blasphemy,* Matt. *12,* 31; *27,* 39-44; Mark *3,* 28 f; Luke *12,* 10; *22,* 65; *23,* 35.39-42; Acts *12,* 22; *13,* 45; *18,* 6; *1* Cor. *12,* 3; Col. *3,* 8; *1* Tim. *1,* 20; Apoc. *13,* 6; *16,* 11; *boasting,* Luke *18,* 10 ff; *cursing, sinful,* Rom. *12,* 14; *by nature grave, 1* Cor. *6,* 10; *some curses worse,* Matt. *15,* 4; *just curse not sinful,* Matt. *21,* 19; *25,* 41; *1* Cor. *12,* 3; *16,* 22; Gal. *1,* 8; *3,* 10; *false testimony,* Matt. *19,* 18; *26,* 60 ff; Acts *6,* 11; *insult,* Matt. *5,* 22; Acts *10,* 28; *1* Pet. *3,* 8 f; Jude 9; *slander,* Rom. *1,* 30; *2* Cor. *12,* 20; Eph. *4,* 31; *1* Tim. *3,* 11; *1* Pet. *2,* 1; *swearing,* Matt. *5,* 33-37; Jas. *5,* 12; *when rash,* Mark *6,* 28; *when unjust,* Acts. *23,* 12; *false ideas on oaths,* Matt. *23,* 16-22; *oaths, properly made, lawful, 2* Cor. *1,* 23; *1* Thess. *2,* 10; Phil. *1,* 8; Heb. *6,* 13-17; *7,* 20 ff; Apoc. *10,* 6; *tale bearing,* Matt. *12,* 36; *1* Tim. *5,* 13; Titus *3,* 2; Jas. *4,* 11; *1* Pet. *2,* 1; *4,* 15.

ACTIONS IN WORKS: *good works,* Matt. *3,* 10; Eph. *2,* 10; Col. *1,* 10; Titus *2,* 14; *3,* 8; Philem. 6; Heb. *10,* 24; *1* Pet. *2,* 12; *2* Pet. *1,* 10; *necessary,* Matt. *3,* 10; *7,* 19-27; *25,* 37-46; Luke *6,* 46-49; John *15,* 16; *2* Cor. *9,* 8; Col. *1,* 10; Jas. *2,* 14-22; *meritorious,* Matt. *6,* 4.18-21; *10,* 41 ff; *16,* 27; Mark *9,* 41; John *5,* 28 ff; *15,* 16; Rom. *2,* 6 ff; *10,* 13; *1* Cor. *3,* 13 ff; *9,* 24; *15,* 58; Gal. *6,* 9 f; Eph. *6,* 8; Col. *3,* 23; *1* Tim. *4,* 8; *2* Tim. *4,* 7 f; Heb. *6,* 10; *10,* 35; *13,* 16; Apoc. *14,* 13; *20,* 12; *22,* 12.

Index of Scripture Texts

ADAM: *first man, who was from God,* Luke *3,* 38; *1* Tim. *2,* 13 ff; Jude 14; *Adam and Christ,* Rom. *5,* 12-21; *1* Cor. *15,* 22.45-48.

ADULTERY: *forbidden,* Matt. *19,* 18; Rom. *13,* 9; Gal. *5,* 19; *nature of,* Matt. *5,* 28.32; *15,* 19; *19,* 9; Mark *7,* 21; *10,* 11; Luke *16,* 18; Rom. *7,* 3; *penalty of,* Rom. *1,* 26-27.32; *1* Cor. *6,* 9; Heb. *13,* 4; Jas. *4,* 4; *woman taken in,* John *8,* 3 ff.

AFFLICTION: *benefits of,* John *15,* 2; Acts *14,* 22; Rom. *5,* 3; *2* Cor. *4,* 8; *12,* 7; Phil. *1,* 12; Heb. *12,* 11; *1* Pet. *2,* 20; *comfort under,* Matt. *5,* 4 ff; *11,* 28; John *16,* 20.33; *2* Cor. *1,* 4; *7,* 6; Heb. *2,* 18; *1* Pet. *4,* 13; Apoc. *3,* 10; *joy in,* Matt. *5,* 12; Acts *5,* 41; Rom. *5,* 3; *2* Cor. *7,* 4; Col. *1,* 24.

ALMSGIVING: *duty of,* Luke *3,* 11; *11,* 41; *12,* 33 f; *14,* 13; Rom. *12,* 13.20; *1* Cor. *16,* 1 f; Gal. *6,* 9 f; *1* John *3,* 17-19; *examples of,* Mark *12,* 41-44; Luke *21,* 1-4; Acts *2,* 45; *4,* 32-37; *6,* 1; *9,* 36; *10,* 2; *2* Cor. *8,* 1-7; *intention in,* Matt. *6,* 3; *10,* 42; Mark *9,* 40; Rom. *12,* 8; *1* Cor. *13,* 3; *2* Cor. *9,* 7; *rewards of,* Matt. *10,* 42; *19,* 21; *25,* 40; Mark *9,* 40; Luke *6,* 38; Acts *20,* 35; *2* Cor. *9,* 6-15; Heb. *13,* 16.

ALTAR: *of the church,* Heb. *13,* 10; *in heaven,* Apoc. *6,* 9; *8,* 3.5; *9,* 13; *14,* 18; *16,* 7.

AMBITION: *inordinate, reproved,* Matt. *18,* 1; *20,* 26 f; *23,* 5-12; Luke *22,* 24-26.

ANGELS: *appearances of,* Matt. *1,* 20; *2,* 13.19 ff; *28,* 2-5; Mark *16,* 5; Luke *1,* 12-20.26-38; *2,* 9-15; *22,* 43; *24,* 4; John *20,* 12; Acts *1,* 10 f; *5,* 19; *8,* 26; *10,* 3-7; *12,* 7-11; *27,* 23; Apoc. *1,* 1; *duties of,* Matt. *4,* 6.11; *24,* 31; Mark *1,* 13; Luke *2,* 13 ff; *16,* 22; *22,* 43; Heb. *1,* 6.14; *1* Pet. *1,* 12; Apoc. *5,* 11 ff; *7,* 11 ff; *8,* 4; *existence of,* Matt. *18,* 10; *22,* 30; *25,* 31; *26,* 53; Mark *12,* 25; Luke *15,* 10; *20,* 36; *1* Tim. *5,* 21; Heb. *1,* 4-14; *12,* 22; Apoc. *1,* 4; *12,* 7; *ranks of,* Rom. *8,* 38; Eph. *1,* 21; Col. *1,* 16; *2,* 10; *1* Thess. *4,* 15; Heb. *9,* 5; *1* Pet. *3,* 22; Jude 9; *spirituality of,* Heb. *1,* 7-14; Apoc. *1,* 4; *3,* 1; *4,* 5; *5,* 6; *not to be worshipped,* Apoc. *19,* 10; *22,* 9.

ANGELS: *wicked,* Acts *19,* 12 ff; *2* Pet. *2,* 4; Jude 6; also see Devils, Evil Spirits, Unclean Spirits.

ANGER: *warning against,* Matt. *5,* 22; Rom. *12,* 19; Gal. *5,* 19-21; Eph. *4,* 26; *6,* 4; Col. *3,* 8; Titus *1,* 7; Jas. *1,* 19 f.

ANGELUS: *beautiful popular devotion,* from Luke, *1,* 35.38; John *1,* 14.

ANOINTED: *The (Christ):* Luke *4,* 18; Acts *4,* 27; *10,* 38.

ANOINTING: *of Christ,* Matt. *26,* 6-13; Mark *14,* 3-9; Luke *7,* 37 ff; John *12,* 3-8; *of the sick,* Jas. *5,* 14 f; *of the spirit,* *2* Cor. *1,* 21; *1* John *2,* 20.

ANTICHRIST: *adversary to Christ,* *2* Thess. *2,* 3-10; *1* John *2,* 18.22; *4,* 3; *2* John 7; Apoc. *19,* 20.

ANXIETY: *to cast it upon God,* Matt. *6,* 25; Luke *12,* 22-24; Phil. *4,* 6; *1* Pet. *5,* 7.

Index of Scripture Texts

APOSTASY: Matt. *13*, 21; Luke *8*, 13; 2 Thess. *2*, 3; *1* Tim. *4*, 1; 2 Tim. *3*, 1 ff; Heb. *3*, 12; *6*, 4 ff; *10*, 26; 2 Pet. *2*, 1.17; *1* John *2*.

APOSTLES: *calling of the*, Matt. *4*, 18-22; *9*, 9-13; Mark *1*, 16-20; *2*, 14; Luke *5*, 10-11.27-28; John *1*, 35-51; *Council of the*, Acts *15*, 6-29; *duties of the*, Matt. *5*, 13-16; *28*, 18-20; Mark *16*, 15-18; *instructions to the*, Matt. *28*, 19 f; Mark *16*, 15; Acts *1*, 8; *ordeals (suffering) of the*, Matt. *10*, 16-26; Mark *13*, 9-13; Luke *12*, 4-12; *21*, 12-17; *powers of the*, Mark *16*, 17 f; Luke *22*, 19; John *20*, 22 f; *1* Cor. *11*, 24 f; *reward of the*, Matt. *10*, 40-42; Luke *10*, 16; John *13*, 20; *selection of the twelve*, Matt. *10*, 1-4; Mark *3*, 14-19; Luke *6*, 13-16; *sending out of the*, Matt. *10*, 5-15; *28*, 16-20; Mark *6*, 7-13; *16*, 15 f; Luke *9*, 1-6.

APOSTLES: Names of the: *Andrew*, brother of Simon Peter, Matt. *4*, 18; *10*, 2; Mark *1*, 16.29; *3*, 18; *13*, 3; Luke *6*, 14; John *6*, 8; Acts *1*, 13;

Bartholomew, Matt. *10*, 3; Mark *3*, 18; Luke *6*, 14; Acts *1*, 13;

James the Greater, or the Elder, son of Zebedee, brother of John, Matt. *4*, 21; *10*, 3; Mark *1*, 19.29; *3*, 17; *5*, 37; *9*, 1; *14*, 33; Luke *6*, 14; *9*, 28; Acts *1*, 13; *12*, 2;

James the Less, or the Younger, son of Alpheus, Matt. *10*, 3; *13*, 55; Mark *3*, 18; *6*, 3; Luke *6*, 15; Acts *1*, 13; *12*, 17; *15*, 13-29; *21*, 18; *1* Cor. *15*, 7; Gal. *1*, 19; *2*, 9.12; Jas. *1*, 1;

John, also Evangelist, brother of James the Greater, Matt. *4*, 21; *10*, 3; Mark *1*, 19.29; *3*, 17; *5*, 37; *9*, 1.37; *14*, 33; Luke *6*, 14; *9*, 28; John *19*, 26 f; *20*, 1-10; *21*, 20-24; Acts *1*, 13; *3*, 4; *8*, 14; Gal. *2*, 9; *1* John *1*, 1-5; *writes the Apocalypse*, Apoc. *1*, 1.4.9.19; *Epistles*, *1*, *2*, and *3* John;

Judas Iscariot, which see.

Jude (Thaddaeus), brother of James the Less, Matt. *10*, 3; *13*, 55; Mark *3*, 18; *6*, 3; Luke *6*, 16; John *14*, 22; Acts *1*, 13; *15*, 22; Jude 1;

Matthew, also Evangelist, Levi, son of Alpheus, Matt. *9*, 9; *10*, 3; Mark *2*, 14; *3*, 18; Luke *5*, 27-29; *6*, 15; Acts *1*, 13;

Matthias, Acts *1*, 23-26;

Paul, as Saul persecutes the young Church, Acts *7*, 58; *8*, 3; *9*, 1 f; *22*, 4; *26*, 9-11; *1* Cor. *15*, 9; Gal. *1*, 13; Phil. *3*, 6; *1* Tim. *1*, 13; *call to the Apostolate*, Acts *9*, 3-19; *22*, 6-21; *26*, 12-18; *missionary travels*, Acts *13*; *14*; *15*; *16*; *17*; *18*; *19*; *20*; *21*, 1-16; *returns to Jerusalem and is seized*, Acts *21*, 17-40; *his defense before the people, the Sanhedrin and the Roman officials*, Acts *22*; *23*; *24*; *25*; *26*; *appeals to Cæsar and is sent to Rome, his voyage and shipwreck*, Acts *25*; *27*; *28*, 1-10; *in Rome*, Acts *28*, 11-31; *his love for the churches*, Rom. *1*, 8.15; *1* Cor. *1*, 4; *4*, 14; 2 Cor. *1*; *2*; *6*; *7*; Phil. *1*; Col. *1*; *1* Thess. *1*; *3*; 2 Thess. *1*; *his sufferings*, 1 Cor. *4*, 9; 2 Cor. *4*, 8 f; *11*, 23-27; *12*, 7; Phil. *1*, 12; 2 Tim. *2*, 10; *his Apostolate divine*, 2 Cor. *11*; *12*; Gal. *1*, 11-24; *2*; *his letters mentioned by Peter*, 2 Pet. *3*, 15 f;

Peter, Simon, called Peter, (the Rock), his call, Matt. *4*, 18; *10*, 2;

Index of Scripture Texts

Mark *1*, 16; *3*, 16; Luke *5*, 4-10; *6*, 14; John *1*, 40-42; *as one of Christ's disciples,* Matt. *14*, 28 f; *16*, 16 f; Mark *1*, 16.29 f. 36; *5*, 37; *8*, 29; *9*, 1.4; *14*, 33; Luke *9*, 20.28; John *13*, 36; *his denial of Christ and his repentance,* Matt. *26*, 69-75; Mark *14*, 66-72; Luke *22*, 55-63; John *18*, 25 ff; *his death foretold by Christ,* John *21*, 18; *2* Pet. *1*, 14 f; *his Apostolate,* Acts *1*, 15; *2*, 14; *3*, 12; *4; 5; 8*, 19-23; *9*, 32.40; *10; 12; 15*, 7; Gal. *1*, 18; *2*, 7-14; *1* and *2* Pet.; *his primacy,* Matt. *16*, 18 f; Luke *22*, 31 f; John *21*, 15 ff; Acts *1*, 13.15 ff; *15*, 7-11;

Philip, Matt. *10*, 3; Mark *3*, 18; Luke *6*, 14; John *1*, 43-48; *12*, 21 f; *14*, 8 f; Acts *1*, 13; *6*, 5; *8*, 5.29-40; *21*, 8 f;

Simon, the Cananean, called Zelotes (the zealous), Matt. *10*, 4; Mark *3*, 18; Luke *6*, 15; Acts *1*, 13;

Thomas, called the Twin, Matt. *10*, 3; Mark *3*, 18; Luke *6*, 15; John *11*, 16; *14*, 5; *20*, 24-29; *21*, 2; Acts *1*, 13.

ARMOR OF GOD: Rom. *13*, 12; *2* Cor. *6*, 7; Eph. *6*, 10-17; *1* Thess. *5*, 8.

ASSEMBLY FOR PRAYER: Matt. *18*, 20; John *20*, 19; Acts *1*, 13 f; *2*, 1; *3*, 1; *16*, 13; *20*, 7.

ATHENS: city in Greece—*Paul preaches there,* Acts *17*, 16-34; *1* Thess. *3*, 1.

ATONEMENT: *made by Christ,* Rom. *3*, 24; *5*, 6; *2* Cor. *5*, 18; Gal. *1*, 4; *3*, 13; Titus *2*, 13-14; Heb. *9*, 28; *1* Pet. *1*, 19; *2*, 24; *3*, 18; *1* John *2*, 2; Apoc. *1*, 5; *13*, 8.

AUTHORITIES: *duties toward,* Rom. *13*, 1-7; *1* Thess. *5*, 12 ff; Titus *3*, 1; Heb. *13*, 17.

AVARICE: Luke *12*, 15; *1* Cor. *5*, 10; *6*, 9 f; Col. *3*, 5; *1* Tim. *6*, 10; Heb. *13*, 5; *2* Pet. *2*, 3; see also Covetousness.

AVE MARIA: *prayer—first part of angelical salutation,* Luke *1*, 28.

BAPTISM: *administration of,* Acts *2*, 41; *8*, 12 f.38; *9*, 18; *10*, 48; *16*, 15.33; *18*, 8; *19*, 4 f; *effects and significance of,* Mark *16*, 16; Acts *2*, 38; *22*, 16; Rom. *6*, 3-23; *1* Cor. *12*, 13-27; Gal. *3*, 27-29; Eph. *5*, 26; Col. *2*, 12; Titus *3*, 4-7; Heb. *10*, 22; *1* Pet. *3*, 21; *institution of,* see under Sacraments; *necessity of,* John *3*, 5; *of infants,* Luke *18*, 16; John *3*, 5; *of Christ,* Matt. *3*, 13-17; Mark *1*, 9-11; Luke *3*, 21 f; John *1*, 31-34; *by John,* Matt. *3*, 6; Mark *1*, 4-8; Luke *3*, 1-18; John *1*, 26.

BARABBAS: a robber, *preferred by the Jews over Jesus,* Matt. *27*, 16.20 f.26; Mark *15*, 7.11.15; Luke *23*, 17-21.25; John *18*, 39 f.

BARNABAS: early disciple of the Apostles, Acts *4*, 36; *9*, 27; *11*, 22.30; *12*, 25; *13; 14; 15; 1* Cor. *9*, 6; Gal. *2*, 1.9.13; Col. *4*, 10.

BEATITUDES: Matt. *5*, 1-12; Luke *6*, 20-23.

BEELZEBUB (BEELZEBUL): prince of the devils, Matt. *12*, 24; Mark *3*, 22; Luke *11*, 15.

BENEVOLENCE: Acts *20*, 35; *1* Cor. *16*, 1 ff; *2* Cor. *8*, 14; *1* Tim. *6*, 17-19; Heb. *13*, 16; Jas. *2*, 15 ff; *1* John *3*, 17 ff.

BETHLEHEM: birthplace of Christ, south of Jerusalem, Matt. *2*, 1.5 f.16; Luke *2*, 4; John *7*, 42.

Index of Scripture Texts

BISHOPS: *duties of,* Acts *20,* 28; *1* Cor. *4,* 1 ff; *2* Cor. *6,* 4; Titus *1, 5; 1* Pet. *5,* 1-4; *ordination of, 1* Tim. *4,* 14; *2* Tim. *1,* 6; *qualities of, 1* Tim. *3,* 1-7; Titus, *1,* 5-9.

BLASPHEMY: see under Actions in words.

BLASPHEMY: *against the Holy Spirit,* Matt. *12,* 31 f; Mark *3,* 29; Luke *12,* 10.

BLINDNESS: *spiritual,* Matt. *6,* 23; *15,* 14; *23,* 16; Luke *6,* 39; *11,* 34 ff; John *1,* 5; *3,* 19; *9,* 39-41; *1* Cor. *2,* 14; Eph. *4,* 18; *2* Pet. *1,* 9; Apoc. *3,* 17 f.

BLOOD OF CHRIST: Acts *20,* 28; Rom. *3,* 25; *5,* 9; *1* Cor. *10,* 16; *11,* 27; Col. *1,* 20; Eph. *2,* 13; Heb. *9,* 12.14; *10,* 19.29; *12,* 24; *13,* 12.20; *1* Pet. *1,* 2.18 f; *1* John *1,* 7; *5,* 6.8; Apoc. *1,* 5; *5,* 9; *7,* 14; *12,* 11.

BOASTING: *condemned,* Rom. *1,* 30; *3,* 27; *11,* 18; *2* Cor. *10;* Eph. *2,* 9; Jas. *3,* 5; *4,* 16.

BODY, HUMAN: *dignity of,* Matt. *6,* 25-32; Luke *12,* 22-31; *1* Cor. *3,* 16 f; *6,* 19 f; Eph. *5,* 29 f; *functions of,* Rom. *6,* 12 ff; *12,* 1; *1* Cor. *3,* 17; *6,* 13-20; *9,* 27; *1* Thess. *5,* 23; *glorification of,* Rom. *8,* 29 ff; *1* Cor. *15,* 42-49; Phil. *3,* 21.

BODY, MYSTICAL: *in Christ,* Rom. *12,* 4 f; *1* Cor. *12,* 12 ff; Eph. *1,* 23; *4,* 12 f; *5,* 23; Col. *1,* 18; *2,* 19; *3,* 15.

BONDAGE: *spiritual,* John *8,* 34; Acts *8,* 23; Rom. *6,* 15 f; *7,* 23; *8,* 2; Gal. *2,* 4; *4,* 3; *2* Tim. *2,* 26; Heb. *2,* 14; *2* Pet. *2,* 19; *deliverance from,* Luke *4,* 18; John *8,* 36; Rom. *8,* 2; Gal. *3,* 13.

BOOK: *of life,* Phil. *4,* 3; Apoc. *3,* 5; *13,* 8; *17,* 8; *20,* 12; *21,* 27; *22,* 19.

BREAD: *breaking of,* Matt. *26,* 26; Mark *14,* 22; Luke *22,* 19; *24,* 30; Acts *2,* 42.46; *20,* 7.11; *1* Cor. *10,* 16; *11,* 24.

BREAD: *of God,* John *6,* 32-34.

BREAD: *of life,* John *6,* 35. See Eucharist.

BREATH: *of life, given by God,* Acts *17,* 25.

BRETHREN: *duty of, towards one another,* Matt. *5,* 22; *18,* 15.21 f; *25,* 40; John *13,* 34 f; *15,* 12 ff; Rom. *12,* 10; *1* Cor. *6; 8; 13;* Gal. *6,* 1 f; *1* Thess. *4,* 9; Heb. *13,* 1; *1* Pet. *1,* 22; *3,* 8; *2* Pet. *1,* 7; *1* John *2,* 9 f; *3,* 17 f.

BRETHREN OF JESUS: Matt. *12,* 46-50; *13,* 55; Mark *3,* 31-35; *6,* 3; Luke *8,* 19-21; John *2,* 12; *7,* 3-5; Acts *1,* 14; *1* Cor. *9,* 5; Gal. *1,* 19.

BROTHERLY CORRECTION: Matt. *18,* 15-17; Luke *17,* 3; *1* Tim. *5,* 20; *2* Tim. *2,* 25; Jas. *5,* 19 f.

BROTHERLY LOVE: John *13,* 33-35; *15,* 9-14; Rom. *13,* 8-10; Gal. *5,* 13-15; Eph. *5,* 1 f; *1* John *3,* 14-16; *4,* 7-21.

CALL OF GOD: *to penance and salvation,* Matt. *3,* 11; John *7,* 37 ff; *12,* 44; Rom. *8,* 28 ff; *9; 10; 11; 2* Cor. *5,* 20; Apoc. *2,* 5; *3,* 3.19; *22,* 17; *penalty of rejecting,* Matt. *22,* 3 ff; Luke *14,* 16-24; John *12,* 48; Acts *13,* 46; *18,* 6; *2* Thess. *2,* 10 f; Heb. *12;* Apoc. *2,* 5; *call to spiritual vocation,* Rom. *11,* 29; *1* Cor. *1,* 26; Eph. *1,* 18; *4,* 1; Phil. *3,* 14;

Index of Scripture Texts

2 Thess. *1*, 11; 2 Tim. *1*, 9; Heb. *3*, 1; *1* Pet. *2*, 9; 2 Pet. *1*, 10; Apoc. *19*, 9. See God's Elect.

CANA: village in Galilee, north of Nazareth, John 2, 1-11; *4*, 46.

CANTICLES: *Benedictus,* Luke *1,* 68-79; *Magnificat,* Luke *1,* 46-55; *Nunc dimittis,* Luke 2, 29-32.

CAPHARNAUM: city by the sea of Galilee, Matt. *4*, 13-17; *8*, 5; *9*, 1; *11*, 23; *17*, 23; Mark *1*, 21; *2*, 1; Luke *10*, 15; John 2, 12; *4*, 46; *6*, 17.

CARE: *worldly, censured:* see Temporal things.

CHARITY: *the greatest of gifts,* Matt. 22, 34-40; *1* Cor. *13;* Phil. *1*, 9; 2 John 4-6; see Almsgiving.

CHASTITY: Matt. *5*, 8.27-32; Rom. *12*, 1; *1* Cor. *3*, 16 f; *6*, 15-20; 7, 7 f. 25 f; Gal. *5*, 16 f. 23 f; *1* Thess. *4*, 3-8; Titus *1*, 15; *1* Pet. *2*, 11; Apoc. *14*, 1-5.

CHILDREN: *blessed by Christ,* Matt. *19*, 13; Mark *10*, 13; Luke *18*, 15; *duties of,* Matt. *15*, 4; *19,* 19; Mark 7, 10; *10*, 19; Luke *18*, 20; Eph. *6*, 1-3; Col. *3*, 20.

CHILDREN OF GOD: Eph. *5*, 1; Heb. *12*, 5; *1* Pet. *1*, 14.

CHILDREN OF LIGHT: Luke *16*, 8; John *12*, 36; Eph. *5*, 8; *1* Thess. *5*, 5.

CHRIST: *divinity of:* (1) *Divine nature:* John *1*, 1.14; *20*, 28; Rom. *9*, 5; Phil. *2*, 6; Col. *2*, 9; Titus 2, 13; Heb. *1*, 8; *1* John *5,* 19 f; *divine origin,* John *3*, 13. 31; *6*, 38. 51; *8*, 23. 42; *16*, 5 ff; *equality with the Father,* Matt. *10*, 32; *11*, 25-27; *28*, 18 ff; Mark *8*, 38; Luke *10*, 22; John *5*, 17 ff; *6*, 46; 7, 28 f; *8*, 18 f; *9*, 35 ff; *10*, 15; *14*, 10 ff; Phil. *2*, 6; *sonship,* Matt. *16*, 16 ff; *25*, 34; *26*, 29; Luke 2, 48 ff; *24*, 49; John *3*, 16; Rom. *8*, 32; Heb. *3*, 3-6; *unity with Father,* John *10*, 30.38; *14*, 9.11. (2) *Divine attributes: eternity and immutability,* John *8*, 58; *17*, 5; Heb. *13*, 8; *omnipotence,* Matt. 7, 21 ff; *9*, 5 ff; *10*, 1.8; *11*, 27; *13*, 41 ff; *25*, 31 ff; Luke *10*, 17; John *1*, 3.10; *5*, 21; *6*, 52.55; *20*, 23; *1* Cor. *8*, 6; Col. *1*, 13 ff; Heb. *1; 2; 3; omniscience,* Col. 2, 3.

(3) *Divine pre-eminence over angels:* Matt. *4*, 11; *13*, 41; *16*, 27; *24*, 31; *26*, 53; Col. *1*, 16; 2 Thess. *1*, 7; Heb. *1*, 4-14; *1* Pet. *1*, 12; *3*, 22; *over men,* Matt. *12*, 6.41-42; *22*, 42 ff; *23*, 34; Mark *12*, 37; Luke *21*, 15; John *4*, 12; *8*, 51 ff; Acts 2, 32-36; Heb. *3*, 1-6.

(4) *Divine honors: faith,* John *14*, 1; *hope, 14,* 13 ff; *charity,* John *15*, 9; Rom. *8*, 35 ff; *1* Cor. *16*, 22; *adoration,* Matt. 2, 11; *28*, 17; John *5*, 23; Phil. *2*, 10; Heb. *1*, 6.

(5) *Divine names: Alpha and Omega,* Apoc. *1*, 8; *21*, 6; *22*, 13; *Emmanuel,* Matt. *1*, 23; *God,* John *1*, 1 f; *20*, 28; Rom. *9*, 5; Heb. *1*, 8; *Lord,* Acts 2, 36; 7, 59 f; *10*, 36; Rom. *10*, 9; *1* Cor. *8*, 6; *16*, 22 f; Phil. 2, 11; Jas. *1*, 1; *1* Pet. *1*, 3; Jude 4; 17; *Word,* John *1*, 1.14; *1* John *1*, 1; *5*, 7; Apoc. *19*, 13.

(6) *Divine works:* John *5*, 17; see also Miracles and Prophecies.

(7) *Testimonials to Christ's divinity: by the Father,* Matt. *3*, 17; Luke *9*, 35; *by James,* Jas. *1*, 1; *by John,* John *20*, 31; *1* John *1*, 1-3; *5*, 20; *by Paul,* Rom. *1*, 1; *1* Cor. *1*, 1.9; 2 Cor. *4*, 5; Gal. *1*, 1.10; 2, 20; Phil. *1*, 21-23; *1* Tim. *1*, 12; *by Peter,* Matt. *16*, 16; Acts *3*, 13; *1* Pet. *3*, 22; 2 Pet. *1*, 1; *by Stephen,* Acts 7, 55 f; *by Thomas,* John *20*, 28.

Index of Scripture Texts

(8) *Credentials:* John 5, 31-40.
(9) *Supreme dignity:* Col. 1, 15-20.
See also Anointed.
CHRIST: *human nature of:* (1) *True humanity:* John 1, 14; 1 Tim. 2, 5; 1 John 1, 1; *union of soul and body,* Matt. 26, 38; Mark 14, 34; Luke 24, 39; John 1, 14; 10, 15; *human acts,* Heb. 5, 7; 10, 9.
(2) *Human soul of Christ:* Matt. 26, 38; 27, 50; Luke 23, 46; John 10, 18; 11, 33; *beatific knowledge,* John 1, 14 ff; 3, 11.34; 8, 55; Col. 2, 3; *infused knowledge,* John 1, 14; Col. 2, 3; Heb. 10, 4-8; *acquired knowledge,* Matt. 8, 10; 15, 34; Mark 5, 30; 8, 5; Luke 2, 52; John 11, 34; Heb. 5, 8; *freedom from ignorance,* Luke 1, 79; John 1, 14; *some knowledge not communicated,* Mark 13, 32; Acts 1, 7; *human will,* Matt. 26, 39; 27, 34; Mark 14, 36; Luke 22, 42; John 5, 30; 6, 38; 7, 1; *free will,* Matt. 27, 34; John 7, 1; 10, 17 f; Phil. 2, 8 f; Heb. 12, 2; *human emotions, anger,* Mark 3, 5; John 2, 15-17; *desire and joy,* Luke 22, 15; John 11, 15; *fear and weariness,* Mark 14, 33; *gentleness,* Matt. 11, 29; 2 Cor. 10, 1; *love,* Mark 10, 21; John 11, 36; *sorrow,* Matt. 26, 37 f; Luke 19, 41; John 11, 35; *wonder,* Matt. 8, 10; *human will conformed to the divine will,* Mark 14, 36; Luke 22, 42; John 10, 18; 12, 49; 14, 31; Phil. 2, 8; Heb. 10, 7; *human holiness of Christ,* John 1, 14.16.32 f; Acts 10, 38; *freedom from sin,* John 8, 46; Heb. 7, 26; 1 Pet. 2, 22; 1 John 3, 5; *fullness of grace and virtue,* Matt. 12, 18 ff; Luke 4, 1; John 1, 14.16.32 ff; Eph. 3, 19; *merits for himself,* Luke 24, 26.46; John 17, 5; Eph. 4, 9; Phil. 2, 7-11; Heb. 2, 9; Apoc. 5; *merits for others,* Matt. 27, 51 ff; Mark 15, 37; Luke 23, 46 f; John 1, 16; 19, 30; Rom. 3, 24; 5, 1 ff; 7, 24 f; 1 Cor. 15, 3; 2 Cor. 5; Eph. 1, 3; 2, 5-10; 2 Tim. 1, 9; Titus 3, 7; *progressive manifestation of grace and virtue,* Luke 2, 52;
(3) *Human body of Christ: real,* Matt. 26, 12-67; 27, 58; Mark, 14, 8-22; 15, 43; Luke 22, 16; 24, 39; John 1, 14; 2, 21; Acts 2, 31; Rom. 1, 3; 9, 5; Gal. 4, 4; Heb. 5, 7; 10, 5; 1 John 4, 2; 2 John 7; *earthly,* Luke 24, 39; *descended from Adam and David,* Matt. 1, 1; 21, 9; Luke 3, 31.38; Gal. 3, 16; Heb. 2, 16; *subject to general infirmities,* Matt. 4, 2; 11, 19; Rom. 8, 3; Phil. 2, 7; Heb. 2, 18; 12, 3; *blood,* Acts 20, 28; Rom. 3, 25; 5, 9; 1 Cor. 10, 16; 11, 27; Col. 1, 20; Heb. 9, 12.14; 10, 19.29; 12, 24; 13, 12.20; 1 Pet. 1, 2.18 ff; 1 John 1, 7; 5, 6.8; Apoc. 1, 5; 5, 9; 7, 14; 12, 11.
CHRIST: *hypostatic union of human and divine in,* Matt. 16, 13-16; John 1, 14; Rom. 8, 32; 9, 5; Phil. 2, 6; Heb. 1, 2 f.
CHRIST: *relations of—to the Father: equal to the Father in His divinity,* see above under Christ, Divinity of; *natural sonship,* Luke 3, 22; John 1, 14.18; Rom. 8, 32; Heb. 5, 7 f; *prayer of Christ to the Father,* Matt. 26, 39; Luke 6, 12; John 11, 41 f; 17; *predestination of,* Rom. 1, 4; Eph. 1, 5; *servant of God in his human nature,* Matt. 12, 18; John 5, 30; 8, 29; 1 Cor. 15, 28; Phil. 2, 8.

Index of Scripture Texts

CHRIST: *relations of—to creatures: as God, Lord of all,* see above under Christ, Divinity of; *adoration of Christ as God-man,* Matt. 2, 11; 28, 17; John 5, 23; 9, 38; Phil. 2, 10; Heb. 1, 6; *as man, head of Angels,* Matt. 4, 11; Col. 2, 10; *head of Church,* Rom. 12, 4 f; Eph. 1, 22 f; 4, 15 f; 5, 23; Col. 1, 18; *head of mankind,* 1 Cor. 11, 3; 1 Tim. 4, 10; 1 John 2, 2; *head of the rulers of the Church,* 2 Cor. 2, 10; 5, 20; *the God-man, mediator,* 1 Tim. 2, 5; Heb. 8, 6; 9, 15; 12, 24; *High-priest,* Heb. 4, 14 f; 5, 4 ff; 9, 11.23 ff; *His sacrifice on the cross,* John 1, 29.36; 3, 14 ff; 10, 11 ff; 15, 13; Rom. 3, 25; 5, 19; 1 Cor. 5, 7; 15, 3; Eph. 5, 2; Phil. 2, 8 ff; Heb. 10, 5 ff; 1 Pet. 1, 18 f; Apoc. 5, 9; 7, 14; *King,* Matt. 27, 11; Mark 15, 2; Luke 1, 32 f; 23, 1-3; John 18, 33-37; *His kingdom not earthly,* Matt. 20, 20 ff; John 18, 36; *His kingdom heavenly,* Matt. 25, 34.40; 28, 18; John 18, 37; 1 Cor. 15, 23 ff; Heb. 1, 2; Apoc. 19, 16; *Lawgiver,* Matt. 5; 6; 7; John 14, 15.23 f; 15, 10; 1 Cor. 9, 21; Gal. 6, 2; *Judge,* Matt. 16, 27; 24, 30 f; 25, 21.31 ff; Mark 13, 26 f; Luke 17, 24; 21, 27; John 5, 22; Acts 1, 11; 10, 42; 17, 31; Rom. 2, 16; 2 Cor. 5, 10; 1 Thess. 4, 16; 2 Thess. 1, 7 ff; 2 Tim. 4, 1; Heb. 9, 28; 1 Pet. 4, 5; Jude 14 f; Apoc. 1, 17 ff; 20, 12; *Redeemer,* Matt. 20, 28; Luke 2, 38; 21, 28; Rom. 3, 24; Gal. 3, 13; Eph. 1, 7; 4, 30; Phil. 2, 7; Col. 1, 14; 1 Tim. 2, 6; Titus 2, 14; Heb. 7, 27; 9, 12-15.25-28; 10, 11-14; 1 Peter 1, 18 f; Apoc. 5, 9; *Teacher,* Matt. 23, 8-10; *others teach in His name,* Matt. 10, 27; 28, 19; John 1, 7 ff; *He teaches with perfect and infallible knowledge,* John 1, 9; 3, 11.31; 8, 12; 14, 6; Col. 2, 2 f; *with supreme authority,* Matt. 17, 5; John 17, 4; *all must accept His teaching,* Matt. 28, 18 ff; Mark 16, 15 ff; *His mode of teaching is perfect,* Mark 1, 22; Luke 24, 32; John 7, 46; 10, 25.38; 13, 15; 14, 14; *Prophet,* Matt. 13, 57; John 4, 44; see also Prophecies; *model to youth,* Luke 2, 46 f.50-52.

CHRISTIAN: *dignity of,* 1 Cor. 3, 16 f; 6, 19 f; 2 Cor. 6, 16; Gal. 3, 28 f; Eph. 2, 19-22; 1 Pet. 2, 9 ff; Apoc. 1, 6; *duties of,* Rom. 6; 12; 13, 11-14; 2 Cor. 6, 14-18; Col. 3, 1-17; 2 Thess. 3, 13-15; *joy of,* John 15, 11; 16, 20-22; 17, 13; Rom. 12, 15; 14, 17; 2 Cor. 8, 2; Gal. 5, 22; Phil. 4, 4; Jas. 1, 2; 1 John 1, 4; *name,* Acts 11, 26; 26, 28.

CHRISTIAN LIVING: 1 Pet. 1, 13—2, 10.

CHURCH: *attributes,* John 10, 16; Eph. 4, 5; 5, 25-27; 1 Tim. 3, 15; *duties of,* Matt. 18, 15-17; Acts 12, 5; Rom. 15, 30; Col. 4, 3; Heb. 13, 17; *expulsion from,* 1 Cor. 5, 2-5; 1 Tim. 1, 20; *foundation of,* John 10, 16; 21, 15-17; *infallibility of,* John 14, 16-18; 16, 13; 1 Tim. 3, 15 f; 1 John 2, 26 f; *members of, and their duties,* Eph. 2; 4; 5, 1-21; *permanence of,* Matt. 16, 18; 28, 20, Luke 1, 33; 22, 32; *unity of,* John 10, 16; Rom. 12, 5; 1 Cor. 10, 17; 12, 13; Gal. 3, 28; Eph. 1, 10; 2, 19-22; 4, 3 ff; 5, 23-30.

CHURCH: *persecution of early,* Acts 8, 1 ff; 12, 1 f.

CHURCH: *loved by Christ,* Eph. 5, 25.29 f.

CHURCH OF GOD: Acts 20, 28; 1 Cor. 1, 2; 10, 32; 11, 22; 15, 9; Gal. 1, 13.

Index of Scripture Texts

CIRCUMCISION: *of Christ*, Luke 2, 21; *of the flesh*, Acts 15, 5-20; Rom. 2, 25-29; 4, 10-12; Gal. 5, 1-11; 6, 12-16; Col. 2, 11; 3, 11; *of the spirit*, Rom. 2, 26-29; Phil. 3, 3; Col. 2, 11.
CIVIL AUTHORITIES: *duties of citizens toward*, Matt. 22, 15-21; Mark 12, 13-17; Luke 20, 20-26; Rom. 13, 6 f.
COMMANDMENTS: *fulfilled by Christ*, Matt. 5, 17-20.
COMMON OWNERSHIP OF PROPERTY: Acts 2, 44 ff; 4, 32-37.
COMMUNION OF THE BLOOD AND BODY OF CHRIST: see Eucharist.
COMMUNION OF SAINTS: Heb. 12, 22 f.
COMPANY: *evil—to be avoided*, 1 Cor. 5, 9 f; 15, 33; Eph. 5, 6-8.
COMPASSION OF JESUS: Matt. 9, 13.22.36; 11, 28; 12, 7; Mark 1, 41; 6, 34; 8, 1 ff; Luke 7, 13.47 ff; 11, 33; 19, 5.9 ff; 23, 42 f; John 8, 11; 10, 11; Titus 3, 5.
COMPASSION: *with one's neighbors*, Luke 6, 27-38; Col. 3, 13 f; *examples of*, Matt. 18, 23 ff; Luke 10, 30-37; 15; Heb. 5, 2; Jude 23.
CONCORD: *among Christians*, Acts 2, 42-47; 4, 32; Rom. 12, 16; 15, 5 ff; 1 Cor. 1, 10; Eph. 4, 2; Phil. 2, 2 ff; 1 Pet. 3, 8.
CONCUPISCENCE: Matt. 5, 28; Rom. 13, 14; Jas. 1, 14; 1 John 2, 16.
CONFESSION OF SINS: see under Sins.
CONFIDENCE IN GOD: Matt. 10, 29-33; Luke 12, 6 f.
CONSCIENCE: *bad*, Matt. 27, 3 ff; *good, and effects of it*, 2 Cor. 1, 12; 1 Tim. 1, 19; 2 Tim. 4, 7 ff; 1 Pet. 3, 15 ff; 1 John 3, 19 ff; *purity of*, 2 Tim. 2, 20-22.
CONSOLATION: *under affliction*, Matt. 5, 5; Rom. 5, 3; 2 Cor. 1, 3 f; 7, 6; 1 Thess. 3, 7.
CONTEMPT OF THE WORLD: Gal. 6, 14; Jas. 4, 4; 1 John 2, 15-17.
CONTENTMENT: Luke 3, 14; Phil. 4, 11; 1 Tim. 6, 8; Heb. 13, 5.
CONVERSION: *of the Gentiles*, Acts 10; 15, 3; Rom. 10; 11; 1 Cor. 1; Eph. 1, 1; 3; 1 Thess. 1; *of the Jews*, Acts 2, 41; 4, 32; 6, 7; *of sinners*, Luke 15, 11-32; 23, 40-43; Jas. 5, 19 f; 1 Pet. 2, 25.
CORINTH: *city in Greece, church founded by Paul*, Acts 18, 1-18; 19, 1; 1 Cor. 1, 2 ff; 2 Cor. 1, 2 ff; 2 Tim. 4, 20.
CORRECTION: See Brotherly Correction.
COURAGE: *models of*, Heb. 10, 32-34; *necessity*, Matt. 10, 28; 1 Cor. 15, 58; 16, 13; Eph. 6, 10; 2 Tim. 2, 1; Jas. 5, 8; *value*, Rom. 12, 21; Jas. 4, 7; 1 Pet. 5, 8; 1 John 2, 14; *human respect for*, Matt. 10, 28.33; Mark 8, 38; Luke 9, 26; 12, 4.9; Acts 4, 19 ff; *in persecution*, Matt. 5, 10-12; 1 Cor. 4, 12; 2 Cor 4, 9; 12, 10; 2 Tim. 3, 12.
COUNCIL OF THE APOSTLES: *at Jerusalem*, Acts 15, 6-29.
COUNSELS: *evangelical*, Matt. 19, 21; Mark 10, 21; Luke 18, 22; 1 Cor. 7, 25 f.32-35.
COURTESY: Col. 4, 6.
COVENANT: *the New*, Rom. 11, 26 f; 2 Cor. 3, 6; Heb. 8, 8-13.
COVETOUSNESS: Matt. 5, 28; Luke 12, 15; Rom. 6, 12 f; 7, 7-25; 13, 9.14; 1 Cor. 5, 10; 10, 6; Gal. 5, 16 f; Eph. 2, 3; 5, 3.5; Col. 3, 5; 2 Tim.

Index of Scripture Texts

2, 22; Titus 2, 12; Jas. *4*, 1-3; *1* Pet. 2, 11; *4*, 2; 2 Pet. 2, 18; *1* John 2, 16. See also Avarice.
COWARDICE: *punished,* Apoc. 21, 8.
CROSS: *enemies of the,* Phil. *3*, 18 f; *power and glory,* 1 Cor. *1*, 17 f; Gal. 6, 14; *a stumbling-block,* 1 Cor. *1*, 23; Gal. *5*, 11.
CROSS: *taking upon oneself.* Matt. *10*, 38; *16*, 24; Mark *8*, 34; Luke *9*, 23; *14*, 27; Gal. 2, 19; Heb. 2, 14.
CROWN: *of glory,* 1 Pet. *5*, 4;—*of life;* Jas. *1*, 12; Apoc. 2, 10;—*of justice,* 2 Tim. *4*, 8; *sign of victory,* 1 Cor. *9*, 25; 2 Tim. 2, 5; *4*, 8; Jas. *1*, 12; *1* Pet. *5*, 4; Apoc. 2, 10.
CURSING: *condemned,* Jas. *3*, 10. See Actions in Words.

DAMNATION: *eternal,* Matt. *3*, 12; *5*, 29 f; *18*, 8 f; *25*, 41.46; Mark *9*, 42-47; Luke *3*, 17; *16*, 23 ff; Heb. *10*, 27; 2 Pet. 2, 4; Jude 6; Apoc. *19*, 20; *20*, 9 ff; *21*, 8.
DANIEL: prophet during Babylonian captivity, Matt. *24*, 15; Mark *13*, 14; Luke *21*, 20.
DARKNESS: *spiritual,* Eph. *4*, 18; *5*, 8; *1* Pet. 2, 9; *1* John *1*, 6; 2, 9; *works of,* Rom. *13*, 12; Eph. *5*, 11.
DAVID: king of the Israelites, *progenitor of Christ,* Matt. *1*, 1.6 ff; *9*, 27; *15*, 22; *21*, 9; *22*, 41-45; Luke *1*, 32; John *7*, 42; Acts 2, 25.29; *13*, 22; *15*, 16; Rom. *1*, 3; *4*, 6; *11*, 9; 2 Tim. 2, 8; Heb. *4*, 7; *11*, 32; Apoc. *5*, 5; *22*, 16.
DEATH: *consequence of sin,* Rom. *5*, 12.17; *6*, 23; *1* Cor. *15*, 21; *time uncertain,* Matt. *24*, 42-45; Mark *13*, 35-37; Luke *12*, 20.39 f; *21*, 34-36; *1* Thess. *5*, 1-6; 2 Pet. *3*, 10; Apoc. *3*, 3; *16*, 15; *victory over,* Rom. *6*, 8-11; *1* Cor. *15*, 25 ff. 54 ff; 2 Tim. *1*, 10.
DEATH: *eternal,* Matt. *25*, 46; Mark *9*, 44 f; John *5*, 29; Rom. *1*, 32; 2, 9; *6*, 23; *9*, 22; 2 Thess. *1*, 8 f.
DEATH: *spiritual,* Matt. *4*, 16; *8*, 22; Luke *1*, 79; John *6*, 54; Rom. *5*, 15; *6*, 13 f; *8*, 6; Eph. 2, 1; *4*, 18; Col. 2, 13; *1* Tim. *5*, 6; Heb. *6*, 1; *9*, 14; *1* John *3*, 14; Apoc. *3*, 1.
DEFILEMENT: Matt. *15*, 10-20; Mark *7*, 14-23.
DEMONS: see Evil Spirits, Satan.
DENIAL OF CHRIST: *consequences of,* Matt. *10*, 33; Mark *8*, 38; Luke *9*, 26; 2 Tim. 2, 12.
DESIRES OF THE FLESH: *battle against the,* Gal. *5*, 17; Col. *3*, 5.
DEVIL, DEVILS: see also Evil Spirits, Prince of this world, Satan.
DEVIL: *adversary of God and Man,* Matt. *13*, 19.25.28; 2 Cor. *4*, 4; *1* Thess. 2, 18; *1* Pet. *5*, 8; Apoc. *16*, 14; *cast out of heaven,* Luke *10*, 18; 2 Pet. 2, 4; Jude 6; *character of the,* Matt. *4*, 5 f; Luke *8*, 29; *9*, 39.42; John *10*, 12; *1* Pet. *5*, 8; *1* John 2, 13; Apoc. *12*, 9; *20*, 2.7; *father of lies and deceit,* John *8*, 44; 2 Cor. *4*, 4; *11*, 14; Apoc. *20*, 7; *punished,* Matt. *25*, 41; Jude 6; Apoc. *20*; *sinner from the beginning,* 1 John *3*, 8; *tempts Christ,* Matt. *4*, 3-10; Mark *1*, 13; Luke *4*, 2-13; *to be resisted by Faithful,* 2 Cor. 2, 11; *11*, 3; Eph. *4*, 27; *6*, 16; 2 Tim. 2, 26; Jas. *4*,

Index of Scripture Texts

7; *1* Pet. *5*, 8 f; *1* John *2*, 13; Apoc. *12*, 11; *vanquished by God,* Rom. *16*, 20; *by Christ,* Matt. *4*, 11.24; Mark *1*, 23-27; *5*, 2-19; *16*, 17; Luke *9*, 1.38-43; *11*, 20; *13*, 32; Acts *16*, 18; *19*, 12; Heb. *2*, 14; *1* John *3*, 8; Apoc. *19*, 11-21; *20*.
DEVILS: *Prince of the,* Matt. *10*, 25; *12*, 24.
DISCIPLES OF CHRIST: Luke *10*, 1-24; Acts *2*, 41; *4*, 4; *11*, 26.
DISCIPLESHIP IN CHRIST: Matt. *5*, 13-16; *10*, 24-28.37-38; *16*, 24-26; Mark *8*, 34-37; Luke *6*, 40; *9*, 23-25; *11*, 33; *14*, 26.33; John *13*, 16; *15*, 20.
DISOBEDIENCE: *consequences of,* Acts *7*, 39 ff; *2* Cor. *10*, 6; Heb. *2*, 2 f.
DIVINE SERVICE: Acts *2*, 41 f.46 f; *1* Cor. *14*, 26; Heb. *10*, 24 f.
DIVORCE: *condemned,* Matt. *5*, 31 f; *19*, 3-9; Mark *10*, 2-12; Luke *16*, 18; *1* Cor. *7*, 10-16.
DOXOLOGIES: Rom. *11*, 36; *16*, 27.
DRESS: *of women,* *1* Cor. *11*, 5-15; *1* Tim. *2*, 9.
DRUNKARDS AND DRUNKENNESS: *warning against,* Luke *21*, 34; Rom. *13*, 13; *1* Cor. *5*, 11; *6*, 10; Gal. *5*, 21; Eph. *5*, 18; *1* Tim. *3*, 3; Titus *1*, 7.

EGYPT: land in Africa, called the land of the Nile River; *Jesus taken there,* Matt. *2*, 13-15.20.
ELECT, GOD'S: Matt. *19*, 30; *20*, 16; *22*, 14; *24*, 22.31; Mark *13*, 20.22.27; Luke *18*, 7; Rom. *8*, 33; *1* Cor. *1*, 27; Eph. *1*, 4; Col. *3*, 12; *1* Thess. *1*, 4; *4*, 15-17; Titus *1*, 1; Jas. *2*, 5; *1* Pet. 2, 9; Apoc. *17*, 14.
ELIAS, OR ELIJAH: Hebrew prophet, Matt. *11*, 14; *16*, 14; *17*, 3 ff.10-13; Mark *9*, 3 ff.10-12; Luke *1*, 17; *4*, 25 ff; *9*, 30-33; John *1*, 21.
ELIZABETH: cousin of Virgin Mary, and mother of John the Baptist, Luke *1*, 5.13.24.40-66.
END OF WORLD: Matt. *24*, 29-31; Mark *13*, 24-27; Luke *21*, 25-28; *2* Pet. *3*, 7-12.
ENEMIES: *command to love your,* Matt. *5*, 44; *6*, 14; *18*, 35; Mark *11*, 25 f; Luke *6*, 26-37; Rom. *12*, 14.17-20; Eph. *4*, 26; *1* Thess. *5*, 15; *1* Pet. *3*, 9; *examples,* Matt. *26*, 52; Luke *23*, 34; Acts *7*, 58 f; *1* Cor. *4*, 12 f.
ENEMIES: *of God,* Rom. *8*, 7; *11*, 28; Jas. *4*, 4.
ENEMIES: *of the Cross,* Phil. *3*, 18 f.
ENVY: Matt. *20*, 9-15; Mark *15*, 10; Luke *15*, 25-30; Rom. *1*, 29; Gal. *5*, 19-21.26; Phil. *1*, 15; Jas. *4*, 2; *1* Pet. *2*, 1.
EPHESUS: city on the coast of Asia Minor, *visited by Paul,* Acts *18*, 19.24; *19*; *20*, 17; *1* Cor. *15*, 32; *16*, 8; Eph. *1*, 1 ff; *1* Tim. *1*, 3; *2* Tim. *4*, 12; Apoc. *1*, 11; *2*, 1-7.
ETERNAL HAPPINESS: *way to,* Matt. *6*, 19-20; Rom. *6*, 8 ff; Gal. *6*, 8; *open to all,* John *3*, 17; *1* Thess. *5*, 9 f; *1* Tim. *2*, 3 f; *greatness of,* Matt. *5*, 12; *13*, 43; *19*, 28; Luke *22*, 30; *2* Tim. *4*, 8; *1* Pet.

Index of Scripture Texts

1, 3-6; *5*, 4; Apoc. *3*, 5; *7*, 13-17; *14*, 13; *21*, 4; *different degrees of*, *1* Cor. *15*, 40-44; 2 Cor. *5*, 10; *9*, 6.

EVANGELISTS: *Matthew*, see under Apostles; *Mark*, which see; *Luke*, which see; *John*, see under Apostles.

EUCHARIST: *promise of*, John *6*, 32-59; *institution of*, Matt. *26*, 17-29; Mark *14*, 12-25; Luke *22*, 7-20; *1* Cor. *11*, 23-26; *real presence in*, Matt. *26*, 27 f; Mark *14*, 22-24; Luke *22*, 19 f; John *6*, 51 f; *1* Cor. *11*, 24 ff; *sacrifice in*, Matt. *26*, 28; Mark *14*, 24; Luke *22*, 19 f; *1* Cor. *11*, 24-26; Heb. *13*, 10-12.

EUCHARIST: *communion of*, John *6*, 53-59; *1* Cor. *10*, 16 f; *11*, 26-29; *preparation for*, Matt. *5*, 8.23 ff; *1* Cor. *5*, 7 f; *11*, 27-31; *fruits of*, John *6*, 56-59; *1* Cor. *10*, 17; *11*, 27-30.

EVE: first woman, *created by God*, 2 Cor. *11*, 3; *1* Tim. *2*, 13.

EVIL: *permission of*, Matt. *8*, 30-32; *13*, 29 f.38 f; Mark *5*, 11-13; John *6*, 71 f; Acts *8*, 1.

EXAMPLE: *good*, Matt. *5*, 14-16; Rom. *12*, 17; *15*, 19; 2 Cor. *8*, 21; Phil. *2*, 15; *1* Thess. *4*, 11 f; *1* Tim. *4*, 12-16; Titus *2*, 6-8; Jas. *3*, 13; *1* Pet. *2*, 12; *bad*, Matt. *18*, 6 f; *1* Cor. *8*, 9-13; *15*, 33; 2 Pet. *2*, 6; Jude 7.

EXAMPLE: *of Christ*, Matt. *11*, 29; John *13*, 15; Rom. *15*, 3.5; Phil. *2*, 5-11; *1* Pet. *2*, 21.

EXTREME UNCTION: Jas. *5*, 14 ff.

EVIL SPIRITS: *existence of*, Luke *10*, 18; Eph. *6*, 12; Jas. 2, 19; *4*, 7; *1* Pet. *5*, 8; 2 Pet. *2*, 4; Jude 6; *works of*, Matt. *4*, 3-10; *8*, 31; *13*, 19-25; *16*, 23; Mark *1*, 13; *4*, 15; *8*, 33; Luke *4*, 2-13; *8*, 12; John *8*, 44; *13*, 2; Acts *5*, 3; *13*, 9 f; *19*, 13; *26*, 18; *1* Cor. *7*, 5; 2 Cor. *2*, 11; *11*, 3.14; Eph. *2*, 1 f; *6*, 11-13; Col. *1*, 13; *1* Thess. *2*, 18; *3*, 5; *1* Pet. *5*, 8; *1* John *3*, 8-10; Apoc. *2*, 10; *20*, 7; see also under Devil, Satan, Unclean Spirits.

FAINTHEARTEDNESS: Matt. *8*, 26; *1* Thess. *5*, 14; 2 Tim. *1*, 7; *1* John *4*, 18.

FAITH: *confession of*, Matt. *10*, 32 ff; Mark *8*, 38; Luke *9*, 26; *12*, 8 ff; Rom. *1*, 16; *10*, 9 ff; *1* Tim. *6*, 12; 2 Tim. *2*, 12; Heb. *4*, 14; *10*, 23; *13*, 15; *1* John *4*, 15; *necessity of*, Mark *16*, 16; John *3*, 36; *6*, 29; *8*, 24; *12*, 48; Acts *4*, 11 f; *16*, 30 ff; Heb. *11*, 6; *1* Pet. *5*, 9; 1 John *3*, 23; *5*, 1.10.13; *fruits of*, Mark *16*, 16; John *6*, 40; *12*, 36; *20*, 31; Acts *10*, 43; *15*, 9; *16*, 31; *26*, 18; Rom. *1*, 17; *3*, 22.28; *4*, 16; *5*, 1; *15*, 13; 2 Cor. *4*, 13; *5*, 7; Gal. *2*, 16; *3*, 14.26; Eph. *1*, 13; *3*, 12.17; *1* Tim. *1*, 4; Heb. *4*, 3; *6*, 12; *10*, 38; Jas. *2*, 17; *1* Pet. *1*, 5.8; Jude 20; *1* John *5*, 4; *power of*, Matt. *9*, 22; *17*, 20; *21*, 21 ff; Mark *9*, 23; *11*, 23; *16*, 17 ff; Luke *8*, 50; *17*, 6; *18*, 41 ff; John *7*, 38; *11*, 15; *14*, 12; Acts *3*, 16; Rom. *1*, 16; Eph. *6*, 16; *apostasy from*, *1* Tim. *1*, 19; *5*, 15; *6*, 21; 2 Tim. *2*, 17 ff; Heb. *3*, 12; *6*, 3 ff; *10*, 26; *denial of*, Matt. *10*, 33; Luke *12*, 9; *1* Tim. *5*, 8; 2 Tim. *2*, 12; *firmness in*,

Index of Scripture Texts

1 Cor. 16, 13; *road to salvation*, Matt. *16*, 16; John *1*, 12; *3*, 16.36; *6*, 40.47; Acts *16*, 31; Gal. *3*, 11; Eph. *2*, 8; Heb. *11*, 6; *1* Pet. *1*, 9; *1* John *5*, 10 ff.

FAITHFULNESS: *in service of God*, Matt. *24*, 45; 2 Cor. *2*, 17; *4*, 2; 2 Tim. *2*, 14 ff; *3* John 5 f; *towards men*, Luke *16*, 10 ff; *1* Cor. *4*, 2; *1* Tim. *3*, 11; *6*, 2; Titus *2*, 10.

FALSE TEACHERS: *warning against*, Matt. *7*, 15; Rom. *16*, 17 ff; Titus *3*, 10 ff; *1* John *4*, 1-6.

FALSE TESTIMONY: Matt. *19*, 18; *26*, 60 ff; Acts *6*, 11.

FAMILY: *husband*, 1 Cor. *11*, 3; Eph. *5*, 25-33; Col. *3*, 19; *1* Pet. *3*, 7; *wife*, Eph. *5*, 22-24.33; Col. *3*, 18; *1* Pet. *3*, 1; *children*, Matt. *15*, 4-6; Mark *7*, 10-13; Eph. *6*, 1-3; Col. *3*, 20; *1* Tim. *5*, 4.

FASTING: *motive for*, Matt. *6*, 16-18; Luke *18*, 12.14; *purpose of*, Matt. *17*, 20; Mark *9*, 28; Luke 2, 37; *18*, 12; Acts *13*, 2 ff; *14*, 22; *examples of*, Matt. *4*, 2; *9*, 14 ff; Mark 2, 18-20; Luke *4*, 2; *5*, 33-35; see also Sobriety.

FATHERS: *duties of*, Luke *11*, 11; Eph. *6*, 4; Col. *3*, 21; Heb. *12*, 9; see also Family.

FEAR OF GOD: Matt. *10*, 28; Luke *1*, 50; *12*, 5; Rom. *11*, 20; 2 Cor. *7*, 1; Eph. *6*, 5; Phil. *2*, 12; Col. *3*, 22; Heb. *4*, 1; *12*, 28; *1* Pet. *2*, 17; Apoc. *11*, 18; *14*, 7; *15*, 4.

FEAR OF MEN: Matt. *10*, 28.33; Mark *8*, 38; Luke *12*, 4; Acts *4*, 19; *1* Pet. *3*, 14.

FELLOWSHIP: *with God*, 1 John *1*, 5-10; —*of Christ*, 1 Cor. *1*, 9; *12*, 27; 2 Cor. *4*, 11; Phil. *3*, 10; —*of the spirit*, Phil. *2*, 1.

FIRE: *of the Holy Spirit*, Matt. *3*, 11; Luke *3*, 16; Acts 2, 3 ff; —*of love*, Luke *12*, 49 ff; *24*, 32; *trial by*, *1* Pet. *1*, 6 ff; *4*, 12 ff; —*of the judgment*, 1 Cor. *3*, 12-15; Jas. *5*, 2 ff; Apoc. *8*, 5; —*of hell*, Matt. *13*, 41 ff; *18*, 8 ff; *25*, 41; Mark *9*, 42-48; Heb. *10*, 26 ff; Jude 7; Apoc. *19*, 20; *20*, 9.

FLESH: *seat of passions*, Rom. *7*, 5-25; *8*, 3-13; Gal. *5*, 16-24; Eph. *2*, 3; Col. *2*, 13; *works of the*, Rom. *8*, 13; Gal. *5*, 19 ff; *battle against desires of the*, Gal. *5*, 17; Col. *3*, 5; *lust of the*, see Concupiscence, Covetousness.

FORGIVENESS: Matt. *18*, 21 f.35; Luke *17*, 3 f; Col. *3*, 13.

FORNICATION: *condemned*, Matt. *15*, 19; Mark *7*, 21; Acts *15*, 20; Rom. *1*, 29; *1* Cor. *5*, 9; *6*, 9; 2 Cor. *12*, 21; Gal. *5*, 19; Eph. *5*, 5; Col. *3*, 5; *1* Thess. *4*, 3; *1* Tim. *1*, 10; Heb. *13*, 4; *1* Pet. *4*, 3; Jude 7; Apoc. *2*, 14; *21*, 8; *22*, 15.

FOUL TALK: *condemned*, Eph. *4*, 29; *5*, 4.

FREEDOM: *in Christ*, Gal. *4*, 31; *5*, 1; Jas. *1*, 25; *2*, 12; *1* Pet. *2*, 16; *false*, Gal. *5*, 13; *1* Pet. *2*, 16; 2 Pet. *2*, 19; —*from law of Moses*, 2 Cor. *3*, 15 ff; Gal. *2*, 4; *5*, 1.13; *1* Pet. *2*, 16.

FREE WILL: Matt. *23*, 37; Luke *13*, 34; Acts *7*, 51; Heb. *12*, 15; 2 Pet. *3*, 9; Apoc. *20*, 4.

Index of Scripture Texts

GABRIEL: Archangel, Luke *1*, 19.26-38.
GALILEE: northern part of Palestine, Matt. 2, 22; *4*, 15; *15*, 29; *26*, 32; *27*, 55; *28*, 7; Mark *1*, 9; Luke *4*, 14; *13*, 1; *23*, 5; *24*, 6; Acts *1*, 11; *2*, 7; *10*, 37; *13*, 31.
GENESARETH: Lake of, in northern part of Palestine, also called the Sea of Galilee, or Sea of Tiberias, Matt. *8*, 24 ff; *15*, 29; *14*, 25; *17*, 26; Mark *6*, 45 ff; Luke *5*, 1; John *6*, 1; *21*, 1 ff.
GENTILES: *call of the,* Matt. *8*, 11; Luke *13*, 29; John *10*, 16; Acts *15*, 7 ff; Rom. *11*, 11-32; Eph. *3*, 1 ff; Col. *1*, 24-27.
GETHSEMANI: Garden of, *place of our Lord's agony,* Matt. *26*, 36; Mark *14*, 32; Luke *22*, 39; John *18*, 1.
GIFTS: *spiritual,* Acts *11*, 17; Rom. *12*, 6 ff; *1* Cor. *1*, 7; *12; 13*, 2; *14;* Eph. *2*, 8; Jas. *1*, 5.17; *4*, 6.
GIFT OF TONGUES: Acts *2*, 4; *10*, 46; *11*, 15; *19*, 6; *1* Cor. *12-14*.
GOD: *existence,* Acts *14*, 16; Rom. *1*, 20; Heb. *11*, 3.6; *personality, Lord God Almighty,* 2 Cor. *6*, 18; Apoc. *1*, 8; *Lord of Heaven and Earth,* Matt. *11*, 25; Luke *10*, 21; *Creator,* Mark *10*, 6; John *1*, 3; Acts *4*, 24; *14*, 14; *17*, 24; Rom. *1*, 25; Col. *1*, 16; Heb. *1*, 10; *3*, 4; *11*, 3; *1* Pet. *4*, 19; Apoc. *4*, 11; *14*, 7; *Preserver,* Matt. *6*, 26.28-30; *10*, 29 ff; Luke *12*, 6; John *5*, 17; Acts *14*, 16; *17*, 25.28; Rom. *11*, 36; Col. *1*, 17; *1* Tim. *6*, 13; *Consuming Fire,* Heb. *12*, 29; *Judge,* Rom. *2*, 16; *2* Tim. *4*, 8; Heb. *12*, 23; Jas. *4*, 12; Apoc. *11*, 18; *18*, 8; *19*, 11 f; *Light,* Jas. *1*, 17; *1* John *1*, 5; *Searcher of hearts,* Acts *1*, 24; Rom. *8*, 27; Apoc. *2*, 23; *the Father,* Matt. *5*, 45; *11*, 25; *28*, 19; Mark *14*, 36; Luke *10*, 21; *22*, 42; *23*, 34.46; John *1*, 14; Acts *1*, 4; *2*, 33; Rom. *6*, 4; *8*, 15; *15*, 6; *1* Cor. *8*, 6; *15*, 24; 2 Cor. *1*, 3; *6*, 18; Gal. *1*, 1.3; *4*, 6; Eph. *1*, 17; Phil. 2, 11; Col. *1*, 19; *2*, 2; *1* Thess. *1*, 1; Heb. *12*, 7.9; Jas. *1*, 27; *3*, 9; *1* Pet. *1*, 2.17; 2 Pet. *1*, 17; *1* John *1*, 2; 2 John 3; 4; 9; Jude 1; *the Son,* Matt. *11*, 27; Mark *13*, 32; Luke *1*, 32; John *1*, 18; Acts *8*, 37; *9*, 20; Rom. *1*, 4; 2 Cor. *1*, 19; Gal. 2, 20; Eph. *4*, 13; Heb. *4*, 14; *1* John 2, 22; Apoc. *2*, 18; also see Christ; the Holy Spirit, which see.
GOD: *attributes of, Almighty,* Luke *1*, 49; *Eternal,* John 5, 26; Rom. *1*, 20; *16*, 26; Eph. *3*, 9; *1* Tim. *1*, 17; *6*, 16; 2 Pet. *3*, 8; Apoc. *1*, 8; *4*, 8; *22*, 13; *Faithful, 1* Cor. *1*, 9; *10*, 13; 2 Cor. *1*, 20; *1* Thess. 5, 24; 2 Tim. 2, 13; *1* Pet. *4*, 19; *1* John *1*, 9; Apoc. *1*, 5; *Good,* Matt. *5*, 45; *19*, 17; Acts *13*, 17; *Holy,* Luke *1*, 49; Acts *3*, 14; Rom. 7, 12; *1* John 2, 20; Apoc. *4*, 8; *Immortal, 1* Tim. *1*, 17; *6*, 16; *Immutable,* Jas. *1*, 17; *Incomprehensible, 1* Tim. *6*, 16; *Incorruptible,* Rom. *1*, 23; *Invisible,* John *1*, 18; *5*, 37; Col. *1*, 15; *1* Tim. *1*, 17; *Jealous, 1* Cor. *10*, 22; *Just,* Matt. *10*, 15; *20*, 13; *23*, 14; Luke *12*, 47; *13*, 27; John 7, 18; *17*, 25; Acts *3*, 14; *10*, 34; *17*, 31; Rom. 2, 2; *1* Cor. *4*, 5; 2 Cor. *5*, 10; Gal. *6*, 7; Eph. *6*, 8; Col. *3*, 25; 2 Tim. *4*, 8; Jas. *1*, 13; *1* John *1*, 9; 2, 1; Apoc. *15*, 3; *16*, 7; *19*, 2; *Merciful,* Luke *1*, 50.54 ff.58.72.78; *6*, 36; Rom. *9*, 15 ff; 2 Cor. *1*, 3; Eph. 2, 4 ff; Col. 2, 13 ff; Titus *3*, 5; *1* Pet. *1*, 3; *Most High,* Acts 7, 48; *Omnipotent,*

Matt. *19*, 26; Mark *10*, 27; Luke *1*, 37; *18*, 27; Eph. *3*, 20; *Omnipresent*, Acts *17*, 27 f; *1* John *1*, 5; *Omniscient*, Matt. *6*, 4; *21*, 2; Mark *2*, 8; *14*, 13; Luke *22*, 10; John *1*, 47; *2*, 24 f; *16*, 30; *21*, 17; Acts *15*, 8; Rom. *8*, 27; *the Only Wise*, Rom. *16*, 27; *Patient*, Luke *13*, 8; Rom. *2*, 4; *2* Pet. *3*, 15; *Perfect*, Matt. *5*, 48; *Truthful*, John *3*, 33; *8*, 26; Rom. *3*, 4; Apoc. *16*, 7; *19*, 11; *Unsearchable*, Rom. *11*, 33.
GOD'S ELECT: Matt. *19*, 30; *20*, 16; *22*, 14; *24*, 22.31; Mark *13*, 20.22.27; Luke *18*, 7; Rom. *8*, 33; *1* Cor. *1*, 27; Eph. *1*, 4; Col. *3*, 12; *1* Thess. *1*, 4; *4*, 15-17; Titus *1*, 1; Jas. *2*, 5; *1* Pet. *2*, 9; Apoc. *17*, 14. See Call of God.
GOD: *enemies of*, Rom. *8*, 7; *11*, 28; Jas. *4*, 4.
GOD: *spiritual, gifts of*, Matt. *7*, 7-11; *11*, 28; John *3*, 16; *4*, 10; *6*, 27.32 f; *16*, 23 f; Acts *11*, 18; Rom. *5*, 15-17; *6*, 23; *11*, 29; *12*, 6; *1* Cor. *7*, 7; Eph. *2*, 8; Phil. *1*, 29; *2* Thess. *1*, 7; Jas. *1*, 5.17; *4*, 6; *1* Pet. *4*, 10; *2* Pet. *1*, 3; *temporal*, Matt. *6*, 11.25-34; *25*, 15-30; Acts *14*, 16; *1* Tim. *4*, 4 f; *6*, 17.
GOD: *glory of*, John *17*, 22; Acts *7*, 55; Eph. *3*, 16; Apoc. *21*, 11.23; *exhibited in Christ*, John *1*, 14; Heb. *1*, 3.
GOD: *knowledge of, natural*, Rom. *1*, 19 ff; *1* Cor. *1*, 21; *supernatural*, *1* Cor. *13*, 12; *1* Tim. *2*, 4 f; Heb. *6*, 4 ff; *2* Pet. *1*, 3; *1* John *2*, 13 ff; *5*, 20; *love of—toward men*, Matt. *5*, 45; John *3*, 16; Rom. *5*, 8; *1* John *3*, 16; *4*, 9 ff; *providence of*, Matt. *5*, 45; Luke *12*, 6 f; John *5*, 17; Acts *17*, 26-28; *1* Cor. *9*, 9; Col. *1*, 17; *1* Tim. *6*, 13; Heb. *1*, 3; Jas. *4*, 15; *1* Pet. *5*, 7; *throne of*, Heb. *1*, 8; *4*, 16; Apoc. *1*, 4; *3*, 21; *4*, 10; *trumpet of*, *1* Thess. *4*, 16; *trust in*, Matt. *6*, 25-34; Heb. *4*, 16; *1* Pet. *5*, 7; *worship of*, Matt. *4*, 10; Luke *4*, 8; John *4*, 22-24; Apoc. *4*, 10; *5*, 14; *7*, 11 f; *word of*, which see.
GOLDEN RULE: Matt. *7*, 12; Luke *6*, 31.
GOLGOTHA: place of crucifixion, Matt. *27*, 33; Mark *15*, 22; John *19*, 17 f.
GOOD EXAMPLE: see Example, good.
GOOD WORKS: *meritorious*, Matt. *6*, 4.18; *10*, 41 ff; Mark *9*, 40; John *5*, 28 ff; Rom. *2*, 6 ff; *10*, 13; *1* Cor. *15*, 58; Eph. *6*, 8; Col. *3*, 23; Heb. *6*, 10; *13*, 16; Apoc. *14*, 13; *20*, 12; *22*, 12; *—necessary*, Matt. *3*, 10; *7*, 19.21.24-27; *25*, 37-46; Luke *6*, 46-49; John *15*, 16; Jas. *2*, 14-22. See Actions in Works.
GOSPEL: *authority of*, Mark *16*, 15 f; *1* Cor. *15*, 1 f; Gal. *1*, 11 f; Eph. *3*, 2 ff; *2* Tim. *3*, 16; *2* Pet. *1*, 20 f.
GRACE: *actual*, John *6*, 44; *15*, 4; Acts *20*, 24; Rom. *8*, 26; *11*, 5 f; Eph. *5*, 14; Apoc. *3*, 20; *sanctifying*, Rom. *5*, 2; *6*, 23; *8*, 17.29; Eph. *4*, 24; Titus *3*, 7; *2* Pet. *1*, 4; *1* John *3*, 1; *necessity of*, Luke *18*, 26 ff; John *6*, 44 ff.66; Rom. *12*, 3; *1* Cor. *2*, 10 ff; *4*, 7; *2* Cor. *3*, 5; *5*, 18; Gal. *1*, 15; Eph. *6*, 23; Phil. *2*, 13; Col. *1*, 12 ff; Jas. *1*, 5; *1* John *5*, 20.
GREAT COMMANDMENT: Matt. *22*, 36 ff; Mark *12*, 28-34; Luke *10*, 25-28.
GREECE: country in southeastern Europe, Acts *16; 17; 18; 20*.

Index of Scripture Texts

GREED: see Avarice and Covetousness.
GRIEF: *pleasing to God,* Matt. *5,* 5; *26,* 75; Rom. *9,* 2; *2* Cor. *7, 9.*
GUARDIAN ANGELS: Matt. *18,* 10; Heb. *1,* 14.

HANDS: *laying on of,* Acts *6,* 5 ff; *8,* 17; *13,* 3; *19,* 6; *1* Tim. *4,* 14; *5,* 22; 2 Tim. *1,* 6; *lifting up—in prayer, 1* Tim. *2,* 8.
HAPPINESS: *eternal,* which see.
HATING, HATRED: *condemned,* Matt. *5,* 23 ff.43; Titus *3,* 3; *1* John *2,* 9-11; *3,* 14 f; *4,* 20.
HEAVEN: *the Father's House,* Matt. *5,* 12; *6,* 9; *18,* 10; Luke *10,* 20; John *14,* 1-3; *17,* 24; 2 Cor. *5,* 1; *12,* 2; Col. *1,* 5; *1* Pet. *1,* 4.12; *the possession of all good,* Matt. *11,* 29; *25,* 46; Luke *22,* 29 f; John *14,* 2 f; *17,* 2; Heb. *4,* 9-11; *1* John *3,* 2; Apoc. *3,* 12.21; *4,* 8-11; *7,* 11 f; *21,* 2.22-24; *22,* 3 f; *the third,* 2 Cor. *12,* 2; *the New Heavens,* 2 Pet. *3,* 13; Apoc. *21,* 1.
HEBREW: *language of the Jews,* John *19,* 17.20; Acts *21,* 40.
HEBREWS: *descendants of Abraham,* 2 Cor. *11,* 22; Phil. *3,* 5; *Jewish Christians of Palestine,* Heb. *1,* 1 ff.
HELL: *existence of,* Matt. *5,* 29 ff; *10,* 28; *11,* 23; *13,* 42; *18,* 9; *25,* 41; Mark *9,* 42-48; Heb. *10,* 27; 2 Pet. *2,* 4; Apoc. *14,* 10; *19,* 20; *20,* 10.15; *21,* 8; *eternity of,* Matt. *3,* 12; *25,* 41.46; Mark *9,* 42-47; Luke *3,* 17; *16,* 19-26; 2 Thess. *1,* 7-9; Jude 7; Apoc. *14,* 11; *20,* 10.
HERESIES: *condemned,* Rom. *16,* 17; *1* Cor. *1,* 10; 2 Pet. *2,* 1; Jude 19.
HEROD THE GREAT: king of Judea, Matt. *2,* 1.3.16; Luke *1,* 5.
HEROD AGRIPPA I: king of Judea, Acts *12.*
HEROD AGRIPPA II: Acts *25,* 13.22 ff.
HEROD ANTIPAS: tetrarch of Galilee, Matt. *14,* 1-12; Mark *6,* 14 ff; *8,* 15; Luke *3,* 1.19; *8,* 3; *9,* 7 ff; *13,* 31; *23,* 7 ff; Acts *4,* 27.
HERODIAS: wife of Philip, brother of Herod Antipas, and taken by Herod as his own wife, mother of Salome, Matt. *14,* 3 ff; Mark *6,* 17.22.
HOLINESS: Luke *1,* 75; Rom. *12,* 1; 2 Cor. *7,* 1; Eph. *1,* 4; *4,* 24; Col. *3,* 12; *1* Thess. *2,* 12; *1* Tim. *2,* 15; Heb. *12,* 14; *1* Pet. *1,* 15; 2 Pet. *3,* 11; Apoc. *22,* 11.
HOLY SCRIPTURE: *divine inspiration of,* Luke *1,* 70; 2 Tim. *3,* 16; Heb. *1,* 1; 2 Pet. *1,* 20 f; *interpretation of,* Acts *8,* 30 ff.35; 2 Pet. *1,* 20; *3,* 16; Apoc. *13,* 18; *17,* 9.
HOLY SPIRIT: God the: *divinity of,* Acts *5,* 3 ff; *1* Cor. *2,* 10; *3,* 16; *12,* 4-11; *procession from Father and Son,* John *15,* 26; *16,* 14 ff; *personality and works,* Matt. *12,* 28; John *3,* 5f; *14,* 26; *15,* 26; *16,* 7 ff; Acts *1,* 16; *8,* 29; *10,* 19; *13,* 2.4; *16,* 6; *20,* 28; Rom. *8,* 26 ff; *15,* 13.17-19; *1* Cor. *2,* 10.13; *6,* 11; *12,* 8.11; Titus *3,* 5; Heb. *10,* 15; *1* Pet. *1,* 12; 2 Pet. *1,* 21; *1* John *5, 6* ff; *appearances,* Matt. *3,* 16; Mark *1,* 10; Luke *3,* 22; John *1,* 32 ff; Acts *2,* 1-4; *gifts,* Luke *11,* 13; John *3,* 8; *7,* 38 f; *14,* 16 f; *15,* 26; *16,* 13; Acts *2,* 6 ff.38; *1* Cor. *2,* 10 ff; *12,* 1-31; Gal. *5,* 22 f; *sins against,* Matt. *12,* 31 ff; Mark *3,* 28 ff; Luke *12,* 10; Acts *7,* 51; Heb. *6,* 4-6; *10,* 26 ff.

Index of Scripture Texts

HOPE: *object of,* Rom. *8,* 24 f; Gal. *5,* 5; Titus *2,* 11-14; Heb. *11,* 1; *1* Pet. *1,* 13; *2* Pet. *3,* 9; *value of,* Rom. *15,* 13; *1* Cor. *9,* 10; *13,* 13; Heb. *3,* 6; *4,* 16; *1* John *3,* 3.
HOSPITALITY: Rom. *12,* 13; Titus *1,* 8; Heb. *13,* 2; *1* Pet. *4,* 9.
HOUSE DIVIDED AGAINST ITSELF: Matt. *12,* 25; Luke *11,* 17.
HUMILITY: *exhortation to,* Matt. *11,* 29; *18,* 1-4; *20,* 26; Mark *9,* 35; *10,* 43; Luke *9,* 46 ff; *14,* 11; *18,* 14; *22,* 26; Rom. *11,* 20; *12,* 3.16; *1* Cor. *4,* 6; Eph. *4,* 2; Phil. *2,* 3; Col. *3,* 12; *value of,* Matt. *5,* 3; *11,* 25; *18,* 4; *19,* 14; Gal. *6,* 2; Jas. *1,* 9 ff; *4,* 6.10; *1* Pet. *5,* 5 ff; *examples of,* Matt. *3,* 11; *11,* 29; *20,* 28; Mark *1,* 7; *10,* 45; Luke *1,* 38.48; *3,* 16; *22,* 26 ff; John *1,* 27; *13,* 1-15; Eph. *3,* 8.
HUSBANDS: Matt. *19,* 4; *1* Cor. *7,* 2 f; Eph. *5,* 23.25.33; Col. *3,* 19; *1* Pet. *3,* 7.
HYPOCRISY: Matt. *6,* 5; *7,* 5; *23,* 28; *24,* 51; Mark *12,* 15; Luke *12,* 1-3; *1* Tim. *4,* 2; *1* Pet. *2,* 1; Apoc. *3,* 1.

IMAGES: Heb. *11,* 21.
IMITATION OF CHRIST: *invitation to,* Matt. *8,* 22; *19,* 21.28; Luke *9,* 59; *reward for,* Matt. *19,* 28; Mark *10,* 29 ff; John *8,* 12.
IMMORTALITY: *to put on, 1* Cor. *15,* 53 ff.
IMPENITENCE: Matt. *11,* 20-24; *12,* 32; *23,* 37; Luke *10,* 13-15; Rom. *2,* 4 ff; Apoc. *2,* 21; *9,* 21; *16,* 9.
IMPURITY: *warning against,* Matt. *15,* 19 ff; *1* Cor. *6,* 18-20; *10,* 8; Gal. *5,* 19; Eph. *5,* 3; *2* Tim. *2,* 22; *penalty of,* Luke *15,* 13 ff; *1* Cor. *6,* 9; Eph. *5,* 5; Heb. *13,* 4; Apoc. *21,* 8; see also Adultery.
INCENSE: *symbol of prayers,* Apoc. *8,* 3 f.
INDULGENCES: *power of Church to grant,* Matt. *16,* 19.
INFALLIBILITY: *of Church,* Matt. *16,* 18; *28,* 20; Luke *22,* 32; John *14,* 16 ff; *16,* 13; *1* Tim. *3,* 15; *1* John *2,* 26 ff; —*of the Pope,* Matt. *16,* 18 ff; Luke *22,* 32.
INGRATITUDE: Luke *17,* 17 ff; *2* Tim. *3,* 2; —*to God,* Rom. *1,* 21.
INHERITANCE: *in Christ,* Eph. *1,* 11.14; Col. *1,* 12; *3,* 24; *1* Pet. *1,* 4.
INJUSTICE: Matt. *23,* 14; Luke *16,* 11 ff; *1* Cor. *6,* 9; *2* Pet. *2,* 9.16.
INNOCENTS: *murder of,* Matt. *2,* 16-18.
INTEMPERANCE: *warning against,* Luke *21,* 34; Rom. *13,* 13; *16,* 18; Eph. *5,* 18; Phil. *3,* 19 ff; *penalty of,* Luke *16,* 19-31; *1* Cor. *10,* 5-10; Heb. *12,* 16. See also Drunkenness.
INTENTIONS: see under Actions of the will.
INTERCESSION: *admonition for,* Matt. *5,* 44; Luke *6,* 28; Rom. *15,* 30-32; *2* Cor. *1,* 11; Eph. *6,* 18 ff; Phil. *1,* 19; Col. *4,* 3; *1* Thess. *5,* 25; *2* Thess. *3,* 1; *1* Tim. *2,* 1 ff; Heb. *13,* 18; Jas. *5,* 16; *examples of,* Luke *22,* 32; *23,* 34; John *17,* 9.11.15.20; Acts *7,* 59; *8,* 24; *12,* 5; Rom. *10,* 1 ff; Phil. *1,* 3-5; Col. *4,* 12; *2* Tim. *1,* 3; —*of Christ,* Luke *23,* 34; Rom. *8,* 34; Heb. *7,* 25; *1* John *2,* 1; —*of the Holy Spirit,* Rom. *8,* 26.

Index of Scripture Texts

ISAIAS: Hebrew prophet, referred to in Matt. *3*, 3; *4*, 14; *8*, 17; *12*, 17; *13*, 14; *15*, 7; Mark *1*, 2 f; Luke *3*, 4; *4*, 17; John *1*, 23; *12*, 38 f; Acts *8*, 32; *28*, 25; Rom. 9, 27.29; *10*, 16; *15*, 12.

JERICHO: city in the lower valley of the Jordan, Matt. *20*, 29; Mark *10*, 46; Luke *10*, 30; *18*, 35; *19*, 1; Heb. *11*, 30.

JERUSALEM: the holy city of the Jews, Matt. *21*, 1; *23*, 37-39; *24;* Mark *11*, 1.27; *13;* Luke 2, 22.42 f; *13*, 33-35; *19*, 41-44; *21*, 20-24; John *12*, 12; Acts 2, 5; 9, 28; *10*, 39; *11*, 2.22.27; *15*, 2.34; *21*, 4.17.31; *25*, 1.9.20; *26*, 10; *the heavenly,* Gal. *4*, 26; Heb. *12*, 22; Apoc. *21*, 10-27; *the new,* Apoc. *21*, 2 ff; *the old,* Gal. *4*, 25.

JESUS CHRIST: the Son of God, see Christ.

JESUS: *name of,* Matt. *1*, 21; Luke *1*, 31; 2, 21; Acts *4*, 10-12; Phil. 2, 5-11.

JEWS: inhabitants of Judea, southern part of Palestine; *Christ's mission to the,* Matt. *15*, 24; Acts *3*, 12 ff; *Christ rejected by the,* Matt. *13*, 15.57 f; John 5, 16.38 ff; 7, 11 ff; Acts *3*, 12 ff; *13*, 45 ff; *1* Thess. 2, 14 ff; *gospel first preached to the,* Matt. *10*, 6; Luke *24*, 47; Acts *1*, 8; *13*, 46; *gospel rejected by the,* Acts *13*, 46; *28*, 23 ff; see also Rom. *9; 10; 11*, 1-10; *2* Cor. *3*, 14 ff.

JORDAN: river flowing from north to south through Palestine, Matt. *3*, 5 f.13; *4*, 25; Mark *1*, 5.9; Luke *3*, 3; John *1*, 28.

JOHN, THE BAPTIST: Matt. *3; 9*, 14-17; *11*, 1-18; *14*, 1-12; Mark *1*, 4-14; *2*, 18; *6*, 14-29; Luke *1*, 5-24.57-80; *3*, 1-20; *5*, 33; *7*, 18-28; *9*, 7-9; John *1*, 15.19-35; *3*, 25-30.

JOSEPH: foster-father of Christ, Matt. *1*, 16-25; *2*, 13-23; Mark *6*, 3; Luke *1*, 27; *2*, 4.16; *3*, 23; *4*, 22; John *1*, 45; *6*, 42.

JOY: *in affliction,* Matt. *5*, 12; Acts *5*, 41; *20*, 24; Rom. *5*, 3; *2* Cor. 7, 4; Col. *1*, 24; *—follows sorrow,* John *16*, 20; *2* Cor. *6*, 6-10; Jas. *1*, 2; *—of the Christian,* Luke *1*, 46-49; *10*, 20; John *15*, 11; *16*, 20-22; *17*, 13; Rom. *12*, 12.15; *14*, 17; *2* Cor. *8*, 2; Gal. *5*, 22 f; Phil. *4*, 4; *1* Thess. *5*, 16; Jas. *1*, 2; *1* John *1*, 4.

JUDAS ISCARIOT: *one of the twelve disciples; he betrayed Christ,* Matt. *10*, 4; *26*, 14-16.21-25.48-50; *27*, 3-10; Mark *3*, 19; *14*, 10 f.43-45; Luke 22, 3-6.47-48; John *6*, 71 f; *12*, 4-7; *13*, 26-30; *18*, 2-5; Acts *1*, 16-25.

JUDGMENT: *general,* Matt. *12*, 36; *13*, 41-43; *16*, 27; *24*, 29-31.35; *25*, 31-45; *26*, 64; Mark *14*, 62; Luke *12*, 2; John *12*, 48; *1* Cor. *3*, 13; *4*, 5; *2* Pet. 2, 4; *1* John *4*, 17; Jude 6; 14 ff; Apoc. *1*, 7; *6*, 15-17; *11*, 18; *20*, 12 ff; *particular,* Matt. *25*, 30; Luke *16*, 22 ff; Rom. 2, 5-8; *14*, 10.12; *1* Cor. *3*, 8; *11*, 31; *2* Cor. *5*, 10; Gal. *6*, 8; Heb. *9*, 27; Jas. 2, 13; *God's final,* Apoc. *17; 18; 19; 20.*

JUDGMENT: *rash and unkind—of fellow men,* Matt. 7, 1-5; Luke *6*, 37 ff; Rom. 2, 1-4; *14*, 4.13; Jas. 2, 4; *4*, 11 ff.

JUSTICE OF GOD: John *17*, 25; Acts *3*, 14; *17*, 31; Rom. 2, 2.6; *1* Cor. *4*, 5; *2* Cor. *5*, 10; *2* Tim. *4*, 8; *1* Pet. *3*, 18; *1* John 2, 1.

Index of Scripture Texts

JUSTIFICATION: *necessity of,* Rom. *1,* 18 — *3,* 20; *condition of,* Matt. 7, 21; Mark *16,* 15 ff; Luke *13,* 3; Rom. *3,* 28; Heb. *11,* 6; Jas. 2, 24; *effect of,* Rom. *3,* 24 ff; *8,* 10.29; *1* Cor. *6,* 11; Eph. 2, 10; Titus *3,* 3 ff; 2 Pet. *1,* 4.

KINGDOM: *of Christ,* Luke *1,* 33; *22,* 30; Col. *1,* 14; —*of God,* Mark *4,* 30-32; Luke *13,* 18-21; *17,* 20 f; —*of heaven,* Matt. *5,* 19 f; *13,* 31-33.
KNOWLEDGE: *of God,* see God, knowledge of.

LABOR: *duty to,* Matt. *20,* 1-16; *25,* 14-30; Eph. *4,* 28; *1* Thess. *4,* 11; 2 Thess. *3,* 10-12; *reward of,* Matt. *10,* 10; *20,* 8-15; Luke *10,* 7; *1* Cor. *3,* 8.14; *9,* 10; *1* Tim. *5,* 18; 2 Tim. *2,* 6; Jas. *5,* 4; *examples,* Matt. *6,* 3; Acts *18,* 3; *20,* 34 f; *1* Thess. *2,* 9; 2 Thess. *3,* 8.
LAITY: *1* Tim. *5,* 3-16; Titus *2,* 1-15; *1* Pet. *5,* 5-11.
LAMB OF GOD: John *1,* 29; *1* Pet. *1,* 19; Apoc. *5,* 6; *13,* 8.
LAMENTS OF CHRIST: Matt. *23,* 37-39; Luke *13,* 34 f; *19,* 42-44.
LAST DAY: see Judgment, general.
LAST DISCOURSE OF CHRIST: John *14; 15; 16.*
LAST SUPPER: Matt. *26,* 17-29; Mark *14,* 12-26; Luke *22,* 7-38; John *13,* 1-30; *1* Cor. *11,* 23-26.
LAW: *Natural,* Acts *14,* 16; Rom. *1,* 19 ff; *2,* 14 f; *Mosaic,* Acts *15,* 10; Rom. *2,* 17 ff; *3,* 20 ff; *4,* 13-16; *5,* 20; *7,* 1.7-13; 2 Cor. *3,* 6-11; Gal. *3,* 1-13.19.23; *1* Tim. *1,* 9; Heb. 7, 18 f.
LAW: *Christians subject to,* Rom. *13,* 1; *1* Pet. 2, 13-15,
LAW OF MOSES: *freedom from,* 2 Cor. *3,* 16 f; Gal. 2, 3 f; *3,* 25; *5,* 1-13.
LAWSUITS: *censured, 1* Cor. *6,* 1-9.
LAWYERS: *reproved,* Luke *11,* 45-52; *14,* 3-6.
LAZARUS: *friend of Christ,* John *11,* 1-44; *12,* 1-11.17.
LEAVEN: *action of,* Matt. *13,* 33; Luke *13,* 20 f; *1* Cor. *5,* 6-8; Gal. *5,* 9.
LAYING ON OF HANDS: Acts *6,* 6; *8,* 17; *13,* 3; *19,* 6; *1* Tim. *4,* 14; *5,* 22; 2 Tim. *1,* 6.
LIFE: *value of,* Matt. *6,* 25-34; Luke *12,* 19-31; Jas. *1,* 10; *4,* 15; *1* Pet. *1,* 24; *purpose of,* Matt. *5,* 48; Eph. *1,* 4; *1* Thess. *4,* 3 ff; Titus *2,* 11 ff; *eternal,* Matt. *5,* 12; *19,* 29; *25,* 46; John *3,* 16; *5,* 24; *10,* 28; *17,* 3; Rom. *2,* 7; *6,* 23; *1* Tim. *1,* 16; Heb. *13,* 14; Jas. *1,* 12; *1* John *1,* 2; *2,* 25; Jude 21; Apoc. *2,* 7; *21,* 6; *spiritual,* Rom. *6,* 4.8; Gal. *2,* 20; Eph. *2,* 1 ff; Col. *3,* 3; *1* John *1,* 1-4; *1* Pet. *4,* 1-6.
LIFE: *Book of,* Phil. *4,* 3; Apoc. *3,* 5; *13,* 8; *21,* 27.
LIGHT OF THE WORLD: John *8,* 12.
LIP SERVICE: Matt. *7,* 21-23; *25,* 11 f; Mark *7,* 6 f; Luke *6,* 46; *13,* 25-27; Rom. *2,* 13; Jas. *1,* 22.
LORD'S PRAYER: Matt. *6,* 9-13; Luke *11,* 2-4.
LORD'S SUPPER: see Eucharist, communion of.
LOVE OF GOD: *motive for,* Matt. *22,* 37; Mark *12,* 30; Luke *10,* 27; John *3,* 16; *15,* 12 ff; Rom. *5,* 8 f; *8,* 32-35; Gal. *2,* 20; *1* John *3,* 16;

Index of Scripture Texts

4, 7 ff; *characteristic of,* John *14*, 15.21-24; *21*, 14-17; Rom. *8*, 35-39; *1* Cor. *16*, 22; Eph. 5, 2; *1* John 2, 5; *3*, 16; *4*, 20; *2* John 6.

LOVE OF NEIGHBOR: *duty,* Matt. 7, 12; *19*, 19; *22*, 39; Mark *12*, 31; Luke *10*, 27; John *13*, 34; *15*, 12.17; Rom. *13*, 8-10; *1* Cor. *13*, 1-8; Gal. *5*, 13 f; Col. *3*, 14; *1* Thess. *4*, 9 ff; Heb. *13*, 1; Jas. 2, 8; *1* Pet. *1*, 22; 2, 17; *4*, 8; *1* John 2, 10 f; *3*, 10-23; *4*, 7 ff; *characteristics of,* Matt. *5*, 46 ff; 7, 12; Luke *6*, 31-34; *14*, 12-14; Rom. *12*, 9 f; *1* Cor. *10*, 24; *13*, 4-8; Eph. *4*, 2.32; Phil. 2, 2-4; *1* Pet. *1*, 22; *1* John *3*, 16-18; *examples,* Luke *10*, 27-37; Rom. *9*, 1-5; *11*, 1; *1* Cor. *9*, 19-22; *1* John *3*, 16.

LUKE: physician, companion of Paul, Evangelist, Col. *4*, 14; *2* Tim. *4*, 11; Philem. 24.

LUKEWARMNESS: *condemned,* Apoc. *3*, 15 f.

LUST: see Covetousness.

LUXURIOUSNESS: *in dress, condemned, 1* Tim. 2, 9; *1* Pet. *3*, 3.

LYING: *warning against,* John *8*, 44; Eph. *4*, 25; Col. *3*, 9; Jas. *3*, 14; *punishment for,* Acts *5*, 2-10; Apoc. *21*, 8.27; *22*, 15.

MAGI: Matt. 2, 1-16.

MAGIC ARTS: Acts *19*, 19.

MALICE: *condemned, 1* Cor. *5*, 8; *14*, 20; Eph. *4*, 31; Col. *3*, 8; Titus *3*, 3; Jas. *5*, 9; *1* Pet. 2, 1.

MAMMON: *service of, condemned,* Matt. *6*, 24; Luke *16*, 9-13; Jas. *5*, 1-6.

MAN: *dignity of,* Acts *17*, 28; *1* Cor. *6*, 15-20; *2* Cor. *6*, 16; Heb. *1*, 14; 2, 6-8; Jas. *3*, 9; *2* Pet. *1*, 4; *destiny of,* Matt. *4*, 10; *18*, 14; Acts *17*, 26 ff; *1* Cor. *6*, 19 f; *1* Thess. *4*, 3 ff; *5*, 9 f; *1* Tim. 2, 4; *1* John *3*, 2.

MARRIAGE: see Matrimony.

MARRIED PEOPLE: *duties of, 1* Cor. 7, 3-5; Eph. *5*, 21-33; *6*, 4; Col. *3*, 18 f; *1* Pet. *3*, 1-7.

MARK: disciple of Peter, Evangelist, Acts *12*, 12; *13*, 5.13; *15*, 37-40; Col. *4*, 10; *2* Tim. *4*, 11; Philem. 24; *1* Pet. *5*, 13.

MARTYR: *first,* see Stephen.

MARY, BLESSED VIRGIN: Mother of Christ, Matt. *1*, 18 – 2, 23; *12*, 46-50; Mark *3*, 31-35; *6*, 3; Luke *1*, 26-56; *2*, 4-52; *8*, 19-21; John 2, 1-12; *6*, 42; *19*, 25-27; Acts *1*, 14.

MARY AND MARTHA: sisters of Lazarus, Luke *10*, 38-42; John *11*, 1-44; *12*, 1-8.

MARY MAGDALENE: Matt. *27*, 56; Mark *15*, 40; Luke *8*, 2; *24*, 10; John *19*, 25; *20*, 1 ff. 11-18.

MATRIMONY: *the Sacrament of,* Eph. 5, 22-33; *sanctity of, 1* Thess. *4*, 3-8; Heb. *13*, 4; *indissolubility,* Matt. *5*, 32; *19*, 4-6; Mark *10*, 4-12; Luke *16*, 18; Rom. 7, 2 ff; *1* Cor. *6*, 16; 7, 10 f.39; Eph. *5*, 31; *intercourse in, 1* Cor. 7, 3-7; *1* Thess. *4*, 3 f; *1* Pet. *3*, 7.

Index of Scripture Texts

MEEKNESS: *exhortation to,* Matt. *11,* 29; Gal. *6,* 1; Eph. *4,* 2; Col. *3,* 12; *1* Tim. *6,* 11; Titus *3,* 2; Jas. *1,* 21; *1* Pet. *3,* 15 ff; *reward for,* Matt. *5, 4; 11,* 29.
MERIT: see Actions in Works and Good Works.
MERCY OF GOD: Luke *1,* 50.54 f.58.72.78; *6,* 36; Rom. *9,* 15 f.18; *11,* 1 ff; 2 Cor. *1,* 3; Eph. *2,* 4 ff; Col. *2,* 13 ff; Titus *3,* 5; *1* Pet. *1, 3.*
MILDNESS: 2 Tim. *2,* 23-26.
MIRACLES OF CHRIST:
1) *Miracles over the elements: calming of the storm,* Matt. *8,* 23-27; Mark *4,* 35-40; Luke *8,* 22-25; *draught of fishes,* Luke *5,* 4-11; John *21,* 1-13; *feeding of four thousand,* Matt. *15,* 32-38; Mark *8,* 1-10; *feeding of five thousand,* Matt. *14,* 13-21; Mark *6,* 35-44; Luke *9,* 12-17; John *6,* 1-15; *Christ's passing through crowd unnoticed,* Luke *4,* 30; *payment of temple tax,* Matt. *17,* 24-26; *turning water into wine,* John 2, 1-11; *walking on the sea,* Matt. *14,* 22-23; Mark *6,* 45-52; John *6,* 16-21; *withering of the fig-tree,* Matt. *21,* 18-22; Mark *11,* 12-14.20-24.
2) *Miracles of healing: curing blind man,* Mark *8,* 22-26; *two blind men,* Matt. *9,* 27-31; *man born blind,* John *9,* 1-7; *blind beggar,* Mark *10,* 46-52; Luke *18,* 35-43 *(two blind men* in Matt. *20,* 29-34); *centurion's son,* Matt. *8,* 5-13; Luke *7,* 1-10; John *4,* 46-53; *deaf-mute,* Matt. *15,* 29-31; Mark *7,* 31-37; *man of dropsy,* Luke *14,* 1-4; *woman bent double,* Luke *13,* 10-17; *paralytic,* John *5,* 1-9; *another paralytic,* Matt. *9,* 1-8; Mark *2,* 1-12; Luke *5,* 17-26; *leper,* Matt. *8,* 1-4; Mark *1,* 40-45; Luke *5,* 12-15; *ten lepers,* Luke *17,* 11-19; *Peter's mother-in-law, and others,* Matt. *8,* 14-17; Mark *1,* 29-34; Luke *4,* 38-41; *woman with hemorrhage,* Matt. *9,* 20-22; Mark *5,* 25-34; Luke *8,* 43-48; *withered hand,* Matt. *12,* 9-13; Mark *3,* 1-5; Luke *6,* 6-10; *replacing of ear of Malchus,* Luke *22,* 49-51.
3) *Casting out devils: demoniacs cured,* Matt. *9,* 32-33; Mark *1,* 23-26; Luke *4,* 33-35; Matt. *12,* 22; *17,* 14-20; Mark *9,* 13-28; *demons sent into herd of swine,* Matt. *8,* 28-34; Mark *5,* 1-15; Luke *8,* 26-35; *daughter of the Syro-phoenician woman cured,* Matt. *15,* 21-28; Mark *7,* 24-30.
4) *Raising the dead: the widow's son,* Luke *7,* 11-16; *Jairus' daughter,* Matt. *9,* 18 f.23-26; Mark *5,* 35-43; Luke *8,* 42.49-56; *Lazarus,* John *11,* 1-44.
MODESTY: Gal. *5,* 23;
MONEY: *love of, censured, 1* Tim. *6,* 10.
MORALS: *general instructions on,* Rom. *12; 13; 14; 15;* Col. *3; 4,* 1-6; *1* Thess. *4,* 1-12.
MORTIFICATION OF THE FLESH: Rom. *6,* 12 ff; *13,* 14; 2 Cor. *4,* 10; Gal. *5,* 24; Col. *3,* 5; *1* Pet. *4,* 1 ff.
MURMURING AGAINST GOD: Phil. *2,* 14; Jude 16.

NARROW WAY: Matt. *7,* 13-14; Luke *13,* 24.
NAZARENE: the, Matt. *2,* 23.

Index of Scripture Texts

NAZARETH: village in Galilee, home of the Holy Family, Matt. *2*, 23; *4*, 13; Mark *6*, 1-6; Luke *1*, 26; *2*, 39.51 f; *4*, 16; John *1*, 45 f; Acts *2*, 22; *3*, 6.
NEIGHBOR: *duties toward the,* see under Brethren, also Forgiveness, and Love for one's neighbor.
NEW BIRTH: John *3*, 3-7.
NEW COMMANDMENT: John *13*, 34; *15*, 12; *1* John *2*, 8; *4*, 21.
NEW HEAVEN: 2 Pet. *3*, 13; Apoc. *21*, 1.
NEW JERUSALEM: Apoc. *21*, 2 ff.
NICODEMUS: Pharisee, John *3*, 1-21; *7*, 50 ff; *19*, 39.

OATH: Matt. *5*, 33-37; *23*, 16-22; Heb. *6*, 13-17; *7*, 20-22; Jas. *5*, 12.
OBEDIENCE: *to God,* Matt. *7*, 24 ff; John *4*, 34; *5*, 30; *6*, 38; *14*, 31; Acts *4*, 18 ff; *5*, 29; Rom. *1*, 5; *15*, 18 ff; Phil. *2*, 8 ff; Heb. *5*, 8 ff; *11*, 8; *1* Pet. *1*, 22; *1* John *3*, 24;—*to the Church,* Matt. *18*, 17; Heb. *13*, 17;—*to the State,* Matt. *22*, 21; Rom. *13*, 1-7; Titus *3*, 1; *1* Pet. *2*, 13-17.
OCCASION OF SIN: Matt *5*, 27-30; *18*, 8; Mark *9*, 42-46.
OPPRESSION: *condemned,* Jas. *5*, 1-6.
ORIGINAL SIN: Rom. *5*, 12-19; *7*, 7-25; *1* Cor. *15*, 21 f; Eph. *2*, 3.

PARABLES OF CHRIST: *Banquet, the great,* Luke *14*, 16-24; *banquet at royal wedding,* Matt. *22*, 1-14; *castle builder, the,* Luke *14*, 28-31; *coin, the lost,* Luke *15*, 8-10; *darnel weed,* Matt. *13*, 24-30.36-43; *debtors, the two,* Luke *7*, 41-43; *defrauding manager,* see unjust steward; *dragnet,* Matt. *13*, 47-50; *faithful and prudent servant,* Luke *12*, 41-48; *fig-tree, and all the trees,* Matt. *24*, 32-35; Mark *13*, 28-31; Luke *21*, 29-31; *fig-tree unfruitful (barren),* Luke *13*, 6-9; *fool, and his wealth,* Luke *12*, 16-21; *gold pieces, the ten,* Luke *19*, 11-27; *good shepherd,* John *10*, 1-16; *good Samaritan,* Luke *10*, 30-37; *heartless debtor,* Matt. *18*, 23-35; *house on rocks and house on sands,* Matt. *7*, 24-27; Luke *6*, 47-49; *hidden treasure* Matt. *13*, 44; *king preparing for war,* Luke *14*, 31-33; *laborers in the vineyard,* Matt. *20*, 1-16; *lamp under the grain measure,* Matt. *5*, 15; Mark *4*, 21 f; Luke *8*, 16-18; *11*, 33-36; *leaven,* Matt. *13*, 33; Luke *13*, 20 f; *lost coin,* Luke *15*, 8-10; *lost sheep,* Matt. *18*, 12-14; Luke *15*, 3-7; *mustard seed,* Matt. *13*, 31 f; Luke *13*, 18 f; *new cloth on old garment,* Matt. *9*, 16; Mark *2*, 21; Luke *5*, 36; *new wine in old skins,* Matt. *9*, 17; Mark *2*, 22; Luke *5*, 37-39; *pearl of great value,* Matt. *13*, 45 f; *persistent friend,* Luke *11*, 5-8; *Pharisee and the publican,* Luke *18*, 9-14; *poor guests,* Luke *14*, 12-14; *prodigal son,* Luke *15*, 11-32; *rich man and Lazarus,* Luke *16*, 19-31; *growing seed,* Mark *4*, 26-29; *servants, faithful and wicked,* Matt. *24*, 45-51; Luke *12*, 42-48; *sheep and goats* (last judgment), Matt. *25*, 31-46; *sower and seed,* Matt. *13*, 3-9; Mark *4*, 3-9; Luke *8*, 4-8.11-15; *talents,* Matt. *25*, 14-30; *treasure, hidden (buried),* Matt. *13*, 44; *two sons,* Matt. *21*, 28-32; *unjust judge,* Luke *18*, 2-8; *unjust steward,* Luke *16*, 1-9; *unprofitable servants,* Luke *17*, 7-10; *vine,* John *15*, 1-8; *virgins, ten,* Matt. *25*, 1-13;

Index of Scripture Texts

watchful porter, Mark *13,* 34-37; *watchful servants,* Luke *12,* 35-40; *wicked tenants,* Matt. *21,* 33-41; Mark *12,* 1-9; Luke *20,* 9-16.
PARADISE: Luke *23,* 43; 2 Cor. *12,* 4; Apoc. *2,* 7; *22,* 1-5.
PARENTS: *duties of,* Matt. *18,* 5-7; Eph. *6,* 4; Col. *3,* 21.
PASCHAL SUPPER (PASSOVER): see Last Supper.
PASTURE: *spiritual,* John *10,* 9.
PATIENCE: *of God,* Luke *13,* 7-9; Rom. *2,* 4; 2 Pet. *3,* 9.15; *—in faith,* Heb. *12,* 1 ff; 2 Pet. *1,* 5-7; *—in suffering,* Luke *21,* 19; Rom. *5,* 3-5; *12,* 12; *1* Cor. *13,* 7; 2 Cor. *6,* 4-6; Gal. *5,* 22; Col. *3,* 12; 2 Thess. *1,* 4 ff; *1* Tim. *6,* 11; 2 Tim. *2,* 12; Titus *2,* 1 f; Heb. *6,* 12.15; *10,* 36; *12,* 1; Jas. *1,* 2-4; *5,* 7 f; *—with one's neighbor,* Rom. *15,* 1; Gal. *6,* 1 ff; Eph. *4,* 1 ff; Col. *3,* 12 ff; *1* Thess. *5,* 14; 2 Tim. *2,* 24; *examples of,* Acts *7,* 59; *2* Thess. *5,* 23; 2 Thess. *3,* 16; *—with one another,* Mark *9,* 49; Rom. *12,*
PEACE: *on earth,* Luke *2,* 14; *in heaven,* Luke *19,* 38; *—of mind,* Luke *2,* 14; *24,* 36; John *14,* 27; Rom. *1,* 7; *5,* 1; *14,* 17; Phil. *4,* 7; Col. *3,* 15; *1* Thess. *5,* 23; 2 Thess. *3,* 16; *—with one another,* Mark *9,* 49; Rom. *12,* 18; Eph. *4,* 3; *1* Thess. *5,* 13; Heb. *12,* 14; *love of,* Matt. *5,* 9; *20,* 24; Mark *9,* 49; Rom. *12,* 18; *14,* 19; Eph. *4,* 3; *1* Thess. *5,* 13; 2 Tim. *2,* 22.
PENANCE: see under Sacraments.
PERFECTION: *striving for,* Matt. *5,* 48; *19,* 21; Rom. *12,* 2; *1* Cor. *15,* 58; 2 Cor. *7,* 1; *13,* 11; Eph. *4,* 15.23 ff; Phil. *1,* 9-11; *3,* 12; *1* Thess. *4,* 1-7; *5,* 23; 2 Tim. *3,* 16; Heb. *12,* 14; *1* Pet. *1,* 13-16.
PERJURERS: *condemned, 1* Tim. *1,* 10.
PERSECUTION: Matt. *5,* 10-12; Rom. *8,* 35; *1* Cor. *4,* 12; 2 Cor. *4,* 9; *12,* 10; *foretold,* Matt. *10,* 17; *23,* 34; *24,* 9; Mark *10,* 30; Luke *11,* 49; John *15,* 20; 2 Cor. *4,* 9; 2 Tim. *3,* 12.
PERSEVERANCE: *exhortation to,* Acts *13,* 43; *14,* 21; *1* Cor. *15,* 58; *16,* 13; 2 Thess. *2,* 15; *1* Tim. *1,* 18; Heb. *6,* 11 f; *10,* 23; *value of,* Matt. *10,* 22; *24,* 13; Mark *13,* 13; Rom. *2,* 6 f; *11,* 22; *1* Cor. *15,* 1 ff.58; Gal. *6,* 9; Col. *1,* 22 ff; *1* Thess. *3,* 12 f; *1* Tim. *4,* 16; Heb. *3,* 14; Apoc. *2,* 10.26.
PHARISEES: an ancient Jewish sect, especially exact in the interpretation and observance of the law, both written and oral; *hypocrisy of,* Matt. *6,* 2.5.16; *15,* 7 ff; *23,* 1-33; Mark *7,* 3; Luke *6,* 2; *11,* 37-44; *18,* 11 ff; *enemies of Christ,* Matt. *9,* 34; *12,* 14; Luke *6,* 11; *16,* 14; John *11,* 45-48; *warning against the,* Matt. *16,* 6; Mark *8,* 15; Luke *12,* 1; *woe upon the,* Matt. *23,* 13-33; Luke *11,* 42-44.
PIETY: *1* Tim. *4,* 7 ff; *6,* 5 ff; 2 Tim. *3,* 12; Jas. *1,* 26 f.
PILATE, PONTIUS: Roman procurator of Judea, Idumea and Samaria, at the time of Christ's public life, Matt. *27,* 1 f.11-26.58; Mark *15,* 1-15.43-45; Luke *3,* 1; *23,* 12-25.50-52; John *18,* 28; *19,* 22.38; Acts *3,* 13; *4,* 27; *13,* 28; *1* Tim. *6,* 13.
PLACE OF HONOR: Matt. *23,* 6; Luke *11,* 43; *20,* 46; Mark *12,* 38 f
POOR: *care of,* Acts *6,* 1-3; 2 Cor. *8,* 2-15; Gal. *2,* 10; *1* Tim. *5,* 16.
POVERTY: *voluntary,* Matt. *8,* 20; *19,* 21; Mark *10,* 21; Luke *9,* 58; *18,* 22; 2 Cor. *8,* 9.

Index of Scripture Texts

PLEASURES: *effects of,* Luke *8,* 14; Jas. *5,* 1-5; 2 Pet. *2,* 13; *warnings against,* 2 Tim. *3,* 1-9; Titus *3,* 3; Heb. *11,* 25; *1* Pet. *4,* 1-6.

POWER OF KEYS: Matt. *16,* 18 ff; Luke *22, 32;* John *21,* 17.

PROPHECIES OF CHRIST: *of His passion and death,* Matt. *16,* 21; *17,* 21 f; *20,* 17-19; *26,* 20-25.30 f; and parallels; John *2,* 19; *13,* 21-26; *of His resurrection,* Matt. *12,* 40; *16,* 21; *17,* 9.22; *20,* 19; *26, 32;* Mark *14,* 28; John *2,* 19; *16,* 16; *on the Apostles and the Church,* Matt. *10,* 16-25; *16,* 17-19; Luke *21,* 8-19; John *21,* 18-23; *on the destruction of Jerusalem and the end of the world,* Matt. *24;* Mark *13;* Luke *21,* 20-36.

PROPHECIES ON CHRIST: *fulfilled,* Matt. *2,* 23.

PRAYER: *necessity of,* Luke *18,* 1; *21,* 36; Eph. *6,* 18; Phil. *4,* 6; Col. *4,* 2; *1* Thess. *5,* 17; *qualities of,* Matt. *6,* 5-8; Mark *7,* 6; *11,* 24 f; Luke *11,* 9-13; *18,* 1-14; John *14,* 13; *15,* 5.7; *16,* 23 f; Rom. *12,* 12; 2 Cor. *12,* 7 ff; Eph. *6,* 18; Col. *4,* 2; *1* Thess. *5,* 17; Heb. *4,* 16; Jas. *1,* 6; *1* Pet. *4, 7; value of,* Matt. *7,* 7-11; *17,* 20; *18,* 20; *21,* 22; Luke *11,* 9-13; *18,* 7; *22,* 40; John *16,* 23; Acts *10,* 4; Rom. *10,* 1; Phil. *1,* 19; Col. *4,* 12; Heb. *4,* 16; Jas. *1,* 5; *5,* 13.16; *examples of,* Luke *1,* 46-55.68-79; *2,* 29-32.37; Acts *1,* 14; *7,* 59; *9,* 40; *10,* 2; *12,* 5; *16,* 25; *20,* 36; *21,* 5; 2 Tim. *1,* 3; Jas. *5,* 17 f; *confidence in,* Matt. *7,* 7-11; Mark *11,* 24; Luke *11,* 9-13; John *14,* 13; *16,* 23 f; Heb. *4,* 16; *1* John *3,* 21 f; *5,* 14 f;— *of Christ,* Matt. *11,* 25 f; *14,* 23; *26,* 36; *39,* 42.44; *27,* 46; Mark *6,* 46; *14,* 32.35 f.39; *15,* 34; Luke *6,* 12; *9,* 28; *22,* 41-43; *23,* 34.46; John *11,* 41 f; *12,* 27 f; *17; the Lord's Prayer,* Matt. *6,* 9-13; Luke *11,* 1-4; *—for others,* see Intercession and Assembly for Prayer.

PRECEDENCE: *question of,* Matt. *20,* 25-28; Mark *10,* 42-45; Luke *22,* 24-30.

PRESUMPTION: Jas. *4,* 13-17.

PRESBYTERS: Acts *11,* 30; *14,* 22; *15,* 2.23; *20,* 17; Titus *1,* 5; Jas. *5,* 14; *1* Pet. *5,* 1; 2 John *1;* 3 John *1;* see also Priests.

PRIDE: Luke *18,* 11 ff; *1* Cor. *4,* 6 ff; Gal. *5,* 26; Phil. *2,* 3; *warning against,* Rom. *1,* 29 ff; *11,* 20; *1* Cor. *4,* 7; Gal. *5,* 26; 2 Tim. *3,* 2; Jude 16; *penalty of,* Matt. *23,* 12; Luke *1,* 51; *14,* 11; *18,* 14; Jas. *4,* 6.16; *1* Pet. *5,* 5; Apoc. *18,* 7 ff.

PRIESTS: *office of,* Acts *14,* 22; *15,* 2; *1* Cor. *12,* 5; *1* Tim. *4,* 14; *5,* 17.19; Titus *1,* 5; Jas. *5,* 14 ff; *task of, 1* Cor. *4,* 1; 2 Cor. *5,* 18-20; *6,* 4; Eph. *4,* 11-13; see also Presbyters.

PRIESTHOOD OF CHRIST: Heb. *7; 8; 9; 10.*

PRIMACY: *bestowal of,* Matt. *16,* 15-19; Luke *22,* 31 f; John *21,* 15-17; *exercise of,* Acts *1,* 15; *2,* 14; *15,* 7.29.

PRINCE: *of this world,* Luke *4,* 5-7; John *12,* 31; *14,* 30; *1* Cor. *2,* 6.8; Eph. *2,* 2.

PROGRESS: *spiritual, 1* Pet. *2.*

PROPHETS: *Christian,* Acts *11,* 27; *13,* 1; *15,* 32; *1* Cor. *12,* 28; Eph. *2,* 20; *3,* 5; *4,* 11.

Index of Scripture Texts

PROPHETS: *false,* Matt. 7, 15; *24,* 5.11; Luke *6,* 26; Acts *13,* 6; *1* Tim. *4,* 1; 2 Pet. 2; *1* John *4,* 1.
PROPITIATION: *for our sins,* Rom. *3,* 25; *1* John 2, 2; *4,* 10.
PROVIDENCE OF GOD: see God, providence of.
PROVIDENCE: *trust in divine,* Matt. *6,* 25-34; *10,* 16; Luke *12,* 22-32.
PRUDENCE: Matt. 7, 6; Luke *16,* 8; Rom. *8,* 6-8; *12,* 16 f; *1* Cor. *4,* 10; 2 Cor. *4,* 2; Eph. *5,* 15-17; *1* Pet. *4,* 7.
PSALMS: *singing of,* Eph. *5,* 19 f; Col. *3,* 16 f; Jas. *5,* 13.
PURGATORY: Matt. *12,* 32; *1* Cor. *3,* 11-15.
PURIFICATION: *of the heart,* Acts *15,* 9; *1* Pet. *1,* 22; *1* John *3,* 3; *of women,* Luke 2, 22.
PURITY: see Chastity.
PURITY: *of the heart,* Matt. *5,* 8; 2 Cor. 7, 1; Eph. *1,* 4; Apoc. 7, 14; 22, 14.
PUTTING ON CHRIST: Rom. *13,* 14; Gal. *3,* 27.
PUTTING ON NEW MANHOOD: Eph. *4,* 24; Col. *3,* 10.

QUARRELS, QUARRELLING: *to be avoided,* Gal. *5,* 20; 2 Tim. 2, 23; see also Strife.

RASH JUDGMENT: Matt. 7, 1-5; Luke *6,* 41 f; Rom. 2, 1; Jas. *4,* 11-13.
RECONCILIATION: *with God,* Rom. *5;* 2 Cor. *5,* 18 f; Eph. 2, 16; Col. *1,* 20; Heb. 2, 17.
REDEEMER: Matt. *1,* 21; *9,* 13; *20,* 28; *26,* 28; Mark 2, 17; *10,* 45; *1* Tim. *1,* 15; 2, 5 f; *1* John *3,* 5.8; Apoc. *1,* 5 f.
REDEMPTION: Rom. *3,* 24; *5,* 9 ff; *1* Cor. *1,* 30; Eph. *1,* 7; Col. *1,* 14.
REJOICING: *spiritual,* Matt. *11,* 25-27; *13,* 16 f; Luke *10,* 21-24.
RELICS: *miraculous,* Matt. *9,* 20 f; Mark *5,* 27 f; Luke *8,* 44-47; Acts *19,* 12.
RENUNCIATION OF FAITH: *warning against,* Matt. *24,* 4 f; Acts *20,* 29-31; 2 Cor. *11,* 3.13 ff; Gal. *1,* 6-9; Col. 2, 8.18; 2 Thess. 2, 2; Heb. *13,* 9; *1* John *3,* 7; *penalty for,* 2 Thess. 2, 8-11; Heb. *6,* 4-8; *10,* 26-29; 2 Pet. 2, 15-17.21; Jude 11 f.
REPENTANCE: *motives for,* Luke *13,* 3-5; *15,* 7.10; Acts *3,* 19; Rom. 2, 4; *operation of,* Matt. *16,* 19; *18,* 18; John *20,* 23; *1* John *1,* 9; *examples of,* Matt. *26,* 75; Luke 7, 37 ff.44-48; *15,* 11-24; *18,* 13; *19,* 8 ff; *23,* 40-43; 2 Cor. 7, 9 ff; *delay of,* Matt. *11,* 20-24; *13,* 15; Luke *13,* 34; *19,* 42-44; Rom. 2, 5; Heb. *3,* 15-19.
RESURRECTION: *of Christ, foretold,* see under Prophecies; *witnesses of,* Matt. *28;* Mark *16;* Luke *24;* John *20; 21;* Acts *1,* 1-9.21 ff; 2, 24.32; *3,* 15; *17,* 31; Rom. *1,* 4; *4,* 24 f; *6,* 4; *8,* 11.34; *1* Cor. *6,* 14; *15,* 4-8; Gal. *1,* 1; Eph. *1,* 20; *1* Thess. *4,* 14; *1* Pet. *1,* 3.21.
RESURRECTION: *of the dead, certainty of,* Matt. 22, 30-32; Mark *12,* 26 ff; Luke *20,* 37 ff; John *5,* 28 ff; *6,* 39; *11,* 24; Acts *17,* 18.31 f; *23,* 6; *24,* 15; *26,* 8; Rom. *8,* 11; *1* Cor. *6,* 14; *15,* 12-22; 2 Cor. *1,* 9; *4,* 14; *1* Thess. *4,* 13-17; Apoc. *20,* 12 ff; *state of the resurrected,* Matt. 22, 24-30;

Index of Scripture Texts

Mark *12,* 19-25; Luke *20,* 28-36; *1* Cor. *15,* 35-44.51 ff; Phil. *3,* 21; Col. *3,* 4; *1* John *3,* 2.

REVELATION: *natural,* Acts *14,* 14-16; *17,* 27 ff; Rom. *1,* 19 ff; *1* Cor. *1,* 21; *supernatural,* Matt. *11,* 25; *16,* 17; John *17,* 6; *1* Cor. *2,* 10 ff; Gal. *1,* 15 f; Col. *1,* 26 f; *1* Tim. *3,* 16; *2* Tim. *1,* 10; Heb. *1,* 1 ff.

REVELRY: *warning against,* Luke *21,* 34; Rom. *13,* 13; *penalty of,* Luke *6,* 25; *16,* 19-25; *17,* 26-29; Gal. *5,* 21; Phil. *3,* 18 ff; Jas. *5,* 5.

REVENGE: *warning against,* Matt. *5,* 39; Rom. *12,* 19; *1* Thess. *5,* 15; Heb. *10,* 30; *1* Pet. *3,* 9.

RICHES: *perils of,* Matt. *6,* 24; *19,* 23-26; Mark *10,* 23-27; Luke *6,* 24; *12,* 16-21; *16,* 13; *18,* 24-27; *1* Tim. *6,* 9 ff; Jas. *5,* 1-6; *duties of, 1* Tim. *6,* 17-19; Jas. *1,* 10.

RIGHTEOUS: *blessing of the,* Matt. *13,* 43; Acts *10,* 35; Rom. *2,* 10; *1* Pet. *3,* 12; *1* John *3,* 7; Apoc. *22,* 11; *rewards to the,* Matt. *5,* 12; *6,* 1; *10,* 41; Luke *6,* 35; *1* Cor. *3,* 8; Col. *3,* 24; Heb. *10,* 35; *11,* 6; Apoc. *22,* 12.

ROME: city in Italy, capital of Roman Empire, Acts *2,* 10; *18,* 2; *28,* 14-31; Rom. *1,* 7 ff.

SABBATH: *observance of,* Matt. *12,* 1-14; Mark *2,* 23-28; *Luke 6,* 1-11; *on the first day of the week (our Sunday),* Acts *20,* 7; *1* Cor. *16,* 2.

SACRAMENTS: institution and administration:
Baptism, Matt. *28,* 19; Mark *16,* 16;
Confirmation, Acts *8,* 14-17; *19,* 1-6; Eph. *1,* 13 f; Heb. *6,* 2-4;
Penance, Matt. *16,* 19; *18,* 18; John *20,* 22 f.
Holy Eucharist, Matt. *26,* 17-29; Mark *14,* 12-25; Luke *22,* 7-23; *1* Cor. *11,* 23-26;
Extreme Unction, Jas. *5,* 14 f.
Holy Orders, Luke *22,* 19; John *20,* 22; Acts *6,* 6; *1* Cor. *11,* 24 f; Col. *4,* 17; *1* Tim. *5,* 22; *2* Tim. *1,* 6 f; *1* Pet. *2,* 9;
Matrimony, Matt. *19,* 4-6; Eph. *5,* 22-33; Col. *3,* 18; *1* Pet. *3,* 1.

SACRIFICE: *of Christ,* Rom. *3,* 24 ff; *1* Cor. *5,* 7; *2* Cor. *5,* 21; Heb. *8,* 1-10; *9,* 11 ff; *10,* 1-18; *13,* 10; *spiritual,* Rom. *12,* 1; Heb. *13,* 15; *1* Pet. *2,* 5.

SADDUCEES: religious and political party in Judea, a kind of freethinkers in matters of religion, Matt. *3,* 7; *16,* 1-12; *22,* 23-34; Mark *12,* 18-27; Luke *20,* 27 ff; Acts *4,* 1; *23,* 7 ff.

SAINTS: *according to God,* Rom. *8,* 27; Eph. *1,* 18; *2,* 19; Col. *1,* 12; Jude 3; Apoc. *5,* 8; *their power and glory, 1* Cor. *13,* 12; Heb. *12,* 22 f; Apoc. *2,* 26 f; *4,* 4; *5,* 8; *6,* 9; *7,* 9; *14,* 1-4; *19,* 4-6; *20,* 4.

SALVATION: *call to,* Matt. *20,* 16; *22,* 14; Gal. *1,* 15; Eph. *1,* 18; *2,* 4-10; *1* Tim. *2,* 4; *1* Pet. *2,* 9; *for the Jews,* Rom. *9,* 6-13.24; *11,* 1-32; *15,* 8; *for the Gentiles,* Acts *15,* 7-9; Rom. *9,* 24; *11,* 11-32; *15,* 9-12; Eph. *3,* 6-9; Col. *1,* 25-27; *of the faithful,* Jude 17-23; *of the soul,* Matt. *6,* 33; *16,* 24-26; Mark *8,* 34-37; Luke *9,* 23-25; *10,* 42; *12,* 13-21.

Index of Scripture Texts

SANHEDRIN: or Council, highest Jewish court in Jerusalem, Matt. 5, 22; *10,* 17; *26,* 59; *27,* 1; Mark *13,* 9; *14,* 55; *15,* 1; Luke 22, 66; John *11,* 47; Acts *4,* 15; *5,* 21.27.34.41; *6,* 12.15; *22,* 30; *23,* 1.6.15.20.28; *24,* 20.

SATAN: *fall of,* Luke *10,* 18; John *12,* 31; *16,* 11; Col. 2, 15; *seduction by,* Mark *4,* 15; Luke 22, 3; John *13,* 27; Acts 5, 3 ff; 2 Cor. 2, 11; 2 Thess. 2, 9; Apoc. *12,* 9; also see Devil, Evil Spirit.

SCANDAL: *warning against,* Matt. 5, 29 ff; *17,* 26; *18,* 6-9; Mark *9,* 41-49; Luke *17,* 1 ff; Rom. *14,* 13-16.20 ff; *16,* 17; *1* Cor. *8,* 9-13; *10,* 23-33; 2 Cor. *6,* 3; *of the Pharisees and Scribes,* Matt. *23,* 1-33; Luke *11,* 37-44.

SCRIBES: transcribers of the Sacred Books and professors of the Law, *enemies of Christ,* Matt. *9,* 3; *12,* 38; *15,* 1 f; *26,* 57; *27,* 41; Luke *11,* 53 f; *warning against the,* Mark *12,* 38-40; Luke *20,* 45-47; *Christ pronounces woe upon the,* Matt. *23,* 13-36; Luke *11,* 45-52.

SECOND COMING OF CHRIST: Matt. *24,* 26 f.29-31; Mark *13,* 24-27; Luke *17,* 22-25; *21,* 25-28; Acts *1,* 11; *1* Thess. *4,* 13-18; *5,* 1-11; 2 Thess. 2, 1-12; Jude 14; Apoc. *1,* 7.

SELF-DENIAL: Matt. *16,* 24; Mark *8,* 34; Luke *9,* 23; *14,* 26; John *12,* 25; *1* Cor. *9,* 25-27; Gal. *5,* 24.

SERMON ON THE MOUNT: Matt. *5; 6; 7;* Luke *6,* 17-49.

SERVANTS: *duties of,* Eph. *6,* 5-8; Col. *3,* 22; *1* Tim. *6,* 1 f; Titus 2, 9; *1* Pet. 2, 18.

SERVICE: *key to greatness in Christ,* Matt. *20,* 24-28; Mark *10,* 41-45; Luke 22, 24-27.

SEVEN WORDS FROM THE CROSS: (1) Luke *23,* 34; (2) *23,* 43; (3) John *19,* 26 f; (4) Matt. *27,* 46 f; Mark *15,* 34; (5) John *19,* 29; (6) John *19,* 30; (7) Luke *23,* 46.

SHEPHERD: *the Good,* John *10,* 1-16; Heb. *13,* 20; *1* Pet. 2, 25; *5,* 4.

SHIELD OF FAITH: Eph. *6,* 16.

SICKNESS: *cause and effect of;* John *5,* 14; *9,* 1-3; *11,* 4; Acts *12,* 21-23; Apoc. 2, 21 ff; *consolation in, 1* Cor. *15,* 42 ff; Jas. *5,* 15; *1* Pet. *4,* 19.

SIGHT OF GOD: *acts in the,* Acts *4,* 19; *10,* 31; 2 Cor. 2, 17; *4,* 2 f; *7,* 12 f; *1* Thess. *1,* 3; *1* Tim. 2, 1-4; *6,* 13 f; *1* Pet. *3,* 4 f.

SIGN FROM HEAVEN REFUSED: Matt. *16,* 1-4; Mark *8,* 11-13; Luke *11,* 16 ff.

SIN: *what is,* Rom. *14,* 23; Jas. *1,* 15; *4,* 17; *1* John *3,* 4; *5,* 17; *origin of,* Matt. *15,* 19; John *8,* 44; *15,* 21 ff; Rom. *5,* 12; Jas. *1,* 13-15; *1* John *3,* 8; *mortal sin,* Matt. *5,* 22; *10,* 14 f; Luke *10,* 10-12; *12,* 47; John *19,* 11; *1* John *5,* 16 f; *crying to Heaven,* Jas. *5,* 4; *—against the Holy Spirit,* Matt. *12,* 31 ff; Mark *3,* 28 ff; Luke *12,* 10; Acts *7,* 51; Heb. *6,* 4-6; *10,* 26 ff; *original,* Rom. *5,* 12-19; *7,* 7-25; *1* Cor. *15,* 21 f; Eph. 2, 3; *occasion of,* Matt. *5,* 27-30; *18,* 8 ff; Mark *9,* 42-48; Rom. 7; *relapse into,* Matt. *12,* 43-45; Luke *11,* 24-26; John *5,* 14; *8,* 11; Heb. *6,* 4-8; *10,* 26-30; 2 Pet. 2, 20-22; *penalty of,* Rom. *1,* 18; 2, 6 ff; *6,* 21-23; *1* Cor. *6,* 9; *15,* 56; Gal. *5,* 19-21; *6,* 7 ff; Eph. *5,* 5; Col. *3,* 25; Heb. *10,* 26-29; 2 Pet. 2, 3-6.9 ff; *confession of,* Matt. *3,* 6; Mark *1,* 5;

Index of Scripture Texts

Jas. 5, 16; *1* John *1,* 9; *forgiven,* Matt. *6,* 12; *9,* 2; Luke 7, 47 f; Jas. *5, 15* f; *1* John *1,* 7-10; *2,* 2; *remission of,* Matt. *26,* 28; Luke *24,* 47; Acts *2,* 38; *10,* 43; Col. *1,* 14; Heb. *9,* 22; *10,* 18; *dominion over,* Rom. *5,* 12-21.

SINGLENESS: *of purpose,* Matt. *6,* 22 f; Luke *11,* 34-36.

SLANDER: Rom. *1,* 30; 2 Cor. *12,* 20; Eph. *4,* 31; *1* Tim. *3,* 11; *1* Pet. *2,* 1.

SOBRIETY: Luke *21,* 34; *1* Thess. *5,* 6-8; *1* Pet. *1,* 13; *5,* 8.

SOLITUDE: Matt. *6,* 6; *14,* 13.15; *26,* 36; Mark *1,* 35; *6,* 32; Acts *10,* 9.

SON OF GOD: Matt. *3,* 17; *14,* 33; *16,* 16 f; *17,* 5; *26,* 63 f; Mark *9,* 6; *14,* 61 f; Luke *9,* 35; John *6,* 30; *20,* 31; 2 Pet. *1,* 17.

SON OF MAN: Matt. *8,* 20; *16,* 13 f.27 f; Mark *8,* 31; Acts *7,* 55.

SONSHIP OF GOD: *1* John 2, 28 f; *3; 4,* 1-6.

SOUL: *immortality of,* Matt. *10,* 28; *22,* 32; John *3,* 16; *12,* 25; 2 Cor. *5,* 1; Gal. *6,* 8; *value of,* Matt. *16,* 26; Mark *8,* 36 ff; 2 Cor. *12,* 15.

SPIRITUAL VOCATION: *call to,* Rom. *11,* 29; *1* Cor. *1,* 26; Eph. *1,* 18; *4,* 1; Phil. *3,* 14; 2 Thess. *1,* 11; 2 Tim. *1,* 9; Heb. *3,* 1; *1* Pet. 2, 9; 2 Pet. *1,* 10; Apoc. *19,* 9.

STRENGTH IN GOD: *1* Cor. *16,* 13; Eph. *6,* 10 ff; 2 Tim. 2, 1; *1* John 2, 14.

STRIFE: Rom. *13,* 14; *1* Cor. *3,* 3; Gal. *5,* 20 f; *1* Tim. *6,* 4 f; 2 Tim. 2, 23 f; Titus *3,* 9; Jas. *3,* 14.

SUFFERING: *value of,* Matt. *5,* 10 ff; Luke *24,* 26; John *16,* 22; Rom. *5,* 3; *8,* 18; 2 Cor. *4,* 17; Phil. *2,* 7-11; 2 Thess. *1,* 7-10; 2 Tim. 2, 12; *1* Pet. *1,* 6 ff; *2,* 20 ff; *3,* 14; *4,* 12-16; *consolation in,* 2 Cor. *1,* 4 ff; Heb. *2,* 9 ff; *1* Pet. *2,* 19-24; see also Affliction; *patience in,* see under Patience.

STEPHEN: deacon of the early Church and first martyr, Acts *6,* 5.8-15; *7; 8,* 1 f.

SUPERSTITION: John *16,* 2.

SWEARING: See Actions in Words, also Oath.

TALE BEARING: See Actions in Words.

TAXES: *question of,* see Civil Authorities.

TEMPORAL GOODS: *use of,* Matt. *19,* 21; Luke *18,* 22; *19,* 8; *1* Cor. *7,* 31; *1* Tim. *6,* 17-19; Heb. *13,* 5; *1* John *3,* 17.

TEMPORAL THINGS: *anxious care over, censured,* Matt. *6,* 25-34; Luke *8,* 14; *10,* 41; *12,* 22-31; John *6,* 27; *1* Cor. 7, 32-34; Phil. *4,* 6; *1* Tim. *6,* 7; 2 Tim. 2, 4; Heb. *13,* 5.

TEMPTATION: *origin of,* Matt. *4,* 1-11; Luke *22,* 31 ff; John *13,* 2; Acts *5,* 3; Rom. *7,* 22 ff; 2 Cor. *12,* 7-9; Eph. *6,* 11-17; *1* Thess. *3,* 5; Jas. *1,* 13 ff; *1* Pet. *5,* 8 ff; *value of,* Jas. *1; 2; 4; 1* Pet. *1,* 6.

TESTIMONY: *false,* Matt. *19,* 18; *26,* 60 ff; Acts *6,* 11.

THANKSGIVING: *thankfulness, to God, 1* Cor. *10,* 30; 2 Cor. *1,* 11; Eph. *5,* 20; Phil. *4,* 6; Col. *1,* 12; *2,* 6 f; *3,* 15.17; *4,* 2; *1* Thess. *5,* 18; *1* Tim. *1,* 12; *examples of,* Matt. *15,* 36; Luke *17,* 15 f; Acts *4,* 23-31;

Index of Scripture Texts

Rom. *1*, 8; *6*, 17; *1* Cor. *1*, 4; Phil. *1*, 3-5; Col. *1*, 3 ff; *1* Thess. *1*, 2 ff; 2 Thess *1*, 3; —*by children,* 1 Tim. *5*, 4.
THEOLOGICAL VIRTUES: *the three,* Rom. *5*, 1-5; *1* Cor. *13*, 13; Col. *1*, 4 ff; *1* Thess. *1*, 3.
THOUGHTS: See Actions of Mind.
THRONE OF GOD: Heb. *1*, 8; *4*, 16; Apoc. *1*, 4; *3*, 21; *4*, 10.
TIME: *making use of,* 2 Cor. *6*, 2; Gal. *6*, 9 f; Eph. *5*, 15 f.
TOLERANCE: Mark *9*, 38-40; Luke *9*, 49 f.
TONGUE: *control of the,* Col. *3*, 8; Jas. *1*, 19.26; *3*, 1-14.
TONGUES: *gift of,* Acts *2*, 3 f; *10*, 46; *19*, 6; *1* Cor. *14*, 1-23.
TIMOTHY: son of a Greek father and a Jewish mother, companion of Paul, Acts *16*, 1-3; *17*, 14 f; Rom. *16*, 21; *1* Cor. *4*, 17; *16*, 10 f; 2 Cor. *1*, 1.19; Phil. *2*, 19-23; *1* Tim.; 2 Tim.; Heb. *13*, 23.
TITUS: disciple and companion of Paul, 2 Cor. *2*, 13; *7*, 6; *8*, 6.16; *12*, 18; Gal. *2*, 1.3; 2 Tim. *4*, 10; Titus *1*, 4 ff.
TRADITION: John *20*, 30; *21*, 25; Acts *16*, 4; *1* Cor. *11*, 2; 2 Thess. *2*, 14; 2 Tim. *1*, 13 ff; *2*, 2; *3*, 14; *1* John *2*, 24.
TREASURES IN HEAVEN: Matt. *6*, 19-21; *19*, 21; Luke *12*, 33 f.
TRINITY: Matt. *3*, 16; *28*, 19; Mark *1*, 10 ff; Luke *3*, 22; John *14*, 16; *15*, 26; Acts *2*, 33; 2 Cor. *13*, 13; Eph. *2*, 18; *1* Pet. *1*, 2; *1* John *5*, 7.
TRUST IN GOD: *1* Tim. *4*, 10; 2 Tim. *1*, 12; *4*, 18; Heb. *13*, 5 f; *1* John *2*, 21 ff; *examples of,* Matt. *9*, 2.21; *15*, 22-28; Luke *5*, 4-9.
TRUTH: *to speak the,* Eph. *4*, 25.
TWELVE, THE: see Apostles.

UNBELIEF: *cause of,* Luke *16*, 27-31; John *3*, 19 ff; *5*, 44-47; *8*, 42-47; *10*, 24-26; *12*, 37-40; Acts *13*, 45 ff; *1* Tim. *1*, 13; *penalty of,* Luke *16*, 16; John *3*, 18.36; *8*, 24; *12*, 48; Rom. *11*, 20; Eph. *5*, 6; Apoc. *21*, 8.
UNCLEAN SPIRITS: Matt. *10*, 1; *12*, 43 ff; Mark *1*, 23-27; *5*, 2 ff; Luke *6*, 18; *9*, 38-43; *11*, 24-26; Acts *5*, 16; *8*, 7.
UNGRATEFULNESS: Luke *17*, 17 ff; 2 Tim. *3*, 2.
UNMERCIFULNESS: Matt. *18*, 28-35; *25*, 41-46; Luke *16*, 20 ff; Jas. *2*, 13.

VENGEANCE: *unlawful,* Matt. *5*, 38-42; Luke *6*, 27-29; Rom. *12*, 19; *1* Thess. *5*, 15; Heb. *10*, 30; *1* Pet. *3*, 9.
VIGILANCE: see Watchfulness.
VIRGINITY: Matt. *19*, 12; Luke *20*, 34-36; Acts *21*, 9; *1* Cor. *7*, 25-40; Apoc. *14*, 4 f.
VIRTUE: *necessity and usefulness of,* Matt. *5*, 6; *11*, 29 f; *19*, 17; *1* Cor. *13*, 13; Phil. *3*, 12-14; *1* Thess. *4*, 3; *1* Tim. *4*, 8; Apoc. *22*, 11 f.
VOCATION: *call to spiritual,* see Spiritual vocation.

WATCHFULNESS: Matt. *24*, 42-44; *25*, 1-13; Mark *13*, 33-37; Luke *12*, 35-40; *21*, 34-36; *1* Cor. *16*, 13; *1* Thess. *5*, 6; 2 Tim. *4*, 5; *1* Pet. *4*, 7; *5*, 8; Apoc. *3*, 2 ff; *16*, 15.
WASTE: *censured,* Matt. *14*, 20; Mark *6*, 43; Luke *9*, 17; John *6*, 12.

Index of Scripture Texts

WAY OF THE CROSS: Matt. *27*, 31 f; Mark *15*, 20-24; Luke *23*, 26-33; John *19*, 16-18.
WEAK IN FAITH: *the (weak Christians),* Rom. *14; 15; 1* Thess. *5,* 14; Heb. *12,* 12 f.
WICKED ANGELS: Acts *19,* 12 ff; *2* Pet. *2,* 4; Jude 6.
WIDOW'S MITE: Mark *12,* 41-44; Luke *21,* 1-4.
WIDOWS: *conduct of, 1* Cor. *7,* 8 ff.39; *1* Tim. *5,* 3-16; *support of,* Acts *6,* 1-3; *9,* 39; *1* Tim. *5,* 16; Jas. *1,* 27.
WILL OF GOD: *doing the,* Matt. *7,* 21-25; Luke *6,* 46-47; John *7, 17; 1* Thess. *4,* 3 f; *5,* 18; *1* Pet. *2,* 15; *4,* 2; *1* John 2, 17; *3,* 23 f.
WISE MEN FROM THE EAST: Matt. *2,* 1-16.
WISDOM: *earthly,* Jas. *3,* 14-18; *heavenly, 1* Cor. *2,* 6-16; Jas. *1,* 5; *3,* 13.
WIVES: *duties of, to husbands,* Rom. *7,* 2; *1* Cor. *7,* 3; Eph. *5,* 22.33; Titus *2,* 4 f; *1* Pet. *3,* 1.5 f.
WOMEN: *duties of, 1* Cor. *11,* 3-15; *1* Tim. *2,* 9 f; Titus *2,* 3-5; *1* Pet. *3,* 3-5; *adornment (apparel) of, 1* Tim. *2,* 9 f; *1* Pet. *3,* 3-5; *permission to teach in Church denied to, 1* Cor. *14,* 34 ff; *1* Tim. *2,* 12; *—to keep silence, 1* Tim. *2,* 11 f.
WORD OF GOD: *dignity of,* Matt. *4,* 4; Luke *4,* 4; Eph. *6,* 17; *1* Thess. *2,* 13; *1* Pet. *1,* 25; *effect of,* Matt. *13,* 18-23; Mark *4,* 13-20; Luke *8,* 11-15; Acts *2,* 37-41; *4,* 4; *13,* 48; *14,* 3; *17,* 11 ff; Heb. *4,* 12; Jas. *1,* 21-25.
WORD ETERNAL: *made flesh,* John *1,* 1-14.
WORK: see Labor.
WORKS: see Action in Works.
WORLD: *end of,* Matt. *24,* 3 ff; Mark *13;* Luke *21,* 5-36; *2* Pet. *3,* 1-14.
WORLD: *judgment of,* Apoc. *20,* 11-15.
WORLD: *love of,* John *15,* 19; *2* Tim. *4,* 9; Jas. *4,* 4; *1* John *2,* 15-17.
WORDLINESS: Jas. *4,* 4-10.
WORLDLY GOODS: *foolish worry over,* Matt. *6,* 25-34; *13,* 22; Mark *4,* 18 f; Luke *8,* 14. See Temporal Goods.
WORSHIP: *to God alone,* Matt. *4,* 10; Luke *4,* 8; Apoc. *19,* 10; *22,* 9.
WORSHIP: *public, 1* Tim. *2,* 1-10.

YOKE OF CHRIST: Matt. *11,* 30.

ZACHARY: priest, and father of John the Baptist, Luke *1,* 5-23.57-80.
ZEAL: Matt. *25,* 14-23; John *2,* 13-17; Titus *2,* 11-15; *1* Pet. *3,* 13.

ADDITIONAL READING

1. THE NEW TESTAMENT IN ENGLISH

The Westminster Version of the Sacred Scriptures. 4 Vols. London: Longmans, Green.

F. A. SPENCER, O.P., *The New Testament of our Lord and Saviour Jesus Christ.* New York: Macmillan, 1937.

HUGH POPE, O.P., *The Layman's New Testament.* (Challoner's text; introductions and notes by Pope.) New York: Sheed and Ward, 1938.

2. HARMONIES, CONCORDANCES

S. HARTDEGEN, O.F.M., *A Chronological Harmony of the Gospels.* Paterson: St. Anthony Guild, 1942.

J. E. STEINMUELLER, *A Gospel Harmony, Using the Confraternity Edition of the New Testament.* New York: Sadlier, 1941

N. THOMPSON, *Verbal Concordance to the New Testament.* Baltimore: Murphy, 1928.

N. THOMPSON and R. STOCK, *Concordance to the Bible.* St. Louis: Herder, 1942.

K. VAUGHAN, *The Divine Armory of Holy Scripture.* St. Louis: Herder, 1931.

T. WILLIAMS, *A Textual Concordance of the Holy Scriptures, Arranged Especially for Use in Preaching.* New York: Benziger, 1908.

T. WILLIAMS, *Concordance of Proper Names in the Holy Scriptures.* St. Louis: Herder, 1923.

3. GENERAL AND SPECIAL INTRODUCTION

P. BATIFFOL, *The Credibility of the Gospel.* London: Longsman, Green, 1912.

J. CHAPMAN, O.S.B., *Matthew, Mark and Luke.* London: Longmans, Green, 1937.

J. DOUGHERTY, *Outlines of Bible Study.* Milwaukee: Bruce, 1934.

W. DOWD, S.J., *The Gospel Guide.* Milwaukee: Bruce, 1932.

The Bible, Its History, Authenticity and Authority. Treated in a series of lectures by Catholic Scholars delivered at Aberdeen, 1924-1925.

L. FILLION, S.S., *The Study of the Bible.* New York: Kenedy, 1926.

F. E. GIGOT, S.S., *General Introduction to the Study of the Holy Scriptures.* New York: Benziger, 1901.

F. E. GIGOT, S.S., *Special Introduction to the Study of the New Testament.* New York: Benziger, 1903.

F. E. GIGOT, S.S., *Special Introduction to the Study of the Old Testament.* 2 vols. New York: Benziger, 1906.

C. GRANNAN, *General Introduction to the Bible.* 4 vols. St. Louis: Herder, 1921.

J. HUBY, *The Church and the Gospels.* N. Y.: Sheed and Ward, 1931.

Additional Reading

E. Jacquier, *History of the Books of the New Testament.* Vol. I: St. Paul and His Epistles. London: Kegan Paul, Trench, Trubner, 1907.

T. Moran, *Introduction to the Scriptures.* N. Y.: Sheed and Ward, 1937.

Hugh Pope, O.P., *Catholic Student's Aids to the Study of the Bible.* 5 vols. London: Burns, Oates and Washbourne, 1926.

H. Schumacher, *A Handbook of Scripture Study.* 3 vols. St. Louis: Herder, 1923.

M. Seisenberger, *Practical Handbook for the Study of the Bible.* New York: Wagner, 1911.

J. Simon, O.S.M., *A Scripture Manual, Directed to the Interpretation of Biblical Revelation.* 2 vols. New York: Wagner, 1924, 1928.

J. E. Steinmueller, *A Companion to Scripture Studies.* 3 vols. New York: Wagner, 1941—1943.

L. Vaganay, *An Introduction to the Textual Criticism of the New Testament.* London: Sands, 1937.

4. HISTORY, TOPOGRAPHY, MAPS

F. E. Gigot, S.S., *Outlines of New Testament History.* New York: Benziger, 1898.

F. E. Gigot, S.S., *Outlines of Jewish History.* New York: Benziger, 1913.

E. Graf, O.F.M., *In God's Own Country.* London: Burns, Oates and Washbourne, 1937.

Mgr. Legendre, *The Cradle of the Bible.* London: Sands, 1929.

B. Meistermann, *Guide to the Holy Land.* New York: Kenedy, 1923.

E. Seraphin, O.F.M., *Maps of the Land of Christ.* Paterson: St. Anthony Guild, 1938.

5. LIVES AND TEACHINGS OF CHRIST

T. Conaty, *New Testament Studies.* New York: Benziger, 1898.

L. De Grandmaison, S.J., *Jesus Christ.* 3 vols. N. Y.: Macmillan, 1930.

L. Fillion, S.S., *The Life of Christ.* 3 vols. St. Louis: Herder, 1928.

C. Fouard, *The Christ the Son of God.* 2 vols. New York: Longmans, Green, 1915.

M. J. Lagrange, O.P., *Gospel of Jesus Christ.* 2 vols. New York: Benziger, 1938-39.

J. Lebreton, S.J., *The Life and Teachings of Jesus Christ our Lord.* Milwaukee: Bruce, 1935.

M. Lepin, *Christ and the Gospel.* Philadelphia: McVey, 1910.

M. Meschler, S. J., *The Life of our Lord Jesus Christ the Son of God.* In meditations. 2 vols. St. Louis: Herder, 1924.

I. O'Brien, O.F.M., *The Life of Christ.* Paterson: St. Anthony Guild, 1937.

F. M. Willam, *The Life of Jesus Christ.* St. Louis: Herder, 1936.

6. THE PARABLES

C. Callan, O.P., *The Parables of Christ.* New York: Wagner, 1941.

P. Coghlan, *Parables of Jesus.* New York: Kenedy, 1919.

Additional Reading

L. Fonck, S.J., *The Parables of the Gospel*. New York: Pustet, 1915.
B. Maturin, *Practical Studies on the Parables*. New York: Longmans, Green, 1906.
M. Ollivier, O.P., *The Parables of Our Lord Jesus Christ*. Dublin: Browne and Nolan, 1927.

7. LITURGICAL PASSAGES

Catholic Biblical Association of America, *The Epistles and Gospels for Sundays and Holydays*. New York: Sadlier, 1941.
C. Callan, O.P., *Gospels and Epistles of Sundays and Feasts, with Outlines for Sermons*. New York: Wagner, 1922.
J. Carey, *The Sunday Gospels for Priests and People*. New York: Kenedy, 1935.
L. Kreciszewski, *Sunday Gospels for the Layman*. Winnipeg: Tonkin, 1934.
C. Ryan, *The Gospels of the Sundays and Festivals*. 2 vols. Dublin: Browne and Nolan, 1914.
C. Ryan, *The Epistles of the Sundays and Festivals*. 2 vols. Dublin: Gill, 1931.

8. THE FOUR GOSPELS

A. Breen, *A Harmonized Exposition of the Four Gospels*. 4 vols. Rochester: Smith, 1908.
C. Callan, O.P., *The Four Gospels with a Practical Critical Commentary for Priests and Students*. New York: Wagner, 1940.
Cornelius a Lapide, S.J., *The Great Commentary: The Holy Gospels*. 6 vols. London: Hodges, 1893.
J. MacEvilly, *An Exposition of the Gospels*. New York: Benziger, 1902.

9. MATTHEW

Madame Cecilia, *The Gospel According to St. Matthew*. London: Kegan Paul, Trench, Trubner, 1906.
H. Cladder, S.J., *In the Fullness of Time: The Gospel of St. Matthew Explained*. St. Louis: Herder, 1925.
A. Maas, S.J., *The Gospel According to St. Matthew with Explanatory and Critical Commentary*. St. Louis: Herder, 1898.
J. Maldonatus, S.J., *Commentary on the Holy Gospels: St. Matthew's Gospel*. 2 vols. New York: Benziger, 1898.
L. Miller, *Gospel According to St. Matthew Explained for Religion Classes*. New York: Wagner, 1937.
J. Rickaby, S.J., *The Gospel According to St. Matthew*. London: Burns and Oates, 1913.

10. MARK

Madame Cecilia, *The Gospel According to St. Mark*. London: Kegan Paul, Trench, Trubner, 1904.
R. Eaton, *Gospel According to St. Mark*. London: Burnes, Oates and Washbourne, 1920.

Additional Reading

J. Kleist, S.J., *Memoirs of St. Peter*. Milwaukee: Bruce, 1932.
J. Kleist, S.J., *Gospel of St. Mark*. Milwaukee, Bruce, 1936.
M. J. Lagrange, O.P., *The Gospel According to St. Mark*. London: Burns, Oates and Washbourne, 1930.
S. F. Smith, S.J., *The Gospel According to St. Mark*. New York: Benziger, 1901.

11. LUKE

Madame Cecilia, *The Gospel According to St. Luke*. London: Burns, Oates and Washbourne, 1930.
R. Eaton, *Gospel According to St. Luke*. London: C.T.S., 1917.
V. Rose, O.P., *Holy Gospel According to St. Luke*. Baltimore: Murphy, 1931.
R. Stoll, *Gospel According to St. Luke*. New York: Pustet, 1931.
B. Ward, *Gospel According to St. Luke*. St. Louis: Herder, 1915.

12. JOHN

F. Bormann, *The Treasure Infinite*. Ipswich, S.D. F. Bormann, 1928.
Madame Cecilia, *The Gospel According to St. John*. New York: Benziger, 1923.
J. Donovan, S.J., *The Authorship of St. John's Gospel*. London: Burns, Oates and Washbourne, 1936.
J. McIntyre, *Holy Gospel According to St. John*. London: C.T.S., 1899.
J. MacRory, *The Gospel of St. John*. Dublin: Browne and Nolan, 1900.
J. Rickaby, S.J., *Gospel According to St. John*. London: Burns and Oates, 1912.

13. ACTS OF THE APOSTLES

T. A. Burge, O.S.B., *The Acts of the Apostles*. London: Burns and Oates, 1914.
C. Callan, O.P., *Acts of the Apostles*. New York: Wagner, 1919.
Madame Cecilia, *The Acts of the Apostles*. London: Kegan Paul, Trench, Trubner, 1908.
D. Lynch, S.J., *The Story of the Acts of the Apostles*. New York: Benziger, 1917.
J. MacEvilly, *An Exposition of the Acts of the Apostles*. New York: Benziger, 1911.
E. Parker, *Introduction to the Acts of the Apostles and the Epistles of St. Paul*. New York: Longmans, Green, 1927.

14. ST. PAUL: GENERAL

C. Callan, O.P., *The Epistles of St. Paul*. 2 vols. N. Y.: Wagner, 1931.
C. Lattey, S.J., *Paul*. Milwaukee: Bruce, 1939.
J. MacEvilly, *An Exposition of the Epistles of St. Paul and of the Catholic Epistles*. 2 vols. Dublin: Kelly, 1875.
B. Piconio, *Exposition of the Epistles of St. Paul*. London: Hodges, 1888.

Additional Reading

F. Prat, S.J., *The Theology of St. Paul.* 2 vols. New York: Benziger, 1927.
F. Prat, S.J., *St. Paul.* New York: Benziger, 1928.

15. ROMANS

P. Boylan, *St. Paul's Epistle to the Romans.* Dublin: Gill, 1934.

16. CORINTHIANS

Cornelius a Lapide, S.J., *Commentary on First Corinthians.* London: Hodges, 1896.
C. Lattey, S.J., *Readings in First Corinthians, Church Beginnings in Greece.* St. Louis: Herder, 1928.
J. MacRory, *Epistles of St. Paul to the Corinthians.* Dublin: Gill, 1935.

17. EPISTLES OF THE CAPTIVITY

E. Cerny, S.S., *Firstborn of Every Creature* (Col. 1, 15.). Baltimore: St. Mary's University, 1938.
R. Eaton, *St. Paul's Epistles to the Colossians and Philemon with Introduction and Notes.* London: C.T.S., 1934.
G. Hitchcock, *Epistle to the Ephesians.* New York: Benziger, 1913.
B. Wilberforce, *Devout Commentary on the Epistle to the Ephesians.* St. Louis: Herder, 1902.

18. THESSALONIANS

R. Eaton, *Epistles of St. Paul to the Thessalonians.* London: Walker, 1939.

19. HEBREWS

W. Leonard, *The Authorship of the Epistle to the Hebrews.* London: Burns, Oates and Washbourne, 1939.

20. CATHOLIC EPISTLES

R. Eaton, *Catholic Epistles of SS. Peter, James, Jude and John.* London: Burns, Oates and Washbourne, 1937.

21. APOCALYPSE

E. S. Berry, *Apocalypse of St. John.* Columbus: Winterich, 1922.
R. Eaton, *Apocalypse of St. John.* St. Louis: Herder, 1930.
C. Martindale, S.J., *Apocalypse of St. John.* New York: Kenedy, 1923.
J. Ratton, *The Apocalypse of St. John.* New York: Benziger, 1915.
J. Ratton, *Revelation of St. John.* New York: Benziger, 1918.

22. VARIA

J. Arendzen, *Priests, Prophets and Publicans.* London: Sands, 1926.
J. Arendzen, *The Gospeis—Fact, Myth or Legend?* London: Sands, 1929.
R. Bandas, *Master Idea of St. Paul's Epistles or the Redemption.* Bruges: Desclee, De Brouwer, 1925.

Additional Reading

R. Bandas, *Catechetics in the New Testament.* Milwaukee: Bruce, 1935.
R. Bandas, *Biblical Questions: New Testament.* Paterson: St. Anthony Guild, 1936.
A. Barnes, *Witness of the Gospels.* St. Louis: Herder, 1906.
M. Bover, *Three Studies from St. Paul.* London: Burns and Oates, 1931.
D. Buzy, *The Life of St. John the Baptist, The Forerunner of Our Lord.* London: Burns and Oates, 1933.
P. Coghlan, *St. Paul,* London: Burns and Oates, 1920.
J. Dupperay, *Christ in Christian Life According to St. Paul.* New York: Longmans, Green, 1927.
C. Fouard, *St. Peter and the First Years of Christianity.* New York: Longmans, Green, 1915.
C. Fouard, *St. Paul and His Missions.* N. Y.: Longmans, Green, 1915.
C. Fouard, *The Last Years of St. Paul.* N. Y.: Longmans, Green, 1915.
C. Fouard, *St. John and the Close of the Apostolic Age.* New York: Longmans, Green, 1915.
F. Gigot, S.S., *Christ's Teaching Concerning Divorce in the New Testament.* New York: Benziger, 1912.
H. Heuser, *From Tarsus to Rome.* New York: Longmans, Green, 1929.
J. Husslein, S.J., *Bible and Labor.* New York: Macmillan, 1924.
C. Lattey, S.J., *Religion of the Scriptures.* St. Louis: Herder, 1922.
C. Lattey, S.J., *The New Testament.* London: Burns, Oates and Washbourne, 1938.
A. Lemonnyer, *Theology of the New Testament.* London: Sands, 1929.
A. Lugan, *Social Principles of the Gospels.* New York: Macmillan, 1928.
J. Maritain, *The Living Thoughts of St. Paul.* New York: Longmans, Green, 1941.
C. Martindale, S.J., *Princes of His People:* I St. John the Evangelist. New York: Kenedy, 1921.
C. Martindale, S.J., *Princes of His People:* II St. Paul. New York: Benziger, 1925.
J. MacRory, *The New Testament and Divorce.* Dublin: Burns and Oates, 1934.
W. McGarry, S.J., *Paul and the Crucified.* N. Y.: America Press, 1939.
W. McGarry, S.J., *Unto the End.* New York: America Press, 1941.
W. McGarry, S.J., *He Cometh.* New York: America Press, 1941.
V. McNabb, O.P., *New Testament Witnesses to Our Blessed Lady.* London: Sheed and Ward, 1930.
J. Rickaby, S.J., *Notes on St. Paul: Corinthians, Galatians, Romans.* New York: Benziger, 1905.
J. Rickaby, S.J., *Further Notes on St. Paul: The Epistles of the Captivity.* London: Burns and Oates, 1911.
V. Rose, O.P., *Studies on the Gospels.* London: Longmans, Green, 1903.
W. Russell, *The Bible and Character.* Philadelphia: Dolphin Press, 1934.
H. Schumacher, *Social Message of the New Testament.* Milwaukee: Bruce, 1937.

GLOSSARY OF TERMS AND NAMES

ANTINOMIAN: (Greek: *anti,* against; *nomos,* law) applied to any doctrine which erroneously minimizes the value and obligation of law. In Apostolic times some exaggerated Christian liberation from sin and the Law to include an abrogation not only of certain Mosaic ceremonial laws but also precepts of the natural law. (Cf. *1* Pet. 2, 16; *2* Pet. 2, 13 f. 17-19.) This error was somewhat revived in the sixteenth century by misrepresenting Paul's doctrine of justification by faith.

APOCRYPHAL: books which illegitimately claim to be sacred, inspired and canonical books of Holy Scripture. Although they are very ancient, the Church never received them into the Canon of Scripture. The Apocryphal books are divided into those of the Old and those of the New Testament. For the most part, the Old Testament apocrypha were written by authors in good faith, who ascribed them to celebrated men in order that their works might be acclaimed publicly. They are useful and valuable in that they portray the Jewish mentality of their times in religious matters. The New Testament apocrypha contain pious legends, but for the most part they were written by heretics, who desired to claim divine authority for their heretical doctrines. According to their literary form they are divided into four groups: Gospels, Acts, Epistles, and Apocalypses.

APOSTOLIC CONSTITUTIONS: in the fourth and fifth centuries, various compilations referring to the discipline and liturgy of the primitive Christian Church were composed. Of these the Apostolic Constitutions is the most important. The early Fathers attributed the work to St. Clement of Rome and through him to the Apostles; but an analysis of the eight books manifests that the work was compiled at the end of the fourth or the beginning of the fifth century, near Antioch, by a single anonymous author.

ATHANASIUS (295-373): Bishop of Alexandria and renowned as a great opponent of Arianism. His apologetic and dogmatic treatises are important for his extensive use of Sacred Scripture in which he insists upon the simple and literal explanation of the text.

AUGUSTINE (354-430): Bishop of Hippo in Africa, known as the Doctor of Grace throughout the universal Church. After his conversion in 387, he commented frequently upon Sacred Scripture both in homilies and in the books and commentaries which he composed.

AUTHENTICITY: that quality or characteristic of a work which manifests it to be the product of the author to whom it is ascribed. In Sacred Scripture books are declared to be authentic generally by criticism and history.

Glossary of Terms and Names

BASILIDES: a heretic who taught a Gnostic doctrine in Alexandria between 120 and 140. He wrote a gospel and a commentary on the same. Only a few quotations from his works remain.

BIBLICAL COMMISSION: this Commission consists of a group of Cardinals and priests versed in theology and biblical knowledge, selected by the Supreme Pontiff for the purpose of discussing and deciding problems that pertain to Sacred Scripture. Their decisions are submitted to the Pope for approval before publication. The faithful are obliged in conscience to accept the decisions of the Commission.

BROTHERS OF THE LORD: see p. 94.

CANON: (Greek, *kanon,* a measuring rod, rule) applied first to the rule or norm of faith and conduct, then to the inspired books which give us this rule. At present the canon is the Church's official list of inspired books. A book is canonical when it is admitted and defined by the Church as inspired.

CANONICAL: see Canon.

CERINTHUS: a Gnostic-Ebionite heretic contemporary to John the Evangelist. He taught that the world was created by angels, that Jesus was a mere man, that Christ descended upon Him at His Baptism and withdrew from Him before the Passion, that at the end of time the spiritual kingdom of heaven will be preceded by a period of abundant joy on earth. Irenæus tells us that St. John wrote his Gospel to confute these errors.

CHRISTOLOGICAL HERESIES: Christology is the branch of theology dealing with the nature and personality of Christ. Early errors on this subject were found among the Ebionites, the Docetists, the Cerinthians, and the Gnostic (which see).

CLEMENT OF ALEXANDRIA (150-220): probably a native of Athens, and a disciple of Pantænus, founder of the Catechetical School at Alexandria. He was well versed in Greek philosophy and his numerous writings contain many references to the teaching of the Church on the canonicity, inspiration, and interpretation of Sacred Scripture.

CLEMENT OF ROME: according to tradition St. Clement was the fourth Bishop of Rome (92-101). He is chiefly known for his Letter to the Corinthians in which he asserts his supreme authority as Bishop of Rome. The letter frequently bears witness to the inspired nature and canonicity of the Bible.

CLEMENTINE VULGATE: see Vulgate.

CODEX: the modern book form with leaves arranged in quires or gatherings has its origin in the ancient codices of the Christians some of which date from the early part of the second century. Papyrus was the material used till superseded by vellum in the fourth century.

CYBELE: the name given to the goddess of nature in Asia Minor, known in Greece as Rhea, the mother of the gods.

DEACON: see p. 379.

Glossary of Terms and Names

DEACONESS: see p. 575.

DIASPORA: (Greek, a scattering) the name given to the Jews scattered throughout the pagan world after the exile. At the time of Christ there were four or five million Jews outside of Palestine as opposed to about one million in Palestine. Segregated groups were to be found in almost every part of the Euphrates and Mediterranean areas, usually enjoying political privileges of partial self-government.

DIDACHE: or "The Doctrine of the Twelve Apostles"; a short treatise purporting to be an abstract of the Apostles' teaching composed by the Apostles themselves. It was probably composed by some unknown author in the East between the years 70 and 90. The treatise is a summary of the moral, individual, and social obligations of the early Christians.

DIOGNETUS: the letter to Diognetus is probably of the second or third century. The author and recipient are unknown. It is a reply to questions of a heathen interested in Christianity. It deals with the differences between Christian, pagan and Jewish worship, and with Christian charity.

DOCETAE or DOCETISTS: (Greek, *dokesis,* appearance) Gnostic or Manichaean heretics who held that Christ had only the appearance of man. He was simply a spirit or higher æon and His body was a mere phantasm.

EBIONITES: Jewish-Christian heretics who denied the divinity and the virgin birth of Christ and the authority of St. Paul. They considered matter to be an emanation of the Deity. Their name (Hebrew, *ebyonim,* poor) indicates their general poverty.

EPIPHANIUS (315-403): a saint, a scholar, a prolific reader, and a firm defender of Christian traditions. In his works he has preserved many documents of great worth for the history of Christian thought. He is best known for his book on Biblical Archaeology in which he also treats of the canon and the Versions of the Old Testament, and Palestinian Geography.

EUSEBIUS (265-340): born in Palestine, probably at Cæsarea. In 311 he became Bishop of Cæsarea. He was an erudite scholar and wrote upon History, Geography, Sacred Scripture, Apologetics, Theology and Sacred Eloquence. His Ecclesiastical History is invaluable for a knowledge of the Church during the first three centuries.

GNOSTICS: (Greek, *gnosis,* knowledge) heretics who claimed salvation by knowledge. This "knowledge" was highly speculative without evidence or proof. Concerned principally with the problem of the origin of evil, it was essentially dualistic. Between the good God and the evil creatures there are numerous intermediary spirits or Aeons emanating from God and diminishing in goodness in proportion to their distance from the source. The lowest aeon was Demiurge (Greek, workman), who was the creator and legislator of the Old Testament. Christ was considered a higher æon, but His importance

GLOSSARY OF TERMS AND NAMES

in the system is so slight that Gnosticism can be considered essentially pagan.

HEBREW: one of the Semitic languages. Arabic, Aramaic, Syriac and Ethiopic belong to the same group. In these languages the consonants play the predominant rôle. The vowels, which might color the meaning of a word or give an entirely new word, were not written. They were added to biblical texts only in the fifth century, A.D., and represent interpretations of that period. In place of the tenses of our languages the Semites use a perfect or imperfect form of the verb according as the action or state of the verb is considered complete or incomplete. Semitic poverty in adjectives, adverbs, conjunctions and particles presents a problem to translators. Abstract terms are rare so that the language is often figurative and usually picturesque.

HEGESIPPUS (110?-180?): according to St. Jerome, the first historian of the Church. He was a convert to Christianity from Judaism. After his conversion he visited various Christian Churches in order to observe the uniformity of the faith in the midst of rising heresies. Fragments of his work entitled "Memoirs" are cited by Eusebius and other Christian authors.

HERMAS: the name or pen name of the author of an early Christian work called "The Shepherd." His book was probably written about the middle of the second century. It is a presentation of pretended revelations received in visions and an exhortation to penance and good works.

IGNATIUS OF ANTIOCH (d. 107): one of the chief figures of the early Church. He was probably a disciple of St. Peter and St. Paul, and became Bishop of Antioch about the year 69. On his way to Rome, where he suffered martyrdom, St. Ignatius wrote seven Epistles. He is the first to use the term Catholic Church.

INERRANCY: a consequence of biblical inspiration. It implies that there is not and that there cannot be any error in the Bible of which God is the author. Inerrancy is found only in the original work as it came from the pen of the inspired writer. It applies to the entire content of the books and not just to religious matters. Each statement must be taken according to the intention of the author. Our present Bibles share in this inerrancy as they agree substantially with the original works.

INSPIRATION: as applied to the Bible a supernatural gift by which God, the principal author, uses man, the instrumental author, to write. Man is not used as a mere lifeless instrument but with all his natural aptitudes and individuality of intellect, and will and executive faculties. In a true though different sense God is the author of the whole book and man is the author of the whole book.

INTEGRITY: this implies that all the passages in a book (e.g., John *21*, 1-25) belonged originally to it and were not added by a different

Glossary of Terms and Names

author. Since each book is usually the composition of a single author, the question of integrity is often intimately linked with that of authenticity. Cf. authenticity (p. 721).

IRENÆUS (120-202): a native of Smyrna in Asia Minor. As a young man he often listened to the disciples of the Apostles, especially St. Polycarp. Later he settled down in the city of Lyons, France, where he succeeded St. Pothinus as Bishop of that city in 177. St. Irenæus is known as the "most learned, most prudent, and most illustrious of the early heads of the Church in Gaul." He professed a strong attachment for tradition and a love of Sacred Scripture which he knew perfectly. He is chiefly known for his work "Against Heresies," a refutation of religious errors in his time, especially Gnosticism.

JEROME (347-420): "The Doctor of Sacred Scripture," was born at Stridon in Dalmatia. He was educated at Rome, and in his early life traveled through the East and the West studying, observing, and transcribing his impressions. In 382, at the suggestion of Pope Damasus, he began to revise the Old Latin translation of the Bible. His revision of the New Testament and a translation of the Psalms made from the Septuagint were completed about 384. In 385 St. Jerome retired to Bethlehem where he spent thirty-five years devoted to the study of Sacred Scripture. During this period he translated the Bible into Latin, which in the course of time was given the name "Vulgate." St. Jerome is renowned also for his commentaries on the Old and New Testament, and several historical, dogmatical, and controversial books.

JOHN CHRYSOSTOM (344-407): a learned and holy Doctor of the Church, born at Antioch. In 398 St. John became Bishop of Constantinople. Throughout his life he was known for his sacred eloquence. Besides his numerous letters, sermons, and homilies, St. John is the author also of treatises on the religious life, the priesthood, education, and of several commentaries on Sacred Scripture.

JOSEPHUS FLAVIUS: born about 37 A.D. He was a Jew and belonged to the sacerdotal class of the Jews. His History of the Jewish Wars and The Jewish Antiquities are important for a knowledge and appreciation of the periods treated by him. Although Josephus is sometimes inclined to exaggeration and erroneous statements, in general he employed good sources and is worthy of credence.

JUDAIZER: a Jewish Christian who advocated the retention of such Jewish laws as circumcision, abstinence from unclean food, the observance of Jewish festivals in the Church. Some considered these practices as necessary to salvation, others as conducive to higher perfection.

JUSTIN (100-165): born at Sichem in Palestine and educated in paganism. His study of philosophy led him to embrace Christianity about the year 130. As a layman he spent his life in defending and propagating the truths taught by Christ. The two Apologies addressed

Glossary of Terms and Names

to the Emperor Antoninus Pius have placed St. Justin among the greatest of the early Christian apologists.

MANUSCRIPTS: books or codices written by hand before the invention of printing in 1450. The word is often abbreviated MSS.

MARCION: (d. c. 170) a heretic who taught that there is opposition between the Old and the New Testament, that these are the work of two beings, the righteous and angry God of the Law and the good and loving God of the Gospels. He had an austere system of morals, forbidding marriage, meat, and wine.

NICOLAITES: heretics who claimed origin from the deacon Nicolas mentioned in Acts 6, 5. They sought to kill concupiscence by indulging the passions at will. They are referred to in Apoc. 2, 6.15.

ORIGEN (185-253): born at Alexandria in Egypt, as a youth a student of Clement of Alexandria. His erudition and tenacity earned for him the surname "Man of Steel." It has been estimated that Origen was the author of about 6000 works on Sacred Scripture, theology, apologetics, and ascetics.

PAPIAS: according to Irenæus, St. Papias, Bishop of Hierapolis in Phrygia, was a disciple of St. John the Evangelist. He composed "An Explanation of the Oracles of the Lord," and was the earliest writer whose extant works attest the Gospels according to St. Matthew and St. Mark. St. Papias died about the year 120.

PAPYRI: Papyrus was one of the chief materials used in ancient times for writing. It was supplied by the pith of the papyrus plant which grew plentifully along the Nile River in Egypt. Sheets were formed and fastened side by side to form a roll. In recent years many biblical papyri have been discovered.

PARALLELISM: a rhythm of thought in which an idea is immediately repeated, reversed or expanded. If the thought is repeated, the parallelism is synonymous, as

O God come to my aid

O Lord make haste to help me. (Ps. *69,* 2.)

When the thought is reversed the parallelism is antithetic, as

They are entangled and fallen

But we are risen and stand upright. (Ps. *19,* 9.)

When the thought is expanded by giving the cause, the effect or a circumstance, the parallelism is called synthetic, as

Thou wilt turn, O God, and bring us to life,

And thy people shall rejoice in thee. (Ps. *84,* 7.)

Parallelism is one of the principal characteristics of Hebrew poetry. It is found to some extent also in Hebrew prose.

PAROUSIA: (Greek, *para,* near; *ousia,* being) the name given to the solemn entry or an official visit of a ruler in a province. This often marked a new era. The word now refers to the Second Coming of Christ at the end of this world and the inauguration of the heavenly kingdom.

Glossary of Terms and Names

PESCHITTO: see Versions.

PHILO (25 B.C.-41 A.D.): a Jewish writer of Alexandria. In his writings he generally explained Jewish history and legislation in an allegorical way. He is the author of two apologetical works directed against the adversaries of the Jews at Alexandria.

PHOTIUS (c. 815-897): a pretender to the Patriarchate of Constantinople and author of the great schism between the East and the West.

POLYCARP: the teacher of St. Irenaeus. He was personally acquainted with St. John, and is considered the last survivor of the Apostolic age. As Bishop of Smyrna he wrote an Epistle to the Philippians. He was martyred for the faith about the year 155.

PRÆTORIUM: see p. 188.

PRESBYTER: see p. 577.

PROSELYTE: (Greek, *proselytos,* a newcomer, a stranger.) In the Septuagint the Greek word is applied to aliens. Even the Jews are described as "proselytes" in Egypt. But by the seventh century, B.C., the proselyte was a convert to Judaism.

RECENSION: an emendation or new edition of the sacred text.

SECOND COMING: see Parousia.

SEPTUAGINT: the most important translation in Greek made from the Hebrew. It takes its name from the seventy translators by whom the work was supposed to have been done. The Pentateuch was translated in the third century before Christ; the remaining books of the Old Testament were translated by various persons at different times. By the middle of the second century before Christ all the books had been translated. The complete translation is known as the Septuagint.

SHEPHERD of HERMAS: see Hermas.

SUETONIUS: a non-Christian Latin historian, best known for his lives of the twelve Cæsars. He wrote about the year 120.

TALMUD: a collection of Jewish commentaries on the *Mishna,* the traditional law of the Jews. Composed by the Rabbis from the second to the sixth centuries, the Palestinian or Jerusalem Talmud was committed to writing about the fifth century, that of Babylon about the sixth century.

TARGUMS: paraphrases of the Hebrew text of the Bible in the Aramaic dialect. These paraphrases were made for the Jewish people after the Babylonian Exile when they no longer understood Hebrew.

TERTULLIAN (150/160-240/250): the first great representative of a series of Latin Christian writers from Africa. He was born at Carthage, and converted to the faith about 195. As a Christian Tertullian firmly defended Christianity in his Apologies, but towards the end of his life he fell into the heresy of the Montanists.

THEODORET (d. 458): a scholar of the Antiochian school of biblical interpretation. He commented extensively on the books of the Old Testament, and is known also as the continuator of the Ecclesiastical History of Eusebius.

Glossary of Terms and Names

VERSIONS: translations of Sacred Scripture from the Hebrew and Greek texts into other languages. From the earliest days of the Church translations were made, e.g., into Syriac (Peschitto), Arabic, Greek, Latin, etc.

VULGATE: the most important ancient translation of the Bible. The work was accomplished by St. Jerome, and by degrees it became the only Latin version of Sacred Scripture used in the Western Church. The Council of Trent has declared the Vulgate to be the authentic, that is, the official Latin version of the Church. But due to the fact that many editions of the Vulgate existed in the sixteenth century, the Council ordered that a corrected edition should be published as soon as possible. The revised text of the Vulgate appeared in 1592 under Pope Clement VIII. For this reason it is very often referred to as the **Clementine Vulgate**.

ENVIRONMENTAL ENGINEERING
SAVING A THREATENED RESOURCE—
IN SEARCH OF SOLUTIONS

Proceedings of the Environmental Engineering sessions at Water Forum '92

Sponsors of Water Forum
ASCE Environmental Engineering Division
ASCE Hydraulics Division
ASCE Irrigation and Drainage Division
ASCE Water Resources, Planning and Management Division

Host
ASCE Baltimore Section

Cooperating Organizations
American Association of Port Authorities
American Geophysical Union
American Society of Agricultural Engineers
American Water Resources Association
Association of Ground Water Scientists
 & Engineers
Canadian Society for Civil Engineering
China Civil Engineering Society
Consiglio Nazionale degli Ingegneri
Danish Society of Civil Engineers
International Ref. Center for Community
 Water Supply
International Water Resources Association
International Association for Hydraulic
 Research
International Association of Hydrogeologists
International Commission on Large Dams
Korean Society of Civil Engineers

National Water Resources Association
Norwegian Society of Chartered Engineers
Permanent International Association of
 Navigation Congresses
Societas Internationalia Limnologiao
Societe des Ingenieurs de France
The Institution of Engineering, Pakistan
The Institution of Engineers of Ireland
The Institution of Engineers, Australia
The Institution of Structural Engineers
The Union of International Technical
 Associations
The Universities Council on Water
 Resources
U.S. Committee on Irrigation and Drainage
U.S. Geological Survey
US Net. Comm. IAWPRC
Verein Deutcher Ingenieure

Baltimore, Maryland
August 2-6, 1992

Edited by F. Pierce Linaweaver

Published by the
American Society of Civil Engineers
345 East 47th Street
New York, New York 10017-2398